TRIGGERED STAR FORMATION
IN A TURBULENT INTERSTELLAR MEDIUM

IAU SYMPOSIUM No. 237

COVER ILLUSTRATION: RCW 79

The Galactic HII region RCW 79 contains a cavity of ionized gas surrounded by a dust and molecular ring. The ring has five fragments, three of which contain very young stars that presumably formed when the ring collapsed by self-gravitational forces. Image from Zavagno et al. 2006, Astronomy and Astrophysics, 446, 171, reprinted with permission.

INTERNATIONAL ASTRONOMICAL UNION

UNION ASTRONOMIQUE INTERNATIONALE

TRIGGERED STAR FORMATION IN A TURBULENT INTERSTELLAR MEDIUM

PROCEEDINGS OF THE 237th SYMPOSIUM OF THE INTERNATIONAL ASTRONOMICAL UNION HELD IN PRAGUE, CZECH REPUBLIC AUGUST 14–18, 2006

Edited by

BRUCE G. ELMEGREEN
IBM Research Division, Yorktown Heights, New York, USA

and

JAN PALOUŠ
Astronomical Institute, Czech Academy of Science, Praha 4, Czech Republic

CAMBRIDGE UNIVERSITY PRESS
The Edinburgh Building, Cambridge CB2 2RU, United Kingdom
32 Avenue of the Americas, New York, NY 10013-2473, USA
477 Williamstown Road, Port Melbourne, VIC 3207, Australia
Ruiz de Alarcón 13, 28014 Madrid, Spain
Dock House, The Waterfront, Cape Town 8001, South Africa

First published 2007

Printed in the United Kingdom at the University Press, Cambridge

Typeset in System LaTeX 2_ε

A catalogue record for this book is available from the British Library

Library of Congress Cataloguing in Publication data

ISBN 13 9780521 86346 9 hardback
ISBN 10 0521 86346 5 hardback
ISSN 1743-9213

Table of Contents

Preface xvii

Organizing committee xviii

Conference photograph xix

Conference participants xx

Presentations

Session I. Turbulence and general structure in the ISM
Chair: S. Oey

Observations of turbulence in the diffuse interstellar medium 1
    J. M. Dickey

Turbulence in the molecular interstellar medium 9
    M. H. Heyer & C. Brunt

Turbulence in high latitude molecular clouds 17
    S. N. Shore, T. N. LaRosa, L. Magnani, R. J. Chastain & F. Costagliola

The turbulent environment of low-mass dense cores 24
    E. Falgarone, P. Hily-Blant, J. Pety & G. Pineau des Forêts

Session II. General ISM in galaxies and Magellanic Clouds
Chair: S. Oey

CO, HI, recent Spitzer SAGE results in the Large Magellanic Cloud 31
    Y. Fukui

Dense molecular clouds and associated star formation in the Magellanic Clouds. 40
    M. Rubio

Dust evolution in the star forming turbulent interstellar medium 47
    F. Boulanger

Session III. ISM Turbulence
Chair: Y. Fukui

New results on the distribution of thermal pressures in the diffuse ISM 53
    E. B. Jenkins & T. M. Tripp

Dynamical evolution of a supernova driven turbulent interstellar medium 57
    D. Breitschwerdt & M. A. de Avillez

ISM turbulence driven by the magnetorotational instability 65
    R. A. Piontek & E. C. Ostriker

Non-stellar sources of turbulence in the interstellar medium 70
    E. C. Ostriker

Session IV. Shells I
Chair: Y. Fukui

Observations of supershells in the interstellar medium of nearby galaxies   76
     E. Brinks, I. Bagetakos, F. Walter & E. de Blok

Shells in the Magellanic System   84
     S. Stanimirovic

An automatic identification of HI shells in the 2nd and 3rd Galactic Quadrants.   91
     S. Ehlerová

The SMC super-shells as probes of the turbulent dynamics of the ISM   96
     I. Goldman

Giant molecular clouds and star formation in the Large Magellanic Cloud   101
     A. Kawamura, T. Minamidani, Y. Mizuno, T. Onishi, N. Mizuno,
     A. Mizuno & Y. Fukui

Session V. Shells II
Chair: E. Ostriker

Towards resolving the evolution of multi-supernova superbubbles.   106
     M. S. Oey

Triggering and the gravitational instability in shells and supershells   114
     J. Palouš

The enigmatic Loop III and the local Galactic structure   119
     M. Kun

Footprints of triggering in large area surveys of the nearby ISM and YSOs.   124
     L. V. Tóth & Z. T. Kiss

NANTEN observations of triggered star formation: from H ii regions to galaxy
   collisions   128
     N. Mizuno, A. Kawamura, T. Onishi, A. Mizuno & Y. Fukui

Session VI. Molecular clouds I
Chair: E. Ostriker

Observations of prestellar cores: Probing the initial conditions for the IMF.   132
     P. André

Magnetic fields and star formation – new observational results.   141
     R. M. Crutcher & T. H. Troland

Physics and chemistry of hot molecular cores   148
     H. Beuther

Dissecting a site of massive star formation: IRAS 23033+5951   155
     M. A. Reid & B. C. Matthews

Anatomy of the S255-S257 complex – triggered high-mass star formation   160
     V. Minier, N. Peretto, S. N. Longmore, M. G. Burton, R. Cesaroni,
     C. Goddi, M. R. Pestalozzi & Ph. André

Session VII. Molecular clouds II
Chair: G. Tenorio-Tagle

Dynamical processes in star forming regions: feedback and turbulence generation   165
  J. Bally

Hypersonic swizzle sticks: jets, fossil cavities and turbulence in molecular clouds   172
  A. J. Cunningham, A. Frank, E. G. Blackman & A. Quillen

The newest twists in protostellar outflows: triggered star formation   177
  M. Barsony

Probing turbulence in OMC1 at the star forming scale: observations and simulations   183
  M. Gustafsson, A. Brandenburg, J.-L. Lemaire & D. Field

Protostars in the Elephant Trunk Nebula   188
  W. T. Reach

Session VIII. Molecular clouds III
Chair: G. Tenorio-Tagle

Star formation in the LMC: gravitational instability and dynamical triggering . .   192
  Y.-H. Chu, R. A. Gruendl & C.-C. Yang

Star formation in the Small Magellanic Cloud: the youngest star clusters   199
  E. Sabbi, A. Nota, M. Sirianni, L. R. Carlson, M. Tosi, J. Gallagher, M. Meixner, M. S. Oey, A. Pasquali, L. J. Smith, M. Vlajic & L. Hawks

Molecular clouds in galaxies from extinction studies   204
  J. Alves

Understanding starbursts through giant molecular clouds in high density environments   208
  E. W. Rosolowsky

Observation of triggering in the Milky Way   212
  L. Deharveng, A. Zavagno, B. Lefloch, J. Caplan & M. Pomarès

SK 1: A possible case of triggered star formation in Perseus   217
  M. Rengel, K. Hodapp & J. Eislöffel

Session IX. Clusters I
Chair: F. Mirabel

The life and death of star clusters   222
  B. C. Whitmore

Star-cluster formation and evolution   230
  P. Kroupa

On the structure of giant HII regions and HII galaxies   238
  G. Tenorio-Tagle, C. Muñoz-Tuñón, E. Pérez, S. Silich & E. Telles

Super massive star clusters: from superwinds to a cooling catastrophe and the re-processing of the injected gas   242
  S. Silich, G. Tenorio-Tagle, C. Muñoz-Tuñón & J. Palouš

Triggered star formation in the environment of young massive stars. 246
  M. Gritschneder, T. Naab, F. Heitsch & A. Burkert

Session X. Clusters II
Chair: F. Mirabel

Triggered formation and collapse of molecular cloud cores. 251
  A. P. Whitworth

Massive star and star cluster formation 258
  J. C. Tan

Strongly triggered collapse model confronts observations. 265
  P. Hennebelle, A. Belloche, Ph. André & A. Whitworth

Sequentially triggered star formation in OB associations 270
  Th. Preibisch & H. Zinnecker

Triggered star formation in OB associations 278
  W. P. Chen, H. T. Lee & K. Sanchawala

Session XI. Clusters III
Chair: S. Shore

The formation of self-gravitating cores in turbulent magnetized clouds 283
  P. Padoan, Åke Nordlund, A. G. Kritsuk, M. L. Norman & P. S. Li

Molecular cloud turbulence and the star formation efficiency: enlarging the scope 292
  E. Vázquez-Semadeni

Simulations of ionisation triggering 300
  C. J. Clarke & J. E. Dale

Protostellar turbulence in cluster forming regions of molecular clouds 306
  F. Nakamura & Z.-Y. Li

Session XII. Galaxies I
Chair: S. Shore

Star formation on galactic scales 311
  R. C. Kennicutt, Jr.

Star formation in mergers and interacting galaxies: Gathering the Fuel 317
  C. Struck

Tidal dwarf galaxies as laboratories of star formation and cosmology. 323
  P.-A. Duc, F. Bournaud & M. Boquien

The global star formation law: from dense cores to extreme starbursts 331
  Y. Gao

Session XIII. Galaxies II
Chair: J. Palouš

Galactic-scale star formation by gravitational instability 336
  M.-M. Mac Low, Y. Li & R. S. Klessen

Spiral arm triggering of star formation 344
  I. A. Bonnell & C. L. Dobbs

Cloud formation from large-scale instabilities 351
  W.-T. Kim

Turbulent structure and star formation in a stratified, supernova-driven, interstellar
  medium 358
  M. K. R. Joung & M.-M. Mac Low

Cold HI in turbulent eddies and galactic spiral shocks 363
  S. J. Gibson, A. R. Taylor, J. M. Stil, C. M. Brunt, D. W. Kavars & J. M. Dickey

Session XIV. Galaxies III
Chair: J. Palouš

Dense gas formation triggered by spiral density wave in M 31 368
  T. Tosaki, Y. Shioya, N. Kuno, K. Nakanishi, T. Hasegawa, S. Matsushita, K. Kohno, R. Miura, Y. Tamura, S. K. Okumura & R. Kawabe

Triggered star formation in the Magellanic Clouds 373
  K. Bekki

Turbulence, feedback, and slow star formation 378
  M. R. Krumholz

Conference summary: triggered star formation in a turbulent ISM 384
  B. G. Elmegreen

Poster papers

High supernova rate and enhanced star-formation triggered in M81-M82 encounter   391
  B. Arbutina, D. Urošević & B. Vukotić

Numerical simulations for the interaction of the NGC 1333 IRAS 4A outflow and
  an ambient cloud 392
  C. H. Baek, J. Kim & M. Choi

An inventory of supershells in nearby galaxies: first results from THINGS 393
  I. Bagetakos, E. Brinks, F. Walter & E. de Blok

Fine–scale structure of the neutral ISM in M81 394
  I. Bagetakos, E. Brinks, F. Walter & E. de Blok

Methanol masers and massive star formation 395
  A. Bartkiewicz, M. Szymczak & H. J. van Langevelde

Sequential star formation in the Sh 254-258 molecular cloud: HHT maps of CO
  J=3-2 and 2-1 emission 396
  J. H. Bieging, W. L. Peters, B. Vila Vilaro, K. Schlottman & C. Kulesa

Star formation thresholds derived From THINGS 397
  F. Bigiel, F. Walter, E. de Blok, E. Brinks & B. Madore

Intergalactic star formation around NGC 5291 398
    M. Boquien, P.-A. Duc, J. Braine, E. Brinks, V. Charmandaris & U. Lisenfeld

Comparing the structure of three dark globules 399
    L. Campeggio, B. M. T. Maiolo, F. Strafella & D. Elia

A multifrequency study of the active star forming region NGC 6357 400
    C. E. Cappa, R. H. Barbá, M. Arnal, N. Duronea, E. F. Lajús, W. M. Goss & J. Vasquez

Triggered massive star formation in the LMC HII complex N44 401
    C.-H. R. Chen, Y.-H. Chu, R. A. Gruendl & F. Heitsch

Star formation activity in the cluster spiral NGC 4254 402
    K. T. Chyży, R. Beck & S. Ryś

The Parkes methanol multibeam survey 403
    R. J. Cohen, J. L. Caswell, K. Brooks, M. G. Burton, A. Chrysostomou, J. Cox, P. J. Diamond, S. Ellingsen, G. A. Fuller, M. D. Gray, J. A. Green, M. G. Hoare, M. R. W. Masheder, N. McClure-Griffiths, M. Pestalozzi, C. Phillips, M. Thompson, M. Voronkov, A. Walsh, D. Ward-Thompson, D. Wong-McSweeney & J. A. Yates

The Mopra DQS survey of the G333 region 404
    M. R. Cunningham, I. Bains, N. Lo, T. Wong, M. G. Burton, P. A. Jones and the DQS Team

Studies of the Perseus Region. I. 405
    S. Datta

Evidence for triggered star formation in the Carina Flare supershell 406
    J. Dawson, A. Kawamura, N. Mizuno, T. Onishi & Y. Fukui

New insights into the nature of mid-infrared emission associated with massive star formation: disks and outflow 407
    J. M. De Buizer

From young massive star clusters to old globulars: long-term survival chances . . 408
    R. de Grijs

Star formation In Bright-Rimmed Clouds: a comparison of wind-driven triggering with millimeter observations 409
    Christopher H. De Vries, G. Narayanan & R. L. Snell

The virial balance of clumps and cores in molecular clouds 410
    S. Dib, E. Vázquez-Semadeni, J. Kim, A. Burkert & M. Shadmehri

Structural analysis of molecular cloud maps: the case of the star forming Vela-D cloud 411
    D. Elia, F. Strafella, F. Massi, M. De Luca, L. Campeggio & B. M. T. Maiolo

Different evolutionary stages in S235A-B 412
    M. Felli, F. Massi & R. Cesaroni

Angular power spectra of Galactic HI 413
   D. A. Green

Gemini VRI data of counterparts associated to X-ray sources in CMa R1 414
   J. Gregorio-Hetem, C. V. Rodrigues & T. Montmerle

MHD simulations of supernova driven ISM turbulence 415
   O. Gressel & U. Ziegler

Bright knots along spiral arms in disk galaxies 416
   P. Grosbøl & H. Dottori

Young stellar clusters triggered by a density wave in NGC 2997 417
   P. Grosbøl, H. Dottori & R. Gredel

A method for detecting preferred scale sizes 418
   M. Gustafsson, J.-L. Lemaire & D. Field

HST emission line images of the Orion HII region: proper motions and possible
   variability 419
   L. Gutiérrez, C. Giammanco & J. E. Beckman

The high latitude low mass star forming region Cometary Globule 12: two compact
   cores and a C18O hot spot 420
   L. K. Haikala, M. Juvela, J. Harju, K. Lehtinen, K. Mattila, M. Olberg &
   M. Dumke

Diffuse X-ray emission from the Carina Nebula observed with Suzaku 421
   K. Hamaguchi, R. Petre and the Suzaku η Carina team

Discovery of large-scale masers in W3(OH) 422
   L. Harvey-Smith & R. J. Cohen

IRSF/SIRIUS near-infrared survey of the Magellanic Clouds: triggered star formation in N11 in the Large Magellanic Cloud 423
   H. Hatano, R. Kadowaki, D. Kato, S. Sato and the IRSF/SIRIUS group

A (Sub)Millimeter survey of massive star-forming regions identified by the ISOPHOT
   Serendipity Survey (ISOSS) 424
   M. Hennemann, S. M. Birkmann, O. Krause & D. Lemke

Non-symmetrical protoplanetary disks destroyed by UV photoevaporation 425
   A. Hetem Jr. & J. Gregorio-Hetem

ASTE Submillimeter observations of a YSO condensation in Cederblad 110 426
   M. Hiramatsu, K. Kamegai, T. Hayakawa, K. Tatematsu, T. Onishi,
   A. Mizuno & T. Hasegawa

Giant Molecular Associations in M51 427
   M. Hitschfeld, C. Kramer, K. Schuster, S. Garcia-Burillo & J. Stutzki

Galaxy open clusters and associations : study of stellar population 428
   A. S. Hojaev

Turbulence in the G333 molecular cloud 429
 P. A. Jones, M. R. Cunningham, I. Bains, E. Muller, T. Wong, M. G.
 Burton and the DQS Team

High-resolution mapping of interstellar clouds with near-infrared scattered light 430
 M. Juvela, V.-M. Pelkonen, P. Padoan & K. Mattila

The distribution of early-type stars in the Mon-CMa-Pup-Vel region of the Milky
 Way 431
 N. T. Kaltcheva

Submillimeter-wave observations of outflow and envelope around the low mass
 protostar IRAS 13036-7644 432
 K. Kamegai, M. Hiramatsu, T. Hayakawa, K. Tatematsu, T. Hasegawa,
 T. Onishi & A. Mizuno

3D Spectroscopy of the Blue Compact Dwarf Galaxies IZw18 and IIZw70 433
 C. Kehrig, J. M. Vílchez, S. F. Sanchez, L. Lopez-Martin, E. Telles &
 D. Martin-Gordón

H I clouds in the Large Magellanic Cloud 434
 S. Kim

The dependence of the IMF on the density-temperature relation of pre-stellar gas 435
 S. Kitsionas, A. P. Whitworth, R. S. Klessen & A.-K. Jappsen

ASTE observations of dense molecular gas in galaxies 436
 K. Kohno, K. Muraoka, K. Nakanishi, T. Tosaki, N. Kuno, R. Miura,
 T. Sawada, K. Sorai, T. Okuda, K. Kamegai, K. Tanaka, A. Endo,
 B. Hatsukade, H. Ezawa, S. Sakamoto, J. Cortes, N. Yamaguchi, H. Matsuo
 & R. Kawabe

Three-dimensional MHD simulations of magnetized molecular cloud fragmentation
 with turbulence and ion-neutral friction. 437
 T. Kudoh & S. Basu

Statistics of initial velocities of open clusters 438
 J. R. D. Lépine, W. S. Dias & Yu. Mishurov

Triggered cluster formation in the RMC 439
 J. Z. Li & M. D. Smith

Southern IRDCs seen with Spitzer/MIPS 440
 H. Linz, Ra. Klein, L. Looney, Th. Henning, B. Stecklum & L.-Å. Nyman

Dependence of radio halos on underlying star formation activity and galaxy mass 441
 U. Lisenfeld, M. Dahlem & J. Rossa

New nearby young star cluster candidates within 200 pc 442
 E. E. Mamajek

Superdiffusion in molecular clouds 443
 G. Marschalkó

Stellar formation in H I interstellar bubbles around massive stars 444
 M. C. Martín, G. A. Romero & C. E. Cappa

Triggered star formation in spiral arms 445
    E. E. Martínez-García, R. A. González-Lópezlira & G. Bruzual-A.

The H_2O super maser emission of Orion KL accretion disk, bipolar outflow, shell 446
    L. I. Matveyenko, V. A. Demichev, S. S. Sivakon, P. D. Diamond & D. A. Graham

SUBARU near-infrared multi-color images of Class II Young Stellar Object, RNO91 447
    S. Mayama, M. Tamura & M. Hayashi

Chemical structure of the massive protobinary-forming hot core, W3(H_2O) 448
    Y. C. Minh & H.-R. Chen

Triggered star formation in bright-rimmed clouds. 449
    L. K. Morgan, J. S. Urquhart, M. A. Thompson & G. J. White

Radiation driven implosion model for star formations near an H II region. 450
    K. Motoyama, T. Umemoto & H. Shang

ASTE CO(3-2) observations of M 83: Correlation between CO(3-2)/CO(1-0) ratios
    and star formation efficiencies 451
    K. Muraoka, K. Kohno, T. Tosaki, N. Kuno, K. Nakanishi, K. Sorai & S. Sakamoto

OH masers and magnetic fields in massive star-forming regions: ON1 452
    S. Nammahachak, K. Asanok, B. Hutawarakorn Kramer, R. J. Cohen, O. Muanwong & N. Gasiprong

Subaru high-dispersion spectroscopy of Hα and [NII] 6584 Å emission in the HL
    Tau jet. 453
    T. Nishikawa, M. Takami & M. Hayashi

Radio observation of molecular clouds around the W5-East triggered star-forming
    region. 454
    T. Niwa, Y. Itoh, K. Tachihara, Y. Oasa, K. Sunada & K. Sugitani

Dust evolution in photoevaporating protoplanetary disks 455
    H. Nomura, Y. Aikawa, S. Inutsuka & Y. Nakagawa

Molecular Hydrogen emission from protoplanetary disks: effects of X-ray irradiation
    and dust evolution 456
    H. Nomura, Y. Aikawa, M. Tsujimoto, Y. Nakagawa & T. J. Millar

Photometric and spectroscopic studies of very low mass YSOs and young Brown
    Dwarfs in S106 457
    Y. Oasa

Luminosity functions of YSO clusters in Sh-2 255, W3 Main and NGC 7538 star
    forming regions. 458
    D. Ojha, M. Tamura and the SIRIUS Team

The conditions for star formation at low metallicity: results from the LMC 459
    J. M. Oliveira, J. Th. van Loon & S. Stanimirović

Star formation in the Eagle Nebula and NGC 6611. 460
    J. M. Oliveira, R. D. Jeffries & J. Th van Loon

Thermal and non thermal components of interstellar medium at sub-kiloparsec scales in galaxies... 461
R. Paladino

Mid-IR images of methanol masers and ultracompact HII regions 462
P. Persi, M. Tapia & A. R Marenzi

Degree scale high resolution mapping of CO J=2-1 and 3-2 in giant molecular clouds 463
W. L. Peters, J. H. Bieging, C. E. Groppi, C. A. Kulesa, C. K. Walker, A. S. Hedden & P. S. Puetz

Triggered star formation in the isolated cluster CB 34?..................... 464
D. E. Peterson, R. A. Gutermuth, M. F. Skrutskie, S. T. Megeath, J. L. Pipher, L. E. Allen & P. C. Myers

Massive star formation in the outer Galaxy: S284 465
E. Puga, C. Neiner, S. Hony, A. Lenorzer, A.-M. Hubert & L. B. F. M. Waters

Interaction/merger-induced starbursts in local very metal-poor dwarfs: link to the common SF in high-z young galaxies 466
S. A. Pustilnik, Ekta, A. Y. Kniazev, J. N. Chengalur & L. Vanzi

Evolutionary sequence of expanding Hydrogen shells....................... 467
M. Relaño, J. E. Beckman, O. Daigle & C. Carignan

Arms pattern speed of galaxies in clusters............................... 468
I. Rodrigues, H. Dottori & D. Reichert

High-frequency carbon recombination line as a probe to study the environment of Ultra-compact HII regions...................................... 469
D. A. Roshi

Extraplanar gas and magnetic fields in the cluster spiral galaxy NGC 4569..... 470
S. Ryś, K.T. Chyży, M. Weżgowiec, M. Ehle & R. Beck

Physical and chemical properties of the AFGL 333 cloud 471
T. Sakai, T. Oka & S. Yamamoto

Infrared properties of ultracompact H II regions in the Galaxy and the LMC... 472
M. Sewiło, Ed Churchwell, Barbara Whitney and the GLIMPSE and SAGE Teams

Globular clusters in NGC147, 185, and 205................................ 473
M. E. Sharina, V. L. Afanasiev & T. H. Puzia

Formation and destruction of clouds and spurs in spiral galaxies 474
R. Shetty & E. C. Ostriker

Interaction between molecular outflows and dense gas in the cluster-forming region OMC-2/FIR4 475
Y. Shimajiri, S. Takahashi, S. Takakuwa, M. Saito & R. Kawabe

UCHII regions in the Antennae 476
L. Snijders, L. J. Kewley, P. P. van der Werf & B. R. Brandl

Study of photon dominated regions in IC 348 477

   K. Sun, C. Kramer, B. Mookerjea, V. Ossenkopf, M. Röllig & J. Stutzki

Dense core evolutions induced by shock triggering and turbulent dissipation. . . . 478

   K. Tachihara, A. Hayashi, T. Onishi, A. Mizuno & Y. Fukui

Survey observations of large-scale molecular outflows associated with intermediate-mass protostar candidates in the OMC-2/3 region 479

   S. Takahashi, Y. Shimajiri, S. Takakuwa, M. Saito & R. Kawabe

Starbursts in isolated galaxies: burst modes in coupled star-gas systems 480

   C. Theis & J. Köppen

Modeling the dust and gas temperatures near young stars 481

   A. Urban & N. J. Evans II

The RMS survey: radio observations of candidate massive YSOs in the southern hemisphere 482

   J. S. Urquhart, A. L. Busfield, M. G. Hoare, S. L. Lumsden, A. J. Clarke, T. J. T. Moore, J. C. Mottram & R. D. Oudmaijer

Triggered star formation within the bright-rimmed cloud SFO 75 483

   J. S. Urquhart, M. A. Thompson, L. K. Morgan & G. J. White

A turbulence study in Dwarf Irregular galaxies. 484

   M. Valdez-Gutiérrez & Ivânio Puerari

The history of star formation in the Galactic young open cluster NGC 6231. . . . 485

   M. E. van den Ancker

The generation of dense cores and substructure within them by MHD waves . . . 486

   S. Van Loo, S. A. E. G. Falle & T. W. Hartquist

Photographic variability survey in the M42 region 487

   L. P. R. Vaz, G. H. R. A. Lima & B. Reipurth

Quiescent high mass cores in Orion region. 488

   T. Velusamy, D. Li, P. F. Goldsmith & W. D. Langer

Integral field observations of distant cluster galaxies 489

   D. Vergani, C. Balkowski, H. Flores, V. Cayatte, F. Hammer, S. Mei & J. P. Blakeslee

3D dust radiative transfer simulations in the inhomogeneous interstellar medium 490

   E. Vidal Perez & M. Baes

Alfvén waves damping in protostellar disks providing a way to expand the action of the magneto-rotational instability 491

   A. A. Vidotto & V. Jatenco-Pereira

CONDOR observations of high mass star formatio in Orion 492
    N. H. Volgenau, M. C. Wiedner, G. Wieching, M. Emprechtinger, F. Bielau,
    U. U. Graf, C. E. Honingh, K. Jacobs, B. Vowinkel, R. Güsten, D. Rabanus,
    J. Stutzki & F. Wyrowski

HST ACS/HRC imaging of the intergalactic HII regions in NGC 1533 493
    J. K. Werk, M. E. Putman, G. R. Meurer, E. V. Ryan-Weber & M. S. Oey

Inferring the nature of turbulence in star-forming regions with polarimetric obser-
    vations 494
    D. Wiebe & W. D. Watson

CONDOR – A heterodyne receiver at 1.25-1.5 THz 495
    M. C. Wiedner, G. Wieching, F. Bielau, M. Emprechtinger, U. U. Graf,
    C. E. Honingh, K. Jacobs, D. Paulussen, K. Rettenbacher & N. H. Volgenau

Chandra and Spitzer observations of young clusters 496
    S. J . Wolk, B. D. Spitzbart & T. L. Bourke

HD simulations of super star cluster winds 497
    R. Wünsch, J. Palouš, G. Tenorio-Tagle & S. Silich

A sample of star forming regions triggered by cloud-cloud collision 498
    B. Xin & J.-J. Wang

Synthetic observations of turbulent flows in diffuse multiphase interstellar medium 499
    M. Yamada, H. Koyama, K. Omukai & S. Inutsuka

Large-scale CO observations of a far-infrared loop in Pegasus; detection of a large
    number of very small molecular clouds possibly formed via shocks 500
    H. Yamamoto, A. Kawamura, K. Tachihara, N. Mizuno, T. Onishi &
    Y. Fukui

Molecular loops in the Galactic centre; evidence for magnetic floatation accelerating
    molecular gas 501
    H. Yamamoto, Y. Fukui[1], M. Fujishita[1], K. Torii, N. Kudo, S. Nozawa,
    K. Takahashi, R. Matsumoto, M. Machida, A. Kawamura, Y. Yonekura,
    N. Mizuno, T. Onishi & A. Mizuno

Author index 502

Object index 507

Subject Index 509

Preface

The Interstellar Medium (ISM) is moved and compressed into clouds, filaments, and shells by the actions of stellar ionization, winds and supernovae. It becomes turbulent as a result of these motions through instabilities and gas interactions. Galactic rotation, stellar waves, and large-scale instabilities from magnetic and self-gravitational forces also drive ISM turbulence. All of this activity is in contrast to the quiescence of the dense shielded regions where individual stars form. Yet there is an important connection between the two ISM states. The compressed regions, if they are massive enough, provide seeds for self-gravity to pull the gas together, and in the cores of these seeds nearly all of the energy the gas had before, including the thermal, turbulent, and magnetic energies, dissipates, leading to collapse.

This Symposium on ISM turbulence and star formation was motivated by the recent abundance of high-quality observations of star-forming regions that have been made with new ground and space-based instruments. It was also driven by the prevalence of detailed computational work in which most of the imagined processes can be simulated. We are not yet at the point in computational astronomy where all of the important physics can be included in a simulation, but we seem to be close. We also cannot observe all we would like in star-forming regions, but with the new instruments planned for the next decade, we should soon be at that point as well. Thus it is a good time to review where we stand in our understanding of the interplay between ISM processes and star formation.

This conference took two and a half years to organize. The first discussions between the conference co-chairs took place in January of 2004 and within a few months, the Scientific Organizing Committee was in place. The SOC included scientists from a wide range of backgrounds and nationalities in order to find the most appropriate speakers, the most interesting topics, and the greatest diversity of participants from all over the world. By the end of 2004, a formal proposal was made to the IAU to have a 3.5 day conference in Prague on Triggered Star Formation in a Turbulent ISM. It was accepted in April of 2005 and a conference web site was in place by May. After several more rounds of discussion among the SOC, the first speakers were invited in July. In February 2006, we advertised the meeting in the Star Formation, Cluster (SCYON) and Dwarftales Newsletters, and broadcast the first announcement for contributed and poster papers, with invitations to apply for IAU Travel Grants. A link to the meeting website was also listed on the websites of the GA IAU and the American Astronomical Society. The deadline for early registration was May 15th, and that was also our deadline for early submission of contributed and poster papers. The final deadline for abstract submission was June 26th. This gave the SOC about two months to read all of the abstracts, assign which were appropriate for talks and which should be posters, and then finalize the speaker list.

Symposium No. 237 was part of the XXVIth General Assembly of the IAU held in August 2006 in Prague. The Symposium was attended by almost 500 participants interested in topics ranging from galaxy evolution on large scales to planetary system formation on a small scales. It is a great pleasure to acknowledge the support of the NOC and LOC of the General Assembly in Prague and to thank all of the sponsors for helping with the success of this event.

Bruce G. Elmegreen and Jan Palouš, co-chairs SOC,
December 2006

THE SCIENTIFIC ORGANIZING COMMITTEE

P. Andre (France)
B. G. Elmegreen (USA) Co-chair
P. Kroupa (Germany)
E. C. Ostriker (USA)
M. Rubio (Chile)
A. P. Whitworth (UK)
H. Zinnecker (Germany)

L. Blitz (USA)
Y. Fukui (Japan)
M. S. Oey (USA)
J. Palouš (Czech Republic) Co-chair
G. Tenorio-Tagle (Mexico)
R. E. Williams (USA) IAU ex officio

Acknowledgements

The symposium was sponsored by the IAU and coordinated by the IAU Division VI (Interstellar Matter), proposed by the IAU Commission No. 34 (Interstellar Matter), and supported by the IAU Commission No. 37 (Star Clusters and Associations).

The cooperation with National and Local Organizing Committees of the XXVIth General Assembly of the IAU is highly appreciated.

Funding by the
International Astronomical Union,
Ministry of Education, Youth and Sports of the Czech Republic,
Academy of Sciences of the Czech Republic,
Astronomical Institute of the Academy of Sciences of the Czech Republic,
and the American Astronomical Society
is gratefully acknowledged.

Participants

Pavel **Abolmasov**, Special Astrophysical Observatory, Moscow, Russian Federation   pasha@sao.ru
Helmut **Abt**, NOAO, Tucson, USA   abt@noao.edu
Mark **Adams**, NRAO, Charlottesville, USA   mtadams@nrao.edu
Alfonso **Aguerri**, Instituto de Astrofisica de Canarias, La Laguna, Spain   jalfonso@iac.es
Andrea **Ahumada**, Obs. Astronomomico, Univ. Nac. Cordoba, Cordoba, Argentina   andrea@oac.uncor.edu
Juan Facundo **Albacete Colombo**, Osservatorio Astronomico di Palermo, Palermo, Italy   facundo@astropa.unipa.it
Mohsen **Alizadeh**, Institute for Astrophysics, Goettingen, Germany   alizadeh@astro.physik.uni-goettingen.de
Christine **Allen**, Instituto de Astronomia, UNAM, Mexico City, Mexico   chris@astroscu.unam.mx
Joao **Alves**, Calar Alto Observatory - Centro Astronomico Hispano Aleman, Almeria, Spain   jalves@caha.es
Eduardo **Amores**, IAG/Universitat de Barcelona/IEEC, Barcelona, Spain   amores@astro.iag.usp.br
Ricardo O. **Amorín**, Instituto Astrofisico de Canarias, La Laguna, Spain   ricardo.amorin@iac.es
Philippe **André**, CEA Saclay, Gif-sur-Yvette, France   pandre@cea.fr
Ruben **Andreasyan**, Byurakan Astrophysical Observatory, Byurakan, Armenia   randrasy@bao.sci.am
Díaz **Angeles I.**, Universidad Autónoma de Madrid, Madrid, Spain   angeles.diaz@uam.es
Francesca **Annibali**, Space Telescope Science Institute, Baltimore, USA   annibali@stsci.edu
Christopher **Añorve**, INAOE, Tonantzintla, Mexico   canorve@inaoep.mx
Bojan **Arbutina**, Astronomical Observatory, Belgrade, Serbia   barbutina@aob.bg.ac.yu
Joana **Ascenso**, Centro de Astrofísica da Universidade do Porto, Porto, Portugal   joanasba@astro.up.pt
Colin **Aspin**, Gemini Observatory, Hilo, USA   caa@gemini.edu
Olga **Atanackovic-Vukmanovic**, Faculty of Mathematics, University of Belgrade, Belgrade, Serbia   olga@matf.bg.ac.yu
Kedar **Badu**, Galileo Astronomical Society of Pokhara, Pokhara, Nepal   kedarbadu@yahoo.com
Chang Hyun **Baek**, ARCSEC, Sejong University, Seoul, Republic of Korea   chbaek@kasi.re.kr
Maarten **Baes**, Universiteit Gent, Gent, Belgium   maarten.baes@ugent.be
Ioannis **Bagetakos**, University of Hertfordshire, Hatfield, UK   ioannis@star.herts.ac.uk
Rafael **Bachiller**, Observatorio Astronomico Nacional, Madrid, Spain   r.bachiller@oan.es
Lajos György **Balazs**, Konkoly Observatory, Hungary   balazs@konkoly.hu
Jean **Ballet**, Service d'Astrophysique, Gif sur Yvette Cedex, France   jballet@cea.fr
John **Bally**, University of Colorado, Boulder, USA   bally@casa.colorado.edu
Cassio **Barbosa**, UNIVAP, Sao Jose dos Campos, Brazil   cassio@univap.br
Graham **Barnes**, NorthWest Research Associates, Boulder, USA   graham@cora.nwra.com
Mary **Barsony**, Space Science Institute and San Francisco State University, San Francisco, USA   mbarsony@stars.sfsu.edu
Anna **Bartkiewicz**, Torun Centre for Astronomy, Torun, Poland   annan@astro.uni.torun.pl
Pierre **Bastien**, Université de Montréal, Montreal, Canada   bastien@astro.umontreal.ca
Manuel **Bautista**, IVIC, Caracas, Venezuela   bautista@kant.ivic.ve
Kenji **Bekki**, University of New South Wales, Sydney, Australia   bekki@bat.phys.unsw.edu.au
Arnold **Benz**, ETH Zurich, Zurich, Switzerland   benz@astro.phys.ethz.ch
Nils **Bergvall**, Uppsala Astronomical Observatory, Uppsala, Sweden   nils.bergvall@astro.uu.se
Henrik **Beuther**, Max-Planck-Institute for Astronomy, Heidelberg, Germany   beuther@mpia.de
John **Bieging**, University of Arizona, Tucson, USA   jbieging@as.arizona.edu
Frank **Bigiel**, MPI for Astronomy, Heidelberg, Germany   bigiel@mpia.de
Hans **Bloemen**, SRON, Utrecht, Netherlands   H.Bloemen@sron.nl
Nikolai **Bochkarev**, Sternberg Astron.Inst., Moscow, Russian Federation   boch@sai.msu.ru
Samuel **Boissier**, Laboratoire d'Astrophysique de Marseille, Marseille Cedex 12, France   samuel.boissier@gmail.com
Wilfried **Boland**, NOVA, Leiden, Netherlands   boland@strw.leidenuniv.nl
Tom **Boles**, Coddenham Observatory, Coddenham, UK   tomboles@coddenhamobservatories.org
Ramaprosad **Bondyopadhaya**, Jadavpur University, Kolkata, India   rpb_jumath@email.com
Ian **Bonnell**, University of St Andrews, St Andrews, UK   iab1@st-and.ac.uk
Roy **Booth**, Hartebeesthoek Radio Astronomy Observatory, Johannesburg, South Africa   roy@oso.chalmers.se
Médéric **Boquien**, CEA, Gif-sur-Yvette cedex, France   mboquien@discovery.saclay.cea.fr
François **Boulanger**, IAS, Orsay, France   francois.boulanger@ias.u-psud.fr
Bernhard **Brandl**, Leiden Observatory, Leiden, Netherlands   brandl@strw.leidenuniv.nl
Hector **Bravo-Alfaro**, Universidad de Guanajuato, Guanajuato, Mexico   hector@astro.ugto.mx
Dieter **Breitschwerdt**, Univerity of Vienna, Wien, Austria   breitschwerdt@astro.univie.ac.at
Elias **Brinks**, University of Hertfordshire, Hatfield, UK   ebrinks@star.herts.ac.uk
Noah **Brosch**, Tel Aviv University, Tel Aviv, Israel   noah@wise.tau.ac.il
John C **Brown**, University of Glasgow, Glasgow, UK   john@astro.gla.ac.uk
Jane **Buckle**, University of Cambridge, Cambridge, UK   j.buckle@mrao.cam.ac.uk
Martin **Bureau**, University of Oxford, Oxford, UK   bureau@astro.ox.ac.uk
Michael **Burton**, University of New South Wales, Sydney, Australia   m.burton@unsw.edu.au
Michael **Cai**, Academia Sinica, Taipei, China Taipei   mike@asiaa.sinica.edu.tw
Brady **Caldwell**, Uppsala Astronomical Observatory, Uppsala, Sweden   brady.caldwell@astro.uu.se
Loretta **Campeggio**, University of Lecce, Lecce, Italy   campeggio@le.infn.it
Matteo **Cantiello**, Utrecht University, Utrecht, Netherlands   M.Cantiello@astro.uu.nl
Cristina Elisabet **Cappa**, Instituto Argentino de Radioastronomia, U.N.La Plata, La Plata, Argentina   ccappa@fcaglp.unlp.edu.ar
Mónica **Cardaci**, Universidad Autónoma de Madrid, Ciudad Universitaria de Cantoblanco, Spain   monica.cardaci@uam.es
Patrick **Carolan**, N.U.I. Galway, Mayo, Ireland   patrick.carolan@nuigalway.ie
Luis **Carrasco**, INAOE, Puebla, Mexico   carrasco@inaoep.mx
James **Caswell**, CSIRO, ATNF, Epping, Australia   James.Caswell@csiro.au
Huseyin **Cavus**, Canakkale Onsekiz Mart Univ, Canakkale, Turkey   h_cavus@comu.edu.tr
Michal **Ceniga**, Faculty of Science of Masaryk University, Brno, Czech Republic   emceniga@physics.muni.cz
William **Cepeda**, National Astronomical Observatory of Colombia, Bogota D.C., Colombia   wecepedap@unal.edu.co
Catherine **Cesarsky**, ESO, Garching, Germany   ccesarsk@eso.org
Diego **Cesarsky**, MPE, Garching, Germany   diego.cesarsky@mpe.mpg.de
Nicolas **Champavert**, CRAL - Observatoire de Lyon, Saint Genis Laval, France   nicolas.champavert@obs.univ-lyon1.fr
Hum **Chand**, Inter University Centre for Astronomy and Astrophysics, Pune, India   hcverma@iucaa.ernet.in
Thyagarajan **Chandrasekhar**, Physical Research Laboratory, Ahmedabad, India   chandra@prl.res.in
Jacqueline **Chapman**, Macquarie University, Sydney, Australia   jchapman@physics.mq.edu.au
Richard **Chappelle**, Astronomical Institute, AS CR, Praha, Czech Republic   rjc@astro.cas.cz
George **Chartas**, Penn State University, University Park, USA   chartas@astro.psu.edu
Tanuka **Chattopadhyay**, Shibpur D. B. College, Calcutta, India   tanuka@iucaa.ernet.in
Luis **Chavarria**, CfA, Cambridge, USA   lchavarr@cfa.harvard.edu
Wen-Ping **Chen**, National Central University, Chung-Li, China Taipei   wchen@astro.ncu.edu.tw

Chang-Hui Rosie **Chen**, University of Illinois at Urbana-Champaign, Urbana, USA — c-chen@astro.uiuc.edu

Roger **Chevalier**, University of Virginia, Charlottesville, USA — rac5x@virginia.edu

Po-Shih **Chiang**, Graduate Institute of Astronomy, NCU, Taiwan, Jhongli City, China Taipei — pschiang@astro.ncu.edu.tw

Yunhee **Choi**, Kyunghee University, Yongin-si, Republic of Korea — yhchoi@ap1.khu.ac.kr

Jørgen **Christensen-Dalsgaard**, University of Aarhus, Aarhus C, Denmark — jcd@phys.au.dk

You-Hua **Chu**, University of Illinois at Urbana-Champaign, Urbana, USA — chu@astro.uiuc.edu

HyunSoo **Chung**, Korea Astronomy and Space Science Institute, Daejeon, Republic of Korea — hschung@kasi.re.kr

Mark **Clampin**, GSFC, Greenbelt, USA — mark.clampin@nasa.gov

Paul **Clark**, University of Heidelberg, Heidelberg, Germany — pcc@st-andrews.ac.uk

Cathie **Clarke**, Institute of Astronomy, Cambridge, UK — cclarke@ast.cam.ac.uk

Maurice **Clement**, University of Toronto, Toronto, Canada — mclement@astro.utoronto.ca

Francoise **Combes**, Observatoire de Paris, Paris, France — francoise.combes@obspm.fr

Claudio **Cremaschini**, University of Trieste / 565 Bassano Bresciano Obs., Pompiano (BS), Italy — claudiocremaschini@gmail.com

Mark **Cropper**, MSSL/UCL, Dorking, UK — msc@mssl.ucl.ac.uk

Richard **Crutcher**, University of Illinois, Urbana, USA — crutcher@uiuc.edu

Attila **Cséki**, Faculty of Mathematics, University of Belgrade, Pancevo, Serbia — astrofizicar@gmail.com

Jorge **Cuadra**, Max-Planck-Institut für Astrophysik, Garching, Germany — jcuadra@mpa-garching.mpg.de

François **Cuisinier**, UFRJ, Rio de Janeiro, Brazil — francois@ov.ufrj.br

Katia **Cunha**, National Optical Astronomy Observatory, Tucson, Brazil — kcunha@noao.edu

Charles **Cunningham**, HIA, Victoira, Canada — charles.cunningham@nrc-cnrc.gc.ca

Andrew **Cunningham**, University of Rochester, Rochester, USA — ajc4@pas.rochester.edu

Maria **Cunningham**, University of New South Wales, Sydney, Australia — maria.cunningham@unsw.edu.au

Jörg **Dabringhausen**, Universität Bonn, Bonn, Germany — joedab@astro.uni-bonn.de

Alexander **Dalgarno**, Harvard-Smithsonian Center for astrophysics, Cambridge, USA — adalgarno@cfa.harvard.edu

Mousumi **Das**, Raman Research Institute, Bangalore, India — mousumi@rri.res.in

Srabani **Datta**, University of Manchester, Manchester, UK — Srabani.Datta@manchester.ac.uk

Joanne **Dawson**, Nagoya University, Nagoya, Japan — joanne@a.phys.nagoya-u.ac.jp

Francisco Xavier **De Araujo**, Observatorio Nacional, Rio de Janeiro, Brazil — araujo@on.br

Miguel **de Avillez**, University of Evora, Evora, Portugal — mavillez@galaxy.lca.uevora.pt

James **De Buizer**, Gemini Observatory, Tucson, USA — jdebuizer@gemini.edu

Elisabete **de Gouveia Dal Pino**, Instituto Astronômico e Geofísico, USP, São Paulo, Brazil — dalpino@astro.iag.usp.br

Richard **de Grijs**, University of Sheffield, Sheffield, UK — R.deGrijs@sheffield.ac.uk

Ana Beatriz **De Mello** / MCT, Rio de Janeiro, Brazil — demello@on.br

Christopher **De Vries**, Cal State University, Stanislaus, Turlock, USA — chris@physics.csustan.edu

Véronique **Dehant**, Royal Observatory of Belgium, Brussels, Belgium — veronique.dehant@oma.be

Lise **Deharveng**, Laboratoire d'Astrophysique de Marseille, Marseille Cedex 4, France — lise.deharveng@oamp.fr

Vasily **Demichev**, Space Research Institute, Moscow, Russian Federation — demichev@iki.rssi.ru

Ralf-Juergen **Dettmar**, Ruhr-University Bochum, BOCHUM, Germany — dettmar@astro.rub.de

Kathryn **Devine**, University of Wisconsin, Madison / NRAO, Socorro, USA — devine@astro.wisc.edu

Philip **Diamond**, Jodrell Bank Observatory, Macclesfield, UK — pdiamond@jb.man.ac.uk

Sami **Dib**, Centro de Radioastronomia y Astrofisica, Morelia, Mexico — dib@astrosmo.unam.mx

John **Dickel**, Univ. of New Mexico, Albuquerque, USA — johnd@phys.unm.edu

John **Dickey**, University of Tasmania, Hobart, Australia — john.dickey@utas.edu.au

Mark Eric **Dieckmann**, Ruhr Universität Bochum, Bochum, Germany — markd@tp4.rub.de

Clare **Dobbs**, St Andrews, UK — cld2@st-and.ac.uk

J. Richard **Donnison**, Queen Mary, University of London, London, UK — r.donnison@qmul.ac.uk

Michael **Dopita**, RSAA, Australian Nat. U., Weston Creek, Australia — Michael.Dopita@anu.edu.au

Horacio **Dottori**, Universidade Federal do Rio Grande do Sul, Porto Alegre, Brazil — dottori.voy@terra.com.br

Natalia **Drake**, Sobolev Astronomical Institute, St. Petersburg State University, St. Petersburg, Russian Federation — drake@on.br

Gloria **Dubner**, Institute of Astronomy and Space Physics (IAFE), Buenos Aires, Argentina — gdubner@iafe.uba.ar

Pierre-Alain **Duc**, CNRS AIM, CEA-Saclay, Gif sur Yvette cedex, France — paduc@cea.fr

Anne **Dutrey**, Observatoire de Bordeaux, Floirac, France — dutrey@obs.u-bordeaux1.fr

John **Dyson**, University of Leeds, Leeds, UK — jed@ast.leeds.ac.uk

Fumi **Egusa**, the Institute of Astronomy, the University of Tokyo, Mitaka, Japan — fegusa@ioa.s.u-tokyo.ac.jp

Soňa **Ehlerová**, Astronomical Institute, AS CR, Praha, Czech Republic — sona@ig.cas.cz

Davide **Elia**, University of Lecce, Lecce, Italy — eliad@le.infn.it

Bruce **Elmegreen**, IBM Research Division, Yorktown Heights, USA — bge@watson.ibm.com

Debra **Elmegreen**, Vassar College, Poughkeepsie, USA — elmegreen@vassar.edu

Eric **Emsellem**, Centre de Recherche Astronomique de Lyon, Saint Genis Laval, France — emsellem@obs.univ-lyon1.fr

Daniel **Epitácio Pereira**, Observatorio Nacional / MCT, Rio de Janeiro, Brazil — dnep@on.br

Andreas **Ernst**, Astronomisches Rechen-Institut, Heidelberg, Germany — aernst@ari.uni-heidelberg.de

Jarken **Esimbek**, National Astronomical Observatories, ChineseAcademy of Sciences, Urumqi, China Nanjing — jarken@ms.xjb.ac.cn

Stewart **Eyres**, University of Central Lancashire, Preston, UK — spseyres@uclan.ac.uk

Romanus **Eze**, University of Nigeria, Nsukka, Enugu State, Nigeria., Enugu, Nigeria — rnceze@yahoo.com

Diego **Falceta-Gonçalves**, IAG/USP, Sao Paulo, Brazil — diego@astro.iag.usp.br

Edith **Falgarone**, ENS & Paris Observatory, PARIS, France — edith@lra.ens.fr

S. Michael **Fall**, STScI, Baltimore, USA — fall@stsci.edu

Davide **Fedele**, ESO & Universita' di Padova, Garching, Germany — dfedele@eso.org

Steven **Federman**, University of Toledo, Toledo, USA — steven.federman@utoledo.edu

Marcello **Felli**, Osservatorio di Arcetri, Firenze, Italy — felli@arcetri.astro.it

Roger **Ferlet**, CNRS, Paris, France — ferlet@iap.fr

Jullio Angel **Fernández**, Facultad de Ciencias, Montevideo, Uruguay — julio@fisica.edu.uy

Elysandra **Figueredo**, IAG-USP, London, UK — lys.figueredo@gmail.com

Miroslav **Filipovic**, University of Western Sydney, Glenmore Park, Australia — m.filipovic@uws.edu.au

Friedrich **Firneis**, Austrian Acad. of Sc., Vienna, Austria — Friedrich.Firneis@oeaw.ac.at

Robert Scott **Fisher**, Gemini Observatory, Hilo, USA — sfisher@gemini.edu

Bernard **Foing**, ESA, Noordwijk, Netherlands — Bernard.Foing@esa.int

Jose **Funes**, Vatican Observatory, Tucson, USA — jfunes@as.arizona.edu

Terrance **Gaetz**, Smithsonian Astrophysical Observatory, Cambridge, USA — gaetz@cfa.harvard.edu

Gazinur **Galazutdinov**, Korea Astronomy and Space Science Institute, Daejeon, Republic of Korea — gala@boao.re.kr

Yu **Gao**, Purple Mountain Observatory, Nanjing, China Nanjing — yugao@pmo.ac.cn

Beatriz **Garcia**, National Technologcal University, Mendoza, Argentina — beatrizgarciautn@gmail.com

Ralph **Gaume**, US Naval Observatory, Washington, USA — rgaume@usno.navy.mil

Elena **Gavryuseva**, Arcetri Astrophysical Observatory, Florence, Italy — elena.gavryuseva@gmail.com

Elsa **Giacani**, Instituto de Astronomia y Fisica del Espacio, Buenos Aires, Argentina — egiacani@iafe.uba.ar

Steven **Gibson**, Cornell University, Arecibo, USA — gibson@naic.edu

Mark **Gieles**, Utrecht University, Utrecht, Netherlands — gieles@astro.uu.nl
Isabella M. **Gioia**, INAF, Bologna, Italy — gioia@ira.inaf.it
Sunetra **Giridhar**, Indian Institute of Astrophysics, Bangalore, India — giridhar@iiap.res.in
Itzhak **Goldman**, Tel Aviv University, Tel Aviv, Israel — goldman@wise.tau.ac.il
Daniel **Gomez**, IAFE, Buenos Aires, Argentina — dgomez@df.uba.ar
Dimitrios **Gouliermis**, Max-Planck-Institut fuer Astronomie, Heidelberg, Germany — dgoulier@mpia.de
Anahí **Granada**, National University of La Plata, La Plata, Argentina — granada@fcaglp.unlp.edu.ar
Anne **Green**, University of Sydney, University of Sydney, Australia — agreen@physics.usyd.edu.au
David **Green**, Cavendish Laboratory, Cambridge, UK — dag@mrao.cam.ac.uk
Jane **Gregorio-Hetem**, University of Sao Paulo, Sao Paulo, Brazil — jane@astro.iag.usp.br
Oliver **Gressel**, Astrophysikalisches Institut Potsdam, Potsdam, Germany — ogressel@aip.de
Fabien **Grise**, Observatoire astronomique de Strasbourg, Strasbourg, France — grise@astro.u-strasbg.fr
Matthias **Gritschneder**, USM Munich, Munich, Germany — gritschm@usm.uni-muenchen.de
Preben **Grosbol**, European Southern Observatory, Garching, Germany — pgrosbol@eso.org
Jose **Guichard**, INAOE, Tonantzintla, Mexico — jguich@inaoep.mx
Pierre **Guillard**, IAS, Orsay, France, Bures sur Yvette, France — pierre.guillard@ias.u-psud.fr
Sergei **Gulyaev**, Auckland University of Technology, City, New Zealand — sergei.gulyaev@aut.ac.nz
Bo **Gustafson**, University of Florida, Gainesville, USA — gustaf@astro.ufl.edu
Maiken **Gustafsson**, University of Aarhus, Aarhus C, Denmark — maikeng@phys.au.dk
Vasilii **Gvaramadze**, Sternberg Astronomical Institute, Moscow, Russian Federation — vgvaram@sai.msu.ru
Guillermo **Hägele**, Universidad Autónoma de Madrid, Ciudad Universitaria de Cantoblanco, Spain — guille.hagele@uam.es
Lauri **Haikala**, University of Helsinki, Helsinki, Finland — haikala@astro.helsinki.fi
Ashot **Hakobyan**, Byurakan Astrophysical Observatory, Byurakan, Armenia — aakopian@bao.sci.am
Kenji **Hamaguchi**, NASA/GSFC, Greenbelt, USA — kenji@milkyway.gsfc.nasa.gov
Michal **Hanasz**, Nicolaus Copernicus University, Torun, Poland — mhanasz@astri.uni.torun.pl
Toshihiro **Handa**, University of Tokyo, Mitaka, Japan — handa@ioa.s.u-tokyo.ac.jp
JingFang **Hao**, Tsinghua University, Beijing, China Nanjing — jessica-hjf@163.com
Lisa **Harvey-Smith**, Joint Institute for VLBI in Europe, Dwingeloo, Netherlands — harvey@jive.nl
Hashima **Hasan**, NASA Headquarters, Washington, USA — hhasan@nasa.gov
Saiyid Sirajul **Hasan**, Indian Institute of Astrophysics, Bangalore, India — diriia@iiap.res.in
Yasuhiro **Hashimoto**, SAAO, Cape Town, South Africa — hashimot@saao.ac.za
Hirofumi **Hatano**, Nagoya University, Nagoya, Japan — hattan@z.phys.nagoya-u.ac.jp
Christian **Henkel**, MPIfR, Bonn, Germany — chenkel@mpifr-bonn.mpg.de
Patrick **Hennebelle**, Observatoire de Paris, Paris, France — patrick.hennebelle@ens.fr
Martin **Hennemann**, Max Planck Institute for Astronomy, Heidelberg, Germany — hennemann@mpia.de
Gerhard **Hensler**, University of Vienna, Vienna, Austria — hensler@astro.univie.ac.at
Jonathan **Hernández-Fernández**, Instituto de Astrofisica de Andalucia-CSIC, Granada, Spain — jonatan@iaa.es
Annibal **Hetem**, Fundacao Santo Andre, Sao Paulo, Brazil — annibal.hetem.jr@usa.net
Mark **Heyer**, University of Massachusetts, Amherst, USA — heyer@astro.umass.edu
Michael **Hilker**, AIfA, Bonn, Germany — mhilker@astro.uni-bonn.de
Tomas **Hillberg**, Solar Physics - Royal Academy, Stockholm, Sweden — hillberg@astro.su.se
Lynne **Hillenbrand**, California Institute of Technology, Pasadena, USA — lah@astro.caltech.edu
Pierre **Hily-Blant**, IRAM, Saint Martin D'Heres, France — hilyblan@iram.fr
Masaaki **Hiramatsu**, The University of Tokyo, Tokyo, Japan — hiramatsu.masaaki@nao.ac.jp
Marc **Hitschfeld**, University of Cologne, Köln, Germany — hitschfeld@ph1.uni-koeln.de
Luis **Ho**, Carnegie Observatories, Pasadena, USA — lho@ociw.edu
Klaus **Hodapp**, University of Hawaii, Hilo, USA — hodapp@ifa.hawaii.edu
Alisher S. **Hojaev**, UBAI, CfSR, UAS, Tashkent, Uzbekistan — ash@astrin.uzsci.net
Takashi **Hosokawa**, NAOJ, Mitaka, Japan — hosokawa@th.nao.ac.jp
Leo **Houziaux**, Academie Royale de Belgium, Bruxelles, Belgium — peckova@cbttravel.cz
Jian **Hu**, Tsinghua University, Beijing, China Nanjing — hujian98@mails.tsinghua.edu.cn
Juei-Hwa **Hu**, Institute of Astronomy, Jhongli City, Taoyuan County, China Taipei — d939003@astro.ncu.edu.tw
Hui-Chun **Huang**, Nat'l Taiwan Normal Univ, Taipei, China Taipei — hspring@alioth.geos.ntnu.edu.tw
Rene **Hudec**, Astronomical Institute, AS CR, Ondřejov, Czech Republic — rhudec@asu.cas.cz
Song **In-Ok**, ARCSEC, Sejong University, Seoul, Republic of Korea — songio@sejong.ac.kr
Rosina **Iping**, Catholic University of America, Washington, USA — iping@milkyway.gsfc.nasa.gov
William **Irvine**, University of Massachusetts, Amherst, USA — irvine@fcrao1.astro.umass.edu
Meguru **Ito**, University of Tokyo, Hilo, USA — itomg@subaru.naoj.org
Pavel **Jáchym**, Astronomical Institute, AS CR, Praha, Czech Republic — jachym@ig.cas.cz
Anne-Katharina **Jappsen**, Astrophysikalisches Institut Potsdam, Potsdam, Germany — akjappsen@aip.de
Vera **Jatenco-Pereira**, University of São Paulo, São Paulo, Brazil — jatenco@astro.iag.usp.br
Atefeh **Javadi**, Alzahra University, Tehran, Islamic Republic of Iran — atefehj@gmail.com
Edward **Jenkins**, Princeton University, Princeton, USA — ebj@astro.princeton.edu
Philip **Jewell**, National Radio Astronomy Observatory, Charlottesville, USA — pjewell@nrao.edu
Chanda **Jog**, Indian Institute of Science, Bangalore, India, Bangalore, India — cjjog@physics.iisc.ernet.in
Katharine **Johnston**, University of St Andrews, St Andrews, UK — kgj1@st-andrews.ac.uk
Justin **Jonas**, Rhodes University, Grahamstown, South Africa — J.Jonas@ru.ac.za
Paul **Jones**, Waterloo, Australia — Paul.Jones@csiro.au
M. K. Ryan **Joung**, Columbia University, New York, USA — moo@astro.columbia.edu
Mika **Juvela**, Helsinki University Observatory, Helsinki, Finland — mika.juvela@helsinki.fi
Norio **Kaifu**, NAOJ (National Astronomical Observatory of Japan), Hachioji, Japan — Kaifunorio@aol.com
Nadia **Kaltcheva**, University of Wisconsin Oshkosh, Oshkosh, USA — kaltchev@uwosh.edu
Kazuhisa **Kamegai**, Institute of Astronomy, Univ. of Tokyo, Tokyo, Japan — kamegai@ioa.s.u-tokyo.ac.jp
Hidehiro **Kaneda**, ISAS/JAXA, Sagamihara, Japan — kaneda@ir.isas.jaxa.jp
Hyesung **Kang**, Pusan National University, Pusan, Republic of Korea — hskang@pusan.ac.kr
Namir **Kassim**, Naval Research Laboratory, Washington, USA — namir.kassim@nrl.navy.mil
Seiichi **Kato**, Hyogo College of Medicine, Nishinomiya, Japan — katosi@hyo-med.ac.jp
Akiko **Kawamura**, Nagoya University, Nagoya, Japan — kawamura@a.phys.nagoya-u.ac.jp
Simon **Kemp**, Universidad de Guadalajara, Guadalajara, Mexico — snk@astro.iam.udg.mx
Robert **Kennicutt**, University of Cambridge, Cambridge, UK — robk@ast.cam.ac.uk
Ali Reza **Khesali**, Mazandaran Univ., Babolsar, Islamic Republic of Iran — khesali@umz.ac.ir
Woong-Tae **Kim**, Seoul National University, Seoul, Republic of Korea — wkim@astro.snu.ac.kr
Sungsoo **Kim**, Kyung Hee University, Yongin, Republic of Korea — sungsoo.kim@khu.ac.kr
Sungeun **Kim**, Sejong University, Seoul, Republic of Korea — sek@sejong.ac.kr
Maria **Kirsanova**, Institute of Astronomy RAS, Moscow, Russian Federation — kirsanova@inasan.ru
Zoltán T. **Kiss**, Eötvös Loránd University, Budapest, Hungary — Z.Kiss@astro.elte.hu
Spyridon **Kitsionas**, Astrophysikalisches Institut Potsdam, Potsdam, Germany — skitsionas@aip.de
Patricia **Knezek**, WIYN Observatory, Tucson, USA — knezek@noao.edu
Kirsten Kraiberg **Knudsen**, Max-Planck-Institut für Astronomie, Heidelberg, Germany — knudsen@mpia-hd.mpg.de
Chung-Ming **Ko**, Institute of Astronomy, National Central University, Taiwan, Jhongli City, China Taipei — cmko@astro.ncu.edu.tw

Kotaro **Kohno**, The University of Tokyo, Mitaka, Japan · kkohno@ioa.s.u-tokyo.ac.jp
LuboÅ **Kohoutek**, Hamburg Observatory, Hamburg, Germany · lkohoutek@hs.uni-hamburg.de
Shinya **Komugi**, University of Tokyo, Mitaka, Japan · skomugi@ioa.s.u-tokyo.ac.jp
Mary **Kontizas**, National & Kapodistrian University of Athens, Athens, Greece · mkontiza@phys.uoa.gr
Daniela **Korčáková**, Astronomical Institute AS CR, Ondřejov, Czech Republic · kor@sunstel.asu.cas.cz
Maarit **Korpi**, Observatory, Helsinki, Finland · Maarit.Korpi@helsinki.fi
Thijs **Kouwenhoven**, University of Amsterdam, Amsterdam, Netherlands · kouwenho@science.uva.nl
Attay **Kovetz**, Tel Aviv University, Tel Aviv, Israel · attay@etoile.tau.ac.il
Busaba **Kramer**, National Astronomical Research Institute, Chiang Mai, Thailand · busaba@nari.or.th
Marko **Krco**, Cornell University, Ithaca, USA · marko@astro.cornell.edu
Pavel **Kroupa**, Argelander Institute for Astronomy, Bonn, Germany · pavel@astro.uni-bonn.de
Mark **Krumholz**, Princeton University, Princeton, USA · krumholz@astro.princeton.edu
Yi-Jehng **Kuan**, National Taiwan Normal University, Taipei, China Taipei · kuan@alioth.geos.ntnu.edu.tw
Takahiro **Kudoh**, National Astronomical Observatory of Japan, Mitaka, Japan · kudoh@th.nao.ac.jp
Andreas **Kuepper**, University of Bonn, Bonn, Germany · akuepper@uni-bonn.de
Maria **Kun**, Konkoly Observatory, Budapest, Hungary · kun@konkoly.hu
Nario **Kuno**, National Astronomical Observatory of Japan, Nagano, Japan · kuno@nro.nao.ac.jp
Arlo **Landolt**, Louisiana State University, Baton Rouge, USA · landolt@rouge.phys.lsu.edu
Mathieu **Langer**, Institut d'Astrophysique Spatiale, Orsay cedex, France · mathieu.langer@ias.u-psud.fr
Ted **LaRosa**, Kennesaw State Univ, Kennesaw, USA · ted@avatar.kennesaw.edu
Olivera **Latkovič**, Astronomical Observatory Belgrade, Belgrade, Serbia · olatkovic@aob.bg.ac.yu
Myung Gyoon **Lee**, Seoul National University, Seoul, Republic of Korea · mglee@astrog.snu.ac.kr
Chang Won **Lee**, Korea Astronomy and Space Science Institute, Daejeon, Republic of Korea · cwl@kasi.re.kr
Dae-Hee **Lee**, Korea Astronomy and Space Science Institute, Daejoen, Republic of Korea · dhlee@kasi.re.kr
Youngung **Lee**, Korea Astronomy and Space Science Institute, Daejeon, Republic of Korea · yulee@kasi.re.kr
Jeewon **Lee**, Kyunghee University, Khung ki do, Republic of Korea · jwlee@ap1.khu.ac.kr
Elisabet **Leitet**, Uppsala Astronomical Observatory, Uppsala, Sweden · elisabet.leitet@astro.uu.se
Claus **Leitherer**, Space Telescope Science Institute, Baltimore, USA · leitherer@stsci.edu
Jacques **Lepine**, University of Sao Paulo, Sao Paulo, Brazil · jacques@astro.iag.usp.br
Jinzeng **Li**, National Astronomical Observatories, Chinese Academy of Sciences, Beijing, China Nanjing · ljz@bao.ac.cn
Hendrik **Linz**, MPIA Heidelberg, Heidelberg, Germany · linz@mpia-hd.mpg.de
Andre **Lipand**, Tartu University, Tartu, Estonia · sicroft@ut.ee
Ute **Lisenfeld**, Universidad Granada, Granada, Spain · ute@ugr.es
Zhiyong **Liu**, NAO-CAS, Urumqi, China Nanjing · liuzy@ms.xjb.ac.cn
Man Kit **Lo**, The University of Hong Kong, Hong Kong, China Nanjing · h0492064@hkusua.hku.hk
Alex **Lobel**, Royal Observatory of Belgium, Brussels, Belgium · alobel@sdf.lonestar.org
Felix **Lockman**, NRAO, Green Bank, USA · jlockman@nrao.edu
Mireille **Louys**, Observatoire de Strasbourg, Strasbourg, France · louys@astro.u-strasbg.fr
Ray **Lucas**, Space Telescope Science Institute, Baltimore, USA · lucas@stsci.edu
A-Ran **Lyo**, ASIAA, Taipei, China Taipei · arl@asiaa.sinica.edu.tw
Mordecai-Mark **Mac Low**, American Museum of Natural History, New York, USA · mordecai@amnh.org
Walter **Maciel**, University of Sao Paulo, Sao Paulo, Brazil · maciel@astro.iag.usp.br
Greg **Madsen**, Anglo-Australian Observatory, Eastwood, Australia · madsen@aao.gov.au
Jean-Pierre **Maillard**, Institut d'Astrophysique de Paris, Paris, France · maillard@iap.fr
Berlinda **Maiolo**, University of Lecce, Lecce, Italy · berlinda.maiolo@le.infn.it
Valery **Malofeev**, Lebedev Physical Institute, Pushchino, Russian Federation · malofeev@prao.psn.ru
Eric **Mamajek**, Harvard-Smithsonian Center for Astrophysics, Cambridge, USA · emamajek@cfa.harvard.edu
Richard **Manchester**, CSIRO, ATNF, Marsfield, Australia · Dick.Manchester@csiro.au
Gábor **Marschalkó**, Eötvös University, Budapest, Hungary · g.marschalko@astro.elte.hu
Crystal **Martin**, Univ. of California, Santa Barbara, USA · cmartin@physics.ucsb.edu
Christopher **Martin**, Oberlin College, Oberlin, USA · Chris.Martin@oberlin.edu
Eric Emmanuel **Martinez Garcia**, UNAM, Morelia, Mexico · e.martinez@astrosmo.unam.mx
Eder **Martioli**, INPE, Sao Jose dos Campos, Brazil · eder@das.inpe.br
Kevin **Marvel**, American Astronomical Society, Washington, USA · marvel@aas.org
Thomas **Maschberger**, Universität Bonn, Bonn, Germany · tmasch@astro.uni-bonn.de
Kalevi **Mattila**, Observatory, Helsinki, Finland · mattila@cc.helsinki.fi
Leonid **Matveyenko**, Space Research Institute, Moscow, Russian Federation · matveen@iki.rssi.ru
William **McCutcheon**, University of British Columbia, Vancouver, Canada · mccutche@phas.ubc.ca
David **McDavid**, University of Virginia, Charlottesville, USA · dam3ma@virginia.edu
Peter **McGregor**, Australian National University, Weston, Australia · Peter.McGregor@anu.edu.au
Ian **McLean**, UCLA, Los Angeles, USA · mclean@astro.ucla.edu
Jorge **Melnick**, ESO, Santiago, Chile · jmelnick@eso.org
Don **Melrose**, University of Sydney, Sydney, Australia · melrose@physics.usyd.edu.au
Karl **Menten**, Max-Planck-Institut fuer Radioastronomie, Bonn, Germany · kmenten@mpifr-bonn.mpg.de
Martin **Meyer**, STScI, Baltimore, USA · martinm@stsci.edu
Giuseppina **Micela**, INAF Osservatorio Astr. Palermo, Palermo, Italy · giusi@astropa.unipa.it
Young **Minh**, Korea Astronomy and Space Science Institute, Daejeon, Republic of Korea · minh@kasi.re.kr
Vincent **Minier**, CEA Saclay, Gif-sur-Yvette, France · vincent.minier@cea.fr
George **Mitchell**, Saint Mary's University, Halifax, Canada · gmitchell@ap.stmarys.ca
Marc-Antoine **Miville-Deschenes**, Institut d'Astrophysique Spatiale, Orsay, France · mamd@ias.u-psud.fr
Norikazu **Mizuno**, Nagoya University, Nagoya, Japan · norikazu@a.phys.nagoya-u.ac.jp
Anthony **Moffat**, Université de Montréal, Montréal, Canada · moffat@astro.umontreal.ca
Francisco Javier **Moldon**, Universitat de Barcelona, Barcelona, Spain · jmoldon@am.ub.es
Larry **Morgan**, NRAO, Green Bank, USA · lmorgan@nrao.edu
Luca **Moscadelli**, Osservatorio Astrofisico di Arcetri, Firenze, Italy · mosca@ca.astro.it
Kazutaka **Motoyama**, Theoretical Institute for Advanced Research in Astrophysics, Hsin-Chu, China Taipei · motoyama@tiara.sinica.edu.tw
Kazuyuki **Muraoka**, IoA, the University of Tokyo, Mitaka, Japan · kmuraoka@ioa.s.u-tokyo.ac.jp
Yoshitsugu **Nakagawa**, Kobe University, Kobe, Japan · yoshi@kobe-u.ac.jp
Fumitaka **Nakamura**, Niigata University, Niigata, Japan · fnakamur@ed.niigata-u.ac.jp
Sadollah **Nasiri**, Zanjan University, Zanjan, Islamic Republic of Iran · nasiri@iasbs.ac.ir
Ralph **Neuhaeuser**, Univ. Jena, Jena, Germany · rne@astro.uni-jena.de
Marie-Helene **Nicol**, Max-Planck Institute für Astronomie, Heidelberg, Germany · nicol@mpia-hd.mpg.de
Silvana **Nikolic**, Universidad de Chile, Santiago, Chile · silvana@das.uchile.cl
Takayuki **Nishikawa**, The Graduate University for Advanced Studies, Hilo, USA · nishkwtk@subaru.naoj.org
Takahiro **Niwa**, Kobe University, Kobe, Japan · niwa@kobe-u.ac.jp
Hideko **Nomura**, Kobe University, Kobe, Japan · hnomura@kobe-u.ac.jp
Colin **Norman**, JHU/STScI, Baltimore, USA · norman@stsci.edu
Antonella **Nota**, STScI/ESA, Baltimore, USA · nota@stsci.edu
Harry **Nussbaumer**, Institute of Astronomy, Zurich, Switzerland · nussbaumer@astro.phys.ethz.ch

David **Nutter**, Cardiff University, Cardiff, UK — david.nutter@astro.cf.ac.uk
Yumiko **Oasa**, Kobe University, Kobe, Japan — yummy@kobe-u.ac.jp
John **O'Byrne**, University of Sydney, Sydney, Australia — j.obyrne@physics.usyd.edu.au
Sally **Oey**, University of Michigan, Ann Arbor, USA — msoey@umich.edu
Seungkyung **Oh**, Kyung Hee University, Yongin-shi, Republic of Korea — skoh@ap1.khu.ac.kr
Devendra **OJHA**, Tata Institute of Fundamental Research, Mumbai, India — ojha@tifr.res.in
Pius. N. **Okeke**, Centre for Basic Space Science, University of Nigeria Nsukka, Nsukka, Nigeria — okekepius@yahoo.com
Katalin Ilona **Olah**, Konkoly Observatory, Budapest, Hungary — hec@sirrah.troja.mff.cuni.cz
Joana **Oliveira**, Keele University, Keele, UK — joana@astro.keele.ac.uk
Kazuyuki **Omukai**, NAOJ, Mitaka, Japan — omukai@th.nao.ac.jp
David **Ondřich**, Nuncius Sidereus III, Praha, Czech Republic — david.ondrich@gmail.com
Jose **Oñorbe**, Universidad Autónoma de Madrid, Madrid, Spain — jose.onnorbe@uam.es
Roberto **Ortiz**, University of Sao Paulo, Sao Paulo, Brazil — ortiz@astro.iag.usp.br
Eve **Ostriker**, University of Maryland, College Park, USA — ostriker@astro.umd.edu
Juergen **Ott**, CSIRO Australia Telescope National Facility, Marsfield, Australia — Juergen.Ott@csiro.au
Rosita **Paladino**, Astronomical Observatory of Cagliari, Capoterra (CA), Italy — rpaladin@ca.astro.it
Francesco **Palla**, INAF-Osservatorio Astrofisico di Arcetri, florence, Italy — palla@arcetri.astro.it
Jan **Palouš**, Astronomical Institute, AS CR, Prague, Czech Republic — palous@ig.cas.cz
Vishambhar Nath **Pandey**, Raman Research Institute, Bangalore, India — vnpandey@rri.res.in
Vernon **Pankonin**, National Science Foundation, Arlington, USA — vpankoni@nsf.gov
Polychronis **Papaderos**, University of Göttingen, Göttingen, Germany — papade@astro.physik.uni-goettingen.de
Genevieve **Parmentier**, University of Cambridge, Cambridge, UK — gparm@ast.cam.ac.uk
Hannu **Parviainen**, University of Helsinki, Helsinki, Finland — hannu@astro.helsinki.fi
Nicolas **Peretto**, University of Manchester, Manchester, UK — nicolas.peretto@manchester.ac.uk
Kala **Perkins**, Australian National University, Saratoga, USA — kala.perkins@anu.edu.au
Paolo **Persi**, INAF, Roma, Italy — paolo.persi@iasf-roma.inaf.it
William **Peters**, University of Arizona, Tucson, USA — wpeters@as.arizona.edu
Dawn **Peterson**, University of Virginia, Charlottesville, USA — dawnp@virginia.edu
Daniel **Pfenniger**, Geneva Observatory, University of Geneva, Sauverny, Switzerland — daniel.pfenniger@obs.unige.ch
Randy **Phelps**, U.S. National Science Foundation, Arlington, USA — rphelps@nsf.gov
Catherine **Pilachowski**, Indiana University, Bloomington, USA — catyp@astro.indiana.edu
Olga **Pintado**, CONICET, San Miguel de Tucuman, Argentina — opintado@tucbbs.com.ar
Robert **Piontek**, Astrophysikalisches Institut Potsdam, Potsdam, Germany — rpiontek@aip.de
Paul **Plucinsky**, Harvard-Smithsonian Center for Astrophysics, Cambridge, USA — plucinsky@cfa.harvard.edu
Lucia Aurelia **Popa**, Institute for Space Sciences, Bucharest, Romania — lpopa@venus.nipne.ro
Bettina **Posselt**, MPI für Extraterrestrische Physik, Garching, Germany — posselt@mpe.mpg.de
Arcadio **Poveda**, National University of Mexico, Mexico City, Mexico — poveda@servidor.unam.mx
Thomas **Preibisch**, Max-Planck-Institute for Radioastronomy, Bonn, Germany — preib@mpifr-bonn.mpg.de
Ivanio **Puerari**, INAOE, Tonantzintla, Mexico — puerari@inaoep.mx
Elena **Puga**, Katholieke Universiteit Leuven, Leuven, Belgium — elena@ster.kuleuven.be
Simon **Pustilnik**, Special Astropysical Observatory RAS, Zelenchuk, Russian Federation — sap@sao.ru
Andreas **Quirrenbach**, Landessternwarte Heidelberg, Heidelberg, Germany — A.Quirrenbach@lsw.uni-heidelberg.de
Iratius **Radiman**, Institut Teknologi Bandung, Bandung, Indonesia — iratius@as.itb.ac.id
Roman **Rafikov**, Canadian Institute for Theoretical Astrophysics, Toronto, Canada — rrr@cita.utoronto.ca
William **Reach**, California Institute of Technology, Pasade, USA — reach@ipac.caltech.edu
Luisa **Rebull**, JPL/IPAC/Caltech, Pasadena, USA — rebull@ipac.caltech.edu
Michael **Reid**, Harvard-Smithsonian Submillimeter Array, Hilo, USA — mareid@sma.hawaii.edu
Monica **Relano**, Granada University, GRANADA, Spain — mrelano@ugr.es
Miriam **Rengel**, Max Planck Institute for Solar System Research, Katlenburg-Lindau, Germany — rengel@mps.mpg.de
Estela **Reynoso**, IAFE, Buenos Aires, Argentina — ereynoso@iafe.uba.ar
Thomas **Robitaille**, University of St Andrews, St Andrews, UK — tr9@st-andrews.ac.uk
Luis **Rodriguez**, UNAM, Morelia, Mexico — l.rodriguez@astrosmo.unam.mx
Jose M. **Rodriguez Espinosa**, Instituto de Astrofisica de Canarias, La Laguna, Spain — jmr.espinosa@iac.es
Loïc **Rolland**, CEA-Saclay, Gif-sur-Yvette, France — rollandl@in2p3.fr
Anish **Roshi**, Raman Research Institute, Bangalore, India — anish@rri.res.in
Erik **Rosolowsky**, Center for Astrophysics, Cambridge, USA — erosolow@cfa.harvard.edu
Ian **Roxburgh**, Queen Mary, University of London, London, UK — i.w.roxburgh@qmul.ac.uk
Monica **Rubio**, Universidad de Chile, Santiago, Chile — mrubio@das.uchile.cl
Adam **Růžička**, Astronomical Institute, AS CR, Praha, Czech Republic — adam.ruzicka@gmail.com
Tatiana **Ryabchikova**, Institute of Astronomy RAS, Moscow, Russian Federation — ryabchik@inasan.ru
Elena **Sabbi**, Space Telescope Science Institute, Baltimore, USA — sabbi@stsci.edu
Kanak **Saha**, Indian Institute of Science, Bangalore, India — kanak@physics.iisc.ernet.in
Kazuya **Saigo**, National Astronomical Obervatory of Japan, Mitaka, Japan — saigo@th.nao.ac.jp
Takeshi **Sakai**, Nobeyama Radio Observatory, Nagano, Japan — sakai@nro.nao.ac.jp
Itsuki **Sakon**, University of Tokyo, Bunkyo-ku, Japan — isakon@astron.s.u-tokyo.ac.jp
Heikki **Salo**, University of Oulu, Oulu, Finland — heikki.salo@oulu.fi
Wilton **Sanders**, NASA Headquarters, Arlington, USA — wilton.t.sanders@nasa.gov
Alfredo **Santillan**, UNAM, Mexico City, Mexico — alfredo@astroscu.unam.mx
Salvatore **Sciortino**, INAF- Osservatorio Astronomico di Palermo, Palermo, Italy — sciorti@astropa.inaf.it
Peter **Serlemitsos**, NASA/GSFC, Greenbelt, USA — pjs@astron.gsfc.nasa.gov
Alfonso **Serrano**, INAOE, San Andrés Cholula, Mexico — aspg@inaoep.mx
Marta **Sewilo**, Space Science Institute, Boulder, USA — sewilo@astro.wisc.edu
Margarita **Sharina**, Special Astrophysical Observatory RAS, N. Arkhyz, Russian Federation — sme@sao.ru
Rahul **Shetty**, University of Maryland, College Park, USA — shetty@astro.umd.edu
Yoshito **Shimajiri**, the university of Tokyo, Mitaka-shi, Japan — yoshito.shimajiri@nao.ac.jp
Steven **Shore**, University of Pisa, Pisa, Italy — shore@df.unipi.it
Boris **Shustov**, Institute of Astronomy of Russian Academy of Sciences, Moscow, Russian Federation — bshustov@inasan.ru
Daniel **Schaerer**, Geneva Observatory, Sauverny, Switzerland — daniel.schaerer@obs.unige.ch
Eva **Schinnerer**, Max Planck Institut für Astronomie, Heidelberg, Germany — schinnerer@mpia.de
Rita **Schulz**, ESA, Noordwijk, Netherlands — rschulz@rssd.esa.int
Sergiy **Silich**, INAOE, Puebla, Mexico — silich@inaoep.mx
Zdislav **Šíma**, Astronomical Institute, AS CR, Praha, Czech Republic — sima@ig.cas.cz
Michal **Simon**, SUNY-Stony Brook, Stony Brook, USA — michal.simon@sunysb.edu
Howard **Smith**, Center for Astrophysics, Cambridge, USA — hsmith@cfa.harvard.edu
Derck **Smits**, University of South Africa, UNISA, South Africa — smitsdp@unisa.ac.za
Vernesa **Smolcic**, MPIA, Heidelberg, Germany — smolcic@mpia.de
Keely **Snider**, Arizona State University, Phoenix, USA — keely.snider@asu.edu
Leonie **Snijders**, Leiden Observatory, Leiden, Netherlands — snijders@strw.leidenuniv.nl

George **Sonneborn**, NASA Goddard Space Flight Center, Greenbelt, USA — george.sonneborn@nasa.gov
Roberto **Soria**, Harvard-Smithsonian Center for Astrophysics, Cambridge, USA — rsoria@cfa.harvard.edu
Anselmo **Sosa**, Instituto de Astrofísica ed Canarias (IAC), La Laguna. Tenerife, Spain — asosa@iac.es
Kazimierz **Stepien**, Warsaw University Observatory, Warszawa, Poland — kst@astrouw.edu.pl
Irena **Stojimirovic**, University Of Massachusetts, Amherst, USA — irena@nova.astro.umass.edu
Ivana **Stoklasová**, Astronomical Institute, AS CR, Praha, Czech Republic — ivana@sirrah.troja.mff.cuni.cz
Guy **Stringfellow**, University of Colorado, Boulder, USA — Guy.Stringfellow@colorado.edu
Curtis **Struck**, Iowa State University, Ames, USA — curt@iastate.edu
Micaela B. **Stumpf**, Max-Planck-Institut fuer Astronomie, Heidelberg, Germany — stumpf@mpia-hd.mpg.de
Kefeng **Sun**, I. Physikalisches Institut, Köln, Germany — kefeng@ph1.uni-koeln.de
Kazuyoshi **Sunada**, Nobeyama Radio Observatory, Minamisaku, Japan — sunada@nro.nao.ac.jp
Hyun-il **Sung**, Korea Astronomy and Space Science Institute, Daejeon, Republic of Korea — hisung@kasi.re.kr
Marian **Szymczak**, Nicolaus Copernicus University, Torun, Poland — msz@astro.uni.torun.pl
Kengo **Tachihara**, Kobe University, Kobe, Japan — tatihara@kobe-u.ac.jp
Toshinobu **Takagi**, Institute of Space and Astronautical Science, Sagamihara, Japan — takagi@ir.isas.jaxa.jp
Satoko **Takahashi**, Gradiate University for Advances Studies, Mitaka, Japan — satoko.takahashi@nao.ac.jp
Yoichi **Tamura**, the University of Tokyo, Mitaka, Japan — yoichi.tamura@nao.ac.jp
Jonathan **Tan**, University of Florida, Gainesville, USA — jt@astro.ufl.edu
Andrea **Taracchini**, University of Trieste, Casalmaggiore (CR), Italy — taracchini@gmail.com
Bruce **Tarter**, Lawrence Livermore national Laboratory, Livermore, USA — tarter1@llnl.gov
Edward **Taylor**, Leiden Observatory, Leiden, Netherlands — ent@strw.leidenuniv.nl
Guillermo **Tenorio-Tagle**, INAOE, Tonantzintla, Mexico — gtt@inaoep.mx
Roberto **Terlevich**, INAOE, Tonantzintla, Mexico — rjt@inaoep.mx
Elena **Terlevich**, INAOE, Tonantzintla, Mexico — eterlevi@inaoep.mx
Christian **Theis**, University of Vienna, Vienna, Austria — theis@astro.univie.ac.at
Holly **Thomas**, University of Manchester, Manchester, UK — h.thomas@postgrad.manchester.ac.uk
Andrea Fabiana **Torres**, National University of La Plata, La Plata, Argentina — andy@carina.fcaglp.unlp.edu.ar
Tomoka **Tosaki**, Nobeyama Radio Observatory, Minamisaku, Japan — tomoka@nro.nao.ac.jp
Nick **Tothill**, Smithsonian Astrophysical Obs, Cambridge, USA — ntothill@cfa.harvard.edu
Ginevra **Trinchieri**, INAF-OABr, Milano, Italy — ginevra.trinchieri@brera.inaf.it
Masato **Tsuboi**, National Astronomical Observatory, Japan, Minamisaku, Japan — tsuboi@nro.nao.ac.jp
Takashi **Tsuji**, University of Tokyo, Mitaka, Japan — ttsuji@ioa.s.u-tokyo.ac.jp
Ilkka **Tuominen**, Observatory, Helsinki, Finland — Ilkka.Tuominen@helsinki.fi
David **Turnshek**, University of Pittsburgh, Pittsburgh, USA — turnshek@pitt.edu
Stephane **Udry**, Geneva University, Versoix, Switzerland — stephane.udry@obs.unige.ch
Munetaka **Ueno**, University of Tokyo, Tokyo, Japan — ueno@chianti.c.u-tokyo.ac.jp
Andrea **Urban**, University of Texas at Austin, Austin, USA — aurban@astro.as.utexas.edu
Dejan **Urosevic**, University of Belgrade, Belgrade, Serbia — dejanu@matf.bg.ac.yu
James **Urquhart**, University of Leeds, Leeds, UK — jsu@ast.leeds.ac.uk
Petri **Vaisanen**, South African Astronomical Observatory, Observatory, South Africa — petri@saao.ac.za
Margarita **Valdez-Gutierrez**, OAN - IA - UNAM, Ensenada, Mexico — mago@astrosen.unam.mx
Jacques P. **Vallee**, National Research Council Canada, Victoria, Canada — jacques.vallee@nrc.gc.ca
David **Valls-Gabaud**, Observatoire de Paris, Meudon Cedex, France — dvg@cfht.hawaii.edu
Mario **Van den Ancker**, European Southern Observatory, Garching, Germany — mvandena@eso.org
Edward **Van den Heuvel**, University of Amsterdam, Amsterdam, Netherlands — edvdh@science.uva.nl
Johan **Van der Walt**, North-West University, Potchefstroom, South Africa — fskdjvdw@puk.ac.za
Ewine **Van Dishoeck**, Leiden Observatory, Leiden, Netherlands — ewine@strw.leidenuniv.nl
Sven **Van Loo**, University of Leeds, Leeds, UK — svenvl@ast.leeds.ac.uk
Jacco **Van Loon**, Keele University, Keele, UK — jacco@astro.keele.ac.uk
Hugo **van Woerden**, Kapteyn Astron. Institute, Groningen, Netherlands — hugo@astro.rug.nl
Luiz Paulo **Vaz**, Federal University of Minas Gerais, Belo Horizonte, Brazil — lpv@fisica.ufmg.br
Enrique **Vazquez-Semadeni**, UNAM, Morelia, Mexico — e.vazquez@astrosmo.unam.mx
Thangasamy **Velusamy**, Jet Propoulsion Laboratory, Pasadena, USA — velusamy@jpl.nasa.gov
Daniela **Vergani**, INAF - IASF Milan, Milan, Italy — daniela@lambrate.inaf.it
Marc **Verheijen**, Kapteyn Astronomical Institute, Groningen, Netherlands — verheyen@astro.rug.nl
Dmitri **Vibe**, Institute of Astronomy of the RAS, Moscow, Russian Federation — dwiebe@inasan.ru
Edgardo **Vidal Perez**, Universiteit Gent, Gent, Belgium — edgardoandres.vidalperez@ugent.be
Alfred **Vidal-Madjar**, Institut d'Astrophysique de Paris, Paris, France — alfred@iap.fr
Sonja **Vidojević**, Mathematical faculty, Belgrade, Serbia — sonja@alas.matf.bg.ac.yu
Jose M. **Vilchez**, Instituto de Astrofisica de Andalucia-CSIC, Granada, Spain — jvm@iaa.es
Emmanuil **Vilkoviskiy**, Fesenkov Astrophysical Institute MON RK, Almaty, Kazakhstan — vilk@aphi.kz
Nikolaus **Volgenau**, Universität zu Köln, Köln, Germany — volgenau@ph1.uni-koeln.de
Frederick **Vrba**, U.S. Naval Observatory, Flagstaff, USA — fjv@nofs.navy.mil
Saeqa **Vrtilek**, Harvard-Smithsonian Center for Astrophysics, Cambridge, USA — svrtilek@cfa.harvard.edu
Branislav **Vukotič**, Belgrade Astronomical Observatory, Belgrade, Serbia — bvukotic@aob.bg.ac.yu
Nolan **Walborn**, Space Telescope Science Institute, Baltimore, USA — walborn@stsci.edu
Stefanie **Walch**, University Observatory, Munich, Germany — swalch@usm.uni-muenchen.de
Fabian **Walter**, Max Planck Institut fuer Astronomie, 69117 Heidelberg, Germany — walter@mpia.de
Jun-Jie **Wang**, National Astronomical Observatories, Chinese Academy of Sciences, Beijing, China Nanjing — wangjj@bao.ac.cn
Shiya **Wang**, University of Illinois at Urbana-Champaign, Urbana, USA — swang9@uiuc.edu
Hongchi **Wang**, Purple Mountain Observatory, Nanjing, China Nanjing — hcwang@pmo.ac.cn
Jessica **Werk**, University of Michigan, Ann Arbor, USA — jwerk@umich.edu
Brad **Whitmore**, Space Telescope Science Institute, Baltimore, USA — whitmore@stsci.edu
Anthony **Whitworth**, Cardiff Unversity, Cardiff, UK — anthony.Whitworth@astro.cf.ac.uk
Robert **Williams**, Space Tel. Sci. Institute, Baltimore, USA — wms@stsci.edu
P. Frank **Winkler**, Middlebury College, Middlebury, USA — winkler@middlebury.edu
Adolf **Witt**, The University of Toledo, Toledo, USA — awitt@dusty.astro.utoledo.edu
Scott **Wolk**, Harvard-Smithsonian Center for Astrophysics, Cambridge, USA — swolk@cfa.harvard.edu
Henry **Wootten**, NRAO, Charlottesville, USA — awootten@nrao.edu
Richard **Wünsch**, Astronomical Institute, AS CR, Prahe, Czech Republic — richard.wunsch@matfyz.cz
Bei **Xin**, The National Astronomical Observatories, Chinese Academy of Sciences, Beijing, China Nanjing — xp@bao.ac.cn
Masako **Yamada**, National Astronomical Observatory of Japan, Mitaka, Japan — ymasako@th.nao.ac.jp
Hiroaki **Yamamoto**, Nagoya University, Nagoya, Japan — hiro@a.phys.nagoya-u.ac.jp
Aili **Yishamuding**, Urumqi observatory, NAOs,CAS, Urumqi, China Nanjing — aliyi@ms.xjb.ac.cn
Joao **Yun**, Lisbon Observatory, Lisboa, Portugal — yun@oal.ul.pt
Shuang Nan **Zhang**, Tsinghua University, Beijing, China Nanjing — zhangsn@mail.tsinghua.edu.cn
Hans **Zinnecker**, Astrophys. Inst. Potsdam, Potsdam, Germany — hzinnecker@aip.de

Triggered Star Formation in a Turbulent ISM
Proceedings IAU Symposium No. 237, 2006
B. G. Elmegreen & J. Palouš, eds.

© 2007 International Astronomical Union
doi:10.1017/S1743921307001147

Observations of turbulence in the diffuse interstellar medium

John M. Dickey[1]

[1]School of Maths and Physics, University of Tasmania, Hobart 7001, Australia
email: John.Dickey@utas.edu.au

Abstract. The warm neutral medium, warm ionized medium, and cool neutral medium all show strong evidence for turbulence as a process dominating their structure and motions on a wide range of scales. The spatial power spectra of density fluctuations in all three phases are consistent with a Kolmogorov slope. Turbulence in the magnetic field in the diffuse medium can also be measured through the structure function of the Faraday rotation measure. With new surveys, new analysis techniques, and new telescopes, in the next few years it will be possible to measure the structure function of the magnetic field over a similarly wide range of scales. This will give a complete picture of the turbulence as a magneto-acoustic process.

Keywords. ISM: structure, ISM: clouds, ISM: evolution, ISM: kinematics and dynamics, ISM: magnetic fields

1. Introduction

One of the great stories of Galactic astronomy is how successive generations of stars are enriched by increasing abundances of heavy elements formed in other stars that lived and died long before. Understanding how long it takes for gas ejected by stellar winds and supernova remnants to return to new star formation regions, what the intermediate stages are, and how the gas mixes through the Galaxy along the way, requires a more detailed model of the interstellar medium (ISM) than we have. Modelling cloud formation and phase changes in the ISM leads to an appreciation that turbulence is a fundamental process that determines how the gas evolves. Many observations show the effects of this turbulence. The challenge to observers is to measure the quantities that are needed for a theoretical understanding of the physics that drives it.

The diffuse ISM, including the warm ionized medium, the warm neutral medium, and the cool neutral (atomic) medium, fills most of the volume of the Galactic plane. Tracers of these phases show the big structures: spiral arms and interarm regions, and the somewhat smaller shells, filaments, and chimneys that mottle the disk of the Milky Way on scales of tens to hundreds of parsecs. Many of these structures are apparently deterministic, as, for example, an expanding shell of gas shows an ordered density and velocity field that suggests a past origin and a predictable future evolution. The onset of turbulence is a breakdown of this deterministic structure into a random or stochastic process, in which only the statistics of the density or velocity field have significance. It is not always easy to tell from the observations where and on what scale this transition happens.

The word turbulence denotes something much more specific than a random or disordered velocity field. Saying that the ISM is turbulent suggests that there is a physical process underway in which kinetic energy is transferred from larger to smaller scale motions (the "cascade"). Measuring the statistics of the structure of the ISM from a survey does not establish that turbulence is the process that generates this structure.

1

Interpretation of the observations is generally based on an assumed paradigm for the turbulent process.

2. Measurements of small scale structure

Whatever the resolution, observations of the interstellar medium show structure on all angular scales accessible to the telescope. This is generally true for all tracers, line and continuum, and in all directions. But the structure may be stronger or weaker, and it may be caused by spatial variations in density, velocity, temperature, ionization, radiation field strength, magnetic field strength, or other physical parameters of the medium. There are many ways to characterize the random fluctuations that appear on maps of a given observable quantity, such as the brightness temperature of a spectral line. One of the simplest to understand is the spatial power spectrum (P_1, P_2, or P_3 depending on the number of dimensions in the calculation), and its Fourier conjugate, the spatial autocorrelation function, which is closely related to the structure function (Lee and Jokipii 1974, Rickett 1977, Crovisier and Dickey 1983, Cordes and Rickett 1998, Lazarian and Pogosyan 2000, Goldman 2000, Miville-Deschenes et al. 2003).

There are many other mathematical functions that have proven useful for describing the small scale structure in ISM surveys. Two that are particularly well suited to spectral line surveys are the spectral correlation function (Rosolowsky et al. 1999, Ballesteros-Paredes et al. 2002, Padoan et al. 2003) and principal component analysis (Heyer and Schloerb 1997). There has been a great deal of work on the fractal geometry of the ISM, mostly directed toward molecular clouds, but with some applications to the diffuse medium (Westpfahl et al. 1999, Stanimirović et al. 1999, Elmegreen et al. 2001). The comprehensive review articles of Elmegreen and Scalo (2004) and Scalo and Elmegreen (2004) give a thorough discussion of turbulence in the ISM in general, including both observational and theoretical perspectives. One compelling approach is to characterize the structure in the density or velocity field in terms of a spectrum of magneto-acoustic waves (Ferriere et al. 1988, Cho et al. 2002, Heitsch et al. 2004). This is well suited to situations where the magnetic field is dynamically dominant over the gas pressure, which may be quite common in the interstellar medium.

Consideration of the spatial power spectrum of the variations in interstellar density over a broad range of scale sizes leads quickly to the hypothesis that a turbulent cascade of energy from large to small scales may be at work. Figure 1 (from Armstrong et al. 1995) is a compendium of data on the diffuse **ionized** medium. The figure shows the three dimensional power spectrum, P_3, as deduced from various observations, mostly based on the propagation of pulsar signals. Over at least ten orders of magnitude in linear size, a slope of $-\frac{11}{3}$ fits the data adequately. This matches the prediction of the Kolmogorov theory of inertial turbulence in an incompressible medium.

An excellent example of the kind of pulsar data that is used to probe structure in the diffuse ionized medium is the twenty year time series of dispersion measure (proportional to the line of sight integral of electron density) toward the millisecond pulsar B1937+21 (Ramachandran et al. 2006). The autocorrelation function of these data provide the structure function, which shows a remarkably straight power law over two orders of magnitude in time lag, which translates directly to distance offset. The scales measured are 0.3 to 100 AU. Converting this structure function to an autocorrelation function and taking the Fourier transform gives the spatial power spectrum.

The neutral atomic gas in the ISM is best traced with the λ21-cm line of HI. In emission, this is easy to map on angular scales of 1′ and larger, as has been done for wide areas of the Galactic plane in three recent mosaic surveys, the Canadian Galactic Plane Survey

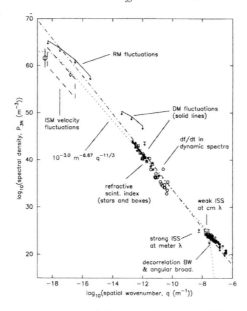

Figure 1. The spatial power spectrum of the ionized medium (Armstrong *et al.* 1996; reproduced by permission of the AAS.)

(Taylor *et al.* 2003), the Southern Galactic Plane Survey (McClure-Griffiths *et al.* 2001, 2005), and the VLA Galactic Plane Survey (Stil *et al.* 2006). From this kind of moderately high resolution data, the spatial power spectrum of the 21-cm emission has been studied for many years (Crovisier and Dickey 1983, Green 1993). In many different directions, observed with many different instruments, the spatial power spectrum consistently gives a power law with spectral index in the range -2.2 to -3. The range of linear scales covered by this kind of observation is typically 0.5 pc to 20 pc. Assuming that a single velocity channel corresponds to a two dimensional slice through the medium (where Galactic rotation translates velocity into distance, whether or not this relation is bi-valued), then the expected power law index is $(-\frac{8}{3})$ for a Kolmogorov cascade, which is very consistent with the observations in most low-latitude directions.

Averaging in velocity steepens the power law, at least in the mosaic survey data, from power law index -2.7, typically, to the range -3 to -4. This is explained by the transition from a thin slice to a thick slice, or three dimensional sample, for which the Kolmogorov slope is $-\frac{11}{3}$ (Stanimirović and Lazarian 2001, Dickey *et al.* 2001, Lazarian and Pogosyan 2000). This is strongly suggestive that the slope of the spatial power spectrum is set by a turbulence process, and not some other random process that modulates the density and/or velocity field. It is interesting also that the neutral medium shows this same power law behavior in very different environments, including the Magellanic Clouds, and even the Magellanic Bridge (Muller *et al.* 2004), where the influence of stellar winds and supernova remnants must be very much weaker than in the solar neighborhood. In the Magellanic Bridge region also the power law steepens, from about -2.25 to -3, with velocity averaging. This similarity of the spatial power spectrum of the turbulence in such different environments raises the question of whether the processes that drive the turbulence have any effect on the spectrum, and the larger question of how much the presence of massive stars is required to drive the turbulence by injecting kinetic energy.

3. Driving the turbulence

There are many possibilities for the source of the kinetic energy on large scales that drives the turbulence in the diffuse medium. Evaluating the role and relative importance of these different processes is an important challenge to Galactic astrophysicists. It is possible that the magnetic field causes an inverse cascade, in which the energy is injected at small scales and propagates to larger and larger patterns of motion (Pouquet *et al.* 1999). But there are some good observational examples of ordered motions on scales of hundreds of parsecs to one kpc that are in the process of breaking down into disordered, smaller scale irregularities in density and velocity. A striking case is the wall of the huge Galactic chimney, GSH 277+00+36 (McClure-Griffiths *et al.* 2000, 2003). As this is a very old supershell, probably 15 to 20 Myr old, Rayleigh-Taylor instabilities are beginning to cause the dense, cool shell to "drip" into the higher pressure, hot, low density interior. As this process continues the ordered expansion velocities are being deflected by density irregularities into a random velocity field. One of these "drips" contains a CO cloud at the same velocity as the atomic gas in the shell wall. The drip has such a narrow line width in HI that the atomic gas must be quite cool (less than 100 K) in the vicinity of the CO cloud. In this case the morphology is very suggestive that the turbulence in the atomic gas is driving turbulent motions in the molecular cloud. It is important to establish how much of the turbulent molecular cloud kinetic energy is imported from the diffuse medium, brought along with the gas during the process of condensation and molecule formation as a cool atomic cloud makes the transition to the molecular phase.

Emission in the λ 21-cm line traces the column density of the atomic medium, without much weighting by the temperature of the gas. In 21-cm absorption the temperature is much more important; the optical depth is primarily due to the cool gas ($T_{kin} \leqslant 100\,\mathrm{K}$) which is about 30% of the total atomic phase. This is an advantage for studying the cool atomic clouds that may be in the process of conversion to molecular gas. It is also much easier to study small scale structure in absorption, since the brightness of the background source may be much higher than the brightness of the line emission. In this way Deshpande (2000) has demonstrated that the spatial power spectrum seen in λ21-cm emission continues to scales of 0.02 pc, from measurements of the absorption toward Cas A. At even much smaller scales, observations with the VLBA show occasional significant variations on angular scales of 10 mas, that correspond to a few hundred AU (Brogan *et al.* 2006). Similar "tiny-scale" structure is seen in Na I absorption in the optical (Lauroesch *et al.* 2000) as is larger scale structure (Meyer and Lauroesch 1999).

An example of a cloud at the borderline between the cool atomic and molecular phases is the Rigel-Crutcher cloud, seen in HI self-absorption (HISA) toward the Galactic center (its distance is 150 to 180 pc, Crutcher and Riegel 1974). Figure 2 shows this cloud as mapped by McClure-Griffiths *et al.* (2006) in the Southern Galactic Plane Survey. The remarkable collection of long filaments is striking. Some have dimensions $\sim$17 pc $\times$ 0.1 pc, for an aspect ratio of 170:1. The filaments are very straight, and very well aligned with the magnetic field direction as deduced from the polarization of starlight (Heiles 2000). McClure-Griffiths *et al.* (2006) estimate the required magnetic field strength if the magnetic energy density is to exceed the gas kinetic energy density, to find a lower limit on B of 30 μG. This is much higher than measurements in the typical diffuse atomic medium (median 6 μG, Heiles and Troland 2005), but similar to field strengths measured in molecular clouds. Assuming that this HISA structure is on the borderline between atomic and molecular clouds, the morphology suggests that one way to carry turbulent energy through this transition is via the magnetic field. This raises the question of what we know about turbulence in the magnetic field elsewhere in the diffuse medium.

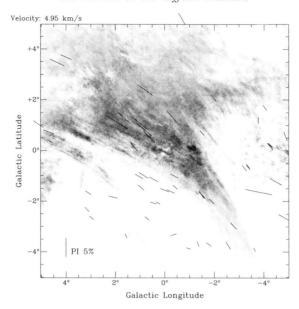

Figure 2. A cold, atomic cloud complex seen in HI self-absorption in the Southern Galactic Plane Survey (McClure-Griffiths *et al.* 2006). Starlight polarization vectors indicating the magnetic field direction are superposed.

4. Turbulence in the magnetic field

There are many ways to estimate the magnetic field strength and direction in the interstellar medium, and a few ways to quantitatively measure the strength of the line of sight component of the field, either in one particular region (using the Zeeman effect, e.g. Heiles and Troland 2005) or integrated along the line of sight (using Faraday rotation, e.g. Weisberg *et al.* 2004). When many measurements are available in the same area, the structure function can be computed. An early study using Faraday rotation measures toward extragalactic sources observed with the VLA was done by Minter and Spangler (1996). They found that the structure functions of the rotation measure and of the emission measure show similar power law behavior on angles of 1° to 10°, but at angles smaller than 0.1^o the structure function of the rotation measure steepens. A similar steepening is found by Haverkorn *et al.* (2004), who also find evidence for anisotropy in the structure function, as would be expected if the magnetic field has a mixture of ordered and random components. In further work, Haverkorn *et al.* (2006) find that the slope of the structure function of the rotation measure changes dramatically between spiral arms and interarm regions, with a flatter slope, i.e. more structure on small scales, in the directions dominated by spiral arms. This result is still preliminary, but it suggests that there will be a lot to learn from future studies of the turbulence of the interstellar magnetic field.

A very promising observational technique for measuring the structure of the magnetic field on a wide range of scales is to map the Faraday depth of the diffuse synchrotron emission from the cosmic rays of the Milky Way. Whenever there is linearly polarized emission interspersed with a magnetized thermal plasma, there will be a superposition of continuum emission with multiple Faraday depths, ϕ,

$$\frac{\phi(r)}{\text{rad m}^2} = 0.81 \int_{los} \frac{n_e}{\text{cm}^{-3}} \frac{\vec{B} \cdot \vec{dr}}{\mu\text{G pc}}$$

where the integral is taken along the line of sight starting from the observer, as in the analogous definition of optical depth. The different regions contributing to the linearly polarized emission can be separated by taking the Fourier transform of the observed linear polarization along an axis constructed to be linear in $x \equiv 2\lambda^2$, since Faraday rotation causes the Stokes Q and U parameters (components of the linear polarization, $P = Q + iU$ expressed as a complex number) to be rotated (i.e. multiplied by $e^{i\phi x}$, giving the Fourier transform relationship

$$P(x) = \int_{-\infty}^{+\infty} F(\phi) \, e^{i\phi x} d\phi$$

where $F(\phi)$ is the intrinsic distribution of linearly polarized emission as a function of Faraday depth along the line of sight (Burn 1966, Brentjens and de Bruyn 2005).

This technique allows the construction of a Faraday cube, similar to a spectral line cube but with radial velocity replaced by Faraday depth, ϕ, the conjugate variable to x. As with the ordinary spectral line cube, this third dimension is related to distance, but in a complicated way. Studying this cube tells us about the three dimensional distribution of the magnetized thermal plasma and the polarized emission (de Bruyn *et al.* 2006). Construction of such cubes from low latitude survey data will allow a much more profound interpretation of the diffuse, polarized Galactic emission, with the goal of separating the structure functions of the electron density and the magnetic field.

5. Future observations of the diffuse medium

Over the next decade radio astronomers will be working toward construction of the Square Kilometer Array telescope (SKA); precursor or "phase 1" projects are underway already in several nations. This telescope will have the power to completely measure the spatial power spectra of the ionized medium, the neutral medium, and the magnetic field, with no gaps, over the full range of scales on figure 1 (Dickey *et al.* 2004). The SKA will revolutionize studies of the interstellar medium, through the λ21-cm line and cm-wave molecular lines, through measurements of the propagation of radiation from pulsars and other compact sources, and through sensitive observations of continuum emission in all Stokes parameters. Surveys of the diffuse ISM with the SKA may show the interaction between the spectrum of magneto-acoustic waves and the gas density and velocity fields. The ultimate goal is to trace the connection between turbulence in all phases of the interstellar medium, from the warm ionized medium, to the cool neutral medium, and on to molecular clouds and star formation.

Acknowledgements

This research was supported in part by the US National Science Foundation under grant AST 03-07603 to the University of Minnesota.

References

Armstrong, J.W., Rickett, B.J. & Spangler, S.R. 1995, *ApJ* 443, 209
Ballesteros-Paredes, J., Vázquez-Semadeni, E. & Goodman, A.A. 2002, *ApJ* 571, 334
Brentjens, M.A. & de Bruyn, A.G. 2005, *A&A* 441, 1217
Brogan, C.L., Zauderer, B.A., Lazio, T.J., Goss, W.M., DePree, C.G. & Faison, M.D. 2005, *AJ* 130, 698
Burn, B.J. 1966, *MNRAS* 133, 67
Cho, J., Lazarian, A. & Vishniac, E.T. 2002, *ApJ* 566, 49

Crovisier, J. & Dickey, J.M. 1983, *A&A* 122, 282

Crutcher, R.M. & Riegel, W.K. 1974, *ApJ* 188, 481

Cordes, J.M & Rickett, B.J. 1998, *ApJ* 507, 846

de Bruyn, A.G., Katgert, P., Haverkorn, M. & Schnitzeler, D.H.F.M. 2006, *Astronomische Nachrichten* 327, 487

Deshpande, A.A. 2000, *MNRAS* 317, 199

Dickey, J.M., McClure-Griffiths, N.M., Stanimirović, S., Gaensler, B.M. & Green, A.J. 2001, *ApJ* 561, 264

Dickey, J.M., McClure-Griffiths, N. & Lockman, F.J. 2004, *New Astron. Revs* 48, 1311

Elmegreen, B.G., Kim, S. & Staveley-Smith, L. 2001, *ApJ* 548, 749

Elmegreen, B.G. & Scalo, J. 2004 *ARAA* 42, 211

Ferriere, K.M., Zweibel, E.G. & Shull, J.M. 1988, *ApJ* 332, 984

Goldman, I. 2000, *ApJ* 541, 701

Green, D.A. 1993, *MNRAS* 262, 327

Haverkorn, M., Gaensler, B.M., McClure-Griffiths, N.M., Dickey, J.M. & Green, A.J. 2004, *ApJ* 609, 776

Haverkorn, M., Gaensler, B.M., Brown, J.C., Bizunok, N.S., McClure-Griffiths, N.M., Dickey, J.M. & Green, A.J. 2006, *ApJ* 637, 33

Heiles, C. & Troland, T.H. 2004, *ApJS* 151, 271

Heiles, C. 2000, *AJ* 119, 923

Heitsch, F., Zweibel, E.G., Slyz, A.D. & Devriendt, J.E.G. 2004, *Ap+SS* 292, 45

Heyer, M.H. & Schloerb, F.P. 1997, *ApJ* 475, 173

Lauroesch, J.T., Meyer, D.M. & Blades, J.C. 2000, *ApJ* 543, 43

Lazarian, A. & Pogosyan, D. 2000, *ApJ* 537, 720

Lee, L.C. & Jokipii, J.R. 1975, *ApJ* 196, 695

McClure-Griffiths, N.M., Dickey, J.M., Gaensler, B.M., Green, A.J., Haynes, R.F. & Wieringa, M.H. 2000, *AJ* 119, 2828

McClure-Griffiths, N.M., Green, A.J., Dickey, J.M., Gaensler, B.M., Haynes, R.F. & Wieringa, M.H. 2001, *ApJ* 551, 394

McClure-Griffiths, N.M., Dickey, J.M., Gaensler, B.M. & Green, A.J. 2003, *ApJ* 594, 833

McClure-Griffiths, N.M., Dickey, J.M., Gaensler, B.M., Green, A.J., Haverkorn, M. & Strasser, S. 2005, *ApJS* 158, 178

McClure-Griffiths, N.M., Dickey, J.M., Gaensler, B.M., Green, A.J. & Haverkorn, M. 2006, *ApJ* in press (astro-ph 0608585)

Meyer, D.M. & Lauroesch, J.T. 1999, *ApJ* 520, 103

Minter, A.H. & Spangler, S.R. 1996, *ApJ* 458, 194

Muller, E., Stanimirović, S., Rosolowsky, E. & Staveley-Smith, L. 2004, *ApJ* 616, 845

Padoan, P., Goodman, A.A. & Juvela, M. 2003, *ApJ* 588, 881

Pouquet, A., Galtier, S. & Politano, H. 1999, in: A.R. Taylor, T.L. Landecker, and G. Joncas (eds.), *New Perspectives on the Interstellar Medium* ASP Conference Series 168 (San Francisco: ASP) p. 417

Ramachandran, R., Demorest, P., Backer, D.C., Cognard, I., Lommen, A. 2006, *ApJ* 645, 303

Rickett, B.J. 1977, *ARAA* 15, 479

Rosolowsky, E.W., Goodman, A.A., Wilner, D.J. & Williams, J.P. 1999, *ApJ* 524, 887

Scalo, J.M. & Elmegreen, B.G. 2004, *ARAA* 42, 275

Stanimirović, S. & Lazarian, A. 2001, *ApJ* 551, 53

Stanimirović, S., Staveley-Smith, L., Dickey, J.M., Sault, R.J. & Snowden, S.L. 1999, *MNRAS* 302, 417

Stil, J.M., Taylor, A.R., Dickey, J.M., Kavars, D.W., Martin, P.G., Rothwell, T.A., Boothroyd, A.I., Lockman, F.J. & McClure-Griffiths, N.M. 2006, *AJ* 132, 1158

Taylor, A.R., Gibson, S.J., Peracaula, M., Martin, P.G. & Landecker, T.L. *et al.* 2003, *AJ* 125, 3145

Weisberg, J.M., Cordes, J.M., Kuan, B., Devine, K.E., Green, J.T. & Backer, D.C. 2004, *ApJS* 150, 317

Westpfahl, D.J., Coleman, P.H., Alexander, J. & Tongue, T. 1999, *AJ* 118, 323

Discussion

BRINKS: You mentioned that shells and bubbles could well, as they disintegrate, be at the basis of the observed Kolmogorov-Smirnov type power spectrum. But then you showed the Magellanic Bridge where no stars or supernovae are around to do the trick. Any idea what might be powering turbulence in such an environment?

DICKEY: It is possible that the turbulence is driven by the large scale dynamics of the clouds and their tidal interaction with each other and with the Milky Way.

GOLDMAN: Don't you think the index is a characteristic of the gas – not the source?

DICKEY: Yes, if we assume that the turbulence is inertial, then conservation of energy might set the spectral index independent of the physical process that drives the cascade on the outer scale.

Y.-H. CHU: There are stars in the Magellanic Bridge. These are blue stars and associations/clusters in the Magellanic Bridge, although there are no current star formation o O stars.

DICKEY: Yes, there are stars. But are there enough to explain the HI structure and motions?

Triggered Star Formation in a Turbulent ISM
Proceedings IAU Symposium No. 237, 2006
B. G. Elmegreen & J. Palouš, eds.

© 2007 International Astronomical Union
doi:10.1017/S1743921307001159

Turbulence in the molecular interstellar medium

Mark H. Heyer[1] and Chris Brunt[2]

[1]Department of Astronomy, University of Massachusetts, Amherst, MA 01003, USA
[2]School of Physics, University of Exeter, Stocker Road, EX4 4QL, United Kingdom
email: heyer@astro.umass.edu,brunt@astro.ex.ac.uk

Abstract. The observational record of turbulence within the molecular gas phase of the interstellar medium is summarized. We briefly review the analysis methods used to recover the velocity structure function from spectroscopic imaging and the application of these tools on sets of cloud data. These studies identify a near-invariant velocity structure function that is independent of the local environment and star formation activity. Such universality accounts for the cloud-to-cloud scaling law between the global line-width and size of molecular clouds found by Larson (1981) and constrains the degree to which supersonic turbulence can regulate star formation. In addition, the evidence for large scale driving sources necessary to sustain supersonic flows is summarized.

Keywords. interstellar turbulence, molecular ISM, star formation

1. Introduction

Turbulent motions are commonly observed within several phases of the interstellar medium (see Elmegreen & Scalo 2005). Within the molecular gas phase, turbulent gas flows are supersonic and possibly, super-Alfvenic, and play a dual role in the dynamics and evolution of these regions. Turbulence can provide a non-thermal, macroscopic pressure that lends support against self-gravity. In addition, compressible, supersonic flows may promote star formation by generating density perturbations within the shocks of colliding gas streams that eventually evolve into self-gravitating or collapsing protostellar cores (Padoan & Nordlund 2002; Mac Low & Klessen 2004).

Spectroscopy of molecular line emission, especially the rotational lines of CO, have long been the primary measurement from which turbulence is defined. In fact, supersonic motions are inferred from the very first CO spectrum observed by Wilson, Jefferts, & Penzias (1970) in which there is a 5 km/s wide line core in addition to the broad 100 km/s wing component that was later attributed to a luminous protostellar outflow. The 5 km/s core is significant broader than the sound speed of molecular hydrogen assuming a temperature of 30 K.

Owing to advancing instrumentation at millimeter and submillimeter wavelengths, our ability to measure the distribution and kinematics of the molecular gas phase of the interstellar medium has greatly expanded since that initial CO spectrum. Sensitive, millimeter wave interferometers routinely probe the circumstellar environments about young stellar objects. Sensitive bolometer imaging arrays identify the sites of protostellar and pre-protostellar cores (see André in these proceedings). Heterodyne focal plane arrays on single dish telescopes enable the construction of high spatial dynamic range imaging of molecular line emission (Heyer 1999). An example of such imaging is displayed in Figure 1. It reveals the varying texture of CO line emission imprinted by the effects of gravity, turbulence, and magnetic fields. A diffuse, low surface brightness component extends across the field and contains localized "streaks" of emission that are aligned along the local magnetic field direction. The sequence of channel images show low column density material moving toward the dense, highly structured filaments that are more apparent in ^{13}CO images and extinction maps. The challenge to the astronomer is to synthesize the information that is resident within these data cubes with suitable analysis tools to place these into a physical context in order to test and constrain model descriptions of turbulence within the molecular interstellar medium.

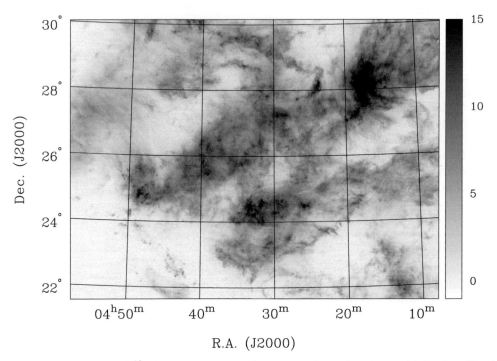

Figure 1. An image of ^{12}CO J=1-0 integrated emission from the Taurus Molecular Cloud observed with the *Five College Radio Astronomy Observatory* 14m telescope and *SEQUOIA* focal plane array. The high spatial dynamic range reveals varying textures across the cloud and clues to the prevailing physical processes.

2. Velocity structure function

A primary goal in the study of ISM physics is to determine the degree of spatial correlation of velocities from observational data. The velocity structure function, $S_q(\tau)$, defined as

$$S_q(\tau) = <|v(r) - v(r+\tau)|^q>$$

provides a statistical measure of the qth order of velocity differences of a field as a function of spatial displacement or lag, τ. For q=2, $S_2(\tau)$, the autocorrelation function, $C(\tau)$, and the power spectrum are equivalent statistical measures of the velocity field. $S_2(\tau)$ is related to the autocorrelation function as

$$S_2(\tau) = 2(C(0) - C(\tau))$$

and $C(\tau)$ is the Fourier transform of the power spectrum.

Within the inertial range of a gas flow, the structure function is expected to vary as a power law with spatial lag,

$$S_q(\tau) = \delta v^q(\tau) \propto \tau^{\zeta_q}.$$

Taking the qth root of the structure function, this expression can be recast into an equivalent linear form, $(S_q(\tau))^{1/q} = <\delta v>_q = v_\circ \tau^{\gamma_q}$ where $\gamma_q = \zeta_q/q$. The power law index, γ_q, measures the degree of spatial correlation and is predicted by model descriptions of turbulence (ex. $\gamma_3 = 1/3$ for Kolmogorov flow). The normalization, $v_\circ$, is the amplitude of velocity fluctuations at a fixed scale and offers a convenient measure of the energy density of a flow.

While the expression for the structure function of a velocity field appears straightforward, the construction of $S(\tau)$ from observational data is, in fact, quite challenging. Observers do not measure velocity *fields*, $v(r)$. Rather, the basic unit of data is a spectrum of line emission that represents a convolution and line of sight integration of density, velocity, and temperature. Furthermore, the effects of chemistry, opacity, and noise can mask or hide contributions to the line profile from features along the line of sight. Despite these limitations, there have been several demonstrated methods to recover the spatial statistics of GMC velocity fields from spectroscopic imaging data.

Analysis of Velocity Centroids: A spectroscopic data cube can be condensed into a 2 dimensional image of centroid velocities determined from the set of line profiles. The spatial statistics of velocity centroids can be formally related to those of the 3 dimensional velocity field (see Ossenkopf *et al.* 2006). With the centroid velocity image, one can assess the power spectrum and hence structure function directly or apply a kernel to calculate the variance of centroid velocities over varying scales (Ossenkopf *et al.* 2006). This method works best under uniform density conditions (Brunt & Mac Low 2004) or with an iterative scheme to account for density fluctuations within the measured power of the observed signal.

Velocity Channel Analysis: Lazarian & Pogosyan (2000) demonstrate a relationship between the power spectra of measured line emission and the respective spectra of the density and velocity fields. The relationship depends on the width of the velocity interval. By calculating the power spectra for both thick and thin velocity windows, one can estimate the power law indices for both the density and velocity fields.

Principal Component Analysis: The spectroscopic data cube is re-ordered onto a set of eigenvectors and eigenimages (Heyer & Schloerb 1997; Brunt & Heyer 2002). The eigenvectors describe the velocity differences in line profiles and the eigenimages convey where those differences occur on the sky. The structure function is constructed from the velocity and angular scales determined from the set of respective eigenvectors and eigenimages that are significant with respect to the noise of the data. To date, the results from PCA have been empirically linked to the velocity structure function parameters based on models under a broad range of physical and observational conditions (Brunt & Heyer 2002; Brunt *et al.* 2003).

2.1. *Universality of turbulence*

The three methods described in the previous section provide valuable tools to determine the velocity structure function for a singular interstellar cloud from a set of spectroscopic imaging data. However, for most observations, the statistical and systematic errors for the derived power law index for a given cloud are large ($\sigma_\gamma/\gamma \sim$ 10-20%) and preclude a designation of a turbulent flow type. Moreover, given the broad diversity of environments and physical conditions within the molecular ISM, any single measurement of a cloud is unlikely to characterize the complete population. Therefore, it is imperative to analyze a large sample of molecular clouds to assess the impact of local effects and to identify trends and differences.

For Velocity Centroid Analysis, Miesch & Bally (1994) analyzed a set of 12 clouds or sub-regions within giant molecular clouds. They determine a mean value of γ to be 0.43 ± 0.15. Using PCA, Heyer & Brunt (2004) studied 28 clouds in the Perseus and local spiral arms and found $\gamma = 0.49 \pm 0.15$. This mean value for γ is consistent with highly supersonic turbulence in which the velocity field is characterized by ubiquitous shocks from converging gas streams. The observed distribution of γ would exclude a Kolmogorov description of incompressible turbulence unless the velocity fields are characterized by strong intermittency. Moreover, they identified the surprising result that the scaling coefficient, v_o, exhibits little variation from cloud to cloud, despite the large range in cloud sizes and star formation activity. Effectively, when the individual structure functions are overlayed onto a single plot, they form a nearly co-linear set of points (see Figure 2).

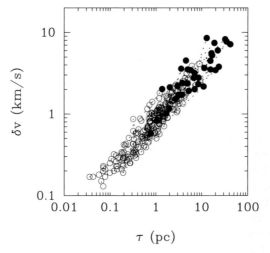

Figure 2. The velocity structure functions for 29 clouds derived from PCA of ^{12}CO J$=$1-0 data cubes (Heyer & Brunt 2004). The nearly co-linear set of points attest to the near-invariant functional form of structure functions despite the large range in size and star formation activity. The filled circles are the upper endpoints for each structure function and are equivalent to the size and global line-width for each cloud.

A necessary consequence of this universality is the Larson (1981) cloud-to-cloud size-linewidth relationship. Basically, the upper endpoint of each individual structure function corresponds to the global size and line width of a cloud (filled points in Figure 2). This set of endpoints are correlated only by the fact that the individual velocity structure functions are described by similar values for γ and v_o. If there were significant variations of these parameters, then the scatter of points on the cloud-to-cloud relationship would be much larger than is observed. Using Monte Carlo simulations to model the scatter of line-width and size for GMCs in the inner Galaxy, Heyer & Brunt (2004) constrain the variation of γ and v_o to be less than 10–15%.

3. Turbulent driving scales

The measurements of velocity structure functions in the molecular ISM point to supersonic turbulent flows in which energy is dissipated in shocks. Unless this energy is replenished within a crossing time, the velocity field would evolve into a Kolmogorov flow comprised exclusively of solenoidal or eddy-like motions. The fact that we observe supersonic turbulence demonstrates that such driving sources must be present in the molecular ISM. Miesch & Bally (1994) summarize candidate sources of energy that could sustain the observed turbulent motions. These include sources that may be resident within the molecular cloud such as protostellar outflows and intermediate and external sources such as HII regions, supernova remnants, and Galactic shear. While all such sources make some contribution, it is important to assess whether any one process is the dominant source.

As first noted by Larson (1981), the universality of velocity structure functions imply a common, *external* source of energy. Otherwise, those regions with significant localized sources would exhibit significant departures from the observed universal relationship. However, GMCs with rich young clusters, and OB stars show the same amplitude of velocity fluctuations as low mass star forming clouds or even those few clouds with negligible star formation activity (Heyer, Williams, & Brunt 2006). Either there are self-regulating processes independent of energy input scales that maintain this amplitude for most interstellar clouds or such internal energy sources contribute only a small fraction of the energy budget of a molecular cloud.

Large scale driving is also implied by the observation that most of the kinetic energy of a cloud is distributed over the largest scales (Brunt 2003). This is illustrated in Figure 3, which displays the first and second PCA eigenimages derived from ^{12}CO J = 1-0 emission from the NGC 7538 molecular cloud. The first eigenimage is similar to an integrated intensity image over the full velocity range of the cloud. The second eigenimage exhibits a dipole-like distribution that identifies the large scale shear across the cloud. All clouds studied by Heyer & Brunt (2004) exhibit this dipole distribution in the second eigenimage. For comparison, we show the first two eigenimages calculated from simulated observations of velocity and density fields produced by computational models that are driven at small, intermediate, and large scales. Using the ratio of characteristic scales determined from each eigenimage, one can quantitatively show that the observations are best described by a large scale driving force (Brunt 2003). Protostellar outflows can have a significant but localized impact on a sub-volume of a cloud and can redistributed energy and momentum to large scales (see Bally in these proceedings). However, it seems quite unlikely that an ensemble of widely distributed outflows within the cloud's volume, can generate the large scale shear that is observed within all molecular clouds analyzed to date.

4. Conclusions

Our understanding of turbulence in the molecular interstellar medium has greatly advanced over the last 10 years owing to more sophisticated computational simulations and ever improving observations. However, there are more critical questions to address to improve these descriptions of turbulence and the role it plays in the star formation process.

• Does the shape of the velocity structure function for a given region change at spatial scales smaller than current resolution limits?

• Does the universality of velocity structure functions extend to the extreme environment of the Galactic Center?

• Are velocity fields of interstellar clouds anisotropic as predicted by the theory of strong, MHD turbulence (Goldreich & Sridhar 1995)?

New telescopes, instrumentation, and analysis methods will be required to address these questions. ALMA will provide both sensitivity and angular resolution to investigate

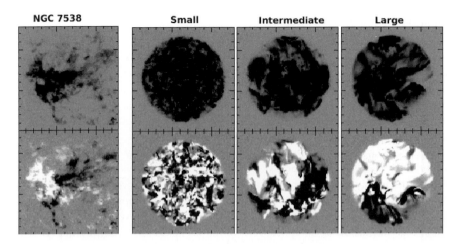

Figure 3. The first two eigenimages derived from ^{12}CO J = 1-0 emission from the NGC 7538 molecular cloud and model spectra from computation simulations driven at small, intermediate, and large scales (Mac Low 1999). Most observations of molecular clouds exhibit a dipole distribution in the second eigenimage that places most of the kinetic energy within the largest scales of a cloud. The observations are more congruent with computational simulations driven at scales larger than or equal to the size of the cloud.

velocity structure functions at the smallest scales and for distant GMCs. The Large Millimeter Telescope will offer the capability to study the low surface brightness component of the molecular ISM to trace the transition from turbulent diffuse material to the dense proto-stellar and proto-cluster cores. These instruments, and others, offer exciting, scientific opportunities to advance our knowledge of interstellar turbulence.

Acknowledgements

This work was supported by NSF grant AST 05-40852 to the Five College Radio Astronomy Observatory. C.B holds a RCUK Academic Fellowship at the University of Exeter. M.H. acknowledges support from the AAS International Travel Grant Program.

References

Brunt, C.M. 2003, *ApJ* 584, 293
Brunt, C.M., Heyer, M.H., Vazquez-Semadeni, E. & Pichardo, B. 2003, *ApJ* 595, 824
Brunt, C.M. & Heyer, M.H. 2002, *ApJ* 566, 276
Elmegreen, B.G. & Scalo, J. 2004, *ARAA*, 42, 211
Goldreich, P. & Sridhar, S. 1995, *ApJ* 438, 763
Heyer, M.H. & Schloerb, F.P. 1997, *ApJ* 475, 173
Heyer, M.H. 1999, in: J.G. Mangum & Simon, J.E. Radford (eds.), *Imaging at Radio through Submillimeter Wavelengths* (ASP-CS), 217, 213
Heyer, M.H. & Brunt, C.M. 2004, *ApJ* 615, L45
Heyer, M.H., Williams J.P. & Brunt, C.M. 2006, *ApJ* 643, 956
Larson, R.B. 1981, *MNRAS* 194, 809
Lazarian, A. & Pogosyan, D. 2000, *ApJ* 537, 720
Mac Low, M. 1999, *ApJ* 524, 169
Mac Low, M. & Klessen, R.S. 2004, *Rev. Mod. Phys.* 76, 125
Miesch, M.S. & Bally, J. 1994, *ApJ* 429, 645
Ossenkopf, V., Esquivel, A., Lazarian, A. & Stutzki, J. 2006, *A&A* 452, 2230

Padoan, P. & Nordlund, A. 2002, *ApJ* 576, 870
Wilson, R.W, Jefferts, K.B. & Penzias, A.A. 1970, *ApJ* 161, L43

Discussion

BLITZ: Have you looked at Maddalena's cloud and the high latitude molecular clouds, both of which have no internal energy sources, and for high latitude clouds, which have a small input scales.

HEYER: Maddalena's cloud is in the sample. Adding the HLCs or diffuse clouds would be interesting in assessing any departure from universality.

MAC LOW: Please amplify on your conclusion that large-scale driving rules out turbulent fragmentation as a control of star formation.

HEYER: It is not large scale driving that rules out turbulent fragmentation. Rather it is the universality of the velocity structure function parameters. In particular, the scaling coefficient is near-invariant. For 2 clouds with extremely different star formation efficiencies, one would expect different turbulent properties to have set these respective star formation properties.

VAZQUEZ-SEMADENI: I don't see the connection between the universality of the turbulence and whether turbulent fragmentation regulates star formation.

HEYER: The universality of structure function parameters excludes turbulent fragmentation as the *exclusive* agent that regulates star formation. Otherwise, I would expect different turbulent properties for regions with different star formation properties.

RICHARDS: The apparent observable scales of turbulence can be affected by other fluctuations, e.g., chemical, outflows, cloud collision - variations are seen on some scales in outer galaxy

HEYER: I agree that there are many ISM processes that can affect turbulent scales and properties. However, the observations demonstrate that the velocity structure function is nearly invariant over the range of resolution of current data (0.1 pc to 50 pc).

CLARK: Is it not possible that the clouds (Rosette, Maddelena) are simply at different stages in their evolution?

HEYER: It is possible! However, we don't rely on single objects within any '1 region of star formation. One would hope that our multiple objects span a range of evolution. Also, the structure functions should also evolve with time - yet, these are nearly invariant.

ELMEGREEN: I think what you are seeing on large scales is a near-uniformity in structure function because of a near-uniformity in ambient pressure. Star formation occurs at the bottom of the spectrum at very small scales. If turbulence triggers star formation in one region and not another, then presumably the structure in the triggered region goes down to stellar mass scales and the clumps are self-gravitating there, whereas in the quiescent clouds, the cascade ends where the clumps are gravitationally stable. So there is another variable that determines whether stars form, and you don't observe that yet because is it at too small a scale.

HEYER: Our scales extend from 0.1 pc to 50 pc that should encompass many of the processes that are forming dense cores. Certainly, turbulent fragmentation promotes density fluctuations in these regime but cannot account for the wide range of star forming properties with our sample of clouds.

Triggered Star Formation in a Turbulent ISM
Proceedings IAU Symposium No. 237, 2006
B. G. Elmegreen & J. Palouš, eds.

Turbulence in high latitude molecular clouds

S. N. Shore[1,2], T. N. LaRosa[3], L. Magnani[4], R. J. Chastain[4] and F. Costagliola[1]

[1]Dipartimento di Fisica "Enrico Fermi", Università di Pisa, Pisa 56127, Italy
email: shore@df.unipi.it, francesco.costagliola@df.unipi.it

[2]INFN-Sezione di Pisa

[3]Department of Biological and Physical Sciences, Kennesaw State University, Kennesaw, GA
USA
ted@avatar.kennesaw.edu

[4]Department of Physics, University of Georgia, Athens, GA USA
email: loris@physast.uga.edu

Abstract. We summarize a continuing investigation of turbulence in high-latitude translucent molecular clouds. These low mass ($\sim$50–100 M$_\odot$), nearby ($\sim$100 pc), non-star forming clouds appear to be condensing out of the atomic cirrus. Unlike star-forming clouds the velocity fields in the clouds must be driven by external processes. Our detailed mapping of the clouds MBM 3,16 and 40 indicates that the dynamics in these clouds result from the combination of shear-flow and thermal instabilities, not shocks. These clouds also show coherent structures, non-Gaussian PDFs but no clear velocity-size relation. Lastly, the energetics of these clouds indicate that radiative loss may terminate the cascade before local heating takes place.

Keywords. turbulence, hydrodynamics, instabilities, ISM: kinematics and dynamics, ISM: clouds, radio lines: ISM

1. Introduction

The general problem of interstellar turbulence has changed its character in recent years from a debate over its existence to a wide variety of studies addressing its diagnosis and role in the dynamics of the large scale medium and the regulation of star formation (e.g. Elmegreen & Scalo 2004, Mac Low & Klessen 2004). For this symposium, the main reason we should care about the turbulence is its feedback into the regulation process for star formation. When the turbulent cascade is in equilibrium, the effect of the turbulent pressure can far outweigh that of the gas in providing dynamical support for the rest of the cloud.

The signatures of turbulence are usually based on spectral line diagnostics and morphology, often using analytic techniques derived from laboratory experience. In many environments, however, experiments cannot serve as an effective guide because of the role played by strong interactions among outflows from recently formed embedded stars, ionization regions, magnetic fields, and self-gravity. These dominate the dynamics of massive, dark clouds: distributed internal sources can easily power the turbulent motions and wipe out the coherence signatures. These flows are too complicated, and previous studies (e.g. Ori, Miesch & Bally 1994) find no correlation scale, evidently because of the long range fluid interactions once jets and winds appear. But collectively the translucent clouds present a very different case study. These are dull, boring, clouds – not forming stars – yet they show all the usual signatures of large scale internal turbulent motions. As we will discuss, even in the absence of self-gravity, the kinematics show an essential element of a turbulent flow, intermittency through the observation of non-Gaussian PDFs, coherent structures, and locally sub-dispersed regions. Finally, we hope the reader will bear in mind that whatever the final mechanism for initiating star formation, the first turbulent dissipative structures to condense in the baryonic component within dynamically stirred dark halos should have looked like translucent clouds *before* star formation started (e.g. the first stars).

The translucent clouds (hereafter TCs) show ordering and structure on many length scales, hence also involving a variety of different scales of mass and time. On the largest size, up to about 10 pc, they are connected with H I clouds that are often parts of larger complexes. All three of the clouds we have studied, MBM 3, 16, and 40 are parts of much larger filamentary and/or sheetlike structures seen with IRAS 100μ images. In these clouds we find quite substantial, large scale systematic velocity gradients, $\sim$(1–3) km s^{-1}pc^{-1}, extending to scales of several pc. For MBM 3, at least, there are several sheets that appear in the channel maps at the same velocity with spacings of about 0.5 pc and also smaller scale (0.1 to 0.5 pc) shear flows that also display sizable velocity gradients but more jetlike (see the movie accompanying Shore *et al.* 2006). On the next scale down, ≈ 1 pc, these clouds display coherent structures and fine structure in velocity and space on typical scale 0.1-0.5 pc. By this we mean regions that are spatially distinct – local column density enhancements that also appear to correspond to volume density peaks – for which the internal centroid velocities are subdispersed relative to the line widths for individual profiles. This is especially true for MBM 16 for which several regions show extremely narrow profiles in both ^{12}CO and H$_2$CO. Finally, these structures show finer scale in the profiles of dense gas tracers such as CS (2-1), CS (1-0), and HCO$^+$. A puzzling feature is the presence of "hairpin structure", commonly observed in terrestrial shear flows (Sreenivasan & Antonia 1997), for the molecular gas within the H I envelope. This may be a red herring, but it is surprisingly frequent (e.g. MBM 16, LaRosa *et al.* 1999; MBM 40, Shore *et al.* 2003, see also Chol Minh *et al.* 2003; Polaris Flare, Falgarone *et al.* 1998) and also appears in numerical simulations (such as those reported at this meeting by Vazquez-Semadeni). In Figs. 1 and 2 we show an example of one such atomic structure, the H I cloud in which MBM 40 is embedded, for which an Arecibo map shows correspondence between the CO, the FIR (which agrees well with the CO, Shore *et al.* 1999) and atomic gas. The 21 cm emission is far more extended than that specifically associated with the velocity range of the CO and the molecular gas is concentrated in a 'hole' in the H I distribution.

2. Analysis methods: probability distribution functions (PDF) and velocity autocorrelation functions (ACF)

The velocity probability distribution function (PDF) is a measure of the frequency distribution of velocities, in effect a histogram of the fluctuations. In laboratory (shear) flows the velocity difference PDFs systematically show non-Gaussian behavior (Minier & Peirano 2001) and in the case of the ISM are a more robust measure of the turbulence (see Miesch, Scalo & Bally 1998 hereafter MSB99; Falgarone & Phillips 1990). Regardless of the parent distribution for the velocity fluctuations, their *PDFs* should approach a Gaussian at sufficiently large lag, when the turbulent motions are completely uncorrelated. In astrophysical studies to date several different schemes have been used to compute this diagnostic, correcting for systematic flows by applying different detrending algorithms to the data. For example, Kitamura *et al.* (1993), Miesch & Bally (1994), Miesch *et al.* (1999), LaRosa, Shore, & Magnani (1999), and Shore *et al.* (2003) removed any large scale trends in their maps by computing a mean fitted, or smoothed map, subtracting this

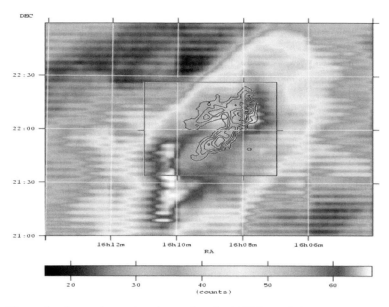

Figure 1. The molecular and atomic relationship for MBM 40. Color represents H I (Arecibo) and contours superposed on the H I map are CO(1-0) emission (FCRAO). The H I spans 2.6–3.5 km s^{-1}, the velocity range covered by the CO. See Shore *et al.* (2003) for details.

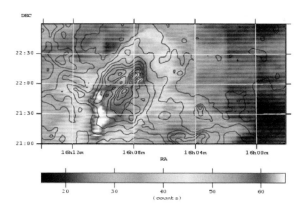

Figure 2. The large-scale environment of MBM 40: H I (Arecibo) and 100μ IRAS emission. Same as Figure 1, except that the contours represent IRAS 100μ emission. The infrared emission is tightly correlated with the total hydrogen column density (atomic plus molecular).

from the original data and analyzing the residual fluctuations. In laboratory turbulence studies this corresponds to the usual Reynolds decomposition that focuses on the fluctuations about the mean flow. Detrending is especially important for correlation analyses since large scale gradients will dominate any correlation and mask correlations generated by turbulence. Alternatively, Ossenkopf & Mac Low (2002) and Pety & Falgarone (2003) did not explicitly remove the large scale trends, arguing that on the scale of their maps such flows are themselves part of the overall cascade and should not be separated. If the shear is powering the cascade, however, it is necessary to remove it to see the smallest scales of the correlation in an ACF but not in a PDF analysis (see Minier & Peirano 2001).

What do PDFs show that the ACF and Principal Component Analysis (PCA) methods don't? Regardless of the parent distribution for the velocity fluctuations, their *PDFs* should approach a Gaussian at sufficiently large lag, when the turbulent motions are completely uncorrelated as required by the Central Limit Theorem. This convergence is also a measure of the correlation length independent of an explicit measurement of the correlation function. The PDF can be described by generalized Lévy processes. Lévy processes are characterized by statistical distributions that have no well defined mean or variance. That is they have long tails indicating that large events can occur with appreciable probability. Although a Gaussian PDF can result from a merely chaotic processes, a Lèvy process cannot. *The smallest scales sampled by the shift PDFs (Lagrangian) for only a few lags, or from those obtained using smoothed centroid map differences (Eulerian) must be detecting the comparatively small regions in which the cascade terminates and the kinetic energy dissipates.*

3. The velocity-size relation in the translucent clouds

First suggested in the 1980s, for a variety of reasons one would expect some sort of dynamical connection between the size and velocity dispersion for a virialized and/or turbulent structure. One was found very early in the work on cloud structures, the first proposal by Larson (1981), for a heterogeneous sample, was $\Delta V \sim L^{0.38}$. As Larson noted, this is suggestive of a turbulent cascade $\epsilon = $ constant with $\Delta V \sim L^{1/3}$ (Richardson-Kolmogorov scaling). Being astrophysicists, it sufficed at first that 0.38 is close enough to $1/3$ to be a reasonable explanation. A similar relation results simply from assuming the clouds are virialized regardless of internal activity or scale. But it's not obvious we are seeing a *single process*, or even a stationary one, in any line of sight and many different environments were combined in this analysis.

Further refinements have changed this simple picture. The signature differs depending on environment. On the largest scale, tens of pc $\alpha = 0.5 \pm 0.05$ (Dame *et al.* 1986, Solomon *et al.* 1987). Multi-cloud, single tracer: $\alpha = 0.31 \pm 0.07$ (Kawamura *et al.* 1998); 0.26 ± 0.09 (Yonekura *et al.* 1997); $\alpha = 0.62 \pm 0.09$ for $3 \leqslant L \leqslant 30$ pc, PCA using FCRAO cubes (Heyer & Brunt 2004). On the smaller scales, the picture is far less clear. High density cores, $\alpha = 0.23 \pm 0.03$ (high mass), 0.53 ± 0.07 (low mass) (Caselli & Myers 1995). In contrast, for the high latitude TCs we don't seem to see anything systematic. For all three clouds, especially our most recent study of MBM 3, the line-widths are *independent* of the region size! Although this result contrasts sharply with those from the FCRAO survey, presented in this meeting by Heyer, the two sets of results may yet be reconcilable. Why should we see this relation on large and not small scales? On the scale of > 1 pc, mixtures of random and systematic motions along any line of sight contribute to the line widths. Systematic motions and gradients are detected on a variety of size scales from velocity centroid maps – when large enough (relative to the intrinsic velocity dispersion) they dominate the line widths. Increasing the surveyed area increases the chance of encountering large gradients, hence ΔV *must* increase. In contrast, for the smaller scales, ΔV is dominated by smaller scale random or turbulent motions, since it is unlikely that small clouds will have more than a single velocity gradient. The velocity dispersion should no longer depend on the size of the sample or its location in the cloud. However, velocity shears and gradients are often unstable, transforming ordered into turbulent motions, so *if* a cascade develops, any large scale gradients are smeared out, leaving only small scale centroid fluctuations.

4. Energetic considerations for these clouds

The bottom line from these studies is that the driving for the high latitude clouds appears to be *external*, by large scale shear flows in the H I medium from which these objects are condensing. The scale for the systematic motions, of order parsecs, may also connect several structures within the same larger HI concentration, two questions naturally arise. Is there evidence for any connection between TCs within the same IRAS or HI structure – e.g. shells and/or filaments – and how do these form as cold molecular clouds embedded within the atomic gas?

Where does turbulent dissipation happen? In the original Kolmogorov picture (also called K41), the cascade proceeds from the injection scale to smallest scales at which some form of smooth distribution of fluctuations mimics viscosity and dissipates the energy into localized heating. Between these extremes, the assumption is a monotonic energy transfer between scales, ending in the intermittent regime above the viscous (Kolmogorov) scale (the so-called K62 "second hypothesis"). We can estimate the heating within the cascade by assuming within the inertial subrange transfer rate is given by $\epsilon_{\text{trans}} \sim \rho(\sigma_{v,\ell})^3 \ell^{-1}$ for a velocity dispersion $\sigma_{v,\ell}$ at the associated length $\ell(\leqslant 0.1\text{pc})$ in the cascade).

Any line profile samples the entire line of sight within a beam (select any ℓ); e.g. MBM 3, $\sigma_v \approx 1.7$ km s^{-1}, independent of the location so for $n_{H_2} \sim 10^3\text{cm}^{-3}$, $\epsilon_{\text{trans}} \approx 3 \times 10^{-23}$ erg s^{-1} cm^{-3}. Assume instead $\sigma_v \approx$ (FWHM 1-lag shift PDF, $\ell \approx 0.03$ pc); thus, for MBM 3, $\sigma_v \approx 0.4$ km s^{-1}, $\epsilon_{\text{trans}} \approx 3 \times 10^{-24}$ erg s^{-1}cm^{-3}. To estimate the rate of energy injection from the shear flow approximate the turbulent viscosity as $\eta_T = \rho\ell\sigma_v$ taking σ_v from the line profiles and estimating ℓ with the correlation length. Since $\epsilon_{\text{inj}} = \eta_T(\Delta V/L)^2$ for a velocity gradient $\Delta V/L$ (for these clouds it is typically ≈ 3 kms^{-1} pc^{-1}), $\epsilon_{\text{inj}} \approx 10^{-24}$ erg s^{-1} cm^{-3}. The magnitude of this injection rate seems similar in at least the clouds observed to date.

Numerical simulations of driven MHD turbulence give, for $T \approx 10$ K and the typical observed column density, 7×10^{20} cm^{-2} (e.g. MBM 3), a CO cooling rate of 10^{-24} ergs s^{-1}cm^{-3} (Juvela, Padoan & Nordlund 2001). *Therefore, CO cooling may keep up with the turbulent energy production.* It thus seems likely that at least in some of these clouds the regions of intense dissipation found in some fine structure surveys may be indicating several channels for the transfer of energy, both by loss and by powering chemistry (i.e. Falgarone, these proceedings). It is not clear that in a medium with radiative loss that the cascade must necessarily proceed to small scales and result in heating.

5. Conclusions

We end with a summary of our conclusions for turbulence in non-star forming, non-self gravitating translucent clouds. MBM 3, MBM 16, and MBM 40 have been mapped with high spatial (0.03 pc) and velocity resolution ($\leqslant 0.08$ km/s) in ^{12}CO (1-0) ^{13}CO (1-0) (NRAO, FCRAO). All show evidence for large-shear flows that we propose powers the turbulent motion. The densest gas is structured into filaments and knots, having a characteristic size of order 0.5 pc. This is the same scale on which we find correlation and also typical of the sub-dispersed regions. Their similarity to laboratory coherent structures may provide another signature of a shear flow origin. The centroid velocity probability distribution function (PDF) is a more precise measure of turbulence. The PDFs exhibit broad wings, consistent with a Lorentzian distribution and showing evidence of non-Gaussian correlated processes. *This is a clear signature of intermittency.* The density field is more likely the result of thermal instability. The line profiles do not change in going across a filament and there are no centroid changes at the boundaries and, evidently, no shocks. And finally, no systematic ΔV-L relation; shears and fine structure can account for the diverse results.

Ultimately the implication for these results for star formation and turbulence is that an external driving, *as well as internal stirring*, can produce a substantial cascade that dynamically structures molecular clouds. Cosmologically this may be important since the first stars formed out of gas that had been dynamically forced by large scale flows that have nothing to do with stellar input.

Acknowledgements

We thank the organizing committee for their kind invitation. SNS thanks the IAU for a travel grant and MIUR/Italy and INFN for additional support of the research presented here. SNS and FC thank the staff of OSO and RadioNet for support. TNL thanks the Kennesaw State University Foundation for a travel grant. We especially want to thank Enrique Vazquez-Semadeni, Bruce Elmegreen, John Black, and Edith Falgarone for discussions.

References

Caselli, P. & Myers, P. C. 1995, *ApJ* 446, 665
Chol Minh, Y. C. Y., Kim, H-G, Lee, Y., Park, H., Kim, K-T & Kim, S. J. 2003, *New Astr.* 8, 795
Dame, T. M., Elmegreen, B. G., Cohen, R. S. & Thaddeus, P. 1986, *ApJ* 305, 892
Elmegreen, B. G. & Scalo, J. 2004, *ARA&A* 42, 211
Falgarone, E., Panis, J. F., Heithausen, A. & Perault, M., *et al.* 1998, *A&A* 331, 669
Falgarone, E. & Phillips, T. G. 1990, *ApJ* 359, 344
Goodman. A. A., Barranco, J. A., Wilner, D. P. & Heyer, M. H. 1998, *ApJ* 504, 223
Heyer, M. H. & Brunt, C. M., 2004, *ApJ* 615, L45
Juvela, M., Padoan, P. & Nordlund, Å. 2001, *ApJ* 563, 853
Kawamura, A., Onishi, T., Yonekura, Y., Dobashi, K., Mizuno, A., Ogawa, H. & Fukui, Y. 1998, *ApJS* 117, 387
LaRosa, T. N., Shore, S. N. & Magnani, L. 1999, *ApJ* 512, 761
Larson, R. B. 1981, *MNRAS* 194, 809
Magnani, L., LaRosa, T. N. & Shore, S. N. 1993, *ApJ* 402, 226
McComb, W. D. 1992, The Physics of Fluid Turbulence, (London: Oxford Univ. Press)
Miesch, M. S. & Bally, J. 1994, *ApJ* 429, 625
Miesch, M. S., Scalo, J. & Bally, J. 1999, *ApJ* 524, 895
Minier, J. & Peirano, E. 2001, *Phys. Rep.* 352, 1
Ossenkopf, V. & Mac Low, M.-M. 2002, *A&A* 390, 307
Pety, J. & Falgarone, E. 2003, *A&A* 412, 417
Robinson, S. K. 1991, *Ann. Rev. Fluid Mech.* 23, 601
Shore, S. N., Magnani, L., LaRosa, T. N. & McCarthy, M. N. 2003, *ApJ* 593, 413
Shore, S. N., LaRosa, T. N., Chastain, R. J. & Magnani, L. 2006, *A&A* 457, 197
Solomon, P. M., Rivolo, A. R., Barret, J. & Yahil, A. 1987, *ApJ* 319, 730
Sreenivasan, K. R. & Antonia, R. A. 1997, *Ann. Rev. Fluid Mech.* 29, 435
Yonekura, Y., Dobashi, K., Mizuno, A., Ogawa, H. & Fukui, Y. 1997, *ApJS* 110, 21

Discussion

MAROV: 1. You have found evidence for dispersion velocities but in your data analysis you use Richardson-Kolmogorov approach applicable for isotropic turbulence only. How do you justify this? Or is it just valid along the line of sight? 2. You invoke shear flow to maintain large eddies formation. What do you think about an other mechanism involving enstrophy acting in the background direction of the cascade energy processes background?

SHORE: Thanks - lovely questions. 1. The approach is because at the dissipation scale, and near the intermittency scale, it should be far more isotropic than at the source scale. 2. To address these flows are fully 3D and the underlying topology of the vorticity is poorly known. I agree it will be interesting to study the enstrophy evolution but I suspect it will be observationally very hard.

Triggered Star Formation in a Turbulent ISM
Proceedings IAU Symposium No. 237, 2006
B. G. Elmegreen & J. Palouš, eds.

© 2007 International Astronomical Union
doi:10.1017/S1743921307001172

The turbulent environment of low-mass dense cores

E. Falgarone[1],
P. Hily-Blant[2], J. Pety[1,2] and G. Pineau des Forêts[3,1]

[1]LERMA/LRA, Ecole Normale Supérieure,
24 rue Lhomond, 75005 Paris, France
email: edith@lra.ens.fr

[2]IRAM, 300 rue de la Piscine, 38406 Saint Martin d'Hères, France
email: hilyblan@iram.fr, pety@iram.fr

[3] IAS, Université Paris-Sud, 91405 Orsay, France
email:guillaume.pineaudesforets@ias.u-psud.fr

Abstract. The signatures of intermittent dissipation of turbulent energy have been sought in the translucent environment of a low-mass dense core. Molecular line observations reveal a network of narrow filamentary structures, found on statistical grounds to be the locus of the largest velocity shears. Three independent properties of these structures make them the plausible sites of intermittent dissipation of turbulence: (1) gas there is warmer and more diluted than average, (2) it bears the signatures of a non-equilibrium chemistry triggered by impulsive heating due to turbulence dissipation, and (3) the power that these structures radiate in the gas cooling lines (mostly H_2) is so large that it balances the total energy injection rate of the turbulent cascade, for a volume filling factor of only a few percents, consistent with other observations in the Solar Neighborhood. These filamentary structures may act as tiny seeds of gas condensation in diffuse molecular gas. They do not exhibit the properties of steady-state low-velocity magneto-hydrodynamic (MHD) shocks, as presently modelled.

Keywords. Turbulence, astrochemistry, MHD, ISM:evolution, ISM:structure, ISM:molecules, ISM:cloud, stars:formation, lines:profiles, radio lines:ISM

1. Introduction

Star formation has at least two (presumably coupled) prerequisites: the growth of a dense core up to a gravitationally unstable state and dissipation of a large fraction of the gas turbulent specific energy (see the detailed reviews of Elmegreen & Scalo 2004 and MacLow & Klessen 2004). The timescales over which these processes occur are critical inputs in the estimates of the star formation efficiency. Yet, little is known on the steps at the origin of the formation of dense cores, nor on those causing turbulence dissipation of interstellar turbulence. Whether it occurs in shocks or in intense velocity shears, turbulent dissipation is intermittent in space and time *i.e.* concentrated in a small fraction of the volume (Kolmogorov 1962, Landau & Lifchitz 1959). Therefore, a scale may exist where, locally, the heating rate due to turbulence dissipation dominates all other heating mechanisms, and induces specific signatures in the gas.

We have sought such signatures in maps of molecular lines in the environment of low-mass dense cores still in the early stages of their evolution. Intermittency manifests itself in the non-Gaussian tails of the probability distribution functions of velocity increments (e.g. Frisch *et al.* 1978, Porter *et al.* 1994, Pety & Falgarone 2000). We followed the statistical method of Lis *et al.* (1996), who showed that the increments of line centroid velocities (CVIs) measured between two different line of sight (LOS) trace a quantity

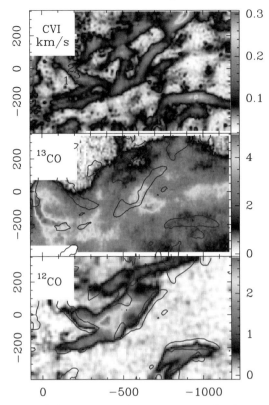

Figure 1. CO observations of the parsec scale environment of a dense core (IRAM-30m observations, resolution 22 arc sec or ~ 0.02 pc). *Top*: map of the largest centroid velocity increments (or velocity shears) for lags of 3 pixels expressed in km s^{-1} (Hily-Blant 2004, Hily-Blant *et al.*, in prep.). The 4 blue numbers give the positions observed in HCO$^+$(1-0). *Middle* : The most opaque regions traced by the ^{13}CO(1-0) line integrated intensity (greyscale in K km s^{-1}) and those of largest velocity shears (blue contours from top panel). The dense core is visible at the eastern edge of the map. *Bottom*: The gas optically thin in ^{12}CO(1-0), thus warmer and more diluted than average (greyscale in K km s^{-1}) and the regions of largest velocity shears (blue contours) (Hily-Blant & Falgarone 2006). The offsets are in arcsec.

related to the LOS average of the plane-of-the-sky projection of the vorticity. It is thus possible, from the CVIs statistics computed in a map of spectra, to approach that of the velocity shears and, most importantly, find the subset of space where the departures from a Gaussian distribution occur.

2. Small-scale velocity structures in a dense core environment

An example is shown in Fig. 1 that displays the parsec scale environment of a low-mass dense core (Heithausen *et al.* 2002) mapped with the IRAM-30m telescope in the ^{12}CO and ^{13}CO J=2-1 and 1-0 transitions (Hily-Blant 2004). The dense core, visible in the East of the central panel, is not isolated but is connected, in space and velocity, to a parsec-scale tail of $\sim 2\ M_\odot$, comparable to its own mass. Longitudinal velocity gradients of a few km s^{-1} pc^{-1} may be interpreted as slow infalling motions onto the core. The upper panel of Fig. 1 shows a new kind of small-scale structures: those traced by the extrema of CVIs, *i.e.* extrema of velocity shear (Pety & Falgarone 2003, Hily-Blant *et al.*, in prep.). They form a conspicuous network of filamentary structures of thickness

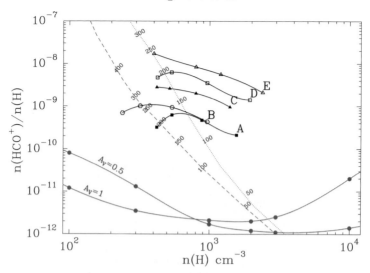

Figure 2. Domain of HCO$^+$ abundances derived from observed spectra (black curves): the HCO$^+$ line intensity increases from curve A to E. The kinetic temperature of the solutions increases from right to left on each curve, and the symbols are located at T_k =20, 40, 100 and 200 K. Predictions of steady-state chemistry computed for two different UV shieldings, bracketing that of the core environment (red curves). Time-dependent HCO$^+$ abundances along two isobaric cooling sequences of gas previously enriched in a burst of dissipation of turbulence (same initial density and two different UV shieldings, $A_v = 0.5$ and 1 mag, blue curves). Labels are gas temperature dropping with time. Models meet observations in the range $T = 100$–200 K, $n_H = 200$–10^3 cm^{-3} (from Falgarone *et al.* 2006).

~ 0.05 pc, unrelated to the tail of dense gas traced by ^{13}CO emission (central panel) and instead clearly associated with gas optically thin in ^{12}CO(1-0) (bottom panel), shown to be warmer ($T_k > 30$ K) and more diluted ($n_{H_2} < 10^3$ cm^{-3}) than the average gas in the core environment (Hily-Blant & Falgarone 2006). Their preferential orientation is close to the direction of the magnetic fields measured one degree away (Heiles 2000).

Observations of one of these filamentary structures with the IRAM Plateau de Bure Interferometer (PdBI) in the ^{12}CO(1-0) line reveal still thinner structures of thickness ~ 800 AU, separated by less than 10 arsec (or 7.5 mpc) in projection and 3.5 km s^{-1} in velocity, all the emission at intermediate velocities being resolved out. Such a pattern may be viewed as a large local velocity shear of ~ 200 km s^{-1} pc^{-1}, or in turn as an impulsive event of timescale $\tau \sim 3000$ yr (Falgarone *et al.*, in prep.).

3. Non-equilibrium chemistry in regions of largest velocity-shear

If large amounts of suprathermal energy are deposited in small regions over such short timescales, the thermal and chemical evolution of the gas there should be different from that of the rest of the volume. We have sought such differences by observing the four positions shown in Fig. 1 in the HCO$^+$(1-0) line. These four positions sample the whole range of velocity-shears in the map from a maximum (#2 and 3) to a minimum (#1).

The observed HCO$^+$ abundances are similar to those obtained by Liszt & Lucas (1996) and Lucas & Liszt (2000) in absorption against extragalactic continuum sources, and exceed by more than one order of magnitude the predictions of steady-state chemistry in gas primarily heated by UV photons and cosmic rays (Fig. 2). In contrast, they can be understood in the framework of the non-equilibrium model described below (Falgarone *et al.* 2006).

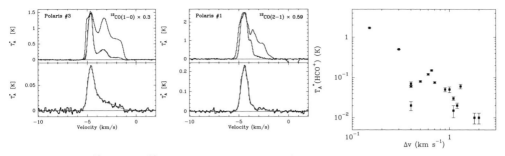

Figure 3. *Left:* ^{12}CO and ^{13}CO(1-0) (top) and HCO$^+$(1-0) (bottom) spectra at positions #3 and #1. *Right:* HCO$^+$ line temperature versus velocity width of all the components identified in the HCO$^+$ line profiles (4 positions). The observed scaling, close to $T_L \propto \Delta v^{-2}$, means $N(\mathrm{HCO}^+) \approx \Delta v^{-1}$ since the lines are almost optically thin.

It is out of reach of present computational capabilities to follow the non-equilibrium chemistry that develops in a turbulent medium at time and size scales close to those of dissipation. An hybrid approach was adopted by Joulain et al. (1998): the time-dependent Lagrangian evolution of a cell of gas is computed as it spirals into a steady-state magnetized Burgers vortex, chosen as a template of turbulent dissipative structure (Moffatt, Kida & Ohkitani 1994). Because of the short timescales, neutrals are only partially coupled to the ions and field. Intense dissipation occurs, both in the layers of largest velocity shear and those of large ion-neutral drift. As the gas cell escapes these *active* layers, it cools down. The temperature excursion therefore includes (1) an impulsive heating over a few 100 yr from $T_k = 80$ K to about 10^3 K, followed by (2) an isobaric relaxation that lasts up to 10^4yr (Falgarone *et al.* 2006).

The main interest of such an hybrid model is to show (1) that chemistry swiftly reacts to sharp temperature variations, triggering a *warm chemistry* by activating a set of endoenergetic reactions that enhance the local production of molecules like CH$^+$, HCO$^+$, CH, OH and H$_2$O by several orders of magnitude, and (2) that molecules formed there survive for more than a thousand years after the gas has escaped the *active* layers, while radiative cooling drives the condensation of the gas. Interestingly, Fig. 2 shows that such non-equilibrium models meet the observational constraints in the range $T = 100$–200 K, $n_{\mathrm{H}} = 200$–10^3 cm^{-3}. Thus, alike MHD shocks, intense and coherent velocity shears induce thermal and chemical excursions in the diffuse gas which eventually turn into high density structures at small scale.

The fact that impulsive heating generates sharp gradients of temperature and molecular abundance over scales much smaller than that of the beam resolution (here, about 0.02 pc) is supported by the following. The four observed HCO$^+$ spectra have multiple velocity components of different linewidths, the broadest present only at positions # 2 and 3 (see Fig. 3). The HCO$^+$ line intensity of these components increases steeply as their linewidth decreases (Fig. 3, right). Since the lines are close to be optically thin, the column density of HCO$^+$ also increases as the linewidth decreases, as $N(\mathrm{HCO}^+) \approx \Delta v^{-1}$. This suggests that: (1) the observed chemical enrichment proceeds at the expense of turbulent energy, traced by the linewidth of each component and (2) that in each beam, gas being enriched, still in *active* layers (the broad and weak components) coexists with chemically enriched gas, already relaxing (the intense and narrower components). As expected, the former is almost absent from the spectrum observed at position #1, an inactive location, and most prominent at positions #2 and 3, located on an extremum of velocity-shear (Fig. 3).

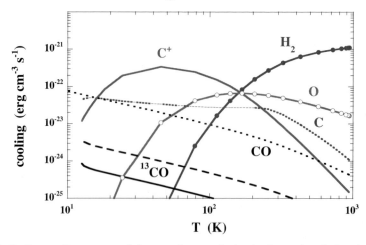

Figure 4. Radiative cooling curves of the gas along an isobaric thermal and chemical relaxation. The time-dependent evolution starts at $T_k = 10^3$ K. H_2 line emission dominates the cooling at high temperatures, and CO lines are never important unless the gas is cold (optically thin (dotted) and thick (dashed) cases).

Alternative scenarios have been proposed to explain the large abundances of CH^+, HCO^+, OH and H_2O in the diffuse ISM. They invoke chemistry developing in warm interfaces between diffuse clouds and the warm intercloud medium (Nguyen, Hartquist & Williams 2001, Lesaffre, Gerin, Hennebelle 2006) or in myriads of low-velocity MHD shocks (e.g. Flower & Pineau des Forêts 1998, Gredel *et al.* 2002).

4. Energetics

The radiative cooling of the gas formerly heated by a dissipation burst and chemically enriched drives its condensation during the relaxation phase. The contribution of all the species contributing is displayed in Fig. 4 as a function of temperature (or time). The pure rotational lines of H_2 dominate the cooling in the warmest stages. At the opposite, between $T_k = 200$ and 100 K, the temperature range of the solutions, the cooling is shared almost equally between H_2 and the fine structure lines of C^+, O I and C I. Note that in that temperature range, CO lines have a negligible contribution since $\Lambda_{tot} = \Lambda_{H_2} + \Lambda_{C^+} + \Lambda_{OI} + \Lambda_{CI} + \Lambda_{CO} = 30$ to $40\,\Lambda_{CO}$.

The radiative cooling rate of the gas in its relaxation phase is much larger than the *average* turbulent heating rate in this dense core environment, estimated to be $\bar{\epsilon}_{turb} = 6 \times 10^{-24}$ erg cm^{-3}s^{-1} (Hily-Blant & Falgarone 2006). For $T_k = 100$–200 K, $\Lambda_{tot} = 2$–3×10^{-22} erg cm^{-3}s^{-1} (Fig. 4). So, even a tiny fraction of warm gas (transient but permanently replenished by dissipation) is able to radiate away the turbulent energy input. The volume filling factor required for the balance is as small as: $f_v = \bar{\epsilon}_{turb}/\Lambda_{tot} = 0.02$ to 0.03. Only a few percent of warm gas in the case of the field under study, are sufficient to radiate the input of energy of the turbulent cascade.

It is all the more interesting that this fraction is also that required, in the Solar Neighborhood, to reproduce the observed abundances of CH^+ (Crane, Lambert & Sheffer 1995, Gredel 1997) and H_2O found by the *SWAS* satellite in diffuse molecular clouds (Neufeld et al. 2002, Plume et al. 2004). Similar findings are inferred from *ISO-SWS* observations of pure rotational lines of H_2 across 15 magnitudes of mostly diffuse gas in the Galaxy (Falgarone *et al.* 2005). The warm H_2 is distributed in a large number of either low velocity MHD shocks or small scale intense velocity-shears, occupying only a few percent of the line of sight. The same fraction of warm H_2 that cannot be heated by the sole UV photons is also derived from far UV spectroscopy with *FUSE* (Gry et al. 2002, Lacour et al. 2005). Warm glitters seem to be ubiquitous, and presumably trace the intermittent dissipation of suprathermal turbulence in the cold gas.

5. Summary and open questions

Signatures of intermittent dissipation of turbulent energy may have been detected for the first time: they appear as extrema of velocity-shears, forming a network of warm and narrow filamentary structures, in the environment of a dense core. These structures are the sites of non-equilibrium chemistry proceeding at the expense of the turbulent energy. Their radiative cooling is powerful and they balance the average turbulent heating while filling only a few percent of the volume. The actual scale at which dissipation of turbulence takes place has not be determined yet. It may be smaller than 800 AU and accessible only to the ALMA interferometer. The full nature of the dissipation process is not understood either: the observational constraints (small sizescale, energy release sufficient to drive the warm chemistry, lack of dense post-shock gas) make the observed structures inconsistent with steady-state low-velocity MHD shocks, as presently modelled.

References

Crane P., Lambert D.L. & Sheffer Y. 1995, *ApJS* 99, 107
Elmegreen B.G. & Scalo J. 2004, *ARAA* 42, 211
Falgarone E., Pineau des Forêts G., Hily-Blant P. & Schilke P. 2006, *A&A* 452, 511
Falgarone E., Verstraete L., Pineau des Forêts G. & Hily-Blant P. 2005, *A&A* 433, 997
Falgarone, E., Pety J. & Hily-Blant, P., in preparation
Flower D. & Pineau des Forêts G. 1998, *MNRAS* 297, 1182
Frisch U., Sulem P.-L. & Nelkin M. 1978, *JFM* 87, 719
Gredel R. 1997, *A&A* 320, 929
Gredel R., Pineau des Forêts G. & Federman S.R. 2002, *A&A* 389, 993
Gry, C., Boulanger, F. Nehmé, C., *et al.* 2002, *A&A* 391, 675
Heiles C. 2000, *AJ* 119, 923
Heithausen A., Bertoldi F. & Bensch F. 2002, *A&A* 383, 591
Hily-Blant, P. 2004, PhD, Université Paris-Sud
Hily-Blant P. & Falgarone E. 2006, *A&A* accepted
Hily-Blant P., Pety J. & Falgarone E., in preparation
Joulain K., Falgarone E., Pineau des Forêts G. & Flower D. 1998, *A&A* 340, 241
Kolmogorov, A.N. 1962, *JFM* 13, 82
Lacour S., Ziskin V., Hébrard G., *et al.* 2005, *ApJ* 627, 251
Lambert D.L. & Danks A.C. 1986, *ApJ* 303, 401
Landau L.D. & Lifchitz E.M. 1959, Fluid Mechanics, Addison-Wesley
Lesaffre P., Gerin M. & Hennebelle P. 2006, *A&A* submitted
Lis D., Pety J., Phillips T.G. & Falgarone E. 1996, *ApJ* 463, 623
Liszt H.S. & Lucas R. 1996, *A&A* 307, 237
Lucas R. & Liszt H.S. 2000, *A&A* 355, 333
MacLow M.-M. & Klessen R. 2004, *Rev.Mod.Phys.* 76, 125
Moffatt H.K., Kida S. & Ohkitani K. 1994, *JFM* 259, 241
Neufeld D.A., Kaufman M.J., Goldsmith P.F., *et al.* 2002, *ApJ* 580, 278
Nguyen, T. K., Hartquist, T. W. & Williams, D. A. 2001, *A&A* 366, 662
Pety, J. & Falgarone E. 2000, *A&A* 356, 279
Pety, J. & Falgarone E. 2003, *A&A* 412, 417
Plume R., Kaufman M.J., Neufeld D.A. *et al.* 2004, *ApJ* 605, 247
Porter D.H., Pouquet A. & Woodward P.R. 1993, *Phys. Fluids* 6, 2133

Discussion

MAC LOW: Why can't these thin, impulsively heated objects be small shocks in a dissipating supersonic turbulent flow?

FALGARONE: Good question! We have no observational evidence for any condensed post-shock layer. But it may well be that the column density of this thin dense layer is still too low to be detected, even with the Plateau de Bure interferometer. If this is true, models will have to explain such thin post-shock layers in the presence of magnetic fields presumably parallel to the shock.

Triggered Star Formation in a Turbulent ISM
Proceedings IAU Symposium No. 237, 2006
B. G. Elmegreen & J. Palouš, eds.

© 2007 International Astronomical Union
doi:10.1017/S1743921307001184

CO, HI, recent Spitzer SAGE results in the Large Magellanic Cloud

Yasuo Fukui

Department of Physics, Nagoya University, Chikusa-ku, Nagoya 464-8602, Japan
email: fukui@a.phys.nagoya-u.ac.jp

Abstract. Formation of GMCs is one of the most crucial issues in galaxy evolution. I will compare CO and HI in the LMC in 3 dimensional space for the first time aiming at revealing the physical connection between GMCs and associated HI gas at a ~ 40 pc scale. The present major findings are 1) [total CO intensity] $\propto$ [total HI intensity]$^{0.8}$ for the 110 GMCs, and 2) the HI intensity tends to increase with the evolution of GMCs. I argue that these findings are consistent with the growth of GMCs via HI accretion over a time scale of a few $\times$ 10 Myrs. I will also discuss the role of the background stellar gravity and the dynamical compression by supershells in formation of GMCs.

Keywords. stars: formation, ISM: atoms, evolution, molecules, Magellanic Clouds, galaxies: star clusters

1. Introduction

The Magellanic system including the LMC, SMC, and the Bridge is an ideal laboratory to study star formation and cloud evolution because of its proximity to the sun. In particular, the LMC offers the best site because of its unrivaled closeness and of the nearly face-on view to us. Among the various objects in the LMC, the molecular clouds which are probed best in the millimetric CO emission, provide a key to understand star formation and galaxy evolution. This is because the molecular clouds are able to highlight the spots of star formation due to their highly clumped distribution both in space and in velocity. The situation should be contrasted with atomic hydrogen gas having lower density and more loosely coupled to the star formation spots.

The key issue I would like to focus on in this talk is the formation of giant molecular clouds (GMC) in the LMC. GMCs are the major site of the star formation and the GMC formation must be a crucial step in the evolution of a galaxy. In order to address this issue, I will make (1) a comparison between HI and CO in 3 dimensional space to understand their physical correlation, (2) a comparison between stellar distribution which is dynamically controlling the HI density, and (3) a comparison with some of the most recent Spitzer SAGE results. The contents of sections 3 and 4 are mainly based on the collaboration with Hinako Iritani and Akiko Kawamura of Nagoya University and Tony Wong of ATNF and will be published elsewhere.

2. Three classes of GMCs

The NANTEN CO survey of the LMC has revealed that there are three classes of GMCs according to the association of young objects (Fukui *et al.* 1999; Yamaguchi *et al.* 2001b; Figure 1). Class I has no apparent sign of star formation and Class II is associated with small HII region(s) only but without stellar clusters. Class III is most actively forming stars as shown by huge HII regions and young stellar clusters. This classification was presented by Fukui *et al.* (1999) based on the first results of the NANTEN survey (see also Mizuno *et al.* 2001; Yamaguchi *et al.* 2001b). The basic scheme of the classification remains valid in the subsequent sensitive survey with NANTEN while the number of GMCs in each class has been increased by a factor of three (Fukui *et al.* 2001; Fukui *et al.* 2006; see also Blitz *et al.* 2006). These Classes are interpreted as an indication of the evolutionary sequence from I to III and the life time of a GMC is estimated to be a few × 10 Myrs in total (Fukui *et al.* 1999). A comparison of physical parameters indicates that size and mass tend to increase from Class I to Class III, and Class III GMC has the largest size and mass among the three. The stage after Class III is perhaps a very violent dissipation of GMCs due to UV photons and stellar winds from formed clusters as seen in the region of 30Dor most spectacularly. More details on this classification are discussed elsewhere in these proceedings (Kawamura *et al.* 2006).

3. HI vs. CO

3.1. *3-D correlation*

Previous studies of star formation in galaxies employed 2-D (2 dimensional) projection of HI intensity at large spatial averaging over ∼ 100 pc – 1 kpc (e.g., Schmidt 1972). We shall here test a 3-D (3 dimensional) comparison of CO and HI in the LMC where the 3-D datacube has a velocity axis in addition to the two axes in the sky. We use the 3-D datacube of CO with NANTEN (Fukui *et al.* 2006) and of HI with ATCA (Kim *et al.* 2003), where CO traces GMCs and HI less dense atomic gas.

Figure 2 shows an overlay of CO and HI and represents that GMCs are located often towards HI filaments. This shows that the CO emission is certainly located towards HI peaks while there are also many HI peaks or filaments without CO emission,

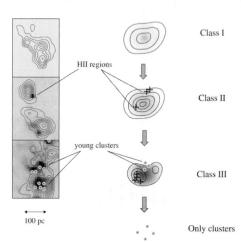

suggesting that HI is the placental site of GMC formation (Blitz *et al.* 2006). We should note that this overlay is of the integrated intensities along the line of sight in velocity space. Figure 2 shows typical CO and HI profiles in the LMC, indicating that the CO emission is highly localized in velocity; the HI emission ranges over 70 km s^{-1} while the CO emission has a width less than 10 km s^{-1}. The large velocity dispersion of HI is most likely due to physically unrelated velocity components in the line of sight and the HI gas associated with the CO gas is a small fraction of HI whose velocity is close to that of CO. Previous studies to correlate HI and CO using such velocity integrated 2-D data obviously overestimates the HI intensity.

Figure 1. Evolutionary sequence of the GMCs in the LMC. An example of the GMCs and illustration at each class are shown in the left panels and the middle column, respectively. The images and contours in the left panels are Hα (Kim *et al.* 1999) and CO integrated intensity by NANTEN (Fukui *et al.* 2006).

The present 3-D analysis is able to pick up the HI gas physically connected to the molecular clouds. Accordingly, we expect it reveals the exact connection of the ambient atomic gas to the dense molecular gas. The present HI and CO datasets have a pixel size of 40 pc × 40 pc × 1.7 km s^{-1} and consist of

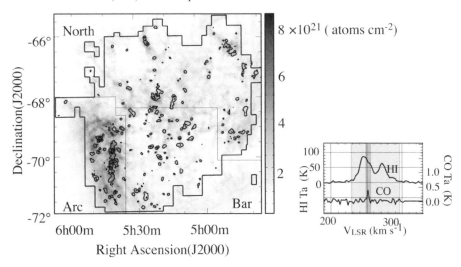

Figure 2. *Left Panel*; HI image (Kim *et al.* 2003) with the CO contours (Fukui *et al.* 2006). The contours are from 1.2 K km s^{-1} with 3.6 K km s^{-1} intervals. The *Right Panel* shows an example of the HI and CO profiles at α(J2000) = $5^h 35^m 42^s$, and δ(J2000) = $-69°11'$.

$\sim$ 2million pixels. The HI and CO intensities are expressed as averaged T_a (K) in a pixel with the lowest value, the 3σ noise level, in each.

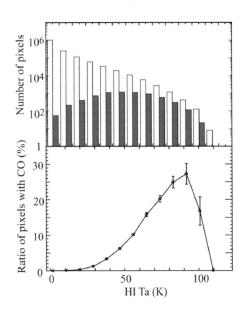

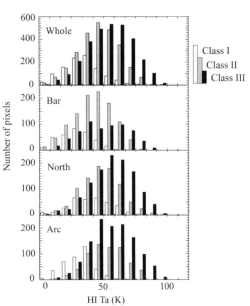

Figure 3. (*Upper Panel*) Distribution of HI integrated intensity in 3-dimension. Distribution of the HI integrated intensity where the significant CO emissions are detected is also shown in gray. (*Lower Panel*) The ratio of the HI pixels with and without CO emission.

Figure 4. Distribution of the HI antenna temperature(T_a) associated with the GMCs in the entire galaxy, the arc , the bar, and the north regions from the top. Each panel indicates the HI intensity distribution associated with the GMC Class I (white), Class II (gray), and Class III (black).

Figure 3 shows a histogram of the HI integrated intensity in 3-D and the pixels with the significant CO emission (greater than 0.7 K km s^{-1}) are marked. This histogram

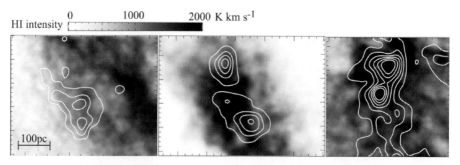

Figure 5. Examples of the HI and CO distributions of the GMC Class I(*Left*), II(*Middle*), and III(*Right*), respectively. Gray images are HI integrated intensity maps; velocity is integrated over the range where the significant CO emissions are detected The contours are CO integrated intensities from 1.2 K km s^{-1} with 1.2 K km s^{-1} intervals(Fukui *et al.* 2006).

shows that the CO fraction increases steadily with the HI intensity, suggesting the HI provides a necessary condition to form GMCs. About one third of the pixels exhibit CO emission near T_a(HI) of ~ 90 K and it seems that there is no sharp threshold value for GMC formation with respect to the HI intensity.

Figure 4 shows a histogram of the HI intensity for the three GMC Classes. This clearly shows that the HI intensity tends to increase from Class I to Class III while the dispersion is not so small. The average HI intensity over the whole LMC is 34 ± 16 K, 47 ± 17 K and 56 ± 19 K for Class I, II and III, respectively. The HI intensity surrounding GMCs becomes greater with the GMC evolution and star formation. In order to test the variation within the galaxy, we shall divide the galaxy into three regions, i.e., Bar, North and Arc (see the left panel of Figure 2). Histograms for each shown in the lower three panels of Figure 4 again indicate the same trend.

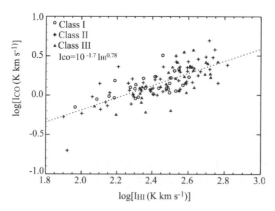

Figure 6. Correlation diagram between HI and CO intensities. Open circles, crosses and open triangles represent GMC Class I, Class II, and Class III, respectively. The dotted line shows the regression line of $I_{CO} = 10^{-1.7} I_{HI}^{0.78}$.

Three images in Figure 5 show the HI distribution associated with GMCs of the three Classes. The CO distribution has detailed structures of ~ 100 pc or less and the HI seems to be associated with the GMC at a scale of 50–100 pc. The HI is not always isotropic with respect to a GMC but indicates close spatial correlations so that the HI is more or less enveloping a GMC. The typical HI velocity width associated is estimated to be ~ 10–14 km s^{-1}, and beyond this velocity span the associated HI peaks generally becomes much weaker or disappear.

We have chosen 110 GMCs with five or more pixels in the two axes in the sky and estimated L_{CO} in the velocity range where CO is detected. Then we divide the L_{CO} by the apparent size of a GMC. This gives an average CO intensity. For each GMC we chose the pixels where the CO emission is significantly detected and tested HI and CO correlation (Figure 6), where only the HI pixels with CO are counted.

The regression shown in Figure 6 is well fitted by a power law with an index of ~0.8, indicating a nearly linear correlation between CO and HI in a GMC. We shall note that this index should become larger as ~1.5 if we use 2-D correlation because the velocity integrated HI intensity has large offsets unrelated with CO.

3.2. *Dressed GMCs; growth of GMCs via HI accretion*

The dependence of HI intensity on Class of GMC indicates that the surroundings of a GMC change appreciably depending on the Class at a length scale of 40–100pc. The HI intensity is generally a product of the spin temperature and the optical depth, and the correlation indicates the temperature and/or density is dependent on Class. The spin temperature of HI is generally estimated to be ~ 100 K in the Milky Way. Since the HI spin temperature may be higher in the LMC where UV is more intense and dust opacity is less than in the Milky Way. The typical HI intensity less than 100 K suggests that the HI emission is optically thin. If so, the HI intensity should represent HI density. We therefore infer that GMCs are "dressed" in HI and that the "HI dress" grows in time.

The correlation between HI and CO is nearly linear (Figure 6). This alone does not provide a strong constraint on the formation of a GMC. Nonetheless, the apparent association of HI with GMCs suggests that the HI is enveloping each GMC and the HI density increases with the cloud evolution. We infer that this represents the enveloping HI gas is accreting onto GMCs and is converted into H_2 due to increased optical depth. This leads to increase the molecular mass of GMC, i.e., the observed mass increase from Class I to III (see section 2). The timescale of the GMC evolution is ~ 10 Myrs and the increased molecular mass is in the order of $10^6 M_\odot$. Namely, a mass accretion rate of $\sim 10^{-1} M_\odot$ yr^{-1} is required. We roughly estimate that this rate is consistent with that calculated for a spherical accretion of the HI gas having $n(HI) \sim 10$ cm^{-3} at an infall speed of ~ 7 km s^{-1}.

4. Stellar gravity and triggering in GMC formation

We shall examine two effects which may be important in converting HI to H_2: one is the stellar gravity and the other supershells driven by OB stars and/or SNRs.

4.1. *Stellar gravity*

We shall here use the K-band image of 2MASS which represents relatively old stellar population dominating the gravity (Sergei *et al.* 2000). This will allow us to test the effects of stellar gravity on GMC formation. The young stars associated with GMCs are small in mass and are not important in such gravity.

Figure 7 shows a histogram of CO clouds vs. K-band stellar density. We find that the number density of CO clouds increases with the increase of the stellar density. It seems that there is a threshold value at ~ 1.0 in log (number of stars/4 arcmin^2) for the HI - H_2 conversion where the number density of CO clouds increases beyond ~ 5 % of the pixel number.

Wong and Blitz (2002) argue that the pressure exerted on HI gas may play a role in converting HI to H_2. They studied several spiral galaxies to show that the HI gas is associated with CO only in the inner part of galaxies where the stellar gravity is strong. The pressure should be basically dominated by the stellar gravity as long as the stars are the dominant source of gravity and the present result on the LMC supports that the pressure plays a role in converting HI into H_2.

4.2. *SAGE results*

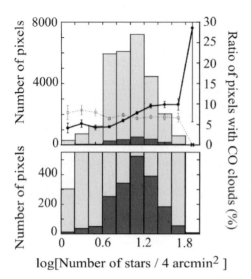

The Spitzer satellite has been used to make an extensive study of the LMC at infrared wavelengths from 3.6–160μm. This is the SAGE project headed by M. Meixner. Figure 8 shows the 3.6–24 μm 3-color composite image (3.6μm, 8.0μm and 24μm in blue, gree, and red) from the SAGE project (NASA/Caltech-JPL/Meixner STScI and the SAGE Legacy Team) with NANTEN CO distribution (Fukui *et al.* 2006). This presents that CO clouds are well correlated with the far-infrared distribution and future comparisons should reveal considerable details of the dust-gas relationship quantitatively.

Figure 7. Distribution of the stellar density derived from the 2MASS K band sources. The histogram in dark gray shows the stellar density distribution where the significant CO emissions are detected. Thick line indicates the ratio of the pixels of the stellar density with and without CO emissions. Thin line shows the ratio as for the thick line but obtained by assuming the CO emissions are distributed randomly. The *Lower Panel* is an enlargement of the *Upper Panel*.

In connection with the GMC formation, we note that the SAGE image indicates numerous good candidates for apparently swept-up matter perhaps by supershells owing to the better angular resolution. Yamaguchi *et al.* (2001a) made a comparison between Hα supershells and CO and concluded that $\sim 30\%$ of the GMCs may be associated with supergiant shells. Since more candidates may be identified from the SAGE image, triggering of GMC formation by supershells may be more important than previously thought by Yamaguchi *et al.* (2001a).

5. Summary

I shall summarize this contribution as follows;

1) 3-D comparison between CO and HI has revealed GMCs are associated with HI gas at a $\sim$ 40pc scale; these are "HI dressed GMCs".

2) There is a clear increase of the HI intensity around GMCs from GMC Class I to Class III. Growth of GMCs in mass via HI accretion has been suggested over a time scale of a few times 10 Myrs.

3) This correlation has a form, [CO intensity] $\propto$ [HI intensity]$^{0.8}$, for selected 110 major GMCs.

4) The background stellar gravity and dynamical compression by supershells may be important in converting HI into H$_2$, i.e. in GMC formation.

Acknowledgements

The NANTEN project is based on a mutual agreement between Nagoya University and the Carnegie Institution of Washington. We greatly appreciate the hospitality of all the staff members of the Las Campanas Observatory. We, NANTEN team, are thankful

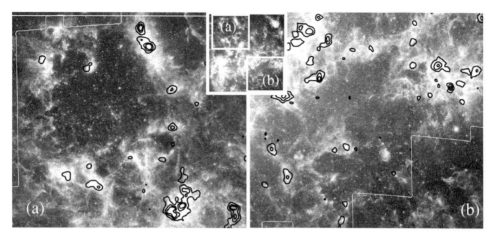

Figure 8. 3.6–24 μm 3-color composite image (3.6μm, 8.0μm and 24μm in blue, gree, and red, respectively) from the SAGE project (NASA/Caltech-JPL/Meixner STScI and the SAGE Legacy Team) of the LMC (*Middle* inset); Overlays with CO contours are shown (a) toward LMC 4, and (b) LMC 7 and 8. The CO contours are from 1.2 K km s^{-1} with 2.4 K km s^{-1} intervals from the CO survey by NANTEN (Fukui *et al.* 2006). The regions of (a) and (b) are indicated by white boxes in the inset.

to many Japanese public donors and companies who contributed to the realization of the project. We would like to acknowledge Drs. L. Staveley-Smith and S. Kim for the kind use of their HI data. We are also thankful to IRAC and MIPS pipeline team to create a beautiful images of the LMC by Spitzer. This work is financially supported in part by a Grant-in-Aid for Scientific Research from the Ministry of Education, Culture, Sports, Science and Technology of Japan (No. 15071203) and from JSPS (No. 14102003, core-to-core program 17004 and No. 18684003).

References

Blitz, L., Fukui, Y., Kawamura, A. Leroy, A., Mizuno, N. & Rosolowsky, E. 2006, in: B. Reipurth, D. Jewitt & K. Keil (eds.), *Protostars and Planets V* (Tucson: Univ. of Arizona) in press

Fukui, Y., Mizuno, N., Yamaguchi, R., Mizuno, A., Onishi, T., Ogawa, H., Yonekura, Y., Kawamura, A., *et al.* 1999, *PASJ* 51, 745

Fukui, Y., Mizuno, N., Yamaguchi, R., Mizuno, A. & Onishi, T. 2001, *PASJ* 53, L41

Fukui, Y., Kawamura, A., Minamidani, T., Mizuno, Y., Kanai, Y., Onishi, T., Mizuno, N., Yonekura, Y., Mizuno, A. & Ogawa, H. 2006, *ApJ* submitted

Kawamura, A. *et al.* 2006, in these proceedings

Kim, S., Dopita, M. A., Staveley-Smith, L. & Bessell, M. S. 1999, *AJ* 118, 2797

Kim, S., Staveley-Smith, L., Dopita, M. A., Sault, R. J., Freeman, K. C., Lee, Y. & Chu, Y.-H. 2003, *ApJS* 148, 473

Meixner, M. *et al.* 2006, *AJ* in press

Mizuno, N., Yamaguchi, R., Mizuno, A., Rubio, M., Abe, R., Saito, H., Onishi, T., Yonekura, Y., Yamaguchi, N., Ogawa, H. & Fukui, Y. 2001, *PASJ* 53, 971

Nikolaev, S. & Weinberg, M. D. 2000, *ApJ* 542, 804

Schmidt, Th. 1972, *A&A* 16, 95

Wong, T. & Blitz, L. 2002, *ApJ* 569, 157

Yamaguchi, R., Mizuno, N., Onishi, T., Mizuno, A. & Fukui, Y. 2001a, *PASJ* 53, 959

Yamaguchi, R., Mizuno, N., Mizuno, A., Rubio, M., Abe, R., Saito, H., Moriguchi, Y., Matsunaga, K., Onishi, T., Yonekura, Y. & Fukui, Y. 2001b, *PASJ* 53, 985

Discussion

KRUMHOLZ: You mentioned that $\sim 1/3$ of GMCs in the LMC show clusters or HII regions, but in the Milky Way we see almost no starless clouds. How can we understand that? Two possibilities: first we have only a poor view of GMCs in the Milky Way, so the statistics may be bad. Second, GMCs in the LMC may evolve more slowly than those in the Milky Way for some unknown reason.

FUKUI: Regarding your first suggestion: the reasonably studied GMCs in the Milky Way are only 30 and are restricted to within a few kpc of the sun because of confusion and extinction in the disk. And for the second point: this remains an interesting possibility.

TAN: If GMCs form quickly from HI, their angular momentum vector directions should be correlated with those of the surrounding HI gas. Do you see this?

FUKUI: This sounds like an interesting work to do. We will try. Thank you for the suggestion.

ZINNECKER: I have not understood what you meant by your concept of "stellar gravity" compression of the GMCs?

FUKUI: The stellar gravity is kpc scale as probed by 2MASS K band image. This gravity confines the HI gas and regulates the pressure in it. This produces higher pressure to help GMC formation.

VAZQUEZ-SEMADENI: How can you tell whether a larger mass of stars is a cause rather than a consequence of the larger gas mass?

FUKUI: The stellar mass I showed is by 2MASS K band data. These stars must be much older and more massive than the gas. So, the stellar gravity potential is a given condition for the HI gas that evolves into Hα in ~ 10 Myr. The newly formed stars must be much less than that K band image shows.

DICKEY: Years ago Ulrich Mebold and I did a survey of 21 cm absorption in the LMC. We found many absorption lines, but none so optically thick that the absorption would obscure any HI emission. For this reason, you can measure the HI column density directly from the 21 cm brightness temperature, without worrying about the excitation temperature of HI. Most of the HI in the LMC is very warm, hundreds to thousands of K.

FUKUI: Thank you, I agree with that.

ELMEGREEN: You are saying there is an evolutionary sequence going from no star formation in low mass clouds to high star formation in high mass clouds, and you interpret this as a result of accretion. But couldn't it also be that there are clouds of all masses and the low mass clouds are not sampling far into the IMF so there are no massive stars and they are not sampling vary far into the cluster mass function so there are no massive clusters? There could be low mass stars and low mass clusters in these low mass clouds.

FUKUI: The lowest mass class of GMCS, Class 1, is already fairly massive, $\sim$ several $\times 10^5$ $M_\odot$ to more than 10^6 $M_\odot$. In the Milky Way, these mass range of GMCs are almost all forming high mass stars. So, a simple different in IMF is not a good explanation for the difference in star formation among the three classes of GMCs.

Triggered Star Formation in a Turbulent ISM
Proceedings IAU Symposium No. 237, 2006
B. G. Elmegreen & J. Palouš, eds.

Dense molecular clouds and associated star formation in the Magellanic Clouds

Mónica Rubio[1]

[1]Departamento de Astronomia, Universidad de Chile, Casilla 36-D, Santiago, Chile
email: mrubio@das.uchile.cl

Abstract. Multiwavelengths studies of massive star formation regions in the LMC and SMC reveal that a second generation of stars is being formed in dense molecular clouds located in the surroundings of the massive clusters. These dense molecular clouds have survived the action of massive star UV radiation fields and winds and they appear as compact dense H_2 knots in regions of weak CO emission. Alternatively, we have found that large molecular clouds, probably remnants of the parental giant molecular clouds where the first generation of stars were formed, are suffering the interaction of the winds and UV radiation field in their surfaces in the direction of the central massive cluster or massive stars. These molecular regions show 1.2 mm continuum emission form cold dust and they show embedded IR sources as determined from deep ground base JHKs imaging. The distribution of young IR sources as determined from their Mid IR colors obtained by SPITZER concentrate in the maxima of CO and dust emission. IR spectroscopy of the embedded sources with high IR excess confirm their nature as massive young stellar objects (MYSO's). Our results are suggestive of contagious star formation where triggering and induced star formation could be taking place.

Keywords. Magellanic Clouds, Star Formation, Molecular Clouds

1. Introduction

Massive Star Formation occurs in giant molecular clouds. In the Magellanic Clouds CO(1-0) surveys have traced the distribution if giant molecular clouds (Fukui *et al.* 1999; Mizuno *et al.* 2001). In these galaxies, the ISM has different conditions such as a lower dust to gas ratio, and thus the shielding of molecules from the external UV radiation field is not as effective as in our Galaxy. Moreover, the amount of gas in GMC's remain unknown since H_2 is almost impossible to observe directly in cold interstellar regions and CO seems to be present only in the densest part of the molecular clouds. Thus, we are faced with the following questions. Is the CO a reliable tracer of the molecular gas? Are the properties of the CO clouds similar or different to those in our Galaxy? How do these issues affect the determination of the molecular mass? These questions are further complicated by the fact that the C and O elements are less abundant and thus CO is more difficult to form. The CO observations reveal that the emission in the Magellanic Clouds is very weak and thus difficult to observe.

We have gathered a large set of observations done with different telescopes and wavelengths (VLT, NTT, LCO, CTIO, ISO, SEST, SIMBA, HST) and we have found evidence of a second generation of massive stars in several of the massive star forming regions in the LMC and SMC. We have investigated their embedded stellar content in the molecular regions by deep NIR imaging, and we are currently doing a spectroscopic follow up of the brightest embedded IR sources detected. In addition, we are doing studies using the millimeter continuum emission obtained with the SIMBA bolometer as an alternative way to determine the amount of the molecular gas. We are combining these data with

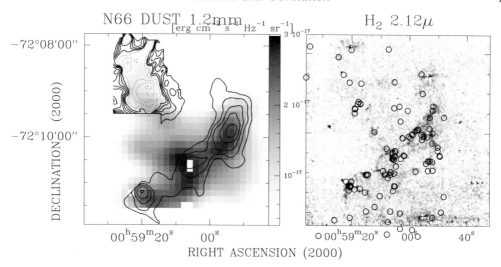

Figure 1. Left) The dust image at 1.2 mm of N66. Superimposed is the CO(2-1) emission line contours. Emission stronger than 2 K km s^{-1} is shown in dotted contours. Right) The 2.12 μm H_2 image of N66. The IR sources with J-Ks > 1.0, plotted as open circles, follow the H_2 emission and tend to concentrate towards the H_2 knots.

SPITZER images (Bolatto *et al.* 2006) to undertake a study of the dust properties in these massive star forming regions. In this contribution we concentrate mainly on results obtained in N66 in the SMC and 30Dor in the LMC.

2. The molecular gas

There is existing evidence supporting that in low metallicity environments, due to an enhance photodissociation of the CO molecule, the gas mass derived from CO measurements could be underestimating the total amount of molecular hydrogen H_2 (Rubio *et al.* 1993; Lequeux *et al.* 1994; Galliano *et al.* 2003). We have used the 1.2 mm millimeter dust emission to determine molecular masses and compared them to those obtained from CO observations. In the SMC, we have done this study in two regions, the bright HII region N66 (Perez 2006) and the SW region of the SMCBAR (Bot 2005; Bot *et al.* 2006).

In N66, we have combined deep NIR imaging with SPITZER 3.6, 4.5, 5.8, 8.0, 24, 70 and 160 μm, and 1.2 mm SIMBA observations, to determine the spectral energy distribution SED of the gas in the region and we have investigated the dust SED (Perez 2006).

2.1. *N66 in the SMC*

N66 is the brightest HII region in the SMC. It is powered by the central cluster NGC 346, which contains 33 O stars, of which 22 are earlier than 06 type. The distribution of the molecular emission as traced by the CO(2-1) emission line (Rubio *et al.* 2000) is shown in the contours in Figure 1a, and the molecular gas as seen in the molecular H_2 2.12 μm emission line in Figure 1b. Two distinct regions are identified in the CO contour map. One is the molecular gas associated to the bright optical filaments depicting the optical HII region and the molecular cloud to the north-east of the HII region. The CO emission of the HII region is very weak, while the northern cloud is bright in CO. In contrast, the 2.12 μm H_2 emission is detected in the HII region with a hint of emission on the north-eastern spur. The molecular gas in the HII region is exposed to a strong

UV radiation field from the O stars in the central core of NGC 346 as well as other O stars distributed in the area.

To determine the contribution of the dust emission to the measured SIMBA flux, we have determined the contribution due to the free-free emission in the region and that of the CO line. The latter is negligible and the free-free contribution represents $\sim 15\%$ to the continuum 1.2 mm emission (Perez 2006). To determine the free-free contribution we calibrated a Brγ 2.16 μm image of the region obtained with the 1.5m Telescope at CTIO Observatory. We derived the free-free emission at 1.2 mm from the 2.16 μm image by adopting a T = 10^4 K, ionized region Case B (Reynolds 1992). We produced a derived free-free image at 1.2 mm at the SIMBA resolution and registered it to the 1.2 mm SIMBA image. We subtracted one from the other to obtain the dust emission image of N66 at 1.2 mm, shown in Figure 1a. In this dust image we choose three different areas, two covering the HII region and one covering the northern-east molecular cloud. In these three areas we obtained the gas mass from the dust emission, assuming a dust to gas ratio of 0.10 for the SMC. In the same areas we obtained the integrated CO emission and determined the N(H_2) column density of the gas using the Galactic conversion factor of 2.8×10^{20} mol cm^{-2} K^{-1} km^{-1}s (Bloemen et al. 1986). The gas mass determined from the dust emission is larger by factor of 10 than that obtained from the CO emission in the massive star forming HII region while it is only a factor of 2.5 that of the quiescent north-east molecular cloud.

Using the N66 SPITZER images, plus the 2.16 μm image and 2.12 μm H_2 image and the ISO CVF spectral images, it was possible to separate the SED due the contribution of the dust emission and that of the HII region. The HII and PDR-dust model (Boulanger, priv. comm.) cannot explain the measured dust emission at 1.2 mm. To explain the 1.2 mm dust contribution to the SED a cold dust component is needed. The spatial extension of the dust is larger than the regions where CO is detected. The 160 μm SPITZER data also shows a larger spatial extension than that of the CO clouds (Leroy et al. 2006). Thus, all these observations are consistent with large amounts of molecular mass in H_2 gas envelopes and CO tracing only the dense region of the molecular cloud.

2.2. The SMCBAR

SIMBA 1.2 mm continuum observations were done toward the SMCBAR region. In this region, the gas mass derived from the dust emission was a factor of 10 or more larger than the mass determined from the virial mass determination using the CO emission for the cold quiescent cloud SMCB1-1 (Rubio et al. 2004). This unexpected result could only be explained if the cloud had a large H_2 gas envelope which the CO emission was not tracing. An extension of this study, where we combined all the observations done towards the SMCBAR region, was done by Bot 2005. The 1.2 mm SIMBA image is shown in Figure 2a. The dust 1.2 mm emission was obtained by subtracting the CO emission line contribution and the free-free emission at 1.2 mm deduced from published continuum radio observations. The dust emission was used to compute gas masses (Bot 2005, Bot et al. 2006). These gas masses were compared to virial masses determined from the CO emission, obtained from the CO(1-0)observations of the region done in ESO/SEST Key Programme: CO in the Magellanic Clouds (Rubio et al. 1993). The CO(1-0) emission line contours are overlaid in the SIMBA image in Figure 2b. The comparison of the gas mass determination for these GMC's in the SMCBAR showed that SMC cloud masses are systematically larger than the virial mass determination, confirming the result obtained in the SMCB1-1 cloud. Bot et al. (2006) conclude that CO is confined in clumps in large envelopes of H_2 and that their motion is not representative of the gravitational potential and they suggest that the GMC's in the SMC are magnetically supported.

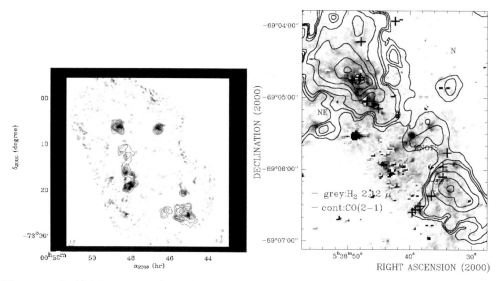

Figure 2. Left) The 1.2 mm SIMBA image of the SMCBAR with CO(2-1) contours superimposed. Right) The 2.12 μm H_2 image of 30Dor. The CO(2-1) contours are superimposed. The IR embedded sources are plotted as black crosses.

3. Massive star formation

Deep JHKs images were obtained with the 2.5m Dupont Telescope at LCO observatory of the central core NGC 346 and southern part of the HII region in N66. These images revealed very bright red sources, with K magnitude brighter than 13 and H-Ks colors larger than 2.0. These embedded IR sources were found in the central NGC 346 cluster and in the 6.7 μm ISO peaks E and H (Contursi *et al.* 2000) and maximum CO emission (Rubio *et al.* 2000). Their IR properties are those of massive young objects with large intrinsic IR excess, most probable Massive Young Stellar Objects (MYSO's) (Bik *et al.* 2006). IR spectra of two bright sources and one H_2 knot were obtained with SOFI/NTT at La Silla Observatory. The spectra show a rising continuum towards longer wavelengths and the series of Brγ lines and H_2 emission lines are detected, confirming their nature as MYSO's (Rubio & Barbá 2006). In two spectra the He lines are present indicating a high temperature of the embedded source. These MYSO's represents a younger generation of massive stars than the 3 Myr assigned to the NGC 346 cluster.

Nota *et al.* 2006. have found a rich population of pre-main sequence stars which are coeval with the massive central core NGC 346. This cluster concentrate the largest number of the pre-main sequence stars detected by them. The distribution of these stars concentrate towards the CO peaks in the HII region. Recent SPITZER observation of N66 evidence a large population of bright, very red infrared sources which are suggested to be YSO's based on their SED (Simon *et al.* 2006).

We have plotted in Figure 1b the IR sources which have colors Ks-J > 1.0 obtained and Ks < 15 from NIR photometry on the distribution of the 2.12 μm molecular hydrogen H_2 gas in N66. The H_2 gas delineates the filaments of the HII region as seen in dark lanes in the HST image of Nota *et al.* (2006). These sources are located in the filaments of the 2.12 μm H_2 molecular gas and tend to concentrate near the H_2 knots. The confirmed MYSO's are located in the H_2 knots.

4. 30 Doradus

30 Doradus is the brightest HII region in the LMC and has no counterpart in the Local Group. It is a giant HII region containing the nearest super star cluster (SSC) R136. The central cluster R136 has more than 65 O stars with several WR concentrated in about 0.5 pc. This massive star forming region has two large giant molecular cloud in its surrounding. Weak CO emission coming from several dense clumps was detected in CO(2-1) between the two GMC's (Figure 2b). These dense clumps have masses of $10^4 M\odot$ and they show a velocity gradient of several tens of km s^{-1}, consistent with expansion due to the kinetic energy of the winds from the massive stars in R136.

A second generation of massive stars was found in the border of the SW molecular cloud and in the NE one (Rubio et al. 1998) from NIR studies. These sources have magnitudes brighter than Ks = 13 and IR colors with H-Ks > 1.6. Their IR properties are consistent with massive embedded sources from their positions in the color-color and color-magnitude IR diagrams. NICMOS/HST images towards these IR sources revealed that they were multiple sources in compact groups with several bright IR stars (Walborn et al. 1999).

Figure 2b shows the distribution of the molecular H_2 2.12 μm gas with the CO(2-1) emission line contours superimposed. The H_2 gas is clumpy and has a completely different morphology with respect to the ionized gas. Combining the 2.12 μm image, a 2.16 Brγ image, the ISO CVF data, and the SPITZER data, Boulanger & Rubio (2006) have related the spatial and spectral distribution of the mid-infrared emission to the radiative and dynamical impact of the super star cluster R136 on the ISM structure and dust composition. They suggest that the spatial/clumpiness of the ionized and molecular gas is a key to understand the mid-IR emission of massive star forming regions. The surface filling factor of the molecular gas fixes the fraction of the stellar radiation which is absorbed in the PDRs. The density and the dust abundance in the HII gas fixes the fraction of Lyman continuum absorbed by dust rather than by H and He atoms. In 30 Doradus, the CO is arising from shielded interiors of dense clumps with a small surface filling factor (Poglitsch et al. 1995). In a clumpy medium, the ionizing photons are not expected to propagate ahead of the wind swept shell in the lower density gas and thus they could ionize the inter-clump medium. If so, then the ionization could increase the pressure at the surface of the dense clumps and may be produce infall, creating the conditions for a new burst of star formation.

In figure 2b we have included the position of the brightest IR sources detected by Rubio et al. (1998). These concentrate towards the peak emission in CO(2-1) in the NE CO Cloud and projected over the bright H_2 gas concentrations. In the SW CO Cloud, the IR sources are located in the outer layer facing the R136 cluster where the CO emission shows a steep density gradient as if being compressed. The IR sources appear associated to H_2 knots where the PDRs are formed in the clumpy medium.

The general picture in this region points that new burst of massive star formation has occurred associated to the dense molecular clumps in the region.

5. Conclusions

We have found evidence from the dust emission in the SMC that the molecular gas H_2 is more spatially extended than the CO. In quiescent regions, the molecular gas mass is similar to a factor of two for CO, mm and FIR determinations. In regions of massive star formation, the molecular gas mass determination can differ by an order of magnitude when derived from the dust mm emission, than when it is determined from CO or virial assumption. The molecular gas in massive star forming regions is clumpy with many PDR's regions associated to the dense molecular clumps. The molecular clouds in low metallicity environments seem to be mainly composed of H_2.

In massive star forming regions, i.e, N66 in the SMC and 30 Dor and N11 in the LMC, studies reveal vigorous star formation in the surrounding of the massive central cluster. Embedded IR sources, H_2 nebular knots, visible optical knots and O stars, and MYSO's are found associated to the dense clumps of molecular gas. The action of the UV radiation field and winds of the massive stars is interacting with remnants of the parental molecular cloud. Either by radiation pressure or by pushing the material through mechanical energy these stars contribute in generating the conditions for new star formation. However, the observations show that no massive cluster as the existing one is being formed. N66 could represent a case of triggered star formation as modeled by Elmegreen (1999).

Studies with SPITZER will be able to address questions such as which is the IMF of these new generation of stars, and the dust properties in low metallicity environments. ALMA, with its high resolution and sensitivity, will be the most suitable instrument to study the dense molecular clouds in the regions of massive star formation in the Magellanic Clouds.

Acknowledgements

We would like to acknowledge Dr. J. Palouš and Dr. B. Elmegreen for an excellent organization of the symposium. M.R. is supported by the Chilean *Center for Astrophysics* FONDAP No. 15010003.

References

Bik, A., Kaper, L. & Waters, L. B. F. M 2006, *A&A* 455, 561

Bloemen, J. B. G. M., Strong, A. W., *et al.* 1986, *A&A* 154, 25

Bolatto, A., Simon, J., Stanimirovic, S., *et al.* 2006, *ApJ* (Submitted)

Bot, C. 2005, PhD thesis, Universite Louis Pasteur, Strasbourg

Bot, C., Boulanger, F., Rubio, M. & Rantakyrö, F. 2006, *ApJ* (Submitted)

Boulanger, F. & Rubio M. 2006, in: L. Infante, M. Rubio & S. Torres-Peimbert (eds.), *Keys to Spitzer Observations of Luminous Star Forming Regions*, RevMexAC (Instituto de Astronomia: UNAM), vol. 26, p. 5

Contursi, A., Lequeux, J., Cesarsky, D., *et al.* 2000, *A&A* 362, 310

Elmegreen, B. G., Kimura, T. & Tosa, M. 1995, *ApJ* 451, 675

Fukui, Y., Mizuno, N., Yamaguchi, R., *et al.* 1999, *PASJ* 51, 745

Galliano, F., Madden, S. C., Jones, A. P., *et al.* 2003, *A&A* 407, 159

Lequeux, J., Le Bourlot, J., Des Forets, G., *et al.* 1994, *A&A* 292, 371

Leroy, A., Bolatto, A., Stanimirovic, S., *et al.* 2006, *In preparation*

Mizuno, N., Rubio, M., Mizuno, A., *et al.* 2001, *PASJ* 53, 971

Nota, A., Sirianni, M., Sabbi, E., *et al.* 2006, *ApJ* 640, L29

Perez, L. 2005, MSc thesis, Universidad de Chile, Santiago, Chile

Poglitsch A., Krabbe, A., Madden, S. C., *et al.* 2006, *ApJ* 454, 293

Reynolds, R. J. 1992, *ApJ* 392, L35

Rubio, M., Lequeux, J., Boulanger, F., *et al.* 1993a, *A&A* 271, 1

Rubio, M., Barbá, R. H., Walborn, N., *et al.* 1998, *A&A* 271, 1

Rubio, M., Contursi, A., Lequeux, J., *et al.* 2000, *A&A* 359, 1139

Rubio, M., Boulanger, F., Rantakyrö, F. & Contursi, A. 2004, *A&A* 425, L1

Rubio, M. & Barbá, R. H. 2006, *ApJL* (Submitted)

Simon, J., Bolatto, A., Whitney, B., *et al.* 2006, in preparation

Walborn, N. R., Barbá, R.H., Brandner, W., *et al.* 1999, *AJ* 117, 225

Discussion

ZINNECKER: Monica, would you agree that the big clusters that you have shown (NGC 346 and R136) do not trigger similar-sized new clusters in their neighboring clouds?

RUBIO: Hans, according to the observational evidence on the new generation of stars forming in the region, there is no similar-sized clusters. The ones detected are compact groups with 3 or 4 massive YSO's but not as NGC 346 or R136.

Triggered Star Formation in a Turbulent ISM
Proceedings IAU Symposium No. 237, 2006
B. G. Elmegreen & J. Palouš, eds.

Dust evolution in the star forming turbulent interstellar medium

François Boulanger[1]

[1]Institut d'Astrophysique Spatiale, Université Paris-Sud 11 and CNRS (UMR 8617), F-91405
Orsay
boulanger@ias.u-psud.fr

Abstract. Understanding interstellar dust evolution is a major challenge underlying the interpretation of Spitzer observations of interstellar clouds, star forming regions and galaxies. I illustrate on-going work along two directions. I outline the potential impact of interstellar turbulence on the abundance of small dust particles in the diffuse interstellar medium and translucent sections of molecular clouds. I present results from an analysis of ISO and Spitzer observations of the central part of 30 Doradus, looking for dust evolution related to the radiative and dynamical impact of the R136 super star cluster on its parent molecular cloud.

Keywords. ISM:Structure, ISM: Evolution, Dust, Stars: Formation

1. Introduction

Interstellar dust has only a distant kinship with star dust. Its composition, abundance and size distribution reflects the action of interstellar processes that contribute to break and re-build grains over time-scales much shorter than the renewal time scale by stellar ejecta. If there is a wide consensus on this statement, observations only provide fragmented evidence of dust evolution and we have yet to understand where in interstellar space and how dust evolution occurs?

This research field is relevant to the symposium topic in several ways. The impact of dust evolution on the interstellar medium (ISM) chemical and physical state and on star formation are wide-ranging and still largely unexplored. On the one hand, dust evolution underlies the interpretation of infrared observations used to trace the condensation of interstellar matter from the diffuse interstellar medium to pre-stellar cores and estimate star formation rates in galaxies. On the other hand, interstellar dust studies need to be set in the general framework of the ISM dynamical evolution including the feedback of newly born stars on their environment.

The physical processes acting on dust leave specific signatures on the dust size distribution and optical properties which in turn affects the dust spectral energy distribution (SED). With spectral bands measuring specifically the emission features from PAHs, the mid-IR emission from stochastically heated very small grains (VSGs) and the far-IR emission from large grains, the Spitzer Space Telescope imaging instruments IRAC and MIPS are particularly well suited to map the relative abundance of dust in these three size bins, while the IRS spectrometer can provide spectroscopic insight for PAHs and VSGs.

The Galactic and nearby galaxies Spitzer surveys and dedicated observing programs open many new perspectives to study dust evolution from small scales in the nearby interstellar medium to Galactic scales. The analysis of these data has just started. Only very few secure results have been obtained. Most of this research is ahead of us. For this paper, I chose to illustrate on-going work along two directions relevant to the symposium topic. First, we outline the potential impact of interstellar turbulence on the dust mass fraction in PAHs and VSGs. Second, we present results from an analysis of dust emission around the R136 super star cluster at the center of 30 Doradus.

2. Small dust particles in the turbulent interstellar medium

A significant fraction of the dust mass is in PAHs and VSGs (e.g. Li & Draine 2001) but their abundance is observed to vary across the diffuse interstellar medium and at the surfaces of molecular clouds (e.g. Miville Deschênes *et al.* 2002). The processes responsible for these abundance variations have yet to be identified.

UV and also X-ray photons that permeate the diffuse ISM and the translucent sections of molecular clouds can alter dust by photo-physical and chemical processing. But far from massive star forming regions gas-grain and grain-grain collisions are thought to be the dominant evolutionary processes. The degree and nature of the processing depends on the rate and the energy of these interactions, both of which are related to the density structure and dynamics of the ISM.

Grain sputtering and shattering in fast shock waves are thought to dominate the overall evolution of interstellar dust in the warm interstellar medium (Jones *et al.* 1996). Metals depletions observed to be lower in the warm interstellar medium than in clouds and changes in the gas-to-dust mass ratio (Bot *et al.* 2004, Borkowski *et al.* 2006, Stanimirovic *et al.* 2005) are evidence for return of dust mass into the gas phase that could be explained by shock destruction.

Turbulence within the cold interstellar medium generates shocks and vortices which might affect the dust evolution more frequently and more deeply on average than fast shocks. In turbulent clouds, the outcome from grain-grain collisions depends on grains relative velocities. Below some size dependent critical velocity, grains are expected to stick to each other (Chokshi *et al.* 1993), while above some threshold collision energy, grain shattering produces smaller fragments (Jones *et al.* 1996). Grain relative velocities are set by turbulent motions and depend on the grain size (i.e. their mass to surface ratio) and their charge state through the coupling with the magnetic field. Yan *et al.* (2004) have calculated the relative grain motions arising from magneto hydrodynamic turbulence as a function of ISM physical conditions. They find that the largest velocities are always those of the large grains but VSGs can also be efficiently decoupled from gas motions by large velocity gradients at small scales, known to exist in turbulence and not included in the Yan *et al.* (2004) work (Falgarone and Puget 1995, Falgarone this volume). Where this occurs, PAHs and VSGs mutually coagulate faster than they stick on large grains.

The gas turbulent velocity which acts as a scaling factor on the grain relative velocities and the gas density are the main physical parameters determining the collision rates and collisions outcome. PAHs and VSGs abundance is locally set by the rates at which grain-grain collisions lead to their production by grain shattering, their growth in size by mutual coagulation or their disappearance when they stick on large grains. Time-scales are uncertain - they depend on empirical factors that await to be determined through observational constraints - but diffuse clouds and the translucent sections of molecular clouds span the critical range of physical conditions where we expect to observe the interplay between dust evolution and the clouds density and velocity structure. Correlation between observed signatures of small dust evolution and physical parameters are presently the missing clues to characterize the evolutionary processes acting on PAHs and VSGs abundance.

3. Dust evolution near the 30 Doradus super star cluster

Spitzer observations are opening a new perspective on massive star formation from the Galaxy to distant galaxies. Mid-IR Spectra of star bursts like those of massive star forming regions are a combination of PAH bands and continuum from stochastically heated VSGs. The key questions raised in many studies is how can the spectral energy distribution be used to investigate the nature and the evolutionary stage of extra galactic IR sources and to estimate star formation rates? To answer these questions it is necessary to understand massive stars impact on dust in star forming regions. PAH bands are widely used as a spectroscopic signature to distinguish between infrared galaxies powered by star bursts rather than an active galactic nucleus. But PAH bands are also weak and even absent in low metallicity star forming galaxies (e.g. Engelbracht *et al.* 2005, Madden *et al.* 2006). Does this reflect PAHs destruction in H II regions (e.g. Giard *et al.* 1994) or a metallicity effect on dust composition?

3.1. *Interstellar matter near the R136 Super Star Cluster*

At the center of 30 Doradus, the R136 star cluster is the only super star cluster sufficiently close to have a detailed view at the cluster impact on the surrounding interstellar medium and dust. Interstellar matter in 30 Doradus has been the topic of many papers over the last decades. There are two CO clouds at the center of the nebula to the North and South West (SW) of R136. These clouds have the same radial velocity and are possibly two pieces of a single parent cloud where the cluster was born.

Spectacular IRAC images of 30 Doradus have been released early in the Spitzer mission. We present results from an analysis of these images on a $3' \times 3'$ field around the R136 cluster covered by an Infrared Space Observatory (ISO) spectroscopic observation. This field encompasses the ionization front and the photo-dissociation region (PDR) at the surface of the SW molecular cloud. A complete account of this work is presented in Boulanger *et al.* (2007).

The interaction of R136 with its parent cloud may be understood within the theoretical framework set by McKee *et al.* (1984) to describe the expansion of photo-ionized wind bubbles in a clumpy medium. Stellar winds have carved out a cavity within the cloud. The winds have burst out of the molecular clouds in many directions and thereafter expanded out faster in the tenuous surrounding medium powering the outer shells seen in wider images of the nebula. In the direction of the CO clouds, the expansion is still occurring within the clouds. The H II filaments observed at the edge of the two clouds delineate the wind-swept shell. Taking the molecular cloud velocity as a reference, in direction of R136 the shell is observed to be moving outwards at a velocity of $10 - 20 \, \mathrm{km \, s^{-1}}$ (Peck *et al.* 1997).

The CO emission has been interpreted as arising form the shielded interiors of dense clumps with a small (5-10%) surface filling factor (Poglitsch *et al.* 1995, Pak *et al.* 1997). These two papers do not set a clear constrain on whether the inter-clump gas is neutral or ionized. At faint levels the Brγ emission extends beyond the shell. This faint emission could arise from the outer surface of the molecular cloud but it may also indicate that the R136 Lyman continuum photons propagate into the molecular cloud, ionizing the clump surfaces and the inter-clump gas. In a clumpy medium one does expect the ionization radiation to propagate ahead of the wind swept shell in the lower density gas (McKee *et al.* 1984). The numerical simulations presented in this volume support and illustrate this view. Within this picture, the expansion of the wind swept shell does not drive a large scale shock in the molecular cloud but ablates matter from the clumps into turbulent

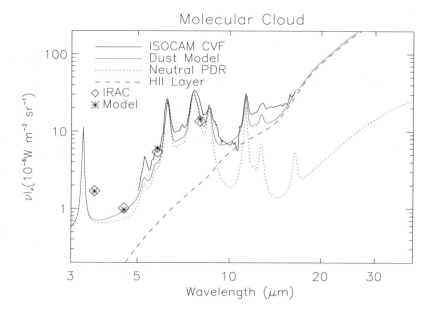

Figure 1. Mid-IR spectrum of the 30 Doradus molecular cloud with proposed decomposition into emission from the ionized and neutral layers of photo-evaporating clumps.

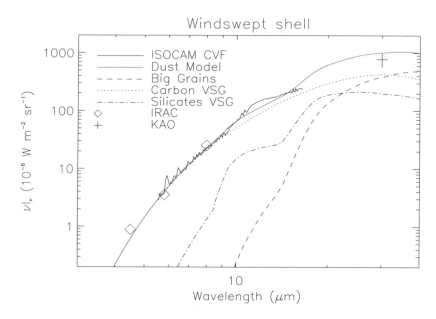

Figure 2. Mid-IR spectrum of ionized shell swept by winds from the 30 Doradus R136 super star cluster.

flows (Nakamura *et al.* 2006). Dust evolution could also here be driven by the turbulent gas dynamics.

3.2. *Dust spectral energy distribution and ISM structure*

The shell geometry permits to spatially separate the infrared emission from the ionized wind-swept shell from that arising from the molecular cloud. The SEDs derived from this

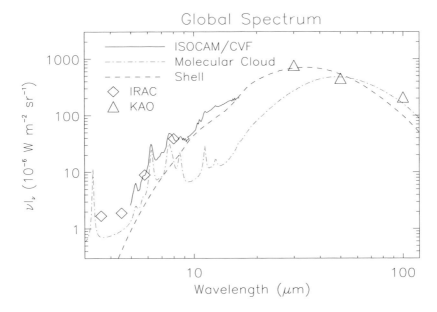

Figure 3. Global spectrum at the far-IR brightness peak to the West of R136 integrated over the 1' aperture of Kuiper Airborne Observatory observations.

correlation analysis are presented in Figs. 1 and 2. In each figure the data are compared to model calculations taking into account differences in the stellar radiation field through the HII/H$_2$ transition.

The fit of the molecular cloud spectrum builds on our view at at its structure. The far-UV radiation field at the surface of the 30 Doradus molecular clumps is comparable to that at the surface of the archetype PDR NGC 7023. But the 30 Doradus molecular cloud spectrum differs from that of NGC 7023 in two significant ways: the PAH bands are weaker by a factor 6 and the continuum is stronger. In the model, the 30 Doradus PAH emission is weak because it is arising from clumps which only fill a small fraction of the telescope beam and the mid-IR continuum represents emission from PAH-free ionized gas evaporating off the clumps. Modeling of the wind-swept shell SED yields insight on dust evolution. The smallest carbon particles including PAHs are destroyed. The mid-IR continuum comes from carbon VSGs with sizes larger than 5000 atoms. The absence of the 9 μm silicate feature in emission sets a lower limit on the small size cut-off of the silicates size distribution a factor 4 larger.

The 30 Doradus SEDs are keys to the interpretation of infrared observations of massive star forming regions and galaxies. The relative contributions of the molecular cloud and the wind-swept shell to the 30 Doradus emission varies with wavelength (Fig. 3). Different infrared wavelengths highlight the emission from different components of star forming regions. The molecular cloud contribution is dominant at short and long wavelengths while emission from dust in HII gas dominates in the mid-IR. The relative scaling between the two contributions depends on the ISM structure. For example, the PAH emission bands contrast depends on the degree of clumping of the molecular matter. In massive star forming regions like in turbulent clouds away from massive stars (Sect. 2), the dust emission spectrum reflects the interplay between dust evolution and the ISM structure. Conclusions on dust composition and evolution may be reached only within an understanding of the radiative and dynamical impact of massive stars on the interstellar medium structure.

Acknowledgements

I thank Edith Falgarone for sharing her insight on interstellar turbulence and many discussions on its potential impact on dust evolution.

References

Borkowski, K. J., Williams, B. J. & Reynolds, S. P., *et al.* 2006, *ApJ* 652, 1259

Bot, C., Boulanger, F. Bot, C., Lagache, G., Cambrésy, L. & Egret, D. 2004, *A&A* 423, 567

Boulanger, F., Rubio, M., Bot, C. & Viallefond, F. 2007, to appear in *A&A*

Chokshi, A., Tielens, A. G. G. M. & Hollenbach, D. 1993 *ApJ* 407, 806

Engelbracht, C. W., Gordon, K. D., Rieke, G. H., Werner, M. W., Dale, D. A. & Latter, W. B. 2005, *ApJ* 628, L29

Falgarone, E. & Puget, J. L. 1995 *A&A* 293, 840

Giard, M., Bernard, J. P., Lacombe, F., *et al.* 1994, *A& A* 291, 239

Jones, A., P., Tielens, A. G. G. M. & Hollenbach, D. J. 1996, *ApJ* 469, 740

Li, A. & Draine, B. T. 2001 *ApJ* 554, 778

Madden S. C., Galliano, F., Jones, A. P. & Sauvage, M. 2006, *A& A* 446, 877

McKee, C. F., Van Buren, D. & Lazareff, B. 1984, *ApJ* 278, L115

Miville-Deschênes, M.-A., Boulanger, F., Joncas, G. & Falgarone, E. 2002, *A&A* 381, 209

Nakamura, F., McKee, C. F., Klein, R. I. & Fisher, R. T. 2006, *ApJS* 164, 477

Pak, S., Jaffe, D. T., van Dishoeck, E. F., Johansson, L. E. B. & Booth, R. S. 1998, *ApJ* 498, 735

Peck, A. B., Goss, W. M., Dickel, H. R., *et al.* 1997, *ApJ* 486, 329

Poglitsch, A., Krabbe, A., Madden, S. C., *et al.* 1995, *ApJ* 454, 293

Stanimirovic, S., Bolatto, A. & Sandstrom, K., *et al.* 2005, *ApJ* 632, L103

Yan, H., Lazarian, A. & Draine, B. T. 2004, *ApJ* 616, 895

Discussion

PADOAN: Is there any reason why large increase in dust/gas ratio with density should not occur in our galaxy, while they occur in the Magellanic Clouds? We have found evidence of that in Taurus (Padoan *et al.* 2006).

BOULANGER: Observed changes in metal depletions show that dust to gas ratio also varies in the Milky Way from the Warm tenuous gas to cold dense gas. But the amplitude of the variation may be smaller than in the SMC. The variation in dust to gas ratio is expected to depend on the ratio between energy injection by stars which trigger dust destruction and mass in dense clouds where dust rebuilding occurs.

GOLDMAN: What processes are dominant in dust destruction: mechanical or radiative? and what are typical time scales for destruction?

BOULANGER: Both mechanical (e.g. shattering in grain-grain collisions) and radiative (e.g., small grain explosion following multiple ionization, photo evaporation) are probably at work but analysis of more data is necessary to characterize destruction processes and identify relevant physical parameters (e.g., radiation field intensity and hardness, x-ray flux, turbulence amplitude, shocks) for dust evolution. This is a main topic to be addressed with the Spitzer LMC and SMC surveys.

Triggered Star Formation in a Turbulent ISM
Proceedings IAU Symposium No. 237, 2006
B. G. Elmegreen & J. Palouš, eds.

New results on the distribution of thermal pressures in the diffuse ISM

Edward B. Jenkins[1] and Todd M. Tripp[2]

[1]Princeton University Observatory, Princeton, NJ, USA
email: ebj@astro.princeton.edu

[2]Department of Astronomy, University of Massachusetts, Amherst, MA, USA
email: tripp@fcrao1.astro.umass.edu

Abstract. The ground electronic state of neutral atomic carbon has three fine-structure levels. In the interstellar medium, the relative populations of the upper two levels are established by collisional excitations (and de-excitations) balanced against spontaneous radiative decay. Consequently, the fractions of C I in the upper two levels indicate acceptable combinations of local temperature and density, which in turn indicate the approximate thermal pressures of the medium. We can measure the values of these fractions and how they vary from one location to the next by observing the multiplets of C I seen in absorption in the ultraviolet spectra of hot stars.

We have identified 102 stars for which the *HST* MAST archive has E140H STIS spectra that are suitable for measuring the absorption features of C I at velocity resolutions of $3\,\mathrm{km\ s^{-1}}$ (or better). A special analysis method developed by Jenkins & Tripp (2001) permits determinations of the amounts of C I in each of the three levels as a function of radial velocity over a wide dynamic range in column density, since several multiplets of vastly different strengths can be considered simultaneously.

The C I data reveal that the much of the diffuse, cold, neutral medium has pressures that are distributed in an approximately log-normal fashion, spread over a range $1000 < p/k < 10^4\,\mathrm{cm^{-3}\,K}$ (FWHM), but with low level tails outside this range. The dispersion of pressures increases slightly for gases that have radial velocities outside the expected range for quiescent material along each line of sight. This link to the kinematics of the gas is consistent with the picture that pressure fluctuations are driven by the dynamics of a turbulent medium. If the gas is a single medium that is being driven by turbulent forces, its barytropic index (slope of $\log p$ vs. $\log n$) is more than 0.9, which is inconsistent with the value 0.72 for material that is expected to be in thermal equilibrium. Slightly less than one part in a thousand of the gas is at pressures of order or greater than $\sim 10^5\mathrm{cm^{-3}\,K}$ and seems to nearly always accompany the gas at normal pressures.

Keywords. ISM: atoms, ISM: general, techniques: spectroscopic, turbulence

1. Introduction

One of the challenges in observing the interstellar medium (ISM) is to sense local physical conditions in space, an exercise which goes beyond the much simpler task of determining the amounts of various constituents integrated along lines of sight. One excellent indicator of local conditions in the diffuse, neutral gas is the neutral carbon atom (C I), which has a ground electronic state that is split into three fine-structure levels, two of which have excitation energies $E/k = 23.6$ and $62.4\,\mathrm{K}$ above the lowest level. These upper two levels, which we designate as C I* and C I**, are populated by collisions with other gas components. Since these excitations compete with spontaneous radiative decay, the equilibrium concentrations of the excited levels indicate local kinetic temperatures and densities. One can measure these relative populations by observing an array of C I multiplets in the ultraviolet region, which show up as absorption features in the spectra of bright background sources. One important feature of these multiplets is that they have line strengths that span 3 orders of magnitude, which offers a large dynamic range in the measurements of column densities.

There have been numerous investigations of the C I excitation in the general ISM, in specific sites of interest (e.g., shocked clouds inside supernova remnants), and for very distant gas systems that are seen in absorption against quasars. The echelle spectrograph of the STIS instrument on the *Hubble Space Telescope*, when used in its highest resolution mode, offers exquisite recordings of the C I multiplets that appear in the spectra of early-type stars. In an earlier survey, we made use of this instrument to observe C I absorption toward 21 stars with a velocity resolution of $1.5 \, \mathrm{km \, s^{-1}}$ (Jenkins & Tripp 2001). We have now built upon this investigation by analyzing the spectra of 81 additional stars obtained from the *HST* MAST archive, and we report the initial conclusions from this new effort here. The resolution of nearly all of these additional stellar spectra, about $3 \, \mathrm{km \, s^{-1}}$, is lower than our original survey because the default slit width of 0.09 arc-sec for the E140H mode was used instead of the extra narrow slit with a width of 0.03 arc-sec.

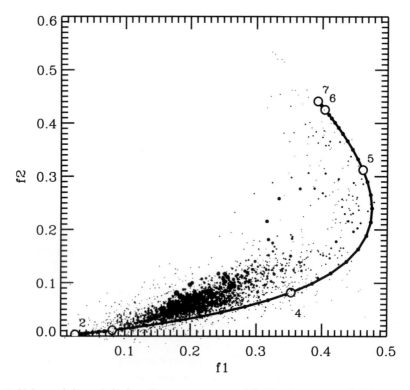

Figure 1. Values of $f1$ and $f2$ for all measurements of C I in the survey of 102 stars. The area of each dot indicates the total column density of C I (in all three of the fine-structure levels). The curve indicates the loci of $f1$ and $f2$ values for homogeneous regions at a temperature $T = 100 \, \mathrm{K}$ for various values of thermal pressure p/k (integer values of $\log[p/k]$ are indicated by open circles with numbers showing these values; bead-like dots on the curve show increments of 0.1 dex.). Other values of T yield curves similar to this one – see Fig. 6 of Jenkins & Tripp (2001). The fact that most of the points fall inside of rather than on the arc indicates that we usually are sensing gas with a small admixture of high pressure material – see Jenkins & Tripp (2001, 2007) for an explanation of how to interpret this effect.

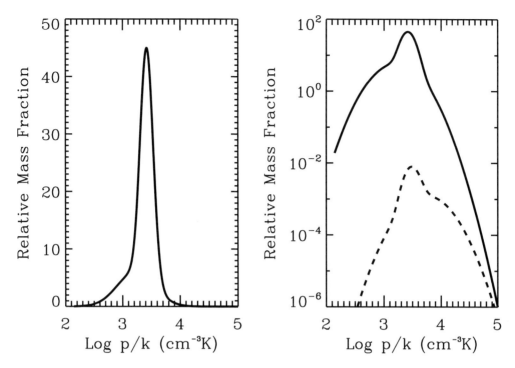

Figure 2. Linear and logarithmic plots of the distribution by mass of thermal pressures of hydrogen gas in the diffuse, neutral medium between us and the target stars in the survey. The dashed curve in the right-hand (logarithmic) display indicates the approximate proportion of gas at $p/k \sim 10^5 \mathrm{cm}^{-3}\,\mathrm{K}$ that accompanies the gas at the indicated value of p/k in the plot. This high pressure component is needed to explain the displacement of the points above the lower portion of the curve shown in Fig. 1.

2. Analysis and results

The normal range of velocity spreads in typical ISM sight lines cause different lines in the C I multiplets to overlap each other. To overcome the confusion caused by these blends, we developed (and described in our 2001 paper) an analysis method that derived solutions for the velocity profiles of each of the three levels that would recreate the patterns seen in the different multiplets. We then calculated for all velocities where measurable absorptions were present two excitation parameters $f1 = N(\mathrm{C\ I^*})/N(\mathrm{C\ I_{total}})$ and $f2 = N(\mathrm{C\ I^{**}})/N(\mathrm{C\ I_{total}})$, where $N(\mathrm{C\ I_{total}}) = N(\mathrm{C\ I_{unexc.}}) + N(\mathrm{C\ I^*}) + N(\mathrm{C\ I^{**}})$.

In Figure 1 we show the determinations of $f1$ and $f2$ for all of the $0.5\,\mathrm{km\ s}^{-1}$ wide velocity bins toward all of the targets in the survey. Most of the points fall within the range $3.0 < \log(p/k) < 4.0$ of the nearby arc that depicts the expected values of $f1$ and $f2$ for $T = 100\,\mathrm{K}$. In Figure 2 we show the distribution of pressures that is consistent with the locations of the points shown in Fig. 1. In deriving this distribution, we had to calculate shifts in the ionization equilibrium between the neutral and ionized forms of C and correct for this effect to obtain the H I-weighted distribution.

We interpret the displacement of the points above the lower portion of the arc in Fig. 2 to arise from either an ever present admixture of a very small amount of gas at very high pressures ($p/k \gtrsim 10^5 \mathrm{cm}^{-3}\,\mathrm{K}$) or excitations by positively charged particles that can more strongly excite the C I** [see Fig. 1 of Silva & Viegas (2002)]. In weakly ionized regions, excitation from shocks or Alfvén waves could create differences in the velocities of ions and neutrals, which in turn could excite the carbon atoms.

As emphasized in our earlier work (Jenkins & Tripp 2001), the points in the higher parts of the $f1 - f2$ diagram represent gas at high pressures ($p/k > 10^4 \mathrm{cm}^{-3}\,\mathrm{K}$), but at temperatures not much below about $80\,\mathrm{K}$. This indicates that the material does not cool appreciably when it is compressed, which would happen if the process were slow enough to allow a thermal equilibrium to be maintained [$d\log p/d\log n = \gamma_{\mathrm{eff}} = 0.72$ (Wolfire, et al. 1995)]. For this reason, we favor the interpretation that the gas is compressed so quickly that the changes in temperature are closer to adiabatic ($\gamma = 5/3$). It should not be difficult for turbulent motions to achieve such compressions if the gas regions have sizes less than about 0.01 pc.

When we examine the kinematics of the gas, we find that material at velocities outside the interval expected for differential galactic rotation along the sight line has a greater tendency to show high pressures. The evidence for this is illustrated in Jenkins & Tripp (2007), and it is consistent with the proposition that shocks and colliding flows of gas are the probably cause for episodic enhancements of the thermal pressure.

Acknowledgements

This research was supported by grant HST-AR-09534.01 from the Space Telescope Science Institute, operated by the Association of Universities for Research in Astronomy, Inc., under NASA contract NAS5-26555.

References

Jenkins, E. B. & Tripp, T. M. 2001, *ApJS* 137, 297

Jenkins, E. B. & Tripp, T. M. 2007, in: M. Goss & M. Haverkorn (eds.), *Small Ionized and Neutral Structures in the Diffuse Interstellar Medium* (ASP-CS), in press

Silva, A. I. & Viegas, S. M. 2002, *MNRAS* 329, 135

Wolfire, M. G., Hollenbach, D., McKee, C. F., Tielens, A. G. G. M. & Bakes, E. L. O. 1995, *ApJ* 443, 152

Discussion

M. Dopita: The distributions you showed have both a broad and a narrow log-normal curve. Are the broad components really separate phases of the ISM?

E. Jenkins: Offhand, I see no easy way to determine if this is the case. Generally, the C I ionization equilibrium strongly enhances our sensitivity to the cold, neutral medium, at the expense of the warm neutral, or partially ionized medium.

B. Elmegreen: Do you see a correspondence between enhancements in the ratio of Ca II vs. Na I and elevated thermal pressures, as indicated by C I?

E. Jenkins: This is an interesting idea. We have not investigated this issue, but perhaps we should do so.

Triggered Star Formation in a Turbulent ISM
Proceedings IAU Symposium No. 237, 2006
B. G. Elmegreen & J. Palouš, eds.

Dynamical evolution of a supernova driven turbulent interstellar medium

Dieter Breitschwerdt[1] and Miguel A. de Avillez[1,2]

[1]Institut für Astronomie, Universität Wien, Türkenschanzstraße 17, A-1180 Vienna, Austria
email: breitschwerdt@astro.univie.ac.at

[2]Department of Mathematics, University of Évora, R. Romão Ramalho 59, 7000 Évora,
Portugal; email: mavillez@astro.univie.ac.at

Abstract. It is shown that a number of key observations of the Galactic ISM can be understood, if it is treated as a highly compressible and turbulent medium, energized predominantly by supernova explosions (and stellar winds). We have performed extensive numerical high resolution 3D hydrodynamical and magnetohydrodynamical simulations with adaptive mesh refinement over sufficiently long time scales to erase memory effects of the initial setup. Our results show, in good agreement with observations, that (i) volume filling factors of the hot medium are modest (typically below 20%), (ii) global pressure is far from uniform due to supersonic (and to some extent super-Alfvénic) turbulence, (iii) a significant fraction of the mass ($\sim 60\%$) in the warm neutral medium is in the thermally unstable regime ($500 < T < 5000$ K), (iv) the average number density of OVI in absorption is 1.81×10^{-8} cm^{-3}, in excellent agreement with Copernicus and FUSE data, and its distribution is rather clumpy, consistent with its measured dispersion with distance.

Keywords. ISM: general, ISM: evolution, ISM: structure, ISM: magnetic fields, hydrodynamics, MHD, turbulence

1. Introduction

Low resolution observations of the interstellar medium (ISM) at various wavelengths reveal a rather smooth spatial distribution of the gas and magnetic fields, and distinctive gas phases can be discerned. Theoretical studies during the last three decades have culminated in a widely accepted multiphase "standard model" (e.g. McKee & Ostriker 1977 (MO-model), McKee 1990), in which the gas is distributed into three phases in global pressure equilibrium, a cold and warm neutral phase (CNM and WNM, respectively), a warm ionized (WIM) and a hot intercloud (HIM) medium. There is global mass balance by evaporation and condensation, and energy balance between supernova (SN) energy injection and radiative cooling. One of its testable predictions is a large volume filling factor of the HIM ($f_h \geqslant 0.5$) for the Galaxy. Observations, however, also in external galaxies ($f_h \sim 0.1$, e.g. Brinks & Bajaja 1986), point to much lower values. This discrepancy can be removed if one allows for a fountain flow due to the break-out of superbubbles into the galactic halo (so-called chimney model, Norman & Ikeuchi 1989) as well as buoyant outflow from supernova remnants (SNRs). OVI absorption line column densities, which were pivotal in establishing the HIM in the first place, are thought to arise in conductive interfaces, yielding systematically too large values by an order of magnitude, when compared to Copernicus observations (Jenkins & Meloy 1974). Furthermore, Jenkins & Tripp (2006, these proceedings) have measured CI absorption lines towards a sample of ~ 100 stars from the HST archive and find a large variation in the CNM pressure of $500 < P/k_B < 4000$ cm^{-3} K, in contrast to what is expected from a model where pressure equilibrium is a key element. Although turbulence is recognized to play an important role in steady-state multiphase models, it is largely treated as an additional pressure source, ignoring its *dynamical* importance.

A fundamentally different and more physical approach to model the structure and evolution of the ISM goes back to the ideas of von Weizsäcker (1951) who suggested that the ISM is essentially a highly turbulent and compressible medium. Indeed, high resolution observations of the ISM show structures on *all scales* down to the smallest resolvable ones, implying a dynamical coupling over a wide range of scales, which is a main characteristic of a turbulent flow with Reynolds numbers of the order of 10^5 – 10^7 (cf. Elmegreen & Scalo 2004). Another characteristic of widespread ISM turbulence is its enhanced mixing of fluid elements, which, unlike thermal conduction, is largely independent of strong temperature gradients and magnetic fields. Recently, the dynamical importance of turbulence in the ISM and in star formation in molecular clouds has been recognized by several groups using different numerical approaches (e.g. Korpi *et al.* 1999, Vázquez-Semadeni *et al.* 2000, Avillez & Breitschwerdt 2004).

Physically the generation of 3D turbulence is intimately related to vortex stretching and its subsequent enhancement, in contrast to 2D where vorticity is conserved. A natural way to generate vorticity is shear flow in which transverse momentum is exchanged between neighbouring fluid elements. This typically occurs when a flow is decelerated at a surface (giving rise to a boundary layer) like wind gushing down a stre*et al.*ng the wall of a high building, or in case of the ISM, colliding gas flows, like e.g hot gas breaking out of an SNR or superbubble (SB). Various sources of turbulence for the ISM have been identified: stellar (jets, winds, HII regions, SN explosions), galactic rotation, self-gravity, fluid instabilities (e.g. Rayleigh-Taylor, Kelvin-Helmholtz), thermal instability, MHD waves (e.g. due to cosmic ray streaming instability), with SNe representing energetically the most importance source (see e.g. MacLow & Klessen 2004).

We will show in the following sections that the new approach of a turbulent SN driven ISM can reproduce many key observations (Avillez 2000, Avillez & Breitschwerdt 2004, 2005a,b, henceforth AB04, AB05a, AB05b), such as a low volume filling factor of the HIM, large pressure fluctuations in the ISM, observed OVI absorption column densities by Copernicus and FUSE, and WNM gas in thermally unstable temperature ranges.

2. Model setup

We have performed both hydrodynamical (HD) and magnetohydrodynamical (MHD, with a total field of 4.5 μG, with the mean and random components of $B_u = 3.1$ and $\delta B = 3.2\,\mu$G, respectively) simulations to study by adaptive mesh refinement simulations the global and local evolution of the SN driven ISM. We use a grid centred at the solar circle with a square disk area of 1 kpc^2 and extending from $z = -10$ to $+10$ kpc in the directions perpendicular to the Galactic midplane. The finest resolution is 1.25 pc (MHD) and 0.625 (HD), respectively. Gravity is provided by the stars in the disk, radiative equilibrium cooling assuming solar and also 2/3 solar abundances (hence, log (O/H)=-3.07 (Anders & Grevesse 1989) and -3.46 (Meyer 2001), respectively), uniform heating due to starlight varying with z (cf. Wolfire *et al.* 1995) and a magnetic field (setup at time zero assuming equipartition) for the case of MHD runs. SNe types Ia and Ib+c+II are the sources of mass, momentum and energy. SNe Ia are randomly distributed, while the other SNe have their high mass progenitors generated in a self-consistent way according to the mass distribution in the simulated disk (with roughly 60% exploding in associations) and are followed kinematically according to the velocity dispersion of their progenitors. In these runs we do not consider heat conduction, as turbulence provides the dominant mixing process. For setup and simulation details see AB04 and AB05a.

Table 1. Average volume filling factors of the different ISM phases for variable SN rate σ (in units of the Galactic rate). The average was calculated using 101 snapshots (of the 1.25 resolution runs) between 300 and 400 Myr of system evolution with a time interval of 1 Myr.

σ^a	$\langle f_{v,cold} \rangle^b$	$\langle f_{v,cool} \rangle^c$	$\langle f_{v,warm} \rangle^d$	$\langle f_{v,hot} \rangle^e$
1	0.171	0.354	0.298	0.178
2	0.108	0.342	0.328	0.223
4	0.044	0.302	0.381	0.275
8	0.005	0.115	0.526	0.354
16	0.000	0.015	0.549	0.436

^{a} SN rate in units of the Galactic SN rate.
^{b} $T < 10^3$ K, ^{c} $10^3 < T \leqslant 10^4$ K, ^{d} $10^4 < T \leqslant 10^{5.5}$ K, ^{e} $T > 10^{5.5}$ K.

3. Results

In the following we focus on those results of our simulations, which show a clear deviation from the aforementioned standard picture of the ISM. It is important to emphasize that the computational box has to be sufficiently large in order to avoid significant mass loss, and that the evolution time is long enough (400 Myr in our runs) in order to be insensitive of the (necessarily artificial) initial setup and to attain global dynamical equilibrium. In addition we have checked that the results are resolution independent by doubling the resolution and found that changes are less than a few percent.

3.1. *Volume filling factors*

A first striking feature of global ISM simulations in a SN driven ISM is the *continuous* distribution of the plasma over temperature, rather than in distinct phases (for a discussion see below). We have therefore specified temperature regimes that correspond to the classical phases as well as to thermally unstable regimes. The second striking feature is the low volume filling factor of the HIM (see Table 1). This is a result of the unavoidable setup of the Galactic fountain, as the overpressured flow always chooses the path of least resistance. Even for a star formation rate, which is 16 times the Galactic value, the HIM covers less than 50% of the disk volume. It should be stressed that this result is fairly robust and does not depend strongly on the magnetic field as our MHD runs show. Even an initially disk parallel field cannot prevent break-out, as in 3D it is easier to push field lines aside than working against tension forces all the way up into the halo as it is in 2D.

3.2. *The myth of pressure equilibrium*

It has been often argued that there should exist global pressure equilibrium between the various stable phases. This hypothesis would be correct, if there would be sufficient time for relaxation for the various processes responsible for mass and energy exchange like collisional heating, radiative cooling, condensation and evaporation etc.. However, due to the large Reynolds number of the flow, turbulent mixing is the dominant exchange process, and a fortiori this occurs supersonically in a compressible medium. Hence there is in general not enough time to establish pressure equilibrium by pressure waves propagating back and forth. There exists though a global *dynamical* equilibrium, depending on the boundary conditions (e.g. SN rate, gravitational and external radiation field), which results in an "average pressure", however with huge fluctuations as can be seen in Fig. 1. The fact that the dynamical evolution of the ISM is indeed governed by turbulence may be appreciated by noting that in Fig. 1 structures occur on *all scales*. This may on the other hand cast some doubt on the results, as surely structures will occur below the resolution limit. As our resolution checks have shown, this does not seem to be an issue here,

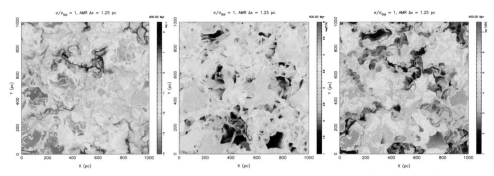

Figure 1. Two dimensional cuts, through the 3D data cube, showing the density n (left panel), the pressure P/k (middle panel) and the temperature T (right panel) distribution in the Galactic plane for an HD simulation with the Galactic SN rate.

since the processes we describe here either dominate on larger scales or do not exhibit any significant energy feedback from smaller to larger scales (as might actually be the case in strong MHD turbulence).

3.3. Does some interstellar gas reside in thermally unstable phases?

HI Arecibo Survey observations by Heiles & Troland (2003) have shown that about 48% of the WNM can be found in the thermally unstable regime between 500 K and 5000 K, and that CNM linewidths are in agreement with supersonic turbulent motions in sheetlike (aspect ratios of up to 280) clouds. Taken at face value this strongly supports a picture in which clouds are immersed in a turbulent medium and are deformed by vortex stretching as well as by shock compression. Our simulations show that ISM turbulence can drive and sustain turbulence inside clouds, which is alleviated to some extent by the fractal structure of clouds, resulting e.g. from colliding gas flows (cf. Burkert 2006). Fig. 2 (left)

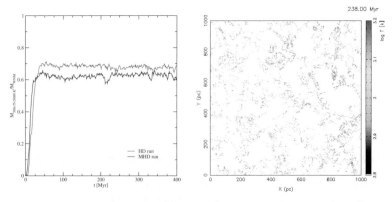

Figure 2. *Left panel:* History of the fraction of mass of the classical thermally unstable gas having $500 < T \leqslant 5000$ K in the disk for the HD (red) and MHD (black) runs. *Right panel:* 2D snapshot of the Galactic plane taken from run HD2a ($\Delta x = 0.625$ pc) at time 238 Myr. The image shows the filamentary structure of the warm neutral gas with $2.8 \leqslant T \leqslant 3.2$ K.

shows the fraction of the WNM derived from our HD and MHD runs in the temperature range between 500 and 5000 K, and the right panel of Fig. 2 demonstrates its filamentary distribution in the narrow band between 630 and 1590 K. The large amount of ISM mass seen in thermally unstable regimes is a direct consequence of SN driven turbulence. Thus, the Field (1965) criterion is necessary, but not sufficient for distributing the ISM gas into stable phases over different temperature ranges. It is essential to realize that turbulence has a stabilizing effect by inhibiting local condensation modes. The reason is that turbulence can be regarded as a diffusion process by which energy is efficiently transferred from large to small scales, thus preventing thermal runaway on scales smaller than the minimal length scale over which thermal instability can overcome turbulent diffusion (note the small scale patchy distribution in Fig. 2 (right)), in much the same way as heat conduction stabilizes the solar chromosphere.

3.4. *Comparison of O*VI *distribution to Copernicus and FUSE data*

The discovery of the widespread OVI absorption line toward background sources led to the discovery of the HIM (e.g. York 1974, Jenkins & Meloy 1974) and identified SNRs as a major source of hot gas. Ever since, starting with the "tunnel network model" of Cox & Smith (1974), to reproduce the observed OVI column densities, N(OVI), has been a touchstone of ISM modeling. In collisional ionization equilibrium OVI is the most abundant ionization stage at $T \sim 3 \times 10^5$ K, a temperature which is typical for transition regions between HIM and cooler gas, like e.g. in conductive interfaces. In the MO model these occur in large numbers between the HIM and embedded clouds and lead to an n(OVI) number density about an order of magnitude larger than the average value of 1.7×10^{-8} cm^{-3}. On the other hand our simulations show, that if turbulent mixing is

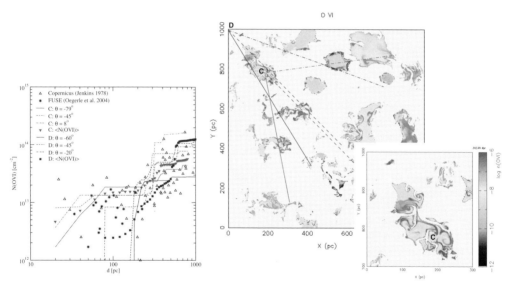

Figure 3. *Left panel:* Comparison of FUSE (stars) and COPERNICUS (open triangles) OVI column densities with spatially averaged (red triangles and blue squares) and single lines of sight (red and blue lines) N(OVI) measurements in the simulated disk at time $t = 393$ Myr. The LOS are taken at positions C (red) and D (blue) that are located inside and outside of a bubble cavity, respectively, as shown in the right panel. *Right panel:* OVI density distribution (in logarithmic scale) in the midplane at time $t = 393$ Myr. The panel also includes a zoom of the bubble located in position C. The colour scale varies between 10^{-12} and 10^{-6} cm^{-3}; grey corresponds to zero OVI. Note the eddy-like structures of OVI inside the bubbles.

the dominant process in energy redistribution, the number of interfaces is reduced, as energy transport is not primarily driven by temperature gradients but by a turbulent cascade from larger to smaller eddies. Fig. 3 (left) shows a comparison between both Copernicus and FUSE data and our simulations, which is remarkably good and yields a time averaged value of n(OVI) $= 1.81 \times 10^{-8}$ cm^{-3} without any tuning. Even the measured

n(Ovi) inside the Local Bubble is in excellent agreement with the FUSE data. Fig. 3 (right) stresses that the Ovi distribution occurs mainly in regions separated by a length scale of about 100 pc. This provides an extreme inherent clumpiness, which ensures the dispersion of the column density to be roughly independent of distance, d, (for $d > 100$ pc; for details see AB05b) rather than declining with the number N_{cl} of interspersed clouds like $1/\sqrt{N_{cl}}$, fully consistent with FUSE observations (Bowen et $al.$ 2005.)

4. Conclusions

Recent high resolution multi-wavelength observations in conjunction with theoretical research have shown that the ISM in star forming galaxies is a highly complex "ecosystem". The key to a better understanding of its nature and evolution lies in the systematic study of compressible HD and MHD turbulence (for more detailed studies see Avillez & Breitschwerdt, these proceedings). Numerical high resolution 3D simulations offer a unique possibility to investigate the nonlinear interaction of the physical processes at work, together with a careful analysis of scaling laws. Since the SN driven ISM model can already explain many important features, as we have shown, we feel encouraged to implement further processes, such as e.g. non-equilibrium cooling, self-gravity and cosmic rays into our bottom-up model. Since these studies require a huge amount of massive parallel computing power, we are just at the beginning.

Acknowledgements

DB thanks Jan Palouš, Bruce Elmegreen and the organizers for financial support.

References

Anders, E. & Grevesse, N. 1989, *Geochim. Cosmochim. Acta* 53, 197
Avillez, M.A. 2000, *MNRAS* 315, 479
Avillez, M.A. & Breitschwerdt, D. 2005, *A&A* 436, 585 (AB05a)
Avillez, M.A. & Breitschwerdt, D. 2005, *ApJ* 634, L65 (AB05b)
Avillez, M.A. & Breitschwerdt, D. 2004, *A&A* 425, 899 (AB04)
Bowen, D.V., Jenkins, E.B., Tripp, T.M., *et al.* 2005, in: G. Sonneborn *et al.* (eds.), Astrophysics in the Far Ultraviolet - Five years of discovery with FUSE (ASP-CS), 348, 412
Brinks, E. & Bajaja, E. 1986, *A&A* 169, 14
Burkert, A. 2006, in: F. Combes & R. Robert (eds.), *Statistical Mechanics of Non-Extensive Systems*, Comptes Rendus Physique, vol. 7, p. 433
Cox, D.P. & Smith, B.W. 1974, *ApJ* 189, L105
Elmegreen, B.G. & Scalo, J. 2004, *ARA&A* 42, 211
Field, G.B. 1965, *ApJ* 142, 531
Heiles, C. & Troland, T.H. 2003, *ApJ* 586, 1067
Jenkins, E.B. & Meloy, D.A. 1974, *ApJ* 193, L121
Korpi, M.J., Brandenburg, A., Shukurov, A., *et al.* 1999, *ApJ* 514, L99
MacLow, M.-M & Klessen, R.S. 2004, *Rev. Mod. Phys.* 76(1), 125
McKee, C.F. 1990, in: L. Blitz (ed.), *The Evolution of the Interstellar Medium* (ASP-CS), 12, p. 3
McKee, C.F. & Ostriker, J.P. 1977, *ApJ* 218, 148
Meyer, D.M. 2001, in: R. Ferlet *et al.* (eds.), *Gaseous Matter in Galaxies and Intergalactic Space* (Paris: Editions Frontiéres), p. 135
Norman, C.A. & Ikeuchi, S. 1989, *ApJ* 345, 372
Vázquez-Semadeni, E., Gazol, A. & Scalo, J. 2000, *ApJ* 540, 271
Wolfire, M.G., McKee, C.F. & Hollenbach, *et al.* 1995, *ApJ* 443, 152
York, D.G. 1974, *ApJ* 193, L127
von Weizsäcker, C.F. 1951, *ApJ* 114, 165

Discussion

ARONS: How do you account for reconnection in the MHD simulations?

BREITSCHWERDT: Reconnection occurs at smaller scales, mediated by numerical resistivity. There is also evidence of a small scale dynamo in the simulations.

PADOAN: You find that for $T < 100$ K magnetic pressure is dominant, but you don't allow the density to grow at the level it should grow in molecular clouds at the scale (0.6 pc) you resolve (no molecular cooling). Furthermore, viscous dissipation of the turbulent velocity (numerical dissipation on the smallest scales) would also reduce the ram pressure.

BREITSCHWERDT: As I have emphasized in my talk, we do not describe the formation of molecular clouds, as we have not yet included molecular cooling and self-gravity. On the smallest scales there should be a transition ala viscous dissipation, so ram pressure will not be the driver at these scales.

DICKEY: I saw in your simulation an effect that is very clear in the 21-cm surveys which is an offset in the gas midplane from the bottom of the gravitational potential.

BREITSCHWERDT: This sounds intriguing. I think that the simulations reflect the effect that the supernova explosions both in the thin and in the extended disk are not distributed symmetrically due to the density/temperature criterion of star formation.

HEYER: Simulations are long enough for the gas to have flowed through spiral potential. Are spiral shocks/gas streaming comparable to turbulent ram pressure component?

BREITSCHWERDT: I agree that spiral density shock waves should be included. My guess would be that SN driven turbulence would dominate thoroughly, as the flow is largely controlled by ram pressure.

MAROV: My question is about conservation laws in 3D turbulence. It is known that 3D Navier-Stokes equations have the second integral of motion which gives rise to spirality preserving in the non-viscous limit. Did you observe spirality in your simulation approach and could it be distinguished on a small eddies background?

BREITSCHWERDT: We see the generation of vorticity and the eddies carry helicity if that's what you mean. But since we do not include galactic rotation there is no net helicity due to Coriolis forces.

MAC LOW: Several groups find $E_{kin} \propto t^{-1}$ for mildly supersonic turbulence (Mac Low *et al.* 1998; Stone, Ostriker & Gammie 1998; Padoan & Nordlund 1999). What is interesting in your results is the slow decay of cold gas $E_{kin} \propto t^{-0.5}$.

BREITSCHWERDT: Yes this is true. The slower decay lasts however only for a few million years and it should be due to the feeding of turbulence from gas cooling down from higher temperatures.

ELMEGREEN: If you were to connect your results with star formation, I imagine you would say stars form in the dense clumps and the clumps form by supernova and other shocks. But then how do you get the observed sensitivity to Toomre Q for star formation, and how do you get the Schmidt law? What is missing from the models?

BREITSCHWERDT: One of the biggest challenges for the next decade is to close the gap between large scale simulations (on which turbulence is driven) and the process of star formation on small scales. I am confident that this is only a matter of increasing computer power. Our present simulations are limited on small scales due to the lack of self-gravity and on the large scales due to the lack of galactic rotation. Both improvements are currently being built in.

FALL: Have you measured the velocity and density correlations in your simulations and compared them with observations (scaling-type relations presented yesterday)? What does the agreement or disagreement between your simulations and observations tell you about the driving mechanisms and or other properties of turbulence in the ISM?

BREITSCHWERDT: We have extracted structure functions of order p from the simulations and have determined the scaling exponent $\xi(p)$ and found that the Hausdorff dimension of the most dissipative structures is 2, implying dissipation by sheets rather than filaments. I agree that it would be useful also to determine the density correlation functions and compare them directly to observations.

Triggered Star Formation in a Turbulent ISM
Proceedings IAU Symposium No. 237, 2006
B. G. Elmegreen & J. Palouš, eds.

© 2007 International Astronomical Union
doi:10.1017/S1743921307001238

ISM turbulence driven by the magnetorotational instability

Robert A. Piontek[1,2] and Eve C. Ostriker[2]

[1]Astrophysikalisches Institut Potsdam, An der Sternwarte 16, D-14482 Potsdam, Germany
email: rpiontek@aip.de

[2]Department of Astronomy, University of Maryland, College Park, MD 20742-2421
email: ostriker@astro.umd.edu

Abstract. We have performed numerical simulations which were designed to further our understanding of the turbulent interstellar medium (ISM). Our simulations include a multi-phase thermodynamic model of the ISM, magnetic fields, and sheared rotation, allowing us to study the effects of the magnetorotational instability (MRI) in an environment containing high density cold clouds embedded in a warm, low density, ambient medium. These models have shown that the MRI is indeed a significant source of turbulence, particularly at low mean densities typical of the outer regions of the Milky Way, where star formation rates are low, but high levels of turbulence persist. Here, we summarize past findings, as well as our most recent models which include vertical stratification, allowing us to self-consistently model the vertical distribution of material in the disk.

Keywords. ISM: general, clouds, kinematics and dynamics, magnetic fields, structure

1. Introduction

In the traditional picture of the ISM (Cox & Smith 1974; McKee & Ostriker 1977; Spitzer 1978) turbulence is driven primarily by supernova explosions. In more recent years, however, some observations of the Milky Way and external galaxies have lead us to believe that another significant source of turbulence must be at work in the ISM. In short, the observed velocity dispersion of neutral hydrogen is found to be relatively constant, while the SNe rate is not. In particular, turbulent velocities of neutral hydrogen are high in the outer regions of galaxies (Dickey *et al.* 1990), and arm/inter-arm regions show no difference in the measured velocity dispersion (Dickey *et al.* 1990). If turbulence is driven by SNe, we would expect that turbulent velocities should be larger near regions of active star formation. These observations motivated Sellwood & Balbus (1999) to propose that turbulence in the ISM may be driven by the magnetorotational instability, and over the past few years we have investigated this theory with numerical simulations.

The MRI has been extensively studied in the context of accretion disks (Hawley & Balbus 1992; Stone *et al.* 1996). The ISM meets the basic requirements to be unstable to the MRI: it is a rotating-shearing system, with a weak magnetic field. However, it is different from typical accretion disk systems in one important aspect. The ISM is structured by radiative heating and cooling processes (Wolfire *et al.* 2003) so that it is found in two distinct "phases". A high density, clumpy, cold phase is thought to be embedded in a warm, diffuse, low density intercloud medium (Field *et al.* 1969). Gas at intermediate temperatures is thermally unstable (Field 1965), and heats or cools to join either the warm or cold stable phases. Our aim has been to study the MRI in the presence of a two-phase medium. Here we briefly summarize results from Piontek &

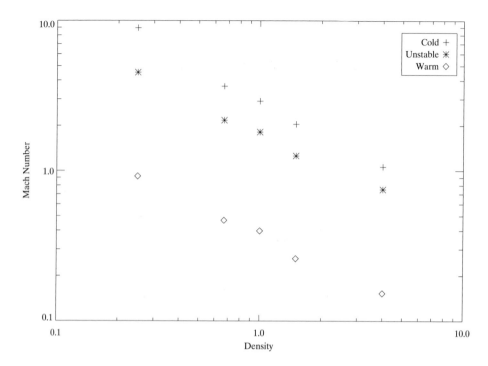

Figure 1. For five different simulations with varying mean density, we plot the time averaged, mass weighted, turbulent Mach number, for the warm, unstable, and cold phases of gas. Our lowest density model has a turbulent velocity of 8 km s^{-1}.

Ostriker (2004), Piontek & Ostriker (2005), and new, unpublished results from a recently submitted paper to *ApJ*.

2. Simulations and results

We performed MHD simulations using an MPI parallel version of ZEUS (Stone & Norman 1992a, 1992b), that includes a radiative cooling function adopted from Sánchez-Salcedo *et al.* (2002), which allows us to simulate a multi-phase ISM. As previously described, there are two stable phases of gas, a cold (100 K), high density, clumpy phase, and a warm (8000 K), low density, intercloud medium. Without a turbulent source, very little gas exists at intermediate temperatures, but turbulence can force gas into the unstable regime. Our simulations are performed in the local, rotating reference frame of the galaxy, and our simulation domain represents a small patch of the ISM, measuring 100-300 pc on each side, depending on the model. The initial magnetic field strength is 0.26 μG, and is vertical. We include shear from the galactic rotation curve using shearing-periodic boundary conditions (Hawley & Balbus 1992; Hawley *et al.* 1995), which in combination with magnetic fields, results in MRI driven turbulence. For details concerning the numerical method, please see Piontek & Ostriker (2004). The primary question we wish to address is whether or not the MRI can produce turbulence at the level which is observed in the Milky Way and other galaxies, which is typically around 7 km s^{-1} Heiles & Troland (2003).

In Piontek & Ostriker (2005), we performed five different simulations, varying only the mean density, from $\bar{n} = 0.25$ cm^{-3} to $\bar{n} = 4$ cm^{-3}. We found that the combination of a

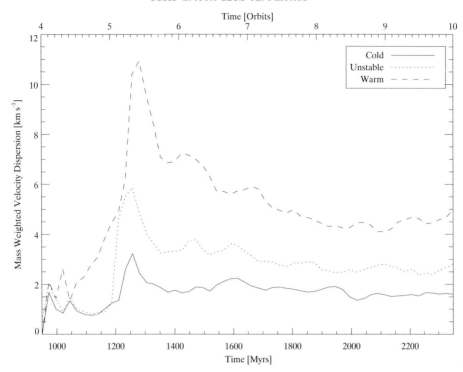

Figure 2. Mass weighted velocity dispersion as a function of time for our inner galaxy stratified model. It can be seen here that the cold component has a much lower velocity dispersion compared to the warm component. This is because the cold component sinks to the mid-plane, increasing the local mean density. In our unstratified models, all three phases were driven at approximately the same velocity.

multi-phase ISM with the MRI results in a scaling of turbulent velocity with mean density that is steeper than what is expected for the MRI in a single-phase model of ISM. As the mean density decreases, MRI turbulence grows more quickly in a multi-phase ISM than it otherwise would. In figure 1 we plot the time averaged, mass weighted, turbulent Mach number as a function of mean density for five different simulations, with each thermal component shown individually. All three components are driven at approximately the same velocity at any particular mean density, but since each phase has a fairly well defined temperature associated with it, the Mach numbers vary significantly. The warm phase is subsonic, and the cold phase is supersonic, and in our model with the lowest mean density, $\bar{n} = 0.25$ cm^{-3}, these Mach numbers are about 0.15, and 9.0, respectively. This corresponds to a velocity of approximately 8 km s^{-1}. This result lead us to conclude that the MRI may indeed be an important source of turbulence in low density environments such as the outer regions of the Milky Way.

We have recently performed simulations which additionally include vertical stratification. In these models the *local* mean density is determined self consistently, whereas previously the mean density was simply a parameter (though the total mass in the model still remains a parameter). Our simulation domain for these stratified models extends 900 pc above and below the mid-plane. In these models we find that the higher density cold clouds tend to sink to the mid-plane. Here they shield themselves from the more turbulent warm medium, and the turbulent velocity of the cold component is lower than what we find for our non-stratified models.

In figure 2 we plot the mass weighted velocity dispersion as a function of time for our standard stratified model, with a surface density of $\Sigma = 10\,M_\odot\,\mathrm{pc}^{-2}$. The warm medium saturates at around 4-5 km/s, approximately the level which is observed, but the cold medium saturates at 1-2 km/s, which is too small to account for observed velocity dispersions. The cold, high density clouds sink to the mid-plane, and the local mean density increases. The increased mean density causes the MRI to be less effective at driving turbulence, as we found in our unstratified models. The MRI is unable to lift the cold gas off of the mid-plane, so that the scale heights we measure in our stratified disks are too small compared to observations. We also performed models of the outer disk with a lower surface density, $\Sigma = 6\,M_\odot\,\mathrm{pc}^{-2}$. This model has a cold mass fraction of only 16%, with a scale height of around 150 pc. In this case the velocity dispersions increase to 6.6, 3.3, and 2.5 km/s for the warm, unstable, and cold phases. We also find that the turbulent magnetic pressure is approximately twice the thermal pressure for both our inner and outer galaxy models, which is consistent with observations.

3. Conclusions

Our unstratified simulations have shown that the MRI may indeed be an important source of turbulence in the ISM, especially when the mean density is low, as is true in the outer galaxy. Our new stratified models, however, have cold component turbulent velocities which are too small compared to observations. A much more detailed analysis of these stratified models will be presented in a forthcoming paper.

Acknowledgements

This work was supported in part by grants NAG5-9167 (NASA), AST 0205972 and AST 0507315 (National Science Foundation). Most of the simulations presented here were performed on the Thunderhead cluster at Goddard Space Flight Center, and the Center for Theory and Computation cluster in the University of Maryland Department of Astronomy. This research has made use of NASA's Astrophysics Data System.

References

Cox, D. P. & Smith, B. W. 1974, *ApJ* 189, L105
Dickey, J. M., Hanson, M. M. & Helou, G. 1990, *ApJ* 352, 522
Field, G. B., Goldsmith, D. W. & Habing, H. J. 1969, *ApJL* 155, L149
Field, G. B. 1965, *ApJ* 142, 531
Hawley, J. F. & Balbus, S. A. 1992, *ApJ* 400, 595
Hawley, J. F., Gammie, C. F. & Balbus, S. A. 1995, *ApJ* 440, 742
Heiles, C. & Troland, T. H. 2003, *ApJ* 586, 1067
McKee, C. F. & Ostriker, J. P. 1977, *ApJ* 218, 148
Piontek, R. A. & Ostriker, E. C. 2004, *ApJ* 601, 905
Piontek, R. A. & Ostriker, E. C. 2005, *ApJ* 629, 849
Sánchez-Salcedo, F. J., Vázquez-Semadeni, E. & Gazol, A. 2002, *ApJ* 577, 768
Sellwood, J. A. & Balbus, S. A. 1999, *ApJ* 511, 660
Spitzer, L. 1978, Physical processes in the interstellar medium (New York Wiley-Interscience), p. 333
Stone, J. M. & Norman, M. L. 1992a, *ApJS* 80, 753
Stone, J. M. & Norman, M. L. 1992b, *ApJS* 80, 791
Stone, J. M., Hawley, J. F., Gammie, C. F. & Balbus, S. A. 1996, *ApJ* 463, 656
Wolfire, M. G., McKee, C. F., Hollenbach, D. & Tielens, A. G. G. M. 2003, *ApJ* 587, 278

Discussion

OMUKAI: (1.) Does your simulation satisfy the so-called Field condition? Namely, if the simulation fails to resolve the Field length, it is know that artificial structures are created. (2.) The conduction depends on the magnetic field orientation. Did you include this effect?

PIONTEK: (1.) Yes, we include thermal conduction at a level which allows us to resolve the most unstable mode of the thermal instability. (2.) This is not accounted for. The level of conduction is constant throughout the simulation domain.

KRUMHOLZ: Can you suggest observational diagnostic to distinguish supernova-driven turbulence from MRI-driven turbulence?

PIONTEK: MRI driven turbulence is stronger in the radial direction than the vertical. It is possible that this could be observed.

CRUTCHER: You get magnetic field strengths of about 3 μG, yet the observed strengths are twice that in the Milky Way, and twice again stronger in some galaxies. Would a stronger initial field still have allowed your MRI to work?

PIONTEK: The saturation of the magnetic field strength is determined by the thermal pressure, which is set by the cooling function we've used, and the thermal pressure is somewhat low in our model. Higher thermal pressures would give higher magnetic field strengths. Very strong fields, however, will stabilize the MRI.

Triggered Star Formation in a Turbulent ISM
Proceedings IAU Symposium No. 237, 2006
B. G. Elmegreen & J. Palouš, eds.

© 2007 International Astronomical Union
doi:10.1017/S174392130700124X

Non-stellar sources of turbulence in the interstellar medium

Eve C. Ostriker

Department of Astronomy, University of Maryland, College Park, MD 20742, USA
email: ostriker@astro.umd.edu

Abstract. Turbulence is believed to be important to star formation both within GMCs (affecting the IMF and the SFE), and on larger scales in the ISM (affecting GMC formation rates and properties). The traditional view of the ISM attributes most of the turbulent driving to stellar sources - in particular, supernovae and HII regions. However, evidence suggests that sources other than star formation must contribute significantly to large-scale turbulent driving in the diffuse ISM, thus also affecting the turbulence that GMCs inherit. I review recent investigations of alternative sources proposed for driving ISM turbulence, including thermal instabilities, the magnetorotational instability, large-scale self-gravitating modes, and instabilities in spiral shocks. I summarize results based on numerical simulations regarding the levels of turbulence that can be driven, and how these amplitudes depend on galactic conditions. This recent work shows that, under certain circumstances, very large-amplitude (trans-sonic with respect to the warm gas) motions and magnetic fields can be driven even without stellar energy inputs. Since turbulence can either trigger or limit star formation, understanding these potentially large non-stellar driving sources is crucial for a developing a theory of star formation regulation in the Milky Way and other disk galaxies.

Keywords. ISM: kinematics and dynamics, ISM: magnetic fields, methods: numerical, turbulence,magnetohydrodynamics

1. Introduction: how does turbulence affect star formation?

The title of this symposium links star formation with turbulence. This link may be positive, with turbulence inducing star formation, but it may also be negative. For example, turbulent *large-scale velocities* concentrate gas at high densities in the post-shock stagnation regions of converging flows, which encourages star formation because the gravitational collapse time decreases with increasing density. Turbulent *large-scale magnetic fields* also encourage star formation by transferring angular momentum out of condensations into their diffuse surroundings, enabling them to contract further. On the other hand, turbulent *small-scale velocities and magnetic fields* may tend to discourage star formation, by contributing to the effective pressure that counteracts self-gravity.

Within **giant molecular clouds** (GMCs), the star formation efficiency and rate likely depend on the overall level of turbulent compression, and the IMF may in part be established by turbulence-induced fragmentation.

In the **large-scale ISM**, the star formation rate and the masses of the largest bound clouds that form may depend on the turbulent velocity dispersion δv and the turbulent Alfvén speed $\delta v_A = \delta B/(4\pi\bar{\rho})^{1/2}$, due to turbulent contributions to an effective pressure. For *isothermal gas* with sound speed c_s and total ISM surface density Σ, the rate of bound cloud formation in spiral arms scales with the Jeans time $t_J \equiv c_s/(G\Sigma)$, while the bound cloud masses scale with the Jeans mass $M_J \equiv c_s^4/(G^2\Sigma)$. Furthermore, whether or not cloud-forming (and ultimately star-forming) gravitational instabilities are possible in a sheared, rotating disk depends on the value of the Toomre parameter $Q \equiv \kappa c_s/(\pi G\Sigma)$.

Here, $\kappa^2 \equiv R^{-3}d(\Omega^2 R^4)/dR$ is the squared epicyclic frequency. It is commonly believed, but has not been proven, that these results on GMC formation time scales, mass scales, and threshold criteria carry over to a turbulent medium if the sound speed c_s is replaced by a suitable effective sound speed c_{eff} that incorporates δv and δv_A. Since δv and δv_A may far exceed c_s for a cold gas component, these effects may be extremely important.

Effects of turbulence may be important, in particular, for defining the outer edges of star-forming disks. Quirk (1972) proposed that only the portions of disks where Q is sufficiently low undergo active star formation, and Kennicutt (1989), Martin & Kennicutt (2001), and others have found that a star formation threshold criterion $Q \lesssim 1.4$ is consistent with observed data, provided that $c_{eff} \approx 6$ km s^{-1} is adopted. The corresponding critical gas surface density at radius R for a galaxy with circular velocity V_c is

$$\Sigma_{crit} \equiv \frac{\kappa c_{eff}}{\pi G Q_{crit}} = 6 M_\odot \text{ pc}^{-2} \left(\frac{V_c}{200 \text{ km s}^{-1}}\right) \left(\frac{c_{eff}}{6 \text{ km s}^{-1}}\right) \left(\frac{R}{15 \text{ kpc}}\right)^{-1} \left(\frac{Q_{crit}}{1.4}\right)^{-1}.$$

$$(1.1)$$

The result that $Q_{crit} \approx 1.4$ is in fact consistent with the results of numerical simulations which use isothermal gas and thus have $c_{eff} = c_s$ (see e.g. the contributions of W.-T. Kim and M. Mac Low to this volume).

An alternative proposal to explain the sharp truncation of star formation in the outer parts of disks is that of Schaye (2004). He argues that star formation thresholds are defined by a transition in the available gas phases as the mean ISM pressure increases moving inward. The proposed star formation criterion is thus $P > P_{min,cold}$ where $P_{min,cold}$ is the minimum pressure at which a cold atomic phase becomes possible. For a self-gravitating gas disk (with no additional stellar or dark matter vertical gravity), the central pressure is related to the surface density by $P = \pi G \Sigma^2 / 2$ so that the surface density at the transition from a pure-warm outer disk to a warm+cold inner disk is given by $\Sigma_{trans} = 3 M_\odot \text{ pc}^{-2} (P_{min,cold}/300 \text{ k cm}^{-3} \text{ K})^{1/2}$. The value of $P_{min,cold}$ depends on UV intensity and metallicity; with a range $P_{min,cold}/k = 100 - 800$ cm^{-3} K (Elmegreen & Parravano 1994, Wolfire *et al.* 1995, Wolfire *et al.* 2003), the transition surface density is in the range $\Sigma_{trans} = 2 - 5 M_\odot$ pc^{-2}.

The numerical similarity between Σ_{trans} and Σ_{crit} makes it difficult to judge between the gravitational instability and phase-change criteria purely based on observed values of the surface density Σ_{th} at the outer star-forming radii R_{th}. However, other evidence suggests that the criterion $\Sigma > \Sigma_{trans}$ is not sufficient for defining a star formation threshold. In particular, Martin & Kennicutt (2001) found that some galaxies are primarily atomic at R_{th}, while other galaxies are primarily molecular. In those galaxies that are primarily molecular at R_{th}, the point at which Σ equals Σ_{trans}, and hence a cold atomic phase becomes possible, must lie at larger radius in the disk.

If the criterion $\Sigma > \Sigma_{trans}$ is necessary but not sufficient for star formation, then what happens in the radial range where $\Sigma_{th} > \Sigma > \Sigma_{trans}$? If $\Sigma > \Sigma_{trans}$ in a hydrostatic (laminar) disk, a cold gas layer with a sound speed ≈ 1 km s^{-1} would form near the disk midplane. If c_{eff} in equation (1.1) included only the thermal sound speed of ≈ 1 km s^{-1}, then Σ_{crit} would be low, the criterion $\Sigma > \Sigma_{crit}$ would easily be met, and the cold layer would be gravitationally unstable. A very low value for Σ_{crit}, however, would be inconsistent with observed values of Σ_{th}. This suggests that *nonthermal* contributions to c_{eff} raise the level significantly above 1 km s^{-1}. If this interpretation is correct, then for self-consistency the turbulence that suppresses gravitational instability in regions with $\Sigma_{th} > \Sigma > \Sigma_{trans}$ must have an origin *not* associated with star formation.

While in the outer parts of galactic disks, non-stellar turbulence sources may dominate, their contributions may also be important in inner-disk regions. In the remainder of this contribution, we outline potential sources of turbulence, and then discuss recent simulations that have quantified the levels of turbulence that can be produced by several different processes.

2. Possible driving sources of ISM turbulence

In the traditional view, turbulence in the diffuse ISM is driven mainly by supernovae. Following Spitzer (1978), the turbulent kinetic amplitude in this scenario may be estimated by equating a driving rate of turbulence with a dissipation rate. The driving rate, per unit ISM mass, is $\dot{E}_{turb,in} = \epsilon_{SN}\epsilon_{SF}E_{SN}/(m_{SN}t_{CF})$. Here $\epsilon_{SN} \approx \delta v/(340 \text{ km s}^{-1})$ is the efficiency of transferring supernova energy to the diffuse ISM (determined by the shell velocity when it becomes radiative), ϵ_{SF} is the efficiency of star formation $(M_{*,tot}/M_{GMC})$ per GMC lifetime, $m_{SN} \sim 200 M_\odot$ is the total mass in stars formed per Type II supernova, $E_{SN} \sim 10^{51}$ ergs is the total initial supernova energy, and the GMC formation time is $t_{CF} \sim t_{orb}\tau$ with $\tau \sim min(Q,1)$ when $Q < Q_{crit}$ and $\tau \to \infty$ when $Q > Q_{crit}$. An estimate of the overall turbulent dissipation rate in a cloudy ISM is given by $\dot{E}_{turb,out} = \delta v^2/t_{coll}$, where the collision time for diffuse ISM clouds is $t_{coll} \sim (H/\delta v)(\Sigma_{ISM}/10M_\odot \text{ pc}^{-2})^{-1}(n_H/10\text{cm}^{-3})(r_{cloud}/10 \text{ pc})$, or $t_{coll} \sim 0.1 t_{orb}Q\Sigma_{cloud}/\Sigma_{ISM}$ in terms of the Toomre parameter and the respective surface densities Σ_{cloud} and Σ_{ISM} of a typical cold atomic cloud and of the ISM as a whole.

Equating $\dot{E}_{turb,in}$ with $\dot{E}_{turb,out}$, one obtains

$$\delta v \sim 7 \text{ km s}^{-1}(\epsilon_{SF}/0.05)max(Q,1)(\Sigma_{cloud}/2\Sigma_{ISM}). \qquad (2.1)$$

Quantitatively, this expression agrees well with observations (e.g. Heiles & Troland 2003) of the mean turbulent H I velocity dispersion. However, other considerations suggest that this agreement may in part be a coincidence. First, star formation is very intermittent – it is associated with GMCs, rather than uniformly distributed throughout the H I itself. Further, these GMCs are highly concentrated in spiral arms. Yet, observations do not indicate a correlation of H I turbulence with star-forming regions, in terms of enhanced velocity dispersions either in arms compared to interarms, or in the inner disk compared to the outer disk (Dickey *et al.* 1990, van Zee & Bryant 1999). Furthermore, as argued in §1, outer disks lack star formation, but without turbulence any cold gas layer that would develop there would be violently gravitationally unstable; thus for self-consistency other *non-stellar* turbulence sources are needed.

A number of potential *non-stellar* sources of turbulence exist; these include thermal instability, magnetic instabilities (either involving vertical magnetic buoyancy, shear and rotation combined with magnetic fields, or cosmic ray coupling to magnetic fields), large-scale self-gravitational instabilities, and unsteady spiral shocks. In the following section, for several of these processes I outline the basic physical effects involved, and discuss recent results from numerical magnetohydrodynamic simulations aimed at estimating the turbulent amplitudes. Production of turbulence by interactions between cosmic rays and magnetic fields is discussed e.g. by Hanasz *et al.* (2004).

3. Recent simulations of non-stellar turbulent driving

3.1. *Thermal instability*

Under the standard heating and cooling processes for diffuse H I in galaxies, a bistable equilibrium curve results (Field, Goldsmith, & Habing 1969, Wolfire *et al.* 1995, Wolfire *et al.* 2003). For typical mean density and pressure conditions, the atomic ISM is thermally unstable and spontaneously separates into a medium consisting of cold dense clouds embedded in an ambient medium of warm, diffuse gas (Field 1965). Since pressure gradients are involved in physically collecting gas as it moves from one phase to another, the thermal instability process creates random motions within the medium. Based on numerical simulations, however, this turbulence has quite low amplitude when the pressure is in the range typical of diffuse HI in the ISM, $P/k = 2000 - 3000$ Kcm^{-3}: Piontek & Ostriker (2004), Koyama & Inutsuka (2006), and Brandenburg *et al.* (2006) all find $\delta v < 0.5$ km s^{-1}. The conclusion is that thermal instability does not, by itself, significantly contribute to turbulence in the warm/cold diffuse ISM.

3.2. *Magnetorotational instability*

Rotating, magnetized disks in which the angular velocity Ω decreases outward are subject to the magnetorotational instability (MRI), which was first explored for its importance to accretion disks by Balbus & Hawley (1991). Sellwood & Balbus (1999) subsequently noted that MRI may be important for driving turbulence in galactic disks as well. In the MRI, magnetic tension forces mediate the transfer of angular momentum from the inner to the outer disk, with a corresponding influx of mass (see contribution of R. Piontek to this volume for further details). In three-dimensional disk models, the MRI has been shown to drive turbulence, with a quasi steady state developing after several orbits.

Galactic disks differ from accretion disks in that the gas is a multi-phase, cloudy medium. In a series of studies, Piontek & Ostriker (2004), Piontek & Ostriker (2005), and Piontek & Ostriker (2006) have initiated explorations of MRI development in realistic ISM gas. One important set of questions concerns how the saturated-state turbulent amplitudes depend on the basic properties of the medium. Due to the bistable thermal equilibrium curve, the medium always evolves to a state in which the cold clouds are a factor ~ 100 denser and colder than the warm intercloud medium. Whereas the physical density in any parcel of gas is set by the equilibrium curve independent of $\bar{n}$, this mean density increases as the number of cold clouds per unit volume – or total ISM surface density – increases.

Based on local, non-stratified disk models, the saturated-state RMS velocity dispersion was found to vary with $\bar{n}$ following $\delta v = 3$ km s$^{-1}(\bar{n}/1\text{cm}^{-3})^{-0.77}$. This suggests that when $\bar{n} \sim 0.2\text{cm}^{-3}$ or less, as is believed to be true in the outer parts of galaxies, turbulent velocity dispersions can become trans-sonic with respect to the warm medium (which has sound speed ~ 7 km s^{-1}). Unlike the turbulent velocity dispersion, however, the turbulent magnetic field amplitudes were found to be independent of $\bar{n}$. Instead, the magnetic field amplifies until the magnetic pressure reaches approximate equipartition with the thermal pressure: in the saturated-state, $\beta \equiv P_{th}/P_{mag} \sim 0.5$. This is very similar to observed magnetization levels (Heiles & Troland (2005)): for $P/k = 3000\text{cm}^{-3}$ K and $B = 6\mu$G, $\beta = 0.3$.

When vertical gravity is included, differential stratification of the phases develops due to the greater buoyancy of warm compared to cold gas. If the gravity level and total gas surface density are both high (corresponding to Solar-neighborhood conditions), then a large fraction of the gas is in the cold phase, and the turbulence that is driven by MRI alone is insufficient to keep the cold clouds aloft; a thin, cold layer develops near the midplane. If the gravity and surface density are both low (corresponding to outer-galaxy conditions), on the other hand, then only a small proportion of the gas is in the cold phase. With low mean density, the MRI amplitudes are large, and a large scale height can be maintained for the cold component. For both inner- and outer-disk models, the ratio of thermal to magnetic pressure reaches realistic levels $\beta = 0.3 - 0.6$ in the saturated state. With the low mean densities of the outer-disk model, the Alfvén speed exceeds

8 km s^{-1} even at the midplane; the velocity dispersion increases from ~ 4 km s^{-1} near the midplane to twice as large at high latitudes.

The high kinetic and magnetic turbulence levels in the outer-disk model are quite interesting because they would be large enough to yield Toomre Q greater than Q_{crit} – and hence suppress star formation – if indeed the gravitational stability of turbulent disks follows a simple prescription based on an effective sound speed c_{eff} (see §1). With self-gravitating simulations (now underway), it is possible to test this idea directly.

3.3. Gravity-driven turbulence

When the value of the Toomre Q parameter is sufficiently small, disks fragment into dense clouds, which presumably further fragment into GMCs. If Q is slightly larger than Q_{crit}, on the other hand, runaway collapse does not occur, but self-gravity still drives large-amplitude motions in the gas. The outer scales involved are quite large, but the turbulence is able to cascade and therefore to contribute ISM turbulence at scales less than the disk thickness as well. Using hydrodynamic simulations that also include the effects of a gravitationally-active stellar disk (modeled with N-body methods), we have found that turbulent amplitudes about half of the warm medium sound speed can be driven when Q is close to Q_{crit} (Kim & Ostriker 2006b). Since observed values of Q typically are close to Q_{crit}, this driving mechanism may at least intermittently – for example, in interarm regions – contribute appreciably to ISM turbulence levels.

3.4. Unsteady spiral shocks

Since the work of Roberts (1969), it has been known that gas entering spiral arms tends to undergo shocks (except in the immediate vicinity of corotation between the flow and the spiral pattern, $\Omega = \Omega_p$). Based on one-dimensional models (with spatial variable perpendicular to the arm), and on two-dimensional models (representing the midplane layer), these shocks are steady as long as compressions are not large enough to make the downstream gas strongly self-gravitating (e.g. Kim & Ostriker 2002, Shetty & Ostriker 2006). Recent work has shown, however, that in *three-dimensional* models the spiral shocks that form are not in general steady (Kim & Ostriker 2006a; see also Martos & Cox 1998, Gómez & Cox 2002, Gómez & Cox 2004).

To explore the shock dynamics more fully, Kim *et al.* (2006) performed a series of high-resolution two-dimensional simulations in the radial-vertical plane. These models tested a range of spiral shock strengths and gas parameters. The typical behavior is for the shock to flap back and forth about the minimum in the spiral gravitational potential, with larger-amplitude flapping at higher z. The basic reason for this flapping is that, when the vertical variations in gravity are taken into account, the shock front must in general be curved in the $R - z$ plane. For a curved shock, vertical motions are excited in the gas, but the frequencies of these oscillations are not the same as the arm-to-arm crossing frequency. Gas streamlines "cross" (i.e. there is no steady flow solution), and turbulence develops and cascades to small scales. Radial and azimuthal velocity amplitudes typically exceed vertical amplitudes by a factor two, and the turbulence decays such that interarm velocities are a factor two lower than those in the post-shock arm region. For effective Mach number exceeding ≈ 4, the total turbulent velocity dispersion is supersonic. This work should be considered preliminary since the models involve several idealizations (including adopting an isothermal equation of state), but the results suggest that spiral shocks may very significantly add to turbulence in the ISM.

4. Conclusion

Stellar sources appear insufficient to power all turbulence in the ISM, while the turbulence itself is key for regulating star formation. Recent work has investigated a number of potential non-stellar turbulence sources, showing that some mechanisms are very important under certain conditions, while others are of moderate or minimal importance. In general, non-stellar turbulence sources must combine with each other and with stellar sources, although there may be some "geographic" complementarity. Understanding these interactions will be necessary for interpreting galactic turbulence observations, and for determining exactly how turbulence can both prompt and suppress star formation.

Acknowledgements

The results reported in this contribution are based on research with W.-T. Kim, C.-G. Kim, and R.A. Piontek. It is a pleasure to acknowledge numerous interesting discussions, and I am grateful for permission to present our joint work in this forum. This research has been financially supported by the National Science Foundation under AST 0507315, and by NASA under NNG05GG43G.

References

Balbus, S. A. & Hawley, J. F. 1991, *ApJ* 376, 214
Brandenburg, A., Korpi, M. J. & Mee, A. J. 2006, astro-ph/0604244
Dickey, J. M., Hanson, M. M. & Helou, G. 1990, *ApJ* 352, 522
Elmegreen, B. G. & Parravano, A. 1994, *ApJ* 435, L121
Field, G. B. 1965, *ApJ* 142, 531
Field, G.B., Goldsmith, D. W. & Habing, H. J. 1969, *ApJ* 155, 149
Gómez, G. C. & Cox, D. P. 2002, *ApJ* 580, 235
Gómez, G. C. & Cox, D. P. 2004, *ApJ* 615, 744
Heiles, C. & Troland, T. H. 2003, *ApJ* 586, 1067
Heiles, C. & Troland, T. H. 2005, *ApJ* 624, 773
Hanasz, M., Kowal, G., Otmianowska-Mazur, K. & Lesch, H. 2004, *ApJ* 605, L33
Kennicutt, R. C., Jr. 1989, *ApJ* 344, 685
Kim, W.-T. & Ostriker, E. C. 2002, *ApJ* 570, 132
Kim, W.-T. & Ostriker, E. C. 2006, *ApJ* 646, 213
Kim, W.-T. & Ostriker, E. C. 2006, *ApJ* submitted
Kim, C.-G., Kim, W.-T. & Ostriker, E. C. 2006, *ApJ* 649, L13
Kim, W.-T., Ostriker, E. C. & Stone, J. M. 2003, *ApJ* 599, 1157
Koyama, H. & Inutsuka, S.-i. 2006, astro-ph/0605528
Martin, C. L. & Kennicutt, R. C., Jr. 2001, *ApJ* 555, 301
Martos, M. A. & Cox, D. P. 1998, *ApJ* 509, 703
Piontek, R. A. & Ostriker, E. C. 2004, *ApJ* 601, 905
Piontek, R. A. & Ostriker, E. C. 2005, *ApJ* 629, 849
Piontek, R. A. & Ostriker, E. C. 2006, *ApJ* submitted
Quirk, W. J. 1972, *ApJ* 176, L9
Roberts, W. W. 1969, *ApJ* 158, 123
Schaye, J. 2004, *ApJ* 609, 667
Sellwood, J. A. & Balbus, S. A. 1999, *ApJ* 511, 660
Shetty, R. & Ostriker, E. C. 2006, *ApJ* 647, 997
Spitzer, L. 1978, New York Wiley-Interscience, 1978
van Zee, Liese & Bryant, J. 1999, *AJ* 118, 2172
Wolfire, M. G., Hollenbach, D., McKee, C. F., Tielens, A. G. G. M. & Bakes, E. L. O. 1995, *ApJ* 443, 152
Wolfire, M. G., McKee, C. F., Hollenbach, D. & Tielens, A. G. G. M. 2003, *ApJ* 587, 278

Triggered Star Formation in a Turbulent ISM
Proceedings IAU Symposium No. 237, 2006
B. G. Elmegreen & J. Palouš, eds.

© 2007 International Astronomical Union
doi:10.1017/S1743921307001251

Observations of supershells in the interstellar medium of nearby galaxies

Elias Brinks[1], Ioannis Bagetakos[1], Fabian Walter[2] and Erwin de Blok[3]

[1]Centre for Astrophysics Research, University of Hertfordshire, UK

[2]Max–Planck–Institut für Astronomie, Heidelberg, Germany

[3]MSSO, Australian National University, Canberra, Australia

Abstract. HI observations at sufficiently high spatial and velocity resolution have revealed a wealth of structures such as shells and bubbles in the ISM of late–type galaxies. These structures are filled with metal–enriched, coronal gas from SNe which, through overpressure, powers their expansion. Material swept up by these expanding shells can go "critical" and form subsequent (secondary or propagating) star formation. Shells that grow larger than the thickness of the gas layer will blow out of the disk, spilling enriched material into the halo (or in the case of violent starbursts, the Intergalactic Medium). We review what has been achieved to date and present some first results of a major project based on THINGS (The HI Nearby Galaxy Survey), which aims to extend studies of the ISM in galaxies to 34 nearby systems (< 10 Mpc), all observed to the same exacting standards (resolution $6'' \times 5\,\mathrm{km\,s}^{-1}$, or better; typical detection threshold of $\sim 5 \times 10^{19}\,\mathrm{atom\,cm}^{-2}$).

Keywords. galaxies: ISM — ISM: bubbles — ISM: structure — radio lines: ISM

1. Introduction

The formation of massive stars in a gas–rich spiral or dwarf galaxy has a dramatic effect on the surrounding Interstellar Medium (ISM). Newly formed massive stars (M> 8M$_\odot$) will have a major impact, first of all through their ionizing flux and stellar winds and, when they eventually have exhausted their fuel supply, as supernovae (SNe). Because massive stars usually form in clusters or associations a large amount of mechanical energy is dumped into the ISM, within a small volume and within a short time span, creating large–scale structures. These go by various descriptions, bubbles, shells or holes, based in part on their appearance at the different wavelengths at which they have been studied. Following Chu *et al.* (2004), bubbles and shells are created by single stars, superbubbles and supergiant shells require multiple stars for their formation. If they grow larger than the scale height of the gas disk, superbubbles break out into the halo. Bubbles and shells appear as "holes" in maps of neutral hydrogen, hence their name.

The study of shells and supershells is relevant to many areas of galaxy research. Supergiant shells are linked to superbubble blow out and might lie at the origin of a galactic fountain (Bregman 1980). If shells do indeed break out of the disk, the halo or in extreme cases even the intergalactic medium can be enriched by metals produced in the massive stars before they exploded as SNe (Tenorio–Tagle 2000). Actively star forming dwarf irregular galaxies are most prone to mass loss (Mac Low & Ferrara 1999; Silich & Tenorio–Tagle 2001). Given that in the currently accepted ΛCDM cosmology dwarf galaxies dominated at large look–back times, this would explain the rapid enrichment and mixing of heavy elements shortly after the first galaxies were formed.

Having said that, X–ray observations of actively star forming dwarfs show that it is exceedingly difficult for material to escape, even from dwarf galaxies (Ott *et al.* 2005a, b). Material can be transported upwards into the dark matter dominated halo, though, increasing the cross section for objects lying between the observer and distant QSOs and going some way to explaining the frequency of Damped Lyα features and associated metal lines as a function of redshift.

On a more local scale, expanding shells compress material on their rims which in turn can reach conditions conducive to star formation (Tenorio–Tagle *et al.* 2005). Examples for this can be found in the LMC where CO observations point at the presence of molecular clouds on the rims of several supergiant shells (Yamaguchi *et al.* 2001; Fukui, this volume). The Magellanic Clouds are not the only example. Leroy *et al.* (2006; see also Blitz, this volume) show similar such accumulations of molecular gas on the interface between two neighbouring shells in IC 10 which seemingly are running into each other. In other words, shells could be driving self–regulating (propagating or stochastic) star formation (Elmegreen *et al.* 2002). On yet smaller scales, the processes leading to shells and supergiant shells provide energy input on the largest turbulent scales which then cascade down to ever smaller scales.

Shells are useful as they provide an independent means to gauge the scale height of the neutral ISM (Brinks *et al.* 2002). As Oey & García–Segura (2004) have shown, they can be used as barometers as well. And finally, the energy input from SNe in the form of expanding shells provides positive feedback as it raises the velocity dispersion in the gas, shutting off further star formation until such time as the gas has cooled down, in a dynamical sense, allowing SF to recommence. This feedback loop is likely what causes the ISM to have a one dimensional velocity dispersion of 6–10 $\mathrm{km\,s^{-1}}$ irrespective of galaxy type (e.g., Dib *et al.* 2006).

2. Observations of supergiant shells

Although the first structures we now describe as supergiant shells were first recognized by Hindman (1967) in the SMC, it was the seminal work by Heiles (1984) in the Milky Way galaxy which gave prominence to this field. In general, observations in the 21–cm line of neutral hydrogen have proven crucial. HI nowadays is easy to observe and aperture synthesis interferometers can reach linear resolutions of 15–50 pc in the Local Group. Maps of the HI surface brightness show a wealth of structure and, assuming that the HI is optically thin, are a direct measure of the surface density. Moreover, as we're dealing with a spectral line, the expansion velocities of giant shells can be measured, allowing estimates to be derived of the age of the structures and energies typically required to produce them. To date, detailed observations of over a dozen galaxies have been published (see Table 1 for a list of published studies of (super)giant shells based on HI observations).

Interestingly, superbubbles are seen in molecular gas as well. I am aware of three galaxies where they have been seen in the line of ^{12}CO: M 82 (Weiß *et al.* 1999), NGC 4666 (Walter *et al.* 2004), and NGC 253 (Sakamoto *et al.* 2006). In the latter galaxy there is a tentative detection in even denser clouds as traced by NH$_3$ (Ott *et al.* 2005c).

Giant and supergiant shells have been seen at other wavelengths as well, of course, notably in Hα. Some recent papers are those by Relaño & Beckman (2005), Ambrocio–Cruz *et al.* (2004), and Valdez–Gutiérrez *et al.* (2002), all reporting Fabry–Perot observations. A lot of the earlier work was based on echelle spectra (cf. early papers on the LMC by Meaburn 1980, and on the 30 Dor region by Chu & Kennicutt 1994).

Most recently, observations with the *Spitzer Space Telescope* have revealed a stunning level of detailed structure in several nearby galaxies, notably the maps at 8 μm and 24 μm

Table 1. Published studies of (super)giant shells based on HI observations

Galaxy	authors	Galaxy	authors
Milky Way	Heiles (1984)	IC 10	Wilcots & Miller (1998)
LMC	Kim *et al.* (2003)	IC 1613	Silich *et al.* (2006)
SMC	Stanimirović *et al.* (1999)	IC 2574	Walter & Brinks (1999)
M 31	Brinks & Bajaja (1986)	Holmberg I	Ott *et al.* (2001)
M 33	Deul & den Hartog (1990)	Holmberg II	Puche *et al.* (1992)
M 101	Kamphuis *et al.* (1991)	DDO 43	Simpson *et al.* (2005b)
NGC 1569	Mühle *et al.* (2005)	DDO 47	Walter & Brinks (2001)
NGC 6946	Boomsma *et al.* (2004)	DDO 88	Simpson *et al.* (2005a)

tracing the emission of PAHs and warm dust, respectively (see, e.g., the recent study of the supergiant shell in IC 2574 by Cannon *et al.* 2005).

These large, coherent structures in the ISM of galaxies have been interpreted in terms of a "standard model" (see Oey, this volume, for an overview) for which there is ample support. X–rays have been detected from the coronal gas filling supergiant shells (Townsley *et al.* 2006, and references therein), many such shells have associated Hα emission in the form of expanding shells (Relaño, this volume). Often a remnant young stellar cluster found to be co–spatial with the shell and UV emission is correlated with these objects (Stewart *et al.* 2000).

On the theoretical side, hydrodynamical and MHD simulations are able to reproduce the observed structures based on energetics commensurate with a SN origin (Oey, this volume). Also, the observed superbubble size distribution fits predictions based on the HII region luminosity function (Oey & Clarke 1997).

Notwithstanding the consensus that star clusters through the dimise of the most massive stars as SNe power supergiant shells, several alternative explanations have been put forward to explain at least some of them, primarily those which are found where star formation is rather low key or for those cases for which the energy requirement to create the structure far surpasses what can be delivered by even a super star cluster (SSC). Among the mechanisms proposed is infall of high velocity clouds (Tenorio–Tagle *et al.* 1987), ram pressure enlarging embryonic shells (Bureau & Carignan 2002), and turbulence, coupled with cooling and gravitational instability (Wada *et al.* 2000; Dib & Burkert 2005).

Based on observations obtained thus far, giant and supergiant shells have been found with diameters ranging from the resolution limit of the telescopes employed (20–100 pc, typically, for nearby galaxies) to kpc size. Energy requirements range from the kinetic energy deposited in the ISM by a single SN to that of several hundred SNe. Estimated ages for the shells can reach of order 10^8 yr for the larger structures.

Walter & Brinks (1999) made a comparison between four objects, two spirals, M 31 and M 33, and two dwarf irregular galaxies, Holmberg II and IC 2574, and reached the following important conclusions. First of all the energy deposited by a young star cluster or stellar association is, to first order, independent of the host galaxy. Secondly, dwarf galaxies have a shallower gravitational potential. This, combined with the fact that they rotate as a solid body over most of their HI disk, results in the counterintuitive result that shells in dwarf galaxies can grow to larger sizes than in large spiral galaxies. Because the neutral gas layer in a spiral galaxy is thinner, in absolute size, than that of a typical dwarf this means that an expanding shell will blow out of the disk once it reaches a diameter which is of order the full width to half density thickness of the gas layer.

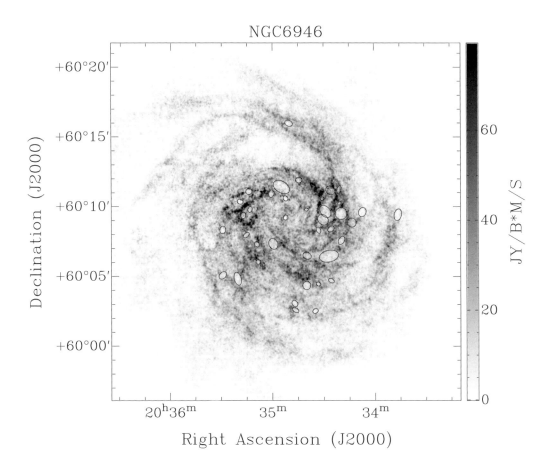

Figure 1. HI surface brightness map of NGC 6946. The ellipses mark the locations of some 40 well–defined HI holes.

3. Future projects

Those galaxies which have been observed at high enough spatial and velocity resolution plus sufficient sensitivity to allow the small scale structure of the ISM to be studied are a rather eclectic collection (see Table 1). In order to put the tentative conclusions reached by Walter & Brinks (1999) on a much firmer footing, we initiated an investigation of the galaxies which form part of THINGS, The HI Nearby Galaxy Survey (Walter *et al.* 2005). THINGS targets 34 nearby (< 10 Mpc) late–type spiral and irregular galaxies, covering a range of metallicities and star formation rates. The resolution of the maps obtained as part of THINGS is $6'' \times 5$ km s^{-1}, or better; typical detection thresholds are $\sim 5 \times 10^{19}$ atom cm^{-2}.

The galaxy M 81 was recently analysed by us (Bagetakos *et al.*, this volume) and several more systems are currently being worked on. Fig. 1 presents the HI surface density map of NGC 6946 showing of order 40 well–defined HI holes. Some HI holes can readily be identified in the HI surface brightness map, others stand out when going through the

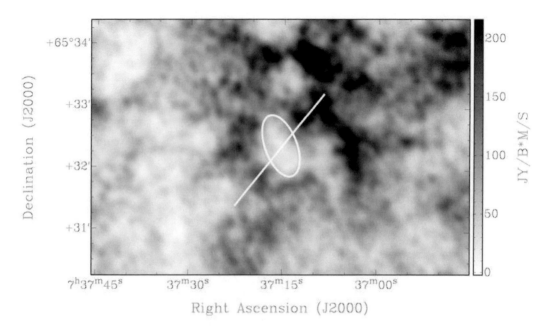

Figure 2. HI surface brightness map of a $4' \times 5'$ area of NGC 2403, showing a wealth of structure. The location of one HI hole is indicated by the ellipse.

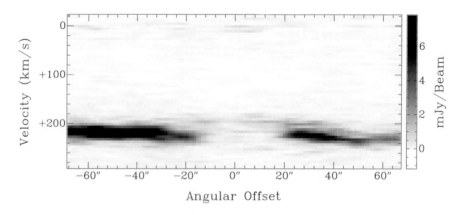

Figure 3. Position–velocity map along the line cutting through the HI hole shown in Fig. 2. Virtually no HI is left at the location of the hole, suggesting that blow out has occurred.

individual velocity channels or upon inspection of the HI data cube in position–velocity space. Despite efforts by various authors to devise an automated procedure to identify HI holes (see Ehlerová & Palouš 2005, and references therein), no single method has emerged yet that manages to match the human eye/brain in picking out these structures. Once a potential HI hole has been identified, a position–velocity cut readily confirms its reality, as illustrated in Figs. 2 and 3 which shows a typical example of an HI hole in yet another object, NGC 2403.

Work has now started to analyse in a consistent and uniform manner all galaxies within THINGS (Bagetakos *et al.*, this volume). The idea is to prepare for each galaxy a catalogue of HI holes. The results for each galaxy will be compared in order to investigate how the diameter distribution of HI holes, their ages and energy requirements correlate with Hubble type, metallicity, star formation rate, etc. The structures found in HI will be compared with mid–IR data from the *Spitzer* SINGS survey, UV maps obtained with GALEX, and optical broad– and narrow–band (Hα) imaging.

4. Summary

Maps in the line of HI at sufficiently high linear resolution (100 pc or better) are dominated by giant and supergiant shells and filaments. Many of the structures are coherent and expand. The dimensions, energetics and general characteristics are compatible with a SN origin. The energy input of a young star cluster is to first order independent of Hubble type. Dwarf galaxies have as a result larger diameter shells. Expanding shells are not only seen in HI but also in Hα and molecular line emission (CO, NH_3). Sites of secondary SF can be seen on the rims of some shells and CO emission is detected on the interface where neighbouring shells run into each other compressing, between them, the surrounding ISM. Shells are at the top of the turbulence cascade and some of the smaller shells could be turbulence driven.

Most of the results and ideas we have about giant and supergiant shells are based on only a handful of galaxies, many of them of the dwarf variety and some clearly pathological. Given the importance of shells, from fields as diverse as the enrichment of the IGM in the early universe to star formation triggered by pressure–driven, expanding bubbles, a major project has been conceived by us, THINGS, which will produce maps of the highest quality of the ISM of 34 nearby galaxies. Eventually, THINGS will firm up many of the tentative results arrived at thus far and go a long way to addressing some of the key questions of this fascinating field.

References

Ambrocio–Cruz, P., Laval, A., Rosado, M., Georgelin, Y. P., Marcelin, M., Comeron, F., Delmotte, N. & Viale, A. 2004, *AJ* 127, 2145

Boomsma, R., van der Hulst, T., Oosterloo, T., Fraternali, F. & Sancisi, R. 2004, in: P.–A. Duc, J. Braine & E. Brinks (eds.), *Proc. IAU Symp. No. 217* (San Francisco: ASP), p.142

Bregman, J. N. 1980, *ApJ* 236, 577

Brinks, E. & Bajaja, E. 1986, *A&A* 169, 14

Brinks, E., Walter, F. & Ott, J. 2002, *ASP–CS* 275, 57

Bureau, M. & Carignan, C. 2002, *AJ* 123, 1316

Cannon, J. M., *et al.* 2005, *ApJ* 630, L37

Chu, Y.–H. & Kennicutt, R. C., Jr. 1994, *ApJ* 425, 720

Chu, Y.–H., Guerrero, M. A. & Gruendl, R. A. 2004, in: E. J. Alfaro, E. Pérez & J. Franco (eds.), *How Does the Galaxy Work? A Galactic Tertulia with Don Cox and Ron Reynolds*, *Ap&SS* (Kluwer, Dordrecht), vol 315, p. 165

Deul, E. R. & den Hartog, R. H. 1990, *A&A* 229, 362

Dib, S. & Burkert, A. 2005, *ApJ* 630, 238

Dib, S., Bell, E. & Burkert, A. 2006, *ApJ* 638, 797

Ehlerová, S. & Palouš, J. 2005, *A&A* 437, 101

Elmegreen, B. G., Palouš, J. & Ehlerová, S. 2002, *MNRAS* 334, 693

Heiles, C. 1984, *ApJS* 55, 585

Hindman, J. V. 1967, *AuJPh* 20, 147

Kamphuis, J., Sancisi, R. & van der Hulst, T. 1991, *A&A* 244, L29

Kim, S., Staveley–Smith, L., Dopita, M. A., Sault, R. J., Freeman, K. C., Lee, Y. & Chu, Y.–H. 2003, *ApJS* 148, 473

Leroy, A., Bolatto, A., Walter, F. & Blitz, L. 2006, *ApJ* 643, 825

Mac Low, M.–M. & Ferrara, A. 1999, *ApJ* 513, 142

Meaburn, J. 1980, *MNRAS* 192, 365

Mühle, S., Klein, U., Wilcots, E. M. & Hüttemeister, S. 2005, *AJ* 130, 524 (Erratum: 2006, *AJ* 132, 443)

Oey, M. S. & Clarke, C. J. 1997, *MNRAS* 289, 570

Oey, M. S. & García–Segura, G. 2004, *ApJ* 613, 302

Ott, J., Walter, F. & Brinks, E. 2005a, *MNRAS* 358, 1423

Ott, J., Walter, F. & Brinks, E. 2005b, *MNRAS* 358, 1453

Ott, J., Weiß, A., Henkel, Ch. & Walter, F. 2005c, *ApJ* 629, 767

Ott, J., Walter, F., Brinks, E., Van Dyk, S. D., Dirsch, B. & Klein, U. 2001, *AJ* 122, 3070

Puche, D., Westpfahl, D., Brinks, E. & Roy, J.–R. 1992, *AJ* 103, 1841

Relaño, M. & Beckman, J. E. 2005, *A&A* 430, 911

Sakamoto, K., Ho, P. T. P., Iono, D., Keto, E. R., Mao, R.–Q., Matsushita, S., Peck, A. B., Wiedner, M. C., Wilner, D. J. & Zhao, J.–H. 2006, *ApJ* 636, 685

Silich, S. & Tenorio–Tagle, G. 2001, *ApJ* 552, 91

Silich, S., Lozinskaya, T., Moiseev, A., Podorvanuk, N., Rosado, M., Borissova, J., Valdez–Gutiérrez, M. 2006, *A&A* 448, 123

Simpson, C. E., Hunter, D. A. & Knezek, P. M. 2005a, *AJ* 129, 160

Simpson, C. E., Hunter, D. A. & Nordgren, T. E. 2005b, *AJ* 130, 1049

Stanimirović, S., Staveley–Smith, L., Dickey, J. M., Sault, R. J. & Snowden, S. L. 1999, *MNRAS* 302, 417

Stewart, S. G., *et al.* 2000, *ApJ* 529, 201

Tenorio–Tagle, G. 2000, *NewAR* 44, 365

Tenorio–Tagle, G., Franco, J., Bodenheimer, P. & Rozyczka, M. 1987, *A&A* 179, 219

Tenorio–Tagle, G., Silich, S., Rodríguez–González, A. & Muñoz–Tuñón, C. 2005, *ApJ* 628, L13

Townsley, L. K., Broos, P. S., Feigelson, E. D., Brandl, B. R., Chu, Y.–H., Garmire, G. P. & Pavlov, G. G. 2006, *AJ* 131, 2140

Valdez–Gutiérrez, M., Rosado, M., Puerari, I., Georgiev, L., Borissova, J. & Ambrocio–Cruz, P. 2002, *AJ* 124, 3157

Wada, K., Spaans, M. & Kim, S. 2000, *ApJ* 540, 797

Walter, F., Dahlem, M. & Lisenfeld, U. 2004, *ApJ* 606, 258

Walter, F. & Brinks, E. 1999, *AJ* 118, 273

Walter, F. & Brinks, E. 2001, *AJ* 121, 3026

Walter, F., Brinks, E., de Blok, W. J. G., Thornley, M. D. & Kennicutt, R. C. 2005, *ASP–CS* 331, 269

Weiß, A., Walter, F., Neininger, N. & Klein, U. 1999, *A&A* 345, L23

Wilcots, E. M. & Miller, B. W. 1998, *AJ* 116, 2363

Yamaguchi, R., Mizuno, N., Onishi, T., Mizuno, A. & Fukui, Y. 2001, *ApJ* 553, L185

Discussion

TENORIO-TAGLE: A suggestion would be to look for shells also in the gaseous halos of galaxies. they would be much fainter than the ones you are detecting on disks but they are there.

BRINKS: That's indeed a good suggestion, and I hope that you will be asked on the time allocation committee to insure we will be awarded the time needed to pursue this!

MAC LOW: Why do density power spectra appear to extend well beyond the supershell scale if supershells are the driver of the turbulence?

BRINKS: I have no idea! Regarding our data, we haven't yet produced any power spectra.

OEY: I don't think I agree with your suggestion that galaxies with higher star formation show more HI shells. The opposite is seen in the Magellanic clouds: the LMC has only 1/4 the number of HI shell as the SMC. I think this may be because if the SFR is high enough, then the shells will merge and shred the ISM, etc. This may also be related to the earlier question about the large extension of the power spectrum.

BRINKS: Yes, I agree that the LMC looks "shredded". We will look, though, across our sample if we find trends with star formation rate, Hubble type, etc., in the hope to be able to increase our underestimate of the shells.

ZINNECKER: Is the secondary star formation on the rim of the shells as violent as that of the first star formation which gave rise to the shells?

BRINKS: This is hard to tell. We can estimate the kind of association that was responsible for creating the holes, but with what do you compare this, with a single star forming region on the rim, or with all such regions averaged over the entire rim?

Proceedings of the IAU Symposium NO. 237
Proceedings IAU Symposium No. 237, 2006
Bruce G. Elmegreen & Jan Palouš, eds.

Shells in the Magellanic System

Snežana Stanimirović†

Radio Astronomy Lab, UC Berkeley, 601 Campbell Hall, Berkeley, CA 94720, USA

Abstract. The Magellanic System harbors > 800 expanding shells of neutral hydrogen, providing a unique opportunity for statistical investigations. Most of these shells are surprisingly young, 2–10 Myr old, and correlate poorly with young stellar populations. I summarize what we have learned about shell properties and particularly focus on the puzzling correlation between the shell radius and expansion velocity. In the framework of the standard, adiabatic model for shell evolution this tight correlation suggests a coherent burst of star formation across the whole Magellanic System. However, more than one mechanism for shell formation may be taking place.

Keywords. ISM: bubbles, ISM: structure, Magellanic Clouds

1. Introduction

Numerous studies over the past three decades have shown that shell-like structures dominate the interstellar medium (ISM) in many galaxies. This sculpturing of the ISM, mainly assumed to be due to star-formation activity, must have an imprint on many physical processes (e.g. the transport of radiation, heating and cooling etc.). In the traditional scenario, shells are expected to be reservoirs of hot gas and are the sole-source of the hot, intercloud medium. At the same time, these dynamic features are sites of local energy depositions in the ISM which contribute significantly to the total energy budget. Another important role shells have is in providing the connection between galactic disks and halos. Shells grow in the disks, expand, and the largest ones can reach sizes larger than the disk scale height. When this happens shells open up and vent hot gas into the halo.

Despite significant observational efforts, the exact mechanism(s) for shell formation is still not fully understood. Similarly, the late stages of shell evolution have not been explored. The most accepted model for shell formation, the "Standard Model" (Weaver *et al.* 1977; McCray & Kafatos 1987), views shells as products of combined effects of stellar winds and supernovae. Numerous observational puzzles however motivated other suggestions. For example, several other types of powering sources were suggested, the most exotic ones being pulsars and gamma-ray bursts. Several proposed mechanisms do not even require the existence of a central energy source. These range from the collision of high-velocity clouds (HVCs) with a galactic disk, through results of the general ISM turbulence, with or without gravitational and thermal instabilities, the ram pressure stripping, flaring of radio lobes, to the complex new conceptual designs of the ISM in the form of an elastic polymer interwoven with magnetic field lines (Cox 2005). This list is obviously long, and surprisingly keeps growing steadily.

In this paper we focus on shells discovered in the Magellanic System: the Small Magellanic Cloud (SMC), the Large Magellanic Cloud (LMC), and the Magellanic Bridge (MB). More than 800 shell-like structures were found and cataloged in high-resolution neutral hydrogen (HI) observations of these three environments. All observations were conducted

† Present address: Department of Astronomy, University of Wisconsin, 475 North Charter Street, Madison, WI 53706, USA email: sstanimi@astro.wisc.edu

Table 1. Summary of shell properties: Number of HI shells, the min/max range of shell radii, the min/max range of shell expansion velocities, the estimated mean dynamic age, and power-law slopes of the shell size and expansion velocity distribution functions.

	N	R_s (pc)	V_{exp} (km s^{-1})	$\langle T_s \rangle$ (Myr)	α_r	α_v
SMC	509	20–800	2–33	5.7	-2.5 ± 0.2^a	-2.2 ± 0.3^a
MB	163	10–200	2–20	6.2	-3.6 ± 0.4	-2.6 ± 0.6
LMC	124 $(54)^b$	50–620	6–36	4.8	-2.5 ± 0.4	?

[a] These slopes were derived using all SMC shells.
[b] The number in brackets refers to stalled shells.

with the Australia Telescope Compact Array (ATCA) and were complemented with the short-spacing data from surveys with the Parkes telescope. These data sets sample a wide and continuous range of spatial scales, providing unique opportunities for finding and studying large samples of expanding shells. It is important to stress that the three systems we investigate here probe very different interstellar environments. The SMC is a dwarf irregular, gas-rich galaxy with a large line-of-sight depth, the LMC is a dwarf disk galaxy with traces of spiral structure and a higher star formation rate, while the MB is a column of gas between the two galaxies that was formed as a result of tidal interactions. In Section 2 we summarize the most important properties of HI shells in the Magellanic System, and address their implications in Section 3. In Section 4 we point out recent observational work in the domain of the late stages of shell evolution, and then summarize in Section 5.

2. Summary of observational properties

2.1. *Shells in the SMC*

More than 500 expanding shells have been cataloged in the SMC (Staveley-Smith *et al.* 1997; Stanimirović *et al.* 1999; Hatzidimitriou *et al.* 2005). Table 1 lists the typical shell radius (R_s) and the expansion velocity (V_{exp}). Curiously, as noted first by Staveley-Smith *et al.* (1997), all SMC shells appear to have a very similar dynamic age, $T_s \sim 5$ Myr. This was interpreted as evidence for a single, coherent, and global burst of star formation in the SMC. The volume occupied by all these shells is large, about 40% of the whole SMC, implying a very bubbly, or a 'Swiss Cheese'-like morphology. Estimating the fraction of HI mass occupied by shells is more difficult as we need an estimate of the local ambient density. If we assume that the local ambient surface density for *all shells* is 0.01 M$_\odot$ pc^{-2}, we arrive at a total mass fraction that is about 20%.

Recently, Hatzidimitriou *et al.* (2005) searched for the remnant stellar population associated with the HI shells by using all available catalogues of the young stellar populations. About 450 shells were found to correlate with one or more stellar objects, while 59 shells do not correlate with any of the known stellar objects. We will refer further to these two classes of shells as "non-empty" and "empty", respectively. The surprising result is that properties of "non-empty" and "empty" are almost indistinguishable. There are no morphological differences between the two groups. "Empty" shells appear smaller and with a lower expansion velocity than "non-empty" ones, however this is primarily a selection effect. Both types also show an almost linear correlation between the shell radius and expansion velocity. Spatially, "empty" shells are found primarily in remote places on the outskirts of the HI distribution. Of course, finding similar shells in the central parts would be impossible. Several "empty" shells with high luminosity appear loosely clustered and possibly connected; these objects may belong to an old chimney. The rest of the empty shells, however, do not have any special location or association.

Hatzidimitriou *et al.* (2005) also derived the shell size and expansion velocity distribution functions, $N(R_s)$ and $N(V_{exp})$. For both types of shells $N(R_s)$ and $N(V_{exp})$ can be fitted with a power-law function, $N(R_s) \propto R_s^{\alpha_r}$ and $N(V_{exp}) \propto V_{exp}^{\alpha_v}$. The slopes α_r and α_v agree within their uncertainties for the two types of shells. In the framework of the standard model, based on whether all shells were formed in a single burst or in a continuous manner, and by assuming either a single input mechanical luminosity function (MLF) for all shells or a power-law function, we can predict α_r and α_v and then compare these values with what we get from observations. This was first shown by Oey & Clark (1997). The positive slope of the $R_s - V_{exp}$ correlation, and the fact that $\alpha_r \approx \alpha_v$ ($\approx -2.2 \pm 0.2$ for "non-empty shells), point to the case of a single burst of shell formation and a power-law input MLF. It is puzzling though that the same arguments apply to "empty" shells as well. If the "empty" shells are > 10 Myr old, and this could be the reason why we do not find their corresponding stellar population, then for them to fit on the same $R_s - V_{exp}$ relation would require a significant and concerted re-acceleration.

2.2. Shells in the Magellanic Bridge

Muller *et al.* (2003) cataloged 163 shells in the MB, applying criteria somewhat tighter than in the case of the SMC shells (for example, they do not include incomplete large shells in their catalog; this results in shell sizes being biased towards smaller shells). Shell sizes and expansion velocities are given in Table 1. The estimated mean dynamic age is $T_s = 6.2$ Myr, with a standard dispersion of 3.4 Myr. Muller *et al.* (2003) cross-correlated their shell catalog with the catalog of OB associations by Bica & Schmitt (1995) and found that about 60% of shells do not have corresponding OB associations. Also, while the mean dynamic shell age is about 6 Myr, the mean age of OB associations is several times larger, 10-25 Myr. Although the MB is a tidal remnant of the interactions between the LMC and the SMC, shells in the MB are primarily spherical and without obvious signs of distortions or tidal stretching.

2.3. Shells in the LMC

There are 101 giant ($R_s < 360$ pc) and 23 super-giant ($R_s > 360$ pc) shells in the LMC (Kim *et al.* 1999). Shell radii are in the range 50–620 pc, while shell expansion velocities are in the range 6–36 km s^{-1}. Interestingly, while the expansion velocity is systematically higher in the LMC than in the SMC, about one half of all LMC shells appear to have stalled, with $V_{exp} = 0$. The mean dynamic age is 4.9 Myr, which is again younger than the age of corresponding OB associations (> 10 Myr). Giant shells in the LMC also follow the almost linear $R_s - V_{exp}$ relation, while the super-giant shells deviate from this trend. Kim *et al.* (1999) also found a poor spatial correlation between shells and OB associations. The shell size distribution is $N(R_s) \propto R_s^{-2.5}$, and is similar to that for the SMC shells.

3. Putting it all together

To summarize, the properties of all shells in the SMC, the MB, and the LMC show striking similarities: the dynamic age, tight $R_s - V_{exp}$ relation, statistical properties, poor correlation with stellar populations, and dynamic ages being younger than those of OB associations. And yet, the three environments in which these shells formed and evolved are drastically different! To emphasize this, we plot the dynamic shell age as a function of Right Ascension in Figure 1 and include all (> 800) shells found in the Magellanic System.

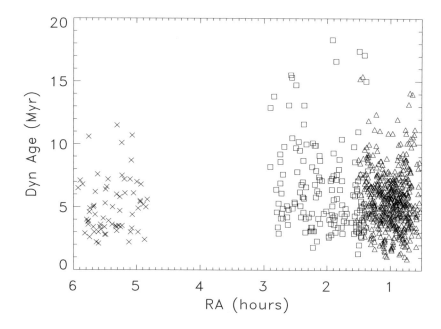

Figure 1. The estimated dynamic age for shells in the SMC (triangles), the MB (squares), and the LMC (crosses).

This distribution appears uniform and there are no obvious discontinuities between the SMC and the MB. It is also apparent that the SMC and the MB have a few shells older than the majority of the LMC shells. Figures 1 and 2 also show that, on average, the shell size increases from the SMC, through the MB, to the LMC. This is primarily a selection effect, however. For example, Muller *et al.* (2003) discuss their lack of large shells in the MB and explain that this is mainly due to their more stringent criteria when identifying shells.

In Figure 2 we plot all shells on the $R_s - V_{exp}$ diagram. It is very curious how all shells appear to follow the same, tight relation, being nested between the solid lines that mark $T = 2$ Myr and $T = 10$ Myr, when viewed within the standard model. In the case of our Galaxy, Ehlerová & Palouš (2005) did not find a correlation between the shell radius and expansion velocity, although even there it looks like larger shells have a larger V_{exp}. In the framework of the standard model, the tight age spread could be interpreted as a result of a recent star formation burst across the whole Magellanic System about 5 Myr ago. However, this is difficult to reconcile with the star formation history of the Clouds. In the case of the SMC, Harris & Zaritsky (2004) estimated ages of $> 5 \times 10^6$ stars and found the closest peak in star formation 60 Myr ago. It is also interesting to note that on the older-age side, the cut-off in the shell distribution (Figure 2) is sharper than on the younger-age side. This may be suggestive of an enhanced shell destruction/fragmentation after ~ 10 Myr. However, this is significantly shorter than the typical predicted shell lifetime of 30–50 Myr (Dove 2000; Wünsch & Palouš 2001).

The $R_s - V_{exp}$ diagram obviously suffers from some selection effects and is biased towards younger objects. Large shells are hard to distinguish observationally as they are often fragmented and also could be seen as a superposition of several smaller shells. In the case of the SMC, Zaritsky & Harris (2004) showed that 10 to 70% of all stars

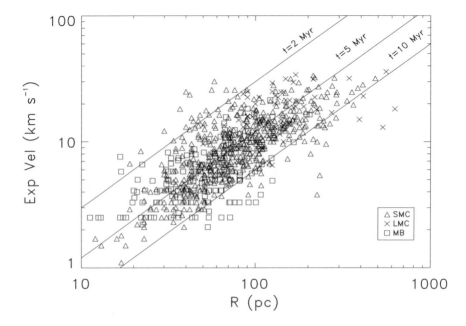

Figure 2. Size and expansion velocity for shells in the SMC (triangles), the LMC (crosses), and the MB (squares). The solid lines represent dynamic ages of 2, 5, and 10 Myrs, respectively, in the framework of the standard model.

could have formed through tidal triggering. Tidal triggering would obviously affect shell formation, and probably even shell evolution. The degree with which these processes are constrained and coordinated across the Magellanic System still needs to be explored. Alternatively, there may be one or more additional shell-formation processes taking place. Obviously, understanding the role of environmental effects (e.g. tidal flows, the turbulent ISM, interactions between shells, magnetic field etc.) is crucial for further advancement.

4. What happens to shells at a very late stage of their evolution?

We would now like to draw attention to numerous HI clouds found recently in the interface region between the disk and the halo of our Galaxy (Lockman 2002; Stil *et al.* 2005; Stanimirović *et al.* 2006). These clouds are small, 5–20 pc in size, cold (< 400 to 1000 K), and often kinematically follow the Galactic disk but at a velocity that is offset by 10–20 km s^{-1} from that of the bulk HI emission. The clouds do not appear to prefer particular regions in the Galaxy, their distribution is most likely radially extended. The origin of these clouds is not clear. There are several possibilities including Galactic fountains and the accretion of extragalactic gas. An alternative possibility, however, is that these clouds are fragments of expanding shells. There are several pieces of evidence that point in this direction. For example, a Galactic chimney, GSH242-03+37 shows small, discrete HI clouds that appear associated with shell caps (McClure-Griffiths *et al.* 2006). Lockman, Pidopryhora, & Shields (2006) found a large plume-like structure which has numerous clouds. Stanimirović *et al..* (2006) found that clouds are embedded in large filamentary structures and morphologically resemble cold clouds that form in simulations of dynamically triggered instabilities (e.g. Audit & Hennebelle 2005). One way of testing the hypothesis that these newly-discovered clouds are shell fragments is to derive the cloud mass spectrum and compare it with theoretical predictions from shell fragmentation. For example, Wünsch & Palouš (2001) predict that the mass spectrum for shell fragments is a power-law with $dN/dM \propto M^{-1.4}$. The mass spectrum of observed clouds will be easily derived in the near future as several large HI surveys are underway with the Arecibo, Parkes and Green Bank telescopes that are particularly suited for finding these clouds.

5. Summary and open questions

There are ~ 800 shells in total in the SMC, the MB and the LMC. At least 1/10 of SMC shells are devoid of stellar counterparts, but surprisingly have properties similar to those of "non-empty" shells. Large similarities in shell properties across the Magellanic System, and especially the tight correlation between the shell radius and expansion velocity, are puzzling and may be highlighting the importance of tidal interactions for both shell formation and evolution. Alternative processes for shell formation and external effects may be also playing important roles. The late stages of shell evolution are finally being addressed observationally, and detailed theoretical attention is highly desirable in this area. And finally, there is the Magellanic Stream, a starless tidal tail that provides the perfect opportunity to quantify the importance of shell formation processes without a powering source. This is an obvious future project!

Acknowledgements

I would like to thank the conference organizers for inviting me to participate in this highly stimulating and enjoyable conference. I would also like to thank Carl Heiles, one of the pioneers of shell exploration, for numerous inspiring discussions, and Jacco van Loon for many insightful comments. This work was partially supported by NSF grants AST 04-06987 and 00-97417.

References

Audit, E. & Hennebelle, P. 2005, *A&A* 433, 1
Bica, E.L.D. & Schmitt, H.R. 1995, *ApJS* 101, 41
Cox, D.P. 2005, *ARAA* 43, 337
Dove, J.B., Shull, J.M. & Ferrara, A. 2000, *ApJ* 531, 846
Ehlerová, S. & Palouš, J. 2005, *A&A* 437, 101
Hatzidimitriou, D., Stanimirović, S., Maragoudaki, F., Staveley-Smith, L., Dapergolas, A. & Bratsolis, E. 2005, *MNRAS* 360, 1171
Kim, S., Dopita, M.A., Staveley-Smith, L. & Bessell, M.S. 1999, *AJ* 118, 2797
Lockman, F.J., Pidopryhora, Y. & Shields, J.C. 2006, this conference.
McClure-Griffiths, N.M., Ford, A., Pisano, D.J., Gibson, B.K., Staveley-Smith, L., Calabretta, M.R., Dedes, L. & Kalberla, P.M.W. 2006, *ApJ* 638, 196
McCray, R. & Kafatos, M. 1987, *ApJ* 317, 190
Muller, E., Staveley-Smith, L., Zealey, W. & Stanimirović, S. 2003, *MNRAS* 339, 105
Stanimirović, et al. 2006, *ApJ* in press (astro-ph/0609137)
Stanimirović, S., Staveley-Smith, L., Dickey, J.M., Sault, R.J. & Snowden, S.L. 1999, *MNRAS* 302, 417
Staveley-Smith, L., Sault, R.J., Hatzidimitriou, D., Kesteven, M.J. & McConnell, D. 1997, *MNRAS* 289, 225
Weaver, R., McCray, R., Castor, J., Shapiro, P. & Moore, R. 1977, *ApJ* 218, 377
Wunsch, R. & Palouš, J. 2001, *A&A* 374, 746

Discussion

Y.-H. CHU: If there was a burst of star formation 5 Myr ago to form all these shells, there should be late O and early B stars remaining in the shells. However, when we searched for such stars from uv observations with IUE/HST/FUSE, we couldn't find any within the larger shells in the LMC. They couldn't have all been formed from a single burst of star formation 5 Myr ago.

STANIMIROVIC: This is just one more argument against interpreting all these shells as being caused by stellar winds/ SNe.

Triggered Star Formation in a Turbulent ISM
Proceedings IAU Symposium No. 237, 2006
B. G. Elmegreen & J. Palouš, eds.

© 2007 International Astronomical Union
doi:10.1017/S1743921307001275

An automatic identification of HI shells in the 2nd and 3rd Galactic Quadrants

Soňa Ehlerová

Astronomical Institute, Academy of Sciences of the Czech Republic,
Boční II 1401, 141 31 Prague, Czech Republic
email: sona@ig.cas.cz

Abstract. We briefly discuss different methods used to identify HI shells in T_B datacubes. Then we give results for our automatic method applied to LDS and LAB HI surveys of the Milky Way (2nd and 3rd quadrants). We fit the radial distribution of HI shells (the exponential profile with the scale length of 3 kpc) and the size distribution (the power law with the index of 2.1). We compare the distribution of identified HI shells with HII regions and study the differences between identifications in the 2nd and 3rd quadrants.

Keywords. ISM: bubbles, methods: data analysis, Galaxy: structure

1. Introduction

HI shells are thoroughly studied in many galaxies, as described by E. Brinks and S. Stanimirovic in this volume. Originally, shells were discovered in the Milky Way Galaxy, which is probably not surprising. In recent years the studies, especially the statistical studies, are done mostly for shells in external galaxies. That might be expected. Our inside view gives us many advantages but also disadvantages. To study the whole Galaxy we need really an all-sky survey. The changing distance to objects makes objects with intrinsically the same dimensions have wildly different angular dimensions. Looking through the galactic disk means you encounter lots and lots of structures. And I do not even mention the problems with kinematical distances.

All that said I still do not think we should abandon the Milky Way and its shells. In this contribution I will give results of a study of shells in the 2nd and 3rd galactic quadrants. Before that I will describe a method used for their identification and since it is not a traditional way, I will also mention different approaches in dealing with the shell identification.

1.1. *Traditional by-eye approach*

The most obvious and historical approach to finding shells is looking at maps and localizing structures in individual velocity channels, probably with the help of velocity spectra. The main problem with that method is not that it is tedious and time-consuming, but that it is subjective and, in my opinion, dependent on the funny things like the observer's favourite colour and intensity scale. Nevertheless, it also has some pluses, like the ability of a human eye (or rather brain) connect disconnected features and disregard 'obviously' disconnected structures. This approach was used for the Milky Way shells by Heiles (1979), Heiles (1984), Hu (1981), and most recently by McClure-Griffiths, Dickey, Gaensler *et al.* (2002).

1.2. *Automatic identification*

An alternative method of identifying shells is some kind of an automatic search. The shell is somehow defined and the pattern is then searched for automatically in datacubes. One

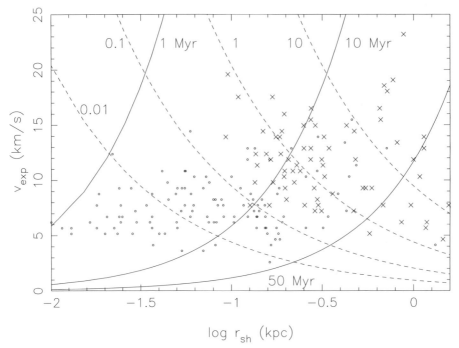

Figure 1. Radius vs expansion velocity of HI shells in the 2nd quadrant. Different symbols stand for different energies (circles $< 10^{51} erg$, crosses $> 10^{51} erg$). Overlaid are contours of constant luminosity and age.

way is to take numerical models of shells. This approach was taken by Thilker, Braun & Walterbos (1998); Mashchenko, Thilker & Braun (1999) and Mashchenko & St-Louis (2002), where (magneto)hydrodynamical and thin-shell models were used.

A different approach is used by Daigle, Joncas, Parizeau *et al.* (2003). They primarily study the velocity spectrum and search for the expanding pattern. Then they look at groups of pixels which contain the expansion and decide if the morphology corresponds to the HI shell.

Our approach (Ehlerová & Palouš (2005)) also differs. We primarily search for local holes in velocity maps and then study consecutive velocity channels and corresponding velocity spectra.

All mentioned methods are not perfect and the ISM is turbulent. Basically, with an automatic identification you have one of extremes: either your models are very restrictive and then you detect nothing or just a small part of real structures, or your models are lax and you have a high number of false identifications.

2. Identification of shells: HOLMES

The detailed description of the code HOLMES can be found in Ehlerová & Palouš (2005). As already told, we search for local minima in T_B velocity channels (step 1 in the procedure), connect them (step 2) and analyse their velocity spectra (step 3). Our method belongs to the type which is prone to find 'false detections': to get rid of them we use the analysis of the spectrum (e.g. step 3).

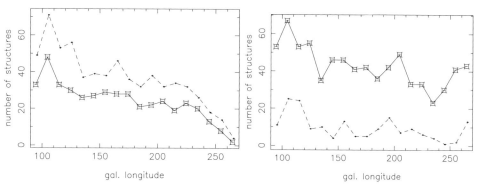

Figure 2. *Left:* number of HI shells in LDS as a function of galactic longitude. Solid line corresponds to the full step 3 in the identification, dashed line to the reduced step 3. *Right:* number of HI shells in the LAB survey (reduced step 3, solid line) and HII regions (dashed line) as a function of longitude.

2.1. *Results for the 2nd Quadrant and LDS*

We applied our searching code HOLMES on the Leiden-Dwingeloo HI survey (Hartmann & Burton (1997)), which covers about 80 % of the sky, in which we identified 628 shells.

For the structures in the 2nd quadrant (fully observed by Hartmann & Burton (1997) and no distance ambiguity) we fitted the exponential radial distribution and found that the radial scale length is about 3 kpc, in agreement with the scale length of the stellar disk.

We also studied the size distribution of shells in the 2nd quadrant using the analysis of Oey & Clarke (1997). We found the power-law index of $\alpha = 2.1$ which is quite shallow but corresponding to values derived for several external galaxies (M31, M33, LMC, SMC, HoII).

The $r - v_{exp}$ diagram for shells in the 2nd quadrant is shown in figure 1. Radii of detected structures range from ~ 10 pc to ~ 1.5 kpc, expansion velocities are between 4 km s^{-1} (an artificial limit set by HOLMES) and ~ 20 km s^{-1}.

2.2. *Application on the LAB survey*

Recently, we began applying our HOLMES code on the LAB all-sky HI survey (Kalberla *et al.* (2005)). So far, we were able to process the 2nd and 3rd quadrants. Step 3 of the procedure (the analysis of the spectra, see above) did not behave quite satisfactorily, therefore we exchange it for the time being by a reduced condition: the identified structure must have $\Delta v_{exp} \geqslant 4$ km s^{-1}. Compared to the full step 3 this increases the number of detections by about 1/2.

Figure 2 (left) shows the number of HI shells as a function of the galactic longitude for the LDS shells using the full and reduced step 3. Absolute numbers differ but the profile is the same. Figure 2 (right) shows the number of HI shells identified in the LAB survey and a number of HII regions from Paladini *et al.* (2003). There are about 3-4 shells for one HII region, but profiles agree well with each other. The discrepancy between profiles of LDS shells and LAB shells in the 3rd quadrant is caused by the fact that the LDS survey does not fully cover this region.

There is a notable difference between numbers of HI shells in the 2nd and 3rd quadrants (as well as HII regions). This is probably caused by the presence of the spiral structure which is dominant in the 2nd quadrant. As studied by McClure-Griffiths, Dickey, Gaensler *et al.* (2002), HI shells prefer interarm regions and avoid arms. This seems to be the case for our identifications as well, even though the absolute number increases with the presence of the spiral structure.

3. Summary

We have briefly summarized advantages and disadvantages of two approaches towards the identification of shells: the traditional 'by-eye' approach and the automatic methods. We then showed results based on one of the automatic searching codes applied to the LDS (Leiden/Dwingeloo HI survey) and to LAB (Leiden/Argentina/Bonn) HI survey. LDS results were used to estimate the radial profile of HI shells (the exponential with the scale length of 3 kpc) and the size distribution (the power law with the index of 2.1). For the LAB shells differences are shown between 2nd and 3rd quadrants; these differences are connected to the spiral structure.

Acknowledgements

This work was supported by the Institutional Research Plan AV10030501 of the Astronomical Institute, Academy of Sciences of the Czech Republic, and the project LC06014 of the Center for Theoretical Astrophysics.

References

Daigle, A., Joncas, G., Parizeau, M. & Miville-Deschênes, M.-A. 2003, *PASP* 115, 662

Ehlerová, S. & Palouš, J. 2005, *A&A* 437, 101

Hartmann, D. & Burton, W.B. 1997, *Atlas of Galactic Neutral Hydrogen* (Cambridge University Press)

Heiles, C. 1979, *ApJ* 229, 533

Heiles, C. 1984, *ApJS* 55, 585

Hu, E. 1981, *ApJ* 248, 119

Kalberla, P.M.W., Burton, W.B., Hartmann, Dap, Arnal, E.M., Bajaja, E., Morras, R. & Pöppel, W.G.L. 2005, *A&A* 440, 775

Mashchenko, S.Y. & St-Louis, N. 2002, in: A.F.J. Moffat & N. St-Louis (eds.), *Interacting Winds from Massive Stars* (ASP-CS), 260, 65

Mashchenko, S.Y., Thilker, D.A. & Braun, R. 1999, *A&A* 343, 352

McClure-Griffiths, N.M., Dickey, J.M., Gaensler, B.M. & Green, A.J. 2002, *ApJ* 578, 176

Oey, M.S. & Clarke, C.J. 1997, *MNRAS* 289, 570

Paladini, R., Burigana, C., Davies, R.D., Maino, D., Bersanelli, M., Cappellini, B., Platania, P. & Smoot, G. 2003, *A&A* 397, 213

Thilker, D.A., Braun, R. & Walterbos, R.A.M. 1998, *A&A* 332, 429

Discussion

MAC LOW: It's remarkable that you also showed a positive slope to the R-v relationship for your shells, like Stanimirovic did for the SMC. Yet, we certainly don't think there was a recent single burst of star formation in the Milky Way, so that cannot be the explanation. Something else must be going on to give a positive slope, rather than the negative one predicted by Oey from standard bubble theory.

EHLEROVÁ: I agree. This kind of R-v diagram seems to be quite frequent among galaxies.

TOTH: What are the velocity limits of your survey? In other term, what is the age limit for the expanding shells?

EHLEROVÁ: There is an artificial lower limit of 4 km s^{-1}.

HEYER: What is the fraction of mass contained within the shells?

EHLEROVÁ: We didn't make precise calculations but the rough estimate is 5–10 % for the outer Galaxy ($R \sim 11$ kpc) and the thickness of the disk of 1–2 kpc.

DEHARVENG: Do you have some statistical results about the morphology of your shells? (spherical shells, half-shells? rings ...?)

EHLEROVÁ: The only morphological quantity we have studied so far is the prolongation of shells in the b direction. The average value of $\frac{\Delta b}{\Delta l cos(b)}$ is 0.8, 70 % of shells are elongated in the l direction rather than in the b one.

Triggered Star Formation in a Turbulent ISM
Proceedings IAU Symposium No. 237, 2006
B. G. Elmegreen & J. Palouš, eds.

© 2007 International Astronomical Union
doi:10.1017/S1743921307001287

The SMC super-shells as probes of the turbulent dynamics of the ISM

Itzhak Goldman[1,2]

[1]Department of Basic Sciences, Afeka Tel Aviv Academic College of Engineering, Tel Aviv, Israel

[2]Department of Astronomy and Astrophysics, School of Physics and Astronomy, Tel Aviv University, Tel Aviv, Israel
email: goldman@wise.tau.ac.il

Abstract. The spatial power spectrum of the HI 21 cm intensity in the Small Magellanic Cloud (Stanimirovic *et al.* 1999) is a power law over scales as large as those of the SMC itself. It was interpreted as due to turbulence by Goldman (2000) and by Stanimirovic & Lazarian (2001). The question is whether the power spectrum is indeed the result of a dynamical turbulence or is merely the result of a structured static density. In the turbulence interpretation of Goldman (2000) the turbulence was generated by the tidal effects of the last close passage of the LMC about 0.2 Gyr ago. The turbulence time-scale was estimated by Goldman to be 0.4 Gyr, so the turbulence has not decayed yet. Staveley-Smith *et al.* (1997) observed in the SMC about five hundreds of HI super shells. Their age is more than an order of magnitude smaller than the turbulence age. Therefore, if the turbulence explanation holds, their observed radial velocities should reflect the turbulence in the gas in which they formed. In the present work we analyze the observed radial velocities of the super shells. We find that the velocities indeed manifest the statistical spatial correlations expected from turbulence. The turbulence spectrum is consistent with that obtained by Goldman(2000).

Keywords. Turbulence, ISM, SMC, Super Shells

1. Introduction

The spatial power spectrum of the HI 21 cm intensity in the Small Magellanic Cloud was obtained by Stanimirovic *et al.* (1999). Interestingly, it is a power law over scales as large as that of the SMC itself. Similar power laws have been observed by Crovisier & Dickey (1983) and by Green (1993) in the galaxy. The outstanding feature in the case of the SMC is the large scale of the observed correlations. The power laws signal underlying long range correlations in what looks like a field of random fluctuations of the intensity. For an optically thin medium along the line of sight, the 21 cm intensity is proportional to the column density. Therefore, the fluctuations in 21 cm intensity represent fluctuations in density.

A natural interpretation of the observed power spectra is that the underlying correlations in density fluctuations are due to a turbulence in which velocity fluctuations, that are coupled to density fluctuations, give rise to the observed power laws. The turbulence interpretation was suggested by Goldman (2000) and Stanimirovic & Lazarian (2001).

Goldman (2000) suggested that this large scale turbulence was generated by instabilities in the bulk flows that resulted from the tidal interaction during the last close passage of the Large Magellanic Cloud (LMC) $\sim$ 2 Gyr ago (Gardiner & Noguchi 1996).

However, since the observations catch a snapshot of the intensity field and since the turbulence timescales are very long (~ 0.4 Gyr) one cannot rule out the possibility of a *static* correlated density field that reflects initial conditions.

In the present paper we propose a test to decide between these two alternatives.

2. Analysis of the supershells radial velocity field

Staveley-Smith *et al.* (1997) observed 501 HI super shells in the SMC. The proposed test relies on the fact that the timescale and age of the turbulence (if indeed there) are typically 1 to 2 orders of magnitude larger than the lifetimes of the super shells. Therefore, they have formed in the turbulent gas and their observed radial velocities should reflect the turbulent velocity field in the gas in which they were formed. We wish to look at them as markers registering the ambient gas velocity. If the radial velocity field exhibits spatial correlations consistent with those of the turbulence, assumed as responsible for the 21 cm intensity spectra, it will strengthen the case for dynamical turbulence as the source of the HI intensity power spectrum.

We use the data of the 501 super shells reported in Table 1 of Staveley-Smith *et al.* (1997). For each super shell, the residual radial velocity was found by subtracting from the observed velocity the large-scale best fit to a linear function of the coordinates,

$$v_i = v_{obs,i} - (c + s_1 x_i + s_2 y_i) \tag{2.1}$$

with shell numbers $1 \leqslant i \leqslant 501$ where

$$c = 155.1\,\mathrm{km\,s^{-1}}, \quad s_1 = 12.34 \times 10^{-3}\,\mathrm{km\,s^{-1}\,pc^{-1}}, \quad s_2 = 4.46 \times 10^{-3}\,\mathrm{km\,s^{-1}\,pc^{-1}}.$$

The coordinates of each shell (x_i, y_i) are in units of pc and were obtained from the angular coordinates by adopting a distance of 60 kpc to the SMC. The velocities are in units of km s^{-1}. The subtracted large scale velocity field is composed of a mean velocity and a term corresponding to a velocity gradient. The magnitude of the latter is consistent with values obtained by Gardiner & Noguchi (1996).

We have computed the second order structure function and the autocorrelation for the residual velocity field along lines parallel to the coordinate axes. Interpolation was used to fit the discrete data along the lines to a continuous function. The different lines yielded similar results.

For simplicity, homogeneous and isotropic velocity field is assumed. In this case, the structure function and the autocorrelation depend only on the distance between the two points, $r = |\vec{r}|$. The structure function is

$$S(r) = < \left(v(\vec{r'} + \vec{r}) - v(\vec{r'}) \right)^2 >. \tag{2.2}$$

Similarly, the autocorrelation function is

$$C(r) = < v(\vec{r'} + \vec{r}) v(\vec{r'}) >. \tag{2.3}$$

The angular brackets denote ensemble averaging. Assuming ergodicity, in addition to homogeneity and isotropy, ensemble averaging equals space averaging. As stated above, we use averages over lines so that

$$S(l) = \frac{1}{L} \int_0^L \left(v(x+l) - v(x) \right)^2 dx \tag{2.4}$$

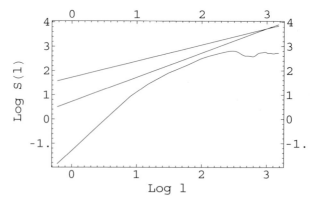

Figure 1. The structure function in units of $(\mathrm{km\,s}^{-1})^2$ as a function of scale in pc. The thin lines have slopes 2/3 and 1. The upper line has a slope of 2/3.

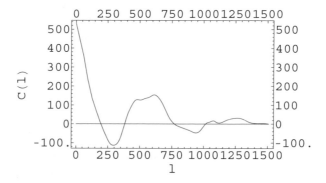

Figure 2. The autocorrelation function in units of $(\mathrm{km\,s}^{-1})^2$ as a function of scale in pc.

where L is the length of the line. Similarly,

$$C(l) = \frac{1}{L} \int_0^L v(x+l)v(x)dx \qquad (2.5)$$

The results of a typical computation are presented in figures 1-2. Figure 1 shows the structure function $S(l)$. For very small values of l, $S(l) \propto l^2$, for larger values of l it varies as $S(l) \propto l^{m-1}$ and then it saturates.

The index m characterizes the inertial range of the turbulent velocity spectral function: $F(k) \propto k^{-m}$. In Kolmogorov turbulence characterizing an incompressible fluid, $m = 5/3$. In the case of turbulence in a compressible medium, $m = 2$. This was also the value deduced by Goldman (2000) on the basis of the 21 cm intensity power spectrum. These two power laws are presented in figure 1. The precision of the data is not enough to decide between them, even though the $m = 2$ line seems to follow better the slope of the computed structure function.

The autocorrelation function is shown in figure 2. It behaves as an autocorrelation function of a turbulent velocity rather than uncorrelated velocity fluctuations.

Figure 3 presents the turbulence spectral function $F(k)$ computed from the autocorrelation function. The curve is noisy but a power law range is clear. Also here the turbulence spectral functions with $m = 5/3$ and $m = 2$ are plotted. The two slopes are compatible with the computed spectral function, although $m = 2$ seems preferable.

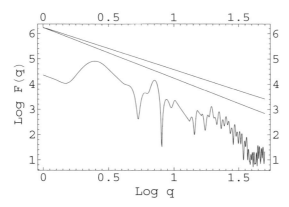

Figure 3. The turbulence spectral function in units of $(\mathrm{km\,s^{-1}})^2\mathrm{pc}$ as function of the normalized wavenumber $q = kL/(2\pi)$. The upper thin line has a slope $-5/3$ and the lower -2.

The wavenumber range shown corresponds to spatial scales between 1500 pc, which is in this case the length of the line L, and 50 pc. Higher wavenumbers correspond to spatial scales that are smaller than the average radius of the shells, and therefore the computed turbulence spectrum is not valid for these scales.

3. Conclusions

The results of the present work strengthen the case for the turbulence interpretation of the 21 cm power spectra of the SMC. The residual radial velocities of the super shells exhibit statistical spatial correlations expected from turbulence. The turbulence spectrum and structure function are consistent with a Kolmogorov spectrum, $m = 5/3$, and with that of incompressible turbulence, $m = 2$. The latter seems preferable. It equals the value deduced by Goldman (2000) from the HI intensity fluctuations.

Acknowledgements

Participation in IAU Symposium 237 was supported by Afeka Engineering College and by the Institute of Astronomy, Department of Astronomy and Astrophysics, Tel Aviv University.

References

Crovisier, J. & Dickey, J. M. 1983, *A&A* 122, 282
Gardiner, L. T. & Noguchi, M. 1996, *MNRAS* 278,191
Goldman, I. 2000, *ApJ* 541, 701
Green, D. A. 1993, *MNRAS* 262, 327
Stanimirovic, S., Stavely-Smith, L., Dickey, J. M., Sault, R. J. & Snowden, S. L. 1999, *MNRAS* 302, 417
Staveley-Smith, L., Sault, R. J., Hatzidimitriou, D., Kesteven, M. J. & McConnell, D. 1997, *MNRAS* 289, 225
Stanimirovic, S. & Lazarian, A. 2001, *ApJ* 551, L53

Discussion

BLITZ: How can you say that your mechanism is anything more than consistent with the observations, since your mechanism doesn't explain the turbulence in the outer regions of disks (beyond the stellar disks) or in spherical HI systems.

GOLDMAN: My talk referred to the particular case of the SMC. Particular in the sense that a tidal interaction with the LMC seems to be the plausible way to generate large-scale shear flows who are unstable and generate turbulence. In any case, the main point was not identifying the energy source but showing that indeed the SMC is pervaded by large-scale velocity turbulence.

PADOAN: It is very difficult, from your velocity power spectrum plot, to evaluate if the power spectrum is k^{-2} or $k^{-5/3}$. The quality of that HI velocity power spectrum is simply too poor to probe the turbulence.

GOLDMAN: I agree. The index is $\sim (-5/3 - 2)$. But the important point is that radial velocities of the supershells exhibit a turbulence spectrum consistent with that deduced from the HI intensity power spectrum.

Triggered Star Formation in a Turbulent ISM
Proceedings IAU Symposium No. 237, 2006
B. G. Elmegreen & J. Palouš, eds.

Giant molecular clouds and star formation in the Large Magellanic Cloud

A. Kawamura[1], T. Minamidani[1], Y. Mizuno[1], T. Onishi[1], N. Mizuno[1], A. Mizuno[2] and Y. Fukui[1]

[1]Department of Astrophysics, Nagoya University, Chikusa-ku, Nagoya 464-8602, Japan
email: kawamura@a.phys.nagoya-u.ac.jp
[2]Solar-terrestrial Environment Lab., Nagoya University, Chikusa-ku, Nagoya 464-8601, Japan

Abstract. In order to elucidate star formation in the Large Magellanic Cloud, a complete survey of the molecular clouds was carried out by NANTEN. In this work, we compare 230 giant molecular clouds (GMCs), whose physical quantities are well determined, with young clusters and H II regions. We find that about 76 % of the GMCs are actively forming stars or clusters, while 24 % show no signs of massive star or cluster formation. Effects of supergiant shells (SGSs) on the formation of GMCs and stars are also studied. The number and surface mass densities of the GMCs are higher by a factor of 1.5–2 at the edge of the SGSs than elsewhere. It is also found that young stellar clusters are more actively formed in the GMCs facing to the center of the SGSs. These results are consistent with the previous studies by Yamaguchi *et al.* and suggest the formation of GMCs and the cluster is triggered by dynamical effects of the SGSs.

Keywords. stars: formation, ISM: clouds, Magellanic Clouds, galaxies: star clusters

1. Introduction

The Large Magellanic Cloud (LMC), at a distance of ~ 50 kpc, is one of the nearest galaxies to our own. Studies of the LMC provide invaluable information to understand the evolution of interstellar matter (ISM) and star formation. The relatively face-on inclination of the LMC also enables us to obtain a sample of astronomical objects with little contamination.

Studies of the distribution and the properties of the young stars and clusters have been carried out by optical, infrared and radio continuum observations for decades. More than 300 H II regions are identified (e.g., Davies *et al.* 1976). Stellar clusters called "populous clusters" with masses $\sim 10^4$–$10^5 \, M_\odot$, were found by photometric studies (e.g., Hodge 1961). There are populous clusters significantly younger, i.e., a few Myr–10 Gyr, than the Galactic globular clusters and some populous clusters are still being formed at present (e.g., Brandl *et al.* 1996). Not only the star formation indicators but also the structure of the ISM has been studied, for example, in Hα and H I (e.g., Meaburn 1980; Kim *et al.* 1999). One of the remarkable feature is the existence of various shells and supergiant shells (SGSs) as commonly found in nearby galaxies (e.g., Tenorio-Tagle & Bodenheimer 1988). The SGSs in the LMC consist of long Hα filaments and bright H II regions, and have more than a few 10 to 100 OB stars in their interior. Nine SGSs were cataloged by Meaburn (1980). Since all of the nine Hα SGSs contain numerous OB stars inside the SGSs, they are likely to be formed due to the effects of stellar winds and supernovae. In this work, we compare the molecular clouds and young clusters with the nine Hα SGSs.

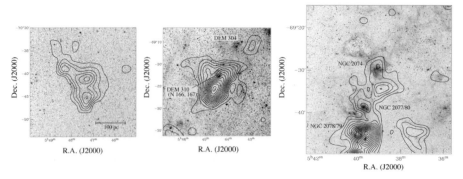

Figure 1. Examples of (*Left*) Class I GMC with no signs of massive star or cluster formation, (*Middle*) Class II GMC associated with H II regions, and (*Right*) Class III GMC associated with H II regions and clusters ($\tau < 10$ Myr). Images are from DSS2 and the contours are from the NANTEN survey. The contour levels are from 1.2 K km s^{-1} with 1.2 K km s^{-1} intervals.

2. Giant molecular clouds

We have carried out CO(1–0) observations toward the LMC with NANTEN, a 4 m telescope of Nagoya University at Las Campanas Observatory, Chile from 1998 April to 2003 August. The resolution is ~ 40 pc at a distance of the LMC. The 3 σ noise level of the velocity-integrated intensity is ~ 1.2 K km s^{-1}. This corresponds to $N(\mathrm{H_2}) \sim 8 \times 10^{20}$ cm^{-2}, by assuming a conversion factor from CO intensity to the hydrogen column density, $X = 7 \times 10^{20}$ cm^{-2}(K km s^{-1})$^{-1}$. We identified 272 molecular clouds by using fitstoprops method (Rosolowsky & Leroy 2006), 230 out of which are detected at more than 2 observed positions (for the details, see Fukui *et al.* 2006).

3. Massive star and cluster formation

We made comparisons of the 230 GMCs with H II regions (e.g., Davies *et al.* 1976, Filipovic *et al.* 1998) and young clusters (e.g., Bica *et al.* 1996). Bica *et al.* (1996) studied the colors of the clusters and classified them into an age sequence from SWB0 to SWB VII. The datasets of H II regions have a detection limit in Hα flux of $\sim 10^{-12}$ ergs cm^{-2} s^{-1} (Kennicutt & Hodge 1986). In order to examine any optically obscured H II regions we have also used the Parkes/ATNF radio continuum survey (Dickel *et al.* 2005; Filipovic *et al.* 2006). The typical sensitivity of the new datasets is good enough to reach the flux limit equivalent to that of Hα. We note that the detection limit of H II regions is quite high, corresponding to one-fourth the luminosity of the Orion Nebula.

A study of the distance of H II regions and clusters measured from the nearest GMCs shows that a large number of the young clusters ($\tau < 10$ Myr) and H II regions are found within 100 pc of the GMCs. On the other hand, older clusters show almost no correlation with the GMCs. We have examined the association between the individual GMCs and the H II regions and young clusters and found that about a half of the H II regions and young clusters are associated with the GMCs. The GMCs are classified into three classes; *Class I*. starless GMCs (no early O stars), *Class II*. GMCs with H II regions only, and *Class III*. GMCs with H II regions and stellar clusters (Fig. 1, see also Fukui *et al.* 1999, Yamaguchi *et al.* 2001b). It is found that ~ 24 % of the GMCs are starless (*Class I*), while 52% (*Class II*) and 24% (*Class III*) are associated with H II regions and young clusters, respectively. Figure 2 shows the mass distribution of the three GMC classes. It is shown that the number ratio of massive GMCs is higher in Class III than those in the other two, and the mass of Class I GMCs tends to be smaller than the rest. We may speculate that Class I, and possibly Class II GMCs, are still growing in mass via mass accretion from their surrounding lower density atomic gas (Fukui 2006 in this proceedings).

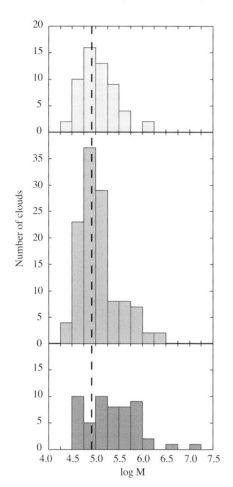

Figure 2. Mass distribution of Class I (*Top*), Class II (*Middle*), and Class III (*Bottom*) GMCs, respectively ($X = 7 \times 10^{20}$ cm^{-2} K km s^{-1} is assumed; Fukui *et al.* 2006).

The completeness of the present GMC sample covering the whole LMC enables us to infer the evolutionary timescales of the GMCs. We assume a steady state evolution and therefore the time spent in each phase is proportional to the number of the GMCs at each class. The absolute time scale is based on the age of young clusters, which is taken to be 10 Myr. The total lifetime of a GMC is then estimated to be ~ 30 Myr.

4. Comparisons with supergiant shells

A comparison of the GMCs and SGSs shows that clumpy GMCs are distributed along the edge of SGSs and only a few clouds are inside the SGSs except for the LMC 2. For simplicity, we assume the SGSs to be circles and introduce a "relative distance" between the GMCs and the SGSs, $d = D_{\mathrm{CO,SGS}}/R_{\mathrm{SGS}}$ where $D_{\mathrm{CO,SGS}}$ is the distance from the peak position of a GMC to the nearest SGS center, and R_{SGS} is the radius of the SGS (Fig. 3a). Here the GMC is just at the edge of the SGS if $d = 1$.

Figure 3b is a histogram of the relative distance with the expected frequency distribution obtained by distributing the same number of GMCs randomly in the observed area. By comparing the observed distribution with a random distribution, the number of GMCs is higher by a factor of ~ 1.5 near/at the edge of SGSs, $0.5 < d < 1.25$ and, particularly, by a factor of ~ 2 at the edge of $0.75 < d < 1.25$. We also found that the surface mass densities of the GMCs are significantly enhanced around the edge of the SGSs by a factor of ~ 2. On the other hand, the number density of the clusters is enhanced by a factor of 4 inside the SGSs compared with the outer parts. The enhancement of the GMCs at the edge suggests the formation of GMCs due to an accumulation of the ISM and/or fragmentation of the SGSs (e.g., Ehlerová *et al.* 1997). The deficiency of GMCs in the central part suggests a dissipation of GMCs due to star-formation activities.

Next, we study where the clusters are formed in the GMC. We introduce here the position angle, θ, as the angle between the two straight lines; one from a CO intensity peak to the center of the nearest SGS, and the other from the CO intensity peak to the associated cluster (Fig. 3a). If the position angle, $| \theta | < 90°$, the cluster is on the side facing the center of the SGSs. We study the position angles for the GMCs at the edge

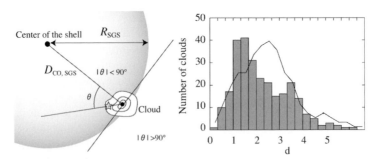

Figure 3. (a) Definition of the relative distance, d, and the position angle, θ. (b) Histogram of the relative distances, d. The solid lines indicate the expected frequency distribution for the same number of samples randomly distributed in the observed area.

of the SGSs, $d = 0.75$–1.25, and found that the number of clusters is by a factor of 2 greater in the GMCs facing to the center of the SGSs than other regions in the GMC.

These results are consistent with Yamaguchi *et al.* (2001a) and suggest that the SGSs play an important role in triggering the formation of molecular clouds and clusters.

Acknowledgements

The NANTEN project is based on a mutual agreement between Nagoya University and the Carnegie Institution of Washington. We greatly appreciate the hospitality of all the staff members of the Las Campanas Observatory. We are thankful to many Japanese public donors and companies who contributed to the realization of the project. We would like to acknowledge Drs. L. Staveley-Smith and M. Filipovic for the kind use of their radio continuum data prior to publication. This work is financially supported in part by a Grant-in-Aid for Scientific Research from the Ministry of Education, Culture, Sports, Science and Technology of Japan (No. 15071203) and from JSPS (No. 14102003, core-to-core program 17004 and No. 18684003).

References

Bica, E., Clariá, J. J., Dottori, H., Santos, J. F. C. Jr. & Piatti, A. E. 1996, *ApJS* 102, 57
Brandl, B., Sams, B. J., Bertoldi, F., Eckart, A., Genzel, R., Drapatz, S., Hofmann, R., Loewe, M. & Quirrenbach, A. 1996, *APJ* 466, 254
Davies, R. D., Elliott, K. H. & Meaburn, J. 1976, *MemRAS* 81, 89
Dickel, J., McIntyre, V. J., Guendl, R. A. & Milne, D. K. 2005, *AJ* 129, 790
Filipovic, M. D., Jones, P. A., White, G. L. & Haynes, R. F. 1998, *AAS* 130, 441
Ehlerová, S., Palouš, J., Theis, Ch. & Hensler, G. 1997, *A&A* 328, 121
Fukui, Y., Kawamura, A., Minamidani, T., Mizuno, Y., Kanai, Y., Onishi, T., Mizuno, N., Yonekura, Y., Mizuno, A. & Ogawa, 2006, *ApJ* submitted
Fukui, Y., Mizuno, N., Yamaguchi, R., Mizuno, A., Onishi, T., Ogawa, H., Yonekura, Y., Kawamura, A., *et al.* 1999, *PASJ* 51, 745
Kennicutt, R. C., Jr. & Hodge, P. W. 1986, *ApJ* 306, 130
Kim, S., Dopita, M. A., Staveley-Smith, L. & Bessell, M. S. 1999, *AJ* 118, 2797
Meaburn, J. 1980, *MNRAS* 192, 365
Rosolowsky, E. & Leroy, A. 2006, *PASP* 118, 590
Tenorio-Tagle, G. & Bodenheimer, P. 1988, *ARAA* 26, 145
Yamaguchi, R., Mizuno, N., Onishi, T., Mizuno, A. & Fukui, Y. 2001a, *PASJ* 53, 959
Yamaguchi, R., Mizuno, N., Mizuno, A., Rubio, M., Abe, R.,Saito, H., Moriguchi, Y., Matsunaga, L., Onishi, T., Yonekura, Y. & Fukui, Y. 2001b, *PASJ* 53, 985

Discussion

Y.-H. CHU: The molecular cloud in the supergiant shell LMC-2 is associated with one of the two sheets of HI gas that confine the supergiant shell. To study the star formation in molecular clouds, YSOs identified by Spitzer observations are much better tracers of current star formation. We have seen star formation in molecular clouds that do not show any HII regions.

KAWAMURA: Yes. We shall present more detailed comparisons of GMCs and SGSs elsewhere. We see a very good correlation of HI and CO. In the present talk, we focussed on massive star formation, but obviously younger or low mass stars are better traced by new IR datasets as you suggest. We have started a comparison of Spitzer and Akari data with GMCs. These comparisons do show some indications of star formation of lower mass stars in the GMCs not associated with HII regions or clusters.

ZINNECKER: Can you comment on the lifetime of the giant molecular clouds in the LMC?

KAWAMURA: About 26%, 52%, and 23% of the GMCs with $M > 10^5$ M$_\odot$ are classified as class I, Class II and Class III. If the star formation proceed uniformly in time, and these classes represent an evolutionary sequence of the GMCs, we can also take this ratio as a ratio of the lifetime of the GMCs at each stage. The GMCs are thought to be dissipated about 5 My after cluster formation, i.e., the life time of the Class III is about 5 My because about a half of the young clusters with lifetimes less than 10 My are associated with the GMCs. Taking these into account, the lifetime of the GMCs is considered to be about a few times 10 My (20-30 My).

WALBORN: In your distinction between Class II and III molecular clouds, do you mean optically visible clusters in the latter? Presumably, the former have embedded or extincted clusters within/behind to power the HII regions. Comment: To refine these correlations, it will be important to discriminate young clusters by actual ages. Two, five, and ten Myr old clusters are very different with respect to stellar content and ISM interactions. Also, some massive clusters trigger second stellar generations around their peripheries, and some IR clusters may not be embedded but viewed through peripheral molecular clouds. (We could be very confused if we have a less fortunate line of sight toward 30 Doradus!)

KAWAMURA: We have compared the GMCs with HII regions seen in visible and/or radio and the optically visible young clusters and associations younger than 10 Myr. It is indeed important to discriminate young clusters according to more accurate age estimations as you point out. Nevertheless, we see a clear difference in mass between the GMCs with and without clusters of < 10 Myrs. This suggests that the Class II and III are different groups of the GMCs. The current information of the age of the young clusters are limited, and it is not yet possible to have a sample of clusters with an accurate age estimation over the entire LMC. A comparison of the GMCs with clusters with more accurate age estimation will be carried out in the near future by using the recent IR studies by SAGE, Akari and SIRIUS.

Triggered Star Formation in a Turbulent ISM
Proceedings IAU Symposium No. 237, 2006
B. G. Elmegreen & J. Palouš, eds.

© 2007 International Astronomical Union
doi:10.1017/S1743921307001305

Towards resolving the evolution of multi-supernova superbubbles

M. S. Oey[1]

[1]Department of Astronomy, University of Michigan, 830 Dennison Building, Ann Arbor, MI 48109-1042, USA

Abstract. Interstellar superbubbles generated by multiple supernova explosions are common in star-forming galaxies. They are the most obvious manifestation of mechanical feedback, and are largely responsible for transferring both thermal and kinetic energy to the interstellar medium from the massive star population. However, the details of this energy transfer remain surprisingly murky when individual objects are studied. I will summarize what we currently know about candidate dominant processes on these scales.

1. Introduction

The previous contributions showed a wealth of H I shells and supershells. How well-established is it, that these structures originate from mechanical feedback, namely, winds and supernovae (SNe) from OB associations? If we are to understand how massive star feedback affects galaxies on global scales, then it is essential that we understand the origin and evolution of superbubbles, which are the direct manifestation of feedback. We refer not only to mechanical energy, but also to photoionization, and dispersal of nucleosynthesis products from the SNe and massive stars. In this presentation, I use the definition introduced by You-Hua Chu in the 1980's: a "superbubble" is a shell structure originating from multiple stars, rather than a definition related to its size.

The default model for superbubbles is the standard, adiabatic evolution that assumes a hot (10^6+ K), shock-heated interior whose pressure drives the outer shell growth (Pikel'ner 1968; Castor *et al.* 1975; Dyson 1977). For this model, the shell parameters are determined by only three input parameters: the mechanical "luminosity" L, the ambient density n, and the age t, e.g.,

$$R \propto (L/n)^{2/5} \, t^{3/5}, \tag{1.1}$$

$$v = \frac{dR}{dt} \propto (L/n)^{2/5} \, t^{-2/5}. \tag{1.2}$$

We also must know conditions in the ambient environment, for example, the ambient pressure and density distribution. The above is the simplest, analytic representation of this model, but there is also a large body of work in hydrodynamic simulations of these objects over the past two decades (e.g., Mac Low *et al.* 1989; Bisnovatyi-Kogan *et al.* 1989; Palouš *et al.* 1990; Tenorio-Tagle *et al.* 1991; Tomisaka 1990; Silich *et al.* 1996; Gazol-Patiño & Passot 1999; Strickland & Stevens 2000).

What evidence do we seek, that "star formation," i.e., massive stars, are responsible for the ubiquitous superbubbles that we see in gas-rich galaxies? Since the adiabatic model is based on the shock-heated interior, we expect gas at multiple temperature phases, including hot, X-ray emitting gas; warm, 10^4 K photoionized gas; and also the neutral gas that we have seen. Second, we should also find that objects whose input parameters are known should follow the prescribed evolution above. Third, the statistical properties of the superbubble populations should be consistent with the statistical properties of the putative parent star-forming regions. Fourth, and perhaps most obvious, we expect a one-to-one correspondence between the superbubbles and the massive star clusters.

2. Multi-phase ISM

As mentioned by Elias Brinks in his talk, we often do see the existence of gas at the different temperatures that are predicted by the adiabatic model. A recent example is the X-ray emission in N51 D (Cooper *et al.* 2004), along with photoionized Hα emission. Superbubbles were already detected in X-rays by *Einstein* and *ROSAT* (e.g., Chu & Mac Low 1990; Wang & Helfand 1991). It has long been known that there are two categories of X-ray emission from these objects: X-ray bright, and X-ray dim (Chu *et al.* 1995). The former show X-ray emission in excess of what is expected from the adiabatic model, and the enhancements are likely caused by secondary SN blastwave impacts to the shell walls (Chu & Mac Low 1990; Oey 1996).

H recombination emission is generally consistent with photoionization by the observed early-type stars, although density-bounding and shock-heating contributions are also often factors (Hunter *et al.* 1995; Oey *et al.* 2000). We usually see Hα emission on the interior of H I shells, consistent with the model that star formation is responsible for the superbubbles. Kim *et al.* (1999) studied the H I shells of the Large Magellanic Cloud (LMC) and were even able to suggest an evolutionary sequence defined by relative Hα and H I morphology.

Intermediate-temperature ions are also present. C IV and Si IV, are often seen in absorption in the lines of sight toward massive stars within superbubbles (Chu *et al.* 1994). Recently, O VI was also detected by *FUSE* in the line of sight toward one LMC superbubble (Danforth & Blair 2006).

3. Dynamics of individual superbubbles

If the input mechanical power, ambient conditions, and age of the superbubbles can be determined, we can test for consistency with the adiabatic evolution. Such studies are possible for objects in the Milky Way and Magellanic Clouds, where the stellar population can be resolved, and these studies invariably show the existence of a growth-rate discrepancy such that the shells are much smaller than predicted by the apparent input parameters (Saken *et al.* 1992; Brown *et al.* 1995; Oey & Massey 1995; Hunter *et al.* 1995; Oey 1996; Cooper *et al.* 2004). The problem has been known for many years, and is even seen in single-star nebulae generated by Wolf-Rayet stars (e.g., Cappa *et al.* 2001, 2005).

What is the current status in resolving this growth-rate discrepancy? There are a number of possible important factors. Those discussed below are the most likely candidates. Combinations of multiple effects may also be at play, including others, such as viscous drag (I. Goldman, private communication), that we do not discuss fully below.

3.1. *Input power overestimated?*

The first possibility for resolving the superbubble growth-rate discrepancy is that the input mechanical luminosity has been overestimated. Stellar wind mass-loss rates $\dot{M}$ are especially suspect, since it has long been suggested that clumping in the winds leads to overestimates in $\dot{M}$ from radio continuum measurements (e.g., Hillier 1991; Nugis *et al.* 1998). In recent years, this appears to be confirmed by X-ray line profile fitting (Cohen *et al.* 2006; Miller *et al.* 2002; Kramer *et al.* 2003). Fullerton *et al.* (2006) also find the same result from P v line profiles observed by *FUSE*. Their analysis of this dominant ion suggests that the overestimates in $\dot{M}$ may be as high as two orders of magnitude in some cases! These overestimates in $\dot{M}$ are relevant primarily to the youngest superbubbles, whose evolution is still dominated by stellar winds instead of SNe, which applies to most dynamical studies of optical or X-ray selected superbubbles.

3.2. *Ambient density underestimated?*

As seen in equations 1.1 and 1.2, an underestimate in the ambient density has an equivalent effect to an overestimate in L. Oey *et al.* (2002) mapped the immediate environment of three Hα-selected LMC superbubbles to determine whether the neutral gas environment was unusually dense. We found an extreme range in conditions for the three objects, with one essentially in an H i void, another nestled amongst a number of H i clouds, and the third in a region with no obvious relation to the observed H i. This lack of any systematic effect suggests that the ambient density may not be the primary source of the growth-rate discrepancy, although it also demonstrates that the ambient environment is more complex than assumed.

We also note that if a superbubble originates in a higher-density cloud, then a miniblowout from the cloud can also accelerate the shell's observed expansion velocity relative to its radius (Oey & Smedley 1998; Mac Low *et al.* 1998). Hence it is possible to reproduce unusual observed kinematics in particular objects.

3.3. *Ambient pressure underestimated?*

Another important environment parameter is the interstellar pressure, which counteracts the growth of the shells. In general, models assume a fiducial ambient P/k on the order of 10^4 cm^{-3} K or less. The various sources of ambient pressure can be broadly described as the thermal pressure, magnetic pressure, turbulent pressure, and cosmic ray pressure. We note that all of these correlate with star-formation rate (SFR). Likewise, superbubbles tend to be prominent in active star-forming galaxies like the LMC and localized active environments. Indeed, the very existence of star-formation in these regions implies higher local pressure (M. Mac Low, private communication). It is therefore plausible that the ambient pressure for superbubbles may have been systematically underestimated.

Oey & García-Segura (2004) discuss this possibility and present 2-D hydrodynamic simulations of six LMC superbubbles for ambient $P/k = 10^4$ and 10^5 cm^{-3} K. The models are generated with the same input mechanical luminosity that is estimated from the observed stellar population by Oey (1996). For the lower pressure, the simulated radial density profiles show an extended photoionized morphology that is inconsistent with the observed Hα data. In contrast, the simulations assuming ambient $P/k = 10^5$ cm^{-3} K show no extended Hα emission, and an Hα morphology that agrees well with the observations. The observed and modeled velocity structures are similarly more consistent with the models for the high-pressure environment in all cases. The H i radial profile predictions also differ, and can be used to further test these models.

3.4. *Radiative cooling?*

If the hot superbubble interiors are in fact losing energy by radiative cooling, then the growth will not keep pace with the adiabatic model. Mass-loading into the hot interior via evaporation from the shell walls or ablation of clouds and clumps can drive radiative cooling (e.g., Hartquist *et al.* 1986; Arthur & Henney 1996). Alternatively, an increase in metallicity due to the injection of nucleosynthesis products from the parent SNe can also enhance the cooling rate. Silich & Oey (2002) show that the X-ray luminosity of a low-metallicity ($Z = 0.05Z_\odot$) superbubble can be increased by almost an order of magnitude, simply by products from 3 – 4 SNe. Silich *et al.* (2001) examine this issue for starburst superwinds. However, the observed X-ray luminosities generally do not appear to suggest anomalous cooling from the superbubble interiors (e.g., Chu *et al.* 1995).

3.5. *Energy transferred to cosmic rays?*

Another factor whose importance has been underemphasized, is the transfer of superbubble energy to cosmic rays. Multi-SN superbubbles are especially efficient at accelerating cosmic rays because the blastwaves expand into a pre-heated environment (e.g., Parizot *et al.* 2004). The superbubble interiors thus harbor strong MHD turbulence and magnetic fields (e.g., Bykov 2001; Bykov & Toptygin 1987), which are needed for cosmic ray acceleration. Because of the multiple SNR shocks, superbubbles also promote multiple accelerations, which can push cosmic rays to higher energies (Parizot *et al.* 2004; Klepach *et al.* 2000). The cosmic ray energy distributions and isotope abundances are broadly consistent with superbubble origins. While the role of superbubbles in explaining cosmic ray properties has been recognized for some time, the effect of energy transfer on the parent objects themselves has only been discussed recently, and rough estimates suggest that a few tenths of superbubble kinetic energy could be lost to cosmic rays (Parizot *et al.* 2004; Bykov 2001). For individual young SNRs, simulations predict growth deviations $\gtrsim 10\%$ and reduced X-ray luminosities (Ellison *et al.* 2004). This energy sink applies more to SN-dominated superbubbles, rather than stellar wind-dominated objects.

4. Properties of global mechanical feedback

Another approach to evaluating the adiabatic model is to compare observations and predictions for the statistical properties of superbubbles in galaxies, based on the known global star-formation properties. Oey & Clarke (1997) derived size distributions and expansion velocity distributions for extremes in star-formation history and star-cluster mass functions. The latter, which produces the H II region luminosity function, is generally a robust power law with a dependence of L^{-2} in differential form (e.g., Efremov & Elmegreen 1997; Oey & Clarke 1998). This yields a differential size distribution,

$$N(R)\, dR \propto R^{-3}\, dR \qquad (4.1)$$

and corresponding distribution in expansion velocities,

$$N(v)\, dv \propto v^{-7/2}\, dv. \qquad (4.2)$$

We find excellent agreement with these relations for the H I shell population found by Staveley-Smith *et al.* (1997) for the SMC. More recently, Hatzidimitriou *et al.* (2005) updated the SMC shell catalog and re-examined these relations, inferring a different star-formation history, but one that is still consistent with a star-formation origin for the structures, plus adiabatic evolution. H I shell size distributions for a few other galaxies also have been examined, for example, the Milky Way (S. Ehlerová, these proceedings), LMC (Kim *et al.* 1999), M31, M33, Holmberg II (Oey & Clarke 1997), and NGC 2403 (Thilker *et al.* 1998; Mashchenko *et al.* 1999). These other catalogs are generally consistent with equation 4.1, but the statistics are more incomplete.

Analysis of the global size distributions leads to the definition of a critical star-formation rate, above which the shells merge and shred the neutral ISM, generating pressure-driven outflows from their parent galaxy disks (Clarke & Oey 2002):

$$\mathrm{SFR_{crit}} = 0.15\, \mathrm{M_{ISM,10}}\, \sigma_{v,10}^2/f_d \quad \mathrm{M_\odot\, yr^{-1}}, \qquad (4.3)$$

where $\mathrm{M_{ISM,10}}$ and $\sigma_{v,10}$ are the mass of the ISM in units of $10^{10}\, \mathrm{M_\odot}$ and the thermal velocity dispersion in units of $10 \ \mathrm{km\,s^{-1}}$, respectively. The geometric correction factor f_d is on the order of unity. Is this expected shredding of the neutral ISM for high SFR morphologically apparent in the H I datasets?

We can compare the H I maps and shell catalogs for the LMC (Kim *et al.* 1999) and SMC (Staveley-Smith *et al.* 1997). The two surveys were both carried out with the Australia Telescope, at similar depth and spatial resolution. Although the LMC is a larger galaxy and has a much higher star-formation rate than the SMC, the number of coherent H I shells identified in the LMC survey is only 126, roughly one-quarter of the 509 (Hatzidimitriou *et al.* 2005) found in the SMC. Ordinarily, we would expect a much *larger* number of shells in the LMC than in the SMC. Morphologically, there is a noticeable contrast between these two galaxies: the LMC H I has a filamentary appearance, consistent with shredding and compression, whereas the SMC has a more quiescent, smoother appearance. It seems surprising that the SMC dataset yields a so much larger shell catalog than the LMC. It turns out that for the LMC, SFR/SFR$_{\mathrm{crit}} \sim 1$, whereas for the SMC, SFR/SFR$_{\mathrm{crit}} \sim 0.1$ (Oey 2001). It will be interesting to examine shell catalogs and properties for larger samples of galaxies, from the THINGS H I survey (E. Brinks, these proceedings), for example.

5. Detailed correspondence with star-forming regions

Last, but not least, we seek a one-to-one correspondence between the observed super-bubbles and parent star clusters. To date, results are not as clear-cut as is desirable, but there does appear to be broad consistency, though somewhat controversial. Kim *et al.* (1999) find enough correspondence between the LMC H I and Hα data to suggest an evolutionary sequence based on the morphological and kinematic relationship between these. Hatzidimitriou *et al.* (2005) examine the spatial relationship between the H I shells in the SMC and existing data for young stellar clusters, and do find broad correspondence, although the quantitative significance of the correlations is ambiguous. They also find that about 10% of the H I shells show no counterparts in the stellar population. In Holmberg II, a targeted search for clusters in the H I shells failed to find the expected correspondence at optical wavelengths (Rhode *et al.* 1999), although Hα and FUV observations seem more consistent with a mechanical feedback origin (Stewart *et al.* 2000).

Finally, in the theme of this Symposium, triggered star-formation is also an association of massive star clusters with the creation of superbubbles. There are many examples of two-stage, sequential star formation; some well-known examples are N11 (Walborn & Parker 1992), N44 (Oey & Massey 1995), and N51 D (Oey & Smedley 1998; Cooper *et al.* 2004) in the LMC, and the Rosette Nebula (Williams *et al.* 1995) in the Galaxy, among others. While it is difficult to establish a *causal* relationship between two-stage star-forming regions, a three-stage sequence is much more convincing. We recently identified the Perseus W3/4 complex in the Galaxy, identified as a triggered system by Thronson *et al.* (1985), as three-stage, hierarchical star formation, whose morphology is difficult to interpret as anything other than causal, triggered star formation (Oey *et al.* 2005).

6. Summary

To summarize, we see that observations are broadly consistent with mechanical feedback from the most massive stars being responsible for the formation of most superbubbles and shell structures seen in star-forming galaxies, and that these objects can be understood in terms of the standard, adiabatic evolution driven by massive star winds and supernovae. There is empirical, multi-wavelength confirmation of the different temperature phases associated with the shock heating of superbubble interiors and also photoionization by OB stars in the youngest objects, as is qualitatively predicted by the adiabatic model. Statistical properties of shell populations are quantitatively consistent with predicted distributions in size and expansion velocity, based on known global properties of the parent star-forming regions and young star clusters. Above a threshold star-formation rate, we predict interaction among the shells and shredding of the ISM, which appears to be observed in the contrasting H I morphology and shell populations between the LMC and SMC. The expected one-to-one correlation between young massive-star clusters and superbubbles remains somewhat ambiguous, although a general consistency is tentatively seen.

When comparing the observed evolution and kinematics of individual objects with standard predictions based on detailed knowledge of the parent stellar population and other input parameters, we find that the objects are also broadly consistent with adiabatic evolution, but that the shells are invariably smaller than expected. There are a number of possible reasons for this growth-rate discrepancy. (1) The assumed input power may be overestimated, especially with respect to measured mass-loss rates for stellar winds. (2) It is possible that the ambient density has been systematically underestimated, although our resolved observations of H I environments for three objects does not especially support this interpretation. (3) There may be a systematic underestimate of the ambient pressure, since the various contributors to interstellar pressure all correlate with star-formation activity. (4) Enhanced radiative cooling may be occurring in the hot superbubble interiors, caused by mass-loading or metallicity enhancements from the parent SNe. (5) Finally, a somewhat-overlooked mechanism is the transfer of mechanical energy from the superbubbles to cosmic rays, which is plausible since superbubbles are an especially effective acceleration environment.

Acknowledgements

Many thanks to the Symposium organizers and participants, with whom I enjoyed many discussions. This work was supported in part by NSF grant AST-0448893.

References

Arthur, S. J. & Henney, W. J. 1996, *ApJ* 457, 752

Bisnovatyi-Kogan, G. S., Blinnikov, S. I. & Silich, S. A. 1989, *Ap&SS* 154, 229

Brown, A. G. A., Hartmann, D. & Burton, W. B. 1995, *A&A* 300, 903

Bykov, A. M. 2001, *Space Sci. Rev.* 99, 317

Bykov, A. M. & Toptygin, I. N. 2001, *Astron. Lett.* 27, 625

Cappa, C., Niemela, V. S., Martín, M. C. & McClure-Griffiths, N. M. 2005, *A&A* 436, 155

Cappa, C. E., Rubio, M. & Goss, W. M. 2001, *AJ* 121, 2664

Castor, J., McCray, R. & Weaver, R. 1975, *ApJ* 200, L107

Chu, Y.-H., Chang, H.-W., Su, Y.-L. & Mac Low, M.-M. 1995, *ApJ* 450, 157

Chu, Y.-H. & Mac Low, M-M. 1990, *ApJ* 365, 510

Chu, Y.-H., Wakker, B., Mac Low, M.-M. & García-Segura, G. 1994, *AJ* 108, 1696

Clarke, C. J. & Oey, M. S. 2002, *MNRAS* 337, 1299

Cohen, D., Leutenegger, M. A., Grizzard, K. T., Reed, C. L., Kramer, R. H. & Owocki, S. P. 2006, *MNRAS* 368, 1905

Cooper, R. L., Guerrero, M. A., Chu, Y.-H., Chen, C.-H. R. & Dunne, B. C. 2004, *ApJ* 605, 751

Danforth, C. W. & Blair, W. P. 2006, *ApJ* 646, 205

Dyson, J. E. 1977, *A&A* 59, 161

Efremov, Y. N. & Elmegreen, B. G. 1997, *ApJ* 480, 235

Ellison, D. C., Decourchelle, A. & Ballet, J. 2004, *A&A* 413, 189

Fullerton, A. W., Massa, D. L. & Prinja, R. K. 2006, *ApJ* 637, 1025

Gazol-Patiño, A. & Passot, T. 1999, *ApJ* 518, 748

Hatzidimitriou, D., Stanimirović, S., Maragoudaki, F., Staveley-Smith, L., Dapergolas, A. & Bratsolis, E. 2005, *MNRAS* 360, 1171

Hillier, D. J. 1991, *A&A* 247, 455

Hunter, D. A., Boyd, D. M. & Hawley, W. N. 1995, *ApJS* 99, 551

Kim, S., Dopita, M. A., Staveley-Smith, L. & Bessell, M. S. 1999, *AJ* 118, 2797

Klepach, E. G., Ptuskin, V. S. & Zirakashvili, V. N. 2000, *Astroparticle Phys.* 13, 161

Kramer, R. H., Cohen, D. H. & Owocki, S. P. 2003, *ApJ* 592, 532

Mac Low, M.-M., Chang, T. H., Chu, Y.-H., Points, S. D., Smith, R. C. & Wakker, B. P. 1998, *ApJ* 493, 260

Mac Low, M.-M., McCray, R. & Norman, M. L. 1989, *ApJ* 337, 141

Mashchenko, S. Y., Thilker, D. A. & Braun, R. 1999, *A&A* 343, 352

Miller, N. A., Cassinelli, J. P., Waldron, W. L., MacFarlane, J. J. & Cohen, D. H. 2002, *ApJ* 577, 951

Nugis, T., Crowther, P. A. & Willis, A. J. 1998, *A&A* 333, 956

Oey, M. S. 1996, *ApJ* 467, 666

Oey, M. S. & Clarke, C. J. 1997, *MNRAS* 289, 570

Oey, M. S. & Clarke, C. J. 1998, *AJ* 115, 1543

Oey, M. S., Clarke, C. J. & Massey, P. 2001, in: K. S. de Boer, R.-J. Dettmar & U. Klein (eds.), *Dwarf Galaxies and Their Environment* (Shaker Verlag), P. 181

Oey, M. S., Dopita, M. A., Shields, J. C. & Smith, R. C. 2000, *ApJS* 128, 511

Oey, M. S. & García-Segura, G. 2004, *ApJ* 613, 302

Oey, M. S., Groves, B., Staveley-Smith, L. & Smith, R. C. 2002, *AJ* 123, 255

Oey, M. S. & Massey, P. 1995, *ApJ* 452, 210

Oey, M. S. & Smedley, S. A. 1998, *AJ* 116, 1263

Oey, M. S., Watson, A. M., Kern, K. & Walth, G. L. 2005, *AJ* 129, 393

Palouš, J., Franco, J. & Tenorio-Tagle, G. 1990, *A&A* 227, 175

Parizot, E., Marcowith, A., van der Swaluw, E., Bykov, A. M. & Tatischeff, V. 2004, *A&A* 424, 747

Pikel'ner, S. B. 1968, *Astrophys. Lett.* 2, 97

Rhode, K. L., Salzer, J. J., Westpfahl, D. J. & Radice, L. A. 1999, *AJ* 118, 323

Saken, J. M., Shull, J. M., Garmany, C. D., Nichols-Bohlin, J. & Fesen, R. A. 1992, *ApJ* 397, 537

Silich, S. A., Franco, J., Palouš, J. & Tenorio-Tagle, G. 1996, *ApJ* 468, 722

Silich, S. A. & Oey, S. 2002, in: D. Geisler, E. K. Grebel & D. Minniti (eds.), *Extragalactic Star Clusters* (ASP-CS), p. 459

Silich, S. A., Tenorio-Tagle, G., Terlevich, R., Terlevich, E., & Netzer, H. 2001, *MNRAS* 324, 191

Staveley-Smith, L., Sault, R. J., Hatzidimitriou, D., Kesteven, M. J. & McConnell, D. 1997, *MNRAS* 289, 225

Stewart, S. G., *et al.* 2000, *ApJ* 529, 201

Strickland, D. K. & Stevens, I. R. 2000, *MNRAS* 314, 511

Tenorio-Tagle, G., Różyczka, M., Franco, J. & Bodenheimer, P. 1991, *MNRAS* 251, 318

Thilker, D. A., Braun, R. & Walterbos, R. A. M. 1998, *A&A* 332, 429

Thronson, H. A., Lada, C. J. & Hewagama, T. 1985, *ApJ* 297, 662

Tomisaka, K. 1990, *ApJ* 361, L5

Walborn, N. R. & Parker, J. W. 1992, *ApJ* 399, L87

Wang, Q. & Helfand, D. J. 1991, *ApJ* 373, 497

Williams, J. P., Blitz, L. & Stark, A. A. 1995, *ApJ* 451, 252

Discussion

BALLY: Have you considered using momentum conservation (instead of energy conservation) to model shell evolution? It seems that mass-loading and radiative cooling would push the shell towards momentum conservation, implying smaller shell radii.

OEY: It's true that momentum conserving bubbles follow an evolution similar to the adiabatic model. The observations of x-ray emission from superbubbles is some of the strongest evidence favoring the adiabatic evolution, but if the objects in fact undergo radiative cooling, then we should certainly consider momentum conserving models.

HENSLER: If you determine the state of superbubbles, then x-ray luminosity, the temperature, metal content, etc., how sensitive is their determination? We found the same growth rate discrepancy e.g., in NGC 1705 (Hensler *et al.*1997, ApJ)

OEY: Because the x-ray fluxes are relatively low for OB superbubbles, we still do not have adequate quantitative confirmation of the x-ray luminosities and metallicities for such objects. I strongly encourage further observations with XMM and Chandra.

BLITZ: One has to be careful when one attributes causality in star formation from shells. In many most (?) cases, the star formation and molecular clouds that precede them may be independent of the shell that now envelopes the region.

OEY: I completely agree, especially for regions with two-stage sequential star formation. For 3-stage sequential star formation, with the morphology seen in W3/4, I find a conclusion of causality hard to avoid.

Triggered Star Formation in a Turbulent ISM
Proceedings IAU Symposium No. 237, 2006
B. G. Elmegreen & J. Palouš, eds.

Triggering and the gravitational instability in shells and supershells

Jan Palouš

Astronomical Institute, Academy of Sciences of the Czech Republic, Boční II 1401,
140 41 Prague 4, Czech Republic
email: palous@ig.cas.cz

Abstract. The gravitational instability of shells and supershells and its relevance to the triggering of the star formation in a turbulent ISM is discussed. The IMF of the self-gravitating clumps formed out of shells is computed and triggering time is compared with the shell dissolution time due to chaotic motions.

Keywords. shock waves, turbulence, ISM: bubbles, ISM: clouds, stars: formation

1. Introduction

Elmegreen & Lada (1978) assume that new stars form in the Cold Post-Shock (CPS) layer separating the shock front from the ionisation front when an HII region created with the energy inserted to the interstellar medium by the previous generation of stars propagates into the dense molecular cloud. The early fast expansion slows down to ~ 5 km s^{-1}, which eliminates the stabilising effect of stretching. In the CPS, the gravitational instability is triggered when it is dense enough ($\geqslant 10^5$ cm^{-3}) at the temperature $\sim 10^2$ K. The gravitational instability of the CPS layer reorganises the accumulated matter forming the clumps, which may be the places of new star formation.

The concept of a turbulent ISM is developed by Elmegreen & Scalo (2004) and Scalo & Elmegreen (2004). The self-similar changing structures cannot be interpreted in terms of simple equilibrium quantities such as velocity dispersion or pressure. Interacting supersonic shock waves create temporary structures including clumps with a range of sizes and masses, which form and dissolve in sheets and filaments. The energy dissipation in supersonic shocks and cooling is compensated with a variety of driving mechanisms including the galactic rotation and galaxy interactions on the large scale, and energy inserted by stars on the small scale.

Here, we discuss if the shell triggering may be important to formation of longer-living star clusters and stars in the ISM clouds.

2. Bubble Bath in the ISM

Shells and supershells in the Milky Way are analysed by Ehlerová & Palouš (2005), who discovered more than 600 structures in the Leiden Dwingeloo HI survey. This analysis is extended with the Leiden/Argentine/Bonn survey, see the contribution by Ehlerová (this volume). There is a lot of structures discovered in other nearby galaxies including LMC, SMC, M31 etc. Reviews of the observational evidence are given by Stanimirovic (this volume) and Brinks (this volume).

The static ISM in two or more fixed components does not exist. The shells are frequently seen in expansion. They move relatively to each other, collide and open to the galactic halo resembling a bubble bath in a hot spring. The high density, low temperature filaments and sheets have 1D or 2D geometry. They form and go showing that the individual structures are far away from an equilibrium situation.

Stars insert energy and mass into this bubble bath. In some cases, we even see the powering OB association, but in the majority of shells the parent stars are not detected. It shows that the average lifetime of shells is longer compared to the lifetime of young and massive stars. Some of the supershells, e.g. those seen in the LMC and SMC, may be connected to galaxy versus galaxy interactions or to galaxy rotation, spiral arms and bars rather than to the star formation. Other structures are due to cold gas infall coming to the galaxy plane in the form of high velocity clouds, which produce incomplete partially open shell-like features. The impacts of high velocity clouds are more destructive and unable to create spherical-like shells. Another mechanism creating or enlarging the existing structures is stripping with ram pressure created by a galaxy high speed motion through the diluted and hot intracluster medium.

3. Shell fragmentation: IMF of clumps

A dispersion relation gives how the frequency $\omega(k) = 2\pi/t_{growth}$ depends on the mode wave number k. Within a volume inside of radius R, the number of fragments produced by an instability is given as $N = \omega \frac{R^3}{(\lambda/4)^3} = \frac{8\omega R^3 k^3}{\pi^3}$, $k = \frac{2\pi}{\lambda}$, where λ is the wave length. The mass of a fragment is $m = \frac{4}{3}\pi(\lambda/4)^3\rho = \frac{1}{6}\pi^4\rho k^{-3}$, and the number of fragments with the wave number in the interval $(k, k+dk)$ is $dN = \frac{24\omega R^3 k^2}{\pi^3}dk$. The mass spectrum $\xi(m) = dN/dm$, or the number of clumps in an interval of mass, is $\xi(m) = -\frac{4}{3}\pi R^3\rho\omega m^{-2}$

The dispersion relation of the Jeans gravitational instability in 3D is

$$\omega^2(k) = -c^2 k^2 + 4\pi G\rho, \tag{3.1}$$

where c is the speed of sound. It transforms to the mass spectrum

$$\xi_{Jeans}(m) = \frac{16}{9}R^3\rho \, m^{-2} \left[-c^2\left(\frac{\pi^4\rho}{6m}\right)^{2/3} + 4\pi G\rho\right]^{1/2}, \tag{3.2}$$

showing for large masses the slope $\alpha = -2$. It flattens when the fragment mass decreases to the Jeans mass $m_{Jeans} = \frac{\pi^{5/2}}{48}G^{-3/2}c^3\rho^{-1/2}$.

In thin CPS layer with the 2D geometry, the dispersion relation is

$$\omega^2(k) = -c_{sh}^2 k^2 + 2\pi G\Sigma_{sh}k, \tag{3.3}$$

where c_{sh} is the speed of sound inside of the CPS and Σ_{sh} is its mass column density. The mass spectrum of CPS fragments is

$$\xi(m) = \frac{1}{4}A\pi^{3/4}\Sigma_{sh}^{3/2}m^{-9/4} \times \frac{-3\pi^{1/2}c_{sh}^2 m^{-1/2} + 2.5G\Sigma_{sh}^{1/2}}{(-\pi^{1/2}c_{sh}^2 m^{-1/2} + G\Sigma_{sh}^{1/2})^{1/2}}, \tag{3.4}$$

where A is the area of the CPS, see Elmegreen & Elmegreen (1978). At the high mass end is the CPS mass spectrum steeper, compared to 3D Jeans mass spectrum, with $\alpha = -2.25$. The expansion would make it even more steep, with the slope $\alpha = -2.35$, see Palouš *et al.* (2003).

Table 1. The time when the gravitational instability sets in a CPS

$c_{sh}\ [kms^{-1}]$ $n\ [cm^{-3}]$	0.3	0.5	1.0
10^{-1}	43	59	91
1	14	19	29
10^2	1.4	1.9	2.9
10^6	0.01	0.02	0.03

4. Velocity – Size Relation from Triggering in CPS Layers

We may ask when the gravitational instability starts in a thin shell expanding into a dense cloud. This triggering time of the gravitational instability of the CPS is given in Table 1 for different densities of the clouds n and for three values of c_{sh}.

The triggering time given in Table 1 sets a velocity - size relation as it is shown in Fig. 1. Having a cloud with expanding shells, it shows an expansion velocity as a function of size under the condition that shells are able to trigger the gravitational instability. When, for given size Δr, is the random velocity Δv in a cloud smaller as derived from velocity - size relation for given density of the ambient medium, the gravitational instability is faster than random motions and the shells in the cloud start to produce fragments out of CPS layers. On the other hand, when the random motions inside of the cloud are larger that corresponding to the velocity - size relation, the gravitational instability is too slow and the cavities are refilled before the gravity sets in.

We compare this theoretical velocity - size relation for triggering in the CPS with the observed relation as given by Larson (1981):

$$\Delta v(\text{km/s}) = 1.1 \Delta r(\text{pc})^{0.38}. \tag{4.1}$$

It is interesting that starting from low $n = 10$ cm^{-3} and sizes $\Delta r = 100$ pc up to the high $n = 10^6$ cm^{-3} and sizes $\Delta r = 0.01$ pc the Larson relation is always just at the theoretical velocity - size relation for triggering. We see only a slightly more gravity dominance for high densities. It shows that in average the clouds are only marginally unstable to gravitational triggering with shells, to almost the same degree on all observed scales. It implies a self-regulation of the level of turbulent random motions: the shell instability increases the level of turbulence leading to faster destruction of shells due to random motions, which decreases the triggering and new star formation.

5. Conclusions

The ISM is described as a bubble bath composed of moving shells expanding inside of clouds. The cold and thin CPS layer, see Elmegreen & Lada (1978), trigger gravitational instability producing clumps with a mass spectrum with a slope $\alpha = -2.25$. The expansion time needed for the initialisation of the instability sets the velocity - size relation of the triggering, which is compared with the observational relation derived by Larson (1981). We conclude that clouds are only marginally unstable due to the gravitational instability of shells. This is valid on all the scales from 0.01 – 100 pc, there is only a slight increase of triggering at high densities. It also shows that there is a self-regulation between the triggering with shells and turbulent motions inside of clouds.

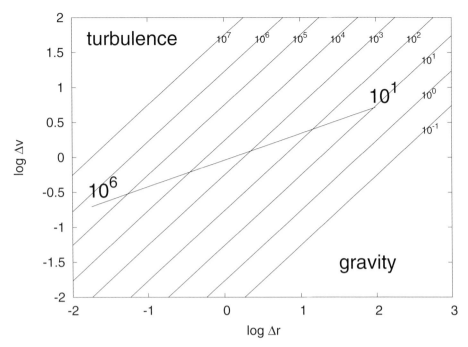

Figure 1. The velocity – size relation for the shells triggering the gravitational instability. The numbers give the density of the medium, we assume $c_{sh} = 1$ km/s. The thick line shows the observational Larson (1981) velocity – size relation.

Acknowledgements

I would like to acknowledge the Institutional Research Plan AV10030501 of the Astronomical Institute, Academy of Sciences of the Czech Republic, and the project LC 06014 Center for Theoretical Astrophysics.

References

Brinks, E. 2006, *this volume*
Ehlerová, S. 2006, *this volume*
Ehlerová, J. & Palouš, J. 2005, *A&A* 437, 101
Elmegreen, B.G. & Elmegreen, D. 1978 *ApJ* 220, 1051
Elmegreen, B.G. & Lada, Ch. 1978, *ApJ* 214, 725
Elmegreen, B.G. & Scalo, J. 2004, *ARA&A* 42, 211
Larson, R.B. 1981, *MNRAS* 194, 809
Palouš, J., Ehlerová, S. & Wünsch, R. 2003 *Astrophys. and Space Sci.* 284, 873
Scalo, J. & Elmegreen, B.G. 2004, *ARA&A* 42, 275
Stanimirovic, S. 2006, *this volume*

Discussion

E. VAZQUEZ-SEMADENI: There is only a small preference for triggering at high compared to small densities. In your comparison of the free-fall time versus the turbulent time, you conclude gravity dominates at high densities. What do you assume to associate a turbulent crossing time to a given density? This seems to implicitly assume a density – size relation.

J. PALOUŠ: Yes, there is a slight increase in the dominance of gravity and shell frag-mentation in the high density medium. But this is a rather tiny tendency, and I like to stress the opposite: the marginal stability of shells holds on all the scales. The differences between high and low density medium are very small as concerns the triggering by shells. I have in mind the average density in a region of a given size. All the substructures like bubbles and sheets are included into an average density. The numbers I have involved in the figure are taken from Larson (1981).

I. GOLDMAN: In the case where gravity dominates – the gravitational instability would generate a turbulence with the   SAME timescale as the gravitational timescale. So the two should be perhaps considered.

J. PALOUŠ: I agree that any mechanism driving the turbulence inserts random mo-tions into smaller scales. Gravitational fragmentation of shells acts in the same way and increases the internal velocity in the thin sheets. I suspect that the same cooling mech-anism, which is responsible for their formation is able to act in the initial phases of the gravitational fragmentation and keep the layer at low velocity dispersion.

K. OMUKAI: How did you convert the dispersion relation to the mass function?

J. PALOUŠ: A finite volume is filled with all the unstable modes of given wavelength with a spatial frequency inversely proportional to their growth-rate. This gives the number of unstable regions as a function of wavelength or as a function of mass. The dispersion relation for a theoretical process where ω does not depend on the wavenumber k gives from purely geometrical reasons the mass spectrum represented with a power law expo-nent equal to $\alpha = -2$. In the case of bubbles and thin layers we get steeper slopes with $\alpha \leqslant -2.25$.

J. BALLY: Sub-mm observations of dusty cores have found mass spectra with a slope, $N(m)\ m^{-\alpha}$ with $\alpha \approx 2.3$, similar to the IMF, and the slope you calculate from the gravitational instability of a shell. Can you speculate about the connection between instabilities and the IMF?

J. PALOUŠ: We see the turbulent, bubble dominated medium, which is out of an equi-librium. Therefore, stars form out of layers via their fragmentation. In my opinion, the dusty cores have been formed in the same process of gravitational fragmentation of thin sheets with a mass spectrum very close to stellar. This shows that all the star forming processes leading from cores to stars probably does not change the mass spectrum, or a similar fraction of the original core mass ends in the star for all the core masses. This may be due to effective cooling at the sub-mm wavelengths, which dominates the ini-tial phases of the evolution of the self-gravitating fragment up to the instant when it is detached from the rest of the ambient medium.

Triggered Star Formation in a Turbulent ISM
Proceedings IAU Symposium No. 237, 2006
B. G. Elmegreen & J. Palouš, eds.

© 2007 International Astronomical Union
doi:10.1017/S1743921307001329

The enigmatic loop III and the Local Galactic Structure

M. Kun[1]

[1]Konkoly Observatory of Hungarian Academy of Sciences,
H-1525 Budapest, P.O. Box 67, Hungary
e-mail: kun@konkoly.hu

Abstract. The aim of the present study, based on literature data, is to find signatures of the giant radio continuum structure Loop III on the nearby interstellar medium, and search for molecular cloud and star formation, possibly triggered by its expansion. The preliminary results are as follows: (1) The 3D map of the Local Bubble, published by Lallement *et al.* (2003) suggests that Loop III is probably more distant than the early models had indicated. (2) The molecular clouds at high galactic latitudes in the 2nd Galactic quadrant are probably associated with the neutral/molecular wall of Loop III. (3) Star formation in Lynds 1333 and Lynds 1082 (GF 9) might have been triggered by the expansion of Loop III. (4) The supernova(e), whose explosion produced Loop III, might have been located in the SU Cas association.

Keywords. ISM: bubbles, ISM: structure, (ISM:) supernova remnants, ISM: individual: Loop III

1. Introduction

The giant galactic radio continuum loops, thought to be nearby, old remnants of single or multiple supernovae, were first studied by Berkhuijsen (1971) and Spoelstra (1972). The basic properties of Loop III, revealed by these early radio continuum surveys, are as follows: *Galactic coordinates of the centre:* $l \approx 125°, b \approx +15.6°$; *Angular diameter:* $65°$; *Estimated distance to the centre:* 150 pc; *Radius:* 85 pc; *Estimated age:* about 10^6 yrs.

Due to its huge angular diameter Loop III affects most of the lines of sight toward the positive latitude part of the second Galactic quadrant. Nevertheless, it has been clearly identified only as a radio continuum source due to the complexity of the interstellar medium over its surface. Loop III is expected to have profound effects on the structure of the local interstellar matter, and possibly on the recent star formation in our Galactic neighbourhood. Nevertheless, only a few data are available on such effects. In particular, Verschuur (1993) pointed out that Loop III has modified the distribution of the high velocity gas at its high-latitude boundaries, demonstrating that the high-velocity hydrogen clouds are local objects.

The aim of the present study, based on literature data, is to find further signatures of Loop III on the nearby interstellar medium, identify interstellar atomic and molecular clouds connected with it, and search for star formation possibly triggered by its expansion.

2. Signatures of Loop III in the distribution of local ISM

Loop III and the Local Bubble. The distance of 150 pc, obtained by Spoelstra (1972), is based on van der Laan's (1962) theory of expanding supernova remnants. In the light of more recent theories of supernova shells (e. g. Asvarov 2006) this value is rather uncertain. If the centre of Loop III were at 150 pc from the Sun, its near wall should be as close as about 65 pc to us, thus a region of its interaction with the Local

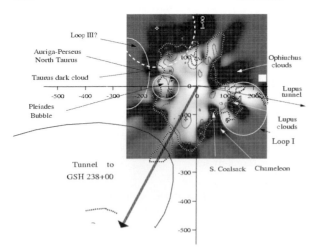

Figure 1. Cross-section of the Local Bubble in the Galactic plane and the outlines of its bordering structures, adopted from Lallement *et al.* (2003). Two hypothetical positions of Loop III are plotted: Solid line: position assumed by Lallement *et al.* (2003); dashed circle: with centre position and angular diameter derived by Spoelstra (1972), and assuming D = 250 pc.

Bubble (LB) could be identified at such a distance. A detailed 3D map of the LB was published by Lallement *et al.* (2003). Their Fig. 9, displaying the cross section of this map in the Galactic plane, demonstrates that the extension of LB is larger than 100 pc toward the 2nd quadrant, suggesting a larger distance for Loop III. The authors also plotted an assumed position of Loop III at larger distance and closer to the Galactic anticentre than indicated by the radio continuum observations. If we accept the position and angular diameter of Loop III, determined by these observations, and assume that the boundary of LB, found by Lallement *et al.* (2003), is defined by its intersection with Loop III in the second quadrant, we obtain 250 pc for the probable distance to the centre and about 140 pc for its radius. Figure 1 shows this hypothetical Loop III, overplotted on Lallement *et al.* 's (2003) Fig. 9 as a dashed circle.

HI and molecular clouds associated with Loop III. Nearby interstellar structures, projected within Loop III, are displayed in the left panel of Fig. 2. The distribution of the neutral hydrogen, $\int T_B dv \sin |b|$, obtained from the Leiden–Dwingeloo survey data base (Hartmann & Burton 1997), is displayed for the radial velocity interval of $-8\,\mathrm{km\,s^{-1}} < v_{\mathrm{LSR}} < 8\,\mathrm{km\,s^{-1}}$. Positions of the most prominent molecular clouds/complexes and the probable members of the SU Cas association (see below), are also indicated.

Large-scale CO maps by Dame, Hartmann & Thaddeus (2001), Heithausen & Thaddeus (1990), and Heithausen *et al.* (1993) show that molecular gas can be found over the whole area of Loop III. Table 1 lists the Galactic positions and distances of these molecular clouds/complexes, and the literature sources of their distances. The most prominent molecular structure projected within the boundary of Loop III is the Cepheus flare. Recently Olano, Meschin & Niemela (2006) established that the interstellar gas in the Cepheus flare is distributed over the surface of an expanding shell (Cepheus flare shell, CFS), centred on $l \sim 124°$, $b \sim +17°$ and at a distance of 300 pc, and having a radius of about 50 pc. According to Olano, Meschin & Niemela (2006), all these quantities are rather uncertain estimates. As most of the molecular clouds in the Cepheus flare region are found at 200 and 300 pc (Kun 1998), the true distance to the centre of this shell may well be about 250 pc, suggesting that Loop III and CFS are nearly concentric and possibly identical.

Another prominent molecular complex, the Polaris flare, is located at a distance of ~ 110 pc, and thus probably is associated with the near boundary of Loop III. At the highest Galactic latitude segment the North Celestial Pole Loop (Meyerdierks, Heithausen &

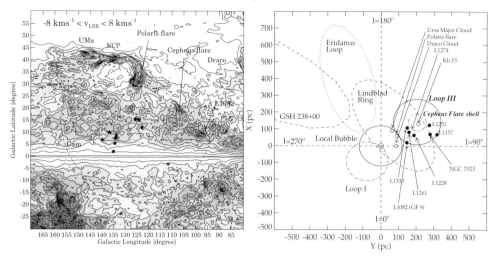

Figure 2. Left: Distribution of the neutral hydrogen over the surface of Loop III in the radial velocity interval $-8\,\mathrm{km\,s^{-1}} < v_{\mathrm{LSR}} < 8\,\mathrm{km\,s^{-1}}$. The position of the centre and boundary are indicated by the cross and dashed line, respectively. Dotted circle indicates the Cepheus flare shell, star symbols show the members of the SU Cas association and other late B–early A type stars located at a distance of about 250 pc. Positions of some molecular clouds are also indicated. Right: Distribution of nearby HI shells, projected onto the Galactic plane. Star forming clouds projected within Loop III are marked by black circles, and non star forming clouds by open circles.

Reif 1991) closely follows the boundary of Loop III. The Camelopardalis clouds are located at the high-longitude, low-latitude boundary of Loop III. According to Zdanavičius *et al.* (1996) absorbing matter can be found between 100 and 300 pc in this direction, demonstrating that the line of sight is parallel to the wall of Loop III.

Star formation possibly triggered by Loop III. In the right panel of Fig. 2 the distribution of the nearby HI shells projected on the Galactic plane, adopted from Heiles (1998), is plotted and supplemented by the position of Loop III and the Cepheus flare shell. The distribution of the molecular clouds of the region is shown as well. Most star forming (possibly older) molecular clouds can be found inside Loop III, whereas the non-star forming (younger) ones are located close to its present surface, suggesting that star formation had been initiated by the expansion of the shell. The trigger by Loop III is conspicuous in the case of L 1333 (see Kun *et al.* 2006). Another candidate is the L 1082 (GF 9, Schneider & Elmegreen 1979), located close to the boundary of Loop III in both projections, in which very early stages of star formation can be observed (Wiesemeyer, Güsten & Wright 1997).

Possible origin of Loop III. The stars plotted in the left panel of Fig. 2 are late B–early A type members of the SU Cas association (Turner & Evans 1984) and a few other stars similar in spectral type, proper motion and distance modulus to the association members. The supernova(e), whose explosion(s) produced Loop III, might have been high-mass member(s) of this association.

3. Conclusion

The preliminary results presented in this paper suggest that careful studies of the available data on the spatial and velocity distribution of the interstellar structures may reveal several connections which combine them into a coherent structure.

Table 1. Molecular cloud distances in the 2nd Galactic quadrant

Cloud	l (°)	b (°)	D (pc)	Reference
Draco	90	38	86	(1)
Lynds 1082 (GF 9)	97	+10	100–150	(2)
Cepheus flare	100–120	10–20	200 and 300	(3), (4), (5)
Polaris flare	118–126	20–30	$105 < D < 125$	(6)
Lynds 1274	118.1	+8.8	200 ± 30	(7)
Khavtassi 15	122.7	+9.6	250 ± 25	(8)
Ursa Major	141.8	+35.9	$100 < D < 120$	(9)
Camelopardalis clouds	140–150	0–10	100–300	(10)
Lynds 1333	127	15	180 ± 30	(11)

References: (1) Lallement *et al.* (2003); (2) Wiesemeyer, Güsten & Wright(1997); (3) Straižys *et al.* (1992); (4) Kun (1998); (5) Olano, Meschin & Niemela (2006); (6) Zagury, Boulanger & Banchet (1999); (7) Nikolić *et al.* (2001); (8) Kiss, Tóth, Sato *et al.* (2000); (9) Penprase (1993); (10) Zdanavičius *et al.* (1996); (11) Obayashi, Kun, Sato *et al.* (1998)

Acknowledgements

This research was supported by the Hungarian OTKA grant T 049082.

References

Asvarov, A. 2006, *A&A* in press (astro-ph/0608079)

Berkhuijsen, E. M. 1971, *A&A* 14, 359

Dame, T. M., Hartmann, D. & Thaddeus, P. 2001, *ApJ* 547, 792

Hartmann, D. & Burton, W. B. 1997, Atlas of Galactic Neutral Hydrogen, Cambridge Univ. Press

Heiles, C. 1998, *ApJ* 498, 689

Heithausen, A. & Thaddeus, P. 1990, *ApJ* 353, L49

Heithausen, A., Stacy, J. G., de Vries, H. W., Mebold, U. & Thaddeus, P. 1993, *A&A* 268, 265

Kiss, Cs., Tóth, L. V., Sato, F., Nikolić, S. & Wouterloot, J. G. A. 2000, *A&A* 363, 755

Kun, M. 1998, *ApJS* 115, 59

Kun, M., Nikolić, S., Johansson, L. E. B., Balog, Z. & Gáspár, A. 2006, *MNRAS* 371, 732

Lallement, R., Welsh, B. Y., Vergely, J. L., Crifo, F. & Sfeir, D. 2003, *A&A* 411, 447

Meyerdierks, H., Heithausen, A. & Reif, K. 1991, *A&A* 245, 247

Nikolić, S., Kiss, Cs., Johansson, L. E. B., Wouterloot, J. G. A. & Tóth L. V. 2001, *A&A* 367, 694

Obayashi, A., Kun, M., Sato, F., Yonekura, Y. & Fukui, Y. 1998, *AJ* 115, 274

Olano, C. A., Meschin, P. I. & Niemela, V. S. 2006, *MNRAS* 369, 867

Penprase, B. E. 1993, *ApJS* 88, 433

Schneider, S. & Elmegreen, B. G. 1979, *ApJS* 41, 87

Spoelstra, T. A. T. 1972, *A&A* 21, 61

Straižys, V., Černis, K., Kazlauskas, A. & Meistas, E. 1992, *Balt. Astr.* 1, 149

Turner, D. G. & Evans, N. R. 1984, *ApJ* 283, 254

van der Laan, H. 1962, *MNRAS* 124, 125

Verschuur, G. L. 1993, *ApJ* 409, 205

Wiesemeyer, H., Güsten, R. & Wright, M. C. H. 1997, in: F. Malbet & A. Castets (eds.), *Low mass Star Formation – from Infall to Outflow*, poster proceedings of IAU Symp. 182, p. 260

Zagury, F., Boulanger, F. & Banchet, V. 1999, *A&A* 352, 645

Zdanavičius, K., Zdanavičius, J. & Kazlauskas, A. 1996, *Balt. Astr.* 5, 563

Discussion

MAC LOW: 1. Has this loop been detected in more recent surveys, such as the sensitive Wisconsin H-α mapper survey, or more recent continuum surveys? 2. Might Loop III actually be a superposition of unrelated objects in a low resolution survey?

BOULANGER: (answered for M. Kun) Loop III is well detected as a coherent structure in emission and polarization in the 23 GHz synchrotron maps provided by the Cosmic Microwave Background Probe WMAP (Page *et al.* 2006). HI gas at $V_{LSR} < -20$ km s^{-1} is seen over the sky area interior to the synchrotron loop.

Triggered Star Formation in a Turbulent ISM
Proceedings IAU Symposium No. 237, 2006
B. G. Elmegreen & J. Palouš, eds.

© 2007 International Astronomical Union
doi:10.1017/S1743921307001330

Footprints of triggering in large area surveys of the nearby ISM and YSOs

L. Viktor, Tóth[1,2] and Zoltán T. Kiss[3,1]

[1]Department of Astronomy, Eötvös University, Pázmány P.s. 1/a H-1117, Budapest, Hungary
email: l.v.toth@astro.elte.hu

[2]Konkoly Observatory, PO Box:67, H-1525, Budapest, Hungary

[3]Baja Astronomical Observatory, PO Box 766, 6500 Baja, Hungary
e-mail: kissz@alcyone.bajaobs.hu

Abstract. Our goal is to evaluate the role of triggering effects on the star formation and early stellar evolution by presenting a statistically large sample of cloud and low-mass YSO data. We conducted large area surveys (ranging from 400 square-degree to 10800 square-degree) in optical, NIR and FIR. The distribution of the ISM and low-mass YSOs were surveyed. A relative excess was found statistically in the number of dense and cold core bearing clouds and low mass YSOs in the direction of the FIR loop shells indicating a possible excess in their formation.

Keywords. stars: early-type, stars: formation, stars: statistics, ISM: bubbles, ISM: clouds, ISM: dust, extinction, Galaxy: structure

1. Introduction

Do we need more motivation for a study of triggered low mass star formation than a better understanding of the formation of our Solar System? We may assume that the formation of the Solar System was triggered by massive stars – based on abundance anomalies of short-lived radio isotopes in meteorites Lee, Papanastassiou & Wasserburg (1977), Cameron & Truran (1977). Either the pre-solar nebula was contaminated by various isotopes including Al^{26} at the time of a close-by supernova explosion, or the YSO disk of the young Sun was "doped" by these radioactive isotopes originating from the atmosphere of a nearby ($d \approx 0.2\,pc$) red supergiant according to Chevalier (2000). The probability of the latter close-massive-star scenario is small but perhaps not negligible. But since massive stars form in associations, it is very likely that other, not so close, but massive YSO members of the association triggered the formation of the proto-Sun by their radiation, stellar and/or supernova winds (see also Hester *et al.* 2004).

The nearby low mass star forming regions are considered as examples for ISM-external effect interactions: Star formation in the outer regions of the ρOph cloud could be triggered by the same event that initiated star formation in Upper Sco (Wilking *et al.* 2005). The dense filamentary structures of the Taurus clouds were produced by large-scale colliding HI streams (Ballesteros-Paredes *et al.* 1999). Cloud collapse is likely triggered by an HII region in the Horsehead nebula, B33 (Ward-Thompson *et al.* 2006).

The low-density intercloud medium is the result of turbulence and overlapping supernova remnants. Clouds are at least partly turbulence-resulted fractals, and as such are also highly clustered, making the mean free path for ionizing photons at least twice as large as in the 'standard cloud' model (Elmegreen 1998). That means however, that a large fraction of dense ISM (i.e. clouds) is affected by either SN or ionizing photons from mid-plane HII regions with a large free path. A large free path means most photons can travel nearly to the next HII region (Elmegreen, 1998). The trigger (by radiation, wind, shock fronts) is then evidently present for a considerable (if not overwhelming) mass fraction of the cold neutral ISM in the Galactic disk.

The footprints of triggered star formation should be observed also in the distribution of few Myrs old YSOs, since those can still be found near their birthplaces.

What is the role of triggering in the formation of low mass stars? The answer is searched in 3 steps:

(1) We study the structure of low and intermediate mass clouds in the obvious presence and assumed absence of external effects.

(2) We survey the distribution of cold dense ISM to locate the interfaces between the very low density intercloud medium, and the dense ISM.

(3) We study the distribution of low mass YSOs, and try to relate their distribution to that of the ISM.

2. Surveys and results

Kiss *et al.* (2006a) performed an unbiased survey of a 256 square-degree area in Cepheus exploring the ISM distribution, cloud and star formation based on optical and NIR data. The nearby ($d \approx 300\,\mathrm{pc}$) Cepheus Flare GMC (Hubble 1934, Lebrun 1986) is known as a site of low- and intermediate-mass star formation with over 200 clouds and with a total ISM mass of $2 \times 10^5 \mathrm{M_\odot}$, of which only 20% is in the form of dark ($A_V > 2.0\,\mathrm{mag}$) clouds (Kiss *et al.* 2006a). The observed projected axis ratios b/a (see Fig. 5) of our clouds correspond to near-prolate 3d shapes (see Jones & Basu 2002). This would be consistent with their formation from large-scale external forcing. While the ISM distribution in the central part of the Cepheus Flare reminds us of turbulent model images (see e.g. examples in this volume), there are high pressure events acting at its outskirts.

The eastern side of the Cepheus Flare GMC is bordered by the Cepheus Flare shell (Olano *et al.* 2006), an expanding shell enclosing an old supernova bubble (Grenier *et al.* 1989) with a radius $r = 50\,\mathrm{pc}$, expansion velocity $v = 4$ km s^{-1} and estimated age $t > 10^6$ yr. The southern border is however "guarded" by an expanding HI shell and FIR loop GIRL G109+11 (Kiss *et al.* 2004). Most of the opaque ($A_V > 5\,\mathrm{mag}$) clouds were found in the loop-boundary area, i.e. 70% (19 out of 27) of these clouds are located at 20% of the total area. Most of the star formation occurs in these clouds as seen from the distribution of Hα, and infrared excess point sources. The immediate conclusion is that an effect results in cloud restructuring and star formation at the GMC-intercloud surface. A straightforward explanation is triggering by slow shock fronts ($v \approx 10$ km s^{-1}) which travel into the Cepheus Flare GMC. Clouds in the walls of the voids are exposed to external pressure and radiation. Triggering by a slow increase in external pressure has been recently discussed by Lesaffre *et al.* (2006). Radiation-driven implosion may result triggered star formation in bright-rimmed clouds as seen e.g. in recent observations by Urquhart *et al.* (2006) and modeling by Miao *et al.* (2006).

Cloud-intercloud interfaces could be found e.g. locating expanding HI shells, of which more than 700 were cataloged by Ehlerova (see also in this volume). The voids however do not always expand, and even when they do, the expansion velocity may decrease to a few km s^{-1} as the expanding shell is aging.

We wanted to search for the cloud-intercloud interfaces without a restriction on the kinematics. The column density of the dusty ISM is seen in the FIR optical depths. With this in mind, the ISSA IRAS all-sky 60μm and $100\,\mu$m surface brightness data were examined and 462 loops were located (Kiss *et al.* 2004 and Könyves *et al.* 2006). Loops are at least 60 percent complete arcs, where an excess $100\,\mu$m surface brightness occurs over the central parts of at least 3 times the local surface brightness fluctuation level. The

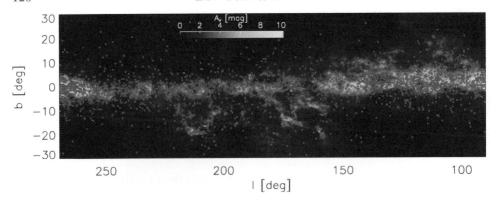

Figure 1. Reddening by Schlegel in the outer galaxy, with CTT candidate point sources over-plotted, marked with (red) dots.

FIR colour of the loops is the same as the cirrus, with a local hydrogen column density excess of 10% in average over their surroundings. The loops appear as typical structure of galactic cirrus in the few pc to a few times 10 pc scale. Large chains and super-loops are seen. Most of our FIR defined loops can be identified on NIR reddening maps as well.

How do the distribution of low mass YSOs compare to the pattern drawn by the projected image of interfaces of the dense cloudy ISM and the low density intercloud medium? We tested it in the outer galaxy ($-30 < b < 30$, $90 < l < 270$). The ISM column density, as derived from the IRAS based FIR optical depth (Schlegel *et al.* 1998) is shown in Figure 1 with all 2MASS point sources over-plotted which have classical TTauri-like (CTT) NIR colours and a good photometry.

A remarkable correlation is seen between the CTT number density and reddening in Orion, i.e. the CTTs follow the loop structure. A careful statistical comparison of the CTT distribution and FIR loops showed that the on-loop CTT number density $N(\text{CTT})$ is highly non-random, where $N(\text{CTT})$ is the number of CTTs seen projected on the loop shell divided by the loop shell area. In order to account for the ISM column density variations we divided $N(\text{CTT})$ by the average NIR based A_V extinction of the loop shell. The observed average $N(\text{CTT})/A_V$ value of 0.2 is far from the results of our Monte Carlo simulations. That we understand as a statistically proved excess star formation on the shells. This suggests triggered star formation. Further details are given in Kiss, Tóth & Elmegreen 2006).

3. Conclusion and outlook

Triggering mechanisms, which may induce, hasten or at least modify the low mass star formation process, are well known. We pointed out that cloud restructuring and YSO distribution indicate an overall importance of triggering in low mass star formation. The low mass clouds do form spontaneously and some of them would indeed condense in gravitationally bound cores. However the opaque, and thus well cooling, and so easily unstable cores form more likely at the outskirts of GMCs (Cepheus) or in filaments (Taurus, Ophiuchus). Those are the locations where the clouds are well exposed to external effects, and triggering acts. It would be important to have an all-sky deep and unbiased census of the nearby low mass young and very young stars. Recent surveys by ISO or Spitzer confined the search for the low mass YSOs to a few star forming regions only.

The ASTRO-F (Akari) (see e.g. Shibai 2005) is in duty scanning the sky like IRAS (Beichman *et al.* 1988) did, but with more spectral filters with at least a factor of 10 better sensitivity and resolution. We expect to see a new all-sky catalogue of infrared point sources. As a part of the Star Formation Mission Programme also the question of low mass star formation triggering will be revisited.

Acknowledgements

This work was partly supported by the OTKA grants No. T-043773, and by the NKTH "Öveges József Grant" TothLVsp.

References

Ballesteros-Paredes, J., Hartmann, L. & Vazquez-Semadeni, E. 1999, *ApJ* 527, 285
Beichman C., Neugebauer, G., Habing, H.J., *et al.* (eds.) 1988, *Infrared astronomical satellite (IRAS) Catalogs and Atlases, Vol. 1, Explanatory Supplement*
Cameron, A.G.W. & Truran,J.W. 1977, *Icar.* 30, 447
Chevalier, R.A. 2000, *ApJ* 538, L151
Elmegreen, B.G. 1998, *Publ. Astron. Soc. Aust.* 15, 74
Grenier, I.A., *et al.* 1989, *ApJ* 347, 231
Hester, J.J., *et al.* 2004, *Science* 304, 1116
Hubble, E. 1934, *ApJ* 79, 8
Jones,C.E. & Basu,S. 2002, *ApJ* 569, 280
Kiss Cs., Moór, A. & Tóth, L.V. 2004, *A&A* 418, 131
Kiss, Z.T., Tóth, L.V. & Krause, O. 2006, *A&A* 453, 923
Kiss, Z.T., Tóth , L.V., Elmegreen, B.G., *et al.* 2006, *submitted to ApJ*
Könyves, V., *et al.* 2006, *A&A in press*
Lebrun, F. 1986, *ApJ* 306, 16
Lee,T., Papanastassiou,D.A. & Wasserburg,G.J. 1977, *ApJ* 211, 107
Lesaffre, P., *et al.* 2006, *A&A* 443, 961
Miao, J., White, G.J., Nelson, R., *et al.* 2006, *MNRAS* 369, 143
Olano ,C.A., Meschin,P.I. & Niemela,V.S. 2006, *MNRAS* 369, 867
Schlegel, D.J., Finkbeiner, D.P. & Davis, M. 1998, *ApJ* 500, 525
Shibai,H. 2005, *IAUS* 216, 347
Urquhart, J.S., *et al.* 2006, *A&A* 450, 625
Ward-Thompson, D., *et al.* 2006, *MNRAS* 369, 1201
Wilking, B.A., *et al.* 2005, *AJ* 130, 1733

Discussion

SHORE: (1.) Is it possible (plausible) that the star forming clouds may be compressed (pre-existing) clouds while the other lower density regions may be formed by dynamics and thermal instability (transient) effects along the shell? (2.) this problem is another reason for going to a LOFAR/SKA type survey – to look at these objects at low frequency.

TOTH: In my view, pre-existing clouds are being re-structured by external effects which turn those to star-forming regions faster on the shells or loops than the clouds would evolve in a spontaneous process inside the relaxed regions. Without being exposed to winds and ionising radiation of the inter-GMC space, low mass star formation is slower and so less effective in the clouds. High resolution (below 0.1 pc) large area surveys of the diffuse ISM will help us understand the physics of transient density fluctuations and their relation to virialized clouds.

Triggered Star Formation in a Turbulent ISM
Proceedings IAU Symposium No. 237, 2006
B. G. Elmegreen & J. Palouš, eds.

© 2007 International Astronomical Union
doi:10.1017/S1743921307001342

NANTEN observations of triggered star formation: from H II regions to galaxy collisions

N. Mizuno[1], A. Kawamura[1], T. Onishi[1], A. Mizuno[2], and Y. Fukui[1]

[1]Department of Astrophysics, Nagoya University, Chikusa-ku, Nagoya 464-8602, JAPAN
email:norikazu@a.phys.nagoya-u.ac.jp

[2]Solar-Terrestrial Environment Laboratory, Nagoya University,
Furo-cho, Chikusa-ku, Nagoya 464-8601, JAPAN

Abstract. In this contribution, we will overview the NANTEN observations of molecular clouds faced to H II regions, supershells, and interacting galaxies, which demonstrate that star/molecular cloud formation is being triggered by young OB associations, supershells, and collisions between galaxies. The large volume filling factor of explosive events like supernovae, ultraviolet radiation fields and stellar winds of massive stars suggest that most of the interstellar medium has been agitated by such strong impacts and triggered star formation is a common event at all scales from small molecular clouds to large galaxy-galaxy mergers. The consequence is the increase of star formation efficiency in many cases, and that more massive stars or clusters of more member stars tend to be formed by triggering than in spontaneous star formation.

Keywords. stars: formation, ISM: bubbles, ISM: clouds, radio lines: ISM, Magellanic Clouds

1. Introduction

The basic process of star formation is gravitational collapse of an interstellar cloud core that mainly consists of molecular hydrogen. This collapse can be accelerated if the core is exposed to high external pressure. High pressure can be caused by the various reasons; ultraviolet photons or stellar winds of OB stars, supernovae, large relative motions between clouds, the galactic density waves and dynamical interactions between galaxies. Most of these actions enhance the star formation efficiency, and influence the evolution of galaxies.

2. Triggering by H II regions

The OB associations are formed in giant molecular clouds of typically 10^5 $M_\odot$ and UV photons of OB stars produce regions of fully ionized hydrogen, H II regions. The H II regions drive the ionization-shock front to compress the molecular clouds into post-shock layers that become gravitationally unstable to form dense fragments. The density increase so-induced should lead to more efficient cooling of the molecular gas since the radiative cooling rate is proportional to $(\text{density})^2$ and these fragments likely lead to star formation when the gravity overcomes the internal pressure. Cloud-cloud collisions in the arm or molecular outflows may also work in triggering to some minor effect. The prevailing occurrence and time-wise persistence make H II regions as important in triggering, although. It has been known that OB associations as well as H II regions and molecular clouds are organized in a galactic spiral pattern, although it is not clear if the spiral arms are enhancing star formation efficiency. We shall focus on the interface between H II regions and molecular clouds to see how triggering is working under the effects of H II regions. A survey for dense molecular gas interacting with H II regions has been made towards 23 southern H II regions within 4 kpc of the sun over 40% of the galactic longitude, from $l = 230°$ to $l = 20°$, including the galactic center (Yamaguchi *et al.* 1999). This study detected 57 molecular clouds of density around 10^3 cm^{-3} whose average mass is about 1,000 M$_\odot$. They show broad velocity dispersion of 5-10 km s^{-1} typical of interacting clouds and the clouds are forming stars as represented by more than 120 associated protostellar FIR sources.

This dataset was used to estimate how star formation is different between the molecular gas interacting with H II regions and the rest of the molecular cloud. The interacting region in a cloud was chosen by dividing the cloud into the two regions; the region adjacent to the H II regions, the interacting region, and the rest. This analysis for the 57 clouds leads to findings that the number of protostellar sources is by a factor of 2 increased in the interacting regions compared to the rest, and that these sources tend to be ten times more luminous than those in the rest. These differences are likely caused by the effects of the H II regions, and cannot be ascribed to the molecular properties since the molecular column density and mass are not significantly different between the both regions. The FIR luminosity of the protostellar sources is significantly enhanced in the interacting regions than in the remaining; e.g., the average FIR luminosity is $\sim 10^4$ L$_\odot$ and $\sim 10^3$ L$_\odot$ for the interacting regions and for the rest, respectively, at a cloud mass of $\sim$10 M$_\odot$. Star formation efficiency defined as a ratio between the formed stars and the combined total mass of the clouds and stars is then estimated to be a factor of 3.5 enhanced in the interacting regions if typical stellar mass-to-luminosity relation is assumed. The total star formation rate is 4 M$_\odot$/yr over the Galaxy (e.g., McKee & Williams 1997). The H II-triggered star formation may account for about 10-30% of this value after corrected for low-mass stars by using a Salpeter IMF. Enhanced massive star formation towards H II regions in fact appears consistent with the flatter initial mass function of galactic OB associations within 2 kpc of the sun for the high-mass part of the IMF (Massey *et al.* 1995).

3. Triggering by supershells

OB stars trigger star formation when they are young and close to parental molecular gas over a time scale of 1 Myr. After this phase, the massive members whose mass is greater than 8 M$_\odot$ in an OB association evolve to a supernova which releases total energy of $\sim 10^{51}$ ergs at an explosion, and can be very effective in accelerating interstellar material at scales of 100–1000 pc. The accumulated mass due to the multiple blast waves and the pressure gradient of the hot gas from supernovae form a nearly spheroidal shell of dense interstellar matter that becomes gravitationally unstable to form stars in them (Palouš & Ehlerova 2000). A typical lifetime of a supershell is up to 30 Myrs, significantly longer than that of H II triggering and after that it may become difficult to identify a spherical shape with no further energy input. Supershell is not a separate phenomenon from H II-triggering phase but is a continuing process subsequent to it. The two phases can be even coexistent as in case of the OB associations in Orion. Figure 1 shows a large field in the Orion-Eridanus region including an extended supershell, the Orion-Eridanus superbubble, one of the typical supershells (superbubbles) in the solar vicinity. It is however to be noted that the role of these shells in triggering star formation was not clear until recently since the expanding gas is all of low density having gas column density of $\sim 10^{20}$ cm^{-2} that are not related to active star formation. The Orion-Eridanus superbubble either shows very little sign of triggered star formation except for a few small molecular clumps (e.g., Kun *et al.* 2001), making a sharp contrast to the active

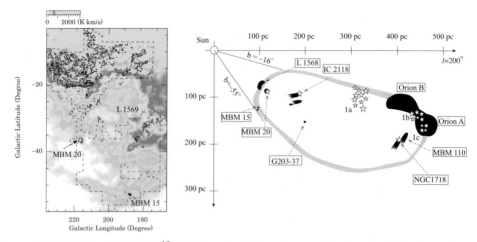

Figure 1. *Left*) The NANTEN ^{12}CO integrated intensity map (contour) and H I intensity map (pseudo-color) taken from Leiden/Dwingeloo Atlas of galactic neutral hydrogen (Hartmann & Burton 1997). *Right*) Schematic cross sections of the Orion-Eridanus superbubble. The x-axis indicates the Galactic plane, and the y-axis indicates the distance from the Galactic plane.

massive star formation in the Orion molecular clouds themselves where star formation is predominantly influenced by the H II regions.

A recent discovery of the Carina Flare is the first case which exhibits massive star formation triggered by a supershell in the Galaxy (Fukui *et al.* 1999, Dawson *et al.* 2006). Triggering in the Carina Flare naturally invokes a question, how the star formation efficiency of supershells is significant all over the Galaxy. Many supershells appear not triggering star formation. The previous sample may be biased towards low density features, a natural consequence of the past extensive usage of the H I data, Hα emission and the FIR dust emission. We need to observe dense molecular gas that is directly connected to on-going star formation over a reasonable large height (z) coverage in order to better understand the impact of supershells. Subsequently, a first attempt to search for molecular supershells in higher resolution with NANTEN has revealed 9 shells including the Carina Flare, 7 of which are new discoveries, where the identification was made primarily based on the arm-like morphology of the molecular gas. Their radius ranges from 50 pc to 230 pc, with ages between 2×10^6 yrs to 1×10^7 yrs (Matsunaga *et al.* 2001). The area covered corresponds to 1/5 of the galactic disk within a distance of $\sim$4 kpc. The 8 molecular supershells are positionally well correlated with galactic spiral arms at a frequency of about $\sim$0.4 kpc^{-2} in the galactic disk. The association with the spiral arms is as expected from the relation with OB associations, while the frequency may be somewhat smaller than expected, implying that only a fraction of OB associations surrounded by rich H I envelope may be able to form molecular supershells. The total mass included in these molecular supershells at z above 150 pc, $\sim$10^{6}M$_\odot$, is very small, less than 1%, compared to 15% of the molecular mass included in the disk of the surveyed area, suggesting that only a small fraction of the high z gas can be identified morphologically as a supershell, probably because the arc-like shape of a supershell is soon lost after $\sim$10 Myrs due to differential rotation etc. It is also to be noted that the identification of associated stars becomes very difficult at distances larger than 2 kpc near the galactic plane, making it hard to study second-generation stars of triggered formation. More insights into the role of supershells are to be obtained in the nearby galaxies which are not heavily obscured. In the Large Magellanic Cloud (LMC) having no clear strong spiral patterns the super(giant) shells may be more important in forming clouds and stars (Fukui 2006, Kawamura *et al.* 2006).

The other relevant aspect is the role of shells in formation of very massive stars and/or rich stellar clusters; The η Carina cluster is where unusually massive stars in the Milky Way may have been formed under triggering of a supershell. η Carina is located just at the same distance and at the same longitude of the Carina Flare. Considering that η Carina cluster is 3-5 $\times 10^6$ yrs old, it is suggested that triggering by the Carina Flare shell may have played a role to form this cluster by effectively collecting gas over a large volume into a small space which was never possible without the action of a supershell. Similarly, the 30 Doradus region in the LMC is well known for its outstanding nature of a massive and rich stellar cluster R136 and is apparently sandwiched by two super giant shells (LMC2 and LMC3). These suggest that super(giant) shells may play a role in the formation of very massive stars and the richest clusters in the system.

4. Triggering by galaxy collisions

Galaxy-galaxy interactions or collisions between galaxies can significantly enhance the rates of star formation. Recently, we found the molecular clouds in the nearest and brightest tidal structure: the Magellanic Bridge (Mizuno *et al.* 2006). We suggest that CO clouds are formed *after* the tidal encounter, rather than being extracted from the SMC. This is supported by the small typical lifetime of CO clouds which is as short as $\sim 10^7$ yrs, and much less than the estimated 200 Myr age of the Bridge itself. These observations have shown that not only is star formation on-going within the Bridge, but that it is quite widespread throughout its kilo-parsec length. Further efforts to probe the spatial extents of the CO emission regions, and to better establish association/cluster properties and star formation will be extremely valuable. The Magellanic Bridge provides an outstanding opportunity to probe the weakly-understood process of star formation in the interacting system, it is difficult to study to a useful level of detail at distant galaxies.

Acknowledgements

The NANTEN project is based on the mutual agreement between Nagoya University and the Carnegie Institution of Washington. We appreciate the hospitality of all the staff member of the Las Campanas Observatory. We also proudly mention that this project has been realized by contributions from many Japanese public donators and companies.

References

Dawson, J., *et al.* 2006, this volume
Fukui, Y., *et al.* 1999, *PASJ* 51, 751
Fukui, Y. 2006, this volume
Hartmann, D. & Burton, W. B. 1997, *Atlas of Galactic Neutral Hydrogen*, (Cambridge: Cambridge University Press)
Kawamura A., *et al.* 2006, this volume
Kun, M., *et al.* 2001, *PASJ* 53, 1063
Massey, P., Johnson, K. E. & Degioia-Eastwood, K. 1995, *ApJ* 454, 151
Matsunaga, K., *et al.* 2001, *PASJ* 53, 1003
McKee, C. F. & Williams, J. P. 1997, *ApJ* 476, 144
Mizuno, N., Muller, E., Maeda, H., Kawamura, A., Minamidani, T., Onishi, T., Mizuno, A. & Fukui, Y. 2006, *Ap. Lett.* 643, 107
Palouš, J. & Ehlerova, S. 2000, *New Astron. Rev* 44, 363
Yamaguchi, R., *et al.* 1999, *PASJ* 51, 791

Triggered Star Formation in a Turbulent ISM
Proceedings IAU Symposium No. 237, 2006
B. G. Elmegreen & J. Palouš, eds.

© 2007 International Astronomical Union
doi:10.1017/S1743921307001354

Observations of prestellar cores: Probing the initial conditions for the IMF

Philippe André[1]

[1]CEA Saclay, DSM/DAPNIA, Service d'Astrophysique, F-91191 Gif-sur-Yvette Cedex, France
email: pandre@cea.fr

Abstract. Several (sub)millimeter-wave studies of nearby star-forming regions have revealed self-gravitating prestellar condensations that seem to be the direct progenitors of individual stars and whose mass distribution resembles the IMF. In a number of cases, small internal and relative motions have been measured for these condensations, indicating they are much less turbulent than their parent cloud and do not have time to interact before evolving into protostars and pre-main sequence stars. These findings suggest that the IMF is at least partly determined by pre-collapse cloud fragmentation and that one of the keys to understanding the origin of stellar masses lies in the physical mechanisms responsible for the formation and decoupling of prestellar cores within molecular clouds.

Keywords. stars: formation, ISM: clouds, ISM: structure

1. Introduction

One of the main limitations in our present understanding of the star formation process is that we do not know well the initial stages of cloud fragmentation and collapse into protostars. In particular, there is a major ongoing controversy between two schools of thought for the formation and evolution of dense cores within molecular clouds: The classical picture based on magnetic support and ambipolar diffusion (e.g. Shu *et al.* 1987, 2004; Mouschovias & Ciolek 1999) has been seriously challenged by a new, more dynamic picture, which emphasizes the role of supersonic turbulence in supporting clouds on large scales and generating density fluctuations on small scales (e.g. Padoan & Nordlund 2002, Mac Low & Klessen 2004).

Improving our knowledge of the initial stages of star formation is crucially important since there is now good evidence that these stages control the origin of the stellar initial mass function (IMF). Indeed, observations suggest that the effective reservoirs of mass required for the formation of individual stars are already selected at the prestellar core stage. First, detailed (sub)-millimeter emission and infrared absorption mapping of a few nearby sources indicates that the density profiles of prestellar cores typically feature flat inner regions and sharp outer edges, hence are reminiscent of the density structure expected for *finite-size/mass*, self-gravitating isothermal spheroids (such as 'Bonnor-Ebert' spheres) (e.g. Ward-Thompson *et al.* 1994, Bacmann *et al.* 2000, Alves *et al.* 2001). Second, several ground-based (sub)-millimeter continuum surveys of nearby, compact cluster-forming clouds have uncovered 'complete' (but small) samples of prestellar condensations whose mass distributions resemble the stellar IMF (see § 2).

2. Link between the prestellar core mass distribution and the IMF

Wide-field (sub)mm dust continuum mapping is a powerful tool to take a census of dense cores within star-forming clouds. The advent of large-format bolometer arrays on

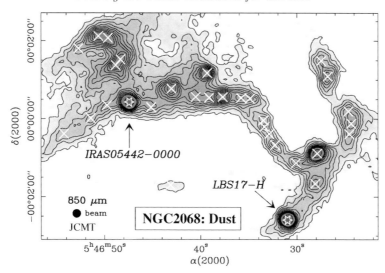

Figure 1. SCUBA 850μm dust continuum map of the NGC 2068 protocluster extracted from the mosaic of NGC 2068/2071 by Motte *et al.* (2001). A total of 30 compact starless condensations (marked by crosses), with masses between $\sim 0.4\,M_\odot$ and $\sim 4.5\,M_\odot$, are detected in this ~ 1 pc $\times$ 0.7 pc field.

(sub)millimeter radiotelescopes such as the IRAM 30m and the JCMT has led to the identification of numerous cold, compact condensations that do not obey the Larson (1981) self-similar scaling laws of molecular clouds and are intermediate in their properties between diffuse CO clumps and infrared young stellar objects (cf. André *et al.* 2000 and Ward-Thompson *et al.* 2006 for reviews). As an example, Fig. 1 shows the condensations found by Motte *et al.* (2001) at 850 μm in the NGC 2068 protocluster (Orion B). Such highly concentrated (sub)millimeter continuum condensations are at least 3 to 6 orders of magnitude denser than typical CO clumps (e.g. Kramer *et al.* 1998) and feature large ($\gg 50\%$) mean column density contrasts over their parent background clouds, strongly suggesting they are self-gravitating. The latter is directly confirmed by line observations in a number of cases (see § 3.1 below). A small fraction of these condensations lie at the base of powerful jet-like outflows and correspond to very young protostars which have not yet accreted the majority of their final masses (Class 0 objects – cf. André *et al.* 2000). However, the majority of them are starless/jetless and appear to be the immediate *prestellar* progenitors of individual protostars or protostellar systems.

In particular, the mass distribution of these prestellar condensations is remarkably similar to the stellar IMF (e.g. Motte, André, Neri 1998 – hereafter MAN98). This is illustrated in Fig. 2 which shows the cumulative mass spectrum of the 57 starless condensations found by MAN98 in their 1.2 mm continuum survey of the ρ Ophiuchi main cloud with the IRAM 30m telescope. These condensations, which were identified using a multi-resolution wavelet analysis (cf. Starck *et al.* 1998), are seen *on the same spatial scales as protostellar envelopes* (i.e., ~ 2500–5000 AU or $\sim 15'' - 30''$ in ρ Oph). Their mass spectrum is consistent with the Salpeter power-law IMF at the high-mass end and shows a tentative flattening below $\sim 0.4\,M_\odot$ (see Fig. 2). The latter is reminiscent of the break observed in the IMF of field stars at $\sim 0.5\,M_\odot$ (e.g. Kroupa 2001, Chabrier 2003), also present in the mass function of ρ Oph pre-main sequence objects (Bontemps *et al.* 2001 – see star symbols in Fig. 2). If real, this flattening occurs at a mass comparable

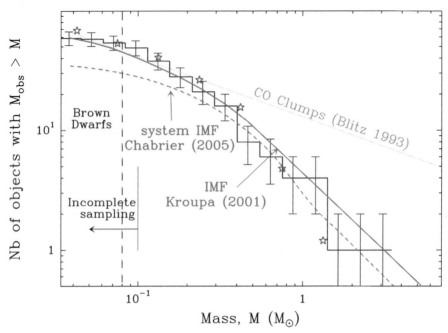

Figure 2. Cumulative mass distribution of a sample of 57 prestellar condensations, complete down to $\sim 0.1\,M_\odot$, in the ρ Oph protocluster (histogram with error bars – from MAN98). For comparison, the solid curve shows the shape of the field star IMF (e.g. Kroupa 2001), while the dashed curve corresponds to the IMF of multiple systems (e.g. Chabrier 2005). The star markers represent the mass function of ρ Oph (primary) pre-main sequence objects as derived from a mid-IR survey with ISOCAM (Bontemps *et al.* 2001). The dotted line shows a $N(> M) \propto M^{-0.6}$ power-law distribution corresponding to the typical mass spectrum found for CO clumps – see Blitz 1993 and Kramer *et al.* 1998).

to the typical Jeans mass in the dense ($n_{\mathrm{H2}} \sim 10^5\,\mathrm{cm}^{-3}$) DCO$^+$ cores of the ρ Oph cloud (cf. Loren *et al.* 1990). Interestingly, the observed mean separation between ρ Oph condensations (~ 6000 AU) is also consistent with the Jeans length at the same density.

Such a close resemblance of their mass spectrum to the IMF in both shape and mass scale suggests that *the starless condensations identified in the (sub)millimeter dust continuum are about to form stars on a one-to-one basis, with a high local efficiency*, i.e., $M_\star/M_{pre} \gtrsim 50\%$. This strongly supports scenarios according to which the bulk of the IMF is at least partly determined by pre-collapse cloud fragmentation (e.g. Larson 1985, 2005; Elmegreen 1997; Padoan & Nordlund 2002).

The results of MAN98 in ρ Oph have been essentially confirmed by independent 850 μm SCUBA and 1.2 mm SIMBA surveys of the same region with JCMT and SEST (Johnstone *et al.* 2000, and Stanke *et al.* 2006, respectively). Similar results have also been found in other nearby cluster-forming clouds such as Serpens (Testi & Sargent 1998), Orion B (Motte *et al.* 2001), and Perseus (Enoch *et al.* 2006). Furthermore, in a recent near-IR extinction imaging study of the Pipe dark cloud complex, Alves *et al.* (2006) also found a dense core mass function closely following the shape of the IMF.

Appealing as a direct connection between the prestellar core mass distribution (CMD) and the IMF might be, several caveats should be kept in mind. First, although core mass estimates based on optically thin (sub)millimeter dust continuum emission are straightforward, they rely on uncertain assumptions about the *dust (temperature and emissivity) properties*. Second, current determinations of the CMD are limited by small-number statistics in any given cloud and may be affected by incompleteness at the low-mass end. With *Herschel*, the future submillimeter space telescope to be launched by ESA in 2008, it will be possible to dramatically improve on the statistics and to largely eliminate the mass uncertainties through direct measurements of the dust temperatures (cf. André & Saraceno 2005). Last but not least, the shape of the CMD agrees better with the IMF of individual field stars (solid curve in Fig. 2) than with the IMF of multiple systems (dashed curve in Fig. 2). This is surprising since current surveys for prestellar cores do not have enough spatial resolution to probe core multiplicity. Furthermore, multiple systems are believed to form *after* the prestellar stage by subsequent dynamical fragmentation during the collapse phase, close to the time of protostar formation (e.g. Goodwin *et al.* 2006). Thus, one would expect the masses of prestellar cores to be more directly related to the masses of multiple systems than to the masses of individual stars. It is possible that a fraction at least of the cores observed below $\sim 0.2-0.3\,M_\odot$ (cf. Fig. 2) are not gravitationally bound, hence not prestellar in nature (but see § 3.1 below). This also implies that other processes besides pure cloud fragmentation are not ruled out and may play an important additional role in generating the low-mass ($M < 0.3\,M_\odot$) end of the IMF (cf. Bate *et al.* 2003, Ballesteros-Paredes *et al.* 2006).

3. Kinematics of protocluster condensations

Investigating the dynamical properties of the prestellar condensations identified in submillimeter dust continuum surveys is of great interest to discriminate between possible theoretical scenarios for core formation and evolution. Interesting results have emerged from recent molecular line studies using tracers such as NH_3, N_2H^+, N_2D^+, and DCO^+, which do not deplete onto dust grains until fairly high densities (e.g. Tafalla *et al.* 2002, Crapsi *et al.* 2005).

3.1. Internal motions

The small-scale (~ 0.03 pc) prestellar condensations observed in the Ophiuchus, Serpens, Perseus, and Orion cluster-forming regions (cf. Fig. 1) are characterized by fairly narrow ($\Delta V_{FWHM} \lesssim 0.5$ km s^{-1}) line widths in optically thin tracers of dense gas such as N_2H^+(1-0) (see, e.g., Fig. 3 and Belloche *et al.* 2001). For instance, the typical nonthermal velocity dispersion observed toward the starless condensations of the ρ Oph protocluster is about half the thermal velocity dispersion of H_2 ($\sigma_{NT}/\sigma_T \sim 0.7$ – Belloche *et al.* 2001). This indicates subsonic or at most transonic levels of internal turbulence and suggests that the initial conditions for individual protostellar collapse are relatively free of turbulence ($\sigma_{NT} < \sigma_T \sim 0.2$ km s^{-1}), even when the parent cluster-forming clouds/cores have supersonic levels of turbulence ($\sigma_{NT} \gtrsim 0.4$ km s^{-1} – cf. Loren *et al.* 1990; Jijina *et al.* 1999). These findings are in qualitative agreement with the 'kernel' picture proposed by Myers (1998), according to which protocluster condensations correspond to zones of minimum turbulence, of size comparable to the cutoff wavelength for MHD waves, developing in turbulent cloud cores.

Importantly, the narrow linewidths measured in N_2H^+(1-0) imply virial masses which generally agree within a factor of ~ 2 with the mass estimates derived from the dust continuum. This confirms that most of the starless condensations identified in the (sub)mm continuum are self-gravitating and very likely *prestellar* in nature. For instance, based on comprehensive N_2H^+(1-0) observations of the ρ Oph protocluster, André *et al.* (2006) conclude that more than $\sim 70\%$ of the starless condensations found by MAN98 at 1.2 mm,

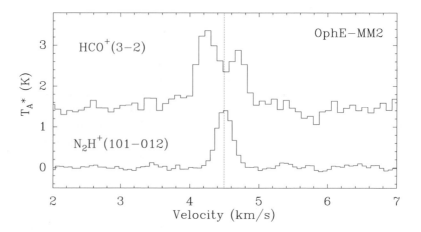

Figure 3. HCO^+(3–2) and N_2H^+(101-012) spectra observed at the IRAM 30 m telescope toward the starless 1.2 mm continuum condensation E-MM2 identified by MAN98 in the ρ Oph protocluster. The optically thick HCO^+ line is self-absorbed and skewed to the blue, which is the classical signature of collapse motions (e.g. Evans 1999), while the optically thin N_2H^+ line is narrow ($\Delta V \lesssim 0.3$ km s^{-1}) indicating a small level of turbulence. (From Belloche *et al.* 2001.)

and essentially all of those more massive than $\sim 0.1\, M_\odot$, are gravitationally bound. The status of the ρ Oph condensations less massive than $\sim 0.1\, M_\odot$ is less clear since these are weaker and often undetected in N_2H^+. Some of them may possibly be unbound transient objects (see also end of § 2 above).

The notion that a majority of the starless condensations identified in the (sub)millimeter dust continuum are prestellar and on the verge of forming protostars is further supported by the detection of infall motions toward some of them. As an example, toward OphE-MM2 in the ρ Oph protocluster, the optically thick HCO^+(3–2) line exhibits a self-absorbed, double-peaked profile with a blue peak stronger than the red peak, while low optical depth lines such as N_2H^+(101–012) are single-peaked and peak in the absorption dip of HCO^+(3–2) (see Fig. 3). This type of blue asymmetry in optically thick line tracers is now accepted as a classical spectroscopic signature of collapse (cf. Evans 1999). The infall speeds derived from radiative transfer modeling are $\sim 0.1 - 0.3$ km s^{-1} (e.g. Belloche *et al.* 2001), consistent with a typical condensation lifetime $\sim 10^5$ yr.

3.2. *Relative motions*

Line observations can also provide information on the relative motions between condensations, as well as on possible global, large-scale motions in the parent protoclusters.

For instance, André *et al.* (2006) have analyzed the distribution of line-of-sight velocities observed toward 45 condensations of the ρ Oph protocluster (see also Belloche *et al.* 2001). The results indicate a global, one-dimensional velocity dispersion $\sigma_{1D} \sim 0.37$ km s^{-1} about the ρ Oph mean systemic velocity. Assuming isotropic random motions, this corresponds to a three-dimensional core–core velocity dispersion $\sigma_{3D} \sim 0.64$ km s^{-1}. With a ρ Oph central cloud diameter of ~ 1.1 pc, such a small velocity dispersion implies a typical crossing time $D/\sigma_{3D} \sim 1.7 \times 10^6$ yr. The crossing times determined for the individual DCO^+ cores or subclusters of ρ Oph are only slightly shorter ($\sim 0.6 \times 10^6$ yr).

Since neither the age of the embedded IR cluster nor the lifetime of the 1.2 mm condensations can be much larger than 10^6 yr (cf. Bontemps *et al.* 2001 and § 3.1 above), it appears that *the ρ Oph pre-stellar condensations do not have time to orbit through the protocluster gas and collide with one another* (even inside individual subclusters) before evolving into pre-main sequence young stars (André *et al.* 2006). (This would require several crossing times – cf. Elmegreen 2001.) Similar results have been obtained for 25 condensations in the NGC 2068 protocluster (Belloche *et al.* in prep.).

While dynamical interactions between protostellar condensations may sometimes occur at the very center of massive, collapsing protoclusters (cf. Peretto *et al.* 2006), the results obtained in ρ Oph and NGC 2068 seem inconsistent with models which resort to interactions and competitive accretion to build up the bulk of the stellar IMF at the protostellar stage (e.g. Bate *et al.* 2003). As the estimated tidal-lobe radius of the ρ Oph condensations is comparable to their observed radius $\lesssim 5000$ AU (cf. MAN98), competitive accretion at the *prestellar* stage may nevertheless play a role in limiting the condensation masses (see Bonnell *et al.* 2001).

4. Conclusions and future prospects

The mass distribution of prestellar cores in star-forming clouds appears to be consistent with the stellar IMF between $\sim 0.1\,M_\odot$ and $\sim 5\,M_\odot$, although large uncertainties remain especially at the low- and high-mass ends (cf. § 2). Small internal and relative motions are measured for these prestellar cores, implying that they are much less turbulent than their parent cloud and generally do not have time to interact before collapsing to (proto)stars (cf. § 3). Taken at face value, these results are at variance with models in which dynamical interactions and competitive accretion play a key role in shaping the distribution of stellar masses (e.g. Bate *et al.* 2003). They strongly support scenarios according to which the IMF is largely determined at the prestellar stage and individual protostellar collapse is initiated in decoupled self-gravitating condensations resulting from turbulent (e.g. Padoan & Nordlund 2002) and/or magnetic (e.g. Shu *et al.* 2004) cloud fragmentation.

In cluster-forming clouds like ρ Oph, the star formation efficiency within each condensation is high (cf. § 2): most (> 50%) of the initial mass at the onset of collapse seems to end up in a star (or a stellar system). There is some evidence that cluster-forming clumps are in a state of global collapse induced by external triggers (e.g. Peretto *et al.* 2006, Nutter *et al.* 2006). A promising mechanism in this case is purely gravitational, Jeans-like fragmentation of compressed cloud layers or thin expanding shells, a process which can produce a fragment mass spectrum approaching the Salpeter IMF (Palouš *et al.* 2003; Palouš, this volume).

In regions of more distributed star formation such as Taurus or the Pipe Nebula, the size of individual cores is larger (e.g. MAN98), the local star formation efficiency is lower ($\sim$ 15%–30% – Onishi *et al.* 2002, Alves *et al.* 2006), and the feedback of protostellar outflows may be more important in limiting accretion and defining stellar masses (e.g. Shu *et al.* 2004).

To fully understand how the IMF comes about, it is crucial to further investigate the processes by which prestellar cores form and evolve in molecular clouds. With present submillimeter instrumentation, observational studies are limited by small-number statistics and restricted to the nearest regions. The advent of major new facilities at the end of the present decade should yield several breakthroughs in this field. With an angular resolution at 75–300 μm comparable to, or better than, the largest ground-based millimeter-wave radiotelescopes, *Herschel*, the Far InfraRed and Submillimeter Telescope to be launched by ESA in 2008 (cf. Pilbratt 2005), will make possible complete surveys for prestellar cores down to the proto-brown dwarf regime in the cloud complexes of the Gould Belt (cf. André & Saraceno 2005). High-resolution $(0.01'' - 0.1'')$ imaging with the 'Atacama Large Millimeter Array' (ALMA, becoming partly available around 2009, fully operational in 2013 – cf. Wootten 2001) at $\sim 450~\mu m - 3$ mm will beat source confusion and allow us to probe individual condensations in distant, massive protoclusters, all the way to the Galactic center and the Magellanic clouds. Complementing each other nicely, *Herschel* and ALMA will tremendously improve our global understanding of the initial stages of star formation in the Galaxy.

References

Alves, J.F., Lada, C.J. & Lada, E.A. 2001, *Nature* 409, 159

Alves, J.F., Lombardi, M. & Lada, C.J. 2006, submitted

André, Ph., Belloche, A., Motte, F. & Peretto, N. 2006, in preparation

André, Ph. & Saraceno, P. 2005, in: A. Wilson (ed.), *The Dusty and Molecular Universe: A Prelude to Herschel and ALMA* (ESA SP), 577, p. 179

André, P., Ward-Thompson, D. & Barsony, M. 2000, in: V. Mannings, A.P. Boss & S.S. Russell (eds.), *Protostars and Planets IV* (Tucson: Univ. of Arizona), p. 59

Bacmann, A., André, P., Puget, J.-L., Abergel, A., Bontemps, S. & Ward-Thompson, D. 2000, *A&A* 361, 555

Ballesteros-Paredes, J., Gazol, A., Kim, J., Klessen, R.S. *et al.* 2006, *ApJ* 637, 384

Bate, M.R., Bonnell, I.A. & Bromm, V. 2003, *MNRAS* 339, 577

Belloche, A., André, P. & Motte, F. 2001, in: T. Montmerle & P. André (eds.), *From Darkness to Light* (ASP-CS), 243, p. 313

Blitz, L. 1993, in: E.H. Levy & J.I. Lunine (eds.), *Protostars & Planets III* (Tucson: Univ. of Arizona), p. 125

Bonnell, I.A., Bate, M. R., Clarke, C. J. & Pringle, J.E. 2001, *MNRAS* 323, 785

Bontemps, S., André, P., Kaas, A.A. *et al.* 2001, *A&A* 372, 173

Chabrier, G. 2003, *ApJ* 586, L133

Chabrier, G. 2005, in: E. Corbelli *et al.* (eds.), *The Initial Mass Function 50 years later* (Dordrecht: Springer), p.41

Crapsi, A., Caselli, P., Walmsley, C.M., Myers, P.C. *et al.* 2005, *ApJ* 619, 379

Elmegreen, B. 1997, *ApJ* 486, 944

Elmegreen, B. 2001, in: T. Montmerle & P. André (eds.), *From Darkness to Light* (ASP-CS), 243, p. 255

Enoch, M.L., Young, K.E., Glenn, J., Evans, N.J. *et al.* 2006, *ApJ* 638, 293

Evans, N.J. 1999, *ARAA* 37, 311

Goodwin, S.P., Kroupa, P., Goodman, A. & Burkert, A. 2006, in: B. Reipurth, D. Jewitt, K. Keil (eds.), *Protostars & Planets V*, in press

Jijina, J., Myers, P.C. & Adams, F.C. 1999, *ApJS* 125, 161

Johnstone, D., Wilson, C. D., Moriarty-Schieven, G., *et al.* 2000, *ApJ* 545, 327

Kramer, C., Stutzki, J., Rohrig, R. & Corneliussen, U. 1998, *A&A* 329, 249

Kroupa, P. 2001, *MNRAS* 322, 231

Larson, R.B. 1981, *MNRAS* 194, 809

Larson, R.B. 1985, *MNRAS* 214, 379

Larson, R.B. 2005, *MNRAS* 359, 211

Loren, R.B., Wootten, A. & Wilking, B.A. 1990, *ApJ* 365, 229

Mac Low, M.-M. & Klessen, R.S. 2004, *Rev. Mod. Phys.* 76, 125

Motte, F., André, P. & Neri, R. 1998, *A&A* 336, 150 – MAN98

Motte, F., André, P. & Ward-Thompson, D. & Bontemps, S. 2001, *A&A* 372, L41

Mouschovias, T.M. & Ciolek, G.E. 1999, in: C.J. Lada & N.D. Kylafis (eds.), *The Origin of Stars and Planetary Systems* (Klewer), p. 305

Myers, P.C. 1998, *ApJL* 496, L109

Nutter, D., Ward-Thompson, D. & André, P. 2006, *MNRAS* 368, 1833

Onishi, T., Mizuno, A., Kawamura, A., Tachihara, K. & Fukui, Y. 2002, *ApJ* 575, 950

Padoan, P. & Nordlund, A. 2002, *ApJ* 576, 870

Palouš, J., Ehlerova, S. & Wünsch, R. 2003, *Ap&SS* 284, 873

Peretto, N., André, Ph. & Belloche, A., 2006, *A&A* 445, 979

Pilbratt, G. 2005, in: A. Wilson (ed.), *The Dusty and Molecular Universe: A Prelude to Herschel and ALMA* (ESA-SP), 577, p. 3

Shu, F.H., Adams, F.C. & Lizano, S. 1987, *ARAA* 25, 23

Shu, F.H., Li, Z.-Y. & Allen, A. 2004, *ApJ* 601, 930

Stanke, T, Smith, M.D., Gredel, R. & Khanzadyan, T. 2006, *A&A* 447, 609

Starck, J.L., Murtagh, F. & Bijaoui, A. 1998, *Image Processing and Data Analysis: The Multiscale Approach* (Cambridge: Cambridge University)

Tafalla, M., Myers, P.C., Caselli, P., Walmsley, C.M. & Comito, C. 2002, *ApJ* 569, 815

Testi, L. & Sargent, A.I. 1998, *ApJL* 508, L91

Ward-Thompson, D., André, P., Crutcher, R. *et al.* 2006, in: B. Reipurth, D. Jewitt & K. Keil (eds.), *Protostars & Planets V*, in press

Ward-Thompson, D., Scott, P.F., Hills & R.E., André, P. 1994, *MNRAS* 268, 276

Wootten, A. 2001, *Science with the Atacama Large Millimeter Array* (ASP-CS), vol. 235

Discussion

ELMEGREEN: You have sampled several regions now where the cloud masses differ and I wonder if the turnover mass, which might be identified with M_J, varies in a systematic way?

ANDRE: A fairly large number of regions have indeed been surveyed in the (sub)millimeter continuum by now, and a possible correlation between the break point in the core mass spectrum and parent cloud density or Jeans mass is an effect we have looked for. The most obvious correlation, however, is not with cloud density but with cloud distance, suggesting the break may be partially due to incompleteness at the low-mass end of the core mass spectrum. For instance, Reid & Wilson (2005, 2006) recently surveyed massive star-forming regions such as M17 at $d = 1.6$ kpc and found a turnover at ~ 4 $M_\odot$ instead of our values of ~ 0.3 $M_\odot$ in ρOph and ~ 0.9 $M_\odot$ in Orion. On the other hand, the high-mass end of the core mass spectrum seems fairly robust.

MACLOW: How do your results compare to the much shallower spectrum found by Velusamy *et al.* in a poster at this meeting?

ANDRE: I must admit that I didn't know the study by Velusamy *et al.* you are referring to. One difference is that they observed at $\lambda = 350$ μm, closer to the peak of the core SEDs, i.e., at a wavelength where temperature effects become important. Our observations were done at $\lambda = 1.2$ mm or $\lambda = 800$ μm, i.e., in the Rayleigh-Jeans regime of the SEDs where in principle we have a more direct proportionality between measured flux and core mass. My other comment is that one has to be very careful when extracting/identifying potential prestellar cores in sub-millimeter star continuum maps. Our team has been using a multi-resolution wavelet analysis to extract starless cloud fragments seen on the same spatial scales as prestellar envelopes observed on the same maps. This is important because if one counts larger structures also visible in our maps as cores, then a flatter mass spectrum is obtained, but these larger structures are too big to qualify as potential progenitors to individual stars or systems (see Motte *et al.* 1998 and Motte & André 2000 for details).

VELUSAMY: We used 51 cores observed in our 350 μm survey of quiescent cores in Orion region about 30 arcmin to the north and south of Orion KL. They have a power law mass function with slope $\alpha = 0.85 \pm 0.21$ which is significantly flatter. I would like to add a note of caution interpreting the multiple slopes in the cumulative mass spectra you showed. We find use of cumulative mass functions can erroneously suggest multiple power law indices particularly if core mass distribution is characterized by a power law index ~ -1 (Discussed in poster S237-230).

ANDRE: Thank you for pointing your new observations of Orion which I was unaware of. The reason for the difference in the derived core mass spectrum may possibly come from the difference in wavelength (350 μm more sensitive to temperature variations than 850 μm or 1.2 mm) or a difference in the method used to identify the cores from the maps. Concerning your note of caution, I am well aware that the cumulative mass function and the differential mass function both have advantages and disadvantages. I only showed cumulative plots for the sake of brevity in my talk but the core mass function originally published in our ρ Oph paper (Motte *et al.* 1998 A&A) was differential (dN/dM) and was entirely consistent with the broken power law core mass spectrum following the Salpeter slope at the high-mass end.

Triggered Star Formation in a Turbulent ISM
Proceedings IAU Symposium No. 237, 2006
B. G. Elmegreen & J. Palouš, eds.

© 2007 International Astronomical Union
doi:10.1017/S1743921307001366

Magnetic fields and star formation – new observational results

Richard M. Crutcher[1] and Thomas H. Troland[2]

[1]Department of Astronomy, University of Illinois, Urbana, IL 61801, USA
email: crutcher@uiuc.edu

[2]Department of Physics and Astronomy, University of Kentucky, Lexington, KY 40502 USA
email: troland@pa.uky.edu

Abstract. Although the subject of this meeting is triggered star formation in a turbulent interstellar medium, it remains unsettled what role magnetic fields play in the star formation process. This paper briefly reviews star formation model predictions for the ratio of mass to magnetic flux, describes how Zeeman observations can test these predictions, describes new results – an extensive OH Zeeman survey of dark cloud cores with the Arecibo telescope, and discusses the implications. Conclusions are that the new data support and extend the conclusions based on the older observational results – that observational data on magnetic fields in molecular clouds are consistent with the strong magnetic field model of star formation. In addition, the observational data on magnetic field strengths in the interstellar medium strongly suggest that molecular clouds must form primarily by accumulation of matter along field lines. Finally, a future observational project is described that could definitively test the ambipolar diffusion model for the formation of cores and hence of stars.

Keywords. Magnetic fields, stars: formation, ISM: clouds, ISM: magnetic fields

1. Introduction

It has become increasingly clear that cosmic magnetic fields are pervasive, ubiquitous, and likely important in the properties and evolution of almost everything in the Universe, from planets to quasars (e.g., Wielebinski & Beck 2005). One area where the role of magnetic fields is far from being understood is star formation. This Symposium is focused on triggered star formation in a turbulent interstellar medium. However, interstellar magnetic fields may play a significant or even dominant role in the star formation process by delaying the collapse of molecular clouds. The reduction of magnetic support through ambipolar diffusion may be an important triggering mechanism for star formation. It is therefore essential to test whether the predictions of models of star formation that include strong magnetic fields meet observational tests. In this paper we discuss how observations of magnetic fields in molecular clouds can test the strong magnetic field model of star formation, present the results of a major new study of magnetic field strengths in dark molecular clouds, summarize the conclusions about the role of magnetic fields in star formation, and suggest what new observations are necessary in order to answer more definitively the question – what triggers star formation?

2. Strong magnetic fields and star formation theory

Until recently, the prevailing view has been that self-gravitating dense clouds are supported against collapse by magnetic fields (e.g., Mouschovias & Ciolek 1999). However, magnetic fields are frozen only into the ionized gas and dust, while the neutral material (by far the majority of the mass) can contract gravitationally unaffected directly by the magnetic field. Since neutrals will collide with ions in this process, there will be support against gravity for the neutrals as well as the ions. But there will be a drift of neutrals into the core without a significant increase in the magnetic flux in the core; this is ambipolar diffusion. Eventually the core mass will become sufficiently large that the magnetic field can no longer support the core, and dynamical collapse and star formation can proceed. The other extreme from the magnetically dominated star formation scenario supposes that magnetic fields are too weak to dominate the star formation process, and that molecular clouds are intermittent phenomena in an interstellar medium dominated by turbulence (e.g., Elmegreen 2000; MacLow & Klessen 2004), and the problem of cloud support for long time periods is irrelevant. Clouds form and disperse by the operation of compressible supersonic turbulence, with clumps sometimes achieving sufficient mass to become self-gravitating. Even if the turbulent cascade has resulted in turbulence support, turbulence then dissipates rapidly, and the cores collapse to form stars.

The ratio of the mass in a magnetic flux tube to the magnitude of the magnetic flux is a crucial parameter for the magnetic support/ambipolar diffusion model. The critical value for the mass that can be supported by magnetic flux Φ is $M_{Bcrit} = \Phi/2\pi\sqrt{G}$ (Nakano & Nakamura 1978); the precise value of the numerical coefficient is slightly model dependent (e.g., Mouschovias & Spitzer 1976). It is convenient to state M/Φ in units of the critical value, and to define $\lambda \equiv (M/\Phi)_{actual}/(M/\Phi)_{crit}$. Inferring λ from observations is possible if the column density N and the magnetic field strength B are measured:

$$\lambda = \frac{(M/\Phi)_{observed}}{(M/\Phi)_{crit}} = \frac{mNA/BA}{1/2\pi\sqrt{G}} = 7.6 \times 10^{-21}\frac{N(H_2)}{B} \qquad (2.1)$$

where $m = 2.8m_H$ allowing for He, A is the area of a cloud over which measurements are made, $N(H_2)$ is in cm^{-2}, and B is in μG.

In the strong field model, clouds are initially subcritical, $\lambda < 1$. Ambipolar diffusion is fastest in shielded, high-density cores, so cores become supercritical, and rapid collapse ensues. The envelope continues to be supported by the magnetic field. Hence, the prediction is that λ must be < 1 in cloud envelopes (models typically have $\lambda \sim 0.3 - 0.8$), while in collapsing cores λ becomes slightly > 1. Hence, this model tightly constrains λ. The turbulent model imposes no direct constraints on λ, although strong magnetic fields would resist the formation of gravitationally bound clouds by compressible turbulence. Also, if magnetic support is to be insufficient to prevent collapse of self-gravitating clumps that are formed by compressible turbulence, the field must be supercritical, $\lambda > 1$.

3. The Zeeman effect

The Zeeman effect provides the only direct method for measuring magnetic field strengths in molecular clouds. Generally only those species with an unpaired electron will have a strong Zeeman splitting. This has limited detections to the the 21-cm line of H I, the 18-cm, 6-cm, 5-cm, and 2-cm Λ-doublet lines of OH, and the 3-mm N=1-0 lines of CN. The sole expectation is the 1.3-cm H_2O maser line, due to very strong line strengths and strong fields in H_2O maser regions.

Except for some OH masers, the Zeeman splitting is a small fraction of the line width, and only the Stokes V spectra can be detected (Crutcher *et al.* 1993); these reveal the sign (i.e., direction) and magnitude of the line-of-sight component B_{los}. By fitting the frequency derivative of the Stokes parameter $I(\nu)$ spectrum $dI(\nu)/d\nu$ to the observed $V(\nu)$ spectrum, B_{los} may be inferred.

It is possible to correct statistically for the fact that only one component of $\mathbf{B}$ is measured, i.e., $B_{los} = |\mathbf{B}| \cos\theta$. For a large number of clouds for which the angle θ between $\mathbf{B}$ and the observed line of sight is randomly distributed,

$$\overline{B}_{los} = \frac{\int_0^{\pi/2} |\mathbf{B}| \cos\theta \sin\theta d\theta}{\int_0^{\pi/2} \sin\theta d\theta} = \frac{1}{2}|\mathbf{B}|. \tag{3.1}$$

If $\mathbf{B}$ is strong, clouds will have a disk morphology with $\mathbf{B}$ along the minor axis (cf, Mouschovias & Ciolek 1999). To properly measure λ, one needs B and N along a flux tube, i.e., parallel to the minor axis. Then, as noted by Crutcher (1999), the path length through a disk will be too long by $1/\cos\theta$ and N will be overestimated, while $|\mathbf{B}|$ will be underestimated by $\cos\theta$. Statistically,

$$\overline{M/\Phi} = \frac{\int_0^{\pi/2} (M/\Phi)_{obs} \cos^2\theta \sin\theta d\theta}{\int_0^{\pi/2} \sin\theta d\theta} = \frac{1}{3}(M/\Phi)_{obs}. \tag{3.2}$$

4. Observing magnetic fields in dark clouds

4.1. *Previous work*

Most previous Zeeman detections in molecular clouds (e.g., Crutcher 1999) have been toward clouds associated with H II regions. Dark clouds offer the possibility of measuring the role of magnetic fields at an earlier stage of the star formation process. However, there have been very few Zeeman detections. Goodman *et al.* (1989) used the Arecibo telescope to observe three dark cloud cores, with one detection (Barnard 1, $B_{los} \approx 27~\mu$G) and two limits of $\sim 10~\mu$G. Crutcher *et al.* (1993) used the NRAO 140-foot telescope to observed 12 clouds and obtained two detections (Barnard 1 and ρ Oph, $B_{los} \approx 10~\mu$G) and 10 upper limits of $\sim 10~\mu$G.

4.2. *The Arecibo Dark Cloud Survey*

In order to improve our knowledge of magnetic field strengths in dark cloud cores, we have used the Arecibo telescope to carry out an extensive program to observe the Zeeman effect in the 1665 and 1667 MHz lines of OH. The project involved ~ 800 hours of allocated telescope time, of which more than 400 hours were actual on-source Zeeman integrations. Thirty-three dark cloud core positions were observed, with integration times ranging from ~ 2 to ~ 50 hours (the limited tracking range of the Arecibo telescope meant that a few positions were the only thing accessible for some periods of the day, so long integration times were accumulated). We achieved 10 detections of B_{los} at the $2-\sigma$ or better level, and sensitive upper limits on the other positions. Full details of this project will be published separately.

Figure 1 shows our Arecibo OH Stokes I and V spectra for L1448. B_{los} was inferred separately for each line, then the two results were weighted averaged to give the final result. L1448 is typical of the results for the detections. We used $2-\sigma$ as the cutoff for a detection, since our experience has shown that the random error computed from the least-squares fitting procedure underestimates the true uncertainty in Zeeman results, probably due to low-level instrumental polarization effects.

In order to compute the mass-to-flux ratio, we need an estimate for the column density of H_2. We obtain this estimate from the OH lines themselves. The Arecibo OH spectra yield N(OH). With OH/H $= 4 \times 10^{-8}$ (Crutcher 1979), we can infer N(H_2). This is not necessarily the total N(H_2) in the telescope beam, for OH does not sample the densest gas. However, N(H_2) inferred from N(OH) is the correct one to use, for it represents the

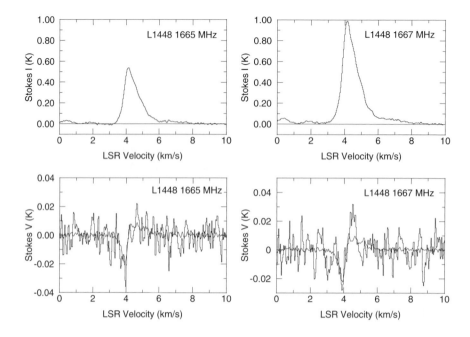

Figure 1. Arecibo Stokes I and V spectra of the OH 1665 and 1667 MHz lines of OH toward L1448. Observed data are histogram plots; fits to Stokes V are the dark lines. The respective results are $B_{los}(1665) = -25 \pm 5~\mu G$, and $B_{los}(1667) = -28 \pm 6~\mu G$. The combined result $\overline{B}_{los}$, together with $N(H_2) \approx 5 \times 10^{21}$ cm^{-2} inferred from the OH lines, yields a mass-to-flux ratio $\lambda \approx 1.6$ (before any geometrical correction), which is nominally slightly supercritical.

total H_2 column density within the telescope beam that is sampled by OH, and B_{los} inferred from OH represents the magnetic field in this same region.

4.3. *Arecibo survey mass-to-flux results*

Figure 2 shows all of the inferred results for B_{los} from the Arecibo survey plotted against $N(H_2)$. The importance of figure 2 is in what it can tell us about the mass-to-flux ratio in dark cloud cores. However, it must be kept in mind that all of the B_{los} results are *lower* limits to the total magnetic field strength. The statistical correction for this is given by equation 3.1. We apply this correction factor of $1/2$ to equation 2.1 and plot the result as the solid line in figure 2. However, if strong magnetic fields result in a disk morphology for cloud cores, then a statistical correction for column densities along flux tubes is also necessary – equation 3.2. This prediction as plotted as the dashed line in figure 2.

5. Discussion

5.1. *The new Arecibo results*

First, note that there are no points in figure 2 that are a factor of 2 above the solid line. If the mean mass-to-flux ratio in these cores were subcritical, one would expect that a few of the magnetic fields would be pointing essentially along the line of sight, and one would see an unambiguous subcritical result without applying any statistical correction.

Second, although the detections scatter roughly equally above and below the solid line (the critical mass-to-flux line with the statistical correction for magnetic field only applied), almost all of the upper limits fall below this line. Even if every upper limit were

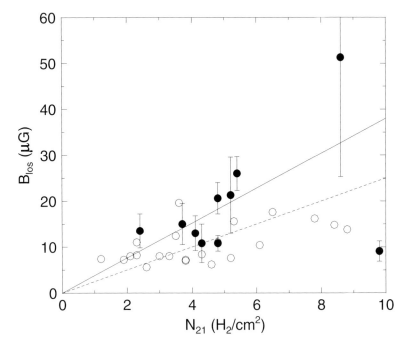

Figure 2. Results for B_{los} from our Arecibo dark cloud survey plotted against the H_2 column density ($N_{21} = 10^{-21} N$). The 10 detections are plotted as filled circles with 1σ error bars, while non-detections are plotted as open circles at the $B_{los} = 2\sigma$ positions. Straight lines are the mean predicted values of B_{los} vs. $N(H_2)$, after geometrical corrections, for a critical mass-to-flux ratio. The solid line applies only the statistical correction for measuring only one component of the magnetic vector $\mathbf{B}$, while the dotted line applies also the correction for the column density in a disk geometry (see equations 3.1, 3.2).

later found to be a detection at the upper limit value, the conclusion would be that with no statistical correction for a disk geometry, the observed mean mass-to-flux ratio would be slightly supercritical ($\overline{\lambda} \approx 1.3$).

Finally, the detections and upper limits scatter roughly equally above and below the dotted line, where both the magnetic field and the disk geometry statistical correction (equation 3.2) have been applied. If all of the $2 - \sigma$ upper limits were detections at that level, the inferred $\overline{\lambda} \approx 0.8$, or slightly subcritical. Hence, the data are consistent with the prediction of the strong magnetic field theory – an approximately critical mass-to-flux ratio in cores with a disk morphology.

5.2. *The bigger picture*

The new data support and extend the earlier conclusion (e.g., Crutcher 1999, 2004) that the data are consistent with the strong magnetic field model of star formation. In addition, the data on magnetic field strengths in the interstellar medium lead to a strong conclusion about the formation of molecular clouds. First, diffuse clouds with $n(H\ I) \sim 50\ cm^{-3}$ are significantly subcritical but not self-gravitating (Heiles & Troland 2005). The change in λ from subcritical values in diffuse clouds to critical ones in molecular clouds probably takes place during the molecular cloud formation process, by material accumulating along flux tubes to form dense clouds (e.g., Hartmann *et al.* 2001). Although this would not actually increase the mass-to-flux ratio in a flux tube, observers of individual H I clouds in the flux tube would infer a lower λ than would be found after H I clouds aggregate to form a single dense molecular cloud. Second, magnetic field strengths have been found to be essentially invariant ($B \approx 5 - 10 \ \mu G$) over the density range $10^{-1} < n(H) < 10^3$ cm^{-3}, and to scale approximately as $B \propto \sqrt{n}$ for $n > 10^3$ cm^{-3}, when clouds may become gravitationally bound. The fact that the magnetic field strength is essentially constant from the lowest densities in the interstellar medium up to self-gravitating molecular clouds provides a very significant clue about the formation of molecular clouds. If densities increased perpendicular to magnetic field lines, field strengths would increase linearly with density. Hence, molecular clouds must form primarily by accumulation of matter along field lines. This process would increase densities but not field strengths. There are possible ways out of this conclusion: magnetic reconnection, turbulence-driven ambipolar diffusion, magneto-rotational instabilities. But in studying triggered star formation, this fact about molecular cloud formation must be explained.

5.3. *The future*

The present situation is that the ambipolar diffusion model of star formation has neither been proved or disproved by observations of magnetic fields. A test that could do this is the measurement of the differential mass-to-flux ratio between the envelope and the core of clouds. The ambipolar diffusion model absolutely requires that mass-to-flux increase from envelope to core. This measurement can now be carried out by using a telescope such as the GBT to measure N(OH) and B_{los} in the envelope regions surrounding the cores where we have achieved Zeeman detections with the Arecibo telescope. Such a differential measurement would eliminate uncertainties due to geometry that we now can only account for statistically. Clear evidence for an increase in the mass-to-flux ratio from envelope to core within individual clouds would then verify the ambipolar diffusion prediction. Alternatively, if the test shows that this is not found, turbulence-driven star formation (although with dynamically important magnetic fields) would be favored.

Acknowledgements

This work was partially supported by NSF grants AST 0205810, 0307642, and 0606822.

References

Crutcher, R. M. 1979, *ApJ* 234, 881

Crutcher, R. M., Troland, T. H., Goodman, A. A., Heiles, C., Kazès, I. & Myers, P. C. 1993, *ApJ* 407, 175

Crutcher, R. M. 1999, *ApJ* 520, 706

Crutcher, R. M. 2004, *Ap&SS* 292, 225

Elmegreen, B. G. 2000, *ApJ* 530, 277

Goodman, A. A., Crutcher, R. M., Heiles, C., Myers, P. C. & Troland, T. H. 1989, *ApJ* 338, L61

Hartmann, L., Ballesteros-Paredes, J. & Bergin, E. A. 2001, *ApJ* 562, 852

Heiles, C. & Troland, T. H., *ApJ* 624, 773

MacLow M.-M. & Klessen R. S. 2004, *Rev. Mod. Phys.* 76, 125

Mouschovias, T. Ch. & Spitzer, L. 1976, *ApJ* 210, 326

Mouschovias, T. Ch. & Ciolek, G. E. 1999, in: C. J. Lada & N. D. Kylafis (eds.), *The Origin of Stars and Planetary Systems* (Kluwer), p. 305

Nakano, T. & Nakamura, T. 1978, *PASJ* 30, 681

Wielebinski, R. & Beck, R. 2005, *Cosmic Magnetic Fields* (Springer: Berlin, Heidelberg)

Discussion

VAZQUEZ-SEMADENI: Two comments: (1) Numerical simulations of MHD turbulence systematically show a lack of correlation of B with ρ. Passot & Vazquez-Semadeni (2003, A&A) gave an explanation, based on the fact that different types of nonlinear MHD waves have different scalings of B with ρ, so in a turbulent medium, where all modes coexist, there is no single preferred scaling at work. So, magnetic reconnection and nonlinear ambipolar diffusion are not the only possibilities. (2) As I'll argue tomorrow, λ is not a fixed parameter of clouds and clumps, but increases in time as the object accretes mass from its environment.

CLARKE: Your important conclusion that dark cloud cores are magnetically critical is based on the fact that both your detections and (more numerous) upper limits to B_{los} are roughly evenly distributed about the critical $\lambda = 1$ line. Higher sensitivity observations would support this conclusion only if the true λ of these non-detections was close to their current lower limit values. Given the importance of this issue, what is the prospect for converting these λ lower limits into detections?

CRUTCHER: It would be practical to improve the sensitivity of some of the non-detections where the integration time was fairly short, but this project has already involved a very large amount of observing time. In any case, some magnetic fields must lie close to the plane of the sky, so even if the total field strength were very large, the line-of-sight component B_{los} that the Zeeman observations can measure could be arbitrarily small. So it would be impossible to convert all upper limits to B_{los} (and hence lower limits to λ) into detections. The important point is that all of the Zeeman results are *lower limits* to the total magnetic field strength, even when a non-detection is an *upper limit* to B_{los}. That is why it is necessary to look at the predicted statistical fraction of detections and non-detections for a given sensitivity limit and an assumed λ in order to infer the most likely value of $\overline{\lambda}$. I think the conclusion that the mass-to-flux ratios in these dark clouds is *approximately* (within a factor of two) critical is solid; the data rule out a *mean* mass-to-flux ratio that is more than a factor of two supercritical, although of course some individual clouds without detections of B_{los} could be significantly supercritical.

Triggered Star Formation in a Turbulent ISM
Proceedings IAU Symposium No. 237, 2006
B. G. Elmegreen & J. Palouš, eds.

© 2007 International Astronomical Union
doi:10.1017/S1743921307001378

Physics and chemistry of hot molecular cores

H. Beuther

Max-Planck-Institute for Astronomy, Königstuhl 17, 69117 Heidelberg, Germany
email: beuther@mpia.de

Abstract. Young massive star-forming regions are known to produce hot molecular gas cores (HMCs) with a rich chemistry. While this chemistry is interesting in itself, it also allows to investigate important physical parameters. I will present recent results obtained with high-angular-resolution interferometers disentangling the small-scale structure and complexity of various molecular gas components. Early attempts to develop a chemical evolutionary sequence are discussed. Furthermore, I will outline the difficulty to isolate the right molecular lines capable to unambiguously trace potential massive accretion disks.

Keywords. accretion disks, astrochemistry, techniques: interferometers, stars: early-type, ISM: kinematics and dynamics, ISM: molecules

1. Introduction

Hot molecular cores (HMCs) are characterized by gas temperatures exceeding $100\,\mathrm{K}$ and a rich chemistry observable in molecular line emission at (sub)mm wavelength. These HMCs are considered to represent an early evolutionary stage in high-mass star formation where the protostars are still actively accreting and ultracompact Hɪɪ regions have not yet formed (e.g., Kurtz *et al.* 2000; Beuther *et al.* 2006). Single-dish observations toward HMCs revealed stunning molecular line forests, but they were not capable to spatially resolve the various molecular components (e.g., Blake *et al.* 1987; Schilke *et al.* 1997; Hatchell *et al.* 1998). Only interferometric high-spatial-resolution observations resolve the spatial complexity in more detail (e.g., Wright *et al.* 1996; Blake *et al.* 1996; Wyrowski *et al.* 1999). For the closest and best known HMC Orion-KL, recent observations with the Submillimeter Array (SMA) dissected its molecular components showing significant spatial differences between, e.g, SiO, oxygen-bearing species like CH_3OH, nitrogen-bearing species like CH_3CN or sulphur-bearing species like SO_2 (Beuther *et al.* 2005a). In the following, I will present recent SMA results toward the HMC in G29.96 as well as a molecular comparison of high-spatial-resolution observation of various massive star-forming regions. Finally, the difficulties of identifying molecular line tracers for massive disks will be discussed.

2. The hot molecular core G29.96

We used the SMA to observe the well-known HMC G29.96 in a broad range of spectral line and continuum emission around $862\,\mu\mathrm{m}$ (Beuther *et al.* in prep.). The achieved angular resolution is exceptional of the order $0.3''$ for the continuum and $0.6''$ for the line emission. The submm continuum data resolved the previously identified HMC (e.g., Cesaroni *et al.* 1994) into four sub-sources within a projected area of $\sim 6900 (\mathrm{AU})^2$. These four source comprise a proto-Trapezium system, and assuming spherical symmetry one can estimate an approximate protostellar density of 2×10^5 protostars/pc^3.

Within the given bandpass of $4\,\mathrm{GHz}$ we detected ~ 80 spectral lines from 18 molecular species, isotopologues or vibrational excited species, with a minor fraction of $\sim 5\%$ of

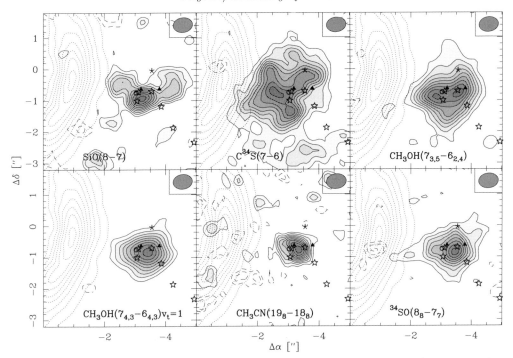

Figure 1. Spectral line maps toward the HMC G29.96 observed with the SMA (Beuther *et al.* in prep.). The dotted contour always show the UCHII region as observed at cm wavelengths (Cesaroni *et al.* 1994). The grey-scale with contours shows the integrated line emission from the species labeled in each panel. The stars mark submm continuum peaks identified during the same observations. The triangles and circles show H_2O and H_2CO maser emission, respectively (Hofner & Churchwell 1996; Hoffman *et al.* 2003), the asterisks mark the mid-infrared peak from De Buizer *et al.* (2002), and the synthesized beam is shown at the top-right of each panel.

unidentified lines. The range of excitation temperatures for the set of lines varies between 40 and 750 K, hence we are capable to study the cold and the warm gas at the same time. Figure 1 presents a compilation of integrated emission maps from a few representative species, and one can already discern from these maps the spatial complexity of the region. A detailed analysis of the whole dataset will be published shortly (Beuther *et al.* in prep.), here I only want to highlight a few characteristics:

• SiO shows some extended emission. Analyzing the spectral data-cube, we identify at least one, potentially two molecular outflows.

• $C^{34}S$ is weak toward the HMC center and the four submm peaks, but it shows strong emission at the edge of the HMC and its interface to the UCHII region. This may be interpreted as chemical evolution: Early-on, at temperatures of the order 30 K CS gets released from the dust grain and the $C^{34}S$ map should have appeared centrally peaked at that time. However, when the HMC heats up to $\geqslant 100$ K, H_2O is released from the grains, this dissociates to OH, and the OH reacts with the S to form SO, which is centrally peaked then (Fig. 1). This leaves significant $C^{34}S$ only at the edges of the HMC.

• No spectral line shows the same morphology as the submm continuum emission (even after smoothing the continuum to the spatial resolution of the line emission). Hence none traces unambiguously the protostellar condensations. In addition to outflow contributions and chemistry effects two other processes are considered to be important for that. On the one hand, many spectral lines are optically thick and therefore only trace the outer envelope of the region without penetrating toward the central protostellar cores. On the other hand, we are likely suffering from confusion because the molecules are not exclusively found in the central protostellar cores but also in the surrounding envelope. Disentangling these components is a difficult task.

• Of the many molecular lines, only a single one exhibits a coherent velocity structure with a velocity gradient perpendicular to the main outflow. Since this structure comprises three of the submm peaks it likely is a larger-scale rotating toroid which may (or may not) harbor accretion disks closer to the protostellar condensations. For the difficulties of massive disk studies see §4.

3. Toward a chemical evolutionary sequence

With the long-term goal in mind to establish chemical sequences – in an evolutionary sense as well as with varying luminosity – over the last few years we observed four massive star-forming regions with the SMA in exactly the same spectral setup around $862\,\mu$m as used originally for the Orion-KL observations (Beuther *et al.* 2005a). These four regions comprise a range of luminosities between $10^{3.8}\,L_\odot$ and $10^5\,L_\odot$, and they cover different evolutionary stages from young pre-HMCs to typical HMCs (Orion-KL: HMC, $L \sim 10^5\,L_\odot$, $D \sim 0.45$ kpc; G29.96: HMC, $L \sim 9 \times 10^4\,L_\odot$, $D \sim 6$ kpc; IRAS 23151, pre-HMC, $L \sim 10^5\,L_\odot$, $D \sim 5.7$ kpc; IRAS 05358: pre-HMC, $L \sim 10^{3.8}\,L_\odot$, $D \sim 1.8$ kpc). Smoothing all datasets to the same linear spatial resolution, we are now capable to start comparing these different regions. Figure 2 presents typical spectra extracted toward the HMC G29.96 and the pre-HMC IRAS 23151 (Beuther *et al.* in prep.).

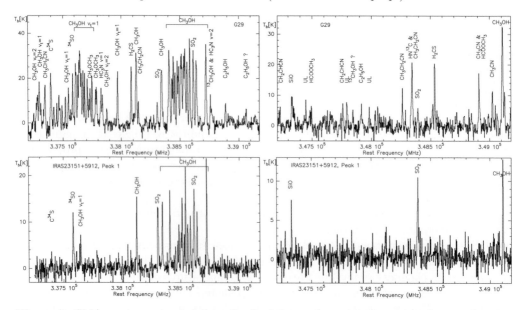

Figure 2. SMA spectra extracted from the final data-cubes toward two massive star-forming regions (G29.96 top row & IRAS 23151+5912 bottom row). The spectral resolution in all spectra is 2 km/s. The left and right column show the lower and upper sideband data, respectively.

A detailed comparison between the four sources will be given in a forthcoming paper (Beuther *et al.* in prep.), here I just outline a few differences in a qualitative manner.

• The HMCs show far more molecular lines than the pre-HMCs. Orion-KL and G29.96 appear similar indicating that the nature of the two sources may be similar as well. Regarding the two pre-HMCs, the higher luminosity one (IRAS 23151) shows still more lines than the lower-luminosity source (IRAS 05358). Since IRAS 05358 is approximately three times closer to us than IRAS 23151, this is not a sensitivity issue but it is likely due to the different luminosity objects forming at the core centers.

• The ground-state CH_3OH lines are detected toward all four sources. However, the vibrational-torsional excited CH_3OH are only strongly detected toward the HMCs Orion-KL and G29.96. Independent of the luminosity, the pre-HMCs exhibit only one CH_3OH $v_t = 1$ line, which can easily be explained by the lower average temperatures of the pre-HMCs.

• A more subtle difference can be discerned by comparing the SO_2 and the $HN^{13}C$ lines near 348.35 GHz (in the upper sideband). While the SO_2 line is found toward all four sources, the $HN^{13}C$ is strongly detected toward the HMCs, but it is not found toward the pre-HMCs. In the framework of warming up HMCs, this indicates that nitrogen-bearing molecules are either released from the grains only at higher temperatures, or they are daughter molecules which need some time during the warm-up phase to be produced in gas-phase chemistry networks. In both cases, such molecules are expected to be found not much prior to the formation of a detectable HMC.

4. Identifying molecules for massive disk studies

There exists ample, however indirect evidence for the existence of massive disks in high-mass star formation (e.g., Cesaroni *et al.* 2006; Beuther *et al.* 2006). Theorists predict that massive accretion disks have to exist (e.g., Jijina & Adams 1996; Yorke & Sonnhalter 2002; Krumholz 2006), and the observations of collimated jet-like molecular outflows from at least B0 stars indicate the presence of underlying accretion disks as well (e.g., Beuther & Shepherd 2005; Arce *et al.* 2006). However, we have not found much observational evidence for massive accretion disks, not to speak that we have not characterized them properly yet (Cesaroni *et al.* 2006). The best known example is the disk in IRAS 20126+4104 which even shows a Keplerian velocity profile, but the mass of the central object is only $\sim 7\,M_\odot$, and it is probably still in its accretion phase (Cesaroni *et al.* 2005). There are more sources observed where we find rotational signatures in the central cores perpendicular to the molecular outflows (see Figure 3 for a small compilation), however, the velocity structure is not Keplerian or they are that large (potentially comprising several sub-sources) that they rather resemble larger-scale rotating toroids than typical accretion disks (e.g., Cesaroni *et al.* 2006; Keto & Wood 2006). Maybe these sources harbor genuine accretion disks at their very centers.

A major observational problem arises because it is difficult to disentangle the spectral line distributions from the various gas components (mainly core-disk, envelope and outflow). Some previously believed good disk-tracers have been shown to be strongly influenced by the molecular outflows (e.g., CN Beuther *et al.* 2004, HCN Zhang priv. comm.). Probably even more difficult, the expected central disks and the close-by surrounding envelopes are both warming up quickly due to the heating of the central accreting source. Therefore, the chemical properties – and hence the molecular emission arising from both components – are similar. This way, one may have to model always both components together. Furthermore, as we have seen in the previous sections, the chemistry varies with evolution. For example, while $C^{34}S$ may be a potentially good disk tracer in young pre-HMCs like IRAS 20126+4104 (Fig. 3), it obviously does not work in more evolved HMCs like G29.96 (Fig. 1). In contrast, nitrogen-bearing molecules like $HN^{13}C$ appear to be a good tracer of rotation in the HMC G29.96, but it remains undetectable in younger sources like IRAS 23151+5912 (Fig. 2). An additional complication arises from varying optical depths: while the 1 mm lines of CH_3CN are good rotation tracers in some sources

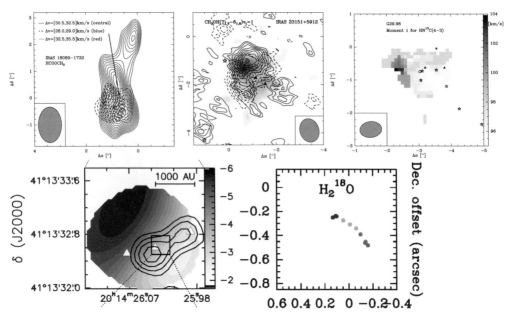

Figure 3. Examples of rotation-tracing molecules: top-left: HCOOCH$_3$ in IRAS 18089-1732 (Beuther *et al.* 2005b), top-middle: CH$_3$OH $v_t = 1$ in IRAS 23151+5912 (Beuther *et al.* in prep.), top-right: HN^{13}C in G29.96 (Beuther *et al.* in prep.), bottom-left: C^{34}S (and also CH$_3$CN) in IRAS 20126+4104 (Cesaroni *et al.* 1999, 2005), bottom-right: H$_2^{18}$O in AFGL2591 (van der Tak *et al.* 2006).

(Cesaroni *et al.* 1999; Beltrán *et al.* 2004), the more excited lines in the submm bands do not always show these signatures. Beuther *et al.* (2005b) interpreted this difference due to increased optical depth at the given high temperatures in the submm bands.

The advent of broad spectral bandpasses in new or upgraded interferometers now allows to observe many spectral lines simultaneously (Fig. 2). Thus, we can identify the best rotation-tracing molecules for individual sources after the observations without the need of strong molecular pre-selection effects. While one would like to observe a large sample of massive disk candidates in the same spectral line to study the kinematic properties as consistently as possible, this may be impossible due to the above discussed physical and chemical difficulties. However, if we are able to observe many sources systematically in a spectral setup covering the most important molecules, we can select the adequate line for each source and still investigate a larger sample in a statistically consistent manner.

5. Conclusions and outlook

Interferometry at (sub)mm wavelengths is the tool of choice if one wants to disentangle the chemical and physical complexity in massive star-forming regions. Only since a couple of years we are capable to spatially map a broad range of molecular lines in selected regions at high-spatial resolution.

One of the first results of these studies is that the spatial diversity of the molecules is extremely complex. For a proper understanding of the given data, it is necessary to enhance the models including physical properties like shocks, outflows, rotation and heating as well as chemical networks containing gas-phase and grain-surface reactions. Furthermore, the physical and chemical models have to be treated with state-of-the art radiative transfer codes to finally produce synthetic images which can be compared with the observations (e.g., Pavlyuchenkov *et al.* 2006). On the observational side, we need to observe larger source-samples consisting of different evolutionary stages as well as different luminosities. Observations and modeling have advanced very much over the last decade, but often the various groups did not interact enough. For a better understanding of the chemical complexity of massive star-forming regions only a concerted effort from theory, modeling and observations is likely to result in significant progress.

Massive accretion disks are considered the holy grail in high-mass star formation research. The availability of broad spectral bandpasses now allows to observe larger source samples in a less pre-selective way since it is likely that one of the observed spectral lines will trace the central rotating structure and hence allow a kinematic analysis. Utilizing the currently available (sub)mm interferometers (mainly PdBI, SMA and CARMA) as well as ALMA in the coming decade, we are expecting to reach a much better understanding of massive accretion disks and thus high-mass star formation in general.

Acknowledgements

Thanks a lot to Hendrik Linz for comments on an early draft of this paper. H.B. acknowledges financial support by the Emmy-Noether-Program of the Deutsche Forschungsgemeinschaft (DFG, grant BE2578).

References

Arce, H., Shepherd, D., Gueth, F., *et al.* 2006, in: B. Reipurth, D. Jewitt & K. Keil (eds.), *Protostars & Planets V*, in press, astro-ph/0603071

Beltrán, M. T., Cesaroni, R., Neri, R., *et al.* 2004, *ApJ* 601, L187

Beuther, H., Churchwell, E., McKee, C. & Tan, J. 2006, in: B. Reipurth, D. Jewitt & K. Keil (eds.), *Protostars & Planets V*, in press, astro-ph/0511294

Beuther, H., Schilke, P. & Wyrowski, F. 2004, ApJ, 615, 832

Beuther, H. & Shepherd, D. 2005, in: M.S. Nanda Kumar, M. Tafalla & P. Caselli (eds.) *Cores to Clusters: Star Formation with Next Generation Telescopes*, Ap&SS 324, 105

Beuther, H., Zhang, Q. Greenhill, L. J., *et al.* 2005a, *ApJ* 632, 355

Beuther, H., Zhang, Q., Sridharan, T. K. & Chen, Y. 2005b, *ApJ* 628, 800

Blake, G. A., Mundy, L. G., Carlstrom, J. E., *et al.* 1996, *ApJ* 472, L49

Blake, G. A., Sutton, E. C., Masson, C. R. & Phillips, T. G. 1987, *ApJ* 315, 621

Cesaroni, R., Churchwell, E., Hofner, P., Walmsley, C. M. & Kurtz, S. 1994, *A&A* 288, 903

Cesaroni, R., Felli, M., Jenness, T., *et al.* 1999, *A&A* 345, 949

Cesaroni, R., Galli, D., Lodato, G., Walmsley, C. & Zhang, Q. 2006, in: B. Reipurth *et al.* (eds.), *Protostars & Planets V*, in press astro-ph/0603093

Cesaroni, R., Neri, R., Olmi, L., *et al.* 2005, *A&A* 434, 1039

De Buizer, J. M., Radomski, J. T., Piña, R. K. & Telesco, C. M. 2002, *ApJ* 580, 305

Hatchell, J., Thompson, M. A., Millar, T. J. & MacDonald, G. H. 1998, *A&AS* 133, 29

Hoffman, I. M., Goss, W. M., Palmer, P. & Richards, A. M. S. 2003, *ApJ* 598, 1061

Hofner, P. & Churchwell, E. 1996, *A&AS* 120, 283

Jijina, J. & Adams, F. C. 1996, *ApJ* 462, 874

Keto, E. & Wood, K. 2006, *ApJ* 637, 850

Krumholz, M. 2006, in: M. Livio & E. Villaver (eds.) *Massive Stars: From Pop III and GRBs to the Milky Way* (Cambridge: Cambridge Univ.), in press, astro-ph/0607429

Kurtz, S., Cesaroni, R., Churchwell, E., Hofner, P. & Walmsley, C. M. 2000, in: V. Mannings, A.P. Boss, S.S. Russel (eds.), *Protostars and Planets IV* (Tucson: Univ. of Arizona), p. 299

Pavlyuchenkov, Y., Wiebe, D., Launhardt, R. & Henning, T. 2006, *ApJ* 645, 1212

Schilke, P., Groesbeck, T. D., Blake, G. A. & Phillips, T. G. 1997, *ApJS* 108, 301

van der Tak, F. F. S., Walmsley, C. M., Herpin, F. & Ceccarelli, C. 2006, *A&A* 447, 1011

Wright, M. C. H., Plambeck, R. L. & Wilner, D. J. 1996, *ApJ* 469, 216

Wyrowski, F., Schilke, P., Walmsley, C. M. & Menten, K. M. 1999, *ApJ* 514, L43

Yorke, H. W. & Sonnhalter, C. 2002, *ApJ* 569, 846

Discussion

ELMEGREEN: Do you see evidence for low mass protostars forming near the high mass protostars associated with hot cores?

BEUTHER: In the case of G29.96, we can identify above the noise gas clumps down to ~ 3 $M_\odot$, but I would expect more sensitive observations to detect even fainter sources. Furthermore, Pratap *et al.* (1999) detected a faint embedded cluster in that region. So, it appears, yes, there have to be low-mass stars around, it's just hard to detect and identify them at the given distances and the large gas column densities.

HANASZ: Is it possible that the hot cores are due to shock heating resulting from e.g., jet propagation?

BEUTHER: Probably, there exist many ways to produce hot-core like spectra. In the case of G29.96, it appears that the major hot-core chemistry driving really stems from embedded protostars, but it is less clear in the case of Orion-KL, there the hot core is $\sim 1''$ offset from the power-house source I. And there may exist other sources where shock heating could contribute as well. But my personal impression is that internal massive protostars are the most likely culprits in many hot-core regions.

Triggered Star Formation in a Turbulent ISM
Proceedings IAU Symposium No. 237, 2006
B. G. Elmegreen & J. Palouš, eds.

Dissecting a site of massive star formation: IRAS 23033+5951

Michael A. Reid[1] and Brenda C. Matthews[2]

[1]Harvard-Smithsonian Center for Astrophysics, Submillimeter Array, 645 North A'ohoku Pl.,
Hilo, HI, 96720, USA
email: mareid@sma.hawaii.edu

[2]Herzberg Institute of Astrophysics, 5071 West Saanich Road, Victoria, British Columbia,
V9E 2E7, Canada
email: brenda.matthews@nrc-cnrc.gc.ca

Abstract. We present new BIMA observations of the massive star-forming region IRAS 23033+ 5951 in Cepheus. 3 mm continuum observations reveal that the source decomposes into at least three dusty clumps, each of which has sufficient mass to form a massive star. The most massive clump has a mass of about 225 $M_\odot$ and appears to house the massive protostar which drives the prominent CO outflow seen in the region. Our $H^{13}CN$ 1–0, N_2H^+ 1–0, and $H^{13}CO^+$ 1–0 maps show that the three continuum sources are all embedded in an elongated structure whose long axis is perpendicular to the outflow. Both $H^{13}CO^+$ and $H^{13}CN$ peak at the geometric center of this structure, which lies between the two prominent continuum peaks. All three lines – $H^{13}CN$, $H^{13}CO^+$, and N_2H^+–show the same velocity gradient along the long axis of their integrated intensity maps. Although the approximately 90,000 AU length of the elongated structure prohibits a disk interpretation, the fact that the dynamical and gas masses of the structure differ by only a factor of a few suggests that the structure may be partially rotationally supported. We also detect a signature of infall toward the center of the structure, seen as an asymmetrically blue HCO^+ line where its optically thin isotope, $H^{13}CO^+$, is symmetric and single-peaked.

1. Introduction

The mechanism of massive star formation is poorly understood. Compelling recent theoretical work has shown that, like low-mass stars, massive stars can and likely do form by disk accretion, rather than by other proposed mechanisms (e.g. Krumholz, McKee & Klein, 2005a,b). The observational evidence is starting to converge on the same conclusion (e.g. Whitney, 2005 and references therein). However, there remain significant uncertainties in the interpretation of the kinematics of massive star-forming objects. For example, several recent studies have claimed to find accretion disks around massive protostars (e.g. Patel *et al.*, 2005; Chini *et al.*, 2004), only to have those claims disputed (Comito *et al.*, 2005; Sako *et al.*, 2005). Other studies show evidence for > 1000 AU rotating structures but without providing direct evidence for an inner accretion disk on outflow-launching scales (~100 AU; e.g. Cesaroni *et al.*, 2005). Moreover, the radii and masses of claimed massive protostellar disks range from 130 AU and 3 $M_\odot$ (Shepherd *et al.*, 2001) to 15,000 AU and 400 $M_\odot$ (Sandell & Sievers, 2004), suggesting that there is either tremendous diversity in the properties of massive accretion disks or ambiguity about what constitutes a "disk". For example, there may be ambiguity among true accretion disks, circumstellar tori, and flattened molecular cloud cores.

Few massive protostellar systems have been analysed kinematically over the wide range of spatial scales necessary to resolve ambiguities among the three aforementioned types of rotating structures. In this paper, we present the results of our BIMA observations of the

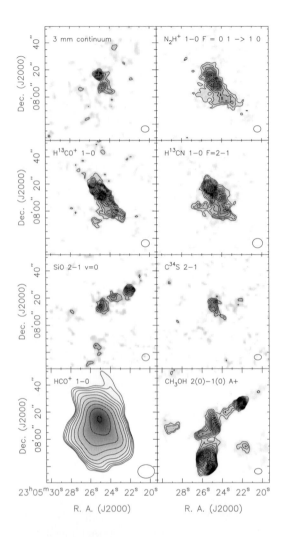

Figure 1. Integrated intensity maps of IRAS 23033+5951 in greyscale and contours. The 3 mm continuum map is shown at the upper left for comparison. Contours begin at 3σ and increase by intervals of 1σ, except for N_2H^+, where the contours interval is 2σ, and HCO^+, where the contours increase from 3σ by factors of 1.5. In each panel, the size of the synthesized beam is represented by a hollow ellipse in the lower right corner.

kinematics of the massive star-forming region IRAS 23033+5951. Previous observations of this region have revealed a massive protostar with a luminosity of about 10^4 $L_\odot$, a collimated CO outflow, and an extended envelope of total mass $\sim$2300 $M_\odot$ (Beuther *et al.* 2002a,b). These results form the first part of a more detailed study of this intriguing region.

2. General morphology of the circumstellar material

In Figure 1, we show 3 mm continuum and integrated intensity maps of IRAS 23033+ 5951 in seven molecular lines. The continuum image shows that the source decomposes

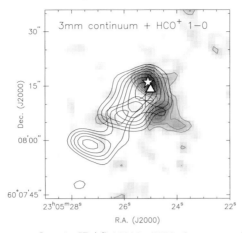

Figure 2. The primary outflow in IRAS 23033+5951. 3 mm continuum emission is shown in greyscale with grey contours. The redshifted and blueshifted HCO$^+$ emission are shown in thick and thin black contours, respectively. The star, triangle, and cross represent the positions of the MSX source, water maser, and IRAS point source, respectively.

into at least three components. Computing their masses from their continuum fluxes, we find that they have masses of 225 M$_\odot$, 205 M$_\odot$, 51 M$_\odot$, assuming a dust temperature of 30 K. All three components are embedded in an elongated structure seen in N$_2$H$^+$, H^{13}CN, and H^{13}CO$^+$. The continuum and N$_2$H$^+$ both show the same prominent double-peaked structure, whereas H^{13}CN and H^{13}CO$^+$ both peak between the continuum peaks. C^{34}S, CH$_3$OH and SiO all show emission coincident with the continuum sources but they also peak between the prominent continuum clumps. The emission from all of the aforementioned molecules is embedded within a much more extended envelope seen in HCO$^+$.

In Figure 2, we show the red and blue lobes of the integrated high-velocity HCO$^+$ emission superposed on the continuum map. Note that SiO and CH$_3$OH also show red-shifted emission along the same axis as the outflow traced by HCO$^+$, though primarily on the opposite side of the continuum sources, suggesting they are tracing a separate outflow. In Figure 2, we have labeled the positions of the water maser, MSX, and IRAS point sources in the region. Clearly at least the MSX and maser sources are associated with the northern continuum source, which also appears to be the source of the HCO$^+$ outflow. The two lobes show significant spatial overlap, suggesting that the outflow is significantly inclined relative to the plane of the sky.

3. A flattened rotating object?

As shown in Figure 3, both H^{13}CO$^+$ and the isolated hyperfine component of N$_2$H$^+$ 1–0 show velocity gradients of several km s^{-1} along the long axis of the elongated structure, which is itself almost exactly perpendicular to the axis of the outflow emission defined by HCO$^+$. The same velocity gradient is seen in H^{13}CN, but the quality of the data is significantly lower than that of either H^{13}CO$^+$ or N$_2$H$^+$. The presence of a velocity gradient of a few km s^{-1} along an axis perpendicular to the outflow axis seen in three separate tracers suggests a rotating disk. However, at an assumed source distance of 3.5 kpc, the $\sim$25$''$ size of the elongated structure in which the continuum sources are embedded translates to $\sim$90,000 AU – far too large for the structure to be a genuine protostellar accretion disk. The dynamical mass of the structure computed from the

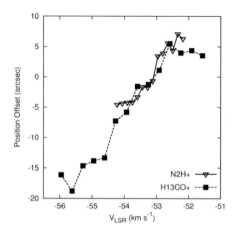

Figure 3. Position-velocity cuts taken along the long axis of the N_2H^+ 1–0 (*solid line and triangles*) and $H^{13}CO^+$ 1–0 (*dashed line and squares*) emission. Only the isolated component of the N_2H^+ 1–0 line is used in making the fit. The points represent the centroid of the emission in each velocity channel. Both lines show the same velocity gradient.

$H^{13}CO^+$ line is $520/\sin^2(i)$ $M_\odot$, which is very close to the sum of the masses computed from the dust continuum (481 $M_\odot$). The interferometric nature of our data means that much of the continuum emission is resolved out, so the continuum masses are effectively lower limits. Thus, the total mass enclosed within the area accounted for by the dynamical mass is likely considerably larger – perhaps approaching the 2300 $M_\odot$ derived from single-dish observations. Nevertheless, the large dynamical mass suggests that, if the elongated structure is rotating, it may derive significant support from rotation. We suggest that this structure may be the remnant of the original molecular cloud from which the massive protostar formed and which has now fragmented into at least two large components.

4. Conclusions

We have shown that the massive protostar in IRAS 23033+5951 is embedded in an elongated structure whose long axis is perpendicular to that of its outflow. We have also shown that all three molecular lines which trace the elongated structure show the same velocity gradient along it. Although the sheer size of the object would seem to prohibit any interpretation of it as an accretion disk, its kinematics are consistent with it being significantly rotationally supported. Planned observations at higher spatial resolution will probe the connection between the kinematics of the large-scale emission shown herein with the kinematics of the material within each continuum peak, perhaps revealing the connection between a large-scale flattened, rotating object and a true accretion disk.

References

Beuther, H., et al., 2002, *ApJ* 566, 945
Beuther, H., et al., 2002, *A&A* 383, 892
Cesaroni, R., et al., 2005, *A&A* 434, 1039
Chini, R., et al., 2004, *Nature* 429, 155
Comito, C., Schilke, P., Jiménez-Serra, I. & Martín-Pintado, J., 2005, in: D. C. Lis, G. A. Blake & E. Herbst (eds.), *Astrochemistry Throughout the Universe: Recent Successes and Current Challenges* (Cambridge: Cambridge Unive.), p. 59
Krumholz, M. R., McKee, C. F. & Klein, R. I., 2005, *ApJ* 618, L33

Krumholz, M. R., McKee, C. F. & Klein, R. I., 2005, *Nature* 438, 332
Patel, N. A. *et al.*, 2005, *Nature* 437, 109
Sako, S., *et al.*, 2005, *Nature* 434, 995
Sandell, G. & Sievers, A. 2004, *ApJ* 600, 269
Shepherd, D. S., Claussen, M. J. & Kurtz, S. E., 2001, *Science* 292, 1513
Whitney, B. A., 2005, *Nature* 437, 7055

Discussion

BONNELL: One potential problem with a "torus" interpretation is that the dynamical time for this object must be of order 10^6 years, which is probably comparable to or longer than its age or formation time. A sheared filament interpretation appears far more straightforward.

REID: We have suggested that the flattened object may be a rotating toroid because it shares so many characteristics with known low-mass disk/jet systems. First, the warm gas and radio continuum emission peak at the geometric center of the flattened structure, with colder gas appearing toward the edges. Second, the driving source of the outflow is roughly coincident with the geometric center of the structure. Third, the outflow axis is almost exactly perpendicular to the long axis of the flattened object. Finally, the rotation curve of the object is consistent with rotation. Collectively, these facts support the interpretation of the flattened structure as a rotating, circumstellar torus, although we acknowledge that they do not conclusively rule out a sheared filament interpretation.

CLARKE: Is this the first of the "giant toroids" that appears to be close to centrifugal balance? Regardless of the issue of whether the rotation curve is "Keplerian", this would appear to be unique in terms of being a toroid where the dynamical mass and measured gas mass are comparable.

REID: The measured dynamical mass is comparable to the sum of the masses of the dust continuum objects embedded within it. However, it remains an open question whether the velocity gradient we see along the long axis of the flattened structure truly represents rotation. We are seeking data with higher sensitivity to study the velocity structure of the apparent toroid.

ELMEGREEN: You shouldn't expect Keplerian rotation anyway. If you flatten an isothermal sphere, you will get something more like a flat rotation curve.

REID: Agreed. The more general question we were trying to raise in referring to the possibility of Keplerian rotation is of the precise nature of the flattened structure: if it truly were a rotationally supported disk, we should expect a Keplerian rotation profile. It is primarily only the size of the object that argued against it being a true rotationally supported disk. If it is merely a flattened molecular cloud core, it would have some other rotation profile. It is also possible that the velocity gradient we see does not represent rotation.

Triggered Star Formation in a Turbulent ISM
Proceedings IAU Symposium No. 237, 2006
B.G. Elmegreen & J. Palouš, eds.

Anatomy of the S255–S257 complex - triggered high-mass star formation

V. Minier[1], N. Peretto[2], S. N. Longmore[3], M. G. Burton[3], R. Cesaroni[4], C. Goddi[4], M. R. Pestalozzi[5] and Ph. André[1]

[1]SAp/DAPNIA/DSM & UMR AIM, CEA Saclay, 91191 Gif-sur-Yvette, France
email: vincent.minier@cea.fr

[2]Department of Physics & Astronomy, University of Manchester, Manchester, UK

[3]School of Physics, University of New South Wales, Sydney, Australia

[4]INAF, Osservatorio Astrofisico di Arcetri, Firenze, Italy

[5]School of Physics, Astronomy & Maths, University of Hertfordshire, Hatfield, UK

Abstract. We present a multi-wavelength (NIR to radio) and multi-scale (1 AU to 10 pc) study of the S255–S257 complex of young high-mass (proto)stars. The complex consists of two evolved HII regions and a molecular gas filament in which new generations of high mass stars form. Four distinct regions are identified within this dusty filament: a young NIR/optical source cluster, a massive protostar binary, a (sub)millimetre continuum and molecular clump in global collapse and a reservoir of cold gas. Interestingly, the massive binary protostellar system is detected through methanol maser and mid-IR emission at the interface between the NIR cluster and the cold gas filament. The collapsing clump is located to the north of the NIR cluster and hosts a young high-mass star associated with an outflow that is observed in mid-IR, methanol maser and radio emission. We interpret this anatomy as the possible result of triggered star formation, starting with the formation of two HII regions, followed by the compression of a molecular gas filament in which a first generation of high-mass stars forms (the NIR cluster), which then triggers the formation of high mass protostars in its near environment (the massive protostellar binary). The global collapse of the northern clump might be due to both the expansion of the HII regions that squashes the filament. In conclusion, we witness the formation of four generations of clusters of high-mass stars in S255–S257.

Keywords. stars: formation, ISM: clouds, HII regions, methods: data analysis

1. Introduction

S255–S257 belongs to the large complex of HII regions, S254, S255, S256, S257, S258, at a distance of 2.4-2.6 kpc (Fig. 1a). These five HII regions are powered by young high-mass stars and clusters of stars. S254 is the most extended HII region. S256 and S258 are still relatively small in extent, and S258 is isolated from the other bubbles of ionized gas. Interestingly, a luminous IR source is observed in between S255 and S257 (e.g. Howard *et al.* 1997). It is surrounded by a cluster of young stars with hundreds of T-Tauri, class I and class II objects (Ojha *et al.* 2006). The NIR source is also associated with a methanol maser that is a signpost of high-mass protostars (Minier *et al.* 2001). A second high-mass young stellar object (HMYSOs) is identified north of the NIR source cluster. It exhibits radio continuum emission that is the signature of an ultra-compact HII (UC HII) region (Kurtz *et al.* 1994).

With signatures of HMYSOs at various stages, the S255–S257 complex is an ideal laboratory to explore possible triggers of star formation. The following section describes

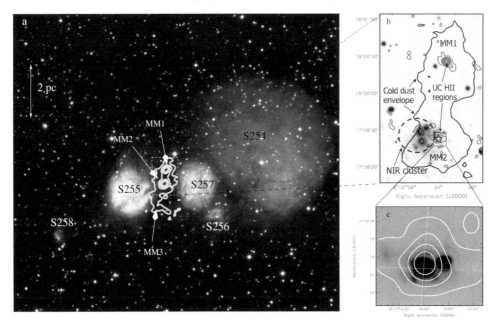

Figure 1. a. Cold dust continuum emission at 1.2 mm (contours) overlaid on the DSS optical image of the S254-S258 HII regions. b. Close-up of MM1 and MM2. Dust emission at 450-μm (single black contour), radio-continuum emission at 15 GHz (grey contours) overlaid on the J-band 2MASS image. The methanol maser site is indicated by a star. c. Close-up of MM2/methanol maser source. Radio-continuum emission overlaid on 18-μm mid-IR emission from hot dust.

the anatomy of the molecular cloud that is squeezed in between S255 and S257, from cloud scale (1 pc) down to protostar scale (1 AU).

2. Anatomy of S255–S257

2.1. *The molecular gas and dust filament*

Observations of S255-S257 were performed with Gemini-North in five mid-IR bands (Longmore *et al.* 2006), with JCMT/SCUBA at 450 and 850 μm and SEST/SIMBA at 1.2-mm (Minier *et al.* 2005), with the IRAM-30m in N_2H^+(1-0) and HCO^+(3-2) spectral lines, with the SEST in the ^{13}CO(1-0) line, with the VLA at 15, 22 and 43 GHz, and with the EVN and VLBA for imaging water and methanol masers (Goddi *et al.* 2006; Minier *et al.* 2001). Complementary data were extracted from the Digitized Sky Survey archives, the Spitzer/IRAC archives, 2MASS, MSX and IRAS IRSA image service, and the VLA and NVSS archives. This work also makes use of other observational results that were obtained in the 2.2-μm H_2 line by Miralles *et al.* (1997) and 44-GHz methanol masers by Kurtz *et al.* (2004).

The observations of cold dust and molecular gas emission reveal a ridge or a filament of dense gas that is squeezed in between S255 and S257 (Fig. 1a,b). It is part of a more extended molecular cloud that links the HII regions from the eastern S258 to the western S254 (Fig. 2a). The molecular gas ridge is observed by Spitzer/IRAC as a dark lane on the bright PAH emission at 8 μm that traces the interface between ionized gas in the HII regions and neutral gas in the cloud (Fig. 3).

Dense molecular clumps are identified in the ridge between S255 and S257. Two clumps are clearly observed in the cold dust emission. MM1 and MM2 present similar physical

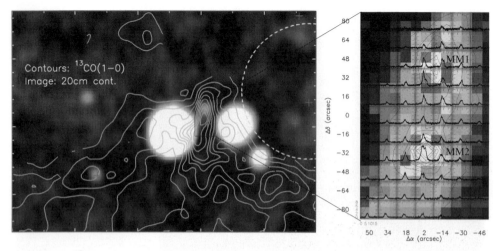

Figure 2. a. ^{13}CO(1-0) line emission (contour) overlaid on the NVSS radio-continuum image at 20 cm. b. Close-up of MM1 and MM2. HCO$^+$(3-2) line spectra overlaid on integrated-line emission.

conditions (see Minier *et al.* 2005). Their mass is ~ 300 M$_\odot$ and their luminosity is $5 - 10 \times 10^4$ L$_\odot$ within ~ 0.3 pc. These clumps are dense (10^5 cm^{-3}) and cold (40 K). They are also observed in HCO$^+$(3-2) (Fig. 2b) and N$_2$H$^+$(1-0). There is a third clump in the N$_2$H$^+$(1-0) emission map that is located in more diffuse dust emission (Fig. 1a).

2.2. *Dust clump MM1*

MM1 hosts a UCHII region at its center (Fig. 1b). The radio-continuum emission is extended in a NE-SW direction as well as the 44-GHz methanol masers (Kurtz *et al.* 2004). Spitzer/IRAC and Gemini-North images also reveal this elongated source. This morphology is likely that of an outflow and its heated-up cavity. At larger scales ($\sim$0.5 pc), MM1 exhibits spatially resolved signatures of infall in HCO$^+$(3-2), which suggest the global collapse of this clump (Fig. 2b; see Peretto *et al.* 2006 for a comprehensive study about global collapse signatures). Simple 1D radiative transfer calculations give an infall velocity of 0.7 km s^{-1}, that translates to a dynamical infall time of 10^5 yr, which is about the free-fall time for such a clump.

2.3. *Dust clump MM2*

MM2 harbours a more complex structure with a NIR source cluster on its East side and neutral molecular gas on its West side (Fig 1b). MM2 hosts two well-developed UC HII regions that are associated with the cluster of NIR sources and a hyper-compact HII region that is coincident with the methanol and water maser source (Fig. 1b; Goddi *et al.* 2006). The NIR cluster is surrounded by a bubble of shocked H$_2$ gas (Miralles *et al.* 1997). The methanol maser is located within this H$_2$ bubble at the interface between the cluster and the neutral molecular gas. A signature of infall in HCO$^+$ is observed at this position (Fig. 2b). Finally, a massive protostellar binary is imaged at the position of the maser (Fig. 1c; Longmore *et al.* 2006).

3. Triggered star formation?

The anatomy study of the molecular gas filament between S255 and S257 reveals two new active sites of high-mass star formation: MM1 and MM2. MM1 is a cold molecular

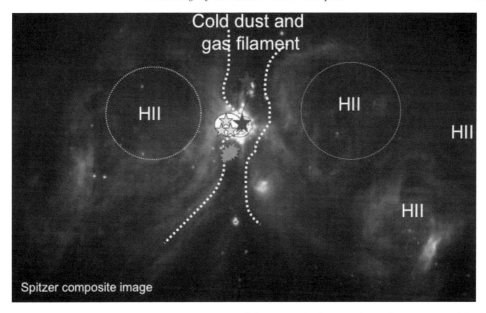

Figure 3. Spitzer composite image in three IRAC bands. Red stars show the position of high–mass protostars. White area and blue stars illustrate the B-star cluster and the H_2 bubble. The grey cloud symbol represents the MM3 cold gas reservoir.

cloud in global collapse. It hosts a high-mass protostar that ionizes its near environment and powers an outflow. It is not clear how the HII regions S255 and S257 have triggered its collapse because the infall time is not greater than the free-fall time of MM1. MM1 is probably a first-generation high-mass star-forming clump. MM2's history is much more complex. MM2 is surrounded by a cluster of low-mass YSOs. Two UCHII regions that are located East from MM2 emission peak are associated with optically visible high-mass stars (Howard *et al.* 1997). They might be part of the cluster. An HII region has developed around them as defined by the shocked H_2 bubble. Interestingly, a massive protostellar binary is born in the shocked interface. A possible interpretation is the formation of high-mass stars through a collect-and-collapse process after the fragmentation of the shocked gas around a HII region (Elmegreen 1998). The massive protostellar binary would therefore be the product of second-generation, triggered high-mass star formation. MM3 is a reservoir of pre-stellar gas that is not yet an active site of star formation.

In summary, this region contains a wide set of star formation mechanisms which might be necessary ingredients to form high-mass stars, such as large scale molecular cloud compression, early global collapse of clumps, or even small scale triggered star formation.

References

Elmegreen, B.G. 1998 in: C.E. Woodward, J.M. Shull & H.A. Thronson (eds.), *Origins* (ASP-CS), 148, p. 150

Goddi, C., Moscadelli, L., Sanna, A., Cesaroni, R. & Minier, V. 2006, *A&A* submitted

Howard, E.M, Pipher, J.L. & Forrest, W.J. 1997, *ApJ* 481, 327

Kurtz, S., Churchwell, E. & Wood, D.O.S. 1994, *ApJS* 91, 659

Kurtz, S., Hofner, P. & Alvarez, C.V. 2004, *ApJS* 155, 149

Longmore, S.N., Burton, M.G., Minier, V. & Walsh, A.J. 2006, *MNRAS* 369, 1196L

Minier, V., Conway, J.E. & Booth, R.S. 2001, *A&A* 369, 278

Minier, V., Burton, M.G., Hill, T., *et al.* 2005, *A&A* 429, 945

Miralles, M.P., Salas, L., Cruz-Gonzalez, I. & Kurtz, S. 1997, *ApJ* 488, 749
Ojha, D., Tamura, M. & Sirius Team 2006, *BASI* 34, 119
Peretto, N., André, Ph. & Belloche, A. 2006, *A&A* 445, 979

Discussion

BIEGING: Poster 84 shows largescale CO maps, which have a 2nd CO(3-2) and IR peak with no obvious triggering mechanism evident.

MINIER: This is also shown in our ^{13}CO(1-0) map.

ZINNECKER: You did not emphasize the presence of the embedded IR-cluster between the S255 and S257 HII regions with hundreds of low-mass stars! This very young cluster could have ejected the two B0-stars that power the S255 and S257 HII regions. These two B0-stars do not have a cospatial low-mass star clusters and were once considered evidence for isolated massive star formation (or bimodal star formation). Taken together with your new data, this region of star formation exhibits a rich variety of complex phenomena that may be characteristic of many other regions and star formation in general.

MINIER: There is a poster by Ohja *et al.* that describes the low-mass star cluster in between S255 and S257. However, this cluster is relatively evolved, 1 Myr or so, and the massive protostars we identified cannot be part of it as they are too young. They must be second-generation stars, especially the massive binary associated with the methanol maser.

Triggered Star Formation in a Turbulent ISM
Proceedings IAU Symposium No. 237, 2006
B. G. Elmegreen & J. Palouš, eds.

Dynamical processes in star forming regions: feedback and turbulence generation

John Bally[1]

[1]Department of Astrophysical and Planetary Sciences, University of Colorado, Boulder, CO 80309, USA
email: John.Bally@colorado.edu

Abstract. Young stellar objects (YSOs) inject large amounts of momentum and kinetic energy into their surroundings. Feedback from low mass YSOs is dominated by their outflows. However, as stellar mass increases, UV photo-heating and ionization play increasingly important roles. Massive stars produce powerful stellar winds and explode as supernovae within 3 - 40 Myr after birth. While low-mass protostellar feedback can drive turbulence in cloud cores and even disrupt the star forming environment, feedback from massive stars plays important roles in the generation of cloud structure and motions in the entire ISM.

Keywords. stars: formation, stars: winds and outflows, ISM: kinematics and dynamics, turbulence

1. Introduction

Chaotic and turbulent motions are ubiquitous in the ISM. On galactic (~ 10 kpc) scales, turbulence can be driven by many processes, including infall from outside the Galaxy, gravitational forces exerted by tidal effects of nearby galaxies or mass concentrations within the galactic disk such as spiral density waves, and processes such as swing amplification, or magneto-rotational instability in the galactic disk (see the various reviews in this volume). While these sources inject energy and momentum into the ISM, convergent flows, vorticity, and dissipation into heat result in the decay of turbulent motions.

It is remarkable that ISM density structure, velocity fields, and turbulence can be characterized by simple power laws on scales ranging from many 10s of kpc to well under 1 pc. On scales larger than the scale-height of the Galactic disk near the Sun, about 100 pc, the disk gas is effectively 2 dimensional while on smaller scales this, disk gas is 3 dimensional. Why is there no break observed in the power-law relationships describing motions and structure in the ISM near 100 pc? Furthermore, why is there no signature of the transition from gravitationally unbound motions in the disk to gravitationally bound molecular clouds and cores? One possible reason is that motions in the ISM may not trace a simple cascade of energy, momentum, and vorticity from a single scale where it is driven to small scales where it is dissipated. Instead, motions may be driven on many scales and such injection may smear out the breaks expected where the flows change dimensionality or become dominated by self-gravity.

In star forming regions, energy and momentum injected by forming stars may dominate all other sources. Most stars are born from dense cloud cores in giant molecular clouds in transient open clusters and in expanding OB associations. Protostellar outflows and UV radiation churn and tend to disrupt parent molecular clouds. UV radiation, winds, and supernovae form superbubbles whose expansion sweeps up the surrounding ISM into shells or rings of denser gas. The roughly 8 $M_\odot$ B3 stars, the least massive to explode as type II supernovae and to produce significant amounts of ionizing radiation, have main-sequence lifetimes of order 30 – 40 Myr. Thus, an OB association or star cluster containing O and early B stars will actively power the expansion of the surrounding ISM and sweep-up shells on this time-scale. After the last supernova, the shell will coast and decelerate. Such superbubbles grow to sizes raging from 100 to over 1,000 pc, comparable to or larger than the typical thickness of the gas layer in galactic disks. They tend to sweep-up rings of dense gas within the galactic plane that become sheared by galactic differential rotation. McCray & Kafatos (1987) found that such rings fragment into self-gravitating clouds with masses or order 10^5 M$_\odot$ in 20 to 50 Myr.

2. Feedback

2.1. *Low-mass star forming regions*

When only low mass stars form, protostellar winds, jets, and bipolar outflows are the only viable form of feedback. Highly collimated jets with steady orientations and velocity tend to blow out of their parent clouds and don't couple energy and momentum efficiently to their surroundings. However, observations show that most YSO jets are surrounded by wide cavities and bipolar outflows that impact a relatively large volume in their surroundings.

Observations of Herbig-Haro (HH) objects show that jets are variable in ejection velocity, degree of collimation, and orientation (e.g. Reipurth & Bally 2001); these processes widen cavities. Ejection velocity variations produce shocks where fast flow elements catch-up to slower moving gas. Pressure gradients in the post-shock layers transform some of this velocity difference into motion orthogonal to the jet axis, resulting in the sideways splashing of ejecta. Examples of such "internal working surfaces" include chains of shocks along the axes of HH1/2 (Bally *et al.* 2002), HH34 (Reipurth *et al.* 2002), HH 26/47 (Hartigan *et al.* 2005), and HH 111 (Reipurth *et al.* 2001; Hartigan *et al.* 1999). Larger bow shocks with higher side-way motions tend to be located downstream from these knotty jets. Low-velocity CO flows and reflection nebulae trace entrained molecular gas and the walls of cavities excavated from the parent cloud by such variable jets. In some flows such as HH 46/47, Spitzer Space Telescope images trace these cavity walls by means of H$_2$ emission (Noriega-Crespo *et al.* 2004).

In the 40 M$_\odot$ L1551 region, a half-dozen outflows criss-cross the cloud. The most luminous YSO, the binary system L1551-IRS5 drives twin jets toward a large cavity lit by reflected starlight and HH objects (Moriarty-Schieven *et al.* 2006; Stojimirović *et al.* 2006). This flow has blown completely out of the L1551 cloud as evidenced by HH objects seen beyond the cloud edge along a line-of-sight containing many background galaxies (Devine *et al.* 1999). The momentum and kinetic energy budgets of the L1551 cloud are dominated by outflows.

While our attention tends to be drawn to the spectacular jets, many YSOs show only marginal evidence for collimated outflows. Forbidden line emission and absorption in many YSO spectra show clear evidence for winds with relatively wide opening angles. Jets also tend to be surrounded by lower density, wide angle winds (e.g. Reipurth & Bally 2001). Wide angle flows tend to generate wide outflow cavities from which ambient material has been displaced and accelerated.

In clustered star forming environments and multiple star systems, outflows can experience forced precession. A companion or passing star in a non co-planar orbit can torque a disk, leading to jet orientation changes. Observationally, such interactions result in outflows that exhibit S-shaped (in case of a uniform precession rate) or Z-shaped point symmetry (if the disk orientation is changed abruptly by a passing star on either a hyperbolic or eccentric orbit). Examples include HH 198 in the L1228 cloud (Bally *et al.* 1995), IRAS 20126+4104 (Cesaroni *et al.* 1997; Shepherd *et al.* 2000), and IRAS 03256+3055 in the NGC 1333 region in the Perseus cloud (Hodapp *et al.* 2005), and Cepheus A (discussed below).

Typical YSOs lose mass at rates ranging from $\dot{M} = 10^{-7}$ to 10^{-5} M$_\odot$ yr^{-1} at velocities ranging from 50 to over 300 km s^{-1}. These flows accelerate the surrounding cloud in momentum-conserving isothermal shocks at rates of order $\dot{P} = 5 \times 10^{-6}$ to 3×10^{-3} M$_\odot$ yr^{-1} km s^{-1}. If a cloud such as L1551 contains 10 outflows at any one time and each lasts 10^5 years, the total momentum transferred to the surrounding ISM is roughly $\dot{M}V\tau \approx 0.5$ to 300 M$_\odot$ km s^{-1}.

The observed ^{13}CO line-widths (away from outflows) in clouds such as the 1 to 2 pc diameter L1551 are around 0.5 to 1.0 km s^{-1}, corresponding to an $M_{1551}V$ product of $20 - 40$ M$_\odot$ km s^{-1} (Larson 1981). Class 0 low-mass YSOs drive flows near the upper end of the mass-loss-rate quoted above for a time-scale of order few 10^4 years, followed by decreasing outflow power over the next 10^6 years as the forming stars evolve through the Class I phase to become classical T Tauri stars (Class II and III YSOs; Reipurth & Bally 2001). Thus, depending on the mass spectrum of YSOs and the star formation history of a particular core, outflows can play roles ranging from insignificant to dominant in the cloud momentum budget. The former case corresponds to clouds with low rates of star formation and little feedback while the latter corresponds to small clusters of stars in which the combined action of multiple YSO outflows tends to disrupt the cloud and terminate star formation. In low-mass star-forming regions, wide-angle winds and outflow cavity widening by jets with variable speeds and orientations increases the coupling of outflow momentum and kinetic energy to the surrounding ISM.

2.2. *Larger clusters and moderate mass star forming regions*

The rate at which random motions are generated by protostellar feedback in molecular clouds is difficult to determine with any degree of reliability. Even rough estimates require complete mapping of entire molecular clouds in tracers such as the optically-thin CO isotopes and dense gas tracers such as CS and/or HCO$^+$, assays of the YSO population in the cloud, sensitive multi-transition CO surveys for bipolar molecular outflows, and deep narrow-band imaging of shock tracers such as Hα, [SII], and the near-IR lines of [FeII] and H$_2$.

For many clouds in the Solar vicinity, surveys are becoming available at visual wavelengths (e.g. Reipurth *et al.* 2004; Walawender *et al.* 2005b; Bally *et al.* 2006b), and in CO (e.g. the "ancient" Bell Labs CO J=1-0 surveys of Orion A, B, and Perseus: Bally *et al.* 1987; Walawender *et al.* 2005; the FCRAO surveys of Perseus: Ridge *et al.* 2006; the Molecular Ring: Jackson *et al.* 2006; and the Outer Galaxy: Brunt & Heyer 2002; Heyer, Carpenter, & Snell 2001).

The CO and visual wavelength surveys of the Perseus molecular cloud (Miesch & Bally 1994; Walawender *et al.* 2004; 2005b) provide an excellent example. The 10^4 M$_\odot$ Perseus cloud, located at a distance of roughly 300 pc contains over a dozen dense cloud cores (Enoch *et al.* 2006; Hatchell *et al.* 2005) several of which are actively forming clusters of several hundred low to intermediate mass stars. The cloud contains the 2 –5 Myr old cluster, IC 348 near its eastern end and the NGC 1333 region, perhaps the most active site of active star formation within 300 pc of the Sun (Quillen *et al.* 2005). About 150 YSOs have formed in NGC 1333 and several dozen are driving outflows into their surroundings. In the Spitzer IRAC near-IR images (Jorgensen *et al.* 2006), and in narrow-band Hα and

[SII] images (Bally *et al.* 1996; Walawender *et al.* 2005b), the cloud is confusion-limited in shocks.

The analyses of Bally *et al.* (1996), Walawender *et al.* (2005b), and Quillen *et al.* (2005) indicate that the energy and momentum released by the dozens of outflows in the NGC 1333 cluster generate a sufficiently large $\dot{M}V$ product to drive the observed cloud motions, perhaps even enough to disrupt the cloud and to eventually stop star formation. Interestingly, Walsh *et al.* (2006) find evidence that gas is falling into this cluster at a rate of about $\dot{M} \approx 10^{-4}$ M$_\odot$ yr^{-1} with a speed of about a half to one km s^{-1}. The accretion rate is comparable to the overall star formation rate. However, for mass ejection rates from stars of order 10% or more of the accretion rate, and outflow speeds of 100 km s^{-1}, the $\dot{M}V$ product associated with this infall is lower than that generated by outflows.

Walawender *et al.* (2005b) present a complete narrow-band survey of the entire Perseus cloud, finding 150 currently active shocks that trace over 50 individual outflows. In addition to the virulent NGC 1333 region, several other cloud cores drive multiple outflows into their surroundings: Barnard 1 (Walawender *et al.* 2005a), the HH 211 region associated with the "Flying Ghost Nebula" located southwest of the older IC 348 cluster of several hundred young stars (Walawender *et al.* 2006), L1448 (Wolf-Chase *et al.* 2001), and L1455 (Bally *et al.* 1997). Additionally, Perseus contains isolated YSOs whose outflows impact adjacent cloud cores (e.g. L1451; Walawender *et al.* 2004).

Walawender's 2005b analysis of the entire Perseus cloud concludes that the turbulent momentum content of the entire Perseus cloud, after subtraction of the overall velocity gradient, is about an order of magnitude larger than the supply rate of momentum generated by outflows. Thus, protostellar outflows may not be able to sustain the observed level of turbulence in the entire cloud. However, on scales of one to a few parsecs, comparable to the size of the cluster forming cores, outflows *do* generate sufficient energy and momentum to self-regulate star formation, stop infall, and disrupt the cloud. However, Walawender *et al.* cautions that most measurements of outflow momentum rely of the detection of shocks, an estimate of the rate at which they process cloud gas, or the analysis of their entrained mass. These measurements are uncertain by about an order of magnitude due to to unknown excitation conditions, poorly constrained extinction corrections, and the non-linear physics of shocks. It is possible that we have not yet detected all of the momentum conveyed by outflows to their surroundings.

The Perseus results provide a natural explanation for why most stars form in clusters: Multiple simultaneous outflows may be required to generate enough outward pressure to stop or even reverse infall. The birth of dozens to hundreds of young stars can produce sufficient feedback to counter the effects of decaying turbulence and self-gravity.

The 2×10^4 L$_\odot$ Cepheus A region (D $\sim$ 700 pc) provides another example of multiple outflows emerging from a region forming several early B stars along with a cluster of low mass YSOs. The most massive YSO, HW2, contains a radio continuum jet (Garay *et al.* 1996; Curiel *et al.* 2006), a luminous cluster of masers (Torrelles *et al.* 2001), a massive cirucmstellar disk (Patel 2005), and a spectacular, collimated outflow that appears to be precessing (Cunningham 2006). Deep H$_2$ images show that over the last 5,000 years, HW2 produced a series of eruptions about every 10^3 years. The orientation of each successive eruption changed by about 15°. The oldest and most distant shock in this system is due east of HW2; the younger and closer shocks are located at smaller position angles. Today, the jet is oriented northeast at PA $\sim$ 45°. Cunningham (2006) proposed that a moderately massive companion star in a highly eccentric, non-coplanar orbit with a period of about 10^3 years triggers episodic accretion/mass-ejection events during periastron passages during which the disk orientation is abruptly changed. The resulting precession provides an excellent example of how a single flow can impact a large portion of a parent cloud.

2.3. *Massive star forming complexes*

The Orion star forming complex illustrates how massive stars dominate star formation feedback. The vicinity of the 10^5 $L_\odot$ Orion Nebula contains several thousand mostly low-mass stars along with about dozen massive ones. The most massive star is the 30 $M_\odot$ θ^1Ori C (see O'Dell 2001 for a review). In addition to the powerful outflows bursting out of the luminous but embedded YSOs in the OMC1 and OMC-1S cloud cores, dozens of low-mass young stars within the HII region also drive jets (Bally 2006a and references therein). However, the momentum injection rate of these outflows pales in comparison to the momentum generated by photo-ionization. The main ionization front at the interface between the Orion A cloud and HII region loses mass at a rate of about 10^{-4} $M_\odot$ yr^{-1} and vents to the southwest at a speed of about 20 km s^{-1}. The advance of the D-type I-front into the molecular cloud must deliver a comparable amount of momentum to the surrounding neutral gas. It is clear that once massive stars are born, their UV radiation, stellar winds, and terminal supernova explosions dominate the energetics of the surrounding ISM.

3. Conclusions

While downward cascades of turbulent energy may dominate the kinematics and structure of the general ISM, in star forming regions, injection by local sources is more important. In dark clouds, and those portions of giant molecular clouds giving birth to only low-mass stars, protostellar winds and jets dominate stellar feedback. A variety of processes such as velocity variability, orientation changes, and wide-angle winds contribute to the widening of outflow cavities to increase the efficiency of the coupling of outflow momentum to the chaotic motions and disruption of the cloud. In clusters, the random orientations of outflows, and dynamical interactions between stars, increases this coupling. However, when massive stars form, their UV radiation, winds, and terminal supernovae explosions dominate momentum and energy injection.

Quantitative measures of the momentum injection rates and feedback from forming stars are just now becoming available. In clustered regions of star formation, these rates appear to dominate all other sources, including the decay of turbulent energy from larger scales. While star formation feedback dominates on the scale of a few parsecs, on the scale of entire GMCs such as the Perseus molecular cloud, turbulent motions must originate form larger scales or be driven by self-gravity. In Orion and Perseus, superbubbles and shells powered by previous generations of stars may feed chaotic motions and structure of molecular gas.

I speculate that multi-scale injection from a variety of processes is responsible for the universality of the power-laws that characterize turbulent motions and density structure in the ISM. The inclusion of momentum injection on scales ranging from one to hundreds of parsecs will tend to flatten these power-laws by adding excess kinetic energy and creating density structures that would otherwise be less evident. Such additional sources may erase the expected breaks where the turbulent cascades transition from two to three dimensions and where self-gravity becomes important.

Acknowledgements

I acknowledge support from NASA grant NCC2-1052 (the University of Colorado Center for Astrobiology) and NSF grant AST 98-19820.

References

Bally, J., Stark, A. A., Wilson, R. W. & Langer, W. D. 1987, *ApJ* 312, L45

Bally, J., Devine, D., Fesen, R. A. & Lane, A. P. 1995, *ApJ* 454, 345

Bally, J., Devine, D. & Reipurth, B. 1996, *ApJ* 473, L49

Bally, J., Devine, D., Alten, V. & Sutherland, R. S. 1997, *ApJ* 478, 603

Bally, J., Heathcote, S., Reipurth, B., Morse, J. & Hartigan, P., Schwartz, R. 2002, *AJ* 123, 2627

Bally, J., Licht, D., Smith, N. & Walawender, J. 2006a, *AJ* 131, 473

Bally, J., Walawender, J., Luhman, K. & Fazio, G. 2006b *AJ* 132, 1923

Brunt, C. M. & Heyer, M. H. 2002, *ApJ* 566, 289

Cesaroni, R., Felli, M., Jenness, T., Neri, R., Olmi, L., Robberto, M., Testi, L. & Walmsley, C. M. 1999, *A&A* 345, 949

Curiel, S., *et al.* 2006, *ApJ* 638, 878

Cunningham, N. 2006, PhD Thesis, University of Colorado, Boulder

Devine, D., Reipurth, B. & Bally, J. 1999, *AJ* 118, 972

Enoch, M. L., *et al.* 2006, *ApJ* 638, 293

Garay, G., Ramirez, S., Rodriguez, L. F., Curiel, S. & Torrelles, J. M. 1996, *ApJ* 459, 193

Hartigan, P., Morse, J. A., Reipurth, B., Heathcote, S. & Bally, J. 2001, *ApJ* 559, L157

Hartigan, P., Heathcote, S., Morse, J. A., Reipurth, B. & Bally, J. 2005, *AJ* 130, 2197

Hatchell, J., Richer, J. S., Fuller, G. A., Qualtrough, C. J. & Ladd, E. F., Chandler, C. J. 2005, *A&A* 440, 151

Heyer, M. H., Carpenter, J. M. & Snell, R. L. 2001, *ApJ* 551, 852

Hodapp, K. W., Bally, J., Eislöffel, J. & Davis, C. J. 2005, *AJ* 129, 1580

Jackson, J. M., *et al.* 2006, *ApJS* 163, 145

Jørgensen, J. K., *et al.* 2006, *ApJ* 645, 1246

Larson, R.B. 1981, *MNRAS* 194, 809

McCray, R. & Kafatos, M. 1987, *ApJ* 317, 190

Miesch, M. S. & Bally, J. 1994, *ApJ* 429, 645

Moriarty-Schieven, G. H., Johnstone, D., Bally, J. & Jenness, T. 2006, *ApJ* 645, 357

Noriega-Crespo, A., *et al.* 2004, *ApJS* 154, 352

O'Dell, C. R. 2001, *ARAA* 39, 99

Patel, N. A., *et al.* 2005, *Nature* 437, 109

Quillen, A. C., Thorndike, S. L., Cunningham, A., Frank, A., Gutermuth, R. A., Blackman, E. G., Pipher, J. L. & Ridge, N. 2005, *ApJ* 632, 941

Reipurth, B., Hartigan, P., Heathcote, S., Morse, J. A. & Bally, J. 1997, *AJ* 114, 757

Reipurth, B. & Bally, J. 2001, *ARAA* 39, 403

Reipurth, B., Heathcote, S., Morse, J., Hartigan, P. & Bally, J. 2002, *AJ* 123, 362

Reipurth, B., Yu, K. C., Moriarty-Schieven, G., Bally, J., Aspin, C. & Heathcote, S. 2004, *AJ* 127, 1069

Ridge, N. A., *et al.* 2006, *AJ* 131, 2921

Shepherd, D. S., Yu, K. C., Bally, J. & Testi, L. 2000, *ApJ* 535, 833

Torrelles, J. M., et al. 2001, *ApJ* 560, 853

Stojimirović, I., Narayanan, G., Snell, R. L. & Bally, J. 2006, *ApJ* 649, 280

Walawender, J., Bally, J., Kirk, H., Johnstone, D., Reipurth, B. & Aspin, C. 2006, *AJ* 132, 467

Walawender, J., Bally, J., Kirk, H. & Johnstone, D. 2005, *AJ* 130, 1795

Walawender, J., Bally, J. & Reipurth, B. 2005, *AJ* 129, 2308

Walawender, J., Bally, J., Reipurth, B. & Aspin, C. 2004, *AJ* 127, 2809

Walsh, A. J., Bourke, T. L. & Myers, P. C. 2006, *ApJ* 637, 860

Wolf-Chase, G. A., Barsony, M. & O'Linger, J. 2000, *AJ* 120, 1467

Discussion

HEYER: 1: Lack of break in structure function is due to resolution and dynamic range of analysis. 2. Outflows need to be widely distributed to affect and drive turbulence in the FULL cloud. It is difficult to imagine that outflows can generate near-coherent large scale shear component observed in CO maps.

BALLY: Josh Walawender demonstrated (in his PhD thesis; see Walawender *et al.* 2005) that in the Perseus molecular cloud energy and momentum injection dominates turbulent decay on scales of about one parsec *in regions where stars are forming*. Outflows may fail by more than an order of magnitude on the scale of the entire cloud (20 pc). I suspect that motions on these large scales may be driven by the nearby Per OB2 association by a combination of shear flow uv-induced photo-ablation, and winds from stars and supernovae.

NAKAMURA: Recently, Li and I did numerical simulations of turbulent magnetized clouds, including effects of protostellar outflows and we found that dynamic interaction between magnetic field and outflows is important to generate supersonic turbulence because outflows generate large amplitude Alfven and MHD waves that transform outflow motions into turbulent motions. So, my question is, is there any evidence showing interaction between magnetic field and outflows in your observations.

BALLY: Possibly, Observations of the L1551 cloud and the outflow from Barnard 5 IRS1 reveal ridges of emission parallel to the outflow axis but located in the molecular cloud well outside the boundaries of the outflow cavity. These density and very low velocity perturbations may have been formed by some sort of perturbation, induced by the outflow, propagating through the cloud. Might these be your MHD or Alfven waves?

ZINNECKER: Since I am blamed to be the bad boy before lunch, I want to make up for it and be the good guy after lunch by advertising that John Bally together with Bo Reipurth, just published a popular book on "The Birth of Stars and Planets", which you can see upstairs in the booth of Cambridge University Press! (Do I get a free copy now?)

BALLY: Yes. Some of the images I used in this presentation are shown in our book.

MAC LOW: Matzner (2002) made a quantitative argument for outflows supporting small molecular clouds, but HII regions supporting larger ones.

BALLY: Outflows are an important feedback mechanism only when massive stars are absent. A and late B stars come to dominate feedback by soft uv photo-heating of PDRs. O stars inject much more energy through the propagation of their ionisation fronts. So I completely agree – massive stars and their uv dominate feedback when they are present – typically in larger, more massive star-forming regions.

Triggered Star Formation in a Turbulent ISM
Proceedings IAU Symposium No. 237, 2006
B. G. Elmegreen & J. Palouš, eds.

© 2007 International Astronomical Union
doi:10.1017/S174392130700141X

Hypersonic swizzle sticks: jets, fossil cavities and turbulence in molecular clouds

Andrew J Cunningham[1], Adam Frank[1], Eric G Blackman[1] and Alice Quillen[1]

[1]Department of Physics and Astronomy, University of Rochester, Rochester, NY 14627, USA
email: ajc4@pas.rochester.edu

Abstract. The ubiquity and high density of outflows from young stars in clusters make them an intriguing candidate for the source of turbulence energy in molecular clouds. In this contribution we discuss new studies, both observational and theoretical, which address the issue of jet/outflow interactions and their ability to drive turbulent flows in molecular clouds. Our results are surprising in that they show that fossil cavities, rather than bow shocks from active outflows, constitute the mechanism of re-energizing turbulence. We first present simulations which show that collisions between active jets are ineffective at converting directed momentum and energy in outflows into turbulence. This effect comes from the ability of radiative cooling to constrain the surface area through which colliding outflows entrain ambient gas. We next discuss observational results which demonstrate that fossil cavities from "extinct" outflows are abundant in molecular material surrounding clusters such as NGC 1333. These structures, rather than the bow shocks of active outflows, comprise the missing link between outflow energy input and re-energizing turbulence. In a separate theoretical/simulation study we confirm that the evolution of cavities from decaying outflow sources leads to structures which match the observations of fossil cavities. Finally we present new results of outflow propagation in a fully turbulent medium exploring the explicit mechanisms for the transfer of energy and momentum between the driving wind and the turbulent environment.

Keywords. ISM: clouds,ISM: jets and outflows

1. Introduction

Star formation occurs within Molecular Clouds (MCs), complex structures whose physical evolution is still not clearly understood (Ballesteros-Paredes *et al.* (2006)). MCs are hierarchical structures with smaller substructures known as clumps and cores. Star formation is believed to occur in cores with larger clusters forming from more massive cores. The expected lifetimes for molecular clouds has become a topic of considerable debate as numerical simulations have shown that MHD turbulence, the nominal means of support for clouds against self-gravity, decays on a crossing timescale (Goldreich & Kwan (1974), Arons & Max (1975), Stone *et al.* (1998), Mac Low & Klessen (2004)). In light of this result the traditional view that MCs are long-lived, quasi-static equilibrium structures has been challenged by a paradigm in which star formation occurs on a timescale comparable to the free-fall time (Ballesteros-Paredes *et al.* (1999), Elmegreen (2000), Hartmann(2003)). In the former case turbulence in the cloud is an important source of support and regulation of the star formation efficiency and it must be re-supplied over time. In the latter case turbulence is produced with the cloud (Yamaguchi *et al.* (2001)) or only needs to be driven up to the point that a cloud is disrupted.

Feedback from protostars forming within a MC has been cited by many authors as a principle means of either re-energizing turbulence or disrupting clouds (Bally &

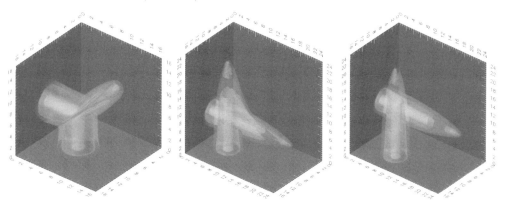

Figure 1. Volume rendered image of density from AMR simulations of two jets undergoing a 90° collision displaced by an impact parameters of 0, 1 jet radius, and non-interacting from left to right.

Reipurth (2001)). When massive stars form their ionization fronts, strong stellar winds and eventual supernova blastwaves are expected to be the major contributor to feedback (Krumholz & McKee (2005)). In lower mass clusters and environments where the effects of massive stars have not been felt protostellar outflows will likely be the dominant form of feedback. In these cases, even if energy is re-supplied from supra-cluster scales, at some wavenumber the outflow injection may come to dominate global dynamics. In fact, energetic outflows associated with low and moderate mass young stellar objects are known to exert a strong effect on their parent molecular clouds (for a recent review see Bally *et al.* (2006)). Young stellar outflows in settings such as NGC 1333 have been shown to contain sufficient kinetic energy to excite a significant fraction of supersonic turbulence in their surrounds and/or unbind and disperse portions of their parent cloud. (Bally & Reipurth (2001), Knee & Sandell (2000), Warin *et al.* (1996)).

In this contribution we explore the role of outflows in returning energy and momentum to the surrounding media from young stars. We focus first on simulation results exploring the role of collisions between active outflows, demonstrating that these types of interaction are not more effective than non-interacting outflows at setting the ambient material in motion or generating randomized isotropic motions which are characteristic of turbulence. Recognizing that the driving properties of star formation outflows (in particular mass loss rates) evolve on timescales shorter than the duration of the outflow, we then focus on fossil outflows first from an observational viewpoint and then in terms of simulations designed to recover and interpret those observations.

2. Active outflow collisions

We estimate $N_{critical}$, the protostellar density that achieves a volume fill ratio of 10% bowshock overlap. Above this density expect the effect of collisions between outflows to become appreciable. Assuming typical values for the protostellar outflow size, bow shock radius, outflow lifetime and cloud lifetime we find $N_{critical} = 500$ pc^{-3} (Cunningham *et al.* (2006A)). This is comparable to the protostellar density of many star forming regions. Outflow interactions are therefore statistically likely to occur in a typical star forming region.

Motivated by the implied likelihood for jet collisions, we explored the efficacy of active outflow interactions in stirring the ambient medium via 3-D AMR simulations of jet collisions (Cunningham *et al.* (2006B)). Our study focused on hydrodynamic simulations

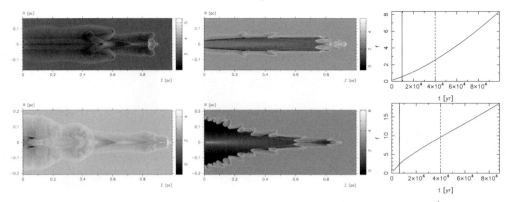

Figure 2. Left: Simulations of fossil cavities driven by winds which are active for 10^4 yr. Number density images are shown at $\sim 10^5$ yr. Note that backfilling via rarefaction waves has occurred. Center: Outflows driven by active winds shown for comparison with frame taken at $\sim 3x104$ yr. Right: fractional difference between the approximate scaling relation of Quillen *et al.* (2005) and the simulation results. Between the two vertical lines delineating the wind shut-off time and the time at which the flows have decelerated to ~ 1 kms^{-1}, the scaling relation and simulation differ by a factor of order unity. The top panel shows the results for a collimated jet driven flow and the lower panel for a wide angle wind driven flow.

of the interaction of two orthogonal outflows at several impact parameters (figure 1). The simulations included the effect of radiative energy loss on the flow and we investigated the role of the impact parameter and degrees of collimation. The simulations were carried out in 3D using the AstroBEAR adaptive mesh refinement (AMR) code.

Surprisingly, our results indicated that the high degrees of compression of outflow material, achieved through radiative shocks near the vertex of the interaction, prevent the redirected outflow from spraying over a large spatial region. Furthermore, the collision reduces the redirected outflow's ability to entrain and impart momentum into the ambient cloud. We note the study by Li & Nakamura (2006) explored outflow interactions with a collapsing turbulent cloud and concluded that outflow activity could re-energize turbulence. These studies do not contradict our results as we agree with their main conclusion however, their simulations were of quite low resolution (128^3 for the entire cluster) and could not adequately resolve the interactions of individual outflows or include the outflow power evolution (a topic we will explore in the next section). Thus combining the results of our simulations with consideration of the probabilities of outflow collisions led us to conclude that individual low velocity fossil outflows, interacting on long timescales, provide the principle coupling between outflows and the cloud.

3. Fossil outflows

Many authors have relied on "0-D" estimates of the energy present in active outflows and compare these with cloud turbulent energy (Bally & Reipurth (2001)). Studies of individual objects however express a more complicated picture. In particular the explicit time-dependent nature of the coupling between cloud material and outflows is not addressed in these estimates. Outflow power evolves rapidly in time in the strongest phases (Class 0 sources) lasting a fraction of the star formation timescale. Thus outflows will continually be turning on and fading across the history of an active star forming region.

The pitfalls of ignoring the temporal domain was highlighted in a recent study of NGC1333 (Quillen *et al.* (2005)). In this work it was found that velocity dispersions, measured in 13CO, did not vary across the cloud. There was no link between active

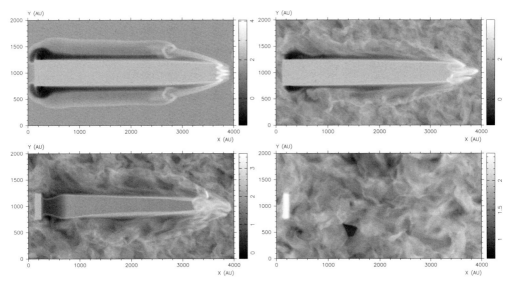

Figure 3. 2-D number density slices of 3-D simulations of jets with different decay times driven into turbulent media. Top Left: Control in which constant jet driven into quiescent media. Top Right: Constant jet driven into turbulent media. Lower Left: Slowly decaying jet driven into turbulent media. Lower Right: Short pulsed jet driven into turbulent media. Note the effect of the turbulent media on then bow shock. In short pulse jet simulations, the outflowing material has been completely subsumed by the turbulence. This is the expected behavior for long-extinct outflows.

outflows and turbulence. Instead a new class of outflow signature was identified in the form of fossil cavities. These fossil cavities proved to be a smoking gun showing strong coupling between outflows and the molecular cloud. Cavities at a range of sizes and velocities were observed in the cloud. Twenty cavities were identified with typical diameter of about 0.1–0.2 pc, and velocity widths 1–3 kms^{-1}. If these cavities were simply empty regions in the cloud, the timescale for them to fill in would be less than a million years implying that they were created relatively recently.

In a recent, more detailed study, AMR simulations of cavity evolution using full H_2 chemistry and cooling (Cunningham *et al.* (2006B)) were performed. In this work jets and wide angle winds were simulated with an injected momentum flux that decreased in time (figure 2). These simulations were compared with runs with constant momentum flux. The decaying flux models exhibited deceleration of the outflow head and backfilling via expansion off of the cavity walls. They also recovered observed morphological signatures including lower density contrasts in comparison to the continuously driven counterparts. Most important, the simulations agree with the scaling relations developed by Quillen *et al.* (2005) to estimate cavity momentum from their readily observable properties (figure 2, right). This work also provided synthetic observations in terms of P-V diagrams which demonstrate that fossil cavities form both jets and wide angle outflows are characterized by linear "Hubble-law" expansions patterns superimposed on "spur" patterns indicative of the head of a bow shock. These should prove useful in future observational work.

4. Conclusion & future work

We have shown that fossil cavities, rather than active outflows, may be the direct link between stellar injection of mechanical energy and turbulence within young clusters.

Thus while turbulent energy may be supplied at larger scales to the clouds as a whole (via supernova or gravitational collapse) which then cascades down, there is also a separate injection of energy at smaller scales which constitutes a feedback from the stars within the cluster. Future work will need to make the interplay between turbulence and energy injection from jets more explicit. Figure 3 shows initial work in this direction in the form of simulations of jets into fully turbulent media and the subsequent evolution of the jet driven cavity (Cunningham *et al.* (2007)). By the time several turbulent crossing times have elapsed, the cavities have slowed via the entrainment of the ambient turbulent gas sufficiently to be subsumed into and become indistinguishable from the turbulent motions of the cloud. Through this process, the jet momentum will act to feed the turbulence that will ultimately destroy it.

References

Arons, J. & Max, C. E. 1975, *ApJ* 196, L77

Ballesteros-Paredes, J., Hartmann, L. & Vázquez-Semadeni, E. 1999, *ApJ* 527, 285

Ballesteros-Paredes, J., Klessen, R., Mac Low, M-M. & Vazquez-Semadeni, E, 2006, in: B. Reipurth, D. Jewitt, & K. Keil (eds.), *Protostars and Planets V* (Tucson: Univ. of Arizona), in press

Bally, J., Reipurth, B. & Davis, C., 2006, in: B. Reipurth, D. Jewitt, & K. Keil (eds.), *Protostars and Planets V* (Tucson: Univ. of Arizona), in press

Bally, J. & Reipurth, B. 2001, *ARA&A* 39, 403

Cunningham, A. J., Frank, A. & Blackman, E. G. 2006A, *ApJ* 646, 1059

Cunningham, A. J., Frank, A., Blackman, E. G. & Quillen, A., 2006B, *ApJ* in press

Cunningham, A. J., Frank, A., Blackman, E. G. & Quillen, A., 2007, in prep

Elmegreen, B. G. 2000, *ApJ* 530, 277

Goldreich, P. & Kwan, J. 1974, *ApJ* 189, 441

Hartmann, L. 2003, *ApJ* 585, 398

Knee, L. B. G. & Sandell, G. 2000, *A&A* 361, 671

Krumholz, M. R. & McKee, C. F. 2005, *ApJ* 630, 250

Li, Z.-Y. & Nakamura, F. 2006, *ApJ* 640, L187

Quillen, A. C., Thorndike, S. L., Cunningham, A., Frank, A., Gutermuth, R. A., Blackman, E. G., Pipher, J. L. & Ridge, N. 2005, *ApJ* 632, 941

Mac Low, M.-M. & Klessen, R. S. 2004, *Rev. Mod. Phys.* 76, 125

Stone, J. M., Ostriker, E. C. & Gammie, C. F. 1998, *ApJ* 508, L99

Warin, S., Castets, A., Langer, W. D., Wilson, R. W. & Pagani, L. 1996, *A&A* 306, 935

Yamaguchi, R., *et al.* 2001, *PASJ* 53, 985

Triggered Star Formation in a Turbulent ISM
Proceedings IAU Symposium No. 237, 2006
B. G. Elmegreen & J. Palouš, eds.

© 2007 International Astronomical Union
doi:10.1017/S1743921307001421

The newest twists in protostellar outflows: triggered star formation

Mary Barsony†

¹Space Science Institute, 4750 Walnut St., Suite 205, Boulder, CO 80301, USA
email: barsony@spacescience.org

Abstract. Winds and outflows powered by the release of gravitational potential energy of infalling matter have been invoked as the triggering agent for further star-formation on scales from dwarf galaxies to individual, low-mass stars. This brief review will touch upon the circumstances under which outflows can serve as star-formation triggers, from specific observations of present-day, nearby protostellar flows.

Keywords. circumstellar matter, stars: formation, stars: winds, outflows, ISM: jets and outflows

1. Observational manifestations of outflows

Protostellar outflows are recognized as such by the radiative cooling of shocked interstellar gas they generate. Since their discovery thirty years ago (Zuckerman *et al.* 1976), bipolar outflows generated by forming stars have been observed via millimeter emission lines of tracer molecules, such as CO, CS, SiO (e.g., Bally & Lada (1983), Fukui *et al.* (1993), Bachiller (1996), Wolf-Chase *et al.* (1998), Chandler *et al.* (2001), Hirano *et al.* (2006)) and molecular ions such as HCO^+ (Aso *et al.* (2000), Rawlings *et al.* (2004), Arce & Sargent (2006)). At optical wavelengths, the most commonly observed outflow tracers used in imaging studies are the emission lines of the [SII] doublet at 6716Å and 6731Å and of Hα at 6563Å (Mundt & Fried (1983), Bally *et al.* (1997), Reipurth & Bally (2001)). The 2.122 μm 1−0 S(1) transition of the H_2 molecule has been used to trace outflow shocks for three decades (Gautier *et al.* 1976). Dramatic improvements in areal coverage and sensitivity of the narrowband H_2 images have been attained with the relatively recent advent of large-format near-infrared detectors, such as in the WIRC instrument at Palomar, IRIS2 at the AAO, ISPI at CTIO, and WFCAM on UKIRT.

The most recent addition to the arsenal of outflow-imaging techniques has been the IRAC on-board the *Spitzer* Space Telescope. Of IRAC's four passbands, centered at 3.6μm [Band 1], 4.5μm [Band 2], 5.8μm [Band 3], and 8.0μm [Band 4], the first three contain 6, 7, and 8 H_2 emission lines, respectively (Smith & Rosen (2005)).

Band 2 has been an especially useful outflow tracer, since it also encompasses the CO v=1-0 fundamental vibrational emission band at 4.7μm, as well as the HI Brα emission line at 4.05μm.

2. Observational evidence for outflow-triggered star-formation

2.1. The case of L1448

Figure 1 shows the *Spitzer* IRAC view of an $11' \times 11'$ field in Perseus (d= 350 pc) harboring four known Class 0 sources and their outflows. The Class 0 sources are L1448C,

---

† Present address: San Francisco State University, Department of Physics and Astronomy, 1600 Holloway Drive, San Francisco, CA 94132 USA

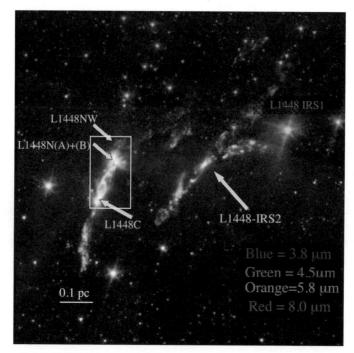

Figure 1. Color-composite image of *Spitzer's* IRAC view of the 11′ × 11′ L1448 field, produced by Dr. Robert Hurt of the Spitzer Science Center. N is up, and E is to the left in this image. See text for details.

the 7″ separation proto-binary, L1448N(A) + L1448N(B), and L1448 IRS2. The nature of L1448NW remains to be determined—it may be a pre-stellar clump or another Class 0 protostar. L1448 IRS1 is known to be a more evolved, Class I object. One of the most remarkable features of this multi-band image is the scattered light appearance of the outflow cavity walls carved out by each outflow. A conical, V-shaped, bluish-tinged scattered light structure is seen opening towards the NW from L1448C, towards the W originating from L1448N(B), and exhibiting a bipolar appearance towards L1448 IRS2. Comparison of this image with the published high-velocity blue- and red-shifted CO gas map of the region (Wolf-Chase *et al.* 2000) reveals the blue-shifted component of the outflow driven by L1448N(B) for the first time. Wolf-Chase *et al.* had already made the case for the possibility of two flows driven by L1448 IRS2, based on the morphology of the blue-shifted high-velocity CO gas in the vicinity. However, the additional information provided by Figure 1, that the scattered light cavity of L1448N(B) points towards the West, calls for a revised interpretation. The shock structures that seem to emanate to the NW from L1448 IRS2, bending in a southerly direction near L1448 IRS1, and are coincident with high-velocity blue-shifted CO, are actually all manifestations of the blue-shifted lobe of the L1448N(B) flow, identified here for the first time. Since L1448 IRS2 has been claimed to be relatively the youngest of these four Class 0 objects (O'Linger *et al.* 2006), there is the possibility that the collision of the blue-shifted flow driven by L1448N(B) may have triggered the collapse of the pre-stellar clump that formed L1448 IRS2.

However, for a clear observational case for outflow-induced star-formation, we refer the reader to Figure 2, which is an enlarged view of the region outlined by the light-blue rectangle surrounding the protostars, L1448C and L1448N(A)+(B), in Figure 1. In Figure 2, red dots indicate the positions of the millimeter continuum sources in this region:

the protostars, L1448C, L1448N(A), L1448N(B), and the pre-stellar clump/protostellar candidate, L1448NW. The blue-shifted, conical outflow cavities dominated by the scattered light from Band 1 (and, therefore, having a light-blue hue in this image) have been outlined by dark-blue lines, pointing to the NW from L1448C and to the W from L1448N(B). The similar structure for L1448N(A), although not as clear, is indicated by the dashed dark blue curve to the NW of L1448N(A). The red-shifted flow/cavity boundaries have been indicated only for L1448C (traced by the solid red curve to the SE), and for L1448N(A) (traced by the dashed red curve towards the SE). The directions of the red-shifted flows powered by L1448C and L1448N(A) are known from other previously published works (Bachiller *et al.* 1995, Barsony *et al.* 1998). The positions of H_2 knots, identified from ground-based imaging observations, are indicated by letters and arrows. The light blue arrows, pointing towards features labeled U-Z, show the locations of low-excitation H_2 shocks, associated with the red-shifted outflow lobe driven by L1448N(A). The red arrows, pointing towards features A-I, mark the locations of high-excitation H_2 emission regions, and are associated with the blue-shifted outflow lobe of L1448C (Davis & Smith 1995). The white, arc-like objects, with their apexes near features A, E, and G, respectively, are pure H_2 emission regions, also detected in narrowband imaging from the ground (Davis *et al.* 1994). The blue-shifted flow driven by L1448C takes a sharp bend (from the high-excitation H_2 shock H to I) right where the flow impacts the gas and dust containing L1448N(A)+L1448N(B). Strong shock emission, detected at 100 μm in the HIRES-processed *IRAS* data of this region, coincides with this bend in the flow, further corroborating the heating and compressional effect the flow is having on the gas structure containing the L1448N binary protostellar system (Barsony *et al.* 1998). This, then, is strong evidence for the L1448C flow triggering the collapse and formation of the binary protostar, L1448N(A)+L1448N(B).

2.2. *Other nearby star-forming regions*

The borderline Class 0/Class I object, L1551NE, lies directly in the red-shifted outflow lobe of its more famous neighbor, the Class I close binary, L1551/IRS5. Both sources are located in the Taurus dark clouds, at 140 pc distance. Combined Nobeyama single-telescope and interferometric maps of various transitions and isopotomers of the CS molecule of the region encompassing L1551NE were acquired by Yokogawa *et al.* (2003). These authors interpret the combined morphology and velocity structure of the CS-emitting gas surrounding the transitional Class 0/Class I object, L1551NE, as evidence for this gas having been swept up, shocked, and dense enough to trigger the formation of L1551NE from the powerful outflow of its neighbor, L1551 IRS5. The fact that L1551 NE is actually physically inside the outflow cavity produced by the combined flows of the Class I L1551 IRS 5 binary, and that L1551NE is at a younger evolutionary stage than L1551/IRS5, both strongly support the contention of outflow-triggered star-formation in this case.

The nearby (310-320 pc) NGC 1333 star-forming cluster may be unique in the sheer wealth of outflows it exhibits. Its *Spitzer* composite color image in all of the IRAC bands was shown by several different speakers at this Symposium. The protostar known as NGC 1333/IRAS 4A is located within the IRAC image. IRAS 4A is known to be dynamically collapsing on the size scale of its protostellar envelope, from detection of red-shifted absorption of molecular lines against its bright, millimeter dust continuum (Di Francesco *et al.*. 2001). IRAS 4A is not directly detected by IRAC—one sees only a very dark absorption feature at its location. The greenish-tinged sources in the vicinity of IRAS 4A in the color-composite IRAC image are shocks associated with various outflows. Two of these, HL 6 and HL 9, are associated with near-infrared 2.12μm shocked H_2 emission first

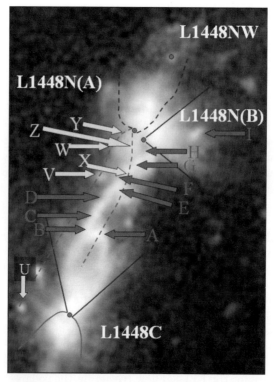

Figure 2. *Spitzer's* IRAC view of the three Class 0 protostars, L1448C and the wide binary, L1448N(A) and L1448N(B)–all marked by red dots. The dust condensation, L1448NW, also marked by a red dot, may be pre-stellar. North is up and East is to the left in this image. The blue-shifted outflow lobes of each protostar, detected via millimeter CO emission, are outlined in blue. Red arrows indicate the locations of high-excitation near-infrared H_2 shocks, A-I, associated with the blue-shifted portion of the L1448C bipolar flow, whereas light blue arrows indicate the locations of low-excitation near-infrared H_2 shocks, U-Z, associated with the red-shifted lobe of the L1448N(A) bipolar flow. Note that the blue-shifted lobe of the L1448C flow veers sharply to the West–from shock H to shock I, right at the position of the proto-binary, whose formation this flow may have triggered. Further north, the blue-shifted flow of L1448N(A) may have helped form the pre-stellar condensation, L1448NW.

detected from the ground (Hodapp & Ladd 1995). The observational evidence for outflow-triggered formation for IRAS 4A is not yet conclusive. A coordinated, multi-wavelength study of both the morphology and the kinematics of the gas and flows in the vicinity of IRAS 4A is required before one could make a well-justified case for outflow-triggered star-formation in this instance.

3. Theoretical considerations regarding outflow-triggered star-formation

Despite the lack of conclusive observational evidence for the triggering of the collapse of IRAS 4A by a single, identifiable outflow, as was the case for L1551NE and L1448N, IRAS 4A has been modelled as having collapsed due to triggering from a sudden increase in external pressure (Belloche *et al.* 2006, and article by Hennebelle in this Volume). This sudden pressure increase was assumed to be caused by the combined flows of NGC 1333. A simple non-rotating, spherical Bonnor-Ebert sphere source model was used as a starting point for the SPH code calculation. The initial model is assumed to be embedded in a hot rarefied external medium. In order to reproduce the envelope's density structure (as inferred from multi-scale mm continuum observations) and velocity structure (derived from multi-line and isotopic spectral line maps), the applied external pressure to IRAS 4A must have doubled in just 3000 years for an envelope with a 10^5 yr. crossing time, in order to be consistent with the densities and velocities deduced from the observations. The resulting model can match the observations over the $1500\text{AU} - 10^4$ AU spatial scale for 10,000 to 30,000 yr after the formation of a central "point" mass. The radial mass distributions calculated from the model are too rarefied to reproduce the high densities observed on spatial scales from a few hundred AU to 1500 AU, however. The primary argument for invoking triggering for the collapse of IRAS 4A is its deduced high mass infall rate of 10^{-4} $M_\odot$ yr^{-1}, which is 2 orders of magnitude above that expected for a collapsing isothermal sphere.

4. Outstanding problems and future work

Outflow triggered star formation is operative on many scales, from intergalactic (e.g., Silk & Rees 1998), to interstellar, to individual stellar systems. Due to space considerations, only the smallest scales were considered here. A single protostellar flow impinging on an ambient interstellar clump may or may not cause it to collapse and form another protostar. Only the clearest cases of outflow-induced star-formation were presented. There are many instances where outflows impact clumps, but these clumps just remain as obstacles in/to the flow, without collapsing. Future systematic studies of known outflows may distinguish the precise range of conditions under which outflow-triggered star formation occurs.

Acknowledgements

Helpful discussions with Josh Walawender in comparing multi-waveband images of the Perseus molecular cloud are gratefully acknowledged. M.B. wishes to acknowledge NASA/JPL Spitzer Cycle 3 funding which made her travel to attend the IAU XXV GA in Prague, and this Symposium 237, possible. Parts of this work were also supported under NSF Grants 95-01788, 97-31797, and 02-06146.

References

Arce, H.G. & Sargent, A.I. 2006, *ApJ* 646, 1070
Aso, Y., Tatematsu, K., Sekimoto, Y., Nakano, T., Umemoto, T., Koyama, K. & Yamamoto, S. 2000, *ApJS* 131, 465
Bachiller, R., Guilloteau, S., Dutrey, A., Planesas, P. & Martin-Pintado, J. 1995, *A&A* 299, 857
Bachiller, R. 1996, *ARA&A* 34, p. 111
Bally, J. & Lada, C.J. 1983, *ApJ* 265, 824
Bally, J., Devine, D., Alten, V. & Sutherland, R.S. 1997, *ApJ* 478, 603
Barsony, M., Ward-Thompson, D., André, P. & O'Linger, J. 1998, *ApJ* 509, 733
Belloche, A., Hennebelle, P. & André, P. 2006, *A&A* 453, 145
Chandler, C.J. & Richer, J.S. 2001, *ApJ* 555, 139
Davis, C.J., Dent, W.R.F., Matthews, H.E., Aspin, C. & Lightfoot, J.F. 1994, *MNRAS* 266, 933
Davis, C.J. & Smith, M.D. 1995, *ApJ* 443, L41
Di Francesco, J., Myers, P.C., Wilner, D.J., Ohashi, N. & Mardones, D. 2001, *ApJ* 562, 770
Fukui, Y., Iwata, T., Mizuno, A., Bally, J. & Lane, A.P. 1993, in E.H. Levy & J.I. Lunine (eds.), *Protostars & Planets III* (Tucson: Univ. of Arizona), p. 603
Gautier, T.N., Fink, U., Larson, H.P. & Treffers, R.R. 1976, *ApJ* 207, L129
Hirano, N., Liu, S.-Y., Shang, H., Ho, P.T.P., Huang, H.-C., Kuan, Y.-J., McCaughrean, M.J. & Zhang, Q. 2006, *ApJ* 636, L141

Hodapp, K.-W. & Ladd, E.F. 1995, *ApJ* 453, 715

Mundt, R. & Fried, J.W. 1983, *ApJ* 274, L83

O'Linger, J.C., Cole, D.M., Ressler, M.E. & Wolf-Chase, G. 2006, *AJ* 131, 2601

Reipurth, B. & Bally, J. 2001, *ARA&A* 39, p. 403

Rawlings, J.M.C., Redman, M.P., Keto, E. & Williams, D.A. 2004, *MNRAS* 351, 1054

Silk, J. & Rees, M. 1998, *A&A* 331, L1

Smith, M.D. & Rosen, A. 2005, *MNRAS* 357, 1370

Wolf-Chase, G.A., Barsony, M., Wootten, H.A., Ward-Thompson, D., Lowrance, P.J., Kastner, J.H. & McMullin, J.P. 1998, *ApJ* 501, L193

Wolf-Chase, G.A., Barsony, M. & O'Linger, J.. 2000, *AJ* 120, 1467

Yokogawa, S., Kitamura, Y., Momose, M. & Kawabe, R. 2003, *ApJ* 595, 266

Zuckerman, B., Kuiper, T.B.H. & Rodriguez-Kuiper, E.N. 1976, *ApJ* 209, L137

Triggered Star Formation in a Turbulent ISM
Proceedings IAU Symposium No. 237, 2006
B. G. Elmegreen & J. Palouš, eds.

© 2007 International Astronomical Union
doi:10.1017/S1743921307001433

Probing turbulence in OMC1 at the star forming scale: observations and simulations†

Maiken Gustafsson[1]‡, Axel Brandenburg[2], Jean-Louis Lemaire[3] and David Field[1]

[1]Department of Physics and Astronomy, University of Aarhus, 8000 Aarhus C, Denmark
email: maikeng@phys.au.dk, dfield@phys.au.dk

[2]NORDITA, Blegdamsvej 17, 2100 Copenhagen Ø, Denmark

[3]Observatoire de Paris & Université de Cergy-Pontoise, LERMA & UMR 8112 du CNRS, 92195 Meudon, France

Abstract. Using radial velocities of vibrationally excited H_2 emission in OMC1 we present the structure functions and the scaling of the structure functions with their order at scales ranging from 70 AU to 30000 AU extending earlier related studies to scales lower by two orders of magnitude. The structure functions for OMC1 show clear deviations from power laws at 1500 AU. The scaling of the higher order structure functions with order deviates from predicted theoretical scalings. Observational results are compared with simulations of supersonic hydrodynamic turbulence. The unusual scaling is explained as a selection effect of preferentially observing the shocked part of the gas. The simulations are unable to reproduce the deviations from power laws of the structure functions.

Keywords. turbulence, shock waves, ISM: kinematics and dynamics, ISM: molecules, ISM: jets and outflows

1. Introduction

The energy spectrum and the related structure functions have been observed to follow power laws in many molecular clouds over a broad range of scales (e.g. Heyer & Brunt (2004)). This feature is usually explained as a manifestation of an energy cascade in a turbulence dominated medium, resulting in self-similar structures without any characteristic scales. However, energy may be injected into or removed from the system by processes such as self-gravity, star formation and supernova explosions and may thus introduce breaks in the energy cascade. In order to understand the evolution of molecular clouds and the importance of turbulence in the star formation process it is essential to identify the presence of these energy modifying processes and the scales at which they occur. This can be achieved by analyzing structure functions of the velocity and should involve a close interplay between observations and numerical simulations. Here we use high spatial resolution (0.15″) infrared observations of velocities in the massive star forming region OMC1. We then compare the observational results with numerical simulations of supersonic hydrodynamic turbulence.

---

† Based on observations obtained at the Canada-France-Hawaii Telescope (CFHT) which is operated by the National Research Council of Canada, the Institut National des Sciences de l'Univers of the Centre National de la Recherche Scientifique of France, and the University of Hawaii.

‡ Present address: Max-Planck-Institute for Astronomy, Königstuhl 17, 69117 Heidelberg, Germany

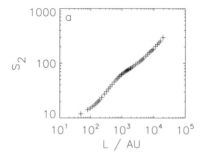 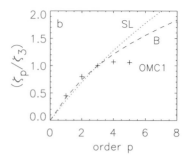

Figure 1. a) Second order structure function of radial velocities in OMC1. **b)** Scaling exponents of structure functions of order p normalized to ζ_3 versus order. For comparison are shown the theoretical scaling of She-Leveque (SL) and Boldyrev(B).

2. Structure functions in OMC1

Near infrared K-band observations of the strongest H_2 emission line in Orion, v= 1-0 S(1) at 2.121 μm, are used as a tracer of radial gas velocities in the BN-KL region of OMC1. The data were obtained at the CFHT with a Fabry-Perot interferometer in conjunction with adaptive optics (the GriF instrument) by spectral scanning of the H_2 line (Gustafsson *et al.* (2003); Nissen *et al.* (2006)). Data consist of a field of $89'' \times 67''$ or $41 \cdot 10^3 \times 31 \cdot 10^3$ AU (assuming a distance to OMC1 of 460 pc, (Bally *et al.* (2000))) centered on the BN object. The AO system achieved a spatial resolution of $0.15''(70\,\text{AU})$. The radial velocity at each spatial position was derived by spectral line fitting to the line profiles provided by the Fabry-Perot and relative velocities are determined with an accuracy of between 1 km s^{-1} (3σ) in the brightest regions and 8-9 km s^{-1} in the weakest regions (Gustafsson *et al.* (2006a)). The emission of excited H_2 observed here traces hot, dense gas, where excitation occurs primarily through shocks. Thus the cold gas goes unobserved.

Using the radial velocity map of OMC1 we obtain the structure functions

$$S_p(L) = \langle B(\vec{r})B(\vec{r}-\vec{\tau}) \mid v(\vec{r}) - v(\vec{r}-\vec{\tau}) \mid^p \rangle = \langle \mid \Delta v \mid^p \rangle. \qquad (2.1)$$

Here v is the line of sight velocity and the average is extended over all spatial positions $\vec{r}$ and all lags $\vec{\tau}$ where $L = |\vec{\tau}|$. $B(\vec{r})$ is the brightness at position $\vec{r}$. We thus weight each velocity difference by the product of the brightness of the two spatial positions involved, thereby giving more weight to the brightest regions which exhibit the highest accuracy in the radial velocity. Using brightness weighting assures that the result is less influenced by noise than if no weighting is used (Gustafsson *et al.* (2006a)). For homogeneous, isotropic turbulence the structure functions are known to follow power laws, $S_p(L) \sim L^{\zeta_p}$, in the inertial range, $\eta \ll r \ll L$, where η is the dissipation scale and L is the integral scale. The scaling exponents, ζ_p, are expected to be characteristic of the turbulence involved – e.g. incompressible or compressible – and universal for all scales, L.

The second order structure function derived from the OMC1 data is shown in Fig. 1a. At lags larger than 2000 AU, that is, over scales varying about an order of magnitude, the structure function is well represented by a power law. However, there is a clear deviation from from a power law at 1500 AU associated with an excess of energy at this scale and below. This indicates a characteristic scale.

For structure functions of order 1 to 5 we use the concept of extended self-similarity introduced by Benzi *et al.* (1993) to obtain power law fits and the normalized scaling

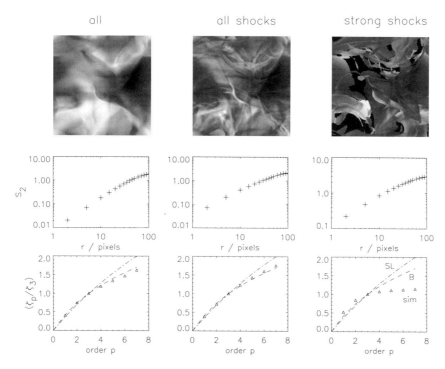

Figure 2. Results for projected maps including all points in the simulation (1st column) and for subsets including all shocks ($\nabla \cdot \vec{u} < 0$, 2nd column) and only strong shocks ($\nabla \cdot \vec{u} < -25$, 3rd column). Top row: projected radial velocity maps. Second row: second order structure functions. Third row: scaling exponents of the structure functions compared to the She-Leveque and Boldyrev scaling.

exponents ζ_p/ζ_3. These are plotted in Fig. 1b as a function of order and are compared with the scaling exponents predicted by the She & Leveque (1994) model of incompressible turbulence and the model of Boldyrev (2002) for supersonic, compressible turbulence. The scaling exponents derived from the velocity in OMC1 are seen to deviate from both the She-Leveque and the Boldyrev scaling at $p \geqslant 4$. The OMC1 scaling exponents show signs of becoming constant at $\zeta_p/\zeta_3 \sim 1$ or even slightly decreasing for $p > 4$, in contrast to the theoretical scalings, which are monotonically increasing.

3. Numerical simulations

In order to understand some of the features found in OMC1 we use data from numerical simulations of supersonic isothermal compressible turbulence. The simulations are purely hydrodynamic and are forced on random scales (Gustafsson *et al.* (2006b)). The root mean square Mach number is 3 in the simulation presented here. The viscosity is enhanced in the neighbourhood of shocks thereby broadening the physical size of shocks. The hydrodynamic equations are solved on a 3-dimensional periodic mesh of size $256 \times 256 \times 256$ using the PENCIL CODE (http://www.nordita.dk/software/pencil-code).

Since the main aim is to test if we can reproduce the features found in the observational data we want the numerical data to resemble best the physical properties of the observations. First, the observed radial velocity at each spatial position represents a projected velocity along the line-of-sight. Accordingly we project the numerical 3D velocity components onto a 2D map of only radial velocity. The projected radial velocity is found by density weighed averaging. Second, we observe a subset of the gas in OMC1 consisting largely of shocked gas. In the simulations we extract regions where shocks occur by selecting regions with strong negative velocity divergence, $\nabla \cdot \vec{u} < C$. Here C is a negative cut-off value and only simulated points satisfying $\nabla \cdot \vec{u} < C$ are included in the calculation of the projected radial velocity.

Figure 2 shows maps of projected radial velocities, the second order structure functions calculated from those maps and related scaling exponents. The maps shown correspond to including all points ($C = \infty$), including all shocks ($C = 0$) and including only stronger shocks ($C = -25$). The structure functions, represented by the 2nd order structure function in Fig. 2, for all values of C are well approximated by power laws. This is in contrast to the structure functions of radial velocities found in OMC1 (Fig. 1) which showed clear deviations from power laws. This suggests that the deviations are not due to the fact that we observe only shocked gas, but arise from physical processes local to the OMC1. It is interesting to note that the preferred scale of 1500 AU detected in the velocities in OMC1 is similar to the size of low-mass protostellar outflows which could in fact re-inject energy into an otherwise turbulent region. If a large number of such outflows are present they could act as a local driving agent for the turbulence and create a bump in the energy cascade similar to that found in OMC1.

The scaling exponents of the structure functions follow the relation predicted by Boldyrev (2002) for supersonic turbulence for the maps including all points and all shocked regions. However, for the map where only strong shocks are included the scaling exponents deviate from the theoretical scalings and resemble the scaling found in OMC1 (Fig. 1). Thus the unusual scaling observed in OMC1 can be explained as an effect of observing only the hot, shocked gas.

4. Conclusion

We have shown that the velocity field of OMC1 is not self-similar but contains a preferred scale size of 1500 AU associated with an excess of energy. This cannot be reproduced by hydrodynamical simulations and more advanced simulations are necessary.

Acknowledgements

MG and DF acknowledge the support of the Instrument Center for Danish Astrophysics (IDA) and the Aarhus Center for Atomic Physics (ACAP) and the natural science funding council (FNU). The Danish Center for Scientific Computing is acknowledged for granting time on the Horseshoe cluster in Odense.

References

Bally, J., O'Dell, C.R. & McCaughrean, M.J. 2000, *AJ* 119, 2919
Benzi, R., Ciliberto, S., Tripiccione, R., Baudet, C., Massaioli, F. & Succi, S. 1993 *Phys. Rev. E* 48, 29
Boldyrev, S. 2002, *ApJ* 569, 841
Gustafsson, M., Kristensen, L.E., Clénet, Y., Field, D., Lemaire, J.L., Pineau des Forêts, G., Rouan, D. & Le Coarer, E. 2003, *A&A* 411, 437
Gustafsson, M., Field, D., Lemaire, J.L. & Pijpers, F.P. 2006, *A&A* 445, 601
Gustafsson, M., Brandenburg, A., Lemaire, J.L. & Field, D. 2006, *A&A* 454, 815
Heyer, M.H & Brunt, C.M. 2004, *ApJ* 615, L45
Nissen, H.D., Gustafsson, M., Lemaire, J.L., Clenet, Y., Rouan, D. & Field, D. 2006, astro-ph/0511226
She, Z. & Leveque, E. 1994, *Phys. Rev. Lett.* 72, 336

Discussion

MAC LOW: OMC-1 is a coherent Hubble outflow, not a region of uniform turbulence, so application of statistical models drawn from uniform turbulence is questionable. A model of an expanding, fragmenting shell is more appropriate (e.g., Xu, Stone & Mundy, or McCaughrean & Mac Low).

GUSTAFSSON: That may be true and yet the statistical models applied have to reproduce some of the observed features. The Fabry-Perot data indicate that the H_2 emission in OMC1 is somewhat more complicated than a Hubble outflow and could suggest that the outflow have triggered turbulence that we now see.

BURTON: 1500 AU at Orion is $\sim$ 3 arcsec, which is the typical scale size of the H_2-emitting HH knots in OMC-1. These are not the sites of star formation in themselves. The knots are part os a Hubble-type expanding flow resulting from an impulsive event $\sim$ 500 years ago.

GUSTAFSSON: A scale of $\sim$ 1500 AU is indeed the size of some of the H_2 emitting clumps in OMC1, and we also see that clearly from the brightness data. That does not necessarily imply that radial velocities are coherent on the same scale. But they are! One more thing is that the Fabry Perot data clearly show that all of these clumps observed cannot be part of a Hubble-type flow with a common origin.

Triggered Star Formation in a Turbulent ISM
Proceedings IAU Symposium No. 237, 2006
B. G. Elmegreen & J. Palouš, eds.

Protostars in the Elephant Trunk Nebula

William T. Reach

Spitzer Science Center, MS 220-6, Caltech, Pasadena, CA, UAS
email: reach@ipac.caltech.edu

Abstract. Extremely red objects were identified in the early *Spitzer* Space Telescope observations of the bright-rimmed globule IC 1396A; they were classified as Class I protostars Class II T Tauri stars with disks based on their colors. New spectroscopic observations covering 5.5–38 μm confirm this identification. The Class I sources have extremely red continua, still rising at 38 μm, with a deep silicate absorption at 9–11 μm, weaker silicate absorption around 18 μm, and weak ice features including CO_2 at 15.2 μm and H_2O at 6 μm. The Class II sources have warm, luminous disks, with a silicate emission feature at 9–11 μm. Optical spectra with the Palomar Hale 200-inch telescope show the Class II sources to be actively accreting, classical T Tauri stars with bright Hα and other emission lines. The Class I sources are located within the molecular globule, while the Class II sources are more widely scattered. This suggests two phases of star formation occurred in the region, the first one leading to the Class II sources including LkHα 349a,c that are located in the center of the globule, and a very recent one (less than 100,000 yr ago) that is occurring within the globule. This second phase was likely triggered by the wind and radiation of the central O star of the IC 1396 H II region, with possible additional contributions from the outflows of LkHα 349a,c and some nearby B stars.

Keywords. stars: formation, stars: pre-main-sequence, ISM: globules

---

1. Introduction

A set of mid-infrared-bright sources was identified from early *Spitzer* (Werner *et al.* 2004) observations of the bright-rimmed globule IC 1396A (Reach *et al.* 2004). These sources were previously unknown, and their nature was determined by comparing broadband colors to those of young stellar objects in nearby star-forming regions. Eight sources were identified based on their spectral energy distributions as 'Class I' protostars, while 30 were identified as 'Class II'. We have now followed up these sources using mid-infrared, optical, and radio spectroscopy to further elucidate their properties. Here we present some mid-infrared spectra obtained with the *Spitzer* Infrared Spectrograph (Houck *et al.* 2004) and optical spectra obtained with the Palomar double-spectroscope (Oke & Gunn 1982).

2. Class I Sources

Figure 1 shows the mid-infrared spectra of the sources whose broad-band colors indicated there were Class I protostars. The spectral shapes are characterized by deep 9–11 μm silicate absorption, moderately deep 15–20 μm silicate absorption, weak 6.0 μm H_2O and 15.2 CO_2 ice absorption, and a dominant, rapidly-rising, infrared continuum. By far the bulk of the luminosity arises in the mid- to far-infrared. These spectra require a source with a centrally-heated core surrounded by a cold envelope that is opaque in the mid-infrared. This is precisely the configuration of a contracting protostar, before reaching the birth-line. Thus we confirm that the photometrically identified Class I sources are indeed protostars.

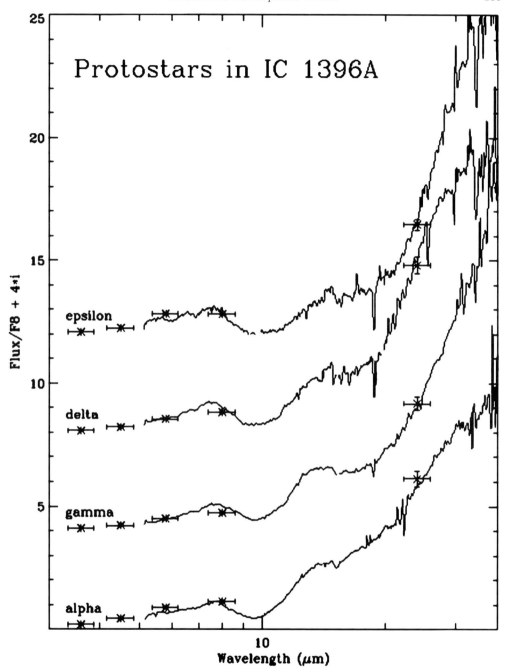

Figure 1. *Spitzer* IRS spectra of the Class I sources in IC 1396A.

3. Class II Sources

Figure 2 shows the mid-infrared spectra of the sources whose broad-band colors indicated there were Class II protostars. The spectral shapes are characterized by bright, rounded 9–11 μm silicate emission and a 'hump' that rises into the mid-infrared then is declining longward of 30 μm. These stars are all optically visible, with optical luminosity

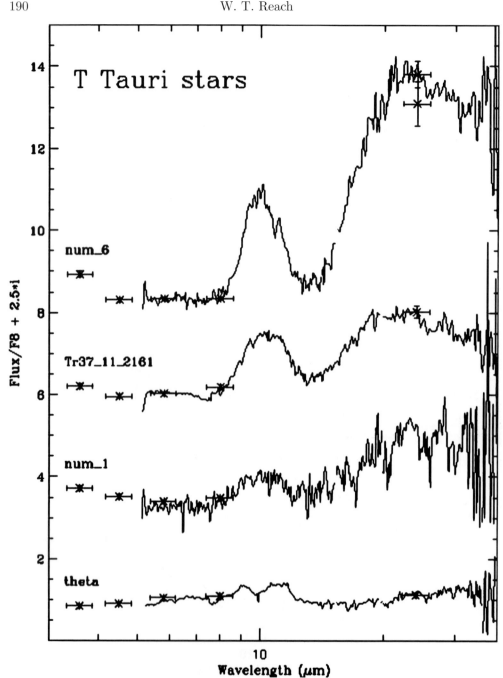

Figure 2. *Spitzer* IRS spectra of the Class II sources in IC 1396A.

comparable to the infrared. The optical spectra show bright emission lines, including Hα with equivalent width > 50 Å and N II, O I, and S II. The infrared emission requires an optically thin region heated far warmer than achieved by the interstellar radiation field, more typical of the environment within 10's of astronomical units of stars. This is precisely the configuration of a young star, already born, with a dusty disk, before planet formation. The optical spectra meet the criteria for 'classical' T Tauri stars. Thus we confirm that the photometrically identified Class II sources are indeed classical T Tauri stars with disks.

4. Star Formation in IC 1396A

The presence of 8 Class I sources and 30 Class II sources within IC 1396A is consistent with their estimated lifetimes of 10^5 and 10^6 yr. The rim of the dense globule is an overdensity due to the pressure of the H II region compressing the dense gas in the globule. The location of the Class I sources near the globule rims suggests they were very recently formed due to that compression, e.g. as predicted for radiative-driven implosion (Lefloch & Lazareff 1994).

Acknowledgements

This work is based in part on observations made with the *Spitzer Space Telescope*, which is operated by the Jet Propulsion Laboratory, California Institute of Technology under NASA contract 1407.

References

Houck, J. R. *et al.* 2004, *ApJS* 154, 18
Lefloch, B. & Lazareff, B. 1994, *A&A* 289, 559
Oke, J. B. & Gunn, J. E. 1982, *PASP* 94, 586
Reach, W. T. *et al.* 2004, *ApJS* 154, 385
Werner, M. W. *et al.* 2004, *ApJS* 154, 1

Discussion

DE VRIES: The presence of dense gas in the globule is confirmed by out observations of N_2H^+ (presented at this conference).

REACH: Observations of the dense gas are important to determine the current state of the globule due to the effect of the H II region and the conditions for potentially ongoing star formation.

Triggered Star Formation in a Turbulent Medium
Proceedings IAU Symposium No. 237, 2006
B. G. Elmegreen & J. Palouš, eds.

Star formation in the LMC: gravitational instability and dynamical triggering

You-Hua Chu[1], Robert A. Gruendl[1] and Chao-Chin Yang[1]

[1]Astronomy Department, University of Illinois, 1002 W. Green Street, Urbana, IL 61801, USA
email: chu@astro.uiuc.edu, gruendl@astro.uiuc.edu, cyang8@astro.uiuc.edu

Abstract. Evidence for triggered star formation is difficult to establish because energy feedback from massive stars tend to erase the interstellar conditions that led to the star formation. Young stellar objects (YSOs) mark sites of *current* star formation whose ambient conditions have not been significantly altered. *Spitzer* observations of the Large Magellanic Cloud (LMC) effectively reveal massive YSOs. The inventory of massive YSOs, in conjunction with surveys of interstellar medium, allows us to examine the conditions for star formation: spontaneous or triggered. We examine the relationship between star formation and gravitational instability on a global scale, and we present evidence of triggered star formation on local scales in the LMC.

Keywords. stars: formation, stars: pre-main-sequence, ISM: evolution, ISM: structure, galaxies: ISM, Magellanic Clouds

1. Introduction

The star formation process is intertwined with the evolution of the interstellar medium (ISM). After the onset of a burst of star formation, massive stars photoionise the ambient ISM into an H II region, and subsequently energize the ambient medium via fast stellar winds and supernova ejecta to form a superbubble. The expansion of H II regions and superbubbles can trigger further star formation. If triggered star formation continues at a high level over an extended period of time, >10 Myr, a kpc-sized supergiant shell may be produced. On the other hand, if a superbubble does not trigger a significant level of star formation, after the O stars have exploded and the remaining stars have dispersed, the superbubble will recombine into an H I shell with no obvious concentration of stars within its boundary.

To illustrate *triggered* star formation, a causal relationship between the prenatal interstellar conditions and the formation of stars need to be convincingly established. Main-sequence and evolved massive stars trace star formation in the past few Myr, but their prenatal interstellar conditions have been significantly altered by the injection of stellar energies. Massive young stellar objects (YSOs), having a short lifetime, trace the current star formation, while their prenatal interstellar conditions are still intact; therefore, massive YSOs and their environment can be used to investigate triggered star formation.

We have chosen the Large Magellanic Cloud (LMC) to investigate the star formation processes because of the following advantages: (1) The LMC is at a small distance, 50 kpc, so that stars can be resolved and the ISM can be mapped with high linear resolution, and it has a nearly face-on orientation so that confusion along the line-of-sight is minimal. (2) The ISM of the LMC has been well mapped – MCELS survey of ionized gas (Smith 1999), ATCA+Parkes survey of the H I gas (Kim *et al.* 2003), and NANTEN survey of molecular gas (Fukui *et al.* 2001). (3) A complete inventory of massive YSOs can be obtained from the *Spitzer* IRAC and MIPS survey of the LMC (Meixner *et al.* 2006).

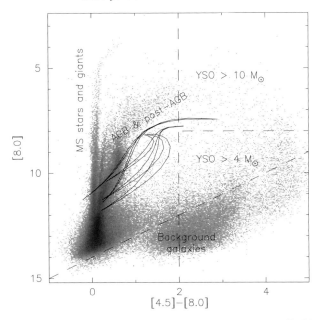

Figure 1. Color-magnitude diagram of all point sources in the LMC. Normal stars are at [4.5]−[8.0] ∼ 0. The curves illustrate locations of O-rich and C-rich AGB stars from Groenewegen (2006). The vertical dashed line roughly separate normal and AGB stars (to the left) and protostars (to the right). The tilted dashed line marks a rough upper boundary of background galaxies determined from the SWIRE data (Evans *et al.* 2005).

We have identified point sources in the *Spitzer* IRAC and 24μm MIPS images and made photometric measurements. Color-color and color-magnitude diagrams are used in conjunction with spectral energy distributions (SEDs) to diagnose YSO candidates from these point sources. Figure 1 shows an example of color-magnitude diagram in the IRAC 4.5 and 8.0 μm bands. The main-sequence stars and giants have zero colors, [4.5]-[8.0] ∼ 0. Contaminations of AGB and post-AGB stars and background galaxies can be largely avoided by employing appropriate color or brightness cutoffs (Groenewegen 2006; Evans *et al.* 2005). Known AGB stars and planetary nebulae are removed from the source list. The final YSO candidates are located in the upper right wedge in Figure 1.

While *Spitzer* observations allow us to identify YSOs in the LMC, the angular resolution severely limits our ability to determine the YSOs' physical properties in detail. Figure 2 illustrates this problem in the H II region N11B. The bright source identified at 8 μm has a SED consistent with a YSO with a substantial envelope. However, a high-resolution 2.1 μm image taken with the ISPI on the CTIO Blanco 4m telescope shows at least three sources within the point-spread-function of the 8 μm source; furthermore, a *Hubble Space Telescope* (*HST*) ACS/WFC Hα image shows multiple sources within a dust pillar whose surface is ionized. It is not clear how much nebular emission contributes to the 8 μm source and whether one or multiple YSOs are present.

The YSOs identified from *Spitzer* observations clearly do not provide unambiguous information about their masses and multiplicity. They are nevertheless excellent markers of sites of current star formation and allow us to examine the star formation process in the LMC both on global and local scales.

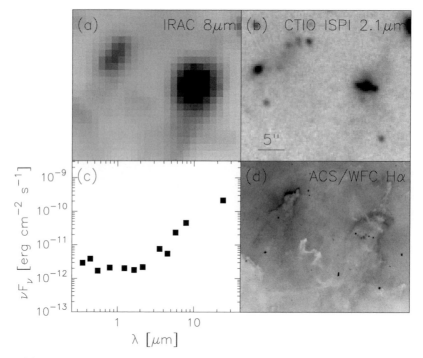

Figure 2. (a) *Spitzer* IRAC 8 μm image of a region in N11B. (b) CTIO Blanco 4m ISPI image of the same region at 2.1 μm. (c) The SED of the brightest source in (a). (d) *HST* ACS/WFC Hα image of the same region.

2. Star formation on a global scale

Star formation activity is seen throughout the disk of the LMC. We will examine the relationship between the distribution of star formation and the gravitational instability on a global scale. In the first case, we consider only a thin gaseous disk, and in the second case we add the stellar contribution to the disk.

A thin, differentially rotating gaseous disk is stable to axisymmetric perturbations, if $Q_g \equiv \frac{\kappa c_g}{\pi G \Sigma_g} > 1$, where Σ_g and c_g are the surface density and the sound speed of the gas, and κ is the epicycle frequency (Binney & Tremaine 1987). The total gas surface density map of the LMC (Fig. 3a) is derived from the the ATCA+Parkes H I survey by Kim *et al.* (2003) and the NANTEN CO survey by Fukui *et al.* (2001). (The total gas mass is $\sim 6.5 \times 10^8 \ M_\odot$.) The epicycle frequency as a function of radial distance is calculated from the H I rotation curve (Kim *et al.* 1998). The sound speed of gas is assumed to be 5 km s^{-1}. The resultant gravitational instability map of the LMC is displayed in Figure 3b, where shaded regions are unstable, with $Q_g < 1$, and the darkness of the shade increases with the degree of instability. The Spitzer sample of massive YSO candidates are also marked in Figure 3b. It is evident that only about 1/2 of the YSO candidates are located in gravitationally unstable regions.

To add stellar contributions to the gravitational instability, we follow Rafikov's (2001) treatment of a disk galaxy consisting of a collisional gas disk and a collisionless stellar disk. The instability condition becomes

$$\frac{1}{Q_{sg}} \equiv \frac{2}{Q_s} \frac{1}{q} \left[1 - e^{-q^2} I_0 \left(q^2 \right) \right] + \frac{2}{Q_g} \frac{1}{q} R \frac{q}{1 + q^2 R^2} > 1, \qquad (2.1)$$

where $Q_s \equiv \frac{\kappa \sigma_s}{\pi G \Sigma_s}$ with Σ_s and σ_s being the surface density and the radial velocity

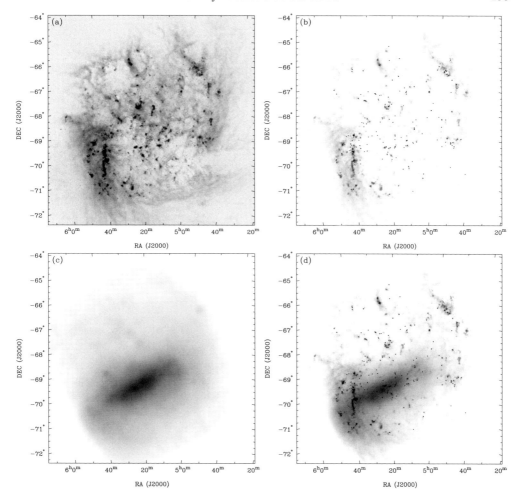

Figure 3. (a) Total gas surface density map of the LMC. (b) The Q_g map. The unstable regions, with $Q_g < 1$, are shaded, and the darkness increases with the degree of instability. YSO candidates are marked. (c) Total stellar surface density map of the LMC, derived from the *Spitzer* 3.6 μm observations. (d) The Q_{sg} map. The unstable regions, with $Q_{sg} < 1$, are shaded, and the darkness increases with the degree of instability. YSO candidates are marked.

dispersion of the stars, I_0 is the Bessel function of order zero, $R \equiv \frac{c_g}{\sigma_s}$, and $q \equiv \frac{k\sigma_s}{\kappa}$ with k being the wavenumber of the axisymmetric perturbations.

To estimate Σ_s, we use a normalized distribution of the 3.6 μm sources detected in the *Spitzer* survey of the LMC (Gruendl *et al.* 2007, in prep) and a total stellar mass of $2 \times 10^9\ M_\odot$ (Kim *et al.* 1998). Note that the 3.6 μm point sources, rather than the direct image, are used to avoid the contamination of diffuse emission from dust, and that each source is weighted by its brightness in the calculation. The resulting stellar surface density map is shown in Figure 3c. We adopt a stellar radial velocity dispersion of 15 km s^{-1}, and use the same Q_g described above. Finally, we calculate the value of Q_{sg} at each pixel by finding its global minimum as a function of the wavenumber k (see Eq. 2.1), following the same approach used by Jog (1996) and Rafikov (2001). The resultant gravitational instability map of the LMC is presented in Figure 3d, where shaded regions are unstable, with $Q_{sg} < 1$, and the darkness of the shade increases with the degree of instability. The

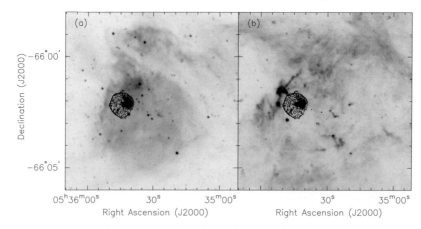

Figure 4. (a) Hα image of N63. Overplotted are X-ray contours extracted from *Chandra* observations to show the spatial extent of the young supernova remnant N63A. (b) *Spitzer* IRAC 8 μm image of N63. YSO candidates are detected along the northeastern rim of the H II region. The star formation is triggered by the expansion of the H II region, but the YSOs and their planetary disks will be enriched by the supernova ejecta.

vast majority of the massive YSO candidates are located within gravitationally unstable regions, in sharp contrast to the first case that includes only the gas disk.

From our analysis of the gravitational instability of the LMC, we conclude that the stellar disk's contribution cannot be ignored. More importantly, the current star formation appears to occur mostly in regions that are unstable against perturbation, implying that star formation can be triggered more easily in the gravitationally unstable regions.

3. Star formation on local scales

Triggered star formation is commonly seen in the LMC at various scales. Below we give three such examples.

Star formation triggered by H II region and enriched by supernova ejecta – N63 shown in Figure 4. The H II region N63 is ionized by the OB association LH 83. A young supernova remnant (SNR), N63A, has been identified in the H II region, as delineated by the X-ray emission. *Spitzer* IRAC 8 μm image shows prominent current star formation on the northeastern rim of the H II region, not in contact with the SNR.

Star formation triggered by photo-implosion of dust globules in a superbubble – N51D shown in Figure 5 (Chu *et al.* 2005). Three YSOs projected within the superbubble N51D are found to be coincident with dust globules. The thermal pressure of the warm photoionised gas on the surface of the dust globules is several times higher than those of the hot gas in the superbubble interior and the cold molecular gas in the dust globules, indicating that the dust globules are compressed by their high external pressure raised by photoionisation and the associated heating.

Star formation triggered by collision of interstellar shells – supergiant shells LMC-4 and LMC-5 shown in Figure 6. These two supergiant shells are expanding into each other, and the colliding region shows a bright ridge of H I and molecular clouds. Ongoing, active star formation is seen in the colliding region.

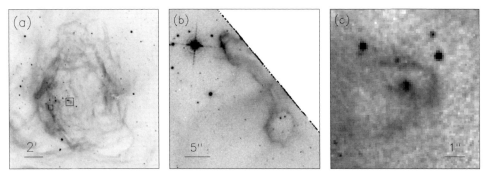

Figure 5. (a) Hα image of the superbubble N51D. (b) & (c) *HST* WFPC2 Hα images of dust globules that host YSOs. The locations of these fields are marked by the two boxes in (a).

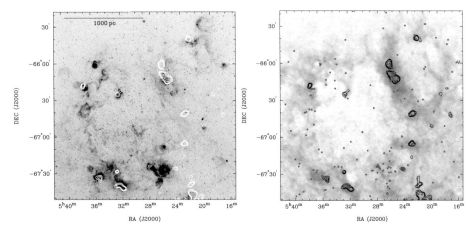

Figure 6. *Left:* Hα image of the supergiant shells LMC-4 and LMC-5 superposed with CO contours (Fukui *et al.* 2001). *Right:* ATCA+Parkes H I map of LMC-4 and LMC-5 in gray-scale. Overplotted are CO contours and positions of YSOs. Star formation is enhanced in the collision zone between the two supergiant shells.

4. Conclusions

The LMC is an excellent site to study star formation processes. Our investigation suggests that global-scale gravitational instabilities prepare the conditions for star formation, but dynamical triggering on local scales determines where to set off the star formation and complete the job.

Acknowledgements

This research is supported by the *Spitzer* grant JPL 1264494.

References

Binney, J. & Tremaine, S. 1987, *Galactic Dynamics* (Princeton University Press)
Chu, Y.-H., *et al.* 2005, *ApJ* 634, L189
Evans, N.J. & c2d Team 2005, *AAS Meeting* 207, #95.02
Fukui, Y., Mizuno, N., Yamaguchi, R., Mizuno, A. & Onishi, T. 2001, *PASJ* 53, L41
Groenewegen, M.A.T. 2006, *A&A* 448, 181
Jog, C.J. 1996, *MNRAS* 278, 209
Kim, S., *et al.* 1998, *ApJ* 503, 674

Kim, S., *et al.* 2003, *ApJS* 148, 473
Meixner, M., *et al.* 2006, *AJ* 132, 2268
Rafikov, R.R. 2001, *MNRAS* 323, 445
Smith, R.C., *et al.* 1999, *IAU Symposium* 190, p.28

Discussion

DOTTORI: You didn't mention but there is a process of star formation triggered by the bar, which has an $m = 1$ perturbation associated, detected through age groups of star clusters. More local are those related to Shapley constellations.

CHU: The role played by the bar needs to be considered. Our preliminary model did not take into account the spheroidal geometry of the bar. The Shapley constellations are older stellar populations formed a few 10^7 yr ago. The gravitational instability map is a snap shot now.

KRUMHOLZ: Many quantities that affect Q are difficult to determine in external galaxies, e.g., the stellar velocity dispersion. How did you get these?

CHU: The stellar radial velocity dispersion is ~ 25 km s^{-1} for the Milky Way. As the LMC is less massive than the Milky Way, we assumed 15 km s^{-1} for the stellar velocity dispersion. We can probe more parameter space later.

ZINNECKER: for clarification: when you say supernova remnants do not trigger star formation (but HII regions do), which stellar masses do you refer to: low mass stars, intermediate mass stars, or high mass stars?

CHU: supernova remnants, when they can be identified as such, are too violent to trigger star formation. For the known supernova remnants that we have examines with Spitzer data, none show any YSOs along their periphery.

Triggered Star Formation in a Turbulent ISM
Proceedings IAU Symposium No. 237, 2006
B. G. Elmegreen & J. Palouš, eds.

© 2007 International Astronomical Union
doi:10.1017/S1743921307001469

Star formation in the Small Magellanic Cloud: the youngest star clusters

E. Sabbi[1]†, A. Nota[1,2], M. Sirianni[1,2], L. R. Carlson[1], M. Tosi[3], J. Gallagher[4], M. Meixner[1], M. S. Oey[5], A. Pasquali[6], L. J. Smith[7], M. Vlajic[8] and L. Hawks[9]

[1]STScI, 3700 San Martin Drive, Baltimore, MD, 21218, USA
email: sabbistsci.edu

[2]ESA: Research & Scientific Support Department;[3]INAF–Osservatorio Astronomico di Bologna, I; [4]Dept. of Astronomy, University of Wisconsin, USA; [5]University of Michigan, USA; [6] MPIA, Heidelberg, Germany; [7]University College London, UK; [8]University of Oxford, UK; [9]Rice University, USA

Abstract. We recently launched a comprehensive ground based (ESO/VLT/NTT) and space (HST & SST) study of the present and past star formation in the Small Magellanic Cloud (SMC), in clusters and in the field, with the goal of understanding how star and cluster formation occur and propagate in an environment of low metallicity, with a gas and dust content that is significantly lower than in the Milky Way. In this paper, we present some preliminary results of the "young cluster" program, where we acquired deep F555W (∼V), and F814W (∼I) HST/ACS images of the four young and massive SMC star clusters: NGC 346, NGC 602, NGC 299, and NGC 376.

Keywords. galaxies: individual (SMC), galaxies: star clusters, stars: pre–main-sequence, stars: formation, techniques: photometric

1. Introduction

The Small Magellanic Cloud (SMC) is an excellent laboratory to investigate the star formation (SF) processes and the associated chemical evolution in dwarf galaxies. Its current sub-solar chemical abundance (Z=0.004) makes it the best local counterpart to the large majority of dwarf irregular (dIrr) and Blue Compact Dwarf (BCD) galaxies, whose characteristics may be similar to those in the primordial universe. The SMC proximity allows us to resolve into single stars the youngest and most compact star clusters, down to the sub solar mass regime.

Here we present some preliminary results of an in–depth study of the stellar content of the four youngest and most massive star clusters (NGC 346, NGC 602, NGC 299 and NGC 376) in the SMC. The clusters differ in age (∼ 3 to 30 Myr), regime of star formation, and location within the galaxy (see Fig. 1). For each cluster we obtained deep ACS/HST images in the F555W and F814W filters.

2. The young clusters

2.1. NGC 346

NGC 346 is an extremely young, (∼ 3 Myr, Bouret *et al.* 2003) moderately compact cluster, that excites the largest and brightest HII region – N66 – in the SMC (Relaño, Peimbert, & Beckman 2002). NGC 346 lies in a very active and interesting region: at a

† Present address: STScI, 3700 San Martin Drive, Baltimore, MD, 21218, USA.

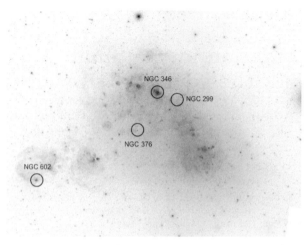

Figure 1. Position of the four clusters in the SMC.

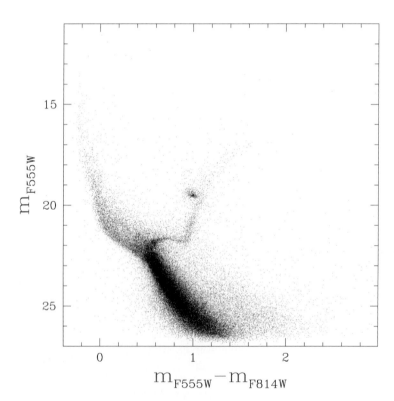

Figure 2. NGC 346 CMD (m_{F555W} vs. $m_{F555W} - m_{F814}$) for all stars with associated photometric errors smaller than 0.1 in both filters.

distance of 2 arc-minutes to the East (35 pc in projection) of the center of the cluster, we find the massive Luminous Blue Variable star HD 5980. N66 contains also at least two known supernova remnants (SNR): SNR 0057-7233 (Ye, Turtle, & Kennicutt 1991), located to the Southwest of N66, and SNR 0056-7226.

In Figure 2 we show the m_{F555W} vs. $m_{F555W} - m_{F814W}$ Color–Magnitude Diagram (CMD) of all the stars with photometric error $\sigma_{DAO} < 0.1$ mag detected in the NGC 346

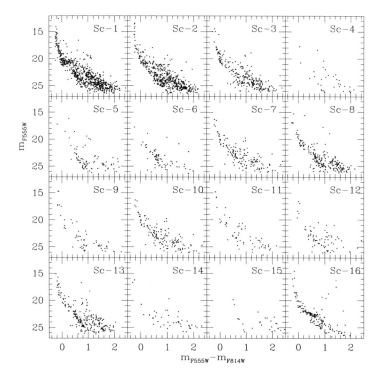

Figure 3. CMDs of the 16 sub–clusters identified within the NGC 346 region.

area. A first inspection of this CMD reveals that different stellar populations are present in the area:

• An old stellar population that belongs to the SMC field. The morphology of the CMD suggests that, in the field, a major episode of SF occurred approximately between 3 and 5 Gyr ago, but stars with ages up to at least 10 Gyr are also present. We also note that at most recent epochs the SF activity in the field has been significantly lower, with a possible moderate enhancement $\sim 150\,\mathrm{Myr}$ ago.

• A young stellar population that belongs to NGC 346. A comparison with Padua isochrones indicates that NGC 346 is $3 \pm 1\,\mathrm{Myr}$ old. The main sequence (MS) of the young population abruptly interrupts at F555W $\simeq 21$. At F555W > 21 we identify hundreds of red and faint stars whose colors and magnitudes are consistent with those predicted for low mass (0.6 - 3 $\mathrm{M}_\odot$) pre–Main Sequence (pre–MS) stars (see also Nota *et al.* 2006), that likely formed at the same time of the central cluster.

The high spatial resolution of our observations shows that the youngest stellar population is not uniformly distributed within the ionized nebula: we identify at least 15 sub–clusters, which differ in size and stellar content (see Fig. 3). Within the uncertainties of the comparison with isochrones, the sub–clusters are likely coeval with each other. However, we also find a relatively older sub–cluster (Sc–16, with an age $15 \pm 2.5\,\mathrm{Myr}$), located at the Northeast periphery of our data. This sub–cluster is likely not related to the star forming episode that originated NGC 346 (see also Sabbi *et al.* 2007).

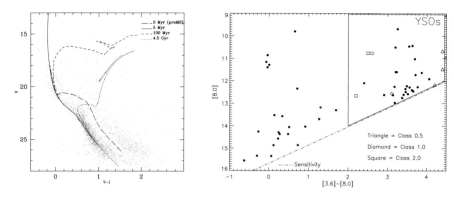

Figure 4. Left Panel: NGC 602 CMD (m_{F555W} vs. $m_{F555W} - m_{F814}$). Padua isochrones have been overlayed to estimate the age of the different stellar populations. Right Panel: NGC 602 IR CMD ([8.0μ], [3.6μ]-[8.0μ]). The expected YSO location is indicated by the box, derived from stellar models (Whitney *et al.* 2004)

2.2. *NGC 602*

Very little is known about the young cluster NGC 602, located in a low stellar density region, the "Bridge" of the SMC, stretching between Shapley's wing (1940) and the main body of the SMC.

Figure 4 (left panel) shows the m_{F555W} vs. $m_{F555W} - m_{F814W}$ CMD of NGC 602. The CMD reveals a rich population of pre–MS low mass (0.6 - 3 $M_\odot$) stars, likely formed coevally with the cluster ∼5 Myr. In the field, a major episode of SF occurred approximately ∼6 Gyr ago, and a second, moderate, episode occurred ∼ 150 Myr ago.

SST data (see Fig. 4–right panel) reveals the presence of a possibly younger generation of young stellar objects (YSOs likely of class 0.5, and I), still embedded in their dust cocoons, likely younger than 1 Myr. We infer that the powerful winds developed by the most massive O stars in the center of the cluster are triggering new episodes of star formation in the periphery, where the dust concentration is higher (Carlson *et al.* 2007, in prep.).

2.3. *NGC 299 & NGC 376*

NGC 299, located in the main body of the SMC, and NGC 376, in the "Wing" of the galaxy, are the oldest star clusters in our survey (∼ 25 Myr). In these clusters the winds from the most massive stars have already dispersed all the gas and dust, and star formation is no longer active (see Fig 5). In these two regions, the field of the SMC had a major episode of SF approximately between 4 and 6 Gyr ago, while at most recent epochs the SF activity has been significantly lower, with a possible enhancement ∼ 150 Myr ago.

It is known that massive star winds at low metallicity are less powerful and efficient. The observations of these two "older clusters" provide an interesting measure of the efficiency of the feedback mechanism at low metallicity. In addition, the clusters show very similar physical characteristics in spite of the very different local environment conditions, in terms of stellar and gas/dust density, suggesting that local condition may have little impact on cluster formation and evolution.

3. Conclusions

Our observation allowed us to follow in detail the evolution of SMC star clusters during their first 30 Myr of lifetime. Our results indicate that star formation is still active in

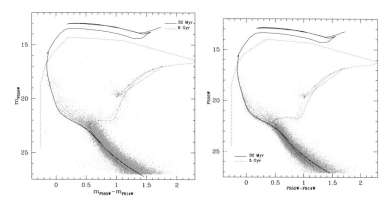

Figure 5. Left Panel: NGC 299 CMD (m_{F555W} vs. $m_{F555W} - m_{F814}$). Right Panel: NGC 376 CMD (m_{F555W} vs. $m_{F555W} - m_{F814}$). Padua isochrones have been overlayed to estimate the age of the different stellar populations.

the SMC, and that metallicity and local conditions do not seem to affect star and cluster formation processes.

Our survey provides a snapshot of the star formation history of the SMC over a look-back time of about 10 Gyr. In all observed fields we can distinguish a major episode of star formation, occurred approximately between 3 and 6 Gyr ago, but stars with ages up to 10 Gyr are also present. We note also that at most recent epochs the star formation activity in the field has been significantly lower, with a possible moderate enhancement 150 Myr ago.

References

Bouret, J.C., Lanz, T., Hillier, D.J., Heap, S.R., Hubeny, I. & Lennon, D. J. 2003, *ApJ* 595, 1182

Hilditch, R.W., Howarth, I.D. & Harries, T.J. 2005, *MNRAS* 125, 336

Nota, A., Sirianni, M., Sabbi, E., Tosi, M., Meixner, M., Gallagher, J., Clampin, M., Oey, S., Smith, L.J., Walterbos, R. & Mack, J. 2006, *ApJ* 640, L29

Relaño, M., Peimbert, M. & Beckman, J. 2002, *ApJ* 564, 704

Rubio, M., Contursi, A., Lequeux, J., Probst, R., Barbà, R.H., Boulanger, F., Cesarsky, D. & Maoli, R. 2000, *A&A* 359, 1139

Sabbi, E., Sirianni, M., Nota, A., Tosi, M., Gallagher, J., Meixner, M., Oey, M.S., Walterbos, R., Pasquali, A., Smith, L.J. & Angeretti, L. 2007, *AJ* in press (astro–ph/0609330)

Walborn, N.R. & Parker, J.W. 1992, *ApJ* 399, L87

Walborn, N.R., Maíz–Apellániz, J. & Barbá, R.H. 2002, *AJ* 124, 1601

Whitney, B.A., Indebetouw, R., Bjorkman, J.E. & Wood, K. 2004, *ApJ* 617, 1177

Ye, T., Turtle, A.J. & Kennicutt, R.C., Jr. 1991, *MNRAS* 249, 722

Discussion

BRANDL: You mentioned an older cluster to the northwest of NGC 346. Can you provide an age estimate for that cluster?

SABBI: Yes, we computed a synthetic CMD for the old cluster. In doing this we derive an age of 4.3 ± 0.1 Gyr.

BLITZ: Is there any remnant molecular cloud within 50-100 pc of NGC 299 or NGC 376?

SABBI: There are not molecular clouds close to these 2 clusters.

Triggered Star Formation in a Turbulent ISM
Proceedings IAU Symposium No. 237, 2006
B. G. Elmegreen & J. Palouš, eds.

© 2007 International Astronomical Union
doi:10.1017/S1743921307001470

Molecular clouds in galaxies as seen from NIR extinction studies

João Alves

Calar Alto Observatory, Almería, Spain
email: jalves@caha.es

Abstract. Near infrared dust extinction mapping is opening a new window on molecular cloud research. Applying a straightforward technique to near infrared large scale data of nearby molecular complexes one can easily construct density maps with dynamic ranges in column density covering, $3\sigma \sim 0.5 < A_V < 50$ mag or $10^{21} < N < 10^{23}$ cm^{-2}. These maps are unique in capturing the low column density distribution of gas in molecular cloud complexes, where most of the mass resides, and at the same time allow the identification of dense cores (n$\sim$10^4 cm^{-3}) which are the precursors of stars. For example, the application of this technique to the nearby Pipe Nebula complex revealed the presence of 159 dense cores (the largest sample of such object in one single complex) whose mass spectrum presents the first robust evidence for a departure from a single power-law. The form of this mass function is surprisingly similar in shape to the stellar IMF but scaled to a higher mass by a factor of about 3. This suggests that the distribution of stellar birth masses (IMF) is the direct product of the dense core mass function and a uniform star formation efficiency of 30%±10%, and that the stellar IMF may already be fixed during or before the earliest stages of core evolution. We are now extending this technique to extragalactic mapping of Giant molecular Clouds (GMCs), and although a much less straightforward task, preliminary results indicate that the GMC mass spectrum in M83 and Centaurus A is a power-law characterized by $\alpha \sim -2$ unlike CO results which suggest $\alpha \sim -1$.

Keywords. ISM: structure, evolution, dust, individual (Pipe Nebula, M83), stars: mass function

1. Introduction

Little is understood about the internal structure of molecular clouds and consequently the initial conditions that give rise to star and planet formation. This is largely due to the fact that molecular clouds are primarily composed of molecular hydrogen, which is virtually inaccessible to direct observation. Therefore, the traditional methods used to derive the basic physical properties of such molecular clouds therefore make use of observations of trace H_2 surrogates, namely those rare molecules with sufficient dipole moments to be easily detected by radio spectroscopic techniques, and interstellar dust, whose thermal emission can be detected by far infrared and radio continuum techniques (e.g., Motte *et al.* 1998, Lehtinen *et al.* 2001). However, the interpretation of results derived from these methods is not always straightforward (e.g., Alves, Lada, & Lada 1999; Chandler & Richer 2000). Several poorly constrained effects inherent in these techniques (e.g., deviations from local thermodynamic equilibrium, opacity variations, chemical evolution, small-scale structure, depletion of molecules, unknown emissivity properties of the dust, unknown dust temperature) make the construction of an unambiguous picture of the physical structure of these objects a very difficult task.

There is then a need for a less complicate and more robust tracer of H_2 to access not only the physical structure of these objects but also to accurately calibrate molecular abundances and dust emissivity inside these clouds. The deployment of sensitive infrared array cameras on large telescopes, however, has fulfilled this need by enabling direct

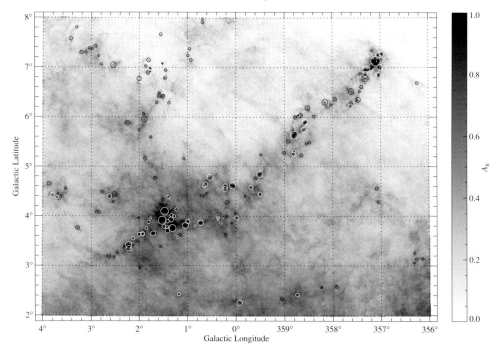

Figure 1. Dust extinction map of the Pipe nebula molecular complex from Lombardi, Alves, & Lada (2006). This map was constructed from near-infrared observations of about 4 million stars in the background of the complex. Approximately 160 individual cores are identified within the cloud and are marked by an open circle proportional to the core radius. Most of these cores appear as distinct, well separated entities. Figure from Lombardi, Alves, & Lada (2006)

and precise measurements of dust extinction toward thousands of individual background stars observed through a molecular cloud (Lada *et al.* 1994, Alves *et al.* 2001, Lombardi & Alves 2001). Such straightforward measurements are free from the complications that plague molecular-line or dust emission data and enable detailed maps of cloud structure to be constructed.

2. Results from the Pipe Nebula

In Figure 1 we present the $8° \times 6°$ extinction map of the Pipe Nebula complex constructed with 4 million stars from the 2MASS database (from Lombardi, Alves, Lada 2006). This molecular complex is one of the closest to Earth (~ 130pc) of this size and mass, and it is particularly well positioned along a relatively clean line of sight to the rich star field of the Galactic bulge, which given the close distance of the Pipe nebula allowed us to achieve spatial resolutions of ~ 0.03 pc, or about 3 times smaller the typical dense core size. The complex also exhibits very low levels of star formation suggesting that its dense cores likely represent a fair sample of the initial conditions of star formation. We present in Figure 2 the probability density function for the 159 core masses identified in the Pipe Nebula complex (marked in Figure 1). The core mass distribution seems to be characterized by two power-laws and a well defined break point at ~ 2 M$_\odot$. For comparison we also present the field star IMF determined by Kroupa (2001) (solid grey) and Chabrier (2003) IMF (dotted grey), and the IMF for the young Trapezium cluster (dashed grey) (Muench *et al.* 2002). All these stellar IMFs were shifted to higher masses by an average scale factor of about 3. The point of this comparison is to simply

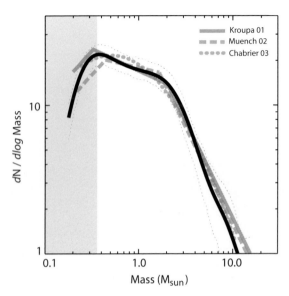

Figure 2. Probability density function of Pipe Nebula core masses (black). The grey region indicates sample incompleteness while the two thin dashed line indicates the 90% confidence limits. The core mass distribution seems to be characterized by two power-laws and a well defined break point at around 2 $M_\odot$. Also plotted are the field star IMFs of Kroupa (2001) (solid grey) and Chabrier (2003) (dotted grey), and the Muench *et al.* (2002) (dashed grey) IMF all scaled up by a factor of ~ 3 in mass. Figure from Alves, Lombardi, and Lada (2007).

illustrate the overall similarity between different stellar IMFs from different environments and constructed in different ways, and the dense core mass function of the Pipe Nebula. Considering the overall uncertainties, apparent differences between the different distributions are likely not significant.

3. Conclusions and Future Work

The close similarity in shape between the dense core mass function and the IMF supports a general concept of a 1-to-1 correspondence between the individual dense cores and soon to be formed stars. In this respect our observations are consistent with and appear to confirm the results of dust continuum surveys in other clouds (e.g., Motte *et al.* 1998). However, our study provides the first robust evidence for a departure from a single power law in the core mass function. We find that the location of the break indicates that there is a factor of about 3 difference in mass scale between the two distributions, which implies that a uniform SFE, that we estimate to be 30%±10%, will likely characterize the star formation in these dense cores, across the entire span of (stellar) mass.

We are currently expanding the technique to map Giant Molecular Clouds (GMCs) in nearby galaxies via seeing or diffraction limited NIR imaging. These clouds can be mapped in a manner analogous to the extinction map in figure 1 and with sub-arcsec resolution (corresponding to few to tens of parsecs in nearby galaxies), although the interpretation of the observed reddening and light attenuation into physical quantities is hampered by effects such as the unknown geometry of the system and non-negligible scattering of NIR light by dust particles. A first investigation on the importance of these effects can be found in Kainulainen, Juvela, and Alves (2007). Preliminary results

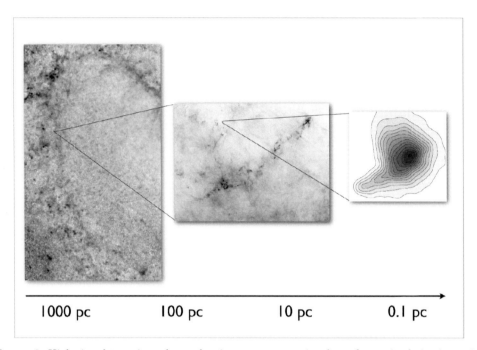

1000 pc   100 pc   10 pc   0.1 pc

Figure 3. High signal-to-noise column density maps across 4 orders of magnitude in size, using one single tracer: extinction by dust grains. From the large to the small: M83, Pipe Nebula, and Barnard 68. Near infrared extinction mapping is opening a new window on molecular cloud research and is very complementary to traditional line and dust emission techniques.

indicate that the GMC mass spectrum in M83 and Centaurus A (Bialetski *et al.* 2007) is a power-law characterized by $\alpha \sim -2$ unlike CO results which suggest $\alpha \sim -1$.

References

Alves, J., Lada, C. & Lada, E. 1999, *ApJ* 515, 265
Alves, J., Lada, C. & Lada, E. 2001, *Nature* 409, 159
Alves, J., Lombardi, M. & Lada, C. 2007, *A&A* in press
Bialetski, Y., Alves, J. *et al.* 2007, *A&A* in preparation
Chabrier, G. 2003, *ApJ* 586, 133
Chandler, C. & Richer, J. 2000, *ApJ* 530, 851
Kainulainen, J., Juvela, M. & Alves, J. 2007, *A&A* in press
Kroupa, P. 2001, *MNRAS*, 322, 231
Lada, C. J., Lada, E. A., Clemens, D. P. & Bally, J. 1994, *ApJ* 429, 694
Lombardi, M. & Alves, J. 2001, *A&A* 377, 1023
Lombardi, M., Alves, J. & Lada, C. 2006, *A&A* 454, 781
Lehtinen, K., Haikala, L. K., Mattila, K. & Lemke, D. 2001, *A&A* 367, 311
Muench, A. A., Lada, E. A., Lada, C. J. & Alves, J. 2002, *ApJ* 573, 366
Motte, F., Andre, P. & Neri, R. 1998, *A&A* 336, 150

Triggered Star Formation in a Turbulent ISM
Proceedings IAU Symposium No. 237, 2006
B. G. Elmegreen & J. Palouš, eds.

© 2007 International Astronomical Union
doi:10.1017/S1743921307001482

Understanding starbursts through giant molecular clouds in high density environments

Erik W. Rosolowsky

Harvard-Smithsonian Center for Astrophysics
60 Garden St., MS-66, Cambridge, MA 02138, USA
email:erosolow[snail]cfa.harvard.edu

Abstract. Starburst galaxies are characterized by uncommonly high star formation efficiencies, but it remains unclear what physical conditions in the molecular gas produce this high efficiency. Invariably, high star formation efficiency is associated with high column densities of molecular material (e.g. the Kennicutt-Schmidt law), but what are the conditions in the molecular clouds in starburst galaxies? Direct observations of starburst are difficult or impossible with current instruments, so I present the properties of GMCs in the Local Group as a starting case and then extend the analysis of GMC properties to nearby systems with surface densities of gas intermediate between the Local Group and starbursts. Rather than being constant, molecular cloud properties follow a continuum with significant variation across the Local Group and the intermediate surface density systems. Concomitant with these variations in the macroscopic properties are significant changes in the internal pressure and densities of molecular clouds, which implies significant variability in the initial conditions of the star formation process.

Keywords. galaxies: evolution, ISM: clouds, stars: formation

1. Introduction

While many phenomena may actively trigger star formation, the common feature of all these physical processes is the collection of gas into a small volume. On galactic scales, the "triggering" of star formation is best represented by the the Kennicutt-Schmidt law (Kennicutt 1998) which states that $\Sigma_{\mathrm{SFR}} \propto \Sigma_{\mathrm{gas}}^{1.4}$ where Σ_{gas} is the total surface density of gas. Subsequent studies have demonstrated that a more accurate relationship relates the surface density of star formation to the molecular gas: $\Sigma_{\mathrm{SFR}} \propto \Sigma_{\mathrm{mol}}^{1.4}$ (Murgia *et al.*, 2002; Wong & Blitz, 2002; Heyer *et al.*, 2004). This star formation law spans over seven orders of magnitude in star formation rate stretching from 10^{-4} to 10^3 $M_\odot$ yr^{-1} kpc^{-2} with relatively little scatter (< 1 dex); for reference, the star formation rate in the solar neighborhood is $\sim 10^{-3}$ $M_\odot$ yr^{-1} kpc^{-2} (Kroupa 1995). For systems with large-scale star formation rates comparable to that of the solar neighborhood, the star formation process likely follows the local pattern with stars forming in dense clumps of gas distributed through molecular clouds. However, for starburst systems where the star formation rate and gas surface density exceeds the solar neighborhood values by orders of magnitudes, star formation likely proceeds along significantly different lines, a conjecture borne out by the ubiquity of super star clusters in starburst systems (e.g. M82; O'Connell *et al.* 1995), which are never found in environments with low star formation rates. In addition, in starburst galaxies, the nature of the ISM must be significantly different. The neutral medium is almost entirely molecular in these systems; and the surface densities are sufficiently high ($> 10^2$ $M_\odot$ pc^{-2}) that if GMCs were like those found locally, they would merge into a single continuous medium. Since the molecular gas in starbursts is the initial

Table 1. Summary of GMC Properties in the Local Group and Transition Starburst Galaxies

Galaxy	σ_0 [km s^{-1}]	$X_{\rm CO}$ [10^{20}cm^{-2}/(K km s^{-1})]	$\Sigma_{\rm H2}$ [$M_\odot$ pc^{-2}]
LMC	0.39	2.7	45
SMC	0.36	6.6	30
M33	0.61	2.0	170
IC10	0.55	1.7	140
M31	0.72	2.6	200
Outer MW	0.40	3.0	50
M64	1.2	2.0	300
M82	>1.9	$\lesssim 2.0$	$\lesssim 500$
MW Center	2.6	<2.0	< 600
Typical Errors	0.5	0.05	10

conditions of their peculiar star and cluster formation, characterizing the properties of this gas is central to understanding the mechanics of the Kennicutt-Schmidt law. In these proceedings, I present some basic tools for characterizing the molecular ISM via scaling laws for giant molecular clouds (GMCs), present a summary of GMC properties in the Local Group galaxies for context, and then examine the GMCs in borderline starburst systems.

2. Using Larson's laws to characterize GMC properties

Larson (1981) first pointed out power law scalings between the macroscopic properties of molecular gas in a variety of systems. There are two independent "laws" expressed in a myriad of algebraically equivalent forms. In this paper, I will focus on the size-line width relationship and the virial parameter (α):

$$\sigma_v = \sigma_0 \left(\frac{R}{1\ {\rm pc}} \right)^\beta, \qquad \alpha = \frac{5R\sigma_v^2}{GM} \qquad (2.1)$$

where σ_v is the one-dimensional velocity dispersion, R is the cloud radius, and M is the cloud mass. For GMCs in the solar neighborhood, $\sigma_0 \approx 0.6$ km s^{-1}, $\beta = 0.5$ and $\alpha \approx 1.5$ (Blitz 1993 and references therein). While the physical interpretation of these relationships continues to be debated, they remain a helpful parameterization of the similarities and differences among GMCs in a variety of environments. Recently, there has been a major effort to re-examine these relationships in Local Group galaxies (Blitz *et al.* 2005) while referencing all their measurements to a common standard, facilitating comparison among systems (Rosolowsky & Leroy 2006). In Table 1, we summarize the size-line width relationship for GMCs in several galaxies across the Local Group. Since there are only a few points in each galaxy and a fit to the relationship is unstable, I only report the constant of proportionality (σ_0) assuming the scaling indices are the same across systems. I also report the derived CO-to-H$_2$ conversion factor ($X_{\rm CO}$) that would be obtained if the virial parameter were $\alpha = 1.5$ for all clouds as well as the mean surface density of the GMCs ($\Sigma_{\rm H2}$). There is significant variation in the GMC properties throughout the Local Group, as has been seen in previous studies (e.g., Rubio *et al.* 1993) and predicted based on environmental effects (Elmegreen 1989). These variations in the macroscopic properties of clouds imply substantial changes in the average internal properties of clouds such as their density and pressure – the average pressure in GMC in M31 is nearly 40 times that of a GMC in the SMC.

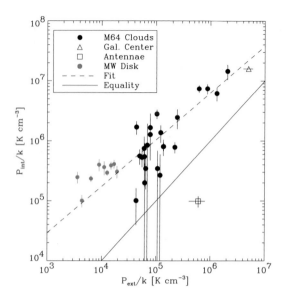

Figure 1. A comparison of internal and external pressures for GMCs in a variety of systems. Points represent pressure estimates in clouds for M64 (Rosolowsky & Blitz 2005), Milky Way (Solomon *et al.* 1987), the Antennae galaxy (Wilson *et al.* 2004), and the Galactic center (Oka *et al.* 2001). Internal pressures are estimated from cloud properties. External pressures are quoted values from X-ray studies (Antennae, Galactic center) or estimated from the mass contributions of galactic components (Milky Way, M64). For most clouds, the internal pressure scales with the external pressure and is consistently above the derived value.

3. GMCs in high density environments

Ideally, we would compare the Local Group GMCs directly to clouds identified in starburst systems. Unfortunately, we cannot directly observe starbursts with the observational facilities currently available, owing to their limited angular resolution or insufficient sensitivity to map optically thin tracers. To make progress we must examine nearby systems that are closer (and therefore better resolved) and have surface densities intermediate between the disks of Local Group galaxies and starbursts. Three candidate systems are available for detailed study: the center of the Milky Way, M82 and M64. In Table 1, I present summary properties for these galaxies based off the analyses of Oka *et al.* (2001) for the galactic center, my own analysis of the data of Shen & Lo (1995) and Keto et al. (2005) for M82, and Rosolowsky & Blitz (2005) for M64. In these systems, the internal velocity dispersions are larger and the surface densities are higher implying an increased internal pressure of the GMCs. The internal pressure is consistently above estimates of the pressure in the ambient ISM and GMCs have internal pressures roughly an order of magnitude above that of the ambient ISM (Figure 1). In the high surface density systems, significantly more massive clouds are found: M64 harbors several GMCs with masses in excess of $10^7 \ M_\odot$ which are unknown in the Local Group. Massive GMCs are necessary to support the formation of massive clusters ($> 10^6 \ M_\odot$) with a reasonable efficiency.

For this limited number of systems, the molecular medium in the (transition) starburst systems supports significantly more massive clouds that are denser and more turbulent than the star-forming clouds seen in the Local Group. This general trend is merely the extension of variations in cloud properties seen throughout the Local Group. With the the parameterization of observational data becoming standardized across galaxies, future work should compare these variations in cloud properties to predictions of star formation based on theoretical models.

Acknowledgements

I am grateful to the National Science Foundation for their generous support of my work through an NSF Astronomy & Astrophysics Postdoctoral Fellowship (AST-0502605).

References

Blitz, L. 1993, in: E.H. Levy & J.I. Lunine (eds.), *Protostars and Planets III* (Tucson: Univ. Arizona), p. 125

Blitz, L., Fukui, Y., Kawamura, A., Leroy, A., Mizuno, N. & Rosolowsky, E. 2006, in: B. Reipurth, D. Jewitt & K. Keil (eds.), *Protostars & Planets V*, in press

Elmegreen, B. G. 1989, *ApJ* 338, 178

Heyer, M. H., Corbelli, E., Schneider, S. E. & Young, J. S. 2004, *ApJ* 602, 723

Kennicutt, R. C. 1998, *ApJ* 498, 541

Keto, E., Ho, L. C. & Lo, K.-Y. 2005, *ApJ* 635, 1062

Kroupa, P. 1995, *MNRAS* 277, 1522

Larson, R. B. 1981, *MNRAS* 194, 809

Murgia, M., Crapsi, A., Moscadelli, L. & Gregorini, L. 2002, *A&A* 385, 412

O'Connell, R. W., Gallagher, III, J. S., Hunter, D. A. & Colley, W. N. 1995, *ApJ* 446, L1

Oka, T., Hasegawa, T., Sato, F., Tsuboi, M., Miyazaki, A. & Sugimoto, M. 2001, *ApJ* 562, 348

Rosolowsky, E. & Blitz, L. 2005, *ApJ* 623, 826

Rosolowsky, E. & Leroy, A. 2006, *PASP* 118, 590

Rubio, M., Lequeux, J. & Boulanger, F. 1993, *A&A* 271, 9

Shen, J. & Lo, K. Y. 1995, *ApJ* 445, L99

Solomon, P. M., Rivolo, A. R., Barrett, J. & Yahil, A. 1987, *ApJ* 319, 730

Wilson, C. D., Scoville, N., Madden & S. C.,Charmandaris, V. 2003, *ApJ* 599, 1049

Wong, T. & Blitz, L. 2002, *ApJ* 569, 157

Discussion

DIB: For the GMCs in the different galaxies is there also a difference in the exponent of the size-line width relationship?

ROSOLOWSKY: We cannot distinguish a change in index from a change in the constant of proportionality with the quality of the data we have. However, we can say that at least one of the index or the constant varies among galaxies and perhaps both.

DIB: It would also be interesting to check the relationship between σ_0 and the star formation rate.

ROSOLOWSKY: Particularly in dwarf galaxies where significant variation in both are observed.

ELMEGREEN: As long as you are looking at virialized clouds, this coefficient depends on pressure as a power law and pressure depends on ambient column density and therefore the star formation rate.

KENNICUTT: In fact, the metal-poor galaxies do have star formation rates that lie systematically above the mean Kennicutt(1998) Schmidt law. The offset is in the direction one would expect if the standard CO-to-H_2 conversion factor underestimated the molecular gas density. So I believe that is consistent with the results presented in this talk.

Triggered Star Formation in a Turbulent ISM
Proceedings IAU Symposium No. 237, 2006
B. G. Elmegreen & J. Palouš, eds.

Observation of triggering in the Milky Way

L. Deharveng[1], A. Zavagno[1], B. Lefloch[2], J. Caplan[1], M. Pomarès[1]

[1]Laboratoire d'Astrophysique de Marseille, 2 Place le Verrier, 13248 Marseille Cedex 4, France
email: lise.deharveng@oamp.fr
[2]Laboratoire d'Astrophysique de l'Observatoire de Grenoble

Abstract. We show how the expansion of classical Galactic H II regions can trigger massive-star formation via the collect & collapse process. We give examples of this process at work. We suggest that it also works in a turbulent medium.

Keywords. stars: formation, stars: early-type, ISM: H II regions

1. Introduction

How do we make a massive star? It has been shown statistically that the luminosities of young stellar objects (YSOs) correlate with the mass of their parental molecular clouds, and that the most luminous YSOs are found in molecular clouds adjacent to H II regions (Dobashi *et al.* 2001). Many physical processes can trigger star formation at the borders of classical H II regions (cf. the review by Elmegreen 1998). These can be divided into two main categories:

• Some of these processes assume the pre-existence of molecular condensations or inhomogeneities. For example, star formation may occur during the radiation-driven implosion of pre-existing molecular globules. This process is often advanced to explain star formation in bright rims; it has been simulated by Lefloch *et al.* (1994) and, more recently, by Miao *et al.* (2006). Also, star formation in EGGS (evaporating gaseous globules) has been proposed by Hester *et al.* (1998) to explain low-mass star formation in the 'pillars' of the Eagle Nebula.

• Some other processes allow formation of molecular condensations out of a nearly homogeneous medium. Among them are various types of instabilities of the ionisation front, e.g. hydrodynamical instabilities as simulated by García-Segura & Franco (1996) and Mizuta *et al.* (2006). The collect & collapse process, proposed by Elmegreen & Lada (1977), is another such process. **This is particularly interesting as it allows us to form massive condensations, and thus possibly massive objects, stars and clusters**. We will now focus on this process.

We also want to stress the advantage of considering objects with a very simple morphology if we want to understand which star formation process is at work at a given location. We have selected spherical H II regions around a central exciting star, surrounded by a shell of neutral material (Deharveng *et al.* 2005). The MSX and the Spitzer-GLIMPSE surveys are particularly well suited for showing the presence of this material, via the emission of the associated dust. The MSX Band A at 12.3 μm and the Spitzer-IRAC band at 8.0 μm are dominated by the emission of polycyclic aromatic hydrocarbons (PAHs). These molecules are destroyed in the ionized gas, but ionized and excited by the far-UV photons leaking from the H II regions. Our selected H II regions are surrounded by a ring of PAH emission. An example of such a region is RCW 79 (Zavagno *et al.* 2006).

2. The collect and collapse process of star formation

2.1. *How it works*

This process was first proposed by Elmegreen & Lada (1977). It has been formulated analytically by Whitworth *et al.* (1994), and simulated by Hosokawa & Inutsuka (2005, 2006).

Due to the high pressure of the warm ionized gas with respect to the low pressure of the cold surrounding neutral material, H II regions expand. During the supersonic expansion of an H II region formed around a first-generation massive star, neutral material accumulates between the ionization front and the shock front which precedes it in the neutral gas. Thus one of the signatures of the collect and collapse process is the presence of a shell of collected neutral material surrounding the H II region. This shell may be massive, up to several thousand solar masses. Later on this shell may collapse, and several massive fragments (up to a few hundred solar masses) are observed along the shell, at the periphery of the H II region. Massive objects – stars or clusters – may form inside these fragments; ultracompact (UC) second-generation H II regions may develop around these second-generation massive stars. As the stars retain the velocity of the material in which they formed, they must be observed, later on, in the direction of the parental shell; this is another signature of the collect and collapse process.

All of these signatures are observed around the Galactic H II regions Sh2-104 (Deharveng *et al.* 2003) and RCW 79 (Zavagno *et al.* 2006), demonstrating that the collect and collapse process is at work in these regions.

2.2. *Examples*

• **Sh2-104** is an optical H II region excited by an O6V star. The mass of the ionized gas is 450 $M_\odot$. The associated molecular material has been observed at IRAM, via its CO emission. The H II region is surrounded by a shell of molecular material formed by the matter collected during the H II region expansion. This swept-up shell has a mass $\sim 6000\ M_\odot$. The dense material is traced thanks to the CS emission. The dense gas is mainly concentrated in four large fragments, almost regularly distributed along the shell. The brightest fragment has a mass $\sim 670\ M_\odot$. A near-IR cluster, which is also an IRAS source of 30000 $L_\odot$, lies in its direction. It ionizes a UC H II region.

Hosokawa & Inutsuka (2005) have developed a model to follow the dynamical evolution of an H II region and its associated photodissociation region. Their model solves the UV and far-UV radiation transfer equations, and takes into account the thermal and chemical processes in a time dependant hydrodynamics code. They have adjusted a model to Sh2-104. Assuming that Sh2-104 evolves in a homogeneous medium of 10^3 cm^{-3}, the model successfully reproduces the size of the H II region and the masses of the H II region and of the collected shell. According to this model, Sh2-104 is 700 000 yr old. The collected shell is predicted to be fully molecular, as observed, and gravitationally unstable.

• **RCW 79** is an optical H II region, excited by a very massive star, according to its radio-continuum flux. Large amounts of dust, traced by the PAH emission at 8.0 μm, surround the H II region. The continuum emission of the cold dust at 1.2 mm has been observed with the SEST at ESO. This dust is distributed in a shell surrounding the ionized gas. This shell is fragmented. The two most massive fragments have masses in the 500–1000 $M_\odot$ range, and are diametrically opposite each other (a configuration observed in a number of regions). The most massive fragment harbours a near-IR cluster exciting a UC H II region. This cluster is also an IRAS point source of 55 000 $L_\odot$. Various maser sources are observed nearby, indicating that star formation is presently occurring in this region.

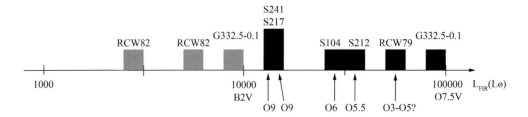

Figure 1. Far-IR luminosity of the second-generation objects formed via the collect & collapse process at the borders of H II regions. B2V and O7.5V are the spectral types of stars with a luminosity of 10000 $L_\odot$ and 100000 $L_\odot$ respectively. The names of the H II regions are indicated, as well as (with arrows) the spectral types of their exciting stars, when known. Black indicates the presence of an associated UC H II region.

The shell of PAH emission surrounding the second-generation cluster is perfectly centred on the shell of PAH emission surrounding the first-generation H II region; this is a very strong argument in favor of the collect and collapse process.

3. What we presently know about this process

3.1. *The second-generation stars*

We have seen that the collect & collapse process forms massive objects – stars and clusters. Fig. 1 gives the far-IR luminosity of the **second-generation** objects, in the few strong cases of formation of such objects by the collect & collapse process. Most of these clusters contain late OB stars exciting UC H II regions.

The spectral type of the first-generation massive star is indicated in Fig. 1. Theoretically, according to Whitworth *et al.* (1994) no correlation is expected between the spectral type of the first- and second-generation massive stars. The dominant physical parameters which influence star formation by this process are the density of the medium into which the H II region evolves and, even more importantly, the turbulence in the swept-up shell.

Two second-generation objects observed at the borders of Sh2-212 and RCW 82 seem to be **isolated massive stars**. This point requires confirmation by high-sensitivity mid-IR observations.

3.2. *Does the collect & collapse process work in a turbulent medium?*

We have seen that shells of molecular material surround, almost entirely, the Sh2-104, and RCW 79 H II regions. This is not the case for all of the H II regions selected as candidates for the collect & collapse process (Deharveng *et al.* 2005), among which Sh2-217, Sh2-212, and Sh2-241 are different.

The CO emission maps of these regions, obtained at IRAM, show that these H II regions are presently associated with molecular filaments; they have probably formed and evolved in a turbulent medium. However, dense molecular condensations are observed at their borders (but only about half way around). Near-IR clusters, stars with near-IR excesses, UC H II regions (Sh2-217 and Sh2-212), and maser sources (Sh2-241) are observed in the direction of these molecular condensations, showing that massive-star formation is presently taking place at their borders.

Dale *et al.* (2005) have simulated the photoionizing feedback of a massive star born at the centre of a turbulent cloud. The simulation relating to a low-density turbulent medium shows that this radiation source is able to stop the accretion in the central region of the cloud, and to compress the neutral material further out in the cloud, forming condensations at the periphery of the ionized gas. The observed distribution of molecular material associated with Sh2-241 (Deharveng *et al.* in preparation) is strikingly similar to the distribution of the neutral material in this simulation. Thus the collect & collapse process probably works in a turbulent medium, if this medium is not too dense and, probably, not too turbulent. These points require further study.

4. Conclusions

The collect & collapse process is an efficient process for triggering massive-star formation at the borders of H II regions. But what is its global impact on star formation? Is this process a common way of forming stars? It may be; Hosokawa & Inutsuka (2005) have shown theoretically that this process alone could sustain the current Galactic star-formation rate. Observationally, we cannot answer at present, as this process is easy to identify only in very simple H II regions.

And finally, we want to stress that even the simplest H II regions can be very complicated! Unsharp-masked images at 8 μm of the photon-dominated regions surrounding H II regions show many low-luminosity structures extending far from the ionization fronts. It shows also that the ionization fronts are very inhomogeneous. This 'leaking' structure of the ionization fronts probably also has some repercussions on star formation triggered by H II regions.

Additional figures are available at http://www.oamp.fr/matiere/Prague2006.html

Acknowledgements

We would like to thank all our collaborators in this long-term program, J. Brand, F. Massi, S. Kurtz, D. Nadeau, and F. Comerón. We also thank A. Whitworth and T. Hosokawa for very helpful and interesting discussions.

References

Dale, J.E., Bonnell, I.A., Clarke, C.J. & Bate, M.R. 2005, *MNRAS* 358, 291
Deharveng, L., Lefloch, B., Zavagno, A., *et al.* 2003, *A&A* 408, L25
Deharveng, L., Zavagno, A. & Caplan, J. 2005, *A&A* 433,565
Dobashi, K., Yonekura, Y., Matsumoto, T., *et al.* 2001, *PASJ* 53, 85
Elmegreen, B.G. & Lada, C.J. 1977, *ApJ* 214, 725
Elmegreen, B.G. 1998, in: C.E. Woodward, J.M., Shull, H.A. Thronson (eds.), *Origins* (ASP-CS), 148, p. 150
García-Segura, G. & Franco, J. 1996, *ApJ* 469, 171
Hester, J.J., Scowen, P.A. & Sankrit, R., *et al.* 1996, *AJ* 111, 2349
Hosokawa, T. & Inutsuka, S.-I. 2005, *ApJ* 623, 917H
Hosokawa, T. & Inutsuka, S.I. 2006, *ApJ* 646, 240
Lefloch, B. & Lazareff, B. 1994, *A&A* 289, 559
Miao, J., White, G.J., Nelson, R., *et al.* 2006, *MNRAS* 369, 143
Mizuta, A., Kane, J.O. & Pound, M.W. 2006, *ApJ* 647, 1151
Whitworth A.P., Bhattal A.S., Chapman S.J., *et al.* 1994, *MNRAS* 268, 291
Zavagno, A., Deharveng, L., Comerón, F., *et al.* 2006, *A&A* 446, 171

Discussion

BREITSCHWERDT: Why are stellar winds less important in triggering star formation than expanding HII regions?

DEHARVENG: Stellar winds may be important in triggering star formation. I did not mention them in my presentation because they are not present or dominant in my regions (their morphologies show no signature of stellar winds).

ZINNECKER: Is it excluded that supernova remnants are present in your HII regions?

DEHARVENG: I cannot answer for all of my regions, but most of them are classical photoionised HII regions. They are thermal radio sources, their [SII]/Hα ratios are not high, we know their exciting stars...

Triggered Star Formation in a Turbulent ISM
Proceedings IAU Symposium No. 237, 2006
B. G. Elmegreen & J. Palouš, eds.

SK 1: A possible case of triggered star formation in perseus

Miriam Rengel[1], Klaus Hodapp[2]
and Jochen Eislöffel[3]

[1]Max Planck Institute for Solar System Research, Katlenburg-Lindau, 37191, Germany
email: rengel@mps.mpg.de

[2]Institute for Astronomy, 640 N. A'hookup Place, Hilo, HI 96720
email: hodapp@ifa.hawaii.edu

[3]Thüringer Landessternwarte Tautenburg, Sternwarte 5, 07778 Tautenburg, Germany
email: jochen@tls-tautenburg.de

Abstract. According to a triggered star formation scenario (e.g. Martin-Pintado & Cernicharo 1987) outflows powered by young stellar objects shape the molecular clouds, can dig cavities, and trigger new star formation. NGC 1333 is an active site of low- and intermediate star formation in Perseus and is a suggested site of self-regulated star formation (Norman & Silk 1980). Therefore it is a suitable target for a study of triggered star formation (e.g. Sandell & Knee 2001, SK 1). On the other hand, continuum sub-mm observations of star forming regions can detect dust thermal emission of embedded sources (which drive outflows), and further detailed structures.

Within the framework of our wide-field mapping of star formation regions in the Perseus and Orion molecular clouds using SCUBA at 850 and 450 μm, we mapped NCG 1333 with an area of around $14' \times 21'$. The maps show more structure than the previous maps of the region observed in sub-mm. We have unveiled the known embedded SK 1 source (in the dust shell of the SSV 13 ridge) and detailed structure of the region, among some other young protostars.

In agreement with the SK 1 observations, our map of the region shows lumpy filaments and shells/cavities that seem to be created by outflows. The measured mass of SK 1 ($\sim$0.07 M$_\odot$) is much less than its virial mass ($\sim$0.2-1 M$_\odot$). Our observations support the idea of SK 1 as an event triggered by outflow-driven shells in NGC 1333 (induced by an increase in gas pressure and density due to radiation pressure from the stellar winds that have presumably created the dust shell). This kind of evidences provides a more thorough understanding of the star formation regulation processes.

Keywords. stars: formation, stars: individual, radio continuum: stars, ISM: jets and outflows, ISM: clouds.

---

1. Introduction

When an area of active or recent star formation is found, a question can be raised: is the star formation there happening spontaneously or is it being triggered? Places where there is no apparent trigger would be a possible evidence of its spontaneous nature, but in areas where there are indications of triggered star formation the determination of the possible causes is not easy. Because star forming areas contain populations of young stellar objects (YSOs), the mentioned question can be addressed by studying the interactions between components of the YSOs and their surrounding Interstellar Medium (ISM).

Powerful outflows are associated with strong accretion activity of the earliest protostellar phase (Class 0 stage) in the formation of a low-mass star. The Class 0 phase was identified by André, Ward-Thompson & Barsony 1993, and consists of a central protostellar object, surrounded by an infalling envelope, and a flattened accretion disk. The later SED classes (1,2 and 3, e.g. Lada 1987) are characterized by progressively diminishing accretion rates and consequently, less outflow power (Bontemps, André, Terebey, *et al.* 1996; Henriksen, André & Bontemps 1997; Davis & Eislöffel 1995).

The Class 0 phase is of special interest, not only because most of the characteristics of a future star are determined during this phase, but outflows can have a profound effect on the surrounding ISM: they shape the molecular clouds, can dig cavities, compress dust shells and trigger new star formation. Observations of this phase are, however, difficult for two reasons: first, the hot, near-stellar core of a Class 0 object is so heavily obscured ($A_V \approx 500$ mag) as to make the object undetectable up into the mid-IR. Second, the Class 0 phase is of short duration; presumably of order of a few 10^5 yr (Visser, Richer & Chandler 2002). Therefore, only a small number of these objects has been found to allow detailed studies. Nevertheless, continuum sub-mm observations of these sources can detect dust thermal emission of the circumstellar envelopes and provide a powerful tool for constraining the distribution of matter in Class 0 objects (Adams 1991).

As part of a more extensive study of star-forming regions and of the physical structure and processes in Class 0 sources (Rengel 2004), we report here the possibility of triggered star formation by outflows driven by YSOs and discuss evidence for this triggering.

2. Imaging of the molecular cloud NGC 1333 and results

2.1. *Target selection*

Together with the regions L1448, L1455, HH211 (in Perseus) and L1634 and L1641 N (in Orion), we map NCG 1333 at 850 and 450 μm using the Submillimetre Common User Bolometer Array (SCUBA) camera at the James Clerk Maxwell Telescope (JCMT). NGC 1333 is an active site of low- and intermediate star formation in Perseus and is a suggested site of self-regulated star formation (Norman & Silk 1980). It has been observed in the sub-mm by Looney, Mundy & Welch 2000, Sandell & Knee 2001, and Chini, Ward-Thompson, Kirk, *et al.* 2001. NGC 1333 IRAS 2 (Jennings; Cameron, Cudlip, *et al.* 1987) is located at the edge of the large cavity (Langer, Castets & Lefloch 1996). It has been resolved into three sources: 2A and 2B, detected by several authors (Sandell, Knee, Aspin, *et al.* 1994; Blake, Sandell, van Dishoeck, *et al.* 1995; Lefloch, Castets, Cernicharo, *et al.* 1998; Rodríguez, Anglada & Curiel 1999; Looney, Mundy & Welch 2000; Sandell & Knee 2001; Jørgensen, Hogerheijde, van Dishoeck, *et al.* 2004), and 2C, detected by Sandell & Knee (2001). IRAS 2 A and B are Class 0 candidates (Sandell & Knee 2001; Motte & André 2001). Observations of IRAS 2 (A,C) in mid-IR are reported by Rebull, Cole, Stapelfeltd, *et al.* 2003.

2.2. *Observations*

NGC 1333 was mapped here with an area of $14' \times 21'$ with both jiggle and scan maps. Data reduction treatment is described in Rengel (2004). These new maps show more structure than previous sub-mm maps of this region. We include SSV 13, south areas, the region surrounded IRAS 1 and NGC 1333 S. This later region was first noted by Rengel, Froebrich, Hodapp, *et al.* (2002), further discussed by Rengel (2004) and independently discovered by Young, Shirley, Evans, *et al.* (2003). Hodapp, Bally, Eislöffel, *et al.* (2005) discussed a subset of the sub-mm data presented here in their relation to NIR observations and pointed out that the driving source of the system HH343A-F–HH340B was likely associated with the easternmost clump in NGC 1333 S. Newly identified structure introduced by Rengel, Hodapp & Eislöffel (2005) strengthen the tentative conclusion reached by Hodapp, Bally, Eislöffel, *et al.* (2005) that NGC 1333 S is the site of secondary,

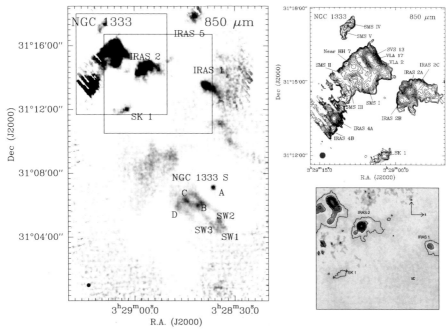

Figure 1. The 850 μm deep field map of NGC 1333 (left) and the 850 μm contour map of the SSV 13 area in NGC 1333 (upper right) (left box in left figure). Contour levels are in log scale with step size 1.25 from 0.0002 Jy beam^{-1}. The filled circles in the bottom left indicate the main beam size. Down right image shows the deep field map of the region containing SK 1.

low-mass star formation triggered by the powerful IRAS 1-9 protostars about 1 pc north of this region.

Furthermore, SK 1 is an embedded source in the dust shell south of the SVS 13 ridge, with an emission (east) detected in the 850 μm map. We unveiled it, and found several dust ridges and shells formed by outflows in the region, among further detailed structure and some other YSOs (Fig. 1). Sandell & Knee (2001) suggest that SK 1 appears to be a case of triggered star formation by outflow-driven shells. Further details of the mapped region are given in Rengel (2004) and Rengel, Hodapp & Eislöffel (2006).

3. Deriving physical parameters

We estimate the gas and dust masses for the sample from the dust emission according to the mass equation of Hildebrand (1983). We find a mass of 3.7 M$_\odot$ for IRAS 2A and $\sim$0.07 M$_\odot$ for SK 1.

We infer the bolometric temperature and luminosity of IRAS 2A and SK 1 by constructing the complete SEDs (we combine the SCUBA fluxes with data from the literature in several wavelengths [Rengel, Froebrich, Wolf, *et al.* 2004]). Ages are further calculated following the Smith protostellar evolutionary scheme (Smith 1999, 2002). For IRAS 2A and SK 1, we find ages of 26 and 15 $\times$ 10^3 yr, respectively.

Within the framework of deriving physical parameters of the YSOs contained our wide-field mapping of star formation regions, we also establish further physical conditions (e.g. temperature distribution, radius, and power-law index of the density) of some sources in the sample, including IRAS 2A. This is performed by using the radiative transfer code MC3D (Wolf, Henning & Stecklum 1999). Further modeling details are given in Rengel (2004) and Rengel, Hodapp & Eislöffel (2006).

4. SK 1: Suspected triggered star formation in NGC 1333

Are the molecular outflows of IRAS 2 (Knee & Sandell 2000; Jørgensen, Hogerheijde, Blake, *et al.* 2004) perhaps the main cause of the formation of SK 1? We report here some evidences that support the idea of modification and disruption of areas by the IRAS 2 outflows, and the formation of SK 1 as a result of the compressed dust shells/cavities.

First, our map of the region shows lumpy filaments and shells/cavities that seem to be created by outflows. Second, the measured mass of SK 1 is much less than its virial mass ($\sim$0.2-1 $M_\odot$). This indicates that SK 1 is not in a stationary state, isolated and self-gravitationally bound. Third, regarding the star formation timing, a hard upper limit is placed on the age: IRAS 2A is older than SK 1. Fourth: which one physical mechanism is responsible of the formation of SK 1? we rule out cluster dissipation as the main mechanism: if a standard cluster dispersion velocity of 1 km s^{-1} is assumed, SK 1 has not traveled far. "Collect-and-collapse" (CAC) and radiation-driven implosion (RDI) scenarios (Elmegreen & Lada 1977; Oort & Spitzer 1955) have been proposed as physical mechanisms for the star formation at the edge of an HII region. Timescales for the RDI mechanism are less than for the CAC scenarios, then it seems that this later one is the most supported. Fifth, IRAS 1 has a long tail pointing away from the HII region, which is an indication of direct interaction between the outflows and the gas and dust in the vicinity.

5. Conclusion

SK 1 provides an interesting study of interactions between outflows and surrounding ISM. In at least one case, we identify this protostellar source whose formation is likely to have been triggered by powerful outflow bow shocks. This kind of evidence provides a more thorough understanding of the star formation regulation processes. Further observations of molecular gas are necessary.

Acknowledgements

We thank the S237 organizers, and T. Jenness for the assistance with SURF.

References

Adams, F. C. 1991, *ApJ* 382, 544
André, P., Ward-Thompson, D. & Barsony, M. 1993, *ApJ* 406, 122
Blake, G. A., Sandell, G., van Dishoeck, E. F., *et al.* 1995, *ApJ* 441, 689
Bontemps, S., André, P., Terebey, S. & Cabrit, S. 1996, *A&A* 311, 858
Chini, R., Ward-Thompson, D., Kirk, J. M., *et al.* 2001, *A&A* 369, 155
Davis, C. J. & Eislöffel, J. 1995, *A&A* 300, 851
Elmegreen, B. G. & Lada, C. J. 1977, *ApJ* 214, 725
Henriksen, R., André, P. & Bontemps, S. 1997, *A&A* 323, 549
Hildebrand, R. H. 1983, *QJRAS* 24, 267
Hodapp, K. W., Bally, J., Eislöffel J. & Davis, C. J. 2005, *AJ* 129, 1580
Jennings, R. E., Cameron, D. H. M., Cudlip, W. & Hirst, C. J. 1987, *MNRAS* 226, 461
Jørgensen, J. K., Hogerheijde, M. R., Blake, G. A., van Dishoeck, E. F., Mundy, L. G. & Schöier, F. L. 2004, *A&A* 415, 1021
Jørgensen, J. K., Hogerheijde, M. R., van Dishoeck, E. F., Blake, G. A. & Schöier, F. L. 2004, *A&A* 413, 993
Knee, L. B. G. & Sandell, G. 2000, *A&A* 361, 671
Lada, C. J. 1987, in: M. Peimbert & J. Jugaku (eds.), *Star Forming Regions* (Dordrecht: Kluwer), p. 1

Langer, W. D., Castets, A. & Lefloch, B. 1996, *ApJ* 471, 111

Lefloch, B., Castets, A., Cernicharo, J. & Loinard, L. 1998, *ApJ* 504, L109

Looney, L. W., Mundy, L. G. & Welch, W. J. 2000, *ApJ* 529, 477

Martin-Pintado, J. & Cernicharo, J. 1987, *A&A* 176, 27

Motte, F. & André, P. 2001, *A&A* 365, 440

Norman, C. & Silk, J. 1980, *ApJ* 238, 158

Oort, J. H. & Spitzer, L., Jr. 1955, *ApJ* 121, 6

Rebull, L. M., Cole, D. M., Stapelfeltd, K. R. & Werner, M. W. 2003, *AJ* 125, 2568

Rengel, M. Froebrich, D., Hodapp, K. & Eislöffel, J. 2002, in: J. F. Alves & M. J McCaughrean (eds.), *The Origins of Star and Planets: The VLT View* (Berlin: Springer), CD-ROM

Rengel, M., Froebrich, D., Wolf, S. & Eislöffel, J. 2004, *BaltA* 13, 449

Rengel, M. 2004, Ph.D. Thesis, Friedrich-Schiller Universität, Jena

Rengel, M., Hodapp, K. & Eislöffel, J. 2005, *AN* 326, 631

Rengel, M., Hodapp, K. & Eislöffel, J. 2006, *A&A* submitted

Rodríguez, L. F., Anglada, G. & Curiel, S. 1999, *ApJS* 125, 427

Sandell, G., Knee, L. B. G., Aspin, C., Robson, I. E. & Russell, A. P. G. 1994, *A&A* 285, L1

Sandell, G. & Knee, L. B. G. 2001, *ApJ* 546, L49

Smith, M. D. 1999, *ApSS* 261, 169

Smith, M. D. 2002, in: J.F. Alves & M.J McCaughrean (eds.), *The Origins of Stars and Planets: The VLT View* (Berlin: Springer), CD-ROM

Visser, A. E., Richer, J. S. & Chandler, C. J. 2002, *AJ* 124, 2756

Wolf, S., Henning, Th. & Stecklum, B. 1999, *A&A* 349, 839

Young, C. H., Shirley, Y. L., Evans, N. J., II & Rawlings, J. M. C. 2003, *ApJS* 145, 111

Discussion

NAKAMURA: Could you make some comment on the physical conditions of SK1 before the outflow shock hit the core?

RENGEL: With our evolutionary models, and knowing the actual physical structure of SK1, we can go back to earlier stages (e.g., before the outflow shock hit the core) and calculate the physical conditions of SK1 (size, T_{bol}, L_{bol}, mass, etc.). As presumably SK1 is a very young protostar, it has more massive envelope, and it was cooler.

Triggered Star Formation in a Turbulent ISM
Proceedings IAU Symposium No. 237, 2006
B. G. Elmegreen & J. Palouš, eds.

The life and death of star clusters

B. C. Whitmore

[1]Space Telescope Science Institute, 3700 San Martin Dr., Baltimore, MD, 21218, USA
email: whitmore@stsci.edu

Abstract. It is generally believed that most stars are born in groups and clusters, rather than in the field. It has also been demonstrated that merging galaxies produce large numbers of young massive star clusters, sometimes called super star clusters. Hence, understanding what triggers the formation of these young massive clusters may provide important information about what triggers the formation of stars in general. In recent years it has become apparent that most clusters do not survive more than ≈10 Myr (i.e., "infant mortality"). Hence, it is just as important to understand the disruption of star clusters as it is to understand their formation if we want to understand the demographics of both star clusters and field stars. This talk will first discuss what triggers star cluster formation in merging galaxies (primarily in the Antennae galaxies), will then demonstrate that most of the faint objects detected in the Antennae are clusters rather than individual stars (which shows that the initial mass function was a power law rather than a Gaussian), and will then outline a general framework designed to empirically fit observations of both star clusters and field stars in a wide variety of galaxies from mergers to quiescent spirals.

1. Introduction

Why are we talking about star clusters in a meeting about the formation of stars? The obvious answer is that most stars are formed in associations, groups and clusters, (e.g., Lada & Lada, 2003). Hence, understanding what triggers the formation of star clusters may provide important clues for understanding the formation of stars in general. It is well known that star formation is enhanced in merging galaxies, so this is a good place to start. The "Antennae Galaxies" (NGC 4038/39; see Whitmore *et al.* 1999 for image) are the youngest and nearest galaxies in the Toomre sequence (Toomre 1977) of prototypical mergers. Hence, they may represent our best chance for understanding the formation of star clusters in interacting galaxies. While other galaxies will occasionally be mentioned in this review, the primary focus will be the Antennae.

2. What triggers the formation of star clusters

2.1. *Kinematic clues*

While the details of star formation are still obscure, one thing that everyone agrees on is that shocks are important (e.g., along spirals arms). In mergers, one popular model (e.g., Kumai *et al.* 1993) has been that high-speed cloud-cloud collisions with velocities ≈50 – 100 km s^{-1} are required. We have used STIS long-slit spectra in three positions angle of the Antennae to test this idea (Whitmore *et al.* 2005).

We find the velocity fields are remarkably quiescent (Figure 1). RMS dispersions are ≈10 km s^{-1}, essentially the same as disks of spiral galaxies. This does not favor high-speed cloud-cloud collision models, but is consistent with models where a high pressure interstellar medium implodes GMCs without greatly altering their velocity distribution (e.g, Jog and Solomon 1992). This also supports the results of Zhang, Fall, & Whitmore

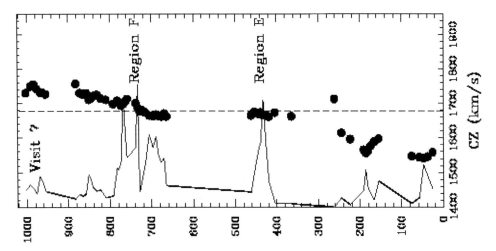

Figure 1. Hα velocities of star clusters in the Antennae based on long-slit observations using the STIS detector on HST. Note how small the velocity dispersion is for a given subregion (e.g., ≈10 km s^{-1} for region F, once the large scale gradient is removed). See Whitmore *et al.* 2005 for details.

(2001) who found essentially no correlation between the positions of young clusters in the Antennae and velocity gradients.

2.2. *Morphological clues*

Another approach to studying triggered star formation is to age-date the stars or star clusters and then look for patterns that suggest sequential star formation. One of the classic example of this is 30 Doradus, where Walborn *et al.* (1999) found evidence for several generations of triggered star formation. Does our age-dating of clusters in the Antennae provide similar evidence for triggered star cluster formation? Figure 2 shows the pattern of ages in Knot S of the Antennae, the Hα image of Knot S, and a plot of the ages vs. the distance from the center of Knot S.

Restricting the range to the inner 50 pixels (≈ 140 pc) provides tentative evidence for some triggering, with a set of five older clusters found near the center of Knot S, and intermediate and young clusters found further out. Note the good correlation between the positions of the youngest clusters (squares in Figure 2a) and the positions of strong Hα flux image (Figure 2b). The Hα shell has apparently met with less resistance traveling to the right side of the image, where it has expanded to larger distance and fewer clusters have been formed. However, we find that the total mass of all the intermediate-age clusters is only about 5 – 10 % of the mass of the central five clusters, hence triggered star formation is a relatively minor effect in this region. Several other knots of star clusters in the Antennae show similar patterns with relatively small fractions of mass being produced in triggered regions of star formation. Perhaps this indicates that most of the gas in the area was used in the original burst of star formation.

However, Knot B (Figure 3) presents both a simpler and a much more dramatic example of triggered star formation. In this case we find a single cluster with an age greater than 10 Myr (and mass ≈ 10^6 M$_\odot$), a smooth gradient in the cluster ages, and a population of younger clusters with a total mass 4×10^6 M$_\odot$, FOUR times greater than the cluster that appears to have triggered their formation. Triggering is not always weak! We note that the Hα supershell is centered on the older cluster, and the young clusters are located where the shell intersects the edge of a major dust lane. This implicates the older

Figure 2. Left figure – I band images of Knot S in the Antennae with older (> 10 Myr: circles),
intermediate-aged (3 – 10 Myr, crosses), and very young (< 3 Myr, squares) clusters. Center figure – H$_\alpha$ image. Right figure – Age profile from center of Knot S. See Whitmore (2007) for a larger image.

cluster as the original spark. An analogy might be that if you throw a match down on a field of dry grass you might only get a few minor brushfires, but if you throw it down next to a tinder box something more impressive is formed.

3. Differentiating stars from clusters

While this might sound like a relatively unimportant technical issue, it is fundamentally important for answering the following two questions. What fraction of stars are formed in clusters vs. the field (i.e., does triggering primarily happen in isolation or in a clustered environment)? Is the initial cluster mass function a power law or a Gaussian (see Fall and Zhang 2001, de Grijs *et al.* 2005, Whitmore, Chandar, & Fall 2006 for background)?

In Whitmore *et al.* (1999), differentiating stars from clusters using the undersampled WFPC2 image was one of our primary difficulties, leading us to conclude that the number of young star clusters in the Antennae was between 800 and 8000; a pretty big range! Our new ACS data provides a better opportunity for making this determination, and for studying the stars in their own right.

Figure 4 shows U-B vs V-I diagrams for 4 luminosity ranges of point-like objects around Knot S in the Antennae. The solid lines are Bruzual & Charlot (2003) solar metalicity models for clusters. The dashed lines in the bottom left panel are Padova models for young stars. The dotted line shows a reddening vector, placed so that the objects above the line are primarily young clusters while the objects below the line are massive stars. Note that this only works when the objects are younger than about 50 Myr, at which point the Bruzual-Charlot models fall below the dotted line. Open squares are objects with PSFs that are indistinguishable from stars while filled circles appear to be slightly resolved (i.e., candidate clusters). The fact that all of the objects in the lower left panel (objects brighter than $M_V = -10$) are in "cluster-space" is reassuring, since the brightest stars are believed to be approximately $M_V = -9$ or fainter. Three candidate stars are found in the next luminosity bin (note that all three have $M_V \approx 9.1$), but most of the other objects still appear to be clusters. As we progress to fainter bins, we find more candidate stars, but the majority of objects still appear to be clusters, based both on their spatial profiles and on their colors. In fact, a combination of size and colors appear to be the best discriminant between stars and clusters (see Whitmore *et al.* 2007 for a

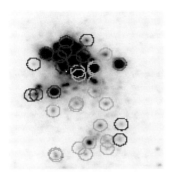 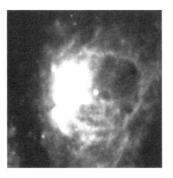 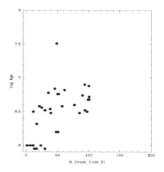

Figure 3. Same as Figure 2, except for Knot B. The center for the profile (right panel) is the set of bright young clusters on the left of the image, rather than the single older cluster in the upper right of Figure 3a. This older cluster is in the middle of the H$_\alpha$ shell, implicating it as the original triggering cluster.

more detailed discussion), since some of the clusters appear to be so compact that they are indistinguishable from stars, even with ACS.

These results provide further evidence that most stars are indeed formed in clusters, and also show that the initial cluster mass function cannot have been a Gaussian, since this would require essentially all of the faint objects to be stars, rather than clusters.

4. A general framework for understanding the demographics of star clusters

Roughly 40 gas-rich mergers have now been observed by the Hubble Space Telescope. All of these show some number of young compact star clusters, similar to what we find in the Antennae. In addition, various authors have found young compact clusters in a variety of other galaxies, including starburst dwarf galaxies, barred galaxies, normal spiral galaxies, and the LMC. In general, the clusters found in non-interacting galaxies have properties similar to those seen in the mergers, but always fewer in number and generally fainter in luminosity. A reasonable hypothesis might therefore be that the most massive super star clusters can only be produced in violent environments rather than relatively quiescent environments such as those found in spiral galaxies? Indeed, various authors have suggested that there are two modes of star formation.

However, Whitmore (2003, originally presented in 2000 as astro-ph/0012546) and Larsen (2002) have shown that there is a smooth, continuous correlation between the brightest cluster in a galaxy and the number of clusters in the galaxy. This suggests that mergers and starburst galaxies may have the brightest clusters only because they have the most clusters. There may be a *universal* luminosity function with the correlation simply being due to statistics, rather than any special physics. This has been dubbed the "size-of-sample" effect, and is been included in a number of recent papers (e.g., Hunter *et al.* 2003). The number of clusters in mergers may be larger because conditions for triggering cluster formation are globally present in mergers but only locally present in spirals (e.g., along spiral arms).

There is growing evidence that the disruption rate of clusters may also be universal. For example, Fall, Chandar & Whitmore (2005) find that 90% of the young clusters in the Antennae are disrupted each factor of 10 of time. This is also true in the Milky Way (Lada & Lada 2003), and the SMC (Rafelski & Zaritsky 2005, Chandar, Fall & Whitmore

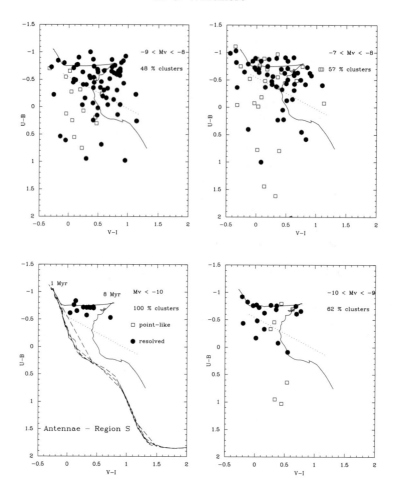

Figure 4. $U - B$ vs. $V - I$ color-color diagram for four luminosity ranges for a sample of objects around Knot S in the Antennae (from Whitmore *et al.* 2006; see text for details).

2006), as shown in Figure 5. Hence, the disruption rate appears to be a power law with index -1.

These two results motivated us to develop a general framework for understanding the demographics of both star clusters and field stars, which we assume are formed as a by-product of the disrupted clusters (Whitmore, Chandar, Fall 2006). The ingredients for the model are: 1) a universal initial mass function (power law, index -2), 2) various star(cluster) formation histories that can be coadded (e.g., constant, Gaussian burst, ...), 3) various cluster disruption mechanisms (e.g., power law, index -1 for < 100 Myr – infant mortality; constant mass loss for > 100 Myr – 2-body relaxation), 4) convolution with observational artifacts and selection effects.

This simple model allows us to predict a wide variety of properties for the clusters, field stars, and integrated properties of a galaxy. Of particular relevance for the present paper is the fact that the model we have developed for the Antennae using this framework predicts that ≈8 % of the UV luminosity should be from clusters rather than field stars, in agreement with the observations (≈10 %). Hence, our data is consistent with the idea

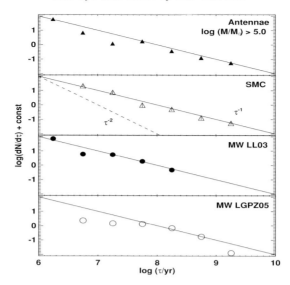

Figure 5. The age distributions for clusters in the Antennae (filled triangles), the SMC (open triangles), and the Milky Way filled and open circles. In each distribution we find that the number of clusters declines as τ^{-1}; i.e., infant mortality – 90 % loss each factor of 10 in time). See Whitmore, Chandar, & Fall (2006) for details.

that all stars are born in clusters. Fall, Chandar & Whitmore (2005) reach a similar conclusion based on Hα flux.

Acknowledgements

The authors wishes to acknowledge various coauthors on various projects discussed in this paper. These include Rupali Chandar, Francois Schweizer, Mike Fall, Qing Zhang, and Barry Rothberg.

References

Bruzual A. G. & Charlot, S. 2003, *MNRAS* 344, 1000
Chandar, R., Fall, S. M. & Whitmore, B. C. 2006, in press
de Grijs, R., Parmentier, G. & Lamers, H. J. G. L. M. 2005, *MNRAS* 364, 1054
Fall, S. M., Chandar, R. & Whitmore, B. C. 2005, *ApJ* 631, L133
Fall, S. M. & Zhang, Q. 2001, *ApJ* 561, 751
Hunter, D. A., Elmegreen, B. G., Dupuy, T. T. & Mortonson, M. 2003, *AJ* 126, 1836
Jog, C. & Solomon, P. M. 1992, *ApJ* 387, 152
Kumai, Y., Hashi, Y. & Fujimoto M. 1993, *ApJ* 416, 576
Lada, C. J. & Lada, E. A. 2003, *ARAA* 41, 57
Larsen, S. S. 2002, *AJ* 124, 1393
Rafelski, M. & Zaritsky, D. 2005, *AJ* 129, 2701
Toomre, A. 1977, in: B. M. Tinsley & R. B. Larson (eds.) *The Evolution of Galaxies and Stellar Populations* (Yale: New Haven), p. 401
Walborn, N. R., Barba, R., H., Brandner, W., Rubio, M., Grebel, E. & Probst, R. 1999, *AJ* 117, 225
Whitmore, B. C. 2003, in: M. Livio, K. Noll, & M. Stiavelli (eds.), *A Decade of HST Science* (Cambridge:Cambridge University), p. 153
Whitmore, B. C. 2007, in: M. Livio (ed.), *Massive Stars: From Pop III and GRBs to the Milky Way* (Cambridge:Cambridge University), in press

Whitmore, B. C., Chandar, R. & Fall, S. M. 2006, *AJ* in press

Whitmore, B. C., Gilmore, D., Leitherer, C., Fall, S. M., Chandar, R., Blair, W. P., Schweizer, F., Zhang, Q. & Miller, B. W. 2005, *AJ* 130, 2104

Whitmore, B. C., Zhang, Q., Leitherer, C., Fall, S. M., Schweizer, F. & Miller, B. W. 1999, *AJ* 118, 1551

Zhang, Q., Fall, M. & Whitmore, B. C. 2001, *ApJ* 561, 727

Discussion

DE GRIJS: (1) I'm a little worried about your distinction between stars and clusters if you're really working at the resolution limit. In a recent paper by Bastian, Lamers, etc., we showed that you need to use the full SED in addition to size estimates. Then all stars fall into the youngest age bin (< 6 Myr). Have you tried to assess your objects this way? (2) I think there is a subtle difference between your argument that the initial cluster luminosity function might be a power law and recent published results advocating lognormal initial CLFs. In your case, you include all objects, in the lognormal case one deals with the bound fraction only, so you would need to include the effects of infant mortality quantitatively (which is hard!) before making definitive statements... I think, therefore, that our results are not at odds with each other.

WHITMORE: (1) We agree that using size alone is only partially successful in separating stars and clusters, but took some pains during the talk to show that in many cases it is actually quite easy using the new ACS data. However, we also use the spectral color information (see figure in paper) to help make the distinction. (2) Of course we would all like to know which clusters are bound and which are not. At present, this is impossible to do for all but the brightest clusters (using stellar velocity dispersions) directly). In your case, you use older clusters which are more likely to be bound, but then you have not direct information about the *initial* cluster mass function, and must make secondary arguments to try to infer the initial mass function. We prefer to work with all the clusters, whether they are bound or not, so that we can directly determine the initial mass function. We then try to estimate the fraction of unbound clusters, which causes the high infant mortality rate.

WALBORN: This is very important work, bridging the very local objects such as in the Magellanic Clouds where we have spectral classifications for all the OB stars to establish ages, and much more distant objects with no spatial resolution. The asymmetry of the triggering is remarkable and agrees with the local archetypes 30 Doradus, N11, NGC 3603, and others, in which the second generation is always in one hemisphere relative to the first. This must be telling us something essential, perhaps that the initial cluster forms near an edge of the GMC in response to an external event, as opposed to a central collapse. As I'm sure you do, one must always bear in mind that a point source is not necessarily a single star. At this distance, 0.1 arcsec is about 10 pc.

WHITMORE: Yes, I agree with everything you say. In particular your point that sometimes a cluster can be so concentrated that it can be indistinguishable from a star at this distance. This is an important sidelight of the color-color diagrams that I showed, that 10-20% of the objects in "cluster-space" appear to be unresolved.

MELNICK: 1. I question the idea that all stars are formed in clusters since in our galaxy up to 70% of the massive stars are not in clusters and are not runaway stars. 2. Don't you find it surprising that the (initial) mass function of clusters has the same slope in the antennae, in the LMC, in the Milky Way, and maybe everywhere else (i.e., is universal)?

WHITMORE: 1. All we can really say is that most of the stars appear to form in groups and clusters. We certainly can't say that 100% do. 2. This just shows that the process that makes clusters is probably pretty universal. I guess Nature has been kind to us.

ZINNECKER: From these observations of the Antennae and other colliding galaxies, what can you infer about the very process of triggering?

WHITMORE: The specific thing we can say is that high speed (50-100 km s^{-1}) cloud-cloud collisions are not directly responsible for triggering star and star cluster formation. The fact that the velocity dispersion between clusters is quite low (< 10 km s^{-1}), and roughly the same as the disk in a normal spiral disk, suggests to me that pre-existing GMCs are triggered into formation by an increase in ISM pressure, without changing the velocity distribution. An alternative possibility is a turbulent cascade of velocities for orbital collision speeds (hundreds of km s^{-1}) all the way down to disk-like velocity dispersions.

ALVES: Just a comment to your result on the mass function of clusters. We find essentially the same mass spectrum for GMCs in both Can A and M83, which ties beautifully to your spectrum of clusters. the fact that you do not see a variation in the cluster mass spectrum between starburst galaxies and more star formation quiet galaxies also suggests a connection to the same underlying mass spectrum.

OEY: In response to Jorge Melnick's comment about field OB stars, we can take the power law mass function that you showed to estimate the fraction of field massive stars by extrapolating to individual stars (Oey, King & Parker 204). But my question to you is: How can we distinguish between cluster dissolution and a recent/on-going burst of star formation? It seems most of the examples you showed are candidates for the latter.

WHITMORE: For a particular galaxy, it is possible that we are catching a galaxy during a recent burst in the last < 10 Myr. However, when nearly everyone finds mean ages for young clusters in any galaxies with ongoing star formation (eg., most spirals) to be < 10 Myr, you have to start wondering. Infant mortality (dissolution of 90% of the clusters each decade of log time) is such a dominant effect that it results in mean cluster ages < 10 Myr even if the star formation is not very constant. Another way to demonstrate how dominant infant mortality is is to break the Antennae into 4 quadrants. It is impossible to coordinate a starburst over the entire galaxy, and in fact there is good evidence that the overlap region is currently where the most recent star formation is. However ,the dN/dT diagram looks very similar in all 4 quadrants, but the mean age is always < 10 Myr, even in regions which had a strong burst ~ 100 Myr ago. I will include this figure in the proceedings.

Triggered Star Formation in a Turbulent ISM
Proceedings IAU Symposium No. 237, 2006
B. G. Elmegreen & J. Palouš, eds.

© 2007 International Astronomical Union
doi:10.1017/S1743921307001524

Star-cluster formation and evolution

Pavel Kroupa

Argelander Institute for Astronomy, University of Bonn, Auf dem Hügel 71, D-53347 Bonn,
Germany
email: pavel@astro.uni-bonn.de

Abstract. Star clusters are observed to form in a highly compact state and with low star-formation efficiencies, and only 10 per cent of all clusters appear to survive to middle- and old-dynamical age. If the residual gas is expelled on a dynamical time the clusters disrupt. Massive clusters may then feed a hot kinematical stellar component into their host-galaxy's field population thereby thickening galactic disks, a process that theories of galaxy formation and evolution need to accommodate. If the gas-evacuation time-scale depends on cluster mass, then a power-law embedded-cluster mass function may transform within a few dozen Myr to a mass function with a turnover near $10^5\,M_\odot$, thereby possibly explaining this universal empirical feature. Discordant empirical evidence on the mass function of star clusters leads to the insight that the physical processes shaping early cluster evolution remain an issue of cutting-edge research.

Keywords. methods: n-body simulations; stars: luminosity function, mass function; stars: formation; binaries: general; globular clusters: general; open clusters and associations: general; Galaxy: structure; Galaxy: evolution; galaxies: star clusters

1. Early cluster evolution

The star-formation efficiency (sfe), $\epsilon \equiv M_{\rm ecl}/(M_{\rm ecl} + M_{\rm gas})$, where $M_{\rm ecl}, M_{\rm gas}$ are the mass in freshly formed stars and residual gas, respectively, is $0.2 \lesssim \epsilon \lesssim 0.4$ (Lada & Lada (2003)) implying that the physics dominating the star-formation process on scales < 10 pc is stellar feedback. Within this volume, the pre-cluster cloud core contracts under self gravity thereby forming stars ever more vigorously, until feedback energy suffices to halt the process (*feedback-termination*, Weidner & Kroupa (2006)). This occurs on one to a few crossing times ($\approx 10^6$ yr), and since each proto-star needs about 10^5 yr to accumulate about 95 per cent of its mass (Wuchterl & Tscharnuter (2003)), the assumption may be made that the embedded cluster is mostly virialised at feedback-termination. Its stellar velocity dispersion,

$$\sigma \approx \sqrt{G\,M_{\rm ecl}/(\epsilon\,R)}, \tag{1.1}$$

may then reach $\sigma = 40\,{\rm pc/Myr}$ if $M_{\rm ecl} = 10^{5.5}\,M_\odot$ which is the case for $\epsilon\,R < 1$ pc. This is easily achieved since the radius of one-Myr old clusters is $R \approx 1$ pc with a weak, if any, dependence on mass (Bastian *et al.* (2005)). Very young clusters (age $\lesssim 10$ Myr) would thus appear super-virial, i.e. with a velocity dispersion too large for the cluster mass.

The above exercise demonstrates that the possibility may be given that a *hot kinematical component* could add to a galactic disk as a result of clustered star formation for reasonable physical parameters. Thickened galactic disks may result. But this depends on

(i) ϵ,

(ii) R (cluster concentration) and

(iii) the ratio of the gas-expulsion time-scale to the dynamical time of the embedded cluster, $\tau_{\rm gas}/t_{\rm cross}$.

1.1. *Empirical constraints*

The first (i) of these is clearly fulfilled: $\epsilon < 40$ per cent (Lada & Lada (2003)). The second (ii) also appears to be fulfilled such that clusters with ages $\lesssim 1$ Myr have $R \lesssim 1$ pc independently of their mass. Some well-studied cases are tabulated and discussed in Kroupa (2005). Finally, the ratio $\tau_{\rm gas}/t_{\rm cross}$ (iii) remains uncertain but critical.

The well-observed cases discussed in Kroupa (2005) do indicate that the removal of most of the residual gas does occur within a cluster-dynamical time, $\tau_{\rm gas}/t_{\rm cross} \lesssim 1$. Examples noted are the Orion Nebula Cluster (ONC) and R136 in the LMC both having significant super-virial velocity dispersions. Other examples are the Treasure-Chest cluster and the very young star-bursting clusters in the massively-interacting Antennae galaxy which appear to have HII regions expanding at velocities such that the cluster volume may be evacuated within a cluster dynamical time.

A simple calculation of the amount of energy deposited by an O star within a cluster crossing time into its surrounding cluster-nebula also suggests it to be larger than the nebula binding energy (Kroupa (2005)). Furthermore, Bastian & Goodwin (2006) note that many young clusters have a radial-density profile signature as expected if they are expanding rapidly.

Thus, the data suggest the ratio $\tau_{\rm gas}/t_{\rm cross}$ to be near one, but much more observational work needs to be done to constrain this number. Measuring the kinematics in very young clusters would be an extremely important undertaking, because the implications of $\tau_{\rm gas}/t_{\rm cross} \lesssim 1$ are dramatic.

To demonstrate these implications it is now assumed that a cluster is born in a very compact state ($R \approx 1$ pc), with a low sfe ($\epsilon < 0.4$) and $\tau_{\rm gas}/t_{\rm cross} \lesssim 1$. As noted in Kroupa (2005), "in the presence of O stars, explosive gas expulsion may drive early cluster evolution independently of cluster mass".

2. Implications

2.1. *Cluster evolution and the thickening of galactic disks*

As one of the important implications, a cluster in the age range of $\approx 1 - 50$ Myr will have an unphysical M/L ratio because it is out of dynamical equilibrium rather than having an abnormal stellar IMF (Bastian & Goodwin (2006)).

Another implication would be that a Pleiades-like open cluster would have been born in a very dense ONC-type configuration and that, as it evolves, a "moving-group-I" is established during the first few dozen Myr comprising roughly 2/3rds of the initial stellar population and expanding outwards with a velocity dispersion which is a function of the pre-gas-expulsion configuration (Kroupa, Aarseth & Hurley (2001)). These computations were in fact the first to demonstrate, using high-precision N-body modelling, that the redistribution of energy within the cluster during the embedded phase and the expansion phase leads to the formation of a substantial remnant cluster despite the inclusion of all physical effects that are disadvantageous for this to happen (explosive gas expulsion, Galactic tidal field and mass loss from stellar evolution).

Thus, in this scenario stars form in very compact clusters that have radii less than about 1 pc and masses larger than a dozen $M_\odot$, and in the presence of O stars the residual gas is removed explosively leading to loss of stars from the cluster which form a moving group I (an expanding population of sibling stars). A cluster re-forms after gas expulsion as a result of energy-equipartition during the embedded phase, or during the expanding phase, and fills its Roche lobe in the Galactic tidal field. Such a cluster appears with an expanded core radius (compare the ONC with about 0.2 pc and the Pleiades with 1.3 pc) and evolves secularly through evaporation until it's demise (Baumgardt & Makino (2003), Lamers *et al.* (2005a), Lamers *et al.* (2005b)). A "moving-group-II" establishes during this stage as the "classical" moving group made-up of stars which slowly diffuse/evaporate out of the re-virialised cluster remnant with relative kinetic energy close to zero. Under unfavorable conditions no cluster re-forms after gas expulsion. Such conditions may arise from time-variable tidal fields (e.g. through nearby cloud formation) or simply through unfavorable density profiles just before gas-expulsion (Boily & Kroupa (2003a), Boily & Kroupa (2003b)), and they may dominate over the favorable ones (§ 2.3). Note that the number of stars in moving-group-I always outnumber the number of stars in moving-group-II, while the unfavourable conditions would lead to no moving-group-II.

Moving-groups-I would be populated by stars that carry the initial kinematical state of the birth configuration into the field of a galaxy. Each generation of star clusters would, according to this picture, produce overlapping moving-groups-I (and II), and the overall velocity dispersion of the new field population can be estimated by adding in quadrature all expanding populations. This involves an integral over the embedded-cluster mass function, $\xi_{ecl}(M_{ecl})$, which describes the distribution of the stellar mass content of clusters when they are born (Kroupa (2002), Kroupa (2005)). It is known to be a power-law (Lada & Lada (2003), Hunter *et al.* (2003), Zhang & Fall (1999)). The integral can be calculated for a first estimate of the effect. The result is that for reasonable upper cluster mass limits in the integral, $M_{ecl} \lesssim 10^5 \, M_\odot$, the observed age–velocity dispersion relation of Galactic field stars in the solar neighbourhood can be reproduced.

This theory can thus explain the "energy deficit", namely that the observed kinematical heating of field stars with age cannot, until now, be explained by the diffusion of orbits in the Galactic disk as a result of scattering on molecular clouds, spiral arms and the bar (Jenkins (1992)). Because the age–velocity-dispersion relation for Galactic field stars increases with stellar age, this notion can also be used to map the star-formation history of the Milky-Way disk by resorting to the observed correlation between the star-formation rate in a galaxy and the maximum star-cluster mass born in the population of young clusters (Weidner, Kroupa & Larsen (2004)).

A possible cosmologically-relevant implication of this "popping-cluster" model emerges as follows (Kroupa (2002), Kroupa (2005)): A thin galactic disk which experiences a significant star burst such that star clusters with masses ranging up to $10^6 \, M_\odot$ can form in the disk, will puff-up and obtain a thick-disk made-up of the fast-moving velocity wings of the expanding clusters after residual gas expulsion. A notable example where this process may have lead to thick-disk formation is the galaxy UGC 1281 (Mould (2005)), which, by virtue of its LSB character, could not have obtained its thick disk as a result of a merger with another galaxy. Elmegreen & Meloy Elmegreen (2006) observe edge-on disk galaxies with a chain-like morphology in the HST ultra-deep field finding many to have thick disks. These appear to be derived from kpc-sized clumps of stars. These clumps are probably massive star-cluster complexes which form in perturbed or intrinsically unstable gas-rich and young galactic disks. The individual clusters in such complexes would "pop" as in the theory above (Fellhauer & Kroupa (2005)), leading naturally to thick disks as a result of vigorous early star formation.

2.2. *Structuring the initial cluster mass function*

Another potentially important implication from this theory of the evolution of young clusters is that *if* the gas-expulsion time-scale and/or the sfe varies with initial (embedded) cluster mass, then an initially featureless power-law mass function of embedded clusters will rapidly evolve to one with peaks, dips and turnovers at "initial" cluster masses that characterize changes in the broad physics involved, such as the gas-evacuation time-scale.

Note that the "embedded" cluster mass is the birth mass in stellar content, while the "initial" cluster mass is the mass of stars the virialised post-gas-expulsion star cluster would have had if it had been born with a SFE of 100 per cent ($\epsilon = 1$).

"Initial" cluster masses are derived by dynamically evolving observed clusters backwards in time to $t = 0$ by taking into account stellar evolution and a galactic tidal field but not the birth process ("classical" evolution tracks, eg. Portegies Zwart *et al.* (2001), Baumgardt & Makino (2003)).

To further quantify this issue, Kroupa & Boily (2002) assumed that the function

$$M_{\mathrm{icl}} = f_{\mathrm{st}}\, M_{\mathrm{ecl}}, \tag{2.1}$$

exists, where M_{ecl} is as above, M_{icl} is the "classical initial cluster mass" and

$$f_{\mathrm{st}} = f_{\mathrm{st}}(M_{\mathrm{ecl}}). \tag{2.2}$$

Thus, for example, for the Pleiades, $M_{\mathrm{cl}} \approx 1000\,M_{\odot}$ at the present time (age: about 100 Myr), and a classical initial model would place the initial cluster mass at $M_{\mathrm{icl}} \approx 1500\,M_{\odot}$ by using standard N-body calculations to quantify the secular evaporation of stars from an initially bound and virialised "classical" cluster (Portegies Zwart *et al.* (2001)). If, however, the sfe was 33 per cent and the gas-expulsion time-scale was comparable or shorter than the cluster dynamical time, then the Pleiades would have been born in a compact configuration resembling the ONC and with a mass of embedded stars of $M_{\mathrm{ecl}} \approx 4000\,M_{\odot}$ (Kroupa, Aarseth & Hurley (2001)). Thus, $f_{\mathrm{st}}(4000\,M_{\odot}) = 0.38$ in this particular case.

By postulating that there exist three basic types of embedded clusters, namely
- clusters without O stars (type I: $M_{\mathrm{ecl}} \lesssim 10^{2.5}\,M_{\odot}$, e.g. Taurus-Auriga pre-main sequence stellar groups, ρ Oph),
- clusters with a few O stars (type II: $10^{2.5} \lesssim M_{\mathrm{ecl}}/M_{\odot} \lesssim 10^{5.5}$, e.g. the ONC), and
- clusters with many O stars and with a velocity dispersion comparable to the sound velocity of ionized gas (type III: $M_{\mathrm{ecl}} \gtrsim 10^{5.5}\,M_{\odot}$),

it can be argued that $f_{\mathrm{st}} \approx 0.5$ for type I, $f_{\mathrm{st}} < 0.5$ for type II and $f_{\mathrm{st}} \approx 0.5$ for type III.

The reason for the high f_{st} values for types I and III is that gas expulsion from these clusters may be longer than the cluster dynamical time because there is no sufficient ionizing radiation for type I clusters, or the potential well is too deep for the ionized gas to leave (type III clusters). Type I clusters are excavated through the cumulative effects of stellar radiation and outflows, while type II clusters probably require multiple supernovae of type II to unbind the gas (Goodwin (1997)). According to the present notion, type II clusters undergo a disruptive evolution and witness a high "infant mortality rate" (Lada & Lada (2003)), therewith being the pre-cursors of OB associations and open Galactic clusters.

Under these conditions and an assumed inverted Gaussian functional form for $f_{\mathrm{st}} = f_{\mathrm{st}}(M_{\mathrm{ecl}})$, the power-law embedded cluster mass function transforms into a cluster mass function with a turnover near $10^5\,M_{\odot}$ and a sharp peak near $10^3\,M_{\odot}$ (Kroupa & Boily (2002)). This form is strongly reminiscent of the initial globular cluster mass function which is inferred by e.g. Vesperini (1998), Vesperini (2001), Parmentier & Gilmore (2005), Baumgardt & Makino (2003) to be required for a match with the evolved cluster mass function that is seen to have a universal turnover near $10^5\,M_{\odot}$. Thus, $\approx 10^{10}$ yr-old cluster populations have log-normal Gaussian mass functions quite independently of the environment.

In the theory presented here, the power-law embedded cluster mass function evolves in the first 10^{7-8} yr to the "initial" cluster mass function with the broad turn-over near $10^5\,M_\odot$. In this context it is of interest to note that de Grijs *et al.* (2003) found a turnover in the cluster mass function near $10^5\,M_\odot$ for clusters with an age near 10^9 yr in the star-burst galaxy M82. This result was challenged, but in their section 2.1 de Grijs *et al.* (2005) counter the criticisms. This age of 1 Gyr is too young to be explainable through secular evolution starting from a power-law cluster mass function. In contrast to this interpretation, the theory outlined here would allow a power-law embedded cluster mass function to evolve to the observed form within 1 Gyr. Evaporation through two-body relaxation in a time-variable tidal field then takes over thereby enhancing the observed log-normal shape of the old cluster mass function (e.g. Vesperini (1998), Vesperini (2001), Parmentier & Gilmore (2005)).

This ansatz may thus bear the solution to the long-standing problem that the initial cluster mass function needs to have this turnover, while the observed mass functions of very young clusters are featureless power-law distributions.

2.3. *Cluster disruption independently of mass*

If, on the other hand, f_{st} = constant such that only about 10 per cent of all clusters survive independently of mass, then the above long-standing problem would remain unsolved. This situation is suggested by the results of Lada & Lada (2003) and Fall *et al.* (2005), among others. Observations show that 90 per cent of all very young clusters dissolve within about 10^4 yr, independently of their mass. This may either be due to a constant f_{st}, or, as pointed out by Fall *et al.* (2005), a result of 90 per cent of all clusters dissolving and only the rest surviving intact. The former case would simply imply a constant shift of the power-law mass function along the mass axis to smaller masses as a result of cluster disruption through stellar feedback (operating in all embedded clusters), while the latter would imply a purely stochastic element to the disruption of clusters.

It would therefore appear that this 'high infant mortality of clusters independently of cluster mass' scenario would be in contradiction with the results of de Grijs *et al.* (2005) on the M82 1 Gyr-old cluster system combined with the observed power-law mass function of very young clusters.

3. Conclusions

Observations show that star clusters of any mass are formed extremely compact ($\lesssim 1$ pc), while older clusters appear well distended with radii of a few-to-many pc. Very young clusters seem to be super-virial, and expanding HII shells also indicate the explosive removal of residual gas. Star-cluster formation thus appears to be rather violent leading to a loss of a large fraction of stars and probably in many cases total cluster disruption.

As this discussion shows, the formation process of star clusters may have significant cosmological implications in that the morphology of galaxies may be shaped by the clusters born within them: the disks of galaxies are probably thickened during periods of high star-formation rates while low-mass dwarf galaxies may attain halos of stars as a result of the violent processes associated with cluster birth as a result of which stars may spread outwards with relatively high velocity dispersions. Particular examples of such events have been noted above. If this is true can be verified by measuring the velocity dispersions of stars in clusters younger than at most a few Myr. This is very important to do because thickened galactic disks are often taken to be evidence for the merging of cold-dark-matter sub-structures. Naturally, in-falling dwarf galaxies will lead to thickened disks, and counter-rotating thick disks would be prime examples of such processes (Yoachim & Dalcanton (2005)), but this discourse shows that sub-pc-scale processes would probably need to be incorporated in any galaxy-evolution study.

The mass function of old globular clusters has a near-universal turnover near $10^5 \, M_\odot$ which has so far defied an explanation, given that the mass function of very young clusters is observed to be a power-law, and that the modelling of classical disruption through two-body relaxation and tidal fields poses a challenge for evolving a power-law to the approximate log-normal form for old clusters. Again, mass-dependent processes related to the physics of residual gas expulsion can explain such a turnover, but in this case the turnover would have to become evident in cluster populations younger than about 100 Myr. The results of de Grijs *et al.* (2005) suggest this to be the case. The results of Zhang & Fall (1999), however, appear to show no such evidence for the cluster population in the Antennae galaxies: the mass function for two age groups of clusters (younger than about 10 Myr, and between 10 and 160 Myr) show the same power-law form, albeit with different normalisations as a result of the high infant mortality rate. If this is true, then orbital anisotropies in the young globular cluster populations as proposed by Fall & Zhang (2001) would be needed to evolve a power-law mass function to the log-normal form.

It therefore appears that the broad principles of early cluster evolution remain quiet unclear. Given that *clusters are the fundamental building blocks of galaxies* (Kroupa (2005)) this is no satisfying state of affairs. On the positive side, the importance of star clusters to stellar populations and other cosmological issues (merging histories of galaxies, galactic morphology) means that much more theoretical and observational effort is needed to clarify which processes act during the early life of a star cluster, and how these affect its appearance at different times.

Acknowledgements

I thank the Jan Palouš and Bruce Elmegreen for organising a great meeting in Prague.

References

Bastian, N. & Goodwin, S. P. 2006, *MNRAS* 369, L9
Bastian, N., Gieles, M., Lamers, H. J. G. L. M., Scheepmaker, R. A. & de Grijs, R. 2005, *A&A* 431, 905
Baumgardt, H. & Makino, J. 2003, *MNRAS* 340, 227
Boily, C. M. & Kroupa, P. 2003a, *MNRAS* 338, 665
Boily, C. M. & Kroupa, P. 2003b, *MNRAS* 338, 673
de Grijs, R., Bastian, N. & Lamers, H. J. G. L. M. 2003, *ApJL* 583, L17
de Grijs, R., Parmentier, G. & Lamers, H. J. G. L. M. 2005, *MNRAS* 364, 1054
Fall, S. M. & Zhang, Q. 2001, *ApJ* 561, 751
Elmegreen, B. G. & Meloy Elmegreen, D. 2006, *ApJ* 650, 644
Fall, S. M., Chandar, R. & Whitmore, B. C. 2005, *ApJL* 631, L133
Fellhauer, M. & Kroupa, P. 2005, *ApJ* 630, 879
Goodwin, S. P. 1997, *MNRAS* 284, 785
Hunter, D. A., Elmegreen, B. G., Dupuy, T. J. & Mortonson, M. 2003, *AJ* 126, 1836
Jenkins, A. 1992, *MNRAS* 257, 620
Kroupa, P. 2002, *MNRAS* 330, 707
Kroupa, P. 2005, in: C. Turon, K. S. O'Flaherty & M. A. C. Perryman (eds.), *The Three-Dimensional Universe with Gaia* (ESA SP-576), 629
Kroupa, P. & Boily, C. M. 2002, *MNRAS* 336, 1188
Kroupa, P., Aarseth, S. J. & Hurley, J. 2001, *MNRAS* 321, 699
Lada, C. J. & Lada, E. A. 2003, *ARA&A* 41, 57
Lamers, H. J. G. L. M., Gieles, M. & Portegies Zwart, S. F. 2005a, *A&A* 429, 173

Lamers, H. J. G. L. M., Gieles, M., Bastian, N., Baumgardt, H., Kharchenko, N. V. & Portegies Zwart, S. 2005b, *A&A* 441, 117

Mould, J. 2005, *AJ* 129, 698

Parmentier, G. & Gilmore, G. 2005, *MNRAS* 363, 326

Portegies Zwart, S. F., McMillan, S. L. W., Hut, P. & Makino, J. 2001, *MNRAS* 321, 199

Vesperini, E. 1998, *MNRAS* 299, 1019

Vesperini, E. 2001, *MNRAS* 322, 247

Weidner, C. & Kroupa, P. 2006, *MNRAS* 365, 1333

Weidner, C., Kroupa, P. & Larsen, S. S. 2004, *MNRAS* 350, 1503

Wuchterl, G. & Tscharnuter, W. M. 2003, *A&A* 398, 1081

Yoachim, P. & Dalcanton, J. J. 2005, *ApJ* 624, 701

Zhang, Q. & Fall, S. M. 1999, *ApJL* 527, L81

Discussion

ELMEGREEN: I think your idea that small massive clusters pop to form a thick disk needs some refinement. It is usually true that things which form by gravitational instabilities and collapse from the ambient ISM have velocity dispersions less than the ambient value. The dispersions also get smaller with smaller size. What we see at high redshift is that kpc-size massive clusters dissolve into a thick disk.

WALBORN: The velocity dispersion of R136 in 30 Dor may be much less than 50 km/s; it has not been determined observationally because of the difficulties of measuring small radial velocities in OB spectra with few broad lines, contaminated by nebular emission and a high fraction of spectroscopic binaries (If that number derived from the nebular emission lines, that is driven by stellar winds and is irrelevant to the stellar velocity dispersion.)

KROUPA: The argument is very simple: before cluster formation the ambient ISM velocity dispersion is ~ 10 km s^{-1}. But once $\sim 10^6$ M$\odot$ are assembled within ~ 1 pc, then the dispersion is greater than 10 km s^{-1} (virial theorem). If residual gas can be removed on a crossing time sale, then a substantial part of this dispersion may be carried into the field (thickened disks). Concerning R136, I agree that the velocity dispersion of say A or F stars needs to be measured to check the high value obtained for O stars.

GOLDMAN: How important is energy dissipation either by dynamical friction or turbulence-hydro drag, in phases I and II?

KROUPA: It ought to be considered because the stars are moving through a dense gaseous medium, but I have not looked at this problem in much detail as it is very complex. A useful reference here would be Just, Kegel, & Weiss (1986, A&A).

MIESKE: Have you made a quantitative calculation with respect to your suggestion that the turnover of the globular cluster luminosity function is caused by the dependence of initial mass loss on mass?

KROUPA: We computed simple models in Kroupa & Boily (2002), and detailed modeling is in progress.

ZINNECKER: You mentioned mass segregation only briefly. Would you care to comment which effect you expect on cluster evolution for initial versus dynamical mass segregation? Which one (initial or dynamical) do you favor?

KROUPA: I personally favor primordial mass segregation because it is intuitive. However, to show this is really true we need to test the hypothesis that it is not. Dynamical mass segregation contracts the core concentrating the massive stars thus maximizing the effects of stellar evolution (cluster expansion). At the same time the cluster expands (to balance the energy of the contracting core) allowing a larger fraction of stars to become unbound compared to the case of primordial mass segregation. However, the violent and disruptive effects of expulsion of residual gas probably mostly dominates the energetics of earlier cluster evolution.

Triggered Star Formation in a Turbulent ISM
Proceedings IAU Symposium No. 237, 2006
B. G. Elmegreen & J. Palouš, eds.

On the structure of giant HII regions and HII galaxies

G. Tenorio-Tagle[1],
C. Muñoz-Tuñón[2], E. Pérez[3], S. Silich[1] and E. Telles[4]

[1] Instituto Nacional de Astrofísica Optica y Electrónica, AP 51, 72000 Puebla, México
email: gtt@inaoep.mx

[2] Instituto de Astrofísica de Canarias, E 38200 La Laguna, Tenerife, Spain
cmt@ll.iac.es

[3] Instituto de Astrofísica de Andalucía (CSIC), Camino bajo de Huetor 50, E 18080 Granada, Spain
eperez@iaa.es

[4] Observatório Nacional, Rua José Cristino 77, 20921-400, Rio de Janeiro, Brazil
etelles@on.br

Abstract. We review the structural properties of giant extragalactic HII regions and HII galaxies based on two dimensional hydrodynamic calculations, and propose an evolutionary sequence that accounts for their observed detailed structure. The model assumes a massive and young stellar cluster surrounded by a large collection of clouds. These are thus exposed to the most important star-formation feedback mechanisms: photoionization and the cluster wind. The models show how the two feedback mechanisms compete with each other in the disruption of clouds and lead to two different hydrodynamic solutions: The storage of clouds into a long lasting ragged shell that inhibits the expansion of the thermalized wind, and the steady filtering of the shocked wind gas through channels carved within the cloud stratum that results into the creation of large-scale superbubbles. Both solutions are here claimed to be concurrently at work in giant HII regions and HII galaxies, causing their detailed inner structure.

Keywords. galaxies: starburst – galaxies: galaxies: HII – galaxies: ISM: HII regions

1. Introduction

Multiple studies during the last decades have addressed the impact of photoionization, stellar winds and supernova explosions on interstellar matter (ISM). Disruptive events restructure the birth place of massive stars and their surroundings, while leading to multiple phase transitions in the ISM (see e.g. Comeron 1997; Yorke *et al.* 1989; García-Segura *et al.* 2004; Hosokawa & Inutsuka 2005 and Tenorio-Tagle & Bodenheimer 1988 and references therein). On the other hand, giant molecular clouds, the sites of ongoing star formation, present a hierarchy of clumps and filaments of different scales whose volume filling factor varies between 10% to 0.1% (see McLow & Klessen, 2004, and references therein). As the efficiency of star formation in molecular clouds is estimated to be $\leqslant 10\%$ (Larson 1988; Franco *et al.* 1994) the implication is that the bulk of the cloud structure remains after a star forming episode. All of these studies are relevant within the fields of interstellar matter and star formation and, in particular, regarding the physics of feedback, a major ingredient in the evolution of galaxies and thus in cosmology.

From the observations, we know that giant extragalactic HII regions and HII galaxies are excellent examples of the impact of massive stars on the ISM. They belong to the same class of objects because they are powered by young massive bursts of star formation, and because of their similar physical size, morphology and inner structure. Detailed studies of 30 Doradus (see Chu & Kennicutt 1994; Melnick *et al.* 1999), NGC 604 (Sabalisck *et al.* 1995; Yang *et al.*1996) and several giant extragalactic HII regions and HII galaxies (Muñoz-Tuñón *et al.* 1996; Telles *et al.* 2001; Maíz-Apellániz *et al.* 1999) have been designed with the aim of unveiling the inner structure and dynamics of the nearest examples. All of them present a collection of nested shells that enclose an X-ray emitting gas and that may extend up to kpc scales. Some of the largest shells have stalled while others present expansion speeds of up to several tens of km s^{-1}. Detailed HST images have also confirmed these issues for HII galaxies. All of these sources, as in the studies of Telles *et al.* (2001) and Cairós *et al.* (2001), present a central bright condensation coincident with the massive burst of stellar formation. In giant HII regions and in some HII galaxies, this is resolved as the brightest filament or a broken ragged shell sitting very close to the exciting cluster.

2. Feedback from massive star clusters

The energy powering these giant volumes comes from the recently found unit of massive star formation: super star clusters (Ho 1997). Massive concentrations of young stars within the range of 10^5 M$_\odot$ to several 10^6 M$_\odot$ are all within a small volume $\sim 3 - 10$ pc. The mechanical energy and the UV photon output from a massive stellar cluster is here confronted with an ISM structured into a collection of clouds. A full description of the following calculations can be found in Tenorio-Tagle *et al.* (2006). Animated versions of the models presented can be found at http://www.iaa.csic.es/~eperez/ssc/ssc.html.

There are two competing events: The pressure acquired by the wind at the reverse shock ($P_{bubble} = \rho_w v_\infty^2$) defines the velocity that the leading shock may have as it propagates into the intercloud medium ($V_S = (P_{bubble}/\rho_{ic})^{0.5}$), and thus, as a first approximation, if the cloudlet distribution has an extent D_{cl}, then the time for the leading shock to travel across it is t_s ($= D_{cl}/V_S$). On the other hand, the pressure gradient between ionized cloudlets and the intercloud medium, established through photoionization, is to disrupt clouds and lead to a constant density medium in a time $t_d = \alpha d_{cs}/(2c_{HII})$; where $d_{cs}/2$ is half the average distance or separation between clouds, c_{HII} is the sound velocity in the ionized medium and α is a small number ($\sim 4 - 6$) and accounts for the number of times that a rarefaction wave ought to travel (at the sound speed) the distance $d_{cs}/2$ to replenish the whole volume with an average even density $\langle \rho \rangle$. In this way if the cloud disruption process promoted by photoionization leads to an average $\langle \rho \rangle$ that could largely reduce the velocity of the leading shock ($V_S = (P_{bubble}/\langle \rho \rangle)^{0.5}$), then t_s could be larger than t_d, leading, as shown below, to an effective confinement of the shocked wind, at least for a significant part of the evolution.

Cases considering both feedback events lead to a rapid evolution of the photoionized cloud stratum, filling almost everywhere the low density intercloud zones, while spreading the density of the outermost cloudlets into the surrounding intercloud medium. This causes the development of an increasingly larger ionized expanding rim around the cloudlet distribution and leads to a rapid enhancement of the intercloud density. On the other hand, the wind – cloudlet interaction leads to a global reverse shock into the wind, that evolves from an initially ragged surface across which the isotropic wind is only partly thermalized, to an almost hemispherical surface that fully thermalizes the wind. Initially, after crossing the reverse shock the hot wind drives the leading shock into the cloudlet distribution and this immediately looks for all possible paths of least resistance in between clouds. This fact, diverts the shocked wind into multiple streams behind every overtaken cloudlet, diminishing steadily its power to reach the end of the cloudlet distribution before the outermost clouds expand and block many of the possible exits into the low density background gas.

In the calculation only one channel, the initially widest channel close to the grid equatorial axis, across which the original cloudlet spacing was set slightly larger than in the rest of the distribution, is successfully crossed by the leading shock and the thermalized wind behind it, after $\sim 10^5$ yr of evolution. A second channel through the cloud stratum is completed (close to the symmetry axis) after a time $t = 5 \times 10^5$ yr. All other possibilities, are blocked by the large densities resultant from the champagne bath.

3. Discussion

We have shown here the effects of feedback from a massive stellar cluster into a selected cloudlet distribution, which although arbitrary, as it could have had a different extent or it could have considered clouds of different sizes, separations and locations, it has allowed us to explore a wide range of possibilities. These however, have lead to the two main possible physical solutions to be expected: partial pressure confinement of the shocked wind, and a stable and long-lasting filtering of the thermalized wind through the cloudlet stratum. Thus despite the arbitrary and simple boundary and initial conditions here assumed, the structures that develop within the flow resemble the structure of giant HII regions and HII galaxies. In particular we refer to the giant and multiple well structured shells evident at optical wavelengths, often referred to as nested shells (see Chu & Kennicutt 1994), that enclose a hot X-ray emitting gas, as in 30 Dor (Wang 1999) and NGC 604 (Maíz-Apellániz *et al.* 2004). The calculations also lead to the slowly expanding brightest filament (or ragged shell) present in all sources close and around the exciting stars, and to the elongated, although much fainter, filaments on either side of the open channels that reassemble the ends of elongated columns into the giant superbubbles.

None of these structural features appear in calculations that assume a constant density ISM, nor in those that allow for a constant density molecular cloud as birth place of the exciting sources, or in calculations where the sources are embedded in a plane stratified background atmosphere. For all of these features to appear, a clumpy circumstellar medium seems to be a necessary requirement. A medium that would refrain the wind from an immediate exit into the surrounding gas and that would also allow for the build up of multiple channels through which the wind energy would flow in a less unimpeded manner. And thus giant HII regions and HII galaxies, both powered by massive star formation events producing an ample supply of UV photons and a powerful wind mechanical energy, all of them seem to process their energy into a stratum of dense cloudlets sitting in the immediate vicinity of the star formation event and this leads to the repeated structure common to all giant nebulae.

Acknowledgements

This study has been partly supported by grants AYA2004-08260-C03-01 and AYA 2004-02703 from the Spanish Ministerio de Educación y Ciencia, grant TIC-114 from Junta de Andalucía and Conacyt (México) grant 47534-F.

References

Cairós, L. M., Caon, N., Vílchez, J. M., González-Pérez, J. N. & Muñoz-Tuñón, C. 2001, *ApJS* 136, 393
Chu, Y. H. & Kennicutt, R. 1994, *ApJ* 425, 720
Comeron, F. 1997, *A&A* 326, 1195

Franco, J., Shore, S. & Tenorio-Tagle, G. 1994, *ApJ* 436,795

Garcia-Segura, G. & Franco, J. 2004, *RMxAC* 22,131

Ho, L. C. 1997, *Rev.Mex.AA, Conf. Ser.* 6, 5

Hosokawa, T. & Inutsuka, S. 2005, *astro-ph/0511165*

Larson, R. 1988, in: R. E. Pudritz & J. M. Fich (eds.), *Galactic and Extra Galactic Star Formation* (Dordrecht: Kluwer), p. 459

Mac Low, M.-M. & Klessen, R. S. 2004, *Rev. Mod. Phys.* 76, N1, 125

Maíz-Apellániz, J., Muñoz-Tuñón, C., Tenorio-Tagle, G., Mas-Hesse, J. M. 1999, *A&A* 343, 64

Melnick, J., Tenorio-Tagle, G. & Terlevich, R. 1999, *MNRAS* 302, 677

Muñoz-Tuñón, C., Tenorio-Tagle, G., Castaneda, H. O. & Terlevich, R. 1996, *AJ* 112, 1636

Sabalisck, N. S. P., Tenorio-Tagle, G., Castaneda, H. O. & Muñoz-Tuñón, C. 1995, *ApJ* 444, 200

Telles, E., Muñoz Tuñón, C. & Tenorio-Tagle, G. 2001, *ApJ* 548, 671

Tenorio-Tagle, G. & Bodenheimer, P. 1988, *ARA&A* 26, 145

Tenorio-Tagle, G. Silich, S. & Muñoz-Tuñón, C. 2005, in: D. Vals-Gabaud & M. Chavez (eds.), *Resolved Stellar Populations* (APS-CS), in press

Tenorio-Tagle, G., Muñoz Tuñón, C., Pérez, E., Silich, S. & Telles, E. 2006, *ApJ* 643, 186

Yang, H., Chu, Y., Skillman, E. D. & Terlevich, R. 1996, *AJ* 112, 146

Yorke, H. W., Tenorio-Tagle, G., Bodenheimer, P. & Rozyczka, M. 1989, *A&A* 216, 207

Discussion

B. ELMEGREEN: What kind of coherent resistance to the wind will an interconnecting magnetic field give if the field connects the clouds with the intercloud medium?

G. TENORIO-TAGLE: A large magnetic pressure would enhance the resistance that the ragged shell presents to the shocked wind. In such cases I expect the ragged shell to have a longer life time despite its nearness to the exciting sources.

S. SHORE: Your calculational scenario has in effect created a "fluffy screen" so your flow is more or less a supersonic grid turbulence (without the ionization effects) so: what is the effective Reynolds number for the large scale? How do your results change with changes in the resolution? Have you tried to use a random and/or fractal structure for the surrounding medium? You might compare your results with laboratory studies.

G. TENORIO-TAGLE: I am sorry but I don't understand what do you mean by a "fluffy screen". We have included in our calculations a realistic computational representation of the feedback from super stellar clusters. Many test calculations reassure us that the results are independent of the adopted numerical resolution. We would like indeed to follow your suggestion and surround our clusters by a random fractal structure. However, I would expect that if a significant degree of clumpiness is adopted as initial condition, then the two solutions here described would be recovered.

J. TURNER: Have you included the effects of gravity in these models? For SSCs, this could be an important effect, and could, like in stars, facilitate a smooth transition to supersonic flow.

G. TENORIO-TAGLE: Gravity is not relevant to super stellar clusters able to drive a stationary wind. This is because of the efficient thermalization of the deposited matter which leads to large temperatures ($T \sim 10^7$ K) and thus to sound velocities that well exceed the escape velocity.

Triggered Star Formation in a Turbulent ISM
Proceedings IAU Symposium No. 237, 2006
B. G. Elmegreen & J. Palouš, eds.

© 2007 International Astronomical Union
doi:10.1017/S1743921307001548

Super massive star clusters: from superwinds to a cooling catastrophe and the re-processing of the injected gas

S. Silich[1], G. Tenorio-Tagle[1], C. Muñoz-Tuñón[2] and J. Palouš[3]

[1] Instituto Nacional de Astrofísica Optica y Electrónica, AP 51, 72000 Puebla, México
email: silich@inaoep.mx

[2] Instituto de Astrofísica de Canarias, E 38200 La Laguna, Tenerife, Spain
cmt@ll.iac.es

[3] Astronomical Institute, Academy of Sciences of the Czech Republic, Boční II 1401, 141 31 Prague, Czech Republic
email: palous@ig.cas.cz

Abstract. Different hydrodynamic regimes for the gaseous outflows generated by multiple supernovae explosions and stellar winds occurring within compact and massive star clusters are discussed. It is shown that there exists the threshold energy that separates clusters whose outflows evolve in the quasi-adiabatic or radiative regime from those within which catastrophic cooling and a positive feedback star-forming mode sets in. The role of the surrounding ISM and the observational appearance of the star cluster winds evolving in different hydrodynamic regimes are also discussed.

Keywords. galaxies: star clusters, galaxies: starburst, ISM: jets and outflows

1. Introduction

In many starburst galaxies, in interacting and merging galaxies, a substantial fraction of star formation is concentrated in a number of compact, young and massive stellar clusters (SSCs) which may represent the earliest stages of globular cluster evolution. In the extreme scenario SSCs represent the dominant mode of star formation in these galaxies (see, for example, McCrady *et al.*, 2003; Smith *et al.*, 2006 and references therein). Powerful gaseous outflows associated with such clusters are now believed to be one of the major agents leading to a large-scale structuring of the ISM in the host galaxies and to the dispersal of heavy elements into the ISM and the IGM.

Analysis of the SSC's outflows led us to realize that radiative cooling may crucially affect the hydrodynamics of the star cluster winds and that the superwind concept proposed by Chevalier & Clegg (1985) required a substantial modification in the case of very massive and compact star clusters. We demonstrated that there exists the threshold line in the mechanical energy input rate vs the cluster size parameter space. This line separates clusters whose outflows evolve in the quasi-adiabatic or radiative regime from those in which catastrophic cooling sets in inside the cluster. In the catastrophic cooling regime (above the threshold line) at least some fraction of the matter reinserted via strong stellar winds and supernovae remains bound within the cluster and is finally re-processed into new generations of stars (see Tenorio-Tagle *et al.*, 2005; Wünsch *et al.*, 2006). Here we review the subject and discuss also how the high pressure in the surrounding medium may prohibit the development of high velocity star cluster winds turning them into low velocity, subsonic outflows. Finally, we make some predictions regarding the observational manifestations of the star cluster outflows evolving in the different hydrodynamic regimes.

2. The threshold mechanical luminosity

In the stationary regime the injection of matter by supernovae and stellar winds, $\dot{M}_{SC}$, is balanced by the mass outflow driven by the large central overpressure which results from the efficient thermalization of the kinetic energy, L_{SC}, deposited by SNe and stellar winds:

$$\dot{M}_{SC} = 4\pi R_{SC}^2 \rho_{SC} a_{SC}, \tag{2.1}$$

where R_{SC} is the radius of the cluster, ρ_{SC} is the density of the out-flowing gas at the star cluster surface, and $a_{SC} \approx V_{A\infty}/2$ is the speed of sound at the star cluster edge. $V_{A\infty} = (2L_{SC}/\dot{M}_{SC})^{1/2}$ is the adiabatic wind terminal speed. Equation 2.1 indicates that the density of the plasma inside more massive clusters is larger, if other parameters $(R_{SC}, V_{A\infty})$ do not change. This implies that the impact of radiative cooling on the star cluster winds becomes progressively more important for more massive clusters.

Silich *et al.*, 2004 and Tenorio-Tagle *et al.*, 2005 have demonstrated that for larger or more massive clusters, the larger the leakage of thermal energy. This leads first to a sharp drop in the wind temperature at some distance from the star cluster surface. When the mechanical energy input rate, L_{SC}, exceeds the critical value, the stationary wind solution is inhibited. The critical power is defined by the condition that the central pressure reaches the maximum value allowed by the radiative cooling. The critical energy crucially depends on the thermalization or heating efficiency, e_t. This parameter characterizes how efficient the transformation of the mechanical energy supplied by supernovae and stellar winds, into thermal energy is (Stevens & Hartwell, 2003; Melioli & Del Pino, 2004).

The threshold line calculated under assumption that $V_{A\infty} = 1500$ km s^{-1} for two cases, $e_t = 1$ and $e_t = 0.1$, together with several massive SSCs, is presented in figure 1a.

3. The impact of the external pressure

Radiative cooling puts important restrictions on the plasma parameters inside the cluster. In particular, the pressure at the stagnation point (R_{st}; the point where the expansion velocity is equal to zero; below the threshold line $R_{st} = 0$), which is the largest across the cluster,

$$P_{st} = kT_{st}q_m^{1/2} \left[\frac{V_{A\infty}^2/2 - a_{st}^2/(\gamma - 1)}{\Lambda(T_{st}, Z)} \right]^{1/2}, \tag{3.1}$$

is restricted by the shape of the cooling function, $\Lambda(T, Z)$, and the cluster's parameters, $q_m = 3\dot{M}_{SC}/4\pi R_{SC}^3$ and $V_{A\infty}$ (Silich *et al.*, 2004). Figure 1b displays P_{st} for clusters with critical mechanical luminosities as presented in figure 1a. Figures 1a and 1b allow one to calculate the pressure at the stagnation point for any cluster above the threshold line without knowing the location of the stagnation radius:

$$P_{st} = P_{thresh}(L_{SC}/L_{thresh})^{1/2}, \tag{3.2}$$

where L_{thresh} is presented in figure 1a, and P_{thresh} is the pressure at the stagnation point along the threshold line (figure 1b). If the pressure in the surrounding interstellar medium exceeds that at the stagnation point, $P_{ISM} \geqslant P_{st}$, the cluster is not able to blow away the inserted matter and drive a high velocity outflow. Such clusters instead of driving a high

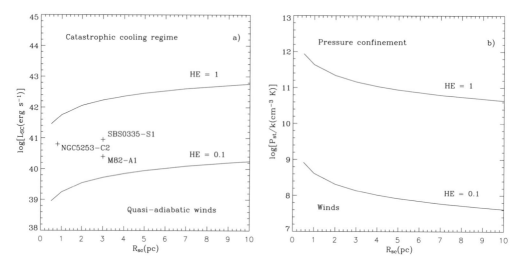

Figure 1. The threshold line. Panel a) presents the critical energy input rates for clusters with different radii. Below the threshold line the stagnation point accommodates at the star cluster center and all mass supplied by massive stars conforms a star cluster wind. Above the threshold line the stagnation point moves to some distance away from the star cluster center (see Wünsch *et al.*, 2006) and only a fraction of the deposited mass and energy leave the cluster. The calculations were done for two values of the heating efficiency, $e_t = 1$ (the upper line), and $e_t = 0.1$ (the lower line). Panel b) presents the pressure at the stagnation point for clusters with critical energy input rates, P_{thresh}. The upper line was calculated for $e_t = 1$ and the lower is for $e_t = 0.1$. Using these two diagrams one can obtain the pressure at the stagnation point for any cluster above the threshold line (see equation 3.2). The locations of several massive SSCs with respect to the threshold lines are indicated in panel a) by the cross symbols.

velocity wind remain buried by the high pressure surrounding medium and effectively reprocess all infalling and reinserted mass into stars. Thus the high pressure ISM may prevent negative feedback and lead to a high star formation efficiency.

4. Observational appearance of SSC's winds

The efficient thermalization of the kinetic energy deposited by stellar winds and supernovae explosions results in a high temperature of the plasma within the cluster. The distributions of density and temperature inside the high velocity outflow defines then the observational manifestations of the star cluster wind in the different energy bands.

Well below the threshold line, in the quasi-adiabatic regime, the temperature rapidly reaches its asymptotic trend, $T \sim r^{-4/3}$, whereas the density drops as $\rho \sim r^{-2}$. Thus well below the threshold line star cluster winds should be associated with the diffuse X-ray sources with the hard component concentrated to the star cluster volume (see, for example, Cantó *et al.* 2000; Silich *et al.*, 2005). In the visible line regime such winds are hardly to be detected due to the fast decrease of density and the negligible emission measure at the distance where the temperature reaches $\sim 10^4$K when the outflowing plasma begins to recombine and produce an emission line spectrum.

However the temperature structure of the outflowing matter changes drastically for clusters approaching or located above the threshold line. For such clusters the temperature falls down and rapidly reaches 10^4 K at much smaller distances from the star cluster surface. This warm gas, photoionised by the star cluster Lyman continuum, moves with velocity around 1000 km s^{-1} and should be detected as a low intensity broad line emission.

When L_{SC} exceeds the threshold value, the catastrophic cooling sets in first at the center. The cooling front and the stagnation point then move from the star cluster center outwards (see Wünsch *et al.*, 2006). The temperature of the initially thermalised material then rapidly drops from $\sim 10^7$ K to approximately 10^4 K where it is balanced by the ionizing radiation from massive stars. The density of the photoionised material grows larger until the gravitational instability sets in and the accumulated gas begins to form new stars (Tenorio-Tagle *et al.*, 2005). The emission line spectra from such clusters should present the central narrow peak associated with the dense gas inside the star forming region which is located inside the stagnation radius, R_{st}, and the lower intensity broad component associated with the warm, photoionised, fast outflow. The two regions are separated by a shell of the hot, thermalised plasma which should be detected as 1.0 – 4.0 keV X-ray emission.

Acknowledgements

We would like to acknowledge the support of this study given by CONACYT – México, research grant 47534-F and AYA2004-08260-CO3-O1 from the Spanish Consejo Superior de Investigaciones Científicas.

References

Chevalier, R.A. & Clegg, A.W. 1985, *Nature* 317, 44
Cantó, J., Raga, A.C. & Rodríguez, L.F. 2000, *ApJ* 536, 896
McCrady, N., Gilbert, A.M. & Graham, J.R. 2003, *ApJ* 596, 240
Melioli, C. & de Gouveia Del Pino E.M. 2004, *A&A* 424, 817
Silich, S., Tenorio-Tagle, G. & Rodríguez-González, A. 2004, *ApJ* 610, 226
Silich, S., Tenorio-Tagle, G. & Añorve Zeferino, G.A. 2005, *ApJ* 635, 1116
Smith, L.J., Westmoquette, M.S., Gallagher, J.S., O'Connell, R.W., Rosario, D.J. & de Grijs, R. 2006, *MNRAS* 370, 513
Stevens, I.R. & Hartwell, J.M. 2003, *MNRAS* 339, 280
Tenorio-Tagle, G., Silich, S., Rodríguez-González A. & Muñoz-Tuñón, C. 2005, *ApJ* 628, L13
Wünsch, R., Palouš, J., Tenorio-Tagle, G. & Silich, S. 2006, this volume

Discussion

S. DIB: Can you please comment on the role of metallicity on your models – for example if the metallicity is subsolar?

S. SILICH: The cooling rate is larger in a plasma with higher metal abundance. Thus the enhanced metallicity will shift the threshold energy towards lower values and vice versa. Note however that we are dealing with matter returned to the ISM through winds and supernovae and thus we expect the metallicities to be large.

HANASZ: Do you observe the thermal instability in your system?

SILICH: We are not able to address this question in the semi-analytical approach. However, it is done in 1D full hydrodynamical calculations (see poster S237-244 by Wünch *et al.*).

Triggered Star Formation in a Turbulent ISM
Proceedings IAU Symposium No. 237, 2006
B. G. Elmegreen & J. Palouš, eds.

Triggered star formation in the environment of young massive stars

Matthias Gritschneder[1], T. Naab[1], F. Heitsch[2] and A.Burkert[1]

[1]Universitäts-Sternwarte München, Scheinerstr. 1, D-81679 München, Germany

[2]Dept. of Astronomy, University of Michigan, Michigan, United States

Abstract. Recent observations with the Spitzer Space Telescope show clear evidence that star formation takes place in the surrounding of young massive O-type stars, which are shaping their environment due to their powerful radiation and stellar winds. In this work we investigate the effect of ionising radiation of massive stars on the ambient interstellar medium (ISM): In particular we want to examine whether the UV-radiation of O-type stars can lead to the observed pillar-like structures and can trigger star formation. We developed a new implementation, based on a parallel Smooth Particle Hydrodynamics code (called IVINE), that allows an efficient treatment of the effect of ionising radiation from massive stars on their turbulent gaseous environment. Here we present first results at very high resolution. We show that ionising radiation can trigger the collapse of an otherwise stable molecular cloud. The arising structures resemble observed structures (e.g. the pillars of creation in the Eagle Nebula, M16, or the Horsehead Nebula, B33). Including the effect of gravitation we find small regions that can be identified as formation places of individual stars. We conclude that ionising radiation from massive stars alone can trigger substantial star formation in molecular clouds.

Keywords. stars: formation, ISM: structure, turbulence, ultraviolet: ISM, methods: numerical

1. Overview

In the surroundings of hot OB-Associations filamentary substructures on different scales are observed (see e.g. Sugitani *et al.* (2002) and references therein). As observational resolution increases, more and more sub-millimeter sources, which could trace the birth of future stars, are detected (e.g. Ward-Thompson *et al.* 2006). It has long been suggested that radiation driven implosion of molecular clouds can explain the morphology and the star formation in these regions (e.g. Elmegreen *et al.* 1995).

Recent simulations (see e.g. Mellema *et al.* (2006), Dale *et al.* (2005), Kessel-Deynet & Burkert (2003)) demonstrate the importance of massive stars for the subsequent evolution of their parental molecular clouds. The ionising radiation is a vital ingredient to understand the disruption of molecular clouds and their star formation efficiency.

Our goal is to investigate the morphology of molecular clouds and the formation of protostars in much greater detail. To do so we use very high resolution simulations of a small region of a molecular cloud ionised by a massive nearby star.

2. Numerical method

We use the prescription for ionising UV-radiation of a young massive star proposed by Kessel-Deynet & Burkert (2000). The ionisation degree x is related to the hydrodynamical quantities by an approximation for the resulting temperature of a partly ionised gas

$$T = T_{ion} \cdot x + T_{cold} \cdot (1 - x). \tag{2.1}$$

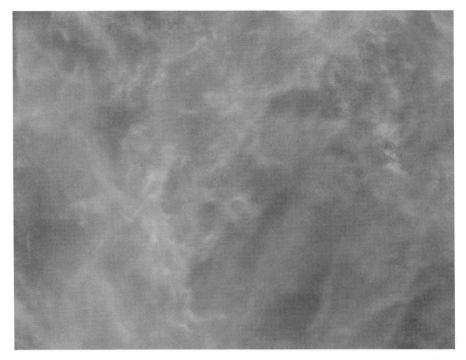

Figure 1. Initial density distribution after the turbulence has decayed to Mach 5. The box dimension are 2 pc in each direction.

T_{cold} is the initial temperature of the cold, un-ionised gas and T_{ion} is the temperature of the ionised gas.

To treat the hydrodynamic and gravitational evolution we use a parallel smoothed particle hydrodynamics (SPH) code called VINE (Wetzstein *et al.*, in prep.). Its Lagrangian nature renders it extremely adept to cover several orders of magnitude in density and time, which is important to follow local gas collapse. We assume plane-parallel UV-irradiation of the simulated area, mimicking a radiation source sufficiently far away such that its distance is larger than the dimensions of the area of infall. To couple ionisation to hydrodynamics we use a flux conserving ray-shooting algorithm. A two dimensional grid is superimposed on the area of interest. Along each of the thereby created bins, the optical depth is calculated. The size of each bin, i.e. the grid resolution, is defined by the volume each SPH-particle occupies. This guarantees that the density information given by the SPH-formalism is transformed to the calculation of radiation correctly.

This implementation is fully parallelised. We call it IVINE (Ionisation+VINE).

3. Numerical tests

A standard test for numerical implementations of ionizing radiation has been proposed by Lefloch & Lazareff (1994). It deals with the steady propagation of an ionisation front: a box of constant density is exposed to a time-dependent ionising source. The initial conditions were chosen to be $n_0 = 100$ cm^{-3} and $T_{cold} = 100$K to enable a direct comparison to their results. The flux increases linearly with time, starting at zero: $\mathrm{d}J/\mathrm{d}t = 5.07 \cdot 10^{-8}$ cm^{-2}s^{-2}. The recombination parameter α_B is set to $\alpha_B = 2.7 \cdot 10^{-13}$ cm^3 s^{-1}. The ionised temperature is $T_{ion} = 10^4$K. At the beginning a small fraction of the box is ionised. Due

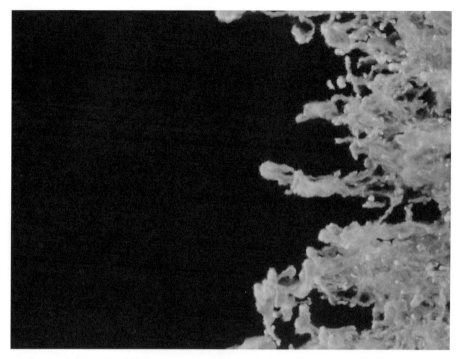

Figure 2. Final stage of evolution after $t \approx 300$ kyrs: the morphology has clearly evolved. The size of the most prominent pillar is roughly 1 pc as it is observed in e.g. M16

Table 1. Comparison of analytical and numerical results for the Lefloch test.

	Analytical	Lefloch & Lazareff (1994)	Kessel-Deynet & Burkert (2000)	IVINE
$n_c(\mathrm{cm}^{-3})$	159	169	155	156.8
$n_i(\mathrm{cm}^{-3})$	0.756	0.748	0.75	0.756
$v_i(\mathrm{km\,s}^{-1})$	3.48	3.36	3.43	3.41
$v_s(\mathrm{km\,s}^{-1})$	3.71	3.51	3.67	3.63

to the higher temperature of the ionised gas a shock front evolves. This shock front moves at a constant speed through the box. The analytical solution provides the exact position and speed of the front as well as of the ionised gas at any given time. We find a very good agreement with the analytical solution, the results are shown in Table 1. n_c and n_i denote the number density of the compressed layer and of the ionised gas, v_i and v_s are the velocities of the ionisation front and the shock front respectively.

4. First application: ionisation of a turbulent ISM

4.1. *Initial conditions*

The first high resolution simulations we performed with the new code address the effect of ionising radiation on a box of turbulent medium. We choose the initial conditions to mimic observed turbulence in the ISM. The cubic simulation domain with a volume of $(2\,\mathrm{pc})^3$ and a mean density of $\bar{n} = 100$ cm^{-3} is set up with a temperature of $T_{cold} = 10$K. The turbulent velocity field is set up adapting a Gaussian random field with a steep power spectrum. The velocity field generates density fluctuations and after a dynamical timescale a turbulent medium with typical velocities of Mach 5 has been generated (see

Fig. 1). At this point the box is exposed to ionising radiation. The UV-radiation is impinging from the negative x-direction with a flux $J = 8.36 \cdot 10^8$ cm^{-2}s^{-1}. This leads to a rapid ionisation of the first $\approx 5\%$ of the cube before the medium reacts to the increased temperature of the ionised gas. The simulations were performed with 2 Million particles on a SGI Altix supercomputer.

4.2. *Results*

The UV-radiation traces the turbulent density distribution, reaching further into the low density regions, and less far in the regions of high density. After a dynamical timescale the hydrodynamics react to the increase in temperature, shock fronts evolve and compress the gas while at the same time increasing the turbulent energy of the cold gas. During this phase a typical morphology evolves. The denser regions shadow regions behind them whereas in lower density regimes the radiation can propagate much further. After the first phase of maximum compression a more quiescent phase of evaporation sets in. The densities are not as high as before, but the structures become even more clear. This leads in the final stage to filamentary, pillar-like substructures, pointing towards the source of radiation as can be seen in Fig. 2. The structures contain high density regions in their tips. These are very likely to become gravitational unstable after a free fall time.

5. Conclusion

We developed a fully parallel treatment for the ionising radiation of young massive stars. Ionising Radiation alone is sufficient to explain the morphology observed in the surroundings of hot OB-clusters. In our simulations of turbulent ISM exposed to UV-radiation characteristic trunks similar to the ones observed in M16 evolve. Further studies including gravity will show whether the UV-radiation from young massive stars is sufficient to trigger gravitational collapse within these pillars.

Acknowledgements

M. Gritschneder is supported by the *Sonderforschungsbereich 375-95 Astro-Particle-Physics* of the Deutsche Forschungsgemeinschaft.

References

Dale, J.E., Bonnell, I.A., Clarke, C.J. & Bate, M.R. 2005, *MNRAS* 358, 291
Elmegreen, B.G., Kimura, T. & Tosa, M. 1995, *ApJ* 451, 675
Kessel-Deynet, O. & Burkert, A. 2000, *MNRAS* 315, 713
Kessel-Deynet, O. & Burkert, A. 2003, *MNRAS* 338, 545
Lefloch, B. & Lazareff, B. 1994, *A&A* 289, 559
Mellema, G., Arthur, S.J., Henney, W.J., Iliev, I.T. & Shapiro, P.R. 2006, *ApJ* 647, 397
Sugitani, K., Tamura, M., Nakajima, Y., Nagashima, C., Nagayama, T., Nakaya, H., Pickles, A.J., Nagata, T.,Sato, S., Fukuda, N. & Ogura, K. 2002, *ApJ* 565, L25
Ward-Thompson, D., Nutter, D., Bontemps, S., Whitworth, A. & Attwood, R. 2006, *MNRAS* 369, 1201

Discussion

DEHARVENG: To compare your simulations with the observations, for example of the pillars in the Eagle Nebula, you must look at: 1. the morphology of your structures, 2. the mass contains in these structures, and 3. the life-time of these structures (you need time to form the young low-mass stars observed inside these structures). Point 1 is very convincing, but what about points 2 or 3?

GRITSCHNEDER: On 2: the average mass per pillar is roughly 1 $M_\odot$, the whole box initially contains 20 $M_\odot$. on 3: the structures evolve at $\sim$ 300 kyr (the end of the movie I showed) and stay stable until the end of our simulation of 600 kyr. It is hard to tell if this is sufficiently long enough since the free fall time at this density is $\sim$ 160 kyr, so it is on the same timescale. The simulations with gravity will show the answer to this question.

Triggered star formation in a turbulent ISM
Proceedings IAU Symposium No. 237, 2006
B. G. Elmegreen & J. Palouš, eds.

© 2007 International Astronomical Union
doi:10.1017/S1743921307001561

Triggered formation and collapse of molecular cloud cores

Anthony P. Whitworth[1]

[1]School of Physics & Astronomy, Cardiff University, Queens Buildings, The Parade, Cardiff
CF24 3AA, Wales, UK
email: A.Whitworth@astro.cf.ac.uk

Abstract. First I discuss the dynamics of core formation in two scenarios relevant to triggered star formation, namely the fragmentation of shock-compressed layers created by colliding turbulent flows and the fragmentation of shells swept up by expanding nebulae. Second I discuss the influence of thermodynamics on the core mass spectrum, on determining which cores are 'pre-stellar' (i.e. destined to spawn stars) and on the minimum mass for a pre-stellar core. Third, I discuss the properties of pre-existing cores whose collapse has been triggered by an increase in external pressure, and compare the results with observations of collapsing pre-stellar cores and evaporating gaseous globules (EGGs).

Keywords. Stars: formation, ISM: clouds, ISM: bubbles, instabilities, turbulence.

1. Introduction

Triggered star formation has several advantages over spontaneous star formation. First it is much easier to understand how nature co-ordinates the approximately coeval formation of stars over an extended region, to form a cluster, if there is a trigger involved. Second, it is much easier to understand why star formation is so patchy, on a galactic scale, if there is a trigger involved. Thirdly it is easier to understand the multiplicity of stars and the great diversity of their properties (separations, eccentricities, mass-ratios), if there is a trigger involved. In spontaneous star formation the approach to instability is normally quite quasistatic, and consequently the material has rather alot of time to organise its subsequent collapse onto a single focus. Triggered star formation tends to bypass the linear stages of instability and launch material directly into the non-linear stages of collapse; the material therefore retains a memory of the asymmetries and substructures generated in its previous history, and can then amplify them gravitationally to produce multiple systems. Other evidence for triggered star formation can be found in the patterns of sequential self-propagating star formation, and in the prodigious bursts of star formation observed in interacting and merging galaxies.

In this contribution I concentrate on three aspects of triggered star formation: (i) the basic dynamics of core formation due to triggered gravitational fragmentation (Sections 2 and 3); (ii) the influence of thermodynamics on the mass function of cores in a turbulent medium (Section 4), on the threshold which must be passed for cores to collapse and form stars (Section 5), and on the minimum mass for a prestellar core (Section 6); and (iii) the evolution, and observable features, of pre-existing cores whose collapse is triggered by a sudden increase in pressure (Section 7), and the fate of pre-existing cores which are overrun by an Hɪɪ Region (Section 8).

2. Fragmentation of shock-compressed layers created by colliding turbulent flows

In a turbulent medium, pre-stellar cores are created wherever two turbulent flows having sufficient density and column-density collide with sufficient ram-pressure to form a self-gravitating condensation. The simplest generic model of this process involves two flows having density ρ_0 colliding at relative speed v_0 to produce a plane-parallel layer, in which the effective isothermal sound speed is a_S and the density is $\rho_S \simeq \rho_0 (v_0/a_S)^2$. Eventually the layer becomes massive enough to fragment gravitationally. If we model a proto-fragment as a disc with radius r (in the plane of the layer) and half-thickness z (perpendicular to the plane of the layer), then the time at which fragmentation becomes non-linear, and the mean initial mass, radius and half-thickness of the resulting fragments, are given by

$$t_{\text{FRAG}} \sim \left(\frac{a_S}{G \rho_0 v_0} \right), \qquad m_{\text{FRAG}} \sim \left(\frac{a_S^7}{G^3 \rho_0 v_0} \right),$$

$$(2.1)$$

$$r_{\text{FRAG}} \sim \left(\frac{a_S^3}{G \rho_0 v_0} \right), \qquad z_{\text{FRAG}} \sim \left(\frac{a_S^5}{G \rho_0 v_0^3} \right)$$

(Whitworth et $al.$, 1994). We stress that the mean fragment mass is not simply the standard Jeans mass evaluated at the shocked density ρ_S and sound speed a_S. This is because at its inception a fragment is flattened: its thickness is determined by ram-pressure, and self-gravity only dominates in the directions parallel to the plane of the layer. The initial aspect ratio of a fragment is effectively equal to the Mach number, $\mathcal{M}$, of the shock bounding the layer, $r_{\text{FRAG}}/z_{\text{FRAG}} \sim v_0/a_S \sim \mathcal{M}$.

These results have been confirmed by numerical simulations (Whitworth et $al.$ 1995). In reality, the layer fragments initially into filaments, and then into cores along the filaments, but it is still the case that the separation between neighbouring filaments, and the separation between cores along a filament, are both $\sim 2r_{\text{FRAG}}$.

We note also that a_S must be defined to represent all contributions to the velocity dispersion of the shocked gas – i.e. thermal motions plus all non-thermal motions which are on sufficiently small scales (i.e. $< r_{\text{FRAG}}$) to contribute to resisting self-gravity. Non-thermal motions on larger scales contribute to fragmentation by creating sub-structure in the shocked gas which then acts to seed fragmentation.

3. Fragmentation of shells swept up by expanding nebulae

A similar mechanism for creating pre-stellar cores involves gravitational fragmentation of a shell swept up by an expanding nebula, i.e. an HII Region (HIIR), a stellar-wind bubble (SWB), or a supernova remnant (SNR). This mechanism is fundamental to the process of sequential self-propagating star formation (Elmegreen & Lada 1977), whereby the massive stars in one generation trigger the formation of the next generation through the compressive action of the HIIRs, SWBs and SNRs which they excite. The phenomenology of shell fragmentation is very similar to that of layer fragmentation, basically because in both cases ram-pressure plays a key role. Indeed, it is the fundamental role of ram-pressure which distinguishes triggered star formation from spontaneous star formation. Again, the fragments are, at their inception, flattened disc-like objects, with an aspect ratio of order the Mach number of the shock bounding the shell on its outer edge. Again the shell fragments first into filaments, and then into cores along the filaments.

As an example, we consider the case of a shell at the edge of a SWB created by a stellar wind with mechanical luminosity L_{WIND} blowing into a medium with density ρ_0. The time at which the shell starts to undergo non-linear fragmentation, the radius of the shell at this juncture, the mean mass of the resulting fragments, their mean radius and their mean half-thickness (i.e. the half-thickness of the shell) are given by

$$t_{\mathrm{FRAG}} \sim \left(\frac{a_{\mathrm{S}}^5}{G^5 \rho_0^4 L_{\mathrm{WIND}}} \right)^{1/8}, \qquad R_{\mathrm{FRAG}} \sim \left(\frac{a_{\mathrm{S}}^3 L_{\mathrm{WIND}}}{G^3 \rho_0^4} \right)^{1/8},$$

$$m_{\mathrm{FRAG}} \sim \left(\frac{a_{\mathrm{S}}^{29}}{G^{13} \rho_0^4 L_{\mathrm{WIND}}} \right)^{1/8}, \tag{3.1}$$

$$r_{\mathrm{FRAG}} \sim \left(\frac{a_{\mathrm{S}}^{13}}{G^5 \rho_0^4 L_{\mathrm{WIND}}} \right)^{1/8}, \qquad z_{\mathrm{FRAG}} \sim \left(\frac{a_{\mathrm{S}}^{23}}{G^7 \rho_0^4 L_{\mathrm{WIND}}^3} \right)^{1/8}$$

(Whitworth & Francis 2002). Here a_{S} is the effective sound speed in the shocked gas of the shell.

4. The core mass function

If we consider a large region of the interstellar medium, i.e. sufficiently large that the overall rate of star formation is a small perturbation to the overall ensemble of cores, there is presumably an approximate statistical equilibrium, in which small cores merge to form larger cores and large cores break up into smaller cores. If the cores subscribe to Larson-type scaling relations,

$$R \propto M^\alpha, \qquad \sigma \propto M^\beta, \tag{4.1}$$

(where M, R, σ are the mass, radius and internal velocity dispersion of a core; Larson 1981), and if their mass spectrum is a power law,

$$\frac{d\mathcal{N}}{dM} \propto M^{-\gamma}, \tag{4.2}$$

then – by implication – there is an inertial range over which the dynamical processes which merge and break up cores are self-similar. Under this circumstance, the amount of mass in equal logarithmic mass intervals (which is proportional to $M^{2-\gamma}$) must be proportional to the dynamical timescale ($t_{\mathrm{DYN}} \sim R/\sigma \propto M^{\alpha-\beta}$), and so

$$\gamma \simeq 2 - \alpha + \beta. \tag{4.3}$$

In high-mass cores ($M > \mathrm{M}_\odot$), σ is dominated by non-thermal motions, $\alpha \simeq 0.5$ and $\beta \simeq 0.25$, so $\gamma \simeq 1.75$, which is essentially what is observed for more diffuse cores and clumps in molecular clouds (e.g. Williams *et al.* 1994). In low-mass cores ($M < \mathrm{M}_\odot$), σ is dominated by thermal motions, $\alpha \simeq 1$ and $\beta \simeq 0$, so $\gamma \simeq 1$, which is significantly flatter. Therefore there should be a knee in the core mass function around $\mathrm{M}_\odot$, *and* the cores below the knee should contribute rather little to the total mass in cores. This knee may be partly responsible for the peak in the core mass function around $\mathrm{M}_\odot$ (e.g. Nutter & Ward-Thompson 2006), and hence for the peak in the initial mass function (IMF) around $0.3\,\mathrm{M}_\odot$ (e.g. Kroupa 2002; Chabrier 2003). However, we should stress that the theory we have presented above is for the mass function of all cores, and most of these cores are presumed to be transient rather than pre-stellar. It is not clear how this mass function would map into the mass function for pre-stellar cores — for example, if it were subjected to a sudden increase in ambient pressure.

5. A pressure threshold for creating prestellar cores

One possibility is that pre-stellar cores are those which are both gravitationally bound *and* sufficiently dense for the gas to couple thermally to the dust (and thereby to avail itself of broadband cooling). The second condition requires that the timescale on which the gas couples to the dust, t_{COUPLE}, is less than the freefall time, t_{FF}. Since we are here concerned with triggered star formation, these timescales should be evaluated at the density, ρ_{S} and sound speed, a_{S}, in a shock-compressed layer or shell:

$$t_{\text{COUPLE}} \simeq \frac{(2\pi)^{1/2} r_{\text{D}} \, \rho_{\text{D}} \, f(a_{\text{S}})}{Z_{\text{D}} \, \rho_{\text{S}} \, a_{\text{S}}} \lesssim t_{\text{FF}} = \left(\frac{3\pi}{32 \, G \, \rho_{\text{S}}} \right)^{1/2} . \tag{5.1}$$

Here r_{D} and ρ_{D} are the radius and internal density of a representative dust grain, and Z_{D} is the fraction by mass in dust. $f(a_{\text{S}})$ is the thermal accommodation coefficient for a gas particle (H_2 molecule) striking a dust grain; for simplicity, we set $f(a_{\text{S}})$ to unity, which is a reasonable assumption for gas kinetic temperatures $T \lesssim 100\,\text{K}$ ($a_{\text{S}} \lesssim 0.6\,\text{km}\,\text{s}^{-1}$). Condition (5.1) then reduces to a condition on the ram pressure creating the layer or shell,

$$P_{\text{RAM}} \equiv \rho_0 \, v_0^2 \simeq \rho_{\text{S}} \, a_{\text{S}}^2 \gtrsim P_{\text{COUPLE}} \simeq \frac{64 \, G}{3} \left(\frac{r_{\text{D}} \, \rho_{\text{D}}}{Z_{\text{D}}} \right)^2 \tag{5.2}$$

In contemporary, local star formation regions, $r_{\text{D}} \sim 10^{-5}\,\text{cm}$, $\rho_{\text{D}} \sim 3\,\text{g}\,\text{cm}^{-3}$ and $Z_{\text{D}} \sim 10^{-2}$, so $P_{\text{COUPLE}} \sim 10^5\,\text{cm}^{-3}\,\text{K}\,k_{\text{B}} \sim 1.4 \times 10^{-21}\,(\text{g}\,\text{cm}^{-3})\,(\text{km}\,\text{s}^{-1})^2$. The corresponding surface-density is $\Sigma_{\text{COUPLE}} \sim (P_{\text{COUPLE}}/G)^{1/2} \sim 1.4 \times 10^{-2}\,\text{g}\,\text{cm}^{-2} \sim 60\,M_{\odot}\,\text{pc}^{-2}$, or equivalently $N_{\text{COUPLE}} \sim 4 \times 10^{21}\,H_2\,\text{cm}^{-2}$. These values are compatible with conditions in observed regions of star formation.

6. The minimum mass for a pre-stellar core (opacity-limited primary fragmentation)

Since the geometry of triggered fragmentation, and the expression for the typical fragment mass, are different from those that obtain in hierarchical three-dimensional fragmentation, it is appropriate to revisit the question of the minimum mass of a pre-stellar core, as determined by the requirement that a core can only undergo *Primary Fragmentation* if it is able to keep cool by radiating away the thermal energy delivered by compression. (*Secondary Fragmentation* may occur at higher densities due to the dissociation of H_2.) If we simply require that the radiative luminosity be greater than the rate of PdV heating of the gas in the condensing fragment, there is an upper limit on the flux of matter flowing into the shock-compressed layer or shell,

$$\rho_0 \, v_0 \lesssim \dot{\Sigma}_{\text{MAX}} \sim \frac{4\pi^2 \, \bar{m}^4 \, a_{\text{S}}^6}{15 \, c^2 \, h^3} , \tag{6.1}$$

and the expression for the minimum mass is essentially the same as for hierarchical 3-D fragmentation,

$$M_{\text{MIN}} \simeq \frac{(30)^{1/2}}{\pi^3} \frac{m_{\text{PLANCK}}^3}{\bar{m}^2} \times \left(\frac{a_{\text{S}}}{c} \right)^{1/2} \tag{6.2}$$

(Whitworth & Stamatellos 2006). Here $m_{\text{PLANCK}} = (hc/G)^{1/2} \simeq 5.5 \times 10^{-5}\,\text{g}$, $\bar{m}$ is the mean gas-particle mass, c is the speed of light, h is Planck's constant and G is the gravitational constant. We note (i) that the part of the above expression preceding '×' is essentially the Chandrasekhar mass (the maximum mass for a non-rotating white dwarf), and (ii) that M_{MIN} depends very weakly on temperature ($\propto T^{1/4}$), as noted by Rees (1976), but quite strongly on the mean gas-particle mass ($\propto \bar{m}^{-9/4}$). For contemporary local star formation in gas with $\bar{m} \simeq 4 \times 10^{-24}$ g and $T \sim 10$ K, Eqn. (6.2) gives $M_{\mathrm{MIN}} \sim 0.001\,\mathrm{M}_\odot$.

In reality the situation is likely to be more complicated. In particular, a core condensing out of a shock-compressed layer (or shell) grows by accreting the material which continues to flow into the layer. Consequently its final mass is larger than its initial mass. Moreover, it must now not only radiate away the PdV heating due to internal contraction, but also the energy dissipated by the accreting material. When this is taken into account in a 2-D semi-analytic model (still with $\bar{m} \simeq 4 \times 10^{-24}$ g and $T \sim 10$ K), the minimum mass becomes $0.0027\,\mathrm{M}_\odot$ (Boyd & Whitworth 2005). In the limiting case, the proto-fragment starts off with a mass of $0.0011\,\mathrm{M}_\odot$ but then more than doubles this by accretion, as it condenses out.

7. Triggered core collapse

Another way in which star formation can be triggered is if a pre-existing core (which would otherwise disperse or persist in a state of approximate hydrostatic equilibrium) experiences a sudden increase in external pressure which causes it to collapse.

7.1. *Velocity fields and accretion rates*

Hennebelle *et al.* (2003) have simulated the response of a stable non-rotating isothermal core to a sudden increase in external pressure. A compression wave propagates into the core setting up an inward velocity field very similar to those which have been inferred from asymmetric line profiles in pre-stellar cores like L1544 (Tafalla *et al.* 1998). When the compression wave reaches the centre, a protostar forms. The accretion rate is initially high, due to the compression wave, but then it declines. Consequently the Class 0 phase is shorter than the Class I phase, in accordance with the observational statistics. We note that this mode of star formation is 'outside-in', in direct contrast to the standard 'inside-out' model involving a singular isothermal sphere.

7.2. *Disc fragmentation and multiplicity*

Hennebelle *et al.* (2004) have simulated the response of a stable rotating core to a sudden increase in external pressure. Again a compression wave propagates into the core triggering the formation of a primary protostar at the centre. However, most of the material has too much angular momentum to accrete directly onto the primary protostar and therefore forms a massive accretion disc around the primary protostar. This disc then fragments to produce one or two companions to the primary protostar. Fragmentation occurs because the compression wave deposits material onto the outer parts of the disc more rapidly than the existing disc is able to stabilise itself by redistributing angular momentum through gravitational torques. As a result, the outer parts of the disc become strongly Toomre unstable. The prediction that a single core spawns only a small number of protostars accords with the observed statistics of binary systems (McDonald & Clarke 1995, Goodwin & Kroupa 2005, Hubber & Whitworth 2005).

8. EGGs and free-floating very low-mass stars

One way in which the external pressure acting on a pre-existing core might increase is if the core is overrun by an H<small>II</small> Region. An ionisation front then eats into the core, preceded by a shock front, which triggers collapse. The final mass of the protostar formed at the centre of the core is determined by a competition between collapse (initiated by the shock front) and photo-erosion (due to the ionisation following close behind). A simple, semi-analytic model (Whitworth & Zinnecker 2004) shows that the final stellar mass is given by

$$M_\star \sim 0.010\,M_\odot \left(\frac{a_{\mathrm{I}}}{0.3\,\mathrm{km\,s^{-1}}}\right)^6 \left(\frac{\dot{\mathcal{N}}_{\mathrm{LyC}}}{10^{50}\,\mathrm{s^{-1}}}\right)^{-1/3} \left(\frac{n_{\mathrm{II}}}{10^3\,\mathrm{cm^{-3}}}\right)^{-1/3}, \qquad (8.1)$$

where a_{I} is the isothermal sound speed in the neutral gas of the core, $\dot{\mathcal{N}}_{\mathrm{LyC}}$ is the output of ionising photons exciting the HII Region, and n_{II} is the number-density in the HII Region. Since the dependence on the parameters which might be expected to have a large range ($\dot{\mathcal{N}}_{\mathrm{LyC}}$ and n_{II}) is weak, and the strong dependence is on a parameter (a_{I}) which cannot vary much, this appears to be a rather robust mechanism for producing free-floating very low-mass stars (brown dwarfs and planetary-mass objects), in the sense of not requiring very special circumstances. Moreover, we are presumably observing it happening in the evaporating gaseous globules (EGGs) seen in M16 and other HII Regions (Hester *et al.* 1996). However, it is also an inefficient way to produce free-floating very low-mass stars, in the sense that only a small fraction of the initial core ends up in the protostar; most of it is photo-eroded. In addition, the mechanism cannot operate in star formation regions like Taurus, where there are no OB stars to excite an HII Region and yet plenty of very low-mass hydrogen-burning stars and brown dwarfs (Luhman 2004). Therefore photoerosion is probably not a major source of very low-mass stars.

Acknowledgements

I acknowledge the support of a PPARC Rolling Grant (PPA/G/O/2002/00497).

References

Boyd, D.F.A. & Whitworth, A.P., 2005, *A&A* 430, 1059
Chabrier G., 2003, *PASP*, 115, 763
Elmegreen, B.G. & Lada, C.J., 1977, *ApJ* 214, 725
Goodwin, S.P. & Kroupa, P., 2005, *A&A* 439, 565
Hennebelle, P., Whitworth, A.P., Gladwin, P.P. & André Ph., 2003, *MNRAS* 340, 870
Hennebelle, P., Whitworth, A.P., Cha, S.-H. & Goodwin, S.P., 2004, *MNRAS* 348, 687
Hester J.J., *et al.*, 1996, *AJ* 111, 2349
Hubber, D.A. & Whitworth, A.P., 2005, *A&A* 437, 113
Kroupa, P., 2002, *Science* 295, 82
Larson, R.B., 1981, *MNRAS* 194, 809
Luhman, K., 2004, *ApJ* 617, 1216
McDonald, J.M. & Clarke, C.J., 1995, *MNRAS* 275, 671
Nutter, D. & Ward-Thompson, D., 2006, *MNRAS* in press
Rees, M.J., 1976, *MNRAS* 176, 483
Tafalla, M., Mardones, D., Myers, P.C., Caselli, P., Bachiller, R., & Benson, P.J., 1998, *ApJ* 504, 900
Whitworth, A.P., Chapman, S.J., Bhattal, A.S., Disney, M.J., Pongracic, H., & Turner, J.A., 1995, *MNRAS* 277, 727
Whitworth, A.P., Bhattal, A.S., Chapman, S.J., Disney, M.J., & Turner, J.A., 1994, *MNRAS* 268, 291
Whitworth, A.P. & Francis, N., 2002, *MNRAS* 329, 641
Whitworth, A.P. & Stamatellos, D., 2006, *A&A* in press
Whitworth, A.P. & Zinnecker, H., 2004, *A&A* 427, 299
Williams, J.P., de Geus, E.J., & Blitz, L., 1994, *ApJ* 428, 693

Discussion

KRUMHOLZ: How are MRI and accretion luminosity likely to change your results on disk fragmentation?

WHITWORTH: I don't think MRI will have an effect. When extra mass is dumped on the outer parts of the disc, instability develops on a dynamical time scale, whereas MRI takes longer to develop and transport angular momentum. However, we will not in the immediate future be able to simulate the MRI. I expect that the accretion luminosity will inhibit fragmentation in the inner disc, but in the outer disc proto-fragments can cool fast enough to condense out.

CLARK: For the core mass spectrum to map to the IMF, each core must contain only one star. If the cores are formed by turbulence, it is unlikely that the flows will deliver just one Jeans mass in the turbulent model, so do you feel the core IMF can ever be related to the star IMF.

WHITWORTH: In my picture, cores form by gravitational fragmentation of a shock-compressed layer or shell, and therefore by definition they contain a Jeans mass. However, they will not necessarily collapse to form just a few stars (e.g., a binary and a couple of ejecta), they may collapse and fragment further to produce a whole cluster of stars.

KROUPA: The idea that photoevaporation may lead to Brown dwarfs that otherwise would have become stars is interesting. In Kroupa *et al.*(2003) we played with this idea and found that a population of photo-evaporated Brown dwarfs would be "substantial" in globular clusters which presumably had 1000's of O stars. But would-be G dwarfs would also have been photo-evaporated to make K- or M- dwarfs, thereby changing the initial stellar mass function. Since globular clusters are found to have very similar IMFs for low-mass stars as the current Galactic field, which mostly stems from low mass clusters (Kroupa 1995, Lada & Lada 2003), this similarity would give constraints on this process.

WHITWORTH: I agree. Even in clusters including OB stars, it can only produce a few very low-mass stars. (Parenthetically, it seems to me that there is a big dynamic range between small cores which end up as EGGs and large cores which end up as bright rimmed clouds. What happens in between?)

Triggered Star Formation in a Turbulent ISM
Proceedings IAU Symposium No. 237, 2006
B. G. Elmegreen & J. Palouš, eds.

© 2007 International Astronomical Union
doi:10.1017/S1743921307001573

Massive star and star cluster formation

Jonathan C. Tan[1]

[1]Department of Astronomy, University of Florida, Gainesville, FL 32611, USA
email: jt @ astro.ufl.edu

Abstract. I review the status of massive star formation theories: accretion from collapsing, massive, turbulent cores; competitive accretion; and stellar collisions. I conclude the observational and theoretical evidence favors the first of these models. I then discuss: the initial conditions of star cluster formation as traced by infrared dark clouds; the cluster formation timescale; and comparison of the initial cluster mass function in different galactic environments.

1. Introduction

Massive stars and star clusters form together as part of a single unified process. All locally-observed massive stars appear to form in star clusters (de Wit *et al.* 2005), particularly in rich star clusters (Massi, Testi & Vanzi 2006). Star clusters make a significant, perhaps dominant, contribution to the total star formation rate of galaxies (Lada & Lada 2003; Fall, Chandar, & Whitmore 2005), so to understand global star formation properties of galaxies (e.g. Kennicutt 1998), one must understand star cluster formation.

2. Massive star formation

There is still some debate about how massive stars form. Do they form from the global collapse of a massive, initially starless gas core, in which a central protostar or binary grows from low to high mass by accretion from a disk (e.g. McKee & Tan 2003)? This is a scaled-up version of the standard model of low-mass star formation (Shu, Adams, & Lizano 1987). Or do they form from favored low-mass protostellar seeds that accrete gas competitively, with the gas being bound to the protocluster potential but not at any stage in a spatially coherent bound core with a mass similar to that of the final massive star (e.g. Bonnell, Vine, & Bate 2004). These latter models typically involve the global collapse of the protocluster gas over a timescale approximately equal to its free-fall time, so the growth of the massive star takes place on the same timescale as the formation of the entire cluster. It has been suggested that protostellar collisions may also be involved in the growth of massive stars (Bonnell, Bate, & Zinnecker 1998; Bally & Zinnecker 2005).

Evidence in support of the core model of massive star formation includes the fact that massive starless cores are observed and the mass function of these cores is similar to the stellar initial mass function (IMF) (Beuther & Schilke 2004; Reid & Wilson 2006). Massive cores tend to have line widths that are much broader than thermal (Caselli & Myers 1995), indicating that other forms of pressure support such as turbulent motions and magnetic fields are important. Indeed observed magnetic field strengths are close to the values needed to support the gas (Crutcher 2005). Known massive protostars tend to be embedded in dense gas cores with masses comparable to the stellar masses (e.g. Source I in the Orion Hot Core; W3(H_2O)). Low-mass protostars, i.e. actively accreting stars, always have relatively massive accretion disks and outflows. A number of claims have been made for disks around massive protostars, although it is usually difficult to determine if these are rotationally supported structures (see Cesaroni *et al.* 2006 for a review).

Powerful outflows from massive protostars with similar degrees of collimation to those from low-mass protostars have been seen (Beuther *et al.* 2002). The expected evolutionary scheme for high-mass star formation from cores has been reviewed in more detail by Beuther *et al.* (2006). Doty, van Dishoeck, & Tan (2006) considered the chemical evolution of this model with particular application to observations of water abundance in hot cores. Kratter & Matzner (2006) investigated the gravitational stability of massive protostellar accretion disks. Krumholz, McKee, & Klein (2007) presented radiation-hydrodynamic simulations of massive star formation from a massive turbulent core.

Massive star formation models involving competitive accretion and stellar collisions face several observational and theoretical hurdles. Edgar & Clarke (2004) showed that Bondi-Hoyle accretion becomes very inefficient for protostellar masses $\gtrsim 10\,M_\odot$ because of radiation pressure on dust in the gas. This feedback has not been included in any of the simulations in which massive stars form by competitive accretion.

To overcome radiation pressure the accretion flow to a massive star must become optically thick, either in a dense core or disk, or in collisions of protostars. The collisional timescale is $t_{\rm coll} = 1.44 \times 10^{10}(n_*/10^4{\rm pc}^{-3})^{-1}(\sigma/2{\rm km\,s}^{-1})(r_*/10R_\odot)^{-1}(m_*/M_\odot)^{-1}$ yr in the limit of strong gravitational focusing, where σ is the 1D velocity dispersion and r_* is the radius of the stellar collisional cross-section. For collisions to occur frequently enough to grow a massive protostar within 10^6 yr (massive zero age main sequence stars are observed) requires protostellar densities of at least 10^6 pc^{-3} and probably closer to 10^8 pc^{-3}, whereas typical observed stellar densities around massive protostars are much smaller. For example from the Orion Nebula Cluster (ONC) x-ray observations of Garmire *et al.* (2000), Tan (2004) estimates a stellar density of about 10^5 pc^{-3} in the KL region. This result is not significantly changed by the deeper x-ray observations of Grosso *et al.* (2005). Hunter *et al.* (2006) find a density of sub-mm cores in the center of protoclusters in NGC 6334 of about 10^4 pc^{-3}. From Fig. 1 and the data of Mueller *et al.* (2002) we see that typical mean densities of the central regions of Galactic protoclusters are $n_{\rm H} \simeq 2 \times 10^5$ cm^{-3}, i.e. 7000 $M_\odot$ pc^{-3}. If all this gas mass formed stars, stellar densities would be about 10^4 pc^{-3}, given a typical IMF. The fiducial core that forms a massive star in the model of McKee & Tan (2003) is also shown in Fig. 1, and has a mean density about one to ten times greater than this. Even if the core fragmented with 100% efficiency into low-mass stars the stellar density would be too low for efficient growth via stellar collisions. In fact numerical simulations show that fragmentation of the core into many stars is impeded by heat input from the forming central massive star (Krumholz 2006). The numerical simulations in which greater degrees of fragmentation are seen (e.g. Dobbs, Bonnell, & Clark 2005) do not include this feedback. Magnetic pressure is also likely to be important for the support of cores more massive than the thermal Jeans mass, but this is also usually not included in simulations of massive star formation.

If collisions are relevant for massive star formation, but not low-mass star formation, then one might expect a change in the slope of the stellar IMF at the mass scale at which the collisional process becomes important. In fact the stellar IMF is reasonably well-fit by a power law from $\sim 1\,M_\odot$ out to the highest observed masses (Massey 1998).

3. Star cluster formation

3.1. *The initial conditions for star cluster formation: infrared dark clouds*

We expect the initial conditions for star clusters to be the densest starless gas clouds. Such clouds reveal themselves by absorption of the Galactic diffuse infrared background and have become known as Infrared Dark Clouds (IRDCs) (Egan *et al.* 1998).

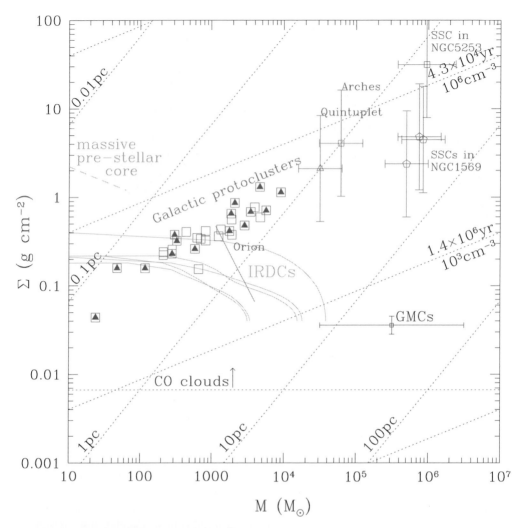

Figure 1. Surface density, $\Sigma \equiv M/(\pi R^2)$, versus mass, M, for star clusters and interstellar clouds. Contours of constant radius, R, and hydrogen number density, n_H, or free-fall timescale, t_{ff}, are shown with dotted lines. The minimum Σ for CO clouds in the local Galactic FUV radiation field is shown, as are typical GMC parameters and the distributions $M(> \Sigma)$ of several IRDCs derived from extinction mapping (Butler *et al.*, in prep.). Open squares are star-forming clumps (Mueller *et al.* 2002): a triangle indicates presence of an HII region. The solid straight line traces conditions from the inner to outer parts of the ONC, assuming equal mass in gas and stars. Several more massive clusters are also indicated. The fiducial massive core in the model of McKee & Tan (2003) is shown by the dashed line.

One way to measure the physical properties of these clouds is through extinction mapping (Fig. 2). Assuming the diffuse Galactic infrared emission behind the cloud is similar to that around it and adopting an infrared extinction law and dust to gas ratio (Weingartner & Draine 2001) allows the measurement of mass surface density, Σ. A kinematic distance can be measured from ^{13}CO line emission (Simon *et al.* 2001), and thus the physical size and mass of the cloud determined. The cumulative distributions of $M(> \Sigma)$, i.e. the mass that is at surface densities greater than or equal to a particular

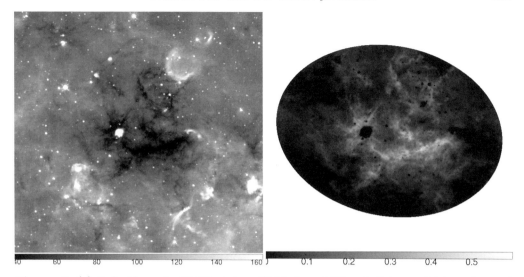

Figure 2. (a) Left: Example IRDC at $l = 28.37$, $b = 0.07$ and distance 4.9 kpc observed by Spitzer at 8 μm. Image is 16.5$'$ across. (b) Right: Σ map of the same cloud derived from extinction of the diffuse background (Butler *et al.*, in prep). Intensity scale is in g cm^{-2}. Note the extinction mapping technique fails where there is a bright source in front of or in the cloud.

Σ, for five typical IRDCs have been measured by Butler, Tan, & Hernandez, in prep and are shown in Fig. 1. The IRDCs span physical properties similar to those of embedded star clusters (Mueller *et al.* 2002), although with somewhat lower surface densities, and so are likely to be representative of the initial conditions of star cluster formation.

3.2. *The timescale for star cluster formation*

Some models of massive star and star cluster formation involve the global collapse of the protocluster in about one free-fall time (e.g. Bonnell, Vine & Bate 2004), while other models that include feedback from the forming stars (e.g. Li & Nakamura 2006) have star formation occurring more gradually over at least several free-fall times. Based on the results of numerical simulations, Krumholz & McKee (2005) argued that supersonically turbulent gas forms stars at a slow rate of only a few percent of the total gas mass per dynamical or free-fall time. Tan, Krumholz, & McKee (2006) extended this analysis to spherical clumps and argued that those clumps that eventually turn a high ($\gtrsim 30\%$) fraction of their mass into stars must do so over at least several ($\gtrsim 7$) free-fall times.

Tan *et al.* (2006) also summarized the observational evidence in support of slow, quasi-equilibrium star cluster formation: (1) The morphologies of CS gas clumps are round (Shirley *et al.* 2003); (2) the spatial distributions of stars in embedded, rich, i.e. high-star-formation-efficiency, star clusters show relatively little substructure; (3) the momentum flux from the combined outflows from protostars in forming clusters is relatively small; (4) the age spreads of stars in rich star clusters are much greater than their current free-fall times; (5) in the ONC a dynamical ejection event associated with the cluster has been dated at 2.5 Myr (Hoogerwerf, de Bruijne, & de Zeeuw 2001), which is much longer than the free-fall time of the present cluster.

Note it is the central, high-star-formation-efficiency region of the cluster where we propose that star formation takes place over several to many free-fall times. These regions have short free-fall times (see Fig. 1), $\sim 10^5$ yr. The outskirts of the cluster have much lower densities, longer free-fall times, and star formation here may occur over just one or two free-fall times, as proposed by Elmegreen (2000), before gas is disrupted by feedback from the newly-formed cluster. The global star formation efficiency here will be relatively low and the young stars will exhibit a greater degree of substructure.

It has been suggested that the star cluster formation process takes only one or two free-fall times when this time is referenced to the pre-cluster conditions at lower density, and that therefore star cluster formation can be regarded as being the result of dynamic collapse of a cloud and is not a quasi-equilibrium process (Hartmann & Burkert 2006). This distinction is important because Krumholz, McKee, & Klein (2005) showed that the process of star formation by competitive accretion cannot be important in virialized, equilibrium clouds. It requires sub-virial conditions associated with global collapse. Several arguments can be made against the global collapse picture: (1) the protocluster gas clouds appear to be approximately virialized (e.g. Shirley *et al.* 2003); (2) the final distributions of the distances of the stars from the cluster center should reflect the locations at which they formed or be even larger because of gas removal, yet we see newly formed rich star clusters with concentrated, dynamically-relaxed distributions. Huff & Stahler (2006) found the star formation history of the ONC showed no dependence on the radial distance from the cluster center; (3) again in the ONC, the 2.5 Myr dynamical ejection event (Hoogerwerf *et al.* 2001) suggests that at this time the cluster was already in a state of high stellar density with a short free-fall time.

3.3. *The initial cluster mass function*

The initial cluster mass function (ICMF) is a fundamental property of the star cluster formation process. If there are external triggers, e.g. cloud collisions, supernova blast waves, that initiate star cluster formation, then these may influence the ICMF. It has been suggested that super star cluster formation may be favored in the low-shear environment of dwarf irregulars (Billett, Hunter, & Elmegreen 2002).

To investigate whether the ICMF depends on galactic environment, Dowell, Buckalew, & Tan (2007) used automated source selection from Sloan Digital Sky Survey (SDSS) data to measured the ICMF at masses $\gtrsim 3 \times 10^4\ M_\odot$ in 13 nearby ($\lesssim 10$ Mpc) dwarf irregular galaxies, which tend to have relatively low metallicity and shear. Cluster ages, masses and reddening were determined by comparing Starburst99 models with the multi-color photometry. Completeness corrections were made, although these are relatively small for massive young clusters at these distances. Foreground stellar and background galactic contamination were assessed and found to be small. The ICMF was assumed to be equal to the mass function of clusters with ages $\leqslant 20$ Myr. The same procedure was repeated on SDSS data of several nearby spiral galaxies at similar distances but with higher metallicity and shear. Several hundred clusters were identified from both the dwarf irregular and spiral galaxy samples.

The main result is that these samples are statistically indistinguishable from each other, suggesting that the ICMF does not depend on galactic shear or metallicity. We find the ICMF is reasonably well fit by a power law $\frac{dN(M)}{dM} \propto M^{-\alpha_M}$ with $\alpha_M \simeq 1.5$ in both dwarf irregular and spiral galaxies. This is somewhat shallower than the power law index of $\alpha_M \simeq 2$ that has been found in spiral galaxies by (Larsen 2002) using HST images. This may be due to the lower resolution of the SDSS observations, which lead to blending of clusters that form within ~ 50 pc of each other. Nevertheless the similarity of the cluster (or association) mass functions between the galaxy samples suggests that a universal process, perhaps turbulent fragmentation inside GMCs (Elmegreen & Efremov 1997), is responsible for star cluster formation.

Acknowledgements

JCT acknowledges support from CLAS, University of Florida.

References

Bally, J. & Zinnecker, H. 2005, *MNRAS* 129, 2281

Beuther, H., Churchwell, E.B., McKee, C.F. & Tan, J.C. 2006, in: B. Reipurth, D. Jewitt & K. Keil (eds.), *Protostars & Planets V*, in press,(astro-ph/0602012)

Beuther, H. & Schilke, P. 2004, *Science* 303, 1167

Beuther, H., Schilke, P., Gueth, F., McCaughrean, M., *et al.* 2002, *A&A* 387, 931

Billett, O.H., Hunter, D.A. & Elmegreen, B.G. 2002, *AJ* 123, 1454

Bonnell, I.A., Bate, M.R. & Zinnecker, H. 1998, *MNRAS* 298, 93

Bonnell, I.A., Vine, S.G. & Bate, M.R. 2004, *MNRAS* 349, 735

Caselli, P. & Myers, P.C. 1995, *ApJ* 446, 665

Cesaroni, R., Galli, D., Lodato, G., Walmsley, C.M. & Zhang, Q. 2006, in: B. Reipurth, D. Jewitt & K. Keil (eds.), *Protostars & Planets V*, in press, (astro-ph/0603093)

Crutcher, R.M. 2005, in: R. Cesaroni, *et al.* (eds.), *Massive star birth: A crossroads of Astrophysics* (Cambridge: Cambridge Univ.), p. 98

de Wit, W.J., Testi, L., Palla, F. & Zinnecker, H. 2005, *A&A* 437, 247

Dobbs, C.L., Bonnell, I.A. & Clark, P.C. 2005, *MNRAS* 360, 2

Doty, S.D., van Dishoeck, E.F. & Tan, J.C. 2006, *A&A* 454, L5

Dowell, J.D., Buckalew, B.A. & Tan, J.C. 2007, *ApJ* submitted.

Edgar, R. & Clarke, C. 2004, *MNRAS* 349, 678

Egan, M.P., Shipman, R., Price, S., Carey, S., Clark, F. & Cohen, M. 1998, *ApJ* 494, L199

Elmegreen, B.G. 2000, *ApJ* 530, 277

Elmegreen, B.G. & Efremov, Y. 1997, *ApJ* 480, 235

Fall, S.M., Chandar, R. & Whitmore, B.C. 2005, *ApJ* 631, L133

Garmire, G., Feigelson, E.D., Broos, P., *et al.* 2000, *AJ* 120, 1426

Grosso, N., Feigelson, E.D., Getman, K.V., *et al.* 2005, *ApJS* 160, 530

Hartmann, L. & Burkert, A. 2006, *ApJ* in press, (astro-ph/0609679)

Hoogerwerf, R., de Bruijne, J.H.J. & de Zeeuw, P.T. 2001, *A&A* 365, 49

Huff, E.M. & Stahler, S.W. 2006, *ApJ* 644, 355

Hunter, T., Brogan, C., Megeath, S., Menten, K., *et al.* 2006, *ApJ* 649, 888

Kennicutt, R.C. 1998, *ApJ* 498, 541

Kratter, K. & Matzner, C.D. 2007, *MNRAS* in press, (astro-ph/0609692)

Krumholz, M.R. 2006, *ApJ* 641, 45

Krumholz, M.R. & McKee, C.F. 2005, *ApJ* 630, 250

Krumholz, M.R., McKee, C.F. & Klein, R.I. 2005, *Nature* 438, 332

Krumholz, M.R., McKee, C.F. & Klein, R.I. 2007, *ApJ* submitted, (astro-ph/0609798)

Lada, C.J. & Lada, E.A. 2003, *ARAA* 41, 57

Larsen, S. 2002, *AJ* 124, 1393

Li, Z-Y. & Nakamura, F. 2006, *ApJ* 640, L187

Massey, P. 1998, in: G. Gilmore, D. Howell (eds.), *The Stellar Initial Mass Function* (ASP-CS), 142, 17

Massi, F., Testi, L. & Vanzi, L. 2006, *A&A* 448, 1007

McKee, C.F. & Tan, J.C. 2003, *ApJ* 585, 850

Mueller, K.E., Shirley, Y.L., Evans, N.J. & Jacobson, H.R. 2002, *ApJS* 143, 469

Reid, M.A. & Wilson, C.D. 2006, *ApJ* 644, 990

Shu, F.H., Adams, F.C. & Lizano, S. 1987, *ARAA* 25, 23

Shirley, Y.L., Evans, N.J., Young, K.E., Knez, C. & Jaffe, D.T. 2003, *ApJS* 149, 375

Simon, R., Jackson, J.M., Clemens, D.P., Bania, T.M. & Heyer, M.H. 2001, *ApJ* 551, 747

Tan, J.C. 2004, in: D. Johnstone, *et al.* (eds.), *Star Formation in the Interstellar Medium: In Honor of David Hollenbach, Chris McKee and Frank Shu* (ASP-CS), 323, 249

Tan, J.C. 2005, in: R. Cesaroni, *et al.* (eds.), *Massive star birth: A crossroads of Astrophysics* (Cambridge: Cambridge Univ.), p. 318

Tan, J.C., Krumholz, M.R. & McKee, C.F. 2006, *ApJ* 641, L121

Weingartner, J.C. & Draine, B.T. 2001, *ApJ* 548, 296

Discussion

LINZ: The speaker mentioned that apparently no current star formation occurs in his sample of IRDCs. Can we be really sure about that, since in most cases, objects are found in IRDCs by Spitzer/MIPS?

TAN: There are two issues here: (1) at each location in the cloud there is a constraint on the embedded luminosity from the lack of flux at $\sim 8\,\mu$m, and, without having done detailed calculations, my impression is that for most of the regions of IRDCs in our sample there is no current, active, luminous star formation, i.e. massive star formation, occurring. There could be embedded lower-luminosity sources and it would be useful to probe this population (either with Spitzer/MIPS or with x-rays). (2) IRDCs are not a particularly well-defined class of objects, and there can in fact be bright sources nearby in adjacent clouds or even in part of the same cloud (the cloud in Fig. 2 has such a source). Still, if one were to measure the total light to mass ratios of IRDCs these should on the average be quite low compared to more evolved star-forming clouds.

LINZ: Still, these objects embedded there might be lower luminosity now but could develop into high-mass YSOs later on?

TAN: I agree that many or most IRDCs, especially the relatively high column density ones that we are studying, are likely to form star clusters and massive stars in the future.

FUKUI: In your turbulent picture, how could you explain the formation of super star clusters?

TAN: Observed super star clusters (SSCs) have $\sim 10^6\,M_\odot$ inside a sphere of radius ~ 3 pc. One basic open question is whether the initial condition is an essentially starless gas cloud with these properties or whether SSCs form more gradually as smaller clouds (perhaps already forming star clusters) merge with the main cluster. In my opinion, it would be difficult to produce the starless initial condition from typical Galactic GMCs without some kind of synchronized, fast trigger. The escape speed from SSCs is greater than the ionised gas sound speed, so they may be forming with very high efficiency from their parent gas clouds (requiring long formation times [and age spreads] in terms of free-fall times) (Tan & McKee 2004, in proc. of Cancun Workshop). This longer formation time may allow more time for infall and merger of surrounding gas clouds, and the higher efficiency means less total gas mass is needed to reach the final stellar mass.

Triggered Star Formation in a Turbulent ISM
Proceedings IAU Symposium No. 237, 2006
B. G. Elmegreen & J. Palouš, eds.

Strongly triggered collapse model confronts observations

Patrick Hennebelle[1], Arnaud Belloche[2], Philippe André[3] and Anthony Whitworth[4]

[1]Laboratoire de Radioastronomie Millimétrique , UMR 8112 du CNRS, École Normale Supérieure et Observatoire de Paris, 24 rue Lhomond 75231 Paris Cedex 05 France
patrick.hennebelle@ens.fr

[2]Max Planck-Institut fur Radioastronomie, Auf dem Hugel 69, 53121 Bonn, Germany
email: belloche@mpifr-bonn.mpg.de

[3]Service d'Astrophysique, CEA/DSM/DAPNIA, C.E. Saclay, 91191 Gif-sur-Yvette Cedex,
France email: pandre@cea.fr

[4]School of Physics & Astronomy, Cardiff University, 5 The Parade, Cardiff CF24 3YB, Wales,
UK email: ant@astro.cf.ac.uk

Abstract. Detailed modelling of individual protostellar condensations is important to test the various theories. Here we present comparisons between strongly induced collapse models with one young class-0 object, IRAS4A, in the Perseus cloud and one prestellar cloud observed in the Coalsack molecular cloud.

Keywords. stars:formation, hydrodynamics, gravitation

1. Introduction

Triggered star formation has been proposed since many years as an important mode of star formation. Indeed many sources of star formation triggering have been proposed and are discussed in this volume, namely turbulence, supernovae remnants, ionisation fronts, stellar outflows or large scale collapse. All of these processes ought to play a role in inducing star formation and probably have their own signatures. In order to be able to identify and to quantify their relative importance, it is necessary to study in great details protostellar condensations in which the collapse has been externally induced, looking for signatures of such violent triggering.

Here we present detailed comparisons between induced collapse models and 2 observed sources in which external triggering may have taken place, namely IRAS4A, a well observed class-0 condensation located in the NGC1333 complex, and the G2 globule located in the Coalsack molecular cloud.

In section 2, we first describe the models with which comparison will be performed. In section 3, comparison with IRAS4A is presented whereas comparison with the G2 globule is described in section 4.

2. Spontaneous versus induced collapse

In order to compare in detail observations of protostellar cores with theoretical models, it is necessary to construct a class of models which on one hand is realistic enough but on the other hand sufficiently simple to be described by few parameters that can be easily varied. It is for example unlikely or at least computationally expensive to find in large scale turbulence simulations, a core which fits sufficiently well an observed object to allow very detailed comparisons. Our approach is as described in Hennebelle *et al.* (2003, 2004).

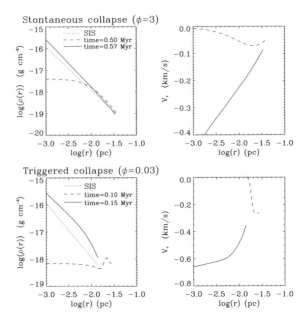

Figure 1. Density and velocity fields in the prestellar and class-0 phase for a spontaneous and a strongly triggered collapse.

We setup initially a stable Bonnor-Ebert sphere and we increase the external pressure at a given rate. In order to quantify the rapidity of the pressure increase, we define $\phi = (R_c/C_s)/(P/\dot{P})$ where R_c is the cloud radius, C_s the sound speed, P the pressure and $\dot{P}$ its time derivative. ϕ represents therefore the ratio of the sound crossing time over the typical pressure increase time. Large values of ϕ correspond to slow pressure increase and describe a spontaneous collapse whereas for small values of ϕ, the collapse is strongly externally triggered. The calculations have been done with the SPH technique.

Figure 1 shows the density and the velocity field in the equatorial plan before and after protostar formation for a spontaneous collapse ($\phi \simeq 3$) and for a strongly triggered collapse ($\phi \simeq 0.03$). In the first case, the density appears to be close to a Bonnor-Ebert sphere density in the prestellar phase, even when the cloud has become unstable, and close to the density of the singular isothermal sphere (SIS) during the class-0 phase. The velocity remains subsonic during the prestellar phase and in the core outer part in the class-0 phase. It becomes more and more supersonic as the collapse proceeds, in the inner part. The situation is much different for $\phi = 0.03$. The velocity is supersonic in the outer part of the core during prestellar phase and everywhere during class-0 phase. The density is also very different from the previous case. It is higher in outer part than in the center. Indeed the strong pressure increase has launched a compression wave that propagates inwards. During the class-0 phase, the density is significantly denser than the SIS density. All of these features appear therefore to be characteristic of externally induced collapse.

3. Comparison with the young class-0 IRAS4A

NGC1333 is a well studied region of the Perseus cloud. Several outflows have been detected (Knee & Sandell 2000), at least one of them is pointing towards IRAS4A making the possibility of induced collapse for this source rather plausible. IRAS4A has been observed in various molecular lines by Di Francesco *et al.* (2001) and in the contin-

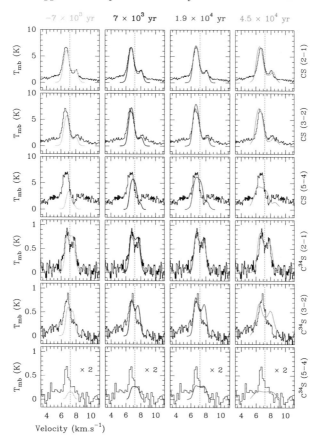

Figure 2. Comparison between observed spectra and synthetic spectra obtained from the strongly induced collapse at various time.

uum by Motte & André (2001). Di Francesco *et al.* (2001) inferred supersonic velocity (0.5-1 km/s) whereas Motte & André (2001) observed density up to 10 times the SIS density. As recalled in previous section, both are signatures of induced collapse. This has been confirmed by observations done recently by Belloche *et al.* (2006).

In order to confirm the scenario of induced collapse for IRAS4A and to set accurate constraints, we have calculated synthetic spectra in the available lines, using the radiative transfer code described in Belloche *et al.* (2002) for the model $\phi = 0.03$ presented in the previous section. Figure 2 shows the comparison between the data and the synthetic profiles at 4 time steps before and after protostellar formation. The comparison reveals that good agreement can be obtained for almost all lines at time 1.9×10^4 yr whereas at earlier times synthetic lines are generally too narrow and at later times, they are generally too broad. As can be seen various features are not well reproduced even at time 1.9×10^4 yr. For example the large wings seen in CS(2-1) are most probably due to the outflow launched by IRAS4A which is not taken into account in our modelling. It is also seen that $C^{34}S(5-4)$ is poorly reproduced. We speculate that the reason for this disagreement is that this line being optically thin, it traces the outer part of the core. The asymmetry of the lines may reveal that the collapse on large scales is indeed not symmetric.

We conclude that IRAS4A can be reasonably well reproduced by our model although further refinements are highly wishable.

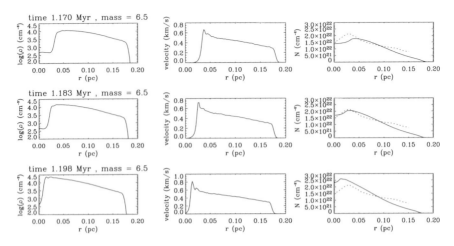

Figure 3. Density, velocity and column density of the collapsing prestellar cloud (solid lines) and column density inferred from observations (dotted line) of Lada *et al.* (2004).

4. Comparison with the prestellar G2 globule of Coalsack

Lada *et al.* (2004) have recently observed in the Coalsack molecular cloud a core which presents an unusual ring-like structure around its centre. The column density appears therefore to decrease inwards when one approaches the core centre. Qualitatively this feature may have similarity with the compression wave shown in figure 1. In our model the compression wave is a dense shell which in projection could be seen as a ring. To test the validity of this scenario, we have compared the column density observed by Lada *et al.* (2004) with the column density of our model at various time steps (Hennebelle *et al.* 2006). Figure 3 displays the density, velocity and column density fields at 3 time steps as well as the column density observed by Lada *et al.* (2004). The column density of the second case appears to be in good agreement with these observations. However no conclusion should be drawn until kinematical data are available for this core (only one spectrum has been observed). Indeed we have calculated synthetic spectra, under optically thin assumption, towards various positions. Our model predicts that toward the centre, the lines should be strongly split. This prediction should be easy to test and we look forward to future observations of this peculiar object.

References

Belloche, A., André P., Despois, D. & Blinder S., 2002, *A&A* 393, 927
Belloche, A., Hennebelle, P. & André P., 2006, *A&A* 453, 145
Di Francesco, J., Myers, P., Wilner, D., Ohashi, N. & Mardones, D., 2001, *ApJ* 562, 770
Hennebelle, P., Whitworth, A., Gladwin, P. & André, P., 2003, *MNRAS* 340, 870
Hennebelle, P., Whitworth, A., Cha, S.-H. & Goodwin, S. P., 2004, *MNRAS* 348, 687
Hennebelle, P., Whitworth, A. & Goodwin, S., 2006, *A&A* 451, 141
Knee, L. & Sandell, G., 2000, *A&A* 361, 671
Lada, C., Huard, T., Crews, L. & Alves, J., 2004, *ApJ* 610, 303
Motte, F. & André, P., 2001, *A&A* 365, 440

Discussion

BARSONY: (1.) Regarding your modeling of NGC 1333 IRAS The initial conditions you use require a very fast timescale for the external pressure change which triggered the collapse, in order to match the density distribution of the infalling envelope on scales of

$1500 < r < 10000$ AU from millimeter continuum observations (Belloche *et al.* 2006). Furthermore, you argue that such a fast external compression is necessary to account for the observed high accretion rate of 10^{-4} M$_\odot$ yr^{-1}. Does this mean that you would expect ALL Class 0 sources to have been triggered (since their accretion rates are so high)? (2.) In your model, you cannot match the density structure of the protostellar envelope at scales of a few $\times 100$ AU $< r < 1500$ AU. Why?

HENNEBELLE: (1.) It depends on whether or not the infall is observed to be supersonic or subsonic. (2.) The model does not take rotation of magnetic fields into account. Including these may lead to better agreement with observations.

Triggered Star Formation in a Turbulent ISM
Proceedings IAU Symposium No. 237, 2006
B. G. Elmegreen & J. Palouš, eds.

Sequentially triggered star formation in OB associations

Thomas Preibisch[1] and Hans Zinnecker[2]

[1]Max-Planck-Institut für Radioastronomie, Auf dem Hügel 69, D–53121 Bonn, Germany
email: preib@mpifr-bonn.mpg.de

[2]Astrophysikalisches Institut Potsdam, An der Sternwarte 16, D–14482 Potsdam, Germany
email: hzinnecker@aip.de

Abstract. We discuss observational evidence for sequential and triggered star formation in OB associations. We first review the star formation process in the Scorpius-Centaurus OB association, the nearest OB association to the Sun, where several recent extensive studies have allowed us to reconstruct the star formation history in a rather detailed way. We then compare the observational results with those obtained for other OB associations and with recent models of rapid cloud and star formation in the turbulent interstellar medium. We conclude that the formation of whole OB subgroups (each consisting of several thousand stars) requires large-scale triggering mechanisms such as shocks from expanding wind and supernova driven superbubbles surrounding older subgroups. Other triggering mechanisms, like radiatively driven implosion of globules, also operate, but seem to be secondary processes, forming only small stellar groups rather than whole OB subgroups with thousands of stars.

Keywords. stars: formation, open clusters and associations: individual (Sco OB2), shock waves, supernovae: general, ISM: clouds, bubbles, evolution

1. Star formation in OB associations

OB associations (Blaauw 1964) are loose, co-moving stellar groups containing O- and/or early B-type stars. As these associations are unstable against galactic tidal forces, they must be young ($\lesssim 30$ Myr) entities, with most of their low-mass members (Briceno *et al.* 2006) still in their pre-main sequence (PMS) phase. OB associations are ideal targets for a detailed investigation of the initial mass function (IMF) and the star formation history, since they allow us to study the *outcome of a recently completed star formation process*. As noted by Blaauw (1964), many OB associations consist of distinct subgroups with different ages, which often seem to progress in a systematic way, suggesting a sequential, perhaps triggered formation scenario.

The massive stars in OB associations affect their environment by ionizing radiation, stellar winds, and, finally, supernova explosions. In their immediate neighborhood, these effects are mostly destructive, since they tend to disrupt the parental molecular cloud and thus terminate the star formation process (e.g. Herbig 1962). A little further away, however, massive stars may also stimulate new star formation. For example, the ionization front from a massive star can sweep up surrounding cloud material into a dense shell, which then fragments and forms new stars (the "collect and collapse" model; see Elmegreen & Lada 1977; Zavagno *et al.* 2006). Also, shock waves from expanding wind- and/or supernova-driven superbubbles can trigger cloud collapse and star formation.

A general problem of scenarios of triggered star formation is that a clear proof of causality is hard to obtain. One can often see Young Stellar Objects (YSOs) and ongoing star formation near shocks caused by massive stars, suggesting triggered star formation. However, such morphological evidence alone does not constitute unequivocal proof for a triggered star formation scenario. It is not clear whether the shock really compressed an empty cloud and triggered the birth of the YSOs, or whether the YSOs formed before the shock wave arrived and the shock appears at the edge of an embedded star formation site simply because the associated dense cloud material has slowed it down. More insight can be gained if one can determine the ages of the YSOs and compare them to the moment in time at which the shock arrived. Agreement of these timings provides much more solid evidence for the triggered star formation scenario than the spatial alignment alone.

Theoretical models for the formation of OB subgroups

The classical model for the sequential formation of OB subgroups was developed by Elmegreen & Lada (1977; see also Lada 1987). Low-mass stars are assumed to form spontaneously throughout the molecular cloud. As soon as the first massive stars form, their ionizing radiation and winds disperse the cloud in their immediate surroundings, thereby terminating the local star formation process. The OB star radiation and winds also drive shocks into other parts of the cloud. A new generation of massive stars is then formed in the dense shocked layers, and the whole process is repeated until the wave of propagating star formation reaches the edge of the cloud. According to this model, one would expect that (1) the low mass stars should be systematically older than the associated OB stars and should show a large age spread, corresponding to the total lifetime of the cloud, and that (2) the youngest OB subgroups should have the largest fraction of low-mass stars (as in these regions low-mass star formation continued for the longest period of time). Another model is based on the mechanism of radiation-driven implosion (e.g., Kessel-Deynet & Burkert 2003). As an OB star drives an ionization shock front into the surrounding cloud, cores within the cloud are triggered into collapse by the shock wave. This model predicts that (1) the low-mass stars should be younger than the OB stars (which initiate their formation), and (2) one may expect to see an age gradient in the low-mass population (objects closer to the OB stars were triggered first and thus should be older than those further away; see, e.g. Chen, these proceedings).

The third model we consider here assumes that a shock wave driven by stellar winds and/or supernova explosions runs though a molecular cloud. Several numerical studies (e.g. Vanhala & Cameron 1998) have found that the effect of the passing shock wave mainly depends on the type of the shock and its velocity: close to a supernova, the shock wave will destroy ambient clouds, but at larger distances, when the shock velocities have decreased to below $\sim$50 km s^{-1}, cloud collapse can be triggered in the right circumstances. The distance from the shock source at which the shock properties are suitable for triggering cloud collapse depend on the details of the processes creating the shock wave (see Oey & Garcia-Segura 2004), the structure of the surrounding medium, and the evolutionary state of the pre-impact core, but should typically range between $\sim$20 pc and $\sim$100 pc. This model predicts that (1) low- and high-mass stars in the triggered subgroup should have the same age, and (2) the age spread in the new subgroup is small (since the triggering shock wave crossed the cloud quite quickly).

These quite distinct predictions of the different models can be compared to the observed properties (i.e. the IMF and the star formation history) of OB associations. The obvious first step of such a study is to identify the complete or a representative sample of the full stellar population of the association. While, at least in the nearby OB associations, the population of high- and intermediate mass stars has been revealed by Hipparcos, the low-mass members (which are usually too faint for proper-motion studies) are quite hard to find. Unlike stellar clusters, which can be easily recognized on the sky, OB associations are generally not so conspicuous because they extend over huge areas in the sky (often several hundred square-degrees for the nearest examples) and most of the faint stars in the area actually are unrelated fore- or background stars. Finding the faint low-mass association members among these field stars is often like finding needles in a haystack. However, the availability of powerful multiple-object spectrographs has now made large spectroscopic surveys for low-mass members possible. The young association members can, e.g., be identified by the strength of their 6708 Å Lithium line, which is a reliable signature for young stars. Studies of the *complete* stellar population of OB associations are now feasible and allow us to investigate in detail the spatial and temporal relationships between high- and low-mass members.

2. Triggered star formation in the Scorpius-Centaurus OB association

At a distance of only ∼140 pc, the Scorpius-Centaurus (ScoCen) association is the OB association nearest to the Sun. It contains at least ∼150 B stars which concentrate in the three subgroups Upper Scorpius (USco), Upper Centaurus-Lupus (UCL), and Lower Centaurus-Crux (LCC). The ages for the B-type stars in the different subgroups, derived from the main sequence turnoff in the HR diagram, were found to be ∼5 Myr for USco, ∼17 Myr for UCL, and ∼16 Myr for LCC (de Geus *et al.* 1989; Mamajek *et al.* 2002).

Upper Scorpius is the best studied part of the ScoCen complex. de Zeeuw *et al.* (1999) identified 120 stars listed in the Hipparcos Catalogue as genuine members of high- and intermediate mass ($\sim 20 - 1.5\,M_\odot$). After the first systematic large-scale search for low-mass members of USco by Walter *et al.* (1994), we have performed extensive spectroscopic surveys for further low-mass members with wide-field multi-object spectrographs at the Anglo-Australian Observatory. These observations are described in detail in Preibisch *et al.* (1998) and Preibisch *et al.* (2002, P02 hereafter), and ultimately yielded a sample of 250 low-mass members in the mass range $\sim 0.1\,M_\odot$ to $\sim 2\,M_\odot$. In combination with the Hipparcos sample of high- and intermediated mass stars, this large sample allowed P02 to study the properties of the full stellar population in USco on the basis of a statistically robust and well defined sample of members. The main results of our detailed analysis of this sample can be summarized as follows: (1) The stellar mass function in USco is consistent with recent field star and cluster IMF determinations. (2) High- as well as low-mass stars have a common mean age of 5 Myr. (3) The spread seen in the HRD, that may seem to suggest an age spread, can be fully explained by the effects of the spread of individual stellar distances, unresolved binary companions, and the photometric variability of the young stars. The observed HRD provides *no evidence for an age dispersion*, although small age spreads of $\sim 1 - 2$ Myr cannot be excluded. (4) The initial size of the association and the observed internal velocity dispersion of the members yield a stellar crossing time of ∼20 Myr.

A very important implication of these results is that the observed age spread of at most ⩽ 1−2 Myr is much smaller than the stellar crossing time of ∼20 Myr. This clearly shows that some external agent must have coordinated the onset of the star formation process over the full spatial extent of the association. In fact, a very suitable trigger is actually available: The structure and kinematics of the large H I loops surrounding the ScoCen association suggest that a shock wave from the older UCL group, driven by stellar winds and supernova explosions, passed through the USco region just about 5 Myr ago (de Geus 1992), which agrees very well with the ages of the USco members. A scenario for the star formation history of USco consistent with the observational results described above is shown in Fig. 1. The shock-wave from UCL initiated the formation of some 2500 stars in USco, including 10 massive stars upwards of $10\,M_\odot$. The new-born massive stars immediately started to destroy the cloud from inside by their ionizing radiation and their strong winds and terminated the star formation process; this explains

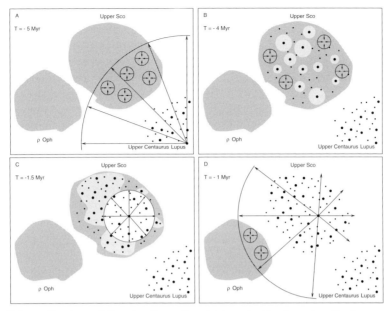

Figure 1. Schematic view of the star formation history in the Scorpius-Centaurus association (from Preibisch & Zinnecker 1999). Molecular clouds are shown as dark regions, high-mass and low-mass stars as large resp. small dots. For further details on the sequence of events see text.

the narrow age distribution. About 1.5 Myr ago, the most massive star in USco exploded as a supernova and created another strong shock wave, which fully dispersed the USco molecular cloud. This shock wave must have crossed the adjacent ρ Oph molecular cloud within the last 1 Myr (de Geus 1992), and thus seems to have triggered the strong star formation activity we witness right now in the L 1688 cloud (see Motte, André, & Neri 1998). Yet another region in which the shock from USco seems to have triggered star formation is the Lupus 1 cloud (see, e.g., Tachihara 2001).

While this scenario provides a good explanation of the star formation history, a potential problem is its implicit assumption that the USco and ρ Oph molecular clouds existed for many Myr without forming stars before the triggering shock waves arrived (otherwise one should see large age spreads in the stellar populations). Elmegreen (2000), Hartmann *et al.* (2001), and other authors provided several convincing arguments that the lifetime of molecular clouds is much shorter than previously thought, and the whole process of molecular cloud formation, star formation, and cloud dispersal (by the feedback of the newly formed stars) occurs on timescales of a few ($\lesssim 5$) Myr. It is now thought that molecular clouds form by the interaction of turbulent convergent flows in the interstellar medium that accumulate matter in some regions. Once the H_2 column density becomes high enough to provide effective self-shielding against the ambient UV radiation field, conversion of atomic H into molecular H_2 quickly follows (e.g., Glover & Mac Low 2006), and star formation may start soon afterwards (e.g., Clark *et al.* 2005).

This new paradigm for the formation and lifetime of molecular clouds may seem to invalidate the idea of a shock wave hitting a pre-existing molecular cloud and triggering star formation. Nevertheless, the basic scenario for the sequence of processes in ScoCen may still be valid. As pointed out by Hartmann *et al.* (2001), wind and supernova shock waves from massive stars are an important kind of driver for ISM flows, and are especially well suited to create *coherent large-scale flows*. Only large-scale flows are able to form large molecular clouds, in which whole OB associations can be born. An updated scenario for ScoCen could be as follows: Initially, the winds of the OB stars created an expanding superbubble around UCL which interacted with flows in the ambient ISM and swept up clouds in some places. When supernovae started to explode in UCL (note that there were presumably some 6 supernova explosions in UCL up to the present day), these added energy and momentum to the wind-blown superbubble and accelerated its expansion. The accelerated shock wave (now with $v \sim 20 - 30$ km s^{-1}) crossed a swept-up cloud in the USco area, and the increased pressure due to this shock triggered star formation.

This scenario does not only explain the temporal sequence of events in a way consistent with the ages of the stars and the kinematic properties of the observed HI shells. The following points provide further evidence: (1) The model (see Fig. 3 in Hartmann et al. 2001) predicts that stellar groups triggered in swept-up clouds should be moving away from the trigger source. A look at the centroid space motions of the ScoCen subgroups (de Bruijne 1999) actually shows that USco is moving away from UCL with a velocity of $\sim 5(\pm 3)$ km s^{-1}. Furthermore, (2) a study by Mamajek & Feigelson (2001) revealed that several young stellar groups, including the η Cha cluster, the TW Hydra association, and the young stars associated with the CrA cloud, move away from UCL at velocities of about 10 km s^{-1}; tracing their current motions back in time shows that these groups were located near the edge of UCL 12 Myr ago (when the supernova exploded). Finally, (3) the model also predicts that molecular clouds are most efficiently created at the intersection of two expanding bubbles. The Lupus I cloud, which is located just between USco and UCL, may be a good example of this process. Its elongated shape is very consistent with the idea that it was swept up by the interaction of the expanding superbubble from USco and the (post-SN) bubble created by the winds of the remaining early B stars in UCL.

3. Triggered star formation in other OB associations

While model and observations agree quite well for ScoCen, it is important to consider how general these findings are. We first note that numerous other regions show similar patterns of sequential star formation, e.g. the superbubbles in W3/W4a (Oey et al. 2005), or the supergiant shell region in IC 2574 (Cannon et al. 2005). Some OB associations, however, show subgroups with similar ages, which cannot have formed sequentially. This could nevertheless fit into the proposed scenario, if we consider the example of Hen 206 in the LMC (Gorjian 2004), where a supernova driven shock wave from a ~ 10 Myr old OB association has created a huge expanding superbubble, at the southern edge of which stars are forming in a giant swept-up molecular cloud. This cloud contains several discrete peaks, in each of which an entire new OB subgroup seems to be forming. This could be an example of the simultaneous formation of several OB subgroups.

A second important aspect is that most nearby OB associations seem to share some key properties: (1) Their mass function is consistent with the field IMF, without much evidence for any IMF variations. (2) High- and low-mass stars generally have the same ages and thus have formed together, not one first and the other later. (3) In most regions, age spreads are remarkably small, often much smaller than the stellar crossing time. These properties are more consistent with the model of large-scale, fast triggering by passing shock waves from wind and supernova driven expanding (super-)bubbles than with the other triggering mechanisms mentioned above.

Nevertheless, other triggering mechanisms seem to operate simultaneously, at least in some regions. For example, the IC 1396 H II region in the Cep OB2 association contains several globules, which are strongly irradiated by the central O6 star. Spitzer observations of the globule VDB 142 (Reach et al. 2004) and Chandra X-ray observations of another globule (Getman et al. 2006) have revealed several very young stellar objects within these globules, all of which are located close to the illuminated edge of the globules, providing evidence for triggered star formation. The triggering mechanism at work here (radiation-driven implosion of globules) is obviously different from the large-scale triggering by expanding wind/supernova driven superbubbles. However, just a handful of stars are formed in these globules and thus this mechanism seems to be a small-scale, secondary process, which does not seem capable of forming whole OB subgroups.

4. Summary and conclusions

OB associations with subgroups showing well defined age sequences and small internal age spreads suggest a large-scale triggered formation scenario, presumably due to supernova/wind driven shock waves. Expanding, initially wind driven superbubbles around OB groups produce coherent large-scale ISM flows that form new clouds; supernova shock waves then compress these newly formed clouds and trigger the formation of whole OB subgroups (several thousand stars) in locations with suitable conditions. Other triggering mechanisms (e.g. radiatively driven implosion) may operate simultaneously, but seem to form only small groups of stars and thus appear to be secondary processes.

References

Blaauw, A. 1964, *ARAA* 2, 213
Briceno, C., Preibisch, Th., Sherry, W., Mamajek, E., Mathieu, R., Walter, F. & Zinnecker, H. 2006, in: B. Reipurth, D. Jewitt, & K. Keil (eds.), *Protostars & Planets V* (Tucson: Univ. of Arizona), in press [astro-ph/0602446]
Cannon, J. M., *et al.* 2005, *ApJ* 630, L37
Clark, P. C., Bonnell, I. A., Zinnecker, H. & Bate, M. R. 2005, *MNRAS* 359, 809
de Bruijne, J. H. J. 1999, *MNRAS* 310, 585
de Geus, E. J. 1992, *A&A* 262, 258
de Geus, E. J., de Zeeuw, P. T. & Lub, J. 1989, *A&A* 216, 44
Elmegreen, B. G. 2000, *ApJ* 530, 277
Elmegreen, B. G. & Lada, C. J. 1977, *ApJ* 214, 725
Getman, K. V., Feigelson, E. D., Garmire, G., *et al.* 2006, *ApJ* in press [astro-ph/0607006]
Glover, S. C. O. & Mac Low, M.-M. 2006, *ApJ* in press [astro-ph/0605121]
Gorjian, V., *et al.* 2004, *ApJS* 154, 275
Hartmann, L., Ballesteros-Paredes, J. & Bergin, E. A. 2001, *ApJ* 562, 852
Herbig, G. H. 1962, *ApJ* 135, 736
Kessel-Deynet, O. & Burkert, A. 2003, *MNRAS* 338, 545
Lada, C. J. 1987, IAU Symp. 115: Star Forming Regions, 115, 1
Mamajek, E. E., & Feigelson, E. D. 2001, in: R. Jayawardhana & T. Greene (eds.), *Young Stars Near Earth: Progress and Prospects* (ASP-CS), 244, 104
Mamajek, E. E., Meyer, M. R. & Liebert, J. 2002, *AJ* 124, 1670
Motte, F., André, P. & Neri, R. 1998, *A&A* 336, 150
Oey, M. S. & Garcia-Segura, G. 2004, *ApJ* 613, 302
Oey, M. S., Watson, A. M., Kern, K. & Walth, G. L. 2005, *AJ* 129, 393
Preibisch, Th. & Zinnecker, H. 1999, *AJ* 117, 2381
Preibisch, Th., Guenther, E., Zinnecker, H., *et al.* 1998, *A&A* 333, 619
Preibisch, Th., Brown, A. G. A., Bridges, T., Guenther, E. & Zinnecker, H. 2002, *AJ* 124, 404
Reach, W. T., *et al.* 2004, *ApJS* 154, 385
Tachihara, K., Toyoda, S., Onishi, T., *et al.* 2001, *PASJ* 53, 1081
Vanhala, H. A. T & Cameron, A. G. W. 1998, *ApJ* 508, 291
Walter, F. M., Vrba, F. J., Mathieu, R. D., Brown, A. & Myers, P. C. 1994, *AJ* 107, 692
Zavagno, A., Deharveng, L., Comerón, *et al.* 2006, *A&A* 446, 171
de Zeeuw, P. T., Hoogerwerf, R., de Bruijne, J. H. J., *et al.* 1999, *AJ* 117, 354

Discussion

KRUMHOLZ: If triggering by OB associations is a dominant mechanism for star formation, why are there spiral arms? Creating clouds by sweeping up gas would not produce a spiral pattern.

PREIBISCH: Our argument for the importance of large-scale triggering by expanding wind- and supernova-driven shocks from OB associations does not imply that this mechanism alone dominates the whole galactic star formation; other mechanisms are at work simultaneously. Triggering by large-scale shocks may be efficient only in suitable locations and in the right circumstances. Spiral shocks increase the density of the interstellar material and create large-scale flows; in this way they may set the stage for wind and supernova shock triggering to work during spiral arm passages.

TAN: You argue that evidence for SN triggering of star formation is the small age spread compared to the crossing time. However, the relevant crossing time is that of the stars and gas at the embedded stage. It is hard to measure this in clusters that have been expanding for more than several Myr and expelled their natal gas. When we look at embedded clusters, like the ONC (Tan *et al.* 2006), we find age spreads that are greater than the crossing time. (The high stellar density and the background obscuration mean that line of sight contamination is not significant for the ONC).

PREIBISCH: We agree that the relevant crossing time is that of the initial configuration, i.e. at the time of star formation and not today. Our determination of the crossing time is based on the *initial size* of Upper Sco of ~25 pc, which we derived from the spatial structure and kinematics of the association (see Preibisch *et al.* 2002 for details). Upper Sco was apparently not born as a dense cluster. With respect to the ONC, we would like to note that the presence of a large age spread (Tan *et al.* 2006) is not fully clear. Individual stellar ages derived from the HRD must not be taken too literally, because they are easily adulterated by effects such as the presence of unresolved binary companions, photometric variability, and errors in the luminosity determination. Thus, the dispersion of the ages determined from the HRD is always only an upper limit to the true age dispersion. We note that HST photometry of the inner ONC by Prosser *et al.* (1994), taking into account these effects, showed that "only a small fraction of the stars (if truly members) have been born outside of a 1 Myr era of star formation." Another factor is contamination: as the ONC coincides with the center of the ~5 Myr old Ori OB1c association, some of the apparently older ONC stars may in fact be Ori OB1c members. Furthermore, star formation does not continue *in* the ONC but in the molecular cloud *behind* the ONC. In view of these arguments, it does not seem obvious that the age spread in the ONC is greater than the crossing time.

BREITSCHWERDT: Some of the stars that belong to UCL and LCC are members of a moving group that was born about 20 Myr ago and its high mass members generated the Local Bubble before entering the Loop I region just 1-3 Myr ago. How would explosions from these stars affect your model of triggering star formation by a cluster with a single age?

PREIBISCH: The suggested star formation history would not be affected by SN explosions during the last 5 Myr (when Upper Sco was formed).

BONNELL: How can you relate your small age spreads to the large age spreads reported for the ONC?

PREIBISCH: The large age spreads that are apparently seen in some regions may in fact, at least in some cases, be caused by projection effects, i.e., we see several stellar groups with different ages projected onto each other. A good example may be the ONC region, with the $\sim 10^5$ yr old BN region, in the background of the 1 Myr Trapezium cluster, which is seen in projection behind the Ori OB1c association (several My old).

CHU: IAU commission 34 (Div VI) includes both ISM and star formation, but the communication between ISM and star formation researchers is lacking. "SN/wind shocks"' and "bubbles" are used by star formation people loosely with different definitions from the ISM people. SNRs, when they are still recognizable as SNRs, are too violent to trigger star formation. SNRs and stellar winds collectively form "superbubbles," and superbubbles can trigger star formation. We really agree on all physical processes, but we need to use terminology consistently.

PREIBISCH: The expanding bubbles in Sco Cen are superbubbles, driven by several stars and SN explosions. I agree that SNR are too violent to trigger star formation, but at large distances from the SN ($> 20 - 30$ pc) the shock wave may be able to trigger star formation.

Triggered Star Formation in a Turbulent ISM
Proceedings IAU Symposium No. 237, 2006
B. G. Elmegreen & J. Palouš, eds.

Triggered star formation in OB associations

W. P. Chen[1,2], H. T. Lee[1] and K. Sanchawala[1]

[1]Institute of Astronomy, National Central University, Chung-Li 32054, Taiwan
email: wchen@astro.ncu.edu.tw, eridan@astro.ncu.edu.tw, kaushar@outflows.astro.ncu.edu.tw

[2]Department of Physics, National Central University, Chung-Li 32054, Taiwan

Abstract. We present causal and positional evidence of triggered star formation in bright-rimmed clouds in OB associations, e.g., Ori OB1, and Lac OB1, by photoionization. The triggering process is seen also on a much larger scale in the Orion-Monoceros Complex by the Orion-Eridanus Superbubble. We also show how the positioning of young stellar groups surrounding the H II region associated with Trumpler 16 in Carina Nebula supports the triggering process of star formation by the collect-and-collapse scenario.

Keywords. stars: formation, stars: pre-main sequence, H II regions, ISM: clouds, ISM: bubbles, open clusters and associations: general

1. Introduction

A massive star has a profound influence on nearby molecular clouds. On the one hand, the stellar radiation and energetic wind could evaporate the clouds and henceforth terminate the star-forming processes. On the other hand, the massive star may provide "just the touch" to prompt the collapse of a molecular cloud which otherwise may not contract and fragment spontaneously. Except perhaps in an environment such as the nucleus of a starburst galaxy, for which triggering predominates the star formation process, triggering in most cases likely plays a constructive albeit auxiliary role. Other than providing additional push for cloud collapse, triggered star formation is self-sustaining (in time) and self-propagating (in space) in comparison to spontaneous cloud collapse.

The extent to which a massive star influences the starbirth in a cloud depends on the amount and proximity of the cloud material, and the positional configuration between the massive star and the cloud. If the massive star is born deep inside a cloud, which is often the case for a giant molecular cloud, the ionization fronts from the H II region created by the massive star push the cloud from within, forming a cavity. The gas and dust hence accumulate to a layer until the critical density is reached for gravitational collapse to form the next generation of stars. This is the so-called "collect-and-collapse" mechanism first proposed by Elmegreen & Lada (1977), and recently demonstrated observationally by Deharveng, Zavagno & Caplan (2005) and Zavagno *et al.* (2006). Any massive stars thus formed may subsequently break out their own cavities. Once a cavity forms, ionization now takes place on the surface of a remnant cloud. Alternatively, the massive star and neighboring clouds could initially be already oriented in this way. The UV photons from the massive star hence ionize the surface layer of the cloud, which illuminates as a bright rim seen prominently in an H-alpha image. The ionization fronts embracing the surface of the cloud then result in a shock compressing into the cloud to cause the dense clumps to collapse. This so-called "radiation-driven implosion" (RDI) process has been proposed to account for triggered star formation in bright-rimmed clouds near H II regions (Bertoldi 1989, Bertoldi & McKee 1990, Hester & Desch 2005). A massive star at the end of its life, with its Wolf-Rayet winds and supernova explosion, may create a superbubble which can

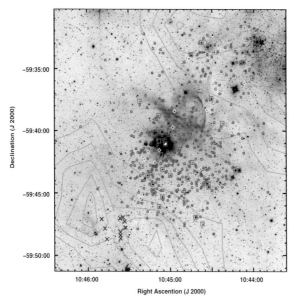

Figure 1. The *Chandra* sources overlaid on the mosaic K_s band image (Sanchawala *et al.* 2006) taken by the IRSF, centered on η Carinae, with the contours showing the $^{12}CO(1-0)$ emission (Brooks *et al.* 1998). Known OB stars are marked as boxes and the candidate OB stars are marked as diamonds. The stars of the embedded X-ray group to the south-east of Trumpler 16 are marked as crosses, whereas all other *Chandra* sources are marked as circles.

have an impact on even larger scales, tens or perhaps hundreds of parsecs away. Here we report on observations to illustrate triggered star formation by the collect-collapse-clear process and by the RDI mechanism in some OB associations.

2. Triggered star formation in Carina Nebula

The Carina Nebula is known to contain the largest number of early-type stars in the Milky Way, with a total of 64 O-type stars (Feinstein 1995). Among the dozen known star clusters in the region, Trumpler 14 and Trumpler 16 are centrally located and are the youngest and the most populous. These two clusters host 6 exceedingly rare main-sequence O3 stars. In particular Trumpler 16 contains a Wolf-Rayet star, HD 93162, and the famous luminous blue variable, η Carinae, which is arguably the most massive star in our Galaxy (Massey & Johnson 1993). The Carina Nebula therefore serves as a unique laboratory to study not only the massive star formation process, but also the interplay among massive stars, interstellar media and low-mass star formation.

Sanchawala *et al.* (2006) studied the X-ray sources detected by *Chandra* in the Carina Nebula (Fig. 1). Of the 454 X-ray sources, 38 coincide with known OB stars. Additionally, 16 anonymous stars have been found to have X-ray and near-infrared properties similar to those of the known OB stars. These candidate OB stars likely escaped earlier optical studies because of their excessive dust extinction. Close to 200 X-ray sources are candidate classical T Tauri stars (CTTSs), judged on the basis of their infrared colors. This sample represents the most comprehensive census of the young stellar population in the Carina Nebula so far and is useful for the study of the star-formation history in this turbulent environment.

In Fig. 1, in addition to Trumpler 16 near the center and Trumpler 14 to the north-west, there is an embedded group of 10 young stars to the south-east of Trumpler 16,

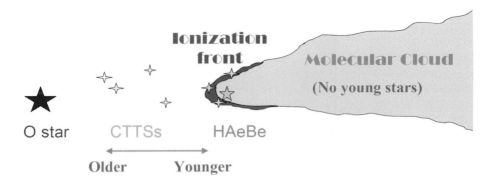

Figure 2. An illustration of a massive star to trigger star formation in a nearby molecular cloud.

'sandwiched' between two cloud peaks (Sanchawala *et al.* 2006). One sees immediately the general paucity of massive stars with respect to molecular clouds; namely Trumpler 16 itself has cast out a cavity and lies between the north-west and south-east cloud complexes, and so has Trumpler 14 to a less extent. The newly identified group suffers a large amount of reddening and also is situated between cloud peaks, apparently in the initial stage to expel the gas. There seems a general tendency for the X-ray sources (i.e., young stars) to be either intervening between clouds or located near the cloud surfaces facing Trumpler 16. The morphology of young stellar groups and molecular clouds peripheral to an H II region (i.e., Trumpler 16 here) fits closely the description of the collect-and-collapse mechanism for massive star formation.

3. Triggered star formation in Orion OB1, near λ Ori, and Lac OB1

An RDI triggering process would leave several imprints that can be diagnosed observationally (Fig. 2): (1) The remnant cloud is extended toward, or pointing to, the massive stars. (2) The young stellar groupings are roughly lined up between the remnant clouds and the luminous star, (3) Stars closer to the cloud, formed later in the sequence, are younger in age, with the youngest distributed at the interacting region, i.e., along the bright rim of a cloud, and (4) No young stars exist far inside the cloud, i.e., leading the ionization front. In particular, the temporal and positional signposts, (3) and (4), are in distinct contrast to the case of spontaneous star formation by a global cloud collapse, which would lead to starbirth spreading throughout the cloud.

Lee *et al.* (2005) and Lee & Chen (2006a) developed an empirical set of criteria, based on the Two Micron All Sky Survey (2MASS) colors, to select CTTSs and Herbig Ae/Be (HAeBe) stars in star-forming regions. The selection criteria prove very effective as follow-up spectroscopy showed that most candidates associated with nebulosity were indeed young stars, but otherwise were carbon stars or M giants. With the sensitivity of 2MASS, young stars within ∼ 1 kpc can be readily recognized. In Ori OB1, λ Ori, and Lac OB1 there is compelling evidence of RDI triggering to form low- and intermediate-mass stars. Fig. 3 shows an example near λ Ori. Analysis of 2MASS colors shows the young stars to be progressively younger toward the clouds, with the youngest near the cloud rim. Furthermore, there is a tendency for HAeBe stars to reside deeper into the cloud, indicating that more massive stars, when prompted to form, appear to favor denser environments where photoevaporation effect is reduced. On the other hand, when a dense core near the ionization layer (i.e., current cloud surface) collapses, the accretion process

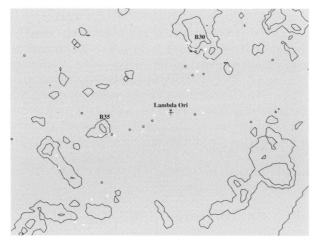

Figure 3. The CTTSs appear to line up between λ Ori and remnant clouds, with the clouds roughly pointing to λ Ori (Lee *et al.* 2005).

has to compete with the mass loss arising from photoevaporation, leading to formation of a less massive star or even substellar objects (Whitworth & Zinnecker 2004). Eventually the remnant cloud would be dispersed completely, and stars of different masses remain in the same volume. On a larger scale, the Wolf-Rayet winds and supernova explosion of a massive star would create a superbubble ramping on one molecular cloud to another (Lee & Chen 2006b). A sequence of such events ("relay star formation") could spread the star formation out to tens or even hundreds of parsec away. Note that the CTTS sample traces only recent star formation, and the bright-rimmed clouds present convenient snapshots to show how triggering leads to formation of low- and intermediate-mass stars once an O star is formed. These processes do not preclude stars formed in earlier epochs, with whatever mechanisms and masses, which already exist in a region.

Acknowledgements

We acknowledge the grant NSC94-2112-M-008-017 from the National Science Council of Taiwan to support this research.

References

Bertoldi, F. 1989, *ApJ*, 346, 735
Bertoldi, F. & McKee, C. F. 1990, *ApJ*, 354, 529
Brooks, K., Whiteoak, J. B. & Storey, J. W. V. 1998, *PASA*, 15, 202
Deharveng, L., Zavagno, A. & Caplan, J. 2005, *A & A*, 433, 565
Elmegreen, B. G. & Lada, C. J. 1977, *ApJ*, 214, 725
Feinstein, A. 1995, *Rev. Mexicana AyA*, 2, 57
Hester, J. J. & Desch, S. J. 2005, in: A. N. Krot *et al.* (eds.), *Chondrites and the Protoplanetary Disk* (ASP-CS) 341, 107
Lee, H.-T., Chen, W. P., Zhang, Z. W. & Hu, J. Y. 2005, *ApJ*, 624, 808
Lee, H. T. & Chen, W. P. 2006, *ApJ*, in press, astro-ph/0509315
Lee, H. T. & Chen, W. P. 2006, *AJ*, in press, astro-ph/0608216
Massey, P. & Johnson, J. 1993, *AJ*, 105, 980
Sanchawala, K., *et al.* 2006, *ApJ*, in press, astro-ph/0603043
Whitworth, A. P. & Zinnecker, H. 2004, *A & A*, 427, 299
Zavagno, A., Deharveng, L., Comerón, F., Brand, J., Massi, F., Caplan, J. & Russeil, D. 2006 *A & A*, 446, 171

Discussion

ZINNECKER: You suggested radiative driven implosion works over tens of parsecs, but have you really shown this? By implication, have you really proven that ionisation fronts can sweep across whole molecular clouds and trigger star formation on tens of parsecs?

CHEN: Triggering by a superbubble from stellar winds or a supernova explosion can have a longer range influence, over tens of parsecs, as we witness in the Orion OB1 association. The RDI works only on nearby clouds, but triggering can be self-propagating, so a sequence of such processes can still go further than just a few parsecs.

TOTH: What is the typical mass of the bright rimmed clouds? Do you see any relation between cloud mass and star forming activity?

CHEN: Measurements of some BRCs suggest $\sim 20 - 30$ M$_\odot$. But this is not really relevant because star formation takes place only near the interacting layer (i.e., the bright rim) of a cloud.

DEHARVENG: Sequential star formation near bright rims at the border of HII regions is also observed at small scales (see for example the observations by the group of Sugitani, which show Hα emission line stars in front of the bright rim – in the direction of the ionised region – then near IR objects near the BR's surface, and then mid-IR sources more deeply embedded inside the BR).

CHEN: Indeed these results have been reported earlier. The use of the 2MASS enables us to study the triggering process on larger distance scales. There has been ambiguity of the configuration you raised. For example, photoevaporation of cloud material and circumstellar disks — instead of triggering — would result in a "cleaned" population toward the massive star, and a "being cleaned" population near a BRC's surface, just as observed. Our studies provide some age discrimination of star groups away or near a BRC, in terms of IR colors and presence of forbidden lines in a star's spectrum. Although this is still ambiguous because photoevaporation of the circumstellar disk would again produce the same outcome, we also found a genuine lack of NIR YSOs inside a BRC (rather than due to dust extinction), which cannot be circumvented in a spontaneous scenario.

Triggered Star Formation in a Turbulent ISM
Proceedings IAU Symposium No. 237, 2006
B. G. Elmegreen & J. Palouš, eds.

The mass distribution of unstable cores in turbulent magnetized clouds

Paolo Padoan[1], Åke Nordlund[2], Alexei G. Kritsuk[1], Michael L. Norman[1] and Pak Shing Li[3]

[1]Department of Physics, University of California, San Diego, CASS/UCSD 0424, 9500 Gilman Drive, La Jolla, CA 92093-0424,
email: ppadoan@ucsd.edu

[2]Astronomical Observatory/NBIfAFG, Juliane Maries Vej 30, DK-2100, Copenhagen, Denmark

[3]Astronomy Department, University of California, Berkeley, CA 94720

Abstract. The predictions of the Padoan and Nordlund IMF model are tested using the largest simulations of supersonic hydrodynamic (HD) and magneto-hydrodynamic (MHD) turbulence to date ($\sim 1000^3$ computational zones). The striking difference between the HD and MHD regimes, predicted by the model, is recovered.

Keywords. star formation, IMF, turbulence

1. Introduction

The mass distribution of prestellar cores was derived by Padoan & Nordlund (2002) assuming that: i) the turbulence has a power law energy power spectrum; ii) cores are formed by shocks in the turbulent flow and have size and density scaling as the postshock layer thickness and density; iii) the number of such shocks scales self-similarly as the inverse of the cube of their size; iv) the condition for the collapse of small cores is that they exceed their Bonnor-Ebert mass, derived from the lognormal probability density function (pdf) of the gas density independently of the core mass. After integrating over the probability of exceeding the Bonnor-Ebert mass, the mass distribution is:

$$N(m)dm = C\left[1 + \mathrm{erf}\left(\frac{4\ln(m) + \sigma^2}{2\sqrt{2}\sigma}\right)\right]m^{-x}dm \qquad (1.1)$$

where the mass $m = m/m_{\mathrm{BE},0}$ is in units of the average Bonnor-Ebert mass,

$$m_{\mathrm{BE},0} = 3.3\mathrm{M}_\odot\left(\frac{n_0}{10^3\mathrm{cm}^{-3}}\right)^{-1/2}\left(\frac{T_0}{10\mathrm{K}}\right)^{3/2}, \qquad (1.2)$$

σ is the standard deviation of the gas density pdf (assumed to be a lognormal) related to the rms Mach number of the turbulence (the sonic or the Alfvénic Mach number in the HD or MHD regime respectively, see below):

$$\sigma = \sqrt{\ln(1 + M_0^2/4)} \qquad (1.3)$$

The coefficient C is not discussed here, but it would be important for modeling the star formation efficiency. The power law slope, x, is determined by the power law slope of the energy spectrum, β ($\beta \approx 5/3$ in incompressible turbulence and $\beta = 2$ in Burgers zero-pressure model), and by the shock jump conditions:

$$x = 3/(4 - \beta) \qquad (1.4)$$

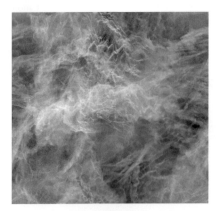 

Figure 1. Logarithm of projected density from a snapshot of the Stagger-Code HD run (left) and MHD run (right).

for $B \geqslant B_{\text{cr}}$ (MHD jump conditions), and

$$x = 3/(5 - 2\beta) \qquad (1.5)$$

for $B < B_{\text{cr}}$ (isothermal HD jump conditions). The critical magnetic field value that separates the two regimes is given by the condition that the postshock gas pressure is of order the postshock magnetic pressure, corresponding to an rms Alfvénic Mach number, M_{A}, of the order of the ratio of the mean gas and magnetic pressures, $M_{\text{A}} \sim P_{\text{g}}/P_{\text{m}}$. This condition gives

$$B_{\text{cr}} \approx 3\,\mu\text{G} \left(\frac{T_0}{10\text{K}} \right) \left(\frac{u_0}{1\text{km/s}} \right)^{-1} \left(\frac{n_0}{10^4\text{cm}^{-3}} \right) \qquad (1.6)$$

Because the Galactic magnetic field strength is locally 6 ± 2 μG (Beck 2001; Han, Ferrière, & Manchester 2004), and most likely larger in prestellar cores (Crutcher 1999; Bourke *et al.* 2001), current star formation in the galactic disk is in the MHD regime. For a value of $\beta = 1.9$, typical of supersonic turbulence, $x = 1.4$, similar to the Salpeter slope of the stellar IMF ($x = 1.35$, Salpeter 1955). For very weak magnetic fields, perhaps in protogalaxies at very large redshifts, the slope is $x = 2.5$, assuming again $\beta = 1.9$. At high redshifts the temperature may also be larger, $T > 100$ K (e.g. Palla, Salpeter, & Stahler 1983; Abel, Bryan & Norman 2000), giving an even larger value of B_{cr}.

The peak of the distribution shifts to smaller masses with increasing Mach number and gas density, increasing the abundance of low mass stars and brown dwarfs as well. This mass distribution matches very well the observed stellar IMF of Chabrier (2003), for reasonable physical parameters, suggesting that the process of turbulent fragmentation may play a major role in the origin of the stellar IMF (Padoan & Nordlund 2002), with only minor effects due to gravitational fragmentation, accretion or merging.

We must stress the statistical nature of the origin of the mass distribution. The present model is based on the statistics of turbulence, and the scaling laws are averages over a large sample. But massive stars originate from shocks on the largest scales, and are rare because such large scale shocks are rare. Being rare, these shocks may deviate from the average properties of the turbulence, resulting in fluctuations of the number of massive stars in different star-forming environments, in excess of the Poisson variance. Large variations of observed mass distributions from place to place are evidence of the stochastic nature of the process of star formation, although this process is universal.

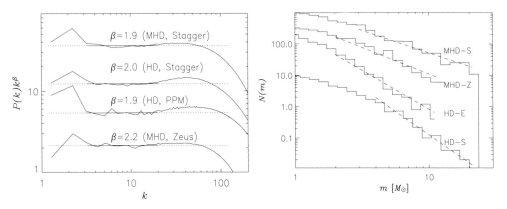

Figure 2. Left: Compensated power spectra of the main four runs and least square fits in the range of wavenumbers $3 \leqslant k \leqslant 20$. **Right:** Mass distributions of gravitationally unstable cores above 1 $M_\odot$, for the main four experiments scaled to a mean density of 10^4 cm^{-3}, a box size of 6 pc, and a clumpfind density resolution $f = 8\%$. The dashed lines show the power law derived from the power spectrum slope and the shock jump conditions of the corresponding simulations, according to the turbulent fragmentation model. The histograms are arbitrarily offset in the vertical direction for clarity.

2. The simulations

Our turbulent fragmentation model outlined in the previous section neglects gravity, so we test it with numerical simulations of supersonic MHD turbulence without self-gravity. The main comparison between the MHD and HD regimes is carried out with the Stagger Code, on a numerical mesh of $1,000^3$ computational zones. We also simulated the HD regime with the Enzo code (Norman & Bryan 1999), and the MHD regime with the Zeus code (Stone & Norman 1992), in both cases on a numerical mesh of $1,024^3$ computational zones.

In all simulations we used periodic boundary conditions, isothermal equation of state, random forcing in Fourier space at wavenumbers $k \leqslant 2$ ($k = 1$ correspond to the computational box size), uniform initial density and magnetic field (in the MHD runs), random initial velocity field with power only at wavenumbers $k \leqslant 2$. The results of this work are for rms Mach number 10, unless otherwise specified . In Figure 1 we show two projections of the density field, from the HD and MHD Stagger Code runs. The density field in the HD run appears to be significantly more fragmented than its MHD counterpart. This is due to the fact that i) The density contrast in the HD shocks is larger than in the MHD shocks, creating thinner postshock layers from shocks with equal sonic Mach number; ii) the HD postshock layers are Kelvin-Helmholtz unstable, due to the strong shear flow that originates in oblique shocks, while in most of the MHD layers the same instability is suppressed by the magnetic field that is amplified in the compression.

Figure 2 (left panel) shows the compensated power spectra of the four main simulations. The power spectra are defined as the squared of the modulus of the Fourier transform of the velocity, integrated over a wave-number shell. If $\hat{\mathbf{u}}_i(\mathbf{k})$ is the Fourier transform of the i velocity component, $\mathbf{u}_i(\mathbf{r})$, the power spectrum is $P_i(k) = \sum \hat{\mathbf{u}}_i\hat{\mathbf{u}}_i^*$, where the sum is over all i and all wave-numbers $\mathbf{k}$ in the shell $k \leqslant |\mathbf{k}| < k + dk$. $P(k)$ is proportional to the contribution to the mean square velocity from all wave-numbers in the shell $k \leqslant |\mathbf{k}| < k + dk$.

The plots in Figure 2 have been arbitrarily shifted in the vertical direction. Deviations of more than a factor of two from a power law fit are found only at wavenumbers $k > 100$, so the turbulence is roughly scale-free for almost two orders of magnitude

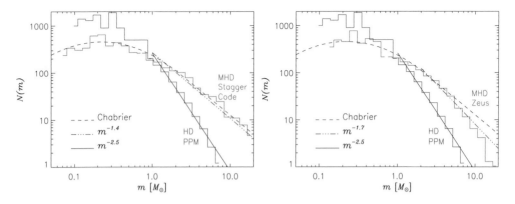

Figure 3. Left: Mass distributions of gravitationally unstable cores from the HD and MHD regimes (Enzo and Stagger Code respectively), computed with $f = 16\%$ and assuming a mean gas density of 10^4 cm^{-3}. Each mass distribution is the result of matching two mass distributions, computed for a computational box size of 1 pc and 6 pc. The Chabrier IMF (Chabrier 2003) and the fragmentation model predictions for the power law mass distributions of each run are also plotted. **Right:** Same as left, but for the Zeus MHD run. Notice how the steeper Zeus mass distribution is recovered. The model predicts the Zeus mass distribution to be steeper than the Stagger-Code one, as a result of the steeper turbulence power spectrum in the more diffusive Zeus run.

in wavenumbers. We have chosen to measure the power spectrum slope in the range $3 \leqslant k \leqslant 20$, because larger wavenumbers are affected by the bottleneck effect (e.g. Falkovich 1994; Dobler *et al.* 2003; Haugen & Brandenburg 2004). We get $\beta = 1.9$ and 2.0 from the Stagger code in the MHD and HD regimes respectively. The Enzo code in the HD regime gives $\beta = 1.9$, and Zeus in the MHD regime $\beta = 2.2$. The corresponding values of the exponent of the power law range of the mass distribution of unstable cores are, according to the model of turbulent fragmentation, $x = 1.4$ and 3 for the MHD and HD regimes of the Stagger code, and $x = 1.7$ and 2.5 for the MHD and HD regimes of Zeus and Enzo respectively.

If the power spectrum is not converged to its correct slope, due to a lack of dynamic range of scales or an excess of numerical diffusivity, the slope of the mass distribution may be strongly affected. For example, in the MHD case we get $x = 1.4$ from the Stagger code, and $x = 1.7$ from the Zeus code. Padoan *et al.* (2006) have recently obtained a measurement of the velocity power spectrum in the Perseus molecular cloud complex. Their result is $\beta = 1.81 \pm 0.10$, which rules out the larger power spectrum slopes generated by more dissipative SPH simulations (see § 4).

3. Mass distributions

We compute the mass distribution of gravitationally unstable cores formed in turbulence simulations without self-gravity in order to learn about the effect of turbulence, and to compare with the predictions of the turbulent fragmentation model. Our results are obtained after the driven turbulence has statistically relaxed, which could not be achieved with self-gravity. The mass distributions derived in this work and the mass distribution predicted by Padoan & Nordlund (2002), should be considered as a guess of the final outcome of more realistic simulations with self-gravity. In simulations including self-gravity the mass distribution of unstable cores may initially vary with time, as the most massive cores are still being assembled by converging turbulent flows while their

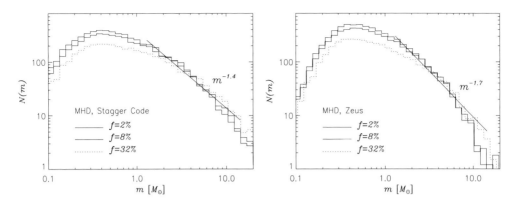

Figure 4. Left: Test of convergence of the mass distribution with decreasing value of the density resolution parameter, f, from the Stagger-Code MHD run. The mass distribution is well converged at $f = 8\%$. **Right:** Same as left, but from the Zeus MHD run.

central part has already collapsed, while in the present study the mass distribution has no time dependence.

Cores are defined as connected overdensities that cannot be split into two or more overdensities of amplitude $\delta\rho/\rho > f$. The unstable cores are the cores with mass larger than their Bonnor-Ebert mass. Our clumpfind algorithm scans the density field with discrete density levels, each of amplitude f relative to the previous one. Only the connected regions above each density level that are larger than their Bonnor-Ebert mass are selected as unstable cores. After this selection, the unstable cores from all levels form a hierarchy tree. Only the final (unsplit) core of each branch is retained.

Different clumpfind algorithms treat the mass surrounding the cores in different ways. The algorithm by Williams, de Geus, & Blitz (1994) uses up all the available mass (see their Figure 2). This results in a core formation efficiency of 100% above the threshold density. Our algorithm, instead, assigns to each core only the mass within the density isosurface that defines the core (below that density level the core would be merged with its next neighbor). With our choice, the smallest possible mass is assigned to each core. With this algorithm, and with conditions typical of molecular clouds, the unstable cores contain a few percent of the total mass, in agreement with the star formation efficiency in molecular clouds.

In the right panel of Figure 2, the mass distributions above 1 $M_\odot$ are plotted for the main four experiments scaled to a mean density of 10^4 cm^{-3}, a box size of 6 pc, and a clumpfind density resolution $f = 8\%$. Overplotted on the corresponding power law section of each mass distribution, the dashed lines show the power law derived from the power spectrum slope and the shock jump conditions of each simulation, according to the turbulent fragmentation model, $x = 3/(4 - \beta)$ in the MHD regime, and $x = 3/(5 - 2\beta)$ in the HD regime. The general trend is recovered well, despite deviations to be expected because this mass distributions are from single snapshots, not time averages.

Figure 3 shows the mass distributions of the HD and MHD regimes, computed with $f = 16\%$ and assuming a mean gas density of 10^4 cm^{-3}. Each mass distribution is the result of matching two mass distributions, computed for a box size of 1 pc and 6 pc. In the 6 pc case we can sample masses in the range $1 - 10$ $M_\odot$, and probe the effect of the turbulence power spectrum and shock jump conditions on the mass distribution, but the mass distribution is incomplete for stars below 1 $M_\odot$. The 1 pc case samples well the turnover region, defining the peak mass for that mean density and rms Mach number,

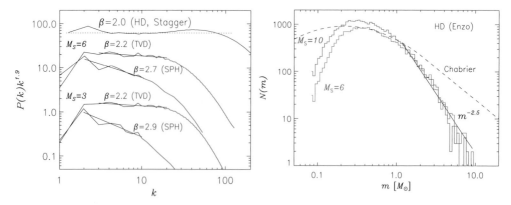

Figure 5. Left: Power spectra compensated for the slope of the Stagger-Code HD run, $\beta = 1.9$. The TVD and SPH power spectra are the same as in Figure 2 of Ballesteros-Paredes *et al.* (2006), for the Mach numbers 3 and 6. **Right:** Mass distributions of gravitationally unstable cores from the Enzo HD runs with $M_S = 6$ and $M_S = 10$, $f = 8\%$ and assuming a mean gas density of 10^4 cm^{-3} and a box size of 6 pc. Each mass distribution contains unstable cores from two snapshots. The Chabrier (2003) IMF (dashed line) and the power law predicted by the turbulent fragmentation model (solid line) are also plotted.

but does not yield intermediate and high mass stars. The numerical mass distributions reproduce the difference between the HD and MHD regimes predicted by the analytical model. The steeper mass distribution expected from the Zeus run, compared with the Stagger Code run, due to the steeper Zeus turbulence power spectrum, is also recovered. Furthermore, the mass distribution of the MHD regime is consistent with Chabrier's stellar IMF (Chabrier 2003).

Once the physical size and mean density of the system are chosen, the clumpfind algorithm depends only on two parameters: i) The spacing of the discrete density levels, f, and ii) the minimum density above which cores are selected, $\rho_{\min}$. Our results do not change significantly for values of $\rho_{\min}$ below the mean gas density, so we scan the density field only above the mean density. We have also verified the convergence of the mass distribution with decreasing values of f. The convergence is typically obtained already at a value of $f \approx 16\%$. In Figure 4 we plot the mass distributions of the MHD regime assuming a mean gas density of 10^4 cm^{-3} and a 6 pc size. Between $f = 32\%$ and $f = 8\%$ there is a tendency to fragment the largest cores and create a larger number of small cores. However, the differences between $f = 8\%$ and $f = 2\%$ are small and the slope of the mass distribution above 2-3 M$_\odot$ is independent of resolution.

4. Discussion

Using TVD and SPH simulations, Ballesteros-Paredes *et al.* (2006) conclude that i) the core mass distribution depends on the rms Mach number and ii) it is not a power law, even at large masses. The first statement is in part correct. The Padoan and Nordlund model contains a Mach number dependence, with the peak of the mass distribution shifting to lower masses as the Mach number increases, in agreement with the numerical results in Ballesteros-Paredes *et al.* (2006). However, in the model the mass distribution above the peak is a power law with slope independent of the Mach number. Based only on the TVD simulations of Ballesteros-Paredes *et al.* (2002), and accounting for their relative low numerical resolution, there is actually no contradiction with our results (see their Figure 4). Significant differences arises only from their SPH simulations (see their Figure 5), but these simulations fail to produce a realistic velocity power spectrum.

The left panel of Figure 5 compares the power spectrum from the Stagger-Code HD run with two TVD and two SPH power spectra from Ballesteros-Paredes *et al.* (2006), for Mach numbers 3 and 6. The inertial range in both the TVD and SPH cases is not very extended, due to the low numerical resolution. The TVD code gives a slope of $\beta \approx 2.2$, as in our Zeus run, for both Mach numbers. The extent of the inertial range in the TVD run is also comparable to that in Zeus at the same resolution. The power spectra of the SPH runs are instead much steeper, and their slope increases with decreasing Mach number, $\beta \approx 2.7$ for $M_S \approx 6$ and $\beta \approx 2.9$ for $M_S \approx 3$. As confirmed by the TVD runs, the power spectrum should not vary much with Mach number between $M_S = 6$ and $M_S = 3$. For even lower Mach numbers, the power spectrum should become shallower, and converge to the Kolmogorov value of $\beta \approx 5/3$ for $M_S < 1$. The dependence of the slope of the SPH power spectrum on the Mach number is therefore both too strong and in the wrong direction. The absence of an inertial range with a reasonable slope or with the correct Mach number dependence, makes the SPH simulations inadequate for testing the turbulent fragmentation model, because the model relies on the scale-free nature of turbulent flows.

The issue of the Mach number dependence, is tested in the right panel of Figure 5, showing the mass distributions from the Enzo HD runs with $M_S = 6$ and $M_S = 10$ and with $f = 8\%$, assuming a mean gas density of 10^4 cm^{-3} and a box size of 6 pc. The Figure shows that the power law part of the mass distribution, above 1-2 M$_\odot$, is independent of the Mach number and matches the prediction of the turbulent fragmentation model, that is $k^{-2.5}$ for the power spectrum slope $\beta = 1.9$ of the HD Enzo runs.

5. Conclusions

We have tested the turbulent fragmentation model of Padoan and Nordlund (2002) with the largest numerical simulations of supersonic MHD and HD turbulence to date. The simulations have confirmed the theoretical prediction that the HD regime should yield a much steeper mass distribution of unstable cores than the MHD regime. This result shows that even rather weak magnetic fields (super-Alfvénic turbulence) can be crucial in setting the initial conditions for star formation. Furthermore, star formation at very high redshift may occur in the HD regime, due to the weak magnetic field and to the larger value of the critical magnetic field strength at larger temperatures. If so, the stellar IMF at high redshift may have a much steeper slope above the peak than in present-day star formation.

Numerical simulations can quantitatively account for the role of the turbulence in setting the initial conditions for star formation only if they generate an inertial range of turbulence. This requires both low numerical diffusivity and large numerical resolution. To model present-day star formation that occurs in the MHD regime, the magnetic field cannot be neglected, even if the turbulence is assumed to be super-Alfvénic. SPH simulations of star formation have too large numerical diffusivity, too low numerical resolution and no magnetic fields. This should cast doubts on the value of comparing their predictions with observational data (see also Agertz *et al.* 2006).

The mass distribution of unstable cores found in the MHD simulations is indistinguishable from the Chabrier stellar IMF (Chabrier 2003) and in agreement with the observed mass distributions of prestellar cores. Such a coincidence may indicate that gravitational fragmentation, competitive accretion or merging, all absent in these turbulence simulations, may not play a major role in the origin of the stellar IMF.

Acknowledgements

This research was partially supported by a NASA ATP grant NNG056601G, by an NSF grant AST-0507768, and by a NRAC allocation MCA098020S. We utilized computing resources provided by the San Diego Supercomputer Center, by the National Center for Supercomputing Applications and by the NASA High End Computing Program.

References

Abel, T., Bryan, G. L. & Norman, M. L. 2000, *ApJ* 540, 39

Agertz, O., *et al.* 2006, astro-ph/0610051

Ballesteros-Paredes, J., Gazol, A., Kim, J., Klessen, R. S., Jappsen, A.-K. & Tejero, E. 2006, *ApJ* 637, 384

Beck, R. 2001, *Space Science Reviews* 99, 243

Chabrier, G. 2003, *PASP* 115, 763

Dobler, W., Haugen, N. E., Yousef, T. A. & Brandenburg, A. 2003, *Phys. Rev. Let.* 68, 026304

Falkovich, G. 1994, *Physics of Fluids* 6, 1411

Han, J. L., Ferrière, K. & Manchester, R. N. 2004, *ApJ* 610, 820

Haugen, N. E. & Brandenburg, A. 2004, *Phys. Rev. E* 70, 026405

Norman, M. L. & Bryan, G. L. 1999, *ASSL Vol. 240: Numerical Astrophysics* 19

Padoan, P. & Nordlund, Å. 2002, *ApJ* 576, 870

Padoan, P., Juvela, M., Kritsuk, A. & Norman, M. L. 2006, *ApJ* in press

Palla, F., Salpeter, E. E. & Stahler, S. W. 1983, *ApJ* 271, 632

Salpeter, E. E. 1955, *ApJ* 121, 161

Stone, J. M. & Norman, M. L. 1992, *ApJS* 80, 791

Williams, J. P., de Geus, E. J. & Blitz, L. 1994, *ApJ* 428, 693

Discussion

BLITZ: The agreement of the slope and mass range of your simulations with observations is impressive. How good is the normalization? That is, how much gas do you need to form a certain integrated mass of stars?

PADOAN: Yes! The typical star formation efficiency we get with the clump find algorithm is ∼5%, similar to the value in molecular clouds. This results from the fact that turbulence puts a fraction of the mass at high density in a small volume (the log normal pdf of gas density), and that mass fraction is a few per cent. The rest of the mass is kept at lower density and higher velocity (until it hits a shock), and cannot take part to the process of star formation.

OSTRIKER: You emphasized that observed structure as well as structure in simulations is hierarchical, in the sense that cores form within clumps within filaments within larger filaments. But your model for the power spectrum does not include any aspects of hierarchy in density structure. Could you comment?

PADOAN: The hierarchical structure is included (albeit in a simple fashion) in the hypothesis of self-similarity. Also, the model does not exclude that a small filament is inside a larger one; it does not take advantage of that, because it more simply assumes self-similarity.

WHITWORTH: (i) How does nature choose the correct combination of M_A and average density to give an approximately universal ratio of brown dwarfs to H-burning stars; (ii) If all massive cores collapse (both compact and extended ones), one would expect high mass stars to be more widely distributed, but this is not what we see. We need some predictions of the distribution of stars as a function of mass.

PADOAN: (i) If you assume Larson's scaling relations you find that the turbulent fragmentation model produces a mass distribution weakly varying with environment. From this point of view, your question may be: what produces the Larson relation? (ii) I don't think this is necessarily the implication of the model but we are starting simulations with sink particles to address this question.

BONNELL: If turbulence gives the upper-mass IMF, then your massive clumps should show the clustering properties of massive stars. This can be used as a strong test of turbulent driven IMFs.

PADOAN: In principle this is a good test. However, observed stellar populations are always dynamically evolved relative to the original star-forming cores, so we will do this comparison as soon as we will follow the collapse of the cores and capture the stars numerically with sink particles. In other words, we will compare stars with stars, not cores with stars.

ANDRE: In a recent paper, Tilly & Pudritz (2004) published purely hydrodynamical simulations of cluster formation and claimed to find a core mass spectrum with a Salpeter slope at the high-mass end. This is in apparent contradiction with your point that turbulent fragmentation leads to a Salpeter core mass spectrum only in the MHD case and to a steeper spectrum in the HD case. Would you care to comment on what might explain this apparent contradiction?

PADOAN: These were much lower resolution simulations than ours. At that resolution, you cannot resolve well the Salpeter range. In fact, you can see from their IMF plots that the statistical significance of the IMF power law fit in the Salpeter range is very low.

Triggered star formation in a turbulent ISM
Proceedings IAU Symposium No. 237, 2006
B. G. Elmegreen & J. Palouš, eds.

© 2007 International Astronomical Union
doi:10.1017/S1743921307001627

Molecular cloud turbulence and the star formation efficiency: enlarging the scope

Enrique Vázquez-Semadeni[1]

[1]Centro de Radioastronomía y Astrofísica, UNAM, Campus Morelia, P.O. Box 3-72, Morelia, Michoacán, 58088, México

Abstract. We summarize recent numerical results on the control of the star formation efficiency (SFE), addressing the effects of turbulence and the magnetic field strength. In closed-box numerical simulations, the effect of the turbulent Mach number $\mathcal{M}_s$ depends on whether the turbulence is driven or decaying: In driven regimes, increasing $\mathcal{M}_s$ with all other parameters fixed decreases the SFE, while in decaying regimes the converse is true. The efficiencies in non-magnetic cases for realistic Mach numbers $\mathcal{M}_s \sim 10$ are somewhat too high compared to observed values. Including the magnetic field can bring the SFE down to levels consistent with observations, but the intensity of the magnetic field necessary to accomplish this depends again on whether the turbulence is driven or decaying. In this kind of simulations, a lifetime of the molecular cloud (MC) needs to be assumed, being typically a few free-fall times. Further progress requires determining the true nature of the turbulence driving and the lifetimes of the clouds. Simulations of MC formation by large-scale compressions in the warm neutral medium (WNM) show that the generation of the clouds' initial turbulence is built into the accumulation process that forms them, and that the turbulence is driven for as long as accumulation process lasts, producing realistic velocity dispersions and also thermal pressures in excess of the mean WNM value. In simulations including self-gravity, but neglecting the magnetic field and stellar energy feedback, the clouds never reach an equilibrium state, but rather evolve secularly, increasing their mass and gravitational energy until they engage in generalized gravitational collapse. However, local collapse events begin midways through this process, and produce enough stellar objects to disperse the cloud or at least halt its collapse before the latter is completed. Simulations of this kind including the missing physical ingredients should contribute to a final resolution of the MC lifetime and the origin of the low SFE problems.

Keywords. ISM: Clouds, stars:formation, turbulence, magnetic fields.

1. Introduction

Molecular clouds (MCs) are the densest regions in the interstellar medium (ISM) and also the site of all present-day star formation in the Galaxy. They are known to have masses much larger than their thermal Jeans mass, a fact that led Goldreich & Kwan (1974) to propose that the clouds should be in a state of generalized gravitational collapse. However, Zuckerman & Palmer (1974) readily noted that this would imply that the MCs should be forming stars at very high rates ($\sim 30 M_{\rm sun}$ yr^{-1}) if all of their mass were to be transformed into stars in roughly one free-fall time $\tau_{\rm ff}$, while the observed rates are much lower ($\sim 5 M_{\rm sun}$ yr^{-1}; see, e.g., Stahler & Palla 2004), suggesting that the star formation efficiency (SFE) is reduced by some mechanism. The observed SFE ranges from a few percent when whole giant molecular complexes are considered (e.g., Myers *et al.* 1986), to 10–30% in cluster-forming cores (e.g., Lada & Lada 2003).

The SFE in MCs, defined as the fraction of the clouds' mass that finally makes it into a star during their lifetime, can be simply written as SFE = SFR $\times \Delta\tau_{\rm c}$, where SFR is the star formation rate, and $\Delta\tau_{\rm c}$ is the cloud lifetime. Thus, a low SFE can be obtained through either a small $\Delta\tau_c$ or a low SFR. Currently, there is an ongoing debate within the community on whether the cloud lifetimes are long, but the SFR is small (or even zero) over a major fraction of the cloud lifetime (e.g. Palla & Stahler 2000; Palla & Stahler 2002; Tassis & Mouschovias 2004; Mouschovias, Tassis & Kunz 2006; Tan, Krumholz & McKee 2006), or else the lifetimes are short, but the SFRs are relatively large (e.g., Ballesteros-Paredes, Hartmann & Vázquez-Semadeni 1999; Klessen, Heitsch & Mac Low 2000; Hartmann, Ballesteros-Paredes & Bergin 2001, hereafter HBB01; Hartmann (2003); Bate, Bonnell & Bromm 2003; Bonnell & Bate 2006; Ballesteros-Paredes & Hartmann 2006; Vázquez-Semadeni *et al.* 2006b).

MCs are also known to have supersonic linewidths, which have been attributed to turbulent motions (e.g., Zuckerman & Evans 1974; Larson 1981; Blitz 1993). Turbulent flows are characterized by a scaling of the typical velocity difference Δv across points separated by a distance ℓ that scales as $\Delta v \propto \ell^\alpha$, with $\alpha > 0$ (e.g., Lesieur 1990), implying that the largest velocity differences occur at the largest spatial scales of a given cloud or clump (Larson 1981). Thus, turbulence is expected to have a *dual* role in the dynamics of MCs (e.g., Vázquez-Semadeni & Passot 1999; Mac Low & Klessen 2004; Ballesteros-Paredes *et al.* 2006): On the one hand, with respect to regions of size L, supersonic compressive turbulent modes of size $\ell > L$ will act mainly as pistons that can form a density peak (a "cloud", "clump" or "core") out of those regions. The timescale for clump formation is essentially the turbulent crossing time across scale ℓ. Since the compressions are supersonic, this is typically shorter than the free-fall or sound-crossing times. On the other hand, turbulent modes with $\ell < L$ will provide support against the self-gravity of a clump of size L.

In this paper we review recent numerical results concerning the effect of the molecular cloud turbulence and the magnetic field on the regulation of the SFE, and discuss how the resolution of certain issues, such as the determination of the most appropriate set of parameters requires studying the formation of the clouds themselves. This review extends the one presented earlier by Vázquez-Semadeni (2005).

2. Effect of the driving scale and turbulent Mach number on the SFE

Results from numerical simulations in the recent past have shown that the effect of the rms Mach number of the turbulence $\mathcal{M}_s$ on the SFE depends on whether the turbulence is driven or decaying. In continuously driven regimes in closed boxes, with periodic boundary conditions and a fixed total mass, Klessen *et al.* (2000) showed that the SFE decreases systematically as either the driving scale of the turbulence λ_d is decreased, or the turbulent Mach number $\mathcal{M}_s$ is increased, and Vázquez-Semadeni, Ballesteros-Paredes & Klessen (2003) subsequently showed that the dependence on $\mathcal{M}_s$ and λ_d could be combined into the dependence with one single parameter, the *sonic scale* λ_s of the turbulence. This is the scale at which the typical turbulent velocity fluctuation (which decreases with scale) equals the sound speed, and is related to $\mathcal{M}_s$ and λ_d by $\lambda_s \approx \lambda_d \mathcal{M}_s^{-1/\alpha}$, where α is the exponent in the velocity dispersion-size relation (cf., §1). At a fixed number of Jeans masses, reducing λ_s leads to a reduction of the fraction of the total mass in scales smaller than λ_s, and the SFE is expected to decrease. Indeed, Vázquez-Semadeni *et al.* (2003) were able to empirically fit a functional dependence of the form SFE $\propto \exp(-\lambda_0/\lambda_s)$, with $\lambda_0 \sim 0.11$ pc in the simulations they studied (fig. 1, *left panel*). From the above relation, this translates into SFE $\approx \exp(-\lambda_0 \mathcal{M}_s^{1/\alpha}/\lambda_d)$, implying that, at fixed λ_d, λ_s decreases with increasing $\mathcal{M}_s$ in driven regimes. This can be understood in terms of the net effect of the turbulent velocity fluctuations, which on

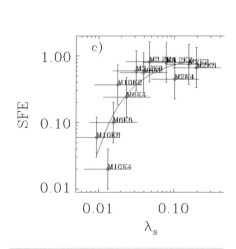
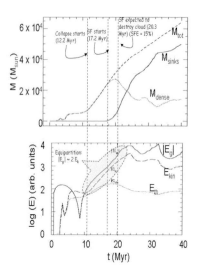

Figure 1. *Left panel:* Star formation efficiency SFE vs. sonic scale λ_s for runs with various rms Mach numbers (M) and turbulence driving wavenumbers K, indicated for each point (from Vázquez-Semadeni *et al.* 2003). *Right panel:* Evolution of the dense gas mass, M_{dense}, mass in stellar objects M_{sinks}, and total mass $M = M_{\mathrm{dense}} + M_{\mathrm{sinks}}$ (*top*) and the gravitational, kinetic and thermal energies (resp. E_g, E_k and E_{th}) for a simulation of colliding WNM streams in the presence of thermal bistability and self-gravity (*bottom*, from Vázquez-Semadeni *et al.* 2006b).

the one hand produce larger-amplitude density fluctuations, but on the other increase the effective "sound" speed in the flow, giving the net result that the effective Jeans mass $M_{\mathrm{J,eff}}$ scales with the rms Mach number as

$$M_{\mathrm{J,eff}} \propto \mathcal{M}_s^2 \tag{2.1}$$

(Mac Low & Klessen 2004, sec. IV.G). At larger $M_{\mathrm{J,eff}}$ it becomes increasingly difficult to collect a core more massive than this mass that can proceed to collapse.

In contrast, in decaying regimes, the SFE appears to *increase* with the rms Mach number of the initial velocity fluctuations (Nakamura & Li 2005). This can be understood because, in this regime, the initial velocity fluctuations can still perform the same fast clump-forming role as in driven regimes. However, at later times, the global decay of the turbulence implies that its supporting action is gradually lost, and $M_{\mathrm{J,eff}}$ decreases, as indicated by eq. (2.1) for $\mathcal{M}_s$ decreasing over time.

In either case, the efficiencies obtained in non-magnetic numerical simulations still appear larger than observational values. For example, Vázquez-Semadeni *et al.* (2003) reported an SFE $\sim 30\%$ in a driven simulation with $\mathcal{M}_s = 10$ and a mass $M = 64 M_{\mathrm{J}}$ ($= 1860 M_{\mathrm{sun}}$), where M_{J} is the thermal Jeans mass, while a two-dimensional decaying simulation with *initial* $\mathcal{M}_s = 10$ and $M = 100 M_{\mathrm{J}}$ reported by Nakamura & Li (2005) reached SFE $\sim 60\%$. Note also that the latter simulation actually had already decayed to $\mathcal{M}_s \sim 2$–3 by the time it was forming stars, a value that appears too low compared with typical turbulent Mach numbers observed in clouds (e.g., Blitz 1993).

3. Effect of the magnetic field strength on the SFE

The magnetic field is an important physical ingredient of interstellar dynamics, and may contribute towards further reducing the SFE obtained in simulations to levels more consistent with observations, even in magnetically supercritical regimes.

In the magnetic case, a fundamental control parameter is the mass-to-magnetic flux ratio μ (in units of the critical value for magnetic support against collapse). Under ideal MHD conditions, supercritical cases ($\mu > 1$) can undergo gravitational collapse, while subcritical cases ($\mu < 1$) are unconditionally supported against it. In this case, collapse can only occur if a Lagrangian fluid parcel loses some of its magnetic flux through some dissipative or diffusive process, such as ambipolar diffusion (AD; e.g., Mestel & Spitzer 1956).

Numerical simulations show that, in magnetically supercritical simulations, collapse is in general delayed with respect to the non-magnetic case (Ostriker, Gammie & Stone 1999; Heitsch, Mac Low & Klessen 2001; Vázquez-Semadeni *et al.* 2005; Nakamura & Li 2005). Recently, the SFE has been measured in simulations of 3D, driven, supercritical simulations of ideal MHD (Vázquez-Semadeni *et al.* 2005) and decaying, 2D simulations including AD (Nakamura & Li 2005). Realistic values of the SFE at the level of whole clouds (SFE $\sim$ a few percent) required moderately subcritical regimes in the decaying cases, but only moderately supercritical regimes in the driven simulations, evidencing again the distinction between driven and decaying regimes. Stronger fields are needed in decaying conditions to compensate for the systematic loss of turbulent support.

In any case, both types of studies show that the SFE is reduced by the presence of a magnetic field even in supercritical regimes, with a tendency to greater reductions at larger mean field strengths. This suggests that the effect of the magnetic field on attenuating the SFE may be gradual rather than dychotomic, as was the case of the distinction between the sub- and supercritical regimes advanced by the "standard" model of magnetic support (e.g., Mouschovias 1976; Shu *et al.* 1987).

4. Discussion: A bigger question

In the previous sections we have summarized results on the SFE in a variety of contexts: driven vs. decaying simulations, and magnetic versus non-magnetic. The main conclusions to be drawn from the existing results are that (1) the very effect of the intensity of the turbulence (measured by the rms Mach number $\mathcal{M}_s$) depends on whether the turbulence is driven or decaying, and (2) the efficiency is reduced as the magnetic field increases from zero to supercritical levels to subcritical levels, but the values of the magnetic field strength needed to attain realistic values of the SFE again depend on whether the turbulence is driven of decaying. Thus, the behavior of the SFE with the parameters $\mathcal{M}_s$ and μ is relatively well understood, but it is necessary to determine what is the true nature of the turbulence driving in molecular clouds (driven, decaying, or somewhere in between) in order to assess the response of the SFE to the parameters.

It is also important to note that in all the simulations described above, it is necessary to define a certain time at which to terminate the accounting of the mass deposited in collapsed objects. This time is typically a few to several free-fall times (a few Myr). If left to run for arbitrarily long times, most of these simulations would eventually turn most of their gas into stars. It is therefore also necessary to address the MC lifetime problem in order to understand the SFE. Presumably, accomplishing both tasks (determining the nature of the driving and the clouds' lifetimes) amounts to addressing the questions of how the clouds themselves form and acquire their properties, and how they are eventually dispersed; that is, their full life cycle.

5. Simulations of cloud formation and evolution

5.1. *Results*

The question of whether the turbulence is driven or decaying is unsettled at present. Arguments in favor of continuous driving include the fact that even nearly starless MCs such as Maddalena's cloud have similar turbulent parameters as clouds with healthy star formation rates (Maddalena & Thaddeus 1985), and that CO clouds fall on a tight velocity dispersion-size relation suggestive of a single cascade process operating at scales ranging from ~ 100 pc to $\lesssim 0.1$ pc in the ISM (e.g., Larson 1981; Heyer & Brunt 2004; see also Breitschwerdt, this volume). Also, the energy feedback from stellar sources once they have started forming is thought to be able to possibly maintain the turbulence in the clouds (e.g., Matzner 2002; Tan, Krumholz & McKee 2006; Krumholz, Matzner & McKee 2006; Li & Nakamura 2006), or even disperse them altogether (e.g., Franco, Shore & Tenorio-Tagle 1994; HBB01; Ballesteros-Paredes 2004).

Recently, it has been proposed that MCs may acquire at least their initial levels of turbulence from the very accumulation process that forms the cloud (Vázquez-Semadeni, Ballesteros-Paredes & Klessen 2003; Heitsch *et al.* 2005; Vázquez-Semadeni *et al.* 2006a, 2006b; Heitsch *et al.* 2006; see also Koyama & Inutsuka 2002; Inutsuka & Koyama 2004), through a combination of the thermal instability and various dynamical instabilities in the compressed layer between converging flows. The precise nature of the instability at work is not yet agreed upon.

These studies have shown that the collision of warm neutral medium (WNM) streams at transonic velocities in the absence of self-gravity produces velocity dispersions of several km s^{-1}, typical of molecular clouds. Furthermore, Vázquez-Semadeni *et al.* (2006a) also showed that the pressure in the dense ($n > 100$ cm^{-3}) gas is larger than the mean WNM pressure by factors 1.5–5, due to the ram pressure of the compressive motion that forms the clouds. These results suggest that cloud formation by WNM stream collisions or passing shocks can produce the observed turbulent velocity dispersions in MCs and at least part of their excess pressure.

Most relevant for our discussion here are the facts that in those studies the turbulence is driven for as long a time as the inflow that forms the cloud persists, and that the rms Mach number of the turbulence in the dense gas depends on the Mach number of the inflow (see also Folini & Walder 2006). This means that, at least during the early epochs of a molecular cloud's existence, the turbulence may be driven, albeit presumably the driving rate itself is decaying, as the inflows that form the cloud subside, and, eventually, the cloud may be left in a decaying state.

This mechanism has been recently investigated including self-gravity and a sink particle prescription for treating collapsed objects by Vázquez-Semadeni *et al.* (2006b). This study has shown that, within its framework and limitations (magnetic fields, stellar energy feedback and chemistry were not included), the clouds evolve secularly, rather than achieving a quasi-stationary state. The collision of WNM streams nonlinearly triggers thermal instability and a transition to the cold neutral medium. Due to the ram pressure of the inflows, densities and temperatures overshoot to values typical of molecular gas. The dense gas (the "cloud") evolves by continuing to incorporate mass, generating an increasingly deep gravitational potential well in the process. Eventually, the gravitational energy $E_{\rm g}$ of the cloud overwhelms the thermal+turbulent energies ($E_{\rm th}$ and $E_{\rm k}$) and the cloud begins to contract gravitationally. This process is illustrated in fig. 1 (*right panel*), which shows the evolution of the dense gas and stellar mass in the simulation, along with the various energies for a simulation in a cubic box of 256 pc per side, in which a cloud is formed by the collision of two oppositely-directed WNM streams at speeds of ± 9.2 km s^{-1}, and each with a length of 112 pc and a radius of 32 pc.

In this simulation, E_g is seen to become dominant at $t \sim 12$ Myr, but the kinetic energy is "dragged along" by the gravitational contraction, with the result that there is near equipartition between the two throughout the collapse, in agreement with observations. After some delay (at $t \sim 17$ Myr for this simulation), local collapse events begin to occur, and within three more Myr ($t \sim 20$ Myr), $\sim 15\%$ of the cloud's mass ($\sim 5000 M_{\rm sun}$) has been converted to stars, at a mean rate $\sim 1.7 \times 10^{-3} M_{\rm sun}$ yr^{-1}. In the simulation, this rate continues for another ~ 5 Myr (to $t \sim 25$ Myr), but already by $t \sim 20$ Myr, the mass that has been converted to stars implies that enough OB stars should be present to destroy the cloud (Franco, Shore & Tenorio-Tagle 1994), assuming a standard IMF. The SFE in this simulation at this time ($\sim 15\%$) is thus comparable to that in the simulations of gravitationally bound clouds discussed in §2. But, as in all those simulations, this is dependent on the assumption that the cloud somehow ceases to form stars some 3–5 Myr after it started.

5.2. *Implications*

Some important consequences of this scenario for MC formation should be noted. First, even though a long delay (~ 15 Myr) occurs between the beginning of the formation process (the time at which the collision between the WNM streams begins), the cloud is expected to remain atomic during most of this time, since the cloud's mean column density is only reaching typical values for molecule formation ($\sim 10^{21}$ cm$^{-2} \sim 8 M_{\rm sun}$ pc^{-2}; see Franco & Cox 1986; HBB01 and references therein; Blitz, this volume) by the time it is beginning to form stars. Thus, even though the cloud as a density enhancement lives ~ 20 Myr, its *molecular* stage is expected to comprise only the last few Myr. That is, there may indeed be a long "dormancy" period before the onset of star formation as suggested by various groups (e.g., Palla & Stahler 2000, 2002; Goldsmith & Li 2005; Mouschovias *et al.* 2006), but most likely it is spent in an atomic, growing state, rather than in a molecular, quasi-equilibrium one.

Second, this scenario of molecular cloud formation implies that the mass-to-flux ratio of the cloud is a variable quantity as the cloud evolves. This ratio is equivalent to the ratio of column density to magnetic field strength (Nakano & Nakamura 1978), with the critical colum density given by $\Sigma \sim 1.5 \times 10^{21} \, [B/5\mu G]$ cm^{-2}. Although in principle under ideal MHD conditions the criticality of a magnetic flux tube involves *all* of the mass contained within it, in practice it is only the mass in the dense gas phase that matters, because the diffuse gas is not significantly self-gravitating at the size scales of MC complexes. As pointed out by HBB01, the above value of the dense gas' column density is very close to that required for gravitational binding, and therefore, the cloud is expected to become magnetically supercritical nearly at the same time it is becoming molecular and self-gravitating. This is consistent with the results of the simulations by Vázquez-Semadeni *et al.* (2006b), in which the column densities of the first four regions to form stars were measured to have column densities within a factor of two of $N = 10^{21}$ cm^{-2} immediately before the first local collapse event occurred there.

Finally, the results from the simulations by Vázquez-Semadeni *et al.* (2006b) would seem to suggest a return to the Goldreich & Kwan (1974) scenario of global gravitational collapse in MCs, except that the criticism by Zuckerman & Palmer (1974) would be avoided in part because the nonlinear turbulent density fluctuations collapse earlier than the whole cloud, involving only a fraction of the total mass, and in part because as soon as the stars form they probably contribute to dispersing the cloud, or at least halting its global collapse. This is consistent with the recent suggestion by Hartmann & Burkert (2006) that the Orion MC may be undergoing global gravitational collapse.

6. Conclusions

We conclude that numerical simulations of isolated clouds up to the present have quantitatively constrained the effect of the turbulent Mach number and the magnetic field strength on the SFE, but in turn this effect depends on the nature of the turbulence production and maintenance, and on the lifetimes of the clouds themselves. Simulations of MC formation within their diffuse environment have begun to shed light on these issues, but much parameter space exploration and inclusion of additional physics (notably, magnetic fields and stellar energy feedback) remain to be done.

Acknowledgements

The author gratefully thanks J. Ballesteros-Paredes for useful comments on the manuscript. This work has received financial support from CRyA-UNAM and CONACYT grant 47366-F. The numerical simulation described in §5.1 was performed on the linux cluster at CRyA-UNAM acquired with funds from CONACYT grant 36571-E.

References

Ballesteros-Paredes, J., Hartmann, L. & Vázquez-Semadeni, E. 1999, *ApJ* 527, 285
Ballesteros-Paredes, J. 2004, *ApSS* 289, 243
Ballesteros-Paredes, J. & Hartmann, L. 2006, *RMAA* submitted (astro-ph/0605268)
Ballesteros-Paredes, J., Klessen, R. S., Mac Low, M.-M. & Vázquez-Semadeni, E. 2006, in: B. Reipurth, D. Jewitt & K. Keil (eds.), *Protostars and Planets V* (Tucson: Univ. of Arizona), in press (astro-ph/0603357)
Bate, M. R., Bonnell, I. A. & Bromm, V. 2003, *MNRAS* 339, 577
Blitz, L. 1993, in: E. H. Levy & J. I. Lunine (eds.), *Protostars and Planets III* (Tucson: Univ. of Arizona), p. 125
Bonnell, I. A. & Bate, M. R. 2006, *MNRAS* 370, 488
Folini, D. & Walder, R. 2006, *A&A* in press (astro-ph/0606753)
Franco, J. & Cox, D. P. 1986, *PASP* 98, 1076
Franco, J., Shore, S. N. & Tenorio-Tagle, G. 1994, *ApJ* 436, 795
Goldreich, P. & Kwan, J. 1974, *ApJ* 189, 441
Goldsmith, P. F. & Li, D. 2005, *ApJ* 622, 938
Hartmann, L., Ballesteros-Paredes, J. & Bergin, E. A. 2001, *ApJ* 562, 852 (HBB01)
Hartmann, L. 2003, *ApJ* 585, 398
Hartmann, L. & Burkert, A. 2006, *ApJ* in press
Heitsch, F., Mac Low, M. M. & Klessen, R. S. 2001, *ApJ* 547, 280
Heitsch, F., Burkert, A., Hartmann, L., Slyz, A. D. & Devriendt, J. E. G. 2005, *ApJ* 633, L113
Heitsch, F., Slyz, A. D., Devriendt, J. E. G., Hartmann, L. W. & Burkert, A. 2006, *ApJ*, submitted (astro-ph/0605435)
Heyer, M. H. & Brunt, C. M. 2004, *ApJ* 615, L45
Inutsuka, S. & Koyama, H. 2004, in *Revista Mexicana de Astronomia y Astrofisica Conference Series*, p. 26
Klessen, R. S., Heitsch, F. & MacLow, M. M. 2000, *ApJ* 535, 887
Koyama, H. & Inutsuka, S.-I. 2002, *ApJ* 564, L97.
Krumholz, M. R., Matzner, C. D. & McKee, C. F., 2006, *ApJ* submitted (astro-ph/0608471)
Lada, C. J. & Lada, E. A. 2003, *ARAA* 41, 57
Larson, R. B. 1981, *MNRAS* 194, 809

M. Lesieur, *Turbulence in Fluids* 2nd. ed. (Kluwer, Dordrecht, 1990)

Li, Z.-Y. & Nakamura, F. 2006, *ApJ* 640, L187

Mac Low, M.-M. & Klessen, R. S. 2004, *Rev. Mod. Phys.* 76, 125

Maddalena, R. J. & Thaddeus, P. 1985, *ApJ* 294, 231

Matzner, C. D. 2002, *ApJ* 566, 302

Mestel, L. & Spitzer, L., Jr. 1956, *MNRAS* 116, 503

Mouschovias, T. C. 1976b, *ApJ* 206, 753

Mouschovias, T. C., Tassis, K. & Kunz, M. W. 2006, *ApJ* 646, 1043

Myers, P. C., Dame, T. M., Thaddeus, P., Cohen, R. S., Silverberg, R. F., Dwek, E. & Hauser, M. G. 1986, *ApJ* 301, 398

Nakamura, F. & Li, Z.-Y. 2005, *ApJ* 631, 411

Nakano, T. & Nakamura, T. 1978, *PASJ* 30, 671

Ostriker, E. C., Gammie, C. F. & Stone, J. M. 1999, *ApJ* 513, 259

Palla, F. & Stahler, S. W., 2000, *ApJ* 540, 255

Palla, F. & Stahler, S. W., 2002, *ApJ* 581, 1194

Shu, F. H., Adams, F. C. & Lizano, S. 1987, *ARA&A* 25, 23.

Stahler, S. W. & Palla, F. 2004, *The Formation of Stars* (New York: Wiley)

Tan, J. C., Krumholz, M. R. & McKee, C. F. 2006, *ApJ* 641, L121

Tassis, K. & Mouschovias, T. Ch. 2004, *ApJ* 616, 283

Vázquez-Semadeni, E. & Passot, T. 1999, in: J. Franco & A. Carramiñana (eds.), *Interstellar Turbulence* (Cambridge: Cambridge Univ.), p. 223

Vázquez-Semadeni, E., Ballesteros-Paredes, J. & Klessen, R. S. 2003, *ApJ* 585, L131

Vázquez-Semadeni, E. 2005, in: E. Corbelli, F. Palla, & H. Zinnecker(eds.), *IMF@50: The Initial Mass Function 50 years later* (Dordrecht: Springer), p. 371

Vázquez-Semadeni, E., Kim, J., Shadmehri, M. & Ballesteros-Paredes, J. 2005, *ApJ* 618, 344

Vázquez-Semadeni, E., Ryu, D., Passot, T., González, R. F. & Gazol, A. 2006, *ApJ* 643, 245

Vázquez-Semadeni, E., Gómez, G. C., Jappsen, K. A., Ballesteros-Paredes, J., González, R. F. & Klessen, R. S. 2006, *ApJ* submitted (astro-ph/0608375)

Zuckerman, B. & Evans, N. J. II 1974, *ApJ* 192, L149

Zuckerman, B. & Palmer, P. 1974, *ARAA* 12, 279

Discussion

CLARK: You say that the final SFE in your simulation is $\sim 15\%$, which is too high. Have you tried unbound clouds (kinetically) to see whether the final SFE goes down?

VÁZQUEZ-SEMADENI: In this simulation the boundedness of the cloud is produced self-consistently by the cloud-formation process, so we do not control it directly. In this particular simulation, the resulting cloud is strongly bound. Nevertheless, one could try to produce a less strongly bound cloud, or even unbound, by decreasing the mass contained in the inflowing streams, or increasing their speed. We are currently performing a parameter study to investigate different cloud masses and inflow velocities.

ROSOLOWSKY: Could you comment on the applicability of your simulations to the formation of GMCs, specifically in the case where the scales over which you have to gather gas become significant on a galactic scale?

VÁZQUEZ-SEMADENI: I am convinced that the process of compression, then cooling with turbulence generation, and finally gravitational collapse, should be representative of GMC formation in spiral arms, although modeling the process more accurately should incorporate the vertical stratification as well.

Triggered star formation in a turbulent ISM
Proceedings IAU Symposium No. 237, 2006
B. G. Elmegreen & J. Palouš, eds.

© 2007 International Astronomical Union
doi:10.1017/S1743921307001639

Simulations of ionisation triggering

C. J. Clarke[1] and J. E. Dale[2]

[1]Institute of Astronomy, Madingley Road, Cambridge, U.K., CB3 OHA
email: cclarke@ast.cam.ac.uk

[2]Department of Physics and Astronomy, University of Leicester, University Road, Leicester,
U.K. LE1 7RH
email: jed20@astro.le.ac.uk

Abstract. We review recent pilot simulations that incorporate feedback from ionising radiation in SPH calculations of star forming clouds. In the case that the ionising radiation source is located within the star forming cloud, the inhomogeneity of the cloud significantly modifies the way that feedback operates compared with spherically symmetric cloud models. Inflow/outflow behaviour develops, combining accretion down dense filaments and thermally driven outflows that can remove many times the binding energy of the parent cloud. If the ionising source is located external to the cloud, we find evidence for triggered star formation but conclude that it is hard to find unambiguous observational signatures that would distinguish "triggered" stars from those created spontaneously.

1. Introduction

The thermal feedback of energy into the ISM via the ionising radiation field of OB stars raises a number of (currently unsolved) questions. An obvious issue is whether the effect of such feedback is net positive or negative (i.e. star formation promoting or disrupting) and how this affects the *efficiency* of star formation. If one decides that the star formation promoting aspect is important, one has then to ask what are the observational signatures of triggered star formation. Another aspect of the problem concerns how the sculpting of the local star forming cloud by ionisation feedback affects the escape of ionising radiation into the larger scale ISM and thus whether star formation is a plausible energy source for sustaining the thermal balance of the warm ionised medium in galaxies.

Historically, there have been two approaches to the numerical study of ionisation feedback. One involves hydrodynamic simulations in smoothly stratified media (Yorke *et al.* 1989; Franco *et al.* 1990; Garcia-Segura and Franco 1996). The other consists of radiative transfer in a realistically clumpy/fractal medium (i.e. with no hydrodynamics: Hobson and Padman 1993; Witt and Gordon 1996; Rollig *et al.* 2002). Evidently one wishes to combine the virtues of both approaches and instead model the hydrodynamic evolution in a realistically clumpy medium.

Such an ambitious task requires considerable simplification of the radiative transfer – specifically neglect of the diffuse field of ionising radiation due to recombinations to the ground state – so that through this 'on the spot' approximation one can compute the instantaneous Stromgren volume (region within which the number of recombinations per second equals the input ionising photon production rate). Determination of this Stromgren volume involves the computation of a recombination integral ($\propto \int n^2 r^2 dr$), a task that is trivial in a grid based code but which requires some thought in a Lagrangian method like SPH. Recently both Dale *et al.* (2005, 2006) and Gritschneder *et al.* (2006) have built on the original scheme of Kessel-Deynet & Burkert (2000) so as to use SPH neighbour lists to compute the recombination integral via jumping along a chain of

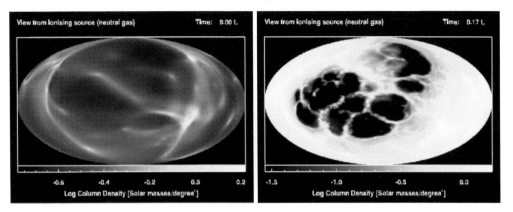

Figure 1. Column density map of sky seen from source: prior to switch on of ionising radiation (left panel) and after sculpting by ionising radiation (right panel).

particles leading to the central star. Once the Stromgren volume is determined in this way, the temperature therein is set to 10^4K. This approach works well in simple geometries and is currently being tested against full Monte Carlo radiative transfer codes in the case of the complex filamentary density fields encountered in realistic hydrodynamic simulations of turbulent molecular clouds (Ercolano *et al.* in prep.).

Here I will report briefly on two projects that use SPH with ionisation feedback included and which explore the effect of radiation from an OB star that is respectively internal (Dale *et al.* 2005) and external (Dale *et al.* 2006) to a star forming cloud. It should be stressed that in neither case has it been possible to explore parameter space and thus at this stage, the conclusions from these pilot studies are mainly qualitative.

2. Internal ionising source

In this case, the OB star is formed at the intersection of a set of dense filaments, and thus the 'sky' as 'seen' from the star is highly inhomogeneous even at the outset of the simulations, there being order of magnitude variations in column density and even larger variations in local density. Once the ionising source is switched on, the density contrasts are exacerbated: ionising radiation can propagate readily through lower density regions where the recombination timescale is long, and pressure gradients in the photoionised gas then drive outflows which reduce the density yet further (see Figure 1). On the other hand, ionising radiation can scarcely penetrate the dense filamentary structures where the recombination timescale is short. Thus an 'inflow-outflow' situation develops: material continues to be channeled onto the central star and yet there are also several regions of loosely collimated yet vigorous outflows. [Indeed, the mass flow rates and opening angles are compatible with the properties of some of the outflows observed in regions of high mass star formation (Churchwell 1997), implying that at least some of these may be environmentally collimated structures. Evidently, the simulations cannot reproduce highly collimated or bipolar structures, which instead demand a, presumably disc related, collimation mechanism close to the star. See Shepherd *et al.* 1997, Beuther *et al.* 2002.]

The present simulations evidence both positive and negative feedback effects. The mass flow rate down the filaments is reduced compared with a control simulation without feedback and yet is not halted in the higher density simulations. At the same time, lateral expansion of hot gas in the outflow channels compresses the gas in the filaments and induces extra star formation which is not occurring in the control simulation. Due

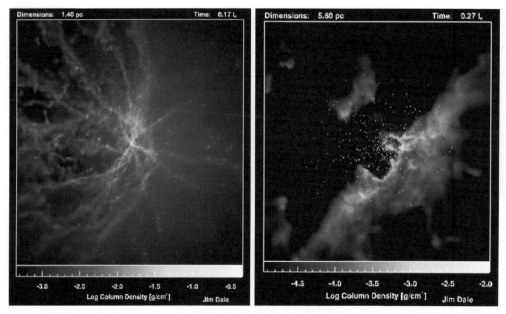

Figure 2. Column density map of high density ($\langle n \rangle \sim 10^4$ cm^{-3} run: left panel) and low density ($\langle n \rangle \sim 10^3$ cm^{-3} run: right panel) after $\sim 2 \times 10^5$ and $\sim 5 \times 10^5$ years respectively.

to limited numerical resolution of the induced star formation, it is not currently possible to assess whether the *net* effect of feedback is positive or negative.

Another noteworthy aspect of this inflow/outflow behaviour is that it is possible for the cluster to absorb a quantity of thermal and kinetic energy that far exceeds the cluster binding energy – and yet, under some circumstances, for the cluster to remain bound. This is simply because a relatively small mass fraction is expelled at ~ 10 km/s, i.e. at several times the escape velocity of the cluster. This result therefore demonstrates that simple criteria based on binding energy may be misleading when assessing the ultimate state (boundedness) of a cluster.

The results are evidently sensitive to mean density (contrast panels of Figure 2, which differ by only an order of magnitude in mean density). Although feedback is much more effective in the low density case (and, in fact, has halted accretion on to the OB star and unbound the cluster), this trend is *less* strong than it would be in the case of equivalent clusters with spherically symmetric density fields. [The spherically symmetric version of the left hand panel would result in essentially no impact from ionising radiation, since the HII region would in this case be confined by ram pressure to deep in the cluster core. In the spherically symmetric version of the low density run, however, the gas would have been completely cleared from the image shown in the right hand panel of Figure 2.]

A final point concerns the escape of ionising radiation. Even in the high density simulation (left hand panel of Figure 2), it is found that around 20% of the ionising photons can escape (along low density channels), whereas none would have escaped in the spherically symmetric equivalent cluster. Lesch *et al.* (1997) estimate that $15 - 20\%$ of the ionising radiation from OB stars would need to escape from their natal regions in order to sustain the thermal state of the diffuse ionised gas in the Galaxy. The simulations demonstrate that, owing to the inhomogeneity of realistic star forming clouds, such escape fractions are credible even when the mean density of star forming clouds is very high ($\sim 10^4$ cm^{-3}).

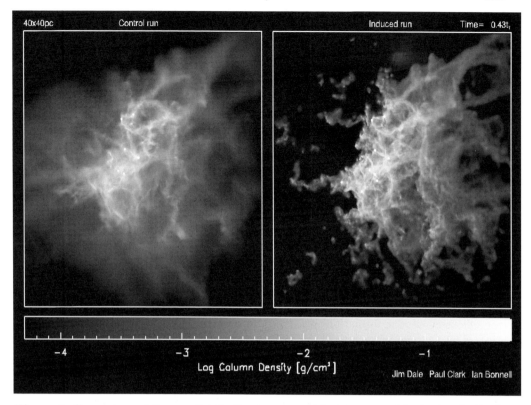

Figure 3. Comparison of the gas distribution in a run subject to external ionisation feedback (right hand panel) with a control run (left panel).

3. The external case

In the following simulations (Dale *et al.* 2006), an OB star is situated next to a turbulent molecular cloud which is initially globally unbound (initial ratio of kinetic to gravitational energy of ∼2). The effect of the ionising source can be carefully assessed through comparison with a control simulation. As can be seen from the left hand panel, the control simulation forms some stars even though it is globally unbound: gas on initially convergent paths undergoes shock compression and dissipation of kinetic energy, producing the filaments in the core of the panel. Gas on initially divergent trajectories can (in the case of this unbound cloud) simply escape in all directions, generating the halo of diffuse gas in the left hand panel.

In the run with the ionising source (located at the middle of the left hand edge of the right hand panel in Figure 3), the gas distribution develops a clear asymmetry. To the right, initially outflowing gas streams freely outwards as before. To the left, however, gas initially flowing outwards has its flow velocity reversed by interaction with the ionising radiation field. When this swept up gas reaches the cloud core (i.e. the region where star formation was proceeding in the control run) it shocks and fragments. It would appear that feedback is triggering some star formation in the cloud (i.e. stars form which are not present in the control run). Analysis of the mass contained in the stars in both simulations shows that, as expected, the stars in the feedback run derive much of their mass from material that was previously outflowing towards the ionisation source. Thus, in a complex environment, these simulations manifest the "collect and collapse" idea first proposed by Elmegreen and Lada 1977 (see also Elmegreen *et al.* 1995 for application of these ideas to a clumpy medium).

Although feedback has triggered some star formation in this case, it turns out to be impossible to find observational diagnostics which distinguish the induced stars from those that would have formed anyway. For example, the two populations are co-spatial since, as noted above, the swept up material only fragments when it encounters dense counter-flowing gas in the core, which is any case the site for star formation. The complexity of the velocity field also removes kinematic signatures of induced star formation, i.e. the net momentum of the gas that fragments following the collision of the swept up gas and turbulent gas in the cluster core can be in either direction. This is because the r.m.s. velocity of the turbulence (a few km s^{-1}) is comparable to that of the swept up gas (which initially attains about 10 km s^{-1} but which is decelerated by mass loading as it ploughs into the cloud core).

The implication of this work is that it is very hard to identify which stars have formed from ionisation induced feedback and that one should therefore be careful about claims in the observational literature about the self-evident hallmarks of triggered star formation.

Finally it is worth asking whether, given that feedback appears to be triggering star formation in this run, one can argue against feedback being the mechanism by which the efficiency of star formation is reduced to observationally acceptable levels? Before discarding the notion that feedback can reduce the star formation efficiency, however, one should stress the fact that this simulation started with an unbound cloud, and so the ionising radiation field played the role of returning material to the cloud core which would otherwise have escaped. It is thus unsurprising that the star formation efficiency is, if anything, enhanced in this case. Further simulations are required in order to discover whether this remains the case if the cloud is bound initially.

References

Beuther, H., Schilke, P., Gueth. F., *et al.* 2002, *A&A* 387, 931
Dale, J., Bonnell, I., Clarke, C. & Bate, M. 2005, *MNRAS* 358, 291
Dale, J., Clark, P.C. & Bonnell, I. 2006, *MNRAS* submitted
Elmegreen, B.G., Kimura, T. & Tosa, M. 1995, *ApJ* 451, 675
Elmegreen, B.G. & Lada, C.J. 1977, *ApJ* 214, 725
Franco, J., Tenorio-Tagle, G. & Bodenheimer, P. 1990, *ApJ* 349, 126
Garcia-Segura, G. & Franco, J. 1996, *ApJ* 469, 171
Gritschneder, M., Naab, T., Heitsch, F. & Burkert, A. 2006, in: B.G. Elmegreen & J. Palouš (eds.), *Triggered Star Formation in a Turbulent ISM* (Cambridge: Cambridge Univ.), in press
Hobson, M. & Padman, R. 1993, *MNRAS* 264, 161
Kessel-Deynet, O. & Burkert, A. 2000, *MNRAS* 315, 713
Lesch H., Dettmar R., Mebold U. & Schlickeiser R. (eds.) 1997, *The Physics of Galactic Halos* (New York: John Wiley & Sons)
Rollig, M., Hegmann, M. & Kegel, W. 2002, *A&A* 392, 1081
Shepherd, D., Churchwell, E. & Wilner, D. 1997, *ApJ* 482, 355

Discussion

DOPITA: A comment and a question: The stellar wind will make an enormous difference to your simulations compared to the photoionization-only simulation. A question: do you include the effect or radiation pressure on grains, which is very important near the stars?

DE GOUVEIA DAL PINO: Just a quick comment (similar to the one made by M. Dopita): C. Melide, A. Raga and I have performed simulations including both the effects of an ionization front and or supersonic winds (or SNR shock fronts) and this later seems to have also relevant influence upon the overall system evolution.

CLARKE: These particular simulations involve only ionising radiation feedback, which is already challenging in the context of fully self-gravitating turbulent hydrodynamic simulations. Work is well under way to add the capability to include stellar winds in such simulations. Radiation pressure on dust is of course the dominant feedback mechanism on the scale of individual stars (i.e. on the $\sim$100 A.U. scale) but is a secondary effect on the cluster scale. Supernova feedback is a potential factor in determining the initial conditions of our star forming clouds but, on the dynamical timescale on which star formation occurs in these simulations, there is of course insufficient time for the stars created in the simulations to undergo supernovae.

KRUMHOLZ: Have you tested against the Spitzer similarity solution, and if so, can you reproduce the result to $\sim$1%.

CLARKE: Yes, agreement with the Spitzer solution is excellent. However, this is not the most challenging aspect of these simulations, which is instead the modeling of the thermal structure of the gas in the case that the gas is distributed in converging filaments. Here there are concerns that the SPH path finder can bias one towards denser regions of the flow, an aspect which we are pursuing through detailed comparison with Monte Carlo radiative transfer calculations.

Triggered Star Formation in Turbulent ISM
Proceedings IAU Symposium No. 237, 2006
B. G. Elmegreen & J. Palouš, eds.

© 2007 International Astronomical Union
doi:10.1017/S1743921307001640

Protostellar turbulence in cluster forming regions of molecular clouds

Fumitaka Nakamura[1] and Zhi-Yun Li[2]

[1]Faculty of Education and Human Sciences, Niigata University, 8050 Ikarashi-2, Niigata 950-2181, Japan,   email: fnakamur@ed.niigata-u.ac.jp

[2]Department of Astronomy, University of Virginia, P.O. Box 400325, Charlottesville, VA 22904, USA,   email: zl4h@virginia.edu

Abstract. We perform 3D MHD simulations of cluster formation in turbulent magnetized dense molecular clumps, taking into account the effect of protostellar outflows. Our simulation shows that initial interstellar turbulence decays quickly as several authors already pointed out. When stars form, protostellar outflows generate and maintain supersonic turbulence that have a power-law energy spectrum of $E_k \sim k^{-2}$, which is somewhat steeper than those of driven MHD turbulence simulations. Protostellar outflows suppress global star formation, although they can sometimes trigger local star formation by dynamical compression of pre-existing cores. Magnetic field retards star formation by slowing down overall contraction. Interplay of protostellar outflows and magnetic field generates large-amplitude Alfven and MHD waves that transform outflow motions into turbulent motions efficiently. Cluster forming clumps tend to be in dynamical equilibrium mainly due to dynamical support by protostellar outflow-driven turbulence (hereafter, protostellar turbulence).

Keywords. ISM: magnetic fields, ISM: clouds, MHD, stars: formation, turbulence

1. Introduction

The majority of stars are thought to form in clusters. Observations show that cluster forming regions are strongly influenced by supersonic turbulence. Also, molecular clouds are known to be magnetized. Observed magnetic energy is comparable to gravitational and turbulent kinetic energies, suggesting that the magnetic field as well as supersonic turbulence is dynamically important in cloud evolution. On the other hand, numerical simulations have demonstrated that supersonic turbulence decays quickly, on a timescale comparable to the turbulence crossing time on the dominant energy-carrying scale, with or without a strong magnetic field (e.g., Stone *et al.* 1998; MacLow *et al.* 1998). Therefore, the supersonic turbulence must be replenished somehow. One promising mechanism of turbulence supply in cluster-forming clumps is protostellar outflows, which are observed in abundance in nearby cluster-forming regions. In fact, there is strong evidence that protostellar outflows strongly affect cloud dynamics and star formation in several regions (e.g., NGC1333: Quillen *et al.* 2005; Circinus cloud: Bally *et al.* 1999). However, it is not well-understood how protostellar outflows affect or control cloud dynamics and star formation in cluster-forming clumps. We have made a start in numerical study of cluster formation including spherical protostellar outflows (Li & Nakamura 2006). Here we present the results of new simulations that take into account the outflow collimation. See Nakamura & Li (2006) for more detail.

2. Numerical model

We consider a centrally condensed self-gravitating spherical isothermal cloud in a cubic box of length $L = 9L_J$, where L_J is the thermal Jeans length. The radius of the central region of nearly constant density is set to $1.5\ L_J$. The initial cloud is threaded by a uniform magnetic field whose strength (B_0) is specified by $\alpha \equiv B_0^2/(8\pi c_s^2 \rho_0)$, where ρ_0 is the central density and c_s is the isothermal sound speed. For the standard model presented below, we set $\alpha = 2.5$. The cloud is magnetically supercritical: the averaged magnetic flux-to-mass ratio is about a half the critical value. At the beginning of the simulation, we impose on the cloud a supersonic velocity field of power spectrum $E_k \propto k^{-1}$. The initial rms turbulent Mach number is set to $\mathcal{M} = 10$. We follow the cloud evolution with a 3D MHD code based on Roe's TVD method.

Initial turbulence generates small self-gravitating cores, from which several stars form. In our simulations, star formation is treated as follows. When the central density of a dense self-gravitating core exceeds a threshold density ($\rho_{\rm th} = 100\rho_0$), a fraction $\epsilon(= 20\%)$ of its mass is converted into a Lagrangian point particle that represents a newborn star. At the same time, the outflow momentum is added in the surrounding gas. The protostellar outflow is assumed to be bipolar, and its direction is parallel to the local magnetic field line. The half opening angle of the protostellar jet component is chosen to be 30 degree. The outflow strength is determined by two parameters: the stellar mass M_* and a dimensionless factor f (Nakamura & Li 2005). The stellar mass is automatically determined when the core mass is identified, $M_* = \epsilon M_{\rm core}$. The factor f is uncertain, and may lie in the range $\sim 0.1 - 1.0$ (see Nakamura & Li 2006 for more detail). We adopt for the standard model presented below $f = 0.5$, corresponding to an outflow momentum of $50\ M_\odot$ km/s per solar mass of stellar material.

3. Numerical results

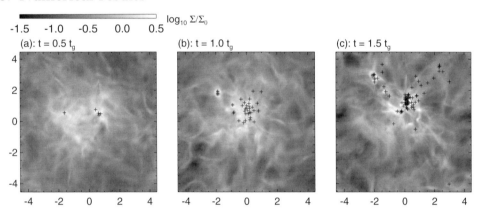

Figure 1. Column density distributions for the model with $\alpha = 2.5$, $\mathcal{M} = 10$, and $f = 0.5$ at three different times: (a) $t = 0.5t_g$, (b) $1.0t_g$, and (c)$1.5t_g$. Crosses indicate the positions of stars.

As a typical example, we show in Figure 1 the evolution of the model with $\alpha = 2.5$, $\mathcal{M} = 10$, and $f = 0.5$. At early times, initial turbulence decays quickly as several authors demonstrated. The first star forms at $t \sim 0.4t_g$, where t_g is the gravitational collapse time defined as $L_J/C_s = (\pi/G\rho_0)^{1/2}$ (roughly 25% longer than the global free fall time). After that, several stars form from the dense fragments generated by the initial turbulent compression [panel (a)]. However, most stars form in high-density regions disturbed strongly by outflows [panels (b) and (c)]. At late times, the velocity dispersion reaches about $\langle \Delta V^2 \rangle^{1/2} \simeq 5c_s \simeq 1.5(T/20\,{\rm K})^{1/2}$ km s^{-1}. This is consistent with the observed values in nearby cluster forming regions such as NGC1333. Our simulation indicates that protostellar outflows influence cloud dynamics and star formation significantly. We note that several stars form by dynamical compression due to outflows. In other words, local star formation can be triggered by outflows.

To assess the effect of protostellar outflows on star formation, we depict in the left panel of Figure 2 the time evolution of star formation efficiency (SFE) for the $\alpha = 2.5$ models with three different outflow strengths: $f = 0.25, 0.5$, and 0.75. Here, the SFE is defined as the ratio of the total stellar mass divided by the total mass of stars and gas. For the model presented in Figure 1, the SFE reaches 6% by $2t_g$. This corresponds to the star formation rate per unit free-fall time of $\mathrm{SFR_{ff}} \simeq 3\%$ (see Krumholz & Tan 2006). For the model with weaker outflows ($f = 0.25$), the SFE reaches about 10.5% by $2t_g$ ($\mathrm{SFR_{ff}} \simeq 5.3\%$). For the stronger outflow case ($f = 0.75$), the SFE reaches 3% by 1.9 t_g ($\mathrm{SFR_{ff}} \simeq 1.5\%$). Thus, we conclude that the outflows tend to suppress global star formation. We note that this agrees with the results of driven-turbulence simulations where driven turbulence tends to reduce star formation efficiency (e.g., Vazquez-Semadeni et al. 2005). To see how SFE depends on initial magnetic field strength, we show in the right panel of Figure 2 three models with $\alpha = 2.5, 0.5$ and 10^{-6}. Clearly, magnetic field also suppresses the global star formation. We conclude that both protostellar outflow and magnetic field are important in regulating the rate of star formation in cluster forming regions.

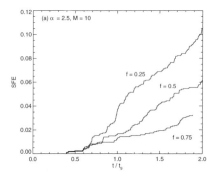 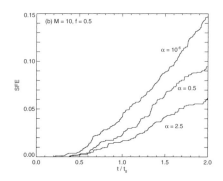

Figure 2. Time evolution of star formation efficiency (SFE). *left*: SFEs for three models with different outflow strength: $f = 0.25, 0.5$, and 0.75. Protostellar outflows tend to suppress global star formation. *right*: SFEs for three models with different magnetic field strength: $\alpha = 10^{-6}, 0.5$, and 2.5. The outflow strength is set to $f = 0.5$. The magnetic field also retards star formation significantly.

To see the dynamical state of the cloud, we define a net radial infall speed from $v_{\mathrm{net}} = \dot{M}_{\mathrm{net}}(r)/4\pi r^2 \tilde{\rho}(r)$, where $\dot{M}_{\mathrm{net}}(r)$ is the net mass flux through a sphere of radius r centered on the minimum of gravitational potential, and $\tilde{\rho}(r)$ is the average mass density at that radius. In the left panel of Figure 3, we plot v_{net} as a function of radius at several representative times: 0.5, 1.0, 1.5, and 2.0 t_g. Note that the net mass flow can either be infall or outflow at a given radius. Its magnitude is, however, generally comparable to, or smaller than the isothermal sound speed. For comparison, the characteristic gravitational speed is $v_g = [GM/(L/2))]^{1/2} = 8.72c_s$, which is much larger than the net infall or outflow speed. Therefore, the average mass redistributes at a speed much less than the characteristic dynamical speed, which is a clear indication that the system has reached a dynamical equilibrium. The radial density profile is approximated by a power-law of $\rho \propto r^{-1.5}$ during the evolution (the right panel of Figure 3).

Our simulations also indicate that protostellar outflows generate large-amplitude MHD waves by strongly distorting magnetic field lines. The waves seem to be important in transforming the outflow motions into local turbulent motions. The kinetic energy power spectrum shows a power-law of $E_k \propto k^{-n}$, with $n \approx 2$. The value of n is more or less independent of the power index of the initial turbulent field. The power index of protostellar

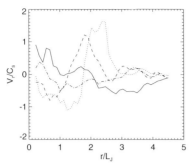 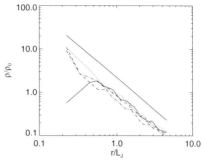

Figure 3. *left*: Net radial velocity normalized by isothermal sound speed. It is measured from the distance from the cluster center. The cluster center is determined as the point where the gravitational potential takes its minimum. *right*: Radial density profile normalized by ρ_0. Straight line denotes the power-law profile of $r^{-1.5}$.

turbulence seems to be somewhat steeper than those of driven MHD turbulence simulations such as Boldyrev *et al.* (2002)'s ($n = 1.74$).

4. Summary

We have carried out 3D numerical simulations of cluster formation taking into account supersonic turbulence, magnetic fields, and protostellar outflows. We found that although initial interstellar turbulence decays very quickly, protostellar outflows can drive and maintain supersonic turbulence in cluster forming clumps. In the model presented above, the velocity dispersion stays around $\langle \Delta V^2 \rangle^{1/2} \sim 5c_s \simeq 1.5(T/20\mathrm{K})^{1/2}$ km s^{-1}, which is in good agreement with the observed values in nearby cluster forming regions. Stars form in high-density regions disturbed strongly by protostellar outflows. We call this outflow-driven turbulence "protostellar turbulence". Protostellar outflows suppress global star formation, although they can sometimes trigger local star formation, e.g., due to dynamical compression of pre-existing dense cores. Our simulations indicate that cluster-forming clumps are close to a dynamical equilibrium. Star formation rate per unit free-fall time (SFR$_{\mathrm{ff}}$) stays around a few percent, which is in good agreement with recent estimates of this number from a range of observations (Krumholz & Tan 2006).

Acknowledgements

Numerical calculations are carried out on NEC SX6 at Niigata University, Hitachi SR8000 at the Center for Computational Science, University of Tsukuba and NEC SX8 at Yukawa Institute, Kyoto University. The research is supported in part by NSF AST-0307368, NASA NNG05GJ49G, and the Grant-in-Aid (18540234) from MEXT of Japan.

References

Bally, *et al.* 1999, *AJ* 117, 410
Boldyrev, S., Nordlund, A. & Padoan, P. 2002, *ApJ* 573, 678
Li, Z.-Y. & Nakamura, F. 2006, *ApJ* 640, L187
Krumholz M. & Tan, J. 2006, *astro-ph/0606277*
MacLow, M.-M., Klessen, R., Burkert, A. & Smith, M.D. 1998, *PRL* 80, 2754
Nakamura, F. & Li, Z.-Y. 2005, *ApJ* 631, 411
Nakamura, F. & Li, Z.-Y. 2006, *in preparation*
Quillen, *et al.* 2005, *ApJ* 632, 941
Stone, J.M., Ostriker, E.C. & Gammie, C.F. 2005, *ApJ* 508, L99
Vazquez-Semadeni, E., *et al.* 2005, *ApJ* 630, 49

Discussion

HILY-BLANT: Since outflows drive turbulence according to your MHD simulations, and since the star formation efficiency decreases with outflow feedback, how do your models compare to Padoan and Vazquez-Semadeni simulations in terms of power spectra?

NAKAMURA: We are now analyzing the power spectrum of our model. Preliminary results show that the energy power spectrum of outflow models follows $E_k \propto k^{-2}$, whereas the power spectrum of no outflow models is steeper than $E_k \propto k^{-2}$. Therefore, the slope of the power spectrum of outflow models is more or less consistent with Padoan's. However, we haven't done any quantitative comparison with Padoan's model yet. As for Enrique's work, our simulation is in good agreement with his results because turbulence driven by outflows tends to suppress star formation.

E. OSTRIKER: (1) Have you quantified the difference in efficiency for transferring outflow energy to turbulence in magnetized vs. unmagnetized models? (2) Have you measured the effective pressure as a function of radius, i.e., $\langle \rho v^2 \rangle(r)$ and $\langle B^2 \rangle(r)$?

NAKAMURA: (1) No, (2) Not yet. I showed that the cloud is in dynamical equilibrium and the thermal pressure is much smaller than turbulent pressure. Therefore, turbulent pressure and magnetic force should contribute to force balance.

Triggered Star Formation in a Turbulent ISM
Proceedings IAU Symposium No. 237, 2006
B. G. Elmegreen & J. Palouš, eds.

© 2007 International Astronomical Union
doi:10.1017/S1743921307001652

Star formation on galactic scales

Robert C. Kennicutt, Jr.[1,2]

[1]Institute of Astronomy, University of Cambridge, Madingley Road, Cambridge CB3 0HA, UK
email: robk@ast.cam.ac.uk

[2]Steward Observatory, University of Arizona, Tucson, AZ 85721, USA

Abstract. New multi-wavelength data on nearby galaxies are providing a much more accurate and complete observational picture of star formation on galactic scales. Here I briefly report on recent results from the Spitzer Infrared Nearby Galaxies Survey (SINGS). These provide new constraints on the frequency and lifetime of deeply obscured star-forming regions in galaxies, the measurement of dust-corrected star formation rates in galaxies, and the form of the spatially-resolved Schmidt law.

Keywords. galaxies: ISM, infrared: galaxies, stars: formation

1. Introduction

One of the least understood facets of the broader problem of star formation has been the triggering and regulation of star formation on galactic scales. The well-defined patterns in star formation within galaxies and along the Hubble sequence provide clear qualitative evidence for the importance of large scale drivers (e.g., Kennicutt 1998a). However our quantitative understanding of these processes has been hampered by the complexity of the ISM on these scales, and until recently by the relative dearth of accurate observations of star formation and the ISM in other galaxies. Now thanks to a flood of new observations from the Galaxy Evolution Explorer (GALEX), the Spitzer Space Telescope, and groundbased telescopes we are constructing accurate maps of star formation in galaxies for the first time. These are enabling statistical studies of star formation rates (SFRs) for well-defined samples of hundreds to thousands of galaxies, along with in-depth spatially-resolved studies of the nearest galaxies, often incorporating high-resolution maps of the atomic and molecular gas.

Space does not allow for a complete discussion of even the most recent work, so this paper will highlight only a few results with emphasis on one project, the Spitzer Infrared Nearby Galaxies Survey (SINGS). SINGS is one of the six original Spitzer Legacy surveys. We have used the Spitzer telescope to image 75 nearby ($d \leqslant 30$ Mpc) galaxies at 3.5 - 160 μm, and obtain spectra maps of the galaxies including their centers and 80 extranuclear star-forming regions. These are supported by ancillary observations ranging from the ultraviolet (deep GALEX imaging) through the visible and near-infrared, to the radio continuum, HI, and CO. Although SINGS was primarily designed to provide a general-purpose multi-wavelength archival resource for studying the structure and evolution of these galaxies, its core science program has been aimed at studying the star formation and ISM properties of the Hubble sequence. A description of the survey and its science goals can be found in Kennicutt *et al.* (2003). Here I highlight a few results that are especially pertinent to the other results presented at this conference, including the statistics of highly-obscured star forming regions and the star formation (Schmidt) law in galaxies.

2. Infrared probes of star formation in galaxies

Dust extinction has long been one of the limiting factors in measurements of SFRs of galaxies, locally and at high redshift. The shape of the cosmic background shows us that approximately half of the ultraviolet and visible starlight in the universe is absorbed and re-emitted in the infrared, so any census of star formation in either the UV/visible or the infrared is bound to be missing half of the star formation on average, with enormous variations from object to object.

The Spitzer telescope has enabled an important breakthrough in this problem, by enabling us to map galaxies in the thermal infrared with an angular resolution that is of the same order as imaging in the ultraviolet and visible. For nearby galaxies ($d < 30$ Mpc) this corresponds roughly to the linear scales over which dust reprocesses the starlight and the scale on which molecular cloud formation is triggered. As a result we have devoted considerable effort in SINGS to using the combination of infrared, ultraviolet, and Hα measurements to derive robust extinction-corrected maps of the star formation.

Figure 1 illustrates the primary diagnostics that are available to us, the photospheric emission of stars in the near-ultraviolet (150 and 250 nm) from the GALEX telescope, the nebular-processed far-ultraviolet emission in Hα, and two tracers of the dust emission at 8 μm and 24 μm. The 8 μm emission is dominated by molecular "PAH" band features, and includes discrete components from the neutral envelopes around HII regions and other star-forming regions, along with a more diffuse component (evident as the arms and filaments in M81) that represents dust in the general interstellar medium being excited by the interstellar radiation field. In contrast the 24 μm emission is dominated by silicate grains, and the positions of the infrared peaks are strongly correlated with optical HII regions, both in position and in flux. There is also a more diffuse, cooler component that dominates at longer wavelengths, the "infrared cirrus" that can now be resolved from the discrete sources directly with Spitzer.

Calzetti *et al.* (2005) and Pérez-González *et al.* (2006) have studied the star formation regions in M51 and M81 in depth, and were able to show that the 24 μm luminosities of the HII regions are strongly correlated with the dust-obscured ionizing fluxes. This encouraged us to calibrate a composite SFR measure based on a weighted sum of the Hα and 24 μm luminosities to provide an extinction-corrected estimate of the Hα and ionizing fluxes. The same approach has already been applied to the combination of UV and total infared fluxes of star-forming regions in the "flux ratio method" of Gordon *et al.* (2000), and we simply have extended the approach to Hα. In the same way an approximate Hα extinction can be estimated from the ratio of 24 μm to Hα fluxes. The promise of this technique is illustrated in Figure 2. In the left panel we have compared extinction-corrected Hα luminosities for 240 HII regions in a sample of SINGS galaxies using this method to independently corrected luminosities based on Pα/Hα ratios. The method is robust for statistical samples of objects, but there is a dispersion of approximately ± 0.2 dex for individual regions, which probably reflects variations in extinction geometry and dust heating along with age and upper mass function variations in the clusters. The right panel shows a similar comparison, but in this case applied to entire galaxies. There we have combined IRAS and Hα fluxes to derive an extinction-corrected SFR, and compared it to SFRs corrected using integrated spectroscopic measurements of the Hα and Hβ lines, from Moustakas & Kennicutt (2006). Again the correlations are not perfect, but the method appears to offer a reasonably accurate means of correcting for extinction for galaxies with attenuations ranging from zero to a few magnitudes.

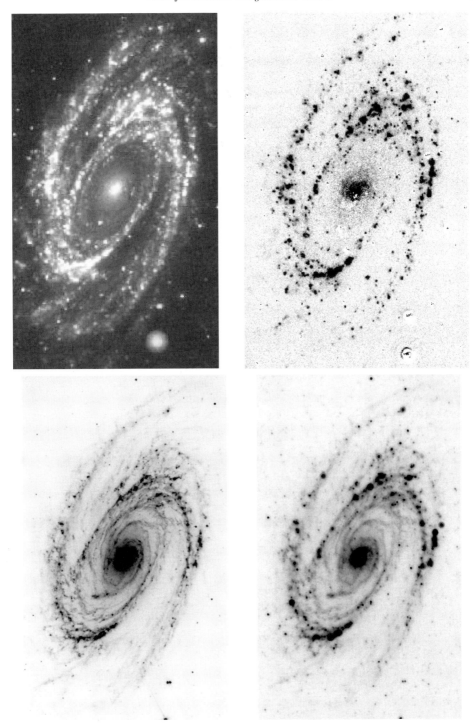

Figure 1. The nearby Sab spiral M81 (NGC 3031) observed at four wavelengths that provide complementary tracers of massive star formation and the associated ISM. Top left: Utraviolet (1350 − 2800 A), imaged with the GALEX satellite. Top right: M81 in Hα, observed with the Burrell Schmidt telescope, courtesy R. Walterbos. Lower Left: 8 μm with the IRAC camera on Spitzer. Lower right: 24 μm, as imaged with the MIPS instrument on Spitzer. Note the dramatic change in the structure and the clumpiness of the infrared emission between 8 μm (PAH dominated) and 24 μm (silicate dust dominated), and the presence of diffuse "cirrus" emission at both infrared wavelengths.

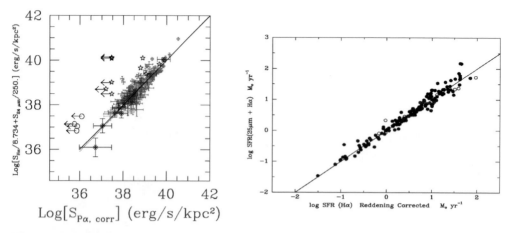

Figure 2. *Left:* Comparison of extinction-corrected Hα luminosities of HII regions based on a combination of 24 μm and Hα fluxes with independently corrected fluxes based on Pα and Hα data. *Right:* A similar comparison for entire galaxies, compared this time to independently corrected SFRs from integrated optical reddenings.

3. Dust-obscured star-forming regions

One of the surprising early results in the SINGS imaging was the rarity of bright infrared sources without Hα and/or ultraviolet counterparts. This can be seen clearly in Figure 1 by comparing the 24 μm and Hα panels in particular; there is nearly a one-to-one correspondence between regions. This contrasts sharply with the results from infrared surveys of star-forming regions on smaller spatial and luminosity scales, where deeply embedded and optically invisible sources represent a significant fraction of the population.

These results have been quantified in a study by Prescott *et al.* (2006, in preparation). Those authors have obtained infrared and Hα photometry for the well-resolved source populations in more than 50 of the SINGS galaxies, and investigated the statistics of the optical to infrared colors of these objects. Using the methods described earlier to estimate extinctions from the ratio of 24 μm to Hα fluxes, Prescott *et al.* find that more than 96% of the infrared sources have Hα counterparts with extinctions (averaged across a beam diameter of 500 pc) of $A_V < 4$ mag. This contrasts sharply with the deeply embedded and even more luminous regions that are often found in the circumnuclear regions of infrared-luminous starburst galaxies. Indeed in our sample the small fraction of highly obscured objects tends to be concentrated in the central regions of the SINGS galaxies, though examples at large radii are not infrequent.

Of course this in no way implies that the familiar population of compact and ultra-compact HII regions with high extinctions are absent in the SINGS galaxies! The average extinction of the star-forming regions is roughly 1 magnitude, so one could contrive a model in which 60% of the young stars in our 500 pc beams were completely invisible, and the other 40% were dust-free, and it would fit our typical observation. I suspect that after more complete study of individual regions in the most well-resolved galaxies that the contribution of deeply embedded stars will turn out to be much lower, but certainly it does not have to be as low as 4%. However our results do demonstrate that very luminous embedded clusters are rare— such objects must be able to clear channels for optical radiation to escape on relatively rapid timescales (order $10^5 - 10^6$ yr) compared to the typical ionizing lifetimes of such cluster. The relative rarity of infrared sources

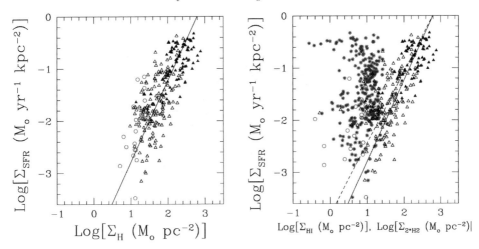

Figure 3. Schmidt law for 257 star-forming regions in M51. *Left:* The SFR surface density is plotted as a function of total gas surface density. *Right:* The SFR density is plotted as functions of the H_2 surface density alone (points) and HI surface density alone (stars).

without HII regions also places interesting constraints on the lifetimes of the natal dust clouds. If the parent clouds remained close to the OB stars for much longer than the lifetimes of HII regions (order ~ 5 Myr), we would expect to see many bright dust regions with bright near-ultraviolet counterparts but little or no $H\alpha$ emission. The rarity of such objects suggests that the dust dissipates on roughly the same timescale as the ionizing lifetime, consistent with results from kinematic measurements of the HI shells about OB associations presented earlier in this symposium.

4. The spatially-resolved star formation law

I conclude with an illustration of what can be learned by applying these multi-wavelength techniques. The ability to measure extinction-corrected SFRs across a galaxy makes it possible to study the behavior of the SFR density vs gas density law (Schmidt law) on a point by point basis in galaxies. It is now known that on global scales the *disk-averaged* SFR and gas densities of disks are well fitted by a Schmidt power law relation with slope $N \simeq 1.4$ (Kennicutt 1998b). However the global relation offers few clues to the physical underpinnings of the star formation law, and other prescriptions, for example a parametrization of the SFR density in terms of the ratio of gas density to local dynamical time fits the disk-averaged properties of galaxies nearly as well (Kennicutt 1998b).

As a pilot study of this application we have combined our SINGS infrared and $H\alpha$ observations of M51 with CO data from the BIMA SONGS survey (Helfer *et al.* 2003) and the THINGS HI survey (Brinks *et al.*, this volume), to correlate the SFR density with the local atomic, molecular, and total cold gas surface densities, on scales ranging from 520 pc (the effective resolution limit of our data) to about 2 kpc. Some of the key results are shown in Figure 3, which shows the SFR vs gas density relations for 257 star-forming regions. The left panel shows the correlation with total gas density, and reveals a Schmidt law with slope $N = 1.55 \pm 0.1$, where our uncertainty includes a conservative allowance for systematic effects. This is similar to the global law observed in disks ($N \sim 1.4$), and certainly rules out a linear dependence, as one might expect if the ratio of stars to gas (SFE) in the regions was constant independent of mass scale. In the right panel we plot the SFR densities against the molecular and atomic gas surface densities separately. The disk of M51 is overwhelmingly molecular, especially in the star-forming clouds, so the correlation of SFR density with H_2 surface density is nearly the same as in the left panel (confirming work on M33 by Heyer *et al.* 2004). Interestingly there is no evidence of a correlation with HI surface density at all; the most prominent feature in the panel is a rather sharp upper cutoff to the observed HI column densities, which may signal the optical depth limit for the efficient conversion to molecular gas. We intend to extend this work to the other galaxies in the SINGS sample with high-quality CO and HI data.

Acknowledgements

Support for the SINGS project, part of the Spitzer Space Telescope Legacy Science Program, is provided by NASA through Contract Number 1224769 issued by JPL, Caltech, under NASA contract 1407. This work was also supported in part by NSF grant AST-0307386.

References

Calzetti, D., *et al.* 2005, *ApJ* 633, 871
Dale, D.A., *et al.* 2005, *ApJ* 633, 857
Gordon, K.D., Clayton, G.C., Witt, A.N. & Misselt, K.A. 2000, *ApJ* 602, 723
Helfer, T.T., *et al.* 2003, *ApJS* 145, 259
Heyer, M.H., Corbelli, E., Schneider, S.E. & Young, J.S. 2004, *ApJ* 533, 236
Kennicutt, R.C. 2003, *PASP* 115, 98
Kennicutt, R.C. 1998a, *ARA&A* 36, 189
Kennicutt, R.C. 1998b, *ApJ* 498, 541
Moustakas, J. & Kennicutt, R.C. 2006, *ApJS* 164, 81
Pérez-Gonzaíez, *et al.* 2006, *ApJ* 648, 987

Discussion

KRUMHOLZ: In evaluating the Σ_g/τ_{dyn} form of the Schmidt law, you are on spiral arms where the epicyclic frequency is doing something interesting. How did you deal with that?

KENNICUTT: We took a simpler approach of setting τ_{dyn} to the local orbit time, to be consistent with Kennicutt (1998b). As you suggest one could examine other formulations that would scale more closely with the local dynamical timescales, but one should bear in mind that these dynamical prescriptions for the star formation law were never intended to be applied on the local scale. There were intended as parametrizations for analytical or numerical models of galaxy formation and evolution. But given the consistency of the Schmidt law on these smaller scales it certainly makes sense to look more closely at the dynamical scaling laws.

BLITZ: You showed a very good improvement in the scatter for Hα-based star formation rates with your extinction corrections applied, but this was for global emission. Have you looked at this on smaller scales, or on annular averages?

KENNICUTT: Yes, to some extent. Within the SINGS project Daniela Calzetti and collaborators have analyzed about 220 individual star-forming regions in a set of SINGS galaxies for which we have infrared, Hα, and Pα measurements, and those results are presented in Figure 2 here. The scatter is considerably larger than for entire galaxies, and I attribute that to object to object variations in dust geometry, cluster ages, and stochasticity in the number of O stars in each object, etc. That sets a physical limit to the precision of these methods for individual HII regions and clusters.

Triggered Star Formation in a Turbulent ISM
Proceedings IAU Symposium No. 237, 2006
B. G. Elmegreen & J. Palouš, eds.

© 2007 International Astronomical Union
doi:10.1017/S1743921307001664

Star formation in mergers and interacting galaxies: gathering the fuel

Curtis Struck[1]

[1]Department of Physics and Astronomy, Iowa State University, Ames, IA 50011, USA
email: curt@iastate.edu

Abstract. Selected results from recent studies of star formation in galaxies at different stages of interaction are reviewed. Recent results from the Spitzer Space Telescope are highlighted. Ideas on how large-scale driving of star formation in interacting galaxies might mesh with our understanding of star formation in isolated galaxies and small scale mechanisms within galaxies are considered. In particular, there is evidence that on small scales star formation is determined by the same thermal and turbulent processes in cool compressed clouds as in isolated galaxies. If so, this affirms the notion that the primary role of large-scale dynamics is to gather and compress the gas fuel. In gas-rich interactions this is generally done with increasing efficiency through the merger process.

Keywords. stars: formation, galaxies: interacting, galaxies: starburst, galaxies: individual (NGC 2207, Arp 82)

1. Introduction

Star formation (SF) in interacting galaxies (IGs) occurs in a vast range of environments, from the dense, turbulent nuclei in major merger remnants to relatively diffuse regions in (literally) far flung tidal tails. In a short paper, it is impossible to review the detailed processes that orchestrate star formation in these different circumstances. Rather, it seems better to try to give an overview of the "big picture," while keeping in mind the key question – what if anything is different about SF in IGs relative to isolated galaxies?

Even with this restriction it is still helpful to break this large topic up into smaller subtopics, and this is easy to do in two parameter dimensions. The first concerns the distribution of the star formation, i.e., it is natural to consider compact SF in galaxy cores separately from extended SF, which occurs primarily in waves. The second classification dimension is the relative time or merger stage of the interaction, which is possible because most significant interactions lead to merger after one or two close encounters. Here it is sufficient to distinguish: 1) early stage encounters, from the onset of the interaction up to the second close approach, 2) intermediate stage encounters, up to the time when the bodies of the two galaxies no longer separate, and 3) late encounter stage, which includes the final merger and continuing relaxation.

2. Compact induced bursts

Compact SF, including nuclear starbursts, are the source of much of the emission in most LIRGs and ULIRGs, which are the most spectacular examples of interaction induced SF. There is, of course, a huge literature on these objects, and on the topic of ULIRGs I would refer the reader to the recent review of Lonsdale, *et al.* (2006). Here I would merely remind the reader that there is now a great deal of evidence confirming that the ULIRG phenomenon generally occurs at the intermediate to late stages of a major merger between

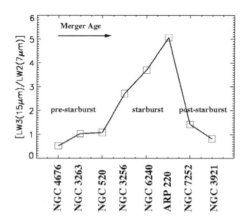

Figure 1. An ISO indicator of star formation rates along the Toomre sequence from
Charmandaris, *et al.* (2001).

two galaxies which both contain a gas-rich disks (very "wet" mergers in the current
jargon). More specifically, the phenomenon is usually the result of a super-starburst
(with perhaps an AGN contribution), triggered by the direct interaction of the two disks
in the core of the developing merger remnant (see e.g., the recent models of Bekki, *et al.*
2006, Hopkins, *et al.* 2006). We can expect that this environment is characterized by
strong shocks and extreme turbulence as a result of high velocity intersecting flows.

This extreme environment has been historically very difficult to observe because it
is compact and deeply buried in dust. The latter fact means that we must observe at
long wavelengths (mid-IR to radio), and until the launching of the Spitzer Space Telescope there have been few long-wavelength instruments with the necessary sensitivity
and resolution. Similarly, with a huge range of spatial scales involved in the dynamics
and a plethora of heating, cooling, and feedback processes, it is very difficult to model
super-starbursts, except in a partial or schematic way. Detailed modeling of individual
systems, like Arp 220, is not yet feasible.

At the same time, we have increasing numbers of observational clues. Among these
is the development of starburst activity with merger age along the Toomre sequence
demonstrated by Charmandaris, *et al.* (2001) using ISO data (see Figure 1). This result
highlights the relation between the large scale dynamics and the SF activity. On the
other hand, the Kennicutt (1998) result that the SFR in ULIRGs follows the Schmidt
Law dependence on gas surface density suggests that the fundamental physics of these
bursts does not differ greatly from SF in isolated late-type disks despite their violent
hydrodynamics. In the coming years the Herschel mission should help us understand this
physics.

The mysteries of LIRGs are no fewer, though their infrared luminosities are less. For
one thing, members of the LIRG class seem to be very heterogeneous. Recent surveys
suggest that a significant fraction are not involved in interactions, though probably the
majority are. Among the latter, we can understand some of the variety just by assuming
that ULIRGs generally pass through a LIRG stage in building up to their full strength,
and another in decaying after reaching their peak. The duration of the former could
be understood as the result of bulk compression of the galaxies from the time that the
central parts of their halos overlap, or the beginnings of disk-disk interaction. The latter

phase may be shortened by prompt feedback effects, but much work is needed to confirm such speculations.

In addition, there are also examples of early stage interactions with core starbursts that reach LIRG levels of intensity. NGC 7469 provides a ring galaxy example, though an active nucleus also contributes in that case (see Weedman, *et al.* 2005).

Finally we note that Hinz & Rieke (2006) have recently pointed out that some LIRGs seem to have SF efficiencies comparable to ULIRGs, but they occur in smaller (less massive) galaxies. Thus, some nearby LIRGs may simply be down-sized ULIRGs.

3. Extended star formation in early stage interactions

At the opposite end of the spectrum of SF activity is the question of how interaction induced SF begins, or alternately, how much is SF enhanced at the early stages of interaction? There has been a good bit of work on these questions recently, and some debate on the answers. I have reviewed this discussion recently in more detail than space allows here, so I will merely note a few highlights, and refer the reader to Struck (2006) for more. Several recent studies of SF in galaxy groups, based on the Sloan or 2dF surveys, have many thousands of objects, and thus, their conclusions are statistically very significant. They reveal modest SF enhancements, especially due to core starbursts, which increase with galaxy-galaxy proximity, and possibly inversely with relative velocity (e.g., Nikolic, *et al.* 2004). The bad news is that these studies may be answering somewhat different questions than those posed above.

For example, many of the galaxies in these groups are not (obviously) interacting, so the proximity effect may be diluted by (possibly nonrandom) projection effects. More seriously, they include interactions at many different stages. Indications of simultaneous proximity and velocity effects suggest systems consummating their merger at intermediate to late stages.

The GO Cycle 1 Spitzer Space Telescope project "Spirals, Bridges and Tails" (SB&T), that my collaborators and I have been working on, takes a different approach to these questions. In this project, we have been attempting to use Spitzer imagery and spectra to study the sites and modes of induced SF in an Arp Atlas sample of quite strongly interacting, but pre-merger galaxies. We have a sample of about three dozen systems, and make comparisons to a control sample of isolated disk galaxies drawn from the SINGS legacy survey. The object identifications and images can be viewed at the website - http://www.etsu.edu/physics/bsmith/research/sbt.html. Complementary GALEX and ground-based Hα observations of a similarly sized and overlapping sample are underway.

In this study we find clear and statistically significant differences in the 3.6μm $- 24\mu$m and 8μm $- 24\mu$m colors of the SB&T systems and non-interacting spirals. The distribution of interacting components has a significant red tail in this color, which in these bands suggests more emission from dust heated by embedded young stellar populations. (For more details see Smith, *et al.* 2006.) This result is supported by the fact that the IRAS fluxes of the interacting sample are about a factor of 2 larger (per galaxy) than those of the comparison sample, in agreement with the results of Bushouse, *et al.* (1988) on a similar sample.

Many of the systems with the reddest [8] − [24] colors appear to be core starbursts in 24μm images, with disturbed disks or mass transfer bridges that may indicate gas transferred to core regions. Thus, the enhancements we see appear to confirm the importance of dynamical triggering from the time of the first close pass. We do not see any evidence of a proximity effect in this relatively small and specialized sample. The ongoing

SF found in tidal features is qualitatively similar to that found in disks, confirming the results of Schombert, *et al.* (1990). Less than 10% of the net SF is found in tidal features.

Every wave formed in a galaxy collision is a unique laboratory of SF processes, and much can be learned from detailed study of individual interacting systems. Given the limited space, I would like to consider one SB&T system in a bit more detail, the Arp 82 system. This system has been studied in detail by SB&T collaborator Mark Hancock (see Hancock, *et al.* 2006).

The primary galaxy in the Arp 82 system has a long tidal tail, a substantial bridge, and an 'ocular' waveform in its disk; all of the features are characteristic of an M51-type fly-by encounter (see Kaufman, *et al.* 1997). The companion galaxy is experiencing a starburst, while the primary and the tidal structures have many star-forming clumps. Considering these facts, it is somewhat surprising to find that the net Spitzer colors of the system are not unusual relative to SINGS spirals, and do not indicate an especially high level of recent star formation.

However, this system has other peculiarities – Kaufman, *et al.* (1997) found that the system contains almost as much HI gas as stellar mass, an unusually high fraction. Hancock, *et al.* (2006) fitted Starburst99 population models to GALEX and optical data, and found moderate extinctions and young ages for the clump sources. More surprisingly, evidence was found that the oldest stellar population has an age of about 2.0 Gyr in the diffuse emission. Although this result should be confirmed with near-infrared observations, and metallicity estimates should be obtained from spectral observations, it has several interesting ramifications. First of all, if the encounter has been of extended duration, most of the visible stars could have been formed in the interaction. A numerical hydrodynamical model presented in Hancock *et al.* suggests that the interaction could have been underway for a long time, with a first close approach about 2.0 Gyr ago. Secondly, the progenitors may have been low surface brightness galaxies with very few old stars. Given that the system is of intermediate mass, and may be forming most of its stars at about the present time, it seems to be a nearby example of "down-sizing" in galaxy formation.

4. Star formation processes in interacting galaxies: universal or exceptional?

Now let us return to the questions posed at the outset. Although not a firm conclusion, it is my impression from detailed modeling and analysis of specific interacting systems over the last decade, that generally – wherever there is compression of cool gas there is SF in IGs. Moreover, the similarity (in colors and rough measures of efficiency) of SF in tidal structures and isolated disks suggests that the physical processes of SF are not very different, in these potentially very different environments. For example, they may have similar, mildly nonlinear, dependences on local mean gas density or pressure, though this point needs confirmation. Conversely, there is very little evidence from detailed studies of IGs for the existence of novel threshold behaviors or nonlinear triggering, except perhaps in the case of flows converging so hypersonically that the underlying turbulent cloud structure is completely destroyed. While we do not observe the formation of (tidal) dwarfs in isolated galaxies, their formation is plausibly explained as an extension of universal processes, e.g., the buildup of gas concentrations, gravitational instability, and more favorable environment of the persistence of marginally bound concentrations than in most galaxy disks (see Duc, this proceedings).

At this symposium Kennicutt has presented evidence from the SINGS survey that the Kennicutt-Schmidt Law, relating SFR to a power n of the gas surface density (generally

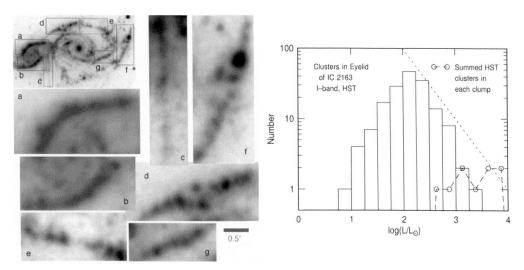

Figure 2. Left panel – Spitzer 8μm logarithmic image (top left), with outlined strings of clumps enlarged below. The scale of 0.5 applies to the enlarged figures and corresponds to 5 kpc. Right panel – luminosity distribution function in the I band of 165 star clusters observed by HST in IC 2163. The dashed line with circles is the luminosity function at I band for the sum of the clusters in each IRAC-defined (8.0μm) star complex. Images from Elmegreen, *et al.* (2006).

$n \simeq 1.4$), extends down to kpc scales in galaxies. Gao argued that when only the densest (HCN) gas is considered, the relation is linear ($n \simeq 1.0$). If the latter conclusion is correct then the extra nonlinearity in the Kennicutt-Schmidt Law may result from a pressure or density dependence in the efficiency of dense core formation in dense clouds (see Blitz & Rosolowsky 2004). If this is the case on small scales in dense clouds, and a nearly scale-free turbulent cascade determines the cloud dynamics on 'intermediate' scales, *then the role of large-scale dynamical processes may be primarily to assemble gas concentrations and set the value of the ambient pressure.*

These tasks can be carried out by several large-scale mechanisms, including shock compression and infall and agglomeration due to gravitational instability. The recent Spitzer study of the NGC 2207/IC 2163 system by Elmegreen, *et al.* (2006) provides evidence of an agglomeration scale much larger than the scale of individual star clusters in the density waves of these galaxies. Figure 2, taken from that paper, illustrates these points. The distribution of star cluster luminosities (from HST I-band observations) is a fairly typical power-law at the more complete high luminosity (mass) end. As in other galaxies, this is probably the result of turbulent dynamics within large clouds and cloud assemblies. Clumps seen in the Spitzer 8μm image have a narrow distribution at high (summed I band) luminosity. These clumps tend to be quite uniform in each wave and are probably the result of large-scale gravitational instabilities and agglomeration. Similar phenomenology can be seen in M51 (see Bastian, *et al.* 2005).

While large-scale shocks can gather and compress the interstellar gas, they may also introduce further nonlinearities. For example, weak shocks may primarily affect the en-velopes of molecular clouds, since they don't have the strength to overcome cloud internal pressures. The result would be increased molecule formation, cooling, and increased star formation efficiency per unit mass of gas. Stronger shocks could compress molecular cloud cores, squeezing subcritical clumps and filaments into star formation, and probably increasing SF efficiency much more. Very strong shocks could shred and strip clouds. The densest clumps within molecular clouds would be violently compressed and heated, and so, would expand strongly afterwards. SF might well be suppressed in such cases. This limiting case might be realized collisions between disks in high speed galaxy encounters. The beautiful intergalactic shock studied recently by Appleton, *et al.* (2006) in Stephan's Quintet might be an example. Thus, very strong shocks probably define a limit to the generalization that shock compression stimulates star formation.

In the cases where shocks or gravitational agglomeration have stimulated SF, feedback effects may ultimately break up the party. There is certainly evidence for this in nuclear starbursts and in dwarf irregular galaxies. In spiral density waves the rarefaction induced by large-scale divergent flow is also important.

We can summarize the large-scale processes for gathering (or dispersing) fuel for SF with the acronym SCAF, for shock compression, agglomeration and feedback. Our understanding of how these processes accomplish large-scale stimulation in detail is still incomplete. However, almost all aspects of that understanding can be tested by observation and models in the next few years. Studies of the effects of large-scale shocks, like the modeling work described by Vazquez-Semadini at this symposium, and related observations will be particularly interesting.

Acknowledgements

I gratefully acknowledge numerous helpful and educational "interactions" with members of the SB&T and Ocular collaborations. I am also grateful for support from NASA Spitzer GO grant 1263961.

References

Appleton, P. N., *et al.* 2006, *ApJ* 639, L51
Bastian, N., *et al.* 2005, *A&A* 443, 79
Bekki, K., Shioya, Y. & Whiting, M. 2006, *MNRAS* 371, 805
Blitz, L. & Rosolowsky, E. 2004, *ApJ* 612, L29
Bushouse, H. A., Werner, M. W. & Lamb, S. A. 1988, *ApJ* 355, 74
Charmandaris, V., Laurent, O., Mirabel, I. F. & Gallais, P. 2001, *Ap Sp Sci Suppl* 277, 55
Elmegreen, D. M., *et al.* 2006, *ApJ* 642, 158
Hancock, M., *et al.* 2006, *AJ* submitted
Hinz, J. L. & Rieke 2006, *ApJ* 642, 872
Hopkins, P. F., *et al.* 2006, *ApJS* 163, 1
Kaufman, *et al.* 1997, *AJ* 114, 2323
Kennicutt, R. C., Jr. 1998, *ApJ* 498, 541
Lonsdale, C. J., Farrah, D. & Smith, H. E. 2006, in: J.W. Mason (ed.), *Astrophysics Update 2* (Chichester: Springer, Praxis), p. 285
Nikolic, B., Cullen, H. & Alexander, P. 2004, *MNRAS* 355, 874
Schombert, J. M., Wallin, J. F. & Struck-Marcell, C. 1990, *AJ* 114, 2323
Smith, B. J., *et al.* 2006, *AJ* submitted
Struck, C. 2006, in: J.W. Mason (eds.), *Astrophysics Update 2* (Chichester: Springer, Praxis), p. 115
Weedman, D. W., *et al.* 2005, *ApJ* 633, 706

Triggered Star Formation in a Turbulent ISM
Proceedings IAU Symposium No. 237, 2006
B. G. Elmegreen & J. Palouš, eds.

© 2007 International Astronomical Union
doi:10.1017/S1743921307001676

Tidal dwarf galaxies as laboratories of star formation and cosmology

Pierre-Alain Duc, Frédéric Bournaud and Médéric Boquien[1]

[1] AIM - Unité Mixte de Recherche CEA - CNRS - Université Paris VII - UMR n° 7158 Service
d'Astrophysique, CEA–Saclay, 91191 Gif-sur-Yvette, France
email: paduc@cea.fr

Abstract. Star formation may take place in a variety of locations in interacting systems: in the
dense core of mergers, in the shock regions at the interface of the colliding galaxies and even
within the tidal debris expelled into the intergalactic medium. Along tidal tails, objects may
be formed with masses ranging from those of super-star clusters to dwarf galaxies: the so-called
Tidal Dwarf Galaxies (TDGs). Based on a set of multi-wavelength observations and extensive
numerical simulations, we show how TDGs may simultaneously be used as laboratories to study
the process of star-formation (SFE, IMF) in a specific environment and as probes of various
cosmological properties, such as the distribution of dark matter and satellites around galaxies.

Keywords. galaxies: dwarf, galaxies: starburst,galaxies: formation,galaxies: interactions

---

1. The various observed types of star–forming tidal objects

Star-formation in colliding galaxies has mostly been studied in their inner most regions.
There the accumulation of gas, previously funneled by various dynamical processes, and
its further collapse may lead to intense circum-nuclear starbursts. A mode of spatially
extended, probably shock–induced, star-formation has also been observed for a long time
at the interface of the interacting galaxies, but was only recently modeled (Barnes 2004).
Further out the detection of HII regions along tidal tails proved that stars may form
well outside the disks of the parent galaxies. The puzzling discovery of even more distant
star-forming regions, located in the intergalactic medium, at more than 100 kpc from
any massive galaxy, has raised the question of their origin. Given these large distances
and related timescales, the young intergalactic stars were born in situ. The relatively
high metallicity measured in the associated ionized gas (typically half solar to solar)
indicate that, without any doubt, the gas fueling these intergalactic SF episodes had
previously been pre-enriched in the disk of parent galaxies and had later been expelled
by a dynamical process: tidal forces or, for systems belonging to dense groups or clusters,
by ram-pressure.

This paper focusses here on the formation of stars and stellar systems in the outermost regions of colliding galaxies. Before presenting how star–formation may take place
far from the parent galaxies, it is worthwhile to briefly enumerate the various types of
external star–forming objects so–far described in the literature:

- *Intergalactic emission–line regions*, with optical spectra typical of star-forming HII
regions. They usually consist of small, often compact – they are also refereed as *EL-
Dots* – blue condensations made of a few OB stars. Their Star Formation Rates are
below 0.01 $M_\odot$ yr^{-1} (Gerhard *et al.* 2002, Ryan-Weber *et al.* 2004, Mendes de Oliveira
et al. 2004, Cortese *et al.* 2006). They will probably not form gravitationally bound
objects but may contribute to the population of Intergalactic Stars, recently found in
nearby clusters of galaxies.

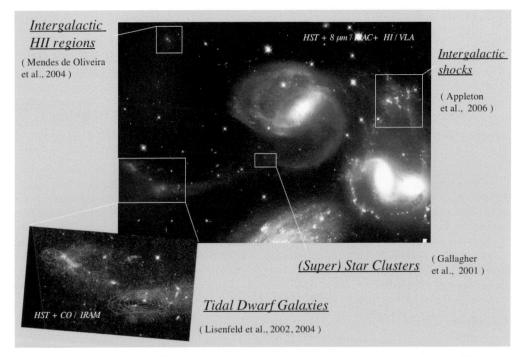

Intergalactic HII regions

(Mendes de Oliveira et al., 2004)

HST + 8 μm / IRAC+ HI / VLA

Intergalactic shocks

(Appleton et al., 2006)

(Super) Star Clusters

(Gallagher et al., 2001)

Tidal Dwarf Galaxies

HST + CO / IRAM

(Lisenfeld et al., 2002, 2004)

Figure 1. Various modes of Intergalactic Star Formation in the Stephan's Quintet. The main image is a montage of the Spitzer/IRAC 8 μm emission (in red), tracing the star–forming regions, the VLA HI emission (in blue), showing the gas reservoirs, superimposed on an optical HST image, showing the stellar populations and dust lanes

- *Young (Super) Star clusters*, born in giant HII complexes and located along tidal tails (e.g., Weilbacher *et al.* 2003, de Grijs *et al.* 2003, López-Sánchez *et al.* 2004). Having typical typical masses of 10^{6-7} $M_\odot$, the latter may evolve into *Globular Clusters* (e.g. Schweizer *et al.* 1996).

- *Tidal Dwarf Galaxies*, i.e. star–forming, gravitationally bound, objects made of tidal material, with apparent masses and sizes of dwarf galaxies. The most massive of them exceed 10^9 $M_\odot$ and are usually found close to the end of tidal tails. They contain large quantities of gas in atomic, molecular and ionized form (Braine *et al.* 2001).

The variety of external star–forming regions/objects is perfectly illustrated by the Stephan's Quintet (HCG92), a well studied compact group of galaxies (see Figure 1). Instances of intergalactic HII regions (Mendes de Oliveira *et al.* 2004), young Star Clusters along tidal tails (Gallagher *et al.* 2001), Tidal Dwarf Galaxies (Lisenfeld *et al.* 2004) and even, probably shock induced Star Formation (Appleton *et al.* 2006), were already reported in the intragroup medium of HCG92.

2. The formation and evolution of sub-structures along tidal tails, according to simulations

Depending on the characteristics of the star–forming regions, various stellar objects will or will not be formed in tidal debris. Numerical simulations may help in understanding the origin and the evolution of sub-structures along tidal tails.

Soon after the study by Mirabel *et al.* (1992) of a TDG candidate in the Antennae system, Barnes & Hernquist (1992) and Elmegreen *et al.* (1993) published numerical models exhibiting bound condensations along tidal tails with typical masses of 10^{6-8} $M_\odot$.

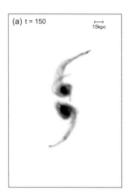 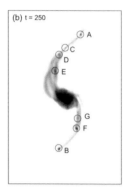 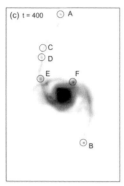 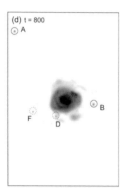

Figure 2. Numerical simulation of colliding galaxies. On each snap shot, the surviving tidal objects are identified (adapted from Bournaud & Duc, 2006).

While in the Barnes & Hernquist (1992) simulations, the condensations formed from gravitational instabilities in the stellar component, those of Elmegreen *et al.* (1993) were produced from gas clouds, the velocity dispersion of which had increased due to the collision. The direct formation of bound stellar objects in models with limited resolution has been questioned by Wetzstein *et al.* (2006). On the other hand, they claimed that the gaseous condensations themselves were not numerical artifacts.

For practical reasons, most simulations of galaxy–galaxy collisions assume that the dark matter halo around the parent galaxies is truncated, since, at first approximation, the latter does not play a major role in the shaping of tidal tails. However, Bournaud *et al.* (2003) found that the internal structure of tidal tails does actually depend on the size of the DM halo. When the latter exceeds ten times the size of the optical disk, prominent concentrations of mostly gaseous matter, with masses exceeding 10^9 M$_\odot$, may form near the end of the tails. Such accumulations of gas resemble those present in the HI maps of several interacting systems. Duc *et al.* (2004) further explained that the shape of tidal forces, within the potential well of an extended DM halo, were such that gas clouds, originally located in the outskirts of the parent galaxies, can be pulled out far away from them without being diluted. In the truncated DM halo case, matter is stretched along the tails, preventing the formation of massive condensations. One should note that the presence of extended DM haloes is a prediction of cosmological simulations which has had sofar only a few rather indirect observational confirmations. However, having an extended DM halo is a necessary but not sufficient condition to form massive tidal objects. As shown by Bournaud & Duc (2006), favorable geometrical parameters for the collision and the initial presence of extended gaseous disks are also key ingredients. The mass ratio between the parent galaxies, which should not be too different, does also matter.

Therefore, depending on their initial conditions, simulations predict the formation along tidal tails of a variety of objects with different origins: local instabilities along tidal tails, forming low or intermediate mass bound objects; a kinematic origin for the most massive ones. How these different types of objects evolve, how long they survive as independent bodies, are key questions that have sofar not had any observational answer, but may again be addressed with numerical simulations.

Based on the analysis of about 100 N–body simulations and 600 tidal objects (see one example on Figure 2), Bournaud & Duc (2006) found that, while most of the intermediate mass tidal objects along the tails vanished in a few 10^8 yrs, those initially located near the end of the tails, with typical masses of 10^9 yrs, could survive up to at least

2 Gyr. They orbit around their parent galaxies like satellite galaxies. The projected distribution of the long–lived tidal objects turns out to be remarkably similar to that of the SDSS satellites around their host (Yang *et al.* 2006). In particular they show a similar anisotropy; the latter has probably a cosmological origin, but at least part of it could be due to a contamination by a population of Tidal Dwarf Galaxies. The simulations by Bournaud & Duc (2006) did not have enough resolution to study the fate of the lowest–mass tidal objects. According to the ad–hoc simulations by Kroupa (1997), those suffer severe gravitational evaporation, loose a significant fraction of their mass, but some of them could still survive as low–mass dwarf spheroidals. Other survivors may turn out to be the progenitors of the Super Star Clusters or even the Young Globular Clusters which were discovered in interacting systems.

3. Physical properties of intergalactic star–forming regions

As shown before, Star Formation in interacting systems may occur in a large variety of environments from the densest ones to the most diffuse ones. It could be expected that the modes of SF and thus the Initial Mass Function (IMF) or the Star Formation Efficiency (SFE) would change from one type of regions to the other. Surprisingly, Gao & Solomon (2004) recently claimed that the SFE may not be so different in the dense core of Ultraluminous Infrared Galaxies and in the more quiescent regions of spiral arms. This result urges the comparison with the other extreme: the outermost regions of spirals and intergalactic star–forming regions.

We have studied the Star Formation processes in the latter environment combining three popular indicators: the ultraviolet emission obtained with Galex, which is sensitive to the SF episodes of time scales of about 100 Myr, the Hα emission, probing SF over time scales of less than 10 Myr and finally the mid–infrared emission obtained with Spitzer, which has a slightly higher timescale. Putting together the values of the Star Formation Rates estimated with each of these tracers, and comparing them with the amount of gas reservoirs, one may determine the dust obscuration and hence the total SFR and SFE as well as speculate on the starburst ages. The IMF and recent star formation history can further be reconstructed from the analysis of the full Spectral Energy Diagram.

The first systematic studies of colliding galaxies with Spitzer indicate that globally the level of external star–formation, in particular along tidal tails, is rather low, contributing for less than 10% to the total SFR (Struck *et al.* in this symposium). The objects we have selected, on the contrary, exhibit in their surroundings collisional debris that are particularly active. For instance, in the interacting system NGC 5291 (see figure 3 and Boquien *et al.*, in these proceedings), more than 80% of the current Star Formation occurs outside the main galaxies, in about 30 individual intergalactic HII regions aligned along a huge 150–kpc long HI ring–like structure. The two most luminous ones have properties similar to TDGs, as defined in the previous section.

4. Conclusions: TDGs as laboratories of star formation and cosmology

Intergalactic SF regions, in general, Tidal Dwarf Galaxies, in particular, appear as attractive laboratories to study the processes of star formation. Indeed, on one hand, they share with galactic SF regions the chemical characteristics – at first order, they have the same ISM, including the molecular gas found in quantity in tidal tails –; on the other hand, they are by nature detached and isolated and are thus much simple to study. Indeed, many of the observed intergalactic star forming regions were formed within

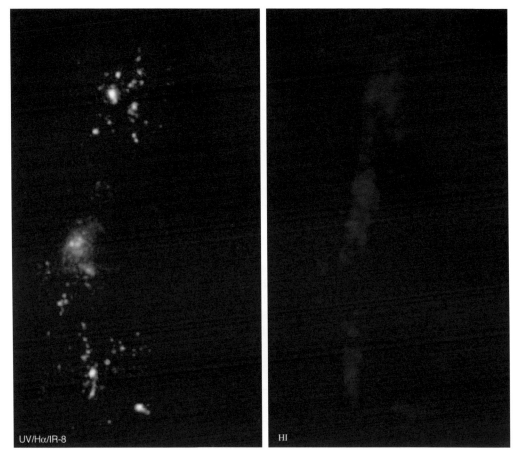

Figure 3. *Left*: composite image of the NGC 5291 system in pseudo-colours: near–ultraviolet Galex (blue), Hα (green) and Spitzer/IRAC 8 μm (red), i.e. a combination of three star—formation indicators. *Right*: VLA-B array HI map with the same scale. The 30 intergalactic star–forming regions along the HI–ring like structure are characterized by a UV excess, best explained by fading, quasi–instantaneous starbursts ignited less than 10 Myr ago (Boquien *et al.*, in prep.). No evidence for an old stellar population has yet been found.

"pure" expelled gas clouds, with no evidence for the presence of a pre-existing stellar component from the parent galaxies. Without such a contamination, the star–formation history (SFH) can be reconstructed with simple Single Star Population models. In galactic disks where several generations of stars coexist, deriving the SFH is much more complex and ambiguous. Beside, for those SF episodes triggered by a tidal interaction, numerical simulations may provide dynamical ages for the system and thus upper limits on the SF onset time, provided it occurred after the formation and development of tidal tails.

Moreover, Tidal Dwarf Galaxies may be used to constrain some parameters related to CDM cosmology, even-though these second–generation galaxies should not contribute much more than 10% to the overall dwarf galaxy population. First of all, TDGs with masses exceeding 10^9 $M_\odot$ may only be formed if their parent galaxies were surrounded by an extended halo of dark matter, or more precisely if they were within a potential well causing flat rotation curves. Cosmological simulations predict that the halos made of non-baryonic DM should be large; however their real sizes are difficult to measure directly. Second, because TDGs are made of material initially located in the disks of their parent galaxies, they should contain little quantities of the halo–type non baryonic dark–matter. Thus, measuring their dynamical mass, and comparing it with the luminous one, should reveal the presence or absence of a yet unknown, baryonic, component of dark matter located in spiral disks. This would be a direct test for the existence of large quantities of, for instance, very cold otherwise undetectable molecular clouds. Obviously deriving dynamical masses in objects as complex and young as tidal tails is still a challenge. This is nevertheless feasible in the most nearby systems (Braine *et al.*, 2006, Bournaud *et al.*, 2006 in prep.). Finally, numerical simulations predict that the long–lived TDGs have orbits resembling those of satellite galaxies around their host. The number and distribution of the latter are critical as they strongly constrain the cosmological, hierarchical, models. The origin for the apparent differences between the observed and predicted numbers of satellites has been actively debated for years. The existence of an additive population of dwarfs of tidal origin may even increase the discrepancies.

One should keep in mind that many of the above conclusions relied on numerical simulations that, even-though they were thoroughly checked (in particular using several codes and numerical resolutions), should be confirmed with observations. So far, only young, still forming, Tidal Dwarf Galaxies have been found and studied. They were classified as such thanks to the tidal tail linking them to their parent galaxies. Older detached TDGs are obviously much more difficult to pinpoint; unambiguous examples of such long-lived objects are still to be found. The absence of a dominating dark matter halo, an unusually high metallicity, are two hints for a tidal origin, which may be checked in nearby objects.

References

Appleton, P. N., *et al.* 2006, *ApJ* 639, L51

Barnes, J. E. & Hernquist, L. 1992, *Nature* 360, 715

Barnes, J. E. 2004, *MNRAS* 350, 798

Bournaud, F., Duc, P.-A. & Masset, F. 2003, *A&A* 411, L469

Bournaud, F. & Duc, P.-A. 2006, *A&A* 456, 481

Braine, J., Duc, P.-A., Lisenfeld, U., *et al.* 2001, *A&A* 378, 51

Cortese, L., Gavazzi, G., Boselli, A., *et al.* 2006, *A&A* 453, 847

de Grijs, R., Lee, J. T., Clemencia Mora Herrera, M., *et al.* 2003, *New Astronomy* 8, 155

Duc, P.-A., Bournaud, F. & Masset, F. 2004, *A&A* 427, 803

Elmegreen, B. G., Kaufman, M. & Thomasson, M. 1993, *ApJ* 412, 90

Gallagher, S. C., Charlton, J. C., Hunsberger, S. D., Zaritsky, D. & Whitmore, B. C. 2001, *AJ* 122, 163

Gao, Y. & Solomon, P. M. 2004, *ApJ* 606, 271

Gerhard, O., Arnaboldi, M., Freeman, K. C. & Okamura, S. 2002, *ApJ* 580, L121

Kroupa, P. 1997, *New Astronomy* 2, 139

Lisenfeld, U., Braine, J., Duc, P.-A., Brinks, E., Charmandaris, V. & Leon, S. 2004, *A&A* 426, 471

López-Sánchez, Á. R., Esteban, C. & Rodríguez, M. 2004, *ApJS* 153, 243

Mendes de Oliveira, C., Cypriano, E. S., Sodré, L. & Balkowski, C. 2004, *ApJ* 605, L17

Mirabel, I. F., Dottori, H. & Lutz, D. 1992, *A&A* 256, L19

Ryan-Weber, E. V., Meurer, G. R., Freeman, K. C., *et al.* 2004, *AJ* 127, 1431

Schweizer, F., Miller, B. W. & Whitmore, B. C. & Fall, S. M. 1996, *AJ* 112, 1839

Weilbacher, P. M., Duc, P.-A. & Fritze-v. Alvensleben, U. 2003, *A&A* 397, 545

Wetzstein, M., Naab, T. & Burkert, A. 2006, astro-ph/0510821

Yang, X., van den Bosch, F. C., Mo, H. J., *et al.* 2006, *MNRAS* 528

Discussion

TAN: Star formation synchronized to 10 Myr along a tidal tail that takes ~ 100 Myr to form is surprising. Do our (or other's) numerical simulations of star formation in tidal tails predict this?

DUC: The time scale to form the tidal tails is about 100-200 Myr. This provides an upper limit for the age of the starbursts ignited in the HI structure. There is besides no hint for an old stellar population along the tail, either produced locally or expelled from the parent galaxies. Within the intergalactic medium, they should be easily detectable.

WHITMORE: The finding that the clusters are < 10 Myr is almost certainly another example of the infant mortality of clusters I talked about this morning rather than the very improbably chance that you are suddenly having a simultaneous burst of young clusters all along the tail some hundred Myr after the initial tidal interaction that pulled the tails out. Whenever you have roughly constant star formation (e.g., spirals), you will always find that the average age is < 10 Myr, since 90% of clusters dissolve each decade of time.

DUC: The star-forming regions studied in this paper were not identified based on the presence of star clusters (possibly suffering a strong infant mortality), but were rather traced by younger HII regions. I nevertheless agree that it may be intriguing to have a quasi-simultaneous onset of star–formation along an HI structure as long as 150 kpc, like in NGC 5291. However, one should note that, for this specific object, the star-formation time scale is not dramatically different than the dynamical time scale for the formation of the HI collisional ring: up to 300 Myr, according to numerical simulations. Besides, in the case of a bulls-eye collision, the star–formation along the ring cannot start just after the impact, unlike in tidal tails. An additional period is required so that the gas can (re)condense. On the observational side, no hint for an old stellar population has yet been found. Given the nature of the object, mostly lying in the intergalactic medium, stars older than 100 Myr should be quite easy to detect.

MIRABEL: You find that dark matter must be large to reproduce the observations. What can you say about baryonic versus non-baryonic dark matter?

DUC: While simulations show that TDGs should contain only a tiny amount of dark matter particles from the cosmological halos, they may have acquired baryonic dark matter from the disk of their parents. Measuring the dynamical mass of TDGs and comparing it with the luminous mass may hence tell about the DM content in the disks of spirals. Our initial measures indicate no strong evidence for large quantities of baryonic DM there. However, observational errors are still large and this result needs to be confirmed.

OEY: I'd like to draw attention to the poster S237.681 by Werk *et al.*in which she has identified some intergalactic HII regions. Her work raises the question of whether the Schmidt law and Kennicutt threshold applies. I realize you don't have CO data for all of your objects, but wonder whether you have any impressions on this issue.

DUC: Unfortunately, we do not have enough objects for which we could map the CO and thus probe the Kennicutt law in that special environment.

STRUCK: Not all tails have dwarfs at the end. If your theory is right, maybe the range of tail types can tell us about the range of halo properties?

DUC: Indeed, this is a promising method. However, other parameters should be taken into account. Having an extended dark matter halo is required to form TDGs, but this is not sufficient. The initial size of the gaseous disk in the parent galaxies, the geometrical parameters of the encounter, the mass ratios of the colliding galaxies are also key parameters.

Triggered Star Formation in a Turbulent ISM
Proceedings IAU Symposium No. 237, 2006
B. G. Elmegreen & J. Palouš, eds.

© 2007 International Astronomical Union
doi:10.1017/S1743921307001688

The global star formation law: from dense cores to extreme starbursts

Yu Gao

Purple Mountain Observatory, 2 West Beijing Road, Nanjing 210008; and National Astronomy
Observatory, Chinese Academy of Sciences, Beijing, P.R. China
yugao@pmo.ac.cn,ygao@nrao.edu

Abstract. Active star formation (SF) is tightly related to the dense molecular gas in the giant molecular clouds' dense cores. Our HCN (measure of the dense molecular gas) survey in 65 galaxies (including 10 ultraluminous galaxies) reveals a tight linear correlation between HCN and IR (SF rate) luminosities, whereas the correlation between IR and CO (measure of the total molecular gas) luminosities is nonlinear. This suggests that the global SF rate depends more intimately upon the amount of dense molecular gas than the total molecular gas content. This linear relationship extends to both the dense cores in the Galaxy and the hyperluminous extreme starbursts at high-redshift. Therefore, the global SF law in dense gas appears to be linear all the way from dense cores to extreme starbursts, spanning over nine orders of magnitude in IR luminosity.

Keywords. stars: formation – ISM: molecules – infrared: galaxies – radio lines: galaxies – galaxies: high-redshift

---

1. Introduction

Schmidt (1959) law of star formation (SF) was first formulated in terms of local SF rate SFR proportional to the HI gas density with a power index n ($SFR \sim \rho^n$, n~1-3) as the atomic gas was then known as the major component of interstellar gas reservoir to possibly form stars. More than a decade later, observations of the CO line emission in the Milky Way and external galaxies enabled by millimeter (mm) astronomy suggest that stars are forming in the giant molecular clouds (GMCs). From 80's, e.g., Kennicutt (1983) found little evidence in parameterizing the global SF law in external galaxies in terms of the total HI gas. But he succeeded in terms of the disk-averaged surface SFR and total surface gas densities of both HI and H_2 (Kennicutt 1989). However, a well determined power index n was still not practical. Other researchers also obtained a wide range of the power index from 1 to 3. Recently, Kennicutt (1998) seemed to obtain a well determined slope of 1.4 though discrepancy exists (e.g., Heyer *et al.* 2004).

We here show that the SF law in terms of the dense gas is a rather simple linear relation (Gao & Solomon 2004a, GS04a) and is straight-forward to understand in what we have learned so far from the physics of SF. The cool atomic gas has the potential to convert into molecular form to possibly provide the fuel to make stars, yet even the bulk of molecular gas and most regions of GMCs are not making stars except for those tiny dense cores. Great Observatories indeed begin to directly link/picture the heavily obscured dusty regions of active SF, embedded in GMCs as mapped by mm/sub-mm telescopes, to the current massive SF.

2. Correlations among FIR, HCN & CO in galaxies

High-dipole moment molecules such as HCN trace more than an order of magnitude higher gas density than that of CO. A major HCN survey in a wide range of 65 galaxies, including 10 ultraluminous infrared galaxies (ULIRGs), tripled the sample of galaxies with global HCN measurements (Gao & Solomon 2004b). Analysis of the various relationships among the global HCN, CO, and FIR luminosities can be statistically conducted for the first time (GS04a).

The strong IR–HCN correlation is linear and extremely tight over 3 orders of magnitude in luminosity, when compared to the non-linear IR–CO correlation (GS04a). While the high luminosity of ULIRGs requires an elevated SF efficiency of the total molecular gas indicated by L_{IR}/L_{CO}, the SFR per unit of *dense* molecular gas, the SF efficiency of the *dense* molecular gas indicated by (L_{IR}/L_{HCN}) is almost constant and independent of the IR luminosity or total SFR. Further, GS04a find the surprising absence of any correlation between L_{IR}/L_{HCN} and L_{CO}/L_{HCN}, yet a still strong correlation between L_{IR}/L_{CO} and L_{HCN}/L_{CO} (Fig. 1). This suggests that the HCN–IR correlation is more physical than the CO–IR correlation and that the global SF efficiency depends on the fraction of the molecular gas in a dense phase (L_{HCN}/L_{CO}). This is somehow reminiscent of the poor HI–IR correlation vs. the better CO–IR correlation discovered more than two decades ago.

The direct consequence of the linear IR–HCN correlation is that the SF law in terms of *dense* gas has a power law index of 1, which is different from the widely used Kennicutt (1998) law of a slope of 1.4 for the disk averaged SFR as a function of the total gas (HI and H_2). As we show in next section that this 1.4 law is not unique, neither valid for normal spiral galaxies nor for extreme starbursts/ULIRGs.

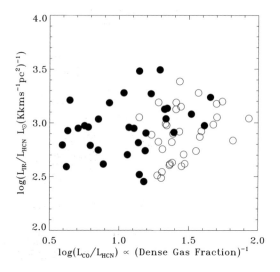

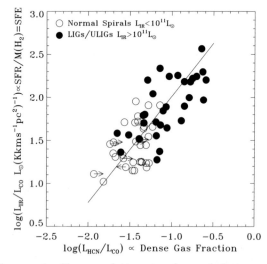

Figure 1. No correlation is observed between L_{IR}/L_{HCN} and L_{CO}/L_{HCN} (top), yet there is still strong correlation between L_{HCN}/L_{CO} and L_{IR}/L_{CO}. The IR–CO correlation can simply be a result of the much better correlations of IR–HCN and HCN–CO.

3. The global SF law in dense gas

The SFR–M_{H_2} (IR–CO) is essentially linear up to $SFR \sim 20M_\odot/yr$ (Fig. 2a). This seems to also be true in terms of the mean surface densities of the SFR and molecular gas mass for the nearest galaxies with spatially resolved observations (e.g., Wong & Blitz 2002). Thus the linear form of the global SF law in terms of total molecular gas density as traced by CO for normal galaxies is due to the constant dense gas mass fraction (HCN/CO) in normal galaxies (GS04a). For normal spirals, the SF law is linear in terms of both the total molecular gas and the dense molecular gas. A fit for the normal galaxies in the IR–CO correlation in HCN sample gives a slope of 1.0. But there is a poor correlation between the SFR and the gas surface density in Kennicutt (1998) normal galaxy sample, difficult to derive a reasonable slope from normal galaxy sample alone.

A direct orthogonal regression fit for all galaxies in HCN sample leads to a slope of 1.4 (the least-squares fit slope is 1.3). These fits are almost identical to the SF law power index in Kennicutt's (1998) 36 circumnuclear starbursts sample. It is obvious from Fig. 2a that only galaxies with SFR > 20 M$_\odot$/yr (mostly ULIRGs) lie above the slope 1. The combination of normal galaxies and ULIRGs leads to a fit of 1.4. Therefore, this slope is not a universal slope at all as it changes according to the sample selection. The Kennicutt's (1998) 1.4 slope is determined mostly from the starburst sample. The circumnuclear starbursts have some of the characteristics of ULIRGs, e.g., a high dense gas fraction indicated by high HCN/CO ratio.

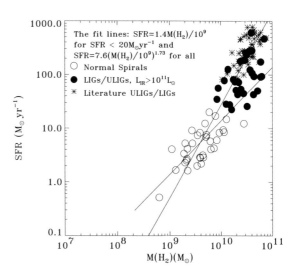

When we add more ULIRGs into the sample, the slope becomes steeper, and the fits lead to a slope of 1.7. It is clear that the ULIRGs steepen the slope of the fit. There also appears to be a trend that some normal spirals with the lowest Σ_{SFR} and Σ_{H_2} in Kennicutt's sample tend to lie below the 1.4 fit. Adding more extreme galaxies, both ULIRGs and low luminosity galaxies, tends to steepen the slope towards 2. Therefore, it is difficult to derive a unique 1.4 power law based upon the total molecular gas or the total gas content.

The SF law in terms of dense gas has a unique slope of 1 (Fig. 2b) since

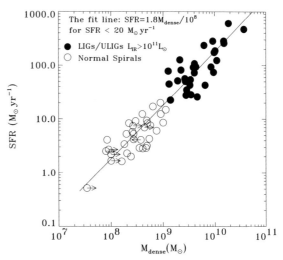

Figure 2. There is no unique SF law in terms of molecular gas or total gas (top) since the power index changes with the sample. But we do observe a linear SF law in dense gas for all star-forming galaxies.

the global SFR is linearly proportional to the mass of the dense molecular gas. Parameterization in terms of observable mean surface densities of the dense molecular gas and the SFR won't change the slope of 1 as both quantities are simply normalized by the same galaxy disk area. This linear SF law in dense gas seems valid for all star-forming galaxies.

Remarkably, the HCN–IR linear correlation is also valid to GMC dense cores implying the same physics drives the active SF in both dense cores and galaxies (Wu *et al.* 2005). New sensitive HCN line observations of four high-redshift submillimeter (sub-mm) galaxies and QSOs with the VLA including first possible HCN detection of a submm galaxy (Gao *et al.* 2006), combined with previous HCN detections and upper limits (e.g., Carilli *et al.* 2005) strongly suggest galaxies at high-redshift follow this same linear law (Fig. 3). The SF law in dense gas appears linear all the way from dense cores to extreme starbursts at high-z, spanning over nine orders of magnitude in SFR or FIR luminosity.

The SFR in a galaxy depends linearly on the dense molecular gas content as traced by HCN, regardless of

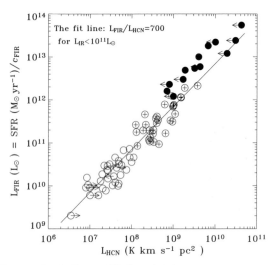

Figure 3. A linear SF law in dense gas appears valid for all star-forming systems including galaxies at high-redshift (filled circles). The plotted fit is for nearby normal spirals, and the dense cores of FIR luminosities of a few times of $10^{3-6} L_\odot$ (not plotted, Wu *et al.* 2005) lie on the same fit.

the galaxy luminosity or the presence of a "starburst", and not the total molecular gas and/or atomic gas traced by CO and/or HI observations respectively. Since dense molecular cloud cores are the sites of high mass SF, it is the physical properties, location and mass of these cores that set the SFR. A detailed SF law can be determined from observations directly probing the Milky Way cloud cores with spatially resolved measurements and the resolved measurements in nearby star-forming galaxies can bridge the dense cores with the galaxies. The global SFR of a star-forming system is best characterized by its mass of dense gas, $SFR \sim M(\text{dense } H_2)$. The gas density traced by HCN emission is apparently near the threshold for rapid SF.

4. Concluding remarks

Although SFR indicated FIR correlates with the various cold neutral gas reservoirs (HI, H_2, HCN–dense H_2), the FIR-HI is the worst among all and the FIR-CO is non-linear and has larger scatter than that of FIR-HCN. The best and tightest FIR-HCN correlation is a linear relation which implies, under the assumption of the constant conversion factors, a linear SF law that SFR is simply proportional to the amount of dense molecular gas available to form high mass stars. Other SF laws based on FIR-CO or FIR-(HI+H_2) won't have a unique power index. This linear FIR-HCN relation appears to be valid all the way from dense cores in GMCs to extreme starbursts at high-z, revealing the same physics that massive SF drives most of the energy output in all these systems.

Acknowledgements

I wish to express my gratitudes to my collaborators, particularly Phil Solomon, for their contributions. China NSF (distinguished young scholars) & Chinese Academy of Sciences (hundred-elite) are thanked for their supports.

References

Carilli, C.L., Solomon., P.M., Vanden Bout, P.A., *et al.* 2005, *ApJ* 618, 586
Gao, Y. & Solomon, P.M. 2004a, *ApJ* 609, 271 (GS04a)
Gao, Y. & Solomon, P.M. 2004b, *ApJS* 152, 63
Gao, Y., Carilli, C.L., Solomon, P.M. & Vanden Bout, P.A. 2006, *ApJ* submitted
Heyer, M.H., Corbelli, E., Schneider, S.E. & Young, J.S. 2004, *ApJ* 602, 723
Kennicutt, R.C. 1983, *ApJ* 272, 54
Kennicutt, R.C. 1989, *ApJ* 344, 685
Kennicutt, R.C. 1998, *ApJ* 498, 541
Schmidt, M. 1959, *ApJ* 129, 243
Wong, T. & Blitz, L. 2002, *ApJ* 269, 157
Wu, J., Evans, N.J.,II, Gao, Y., *et al.* 2005, *ApJ* 635, L173

Discussion

KENNICUTT: The beautiful aspect of your result is that the dense cores do not seem to care about whether they reside in Orion or Arp 220 – they form stars with the same core efficiency. But doesn't this simply reword the problems of explaining why the efficiency of core formation scales with total gas density?

GAO: No. The dense cores do seem to form massive stars with the same efficiency regardless of where they reside. But the core formation and the efficiency of core formation depend upon the location and/or environment the putative cores reside. I'm not exactly sure how the efficiency of core formation scales with total gas density, but I'd think the efficiency of core formation scales with the fraction of gas at high density, rather than total gas or total molecular gas density. For example, a gas-rich low surface brightness galaxy could have nearly same disk-averaged total surface gas density as that of a starburst galaxy, but the starburst has high fraction of molecular gas, particularly dense molecular gas, therefore, a high star formation rate.

ELMEGREEN: I like to interpret these observations in a different way. It seems the extra power of 1.4 in the Kennicutt scaling relation is from the rate at which low density gas evolves toward high density gas. But for observations of only high-density gas, like yours, their evolution is not needed anymore and then the power of density is 1, not 1.4. The really important point of your observations seems to be that there is a universal critical density for star formation. Your observations are close to this density.

GAO: Yes, agree. It seems that a SF law in terms of low density gas needs extra power of 1.4 or higher. But the problem in low density gas (either HI or H_2 or both) is that it is often difficult to derive a SF law with a unique power. As I showed that the SF law even in terms of the total molecular gas does not have a fixed 1.4 power and the situation is even worse in terms of atomic gas or total gas because of the poor correlation between SF rate and HI gas. On the other hand, a SF law in terms of total molecular gas (thus low density) for normal spirals does have a power index of 1, same as that in terms of dense molecular gas probed by HCN observations. But the power index in terms of total molecular gas needs to be increased when extreme (ultraluminous and/or low surface brightness) galaxies are added/considered.

Triggered Star Formation in a Turbulent ISM
Proceedings IAU Symposium No. 237, 2006
B. G. Elmegreen & J. Palouš, eds.

© 2007 International Astronomical Union
doi:10.1017/S174392130700169X

Galactic-scale star formation by gravitational instability

Mordecai-Mark Mac Low[1], Yuexing Li[1,2] and Ralf S. Klessen[3]

[1]Department of Astrophysics, American Museum of Natural History, 79th Street at Central Park West, New York NY 10024-5192, USA. Email: mordecai@amnh.org

[2]Harvard-Smithsonian Center for Astrophysics, Harvard University, 60 Garden Street, Cambridge, MA 02138, USA. Email: yxli@cfa.harvard.edu

[3]Institut für Theoretische Astrophysik, Zentrum für Astronomie der Universität Heidelberg, Alber-Überle-Str. 2, D-69120 Heidelberg, Germany. Email: rklessen@ita.uni-heidelberg.de

Abstract. We present numerical experiments that demonstrate that the nonlinear development of the Toomre instability in disks of isothermal gas, stars, and dark matter reproduces the observed Schmidt Law for star formation. The rate of gas collapse depends exponentially on the (minimum) value of the Toomre parameter in the disk. We demonstrate that spurious fragmentation occurs in the absence of sufficient resolution in our SPH model. Our models also reproduce observed star formation thresholds in disk galaxies. We finally briefly discuss the application of our models to the study of globular cluster formation in merging galaxies.

Keywords. hydrodynamics, instabilities, galaxies: spiral

The tight correlation between gas surface density and star formation rate in galaxies demonstrated, for example, by Kennicutt (1998), suggests that the large-scale star formation rate seen in galaxies may be determined by relatively simple physics. Martin & Kennicutt (2001) find thresholds in Hα emission that traces star formation, and suggest that these are caused by Toomre (1964) instability. (However, note the recent result from GALEX that the radial profiles of UV light emitted by young, massive stars in galaxies do not show the same breaks [e.g. Boissier *et al.* 2006].)

Star formation is a complex process where gravity and radiative cooling are opposed by thermal pressure, magnetic fields, rotation, and turbulent flows (e.g. Shu, Adams, & Lizano 1987, Mac Low & Klessen 2004). Yet, we are suggesting that relatively simple physics dominates the large-scale behavior. In the context of this conference, we are trying to raise the question of whether it is stellar triggering or gravitational instability that determines the large-scale star formation rate. Furthermore, we would like the same mechanism to explain how the very high star formation rates seen in starburst galaxies can be explained.

To address these questions we performed a series of numerical experiments, described in detail by Li, Mac Low, & Klessen (2005b). We set up exponential disks of stars and gas embedded in a dark matter halo with a softened isothermal profile, and simulated their evolution using GADGET, a hybrid N-body/SPH code described by Springel, Yoshida, & White (2001). In our experiments, we control the initial gravitational instability, and measure the properties of the collapsing gas, from which we infer the star formation rate and distribution. We follow regions of collapse using sink particles (Bate, Bonnell, & Price 1995).

We made three major approximations, each of which has been studied in more detail with local computational models. First, we assumed that the gas behaved isothermally for purposes of computing its dynamics. Joung & Mac Low (2006) demonstrated that

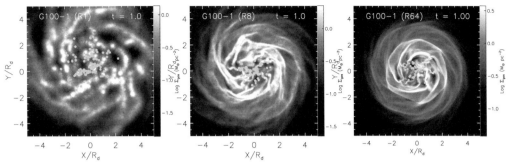

Figure 1. Gas surface density maps from a resolution study, showing models with total particle number of $N_{\rm tot} = 10^5$ (R1), 8×10^5 (R8) and 6.4×10^6 (R64). Model R1, which is underresolved according to the Jeans criterion of Bate & Burkert (1997), shows strong spurious fragmentation. On the other hand model R8, which is marginally resolved, shows only minor differences with the highest-resolution model R64, suggesting the criterion is adequate. From Li, Mac Low, & Klessen (2005b).

a supernova-driven ISM actually is rather *more* compressible than isothermal, so this is actually a conservative assumption. (Note though, that Robertson *et al.* (2004) and others have suggested that use of an isothermal equation of state does not produce stable galactic disks, a result we disagree with, as is discussed below.) Second, we assumed that molecular gas forms quickly once collapse sets in, which has been shown by Glover & Mac Low (2006) using 3D non-equilibrium chemistry simulations to be consistent with the behavior of high-density, turbulent, magnetized, self-gravitating gas. Third, we neglected magnetic fields at these large scales, as they are unlikely to be dynamically important for gas collapsing after reaching the Jeans mass of $\sim 10^6$ M$_\odot$.

The major result that we draw from our experiments is that nonlinear development of gravitational instability quantitatively determines star formation properties.

Before we discuss the results in detail we need to discuss the three numerical criteria that models of gravitational collapse must satisfy. Most important is the Jeans resolution criterion (Bate & Burkert 1997; Truelove *et al.* 1997; Whitworth 1998; Nelson 2006). Also necessary to be satisfied are the gravity-hydro balance criterion for gravitational softening (Bate & Burkert 1997; Nelson 2006), and the equipartition criterion for particle masses (Steinmetz & White 1997).

In Figure 1 we show a resolution study for one of our models, in which we increase the *linear* resolution by a factor of two between models (equivalent to increasing the mass resolution by a factor of 8). The lowest resolution model violates the Jeans criterion, while the two higher resolution models satisfy it. Spurious fragmentation can be clearly seen in the lowest resolution model, demonstrating that SPH simulations are indeed as subject as any other method to this problem, in contradiction to the claim of Hubber, Goodwin, & Whitworth (2006). Robertson *et al.* (2004) and others have presented models of disk galaxies using an isothermal equation of state that collapsed because of the formation of large clumps that lost angular momentum due to dynamical friction, falling to the center of the disk and producing far too high a star formation rate. We believe this occurred due to spurious fragmentation rather than the limitation of the equation of state.

A linear analysis of axisymmetric gravitational instability was performed for collisionless stars by Toomre (1964), giving an instability parameter $Q_s = \kappa \sigma_s / (3.36 G \Sigma_s)$, and for collisional gas by Goldreich & Lynden-Bell (1965), giving $Q_g = \kappa c_g / (\pi G \Sigma_g)$, where κ is the epicyclic frequency, $\Sigma_{s,g}$ are star and gas surface densities, σ_s is the radial stellar velocity dispersion, and c_g is the isothermal gas sound speed. Rafikov (2001) analyzed the

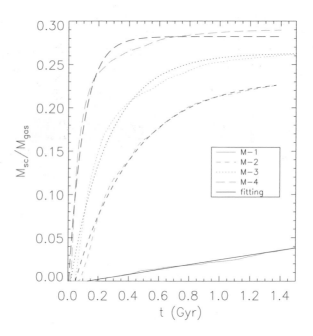

Figure 2. Time history of the mass in star clusters M_{sc} in several sample models, scaled by initial total gas mass $M_{\rm gas}$, and compared to fits to equation (0.1). M_{sc} is taken as 30% of the mass of the sink particles. M-1 to 4 indicate submodels with increasing gas fractions. From Li, Mac Low, & Klessen (2005b).

instability for a combination of collisionless stars and collisional gas, deriving a non-linear equation for a combined instability parameter Q_{sg}. We use this in our analysis.

The history of collapse in our models can be well fit by curves drawn from the single parameter family

$$M_{sc} = M_0(1 - \exp[-t/\tau_{\rm sf}]), \qquad (0.1)$$

as shown in Figure 2, where M_{sc} is the total amount of collapsed mass, M_0 is the initial gas mass, and $\tau_{\rm sf}$ is a collapse or star formation timescale. The collapse timescale $\tau_{\rm sf}$ depends continuously in our model galaxies on the minimum initial value of Q_{sg} in the disk, with an exponential dependence

$$\tau_{\rm sf} \propto \exp(\alpha Q_{sg}), \qquad (0.2)$$

where $\alpha \simeq 3$, as shown in Figure 3. This continuous dependence suggests that nominally stable galaxies can still show low rates of star formation, a conclusion that might extend to stable regions of galaxies as well. The detailed dependence of star formation on the local value of Q_{sg} has been shown in preliminary results presented by Chu in this meeting. She showed a good correlation in the LMC between the location of young stellar objects detected in the Spitzer 8 μm band and the local value of Q_{sg}, computed from the stellar, atomic, and molecular gas densities.

Combining equations (0.1) and (0.2), we can derive a nonlinear equation for the global star formation efficiency $\epsilon_g = M_{sc}/M_0 = f(Q_{sg})$. This predicts that the star formation efficiency in galaxies with $Q_{sg} > 1$ drops continuously to less than 2% (see Li, Mac Low, & Klessen 2006).

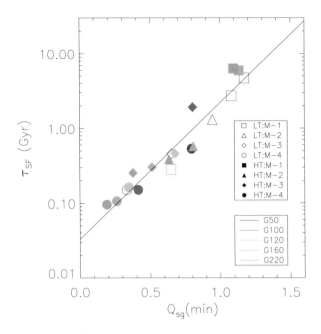

Figure 3. Star formation timescale τ_{sf} correlates with the initial disk instability Q_{sg} The solid line is a least-squares fits to the data. From Li, Mac Low, & Klessen 2005b.

The total collapse rate in our galaxies can be measured at any time, allowing us to derive a quantity related to the star formation rate, and compare it to the gas surface density. As Figure 4 shows, our models fit the observed relation in both slope and normalization. In this plot, we used a constant conversion factor between collapse rate and star formation rate of 30% (a local efficiency for star formation in very unstable regions), but Li, Mac Low, & Klessen (2006) demonstrate that local efficiencies within a reasonable range (i.e. 5–50%) give similar results.

Other groups have recently derived the Schmidt Law in complementary ways. Kravtsov (2003) used an adaptive mesh refinement method to study gas collapse in an ensemble of galaxies evolving from cosmological initial conditions. Unlike many such simulations, he chose a star formation law $\dot{\rho}_* \propto \rho_g$, deliberately chosen to *not* reproduce the Schmidt law by design. Nevertheless, he found that the galaxies reproduced the Schmidt law, in agreement with our results. As he did not measure gravitational instability, he could not make the direct link that we do.

Krumholz & McKee (2005), on the other hand, took the observed distributions of giant molecular clouds and H II regions as input. They assumed the clouds to be in virial equilibrium (an assumption studied more carefully by Krumholz, Matzner, & McKee 2006), and then derived the Schmidt law from the properties of supersonic turbulence in the clouds. We argue that global gravitational instability determines the distribution of clouds and H II regions used in this approach.

Our models also reproduce the star formation thresholds found by Martin & Kennicutt (2001). Li, Mac Low, & Klessen (2005a) show that the radius within which 90% of star formation occurs is tied to the radius at which Q_{sg} passes through unity. However, they also found that the more stable the galaxy, the larger the actual value of Q_{sg} at threshold, over a range from 0.3 for extreme starbursts out to 1.5 for very stable galaxies. The

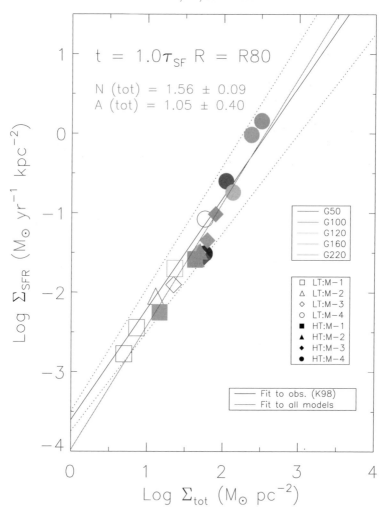

Figure 4. A comparison of the global Schmidt law between our simulations and the observations. The red line is the least-square fit to the total gas of the simulated models, the black solid line is the best fit of observations from Kennicutt (1998), while the black dotted lines indicate the observational uncertainty. The color of the symbol indicates the rotational velocity for each model, while labels from M-1 to M-4 are sub-models with increasing gas fraction; and open and filled symbols represent low and high temperature models, respectively. From Li, Mac Low, & Klessen (2006).

observations of stable, modern galaxies by Martin & Kennicutt (2001) have an average threshold value of $Q_g = 1.4$, which appears consistent with our result.

We have used the same set of approximations to model globular cluster formation in merging galaxies, as described by Li, Mac Low, & Klessen (2004). If we assume that each sink particle represents a cluster formation site, we find clear evidence for large-scale triggering of star formation by galaxy mergers. Our results confirm the original suggestion by Ashman & Zepf (1992) that globular clusters formed during galaxy mergers can explain the factor of three increase in observed specific frequency of globular clusters between spiral and elliptical galaxies.

We conclude by restating our major result: gravitational instability appears to be the dominant mechanism that controls the star formation rate in galaxies. Local triggering by OB stars appears to be roughly a 10–20% effect (see talk by Mizuno in this meeting, and Joung & Mac Low 2006). These conclusions agree with the basic principle laid out by Elmegreen (2002): gravitational instability determines the global star formation rate, but local triggering helps to determine exactly where and when star formation occurs. Finally, our results firmly support the idea that galaxy interactions represent an extremely effective form of global triggering.

Acknowledgements

We thank V. Springel for making both GADGET and his galaxy initial condition generator available, as well as for useful discussions, and A.-K. Jappsen for participating in the implementation of sink particles in GADGET. We are grateful to F. Adams, G. Bryan, J. Dalcanton, B. Elmegreen, D. Helfand, R. Kennicutt, J. Lee, C. Martin, R. McCray, T. Quinn, E. Quataert, M. Shara, C. Struck, and J. van Gorkom for very useful discussions. This work was supported by the NSF under grants AST99-85392 and AST03-07854, by NASA under grant NAG5-13028, and by the Emmy Noether Program of the DFG under grant KL1358/1. Computations were performed at the Pittsburgh Supercomputer Center supported by the NSF, on the Parallel Computing Facility of the AMNH, and on an Ultrasparc III cluster generously donated by Sun Microsystems.

References

Ashman, K. M. & Zepf, S. E. 1992, *ApJ* 384, 50
Bate, M. R., Bonnell, I. A. & Price, N. M. 1995, *MNRAS* 277, 362
Bate, M. R. & Burkert, A. 1997, *MNRAS* 288, 1060
Boissier, S., *et al.* 2006, *ApJ Suppl.* in press (astro-ph/0609071)
Elmegreen, B. G. 2002, *ApJ* 577, 206
Goldreich, P. & Lynden-Bell, D. 1965, *MNRAS* 130, 97
Glover, S. C. O. & Mac Low, M.-M. 2006, *ApJ* submitted (astro-ph/0605121)
Hubber, D. A., Goodwin, S. P. & Whitworth, A. P. 2006, *A&A* 450, 881
Joung, M. K. R. & Mac Low, M.-M. 2006, *ApJ* submitted (astro-ph/0601005)
Krumholz, M. R., Matzner, C. D. & McKee, C. F. 2006, *ApJ* in press (astro-ph/0608471)
Krumholz, M. R. & McKee, C. F. 2005, *ApJ* 630, 250
Kennicutt, R. C., Jr. 1998, *ApJ* 488, 541
Kravtsov, A. V. 2003, *ApJ* 590, L1
Li, Y., Mac Low, M.-M. & Klessen R. S. 2004, *ApJ* 614, L29
Li, Y., Mac Low, M.-M. & Klessen R. S. 2005a, *ApJ* 620, L19
Li, Y., Mac Low, M.-M. & Klessen R. S. 2005b, *ApJ* 626, 823
Li, Y., Mac Low, M.-M. & Klessen R. S. 2006, *ApJ* 639, 879
Mac Low, M.-M. & Klessen, R. S. 2004, *Rev. Mod. Phys.* 76, 125
Martin, C. L. & Kennicutt, R. C., Jr. 2001, *ApJ* 555, 301
Nelson, A. F. 2006, *MNRAS* in press (astro-ph/0609493)
Rafikov, R. R. 2001, *MNRAS* 323, 445
Robertson, B., Yoshida, N., Springel, V. & Hernquist, L. 2004, *ApJ* 606, 32
Shu, F. H., Adams, F. C. & Lizano, S. 1987, *ARAA* 25, 23
Springel, V., Yoshida, N. & White S. D. M. 2001, *New Astron.* 6, 79
Steinmetz, M. & White, S. D. M. 1997, *MNRAS* 288, 545
Toomre, A. 1964, *ApJ* 139, 1217
Truelove, J. K., Klein, R. I., McKee, C. F., Holliman, J. H., II, Howell, L. H. & Greenough, J. A. 1997, *ApJ* 489, L179
Whitworth, A. P. 1998, *MNRAS* 296, 442

Discussion

BONNELL: Do the stars or the gas dominate the gravitational instabilities?

MAC LOW: The large-scale gravitational potential is determined by the total mass, distributed among stars and gas. In model galaxies, the stars dominate the mass, though the cold gas can be locally important in regions where it is concentrated.

ELMEGREEN: You're suggesting Q can be much less than 1 and that starbursts and high efficiencies result from this, but in other global models, Q is self-regulated to stay near 1 and starbursts occur for the same reason as normal star formation, namely in marginally unstable gas with a great sensitivity to column density. Why do you prefer your model?

MAC LOW: We find that low initial Q galaxies form stars with a very short timescale, so that the Q rises quickly towards unity. We need to compare star formation rates with the value of Q observable at that moment, rather than with the initial value. I hope to report on this soon.

TAN: I believe Downes & Solomon (1998) measured $Q \sim 1$ in circumstellar starbursts, in contradiction with your results. I agree that gravitational instability controls where star formation occurs in galaxies (the threshold) but not that it sets the rate. This is because in most of the galaxies and starbursts defining the Kennicutt-Schmidt law, a large fraction (~ 0.5) of the gas is already in molecular clouds that are probably bound (McKee 1999), and that fraction is probably in an approximate steady state. The star formation rate is set by the rate at which these giant molecular clouds form star clusters, i.e., gravitational instabilities on scales much smaller than the galaxy, as in the models of Tan (2000) or Krumholz & McKee (2005). I would thus argue that GMC formation from atomic gas is not the rate limiting step controlling galactic star formation rates.

MAC LOW: I wonder if Downes & Solomon (1998) included the stellar gravitational potential in their Q calculation. A factor of 2 is plenty, as the star formation rate is exponentially dependent on Q. As I understand the models of Tan (2000) or Krumholz & McKee (2005), they depend on internal properties of the clouds (presumably universal) and on the observed distribution of molecular clouds in galaxies. We offer a theoretical grounding for that distribution, which appears to be the property that distinguishes galaxies with different star formation rates.

ZINNECKER: You said triggering would only change the global star formation efficiency by 10-10%. Can I rephrase this by asking: If there were no supernovae whatsoever (just a thought experiment), would the star formation efficiency in galaxies be essentially the same?

MAC LOW: Not necessarily. We assume an isothermal equation of state with an effective temperature of order 10^4K. This implies strong energy input to maintain these temperatures or velocity dispersions. The SNe are a major candidate for this energy input, although as Eve Ostriker showed, there are many other candidates, particularly the magnetorotational instability. If the MRI provides a floor, then the SNe may not make a big difference.

DE GOUVEIA DAL PINO: The detailed calculation of the radiative coding and the implicit evolution of the chemical species have tremendous effects upon the structuring formation at the small scales (e.g., Melioli, de Gouveia Dal Pino, & Raga, A&A, 2005). Thus, to what extent does the adoption of an isothermal equation of state affect the time scales for global (large scale) structuring development and star formation in your simulations?

MAC LOW: I agree that the equation of state, determined by radiative cooling and non-equilibrium chemistry is vital to the details of fragmentation at small scales (e.g., Jappsen *et al.* 2005). However, our isothermal equation of state is primarily designed to simulate the dynamical effects of transonic or supersonic turbulence. The thermal state of the gas is less important at large scales.

CRUTCHER: Without a detailed theory of star formation in simulations such as you present, it seems to me that what you get out may be dominated by your initial assumptions. For example, putting in global magnetic fields could profoundly affect the compression of clouds in the colliding galaxies.

MAC LOW: The theory of gravoturbulent collapse does provide a reasonably well developed theory for what goes on inside sink particles, including magnetic effects (see Heitsch, Klessen, & Mac Low 2001 and Glover & Mac Low 2006). Whether global fields are important isn't directly addressed by our models yet. However, I suspect that the gravitational energy density far exceeds the magnetic energy density.

Triggered star formation in a turbulent ISM
Proceedings IAU Symposium No. 237, 2006
B. G. Elmegreen & J. Palouš, eds.

Spiral arm triggering of star formation

Ian A. Bonnell[1] and Clare L. Dobbs[2]

[1]SUPA, School of Physics and Astronomy, University of St Andrews, KY16 9SS, UK
email: iab1@st-and.ac.uk

[2]Department of Physics, University of Exeter, UK

Abstract. We present numerical simulations of the passage of clumpy gas through a galactic spiral shock, the subsequent formation of giant molecular clouds (GMCs) and the triggering of star formation. The spiral shock forms dense clouds while dissipating kinetic energy, producing regions that are locally gravitationally bound and collapse to form stars. In addition to triggering the star formation process, the clumpy gas passing through the shock naturally generates the observed velocity dispersion size relation of molecular clouds. In this scenario, the internal motions of GMCs need not be turbulent in nature. The coupling of the clouds' internal kinematics to their externally triggered formation removes the need for the clouds to be self-gravitating. Globally unbound molecular clouds provides a simple explanation of the low efficiency of star formation. While dense regions in the shock become bound and collapse to form stars, the majority of the gas disperses as it leaves the spiral arm.

1. Introduction

Star formation has long been known to occur primarily in the spiral arms of disc galaxies (Baade 1963). Spiral arms are denoted by the presence of young stars, HII regions, dust and giant molecular clouds, all signatures of the star formation process (Elmegreen & Elmegreen 1983; Ferguson *et al.* 1998). What is still unclear is the exact role of the spiral arms in inducing the star formation. Is it simply that the higher surface density due to the orbit crossing is sufficient to initiate star formation, as in a Schmidt law, or do the spiral arms play a more active role? Roberts (1969) first suggested that the spiral shock that occurs as the gas flows through the potential minima triggers the star formation process in spiral galaxies. Shock dissipation of excess kinetic energy can result in the formation of bound structures which then collapse to form stars.

Giant molecular clouds (GMCs) are observed to contain highly supersonic motions and a wealth of structure on all length scales (Larson 1981; Blitz & Williams 1999). The supersonic motions are generally thought to be 'turbulent' in nature and to be the cause of the density structure in GMCs (Mac Low & Klessen 2004; Elmegreen & Scalo 2004). We propose an alternative scenario whereby it is the passage of the clumpy interstellar medium through a galactic spiral shock that not only produces the dense environment in which molecular clouds form (Cowie 1981; Elmegreen 1991), but also gives rise at the same time to their supersonic internal motions (Bonnell *et al.* 2006).

2. Global Simulations

Recent global simulations of non-self gravitating gas dynamics in spiral galaxies (Dobbs, Bonnell & Pringle 2006) show that the spiral shocks can account for the formation of molecular gas from cold ($T \lesssim 100$ K) atomic gas and generate the large scale distribution of molecular clouds in spiral arms. Structures in the spiral arms arise due to the shocks that tend to gather material together on converging orbits. Thus, structures grow in time

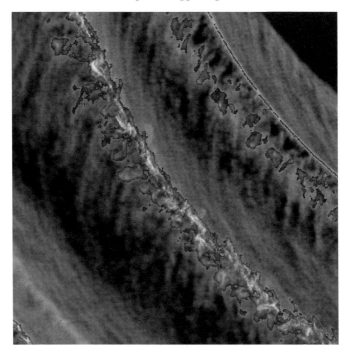

Figure 1. The formation of molecular clouds is shown as the gas passes through a spiral shock (Dobbs *et al.* 2006). Note the spurs and feathering that appear as the dense clumps are sheared away upon leaving the spiral arm.

through multiple spiral arm passages. These structures present in the spiral arms are also found to form the spurs and feathering in the interarm region as they are sheared by the divergent orbits when leaving the spiral arms (Dobbs & Bonnell 2006a). The high gas densities that result from the high Mach number shocks are sufficient for rapid formation of H_2 gas and thus of giant molecular clouds. If, in contrast, the gas is warm ($T \geqslant 1000$ K) when it enters the shock, then H_2 formation cannot occur due to the lower gas densities in the shock. In this model, molecular clouds are limited to spiral arms as it is only there that the gas is sufficiently dense to form molecules. These clouds need not be self-gravitating as their formation is independent of self-gravity. The velocity dispersion in the gas also undergoes periodic bursts during the spiral arm passage as the clumpy shock drives supersonic random motions into the gas (see below). Such bursts in the internal gas motions are likely to be observable and would give support for a spiral shock origin of giant molecular clouds and the triggering of star formation.

3. Triggering of Star Formation in the Spiral Shock

In simulations where self-gravity is included, the passage of gas through a spiral shock can result in the triggering of star formation (Bonnell *et al.* 2006). The evolution, over 34 million years, of 10^6 M$_\odot$ of gas passing through the spiral potential is shown in Figure 3. The initially clumpy, low density gas ($\rho \approx 0.01$ M$_\odot$pc^{-3}) is compressed by the spiral shock as it leaves the minimum of the potential. The shock forms some very dense ($> 10^3$ M$_\odot$pc^{-3}) regions, which become gravitationally bound and thus collapse to form regions of star formation. Further accretion onto these regions, modeled with sink-particles in SPH (Bate *et al.* 1995), raises their masses to that of typical stellar clusters (10^2 to 10^4 M$_\odot$). Star formation occurs within 2×10^6 years after molecular cloud densities are

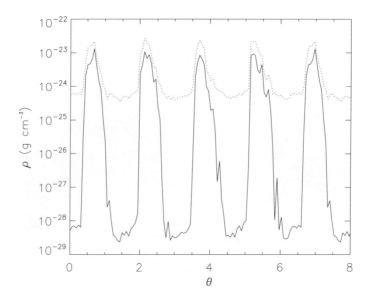

Figure 2. The total gas (dotted) and H_2 (solid) gas densities are plotted against azimuth for an annulus at 5 kpc. The gas is averaged over cells of 50 pc in size (from Dobbs *et al.* 2006). Molecular gas is almost exclusively contained.

reached. The total spiral arm passage lasts for $\approx 2 \times 10^7$ years. The gas remains globally unbound throughout the simulation and re-expands in the post-shock region. The star formation efficiency is of order 10 % and should be taken as an upper limit in the absence of any form of stellar feedback.

The star forming clouds that form in the spiral shocks are generally unbound and thus disperse once they leave the spiral arm. Numerical simulations of unbound clouds (Clark & Bonnell 2004; Clark *et al.* 2005) have recently shown that they can form local subregions which are gravitationally unstable and thus form stars. In contrast, the bulk of the cloud does not become bound and thus disperses without entering into the star formation process. The resultant star formation efficiencies are of order 10 per cent, even without the presence of the divergent flows of gas leaving a spiral arm. In fact, arbitrarily low star formation efficiencies are possible with relatively small deviations from bound conditions. This then offers a simple physical mechanism to explain the low star formation efficiency in our Galaxy.

3.1. *The Generation of the Internal Velocity Dispersion*

In addition to triggering star formation, we must be able to explain the origin of the internal velocity dispersion and how it depends on the size of the region considered (Larson 1981; Heyer & Brunt 2004). The evolution of the velocity dispersion as a function of the size of the region considered is shown in Figure 4 (from Bonnell *et al.* 2006). The initially low velocity dispersion, of order the sound speed $v_s \approx 0.6$ km/s, increases as the gas passes through the spiral shock. At the same time, the velocity dispersion increases more on larger scales, producing a $v_{\mathrm{disp}} \propto R^{0.5}$ velocity dispersion size-scale relation. The basic idea is that when structure exists in the pre-shocked gas, the stopping point of a particular clump depends on the density of gas with which it interacts. Thus some regions will penetrate further into the shock, broadening it and leaving it with a remnant velocity dispersion in the shock direction. Smaller scale regions in the shock are likely to have

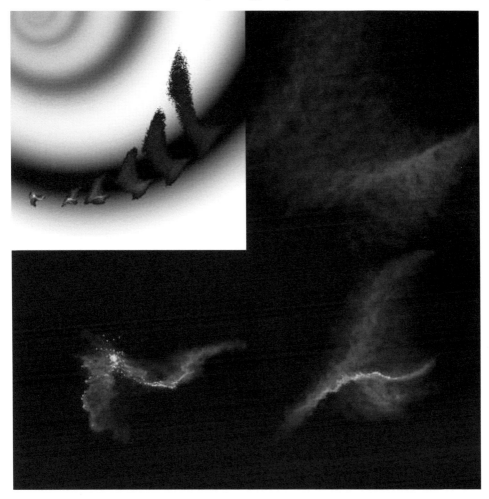

Figure 3. The evolution of cold interstellar gas through a spiral arm is shown relative to the spiral potential of the galaxy (upper left-panel). The minimum of the spiral potential is shown as black and the overall galactic potential is not shown for clarity. The 3 additional panels, arranged clockwise, show close-ups of the gas as it is compressed in the shock and subregions become self-gravitating. Gravitational collapse and star formation occurs within 2×10^6 years of the gas reaching molecular cloud densities. The cloud produces stars inefficiently as the gas is not globally bound.

more uniform momentum injection as well as encountering similar amounts of mass. This then results in small velocity dispersions. Larger regions will have less correlation in both the momentum injection and mass loading such that there will be a larger dispersion in the post-shock velocity. Any clumpy shock can induce such velocity dispersions. Thus the fractal nature of the ISM passing through a spiral shock is a straightforward explanation for how the velocity dispersion size relation arises in molecular clouds (Dobbs & Bonnell 2006b).

3.2. *The clump-mass spectrum*

Clumpy shocks may also be an important role in setting the clump-mass spectrum. Numerical simulations of colliding clumpy flows show that from an initial population of identical clumps, the shocked gas contains a spectrum of clump masses that is consistent

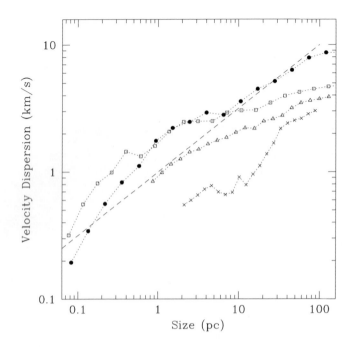

Figure 4. The velocity dispersion is plotted as a function of size at 5 different times during the passage of the gas through a spiral shock. The velocity dispersion is plotted at 4.2, 14, 18, 23, and 27×10^6 years after the start of the simulation. Star formation is initiated at $\approx 23 \times 10^6$ years. The dashed line indicates the Larson relation for molecular clouds where $\sigma \propto R^{-1/2}$ (Larson 1981; Heyer & Brunt 2004).

with a Salpeter-like slope (Clark & Bonnell 2006). This clump mass spectrum arises due to the coagulation and fragmentation of the clumps in the shock, and is very similar to that observed in dense prestellar cores (Motte *et al.* 1998). The relation between this clump-mass spectrum and the stellar IMF is unclear as it arises independently of self-gravity and thus a one-to-one mapping of clump to stellar masses is unlikely. Instead, the produced clumps are likely to be a combination of unbound clumps that will not form stars and clumps that are very bound that will form multiple many stars.

4. Conclusions

The triggering of star formation by the passage of clumpy gas through a spiral arm can explain many of the observed properties of star forming regions. Molecular cloud formation occurs as long as the pre-shock gas is cold ($T \lesssim 100$ K). The shock forms dense structures in the gas which become locally bound and collapse to form stars. The clouds are globally unbound and thus disperse on timescales of 10^7 years, resulting in relatively low star formation efficiencies. In addition, the clumpy shock reproduces the observed kinematics of GMCs, the so-called 'Larson' relation. There is no need for any internal driving of the quasi-turbulent random motions. The internal structure of GMCs can also be understood as being a product of a clumpy shock and even the observed clump-mass spectrum in pre-stellar cores is reproduced. Finally, the clouds are sheared upon leaving the spiral arms, producing the spurs and feathering commonly observed in spiral galaxies.

References

Baade, W., 1963, *Evolution of stars and Galaxies* (Cambridge: Harvard Univ.), p 63

Bate, M.R., Bonnell, I.A. & Price, N.M., 1995, *MNRAS* 277, 362

Blitz, L. & Williams, J., 1999, in: C.J. Lada & N.D. Kylafis (eds.), *The origin of stars and planetary systems* (Kluwer: Dordrecht), p. 3

Bonnell, I.A., Dobbs, C.L., Robitaille, T.P. & Pringle, J.E., 2006, *MNRAS* 365, 37

Clark, P.C. & Bonnell, I.A., 2004 *MNRAS* **347**, L36

Clark, P.C. & Bonnell, I.A., 2006 *MNRAS* **368**, 1787

Clark, P.C., Bonnell, I.A., Zinnecker, H. & Bate, M.R., 2004 *MNRAS* 359, 809

Cowie, L. L., 1981, *ApJ* 245, 66

Dobbs, C.L. & Bonnell, I.A., 2006a, *MNRAS* 367, 873

Dobbs, C.L. & Bonnell, I.A., 2006b, *MNRAS* in press

Dobbs, C.L., Bonnell, I.A. & Pringle, J.E., 2006, *MNRAS* 371, 1663

Elmegreen, B. G., 1991, *ApJ* 378, 139

Elmegreen B.G. & Elmegreen, D.M., 1983, *MNRAS* 203, 31

Elmegreen, B. & Scalo, J., 2004, *ARAA,* 42, 211

Ferguson A.M.N., Wyse R.F.G., Gallagher, J.S., & Hunter, D.A., 1998, *ApJ* 506, L19

Heyer M.H. & Brunt C.M., 2004, *ApJ* 615, 45

Ferguson, A.M.N., Wyse, R.F.G., Gallagher, J.S. & Hunter, D.A., 1998, *ApJ* 506, L19

Larson, R.B., 1981, *MNRAS* 194, 809

Mac Low, M.M. & Klessen, R.S., 2004, *Rev. Mod. Phys* 74, 125

Monaghan, J.J., 1992, *ARAA* 30, 543

Motte, F., André P. & Neri, R., 1998, *A&A* 336, 150

Roberts, W.W., 1969, *ApJ* 158, 123

Discussion

PADOAN: Can you explain what you mean when you say that the motions set up by spiral waves do not generate turbulence? How can you avoid the cloud being turbulent once they are given a velocity dispersion by whatever mechanism?

BONNELL: At present there is no evidence of a turbulent cascade in the simulations but that instead the velocity dispersion is driven at all scales simultaneously. This is not to say that turbulence will not develop in real shocks but just that it is not in the simulations even though the primary kinetic property, the velocity - size scale relation, is attained.

VAZQUEZ-SEMADENI: One question and two comments. Question: In your clumpy simulations, are the clumps at the same temperature as the interclump medium? If so, they are overpressured, and should be dispersing, no? Comment 1: My impression is that the clumpiness you introduce is a substitute, in your isothermal assumption, of the softer than isothermal (even thermally bistable) actual behavior of the clouds. Comment 2: The isothermal assumption may be responsible for several of the effects you observe, in particular the unboundedness of the clouds (cooler clouds would be more bound) and of the very low vorticity of the flow: in the isothermal (or, in general, barotropic) flow, the baroclinic term, $\nabla P \times \nabla \rho$, which is an important source of vorticity, is forced to be zero.

BONNELL: In the single-phase simulations the clumps are unbounded and free to expand. They do so somewhat reducing the gas densities before entering the shock. The isothermal equation of state is an oversimplification of the physics but should not affect the boundedness of the clumps as this is primarily determined by the kinetic energy. I do agree that cooling will help introduce the types of structures we assume in our initial conditions.

BLOCK: A general question on longevity of the feathers, with particular application to NGC 2841 (e.g., Block & Elmegreen, Nature) which is optically flocculent with dense grand-design arms of dust observed at 2.1 μm.

BONNELL: The feathering, which develops in the interarm regions due to the shearing of the spurs, appears to last most of the way through the interarm region for a timescale of several 10's of Myrs.

E. OSTRIKER: I am concerned that in your global isothermal models, the pressure is too low by two orders of magnitude, and that as a consequence your spiral shocks are too strong, leading to unphysical clumping (without self-gravity).

BONNELL: In order to get H_2 formation, a gas temperature of 100K is required in our isothermal simulation. This implies that for these models to be correct requires the presence of cold HI gas or sufficiently fast cooling of the gas as it enters the shock (as reported by Glover & Mac Low 2006).

Triggered Star Formation in a Turbulent ISM
Proceedings IAU Symposium No. 237, 2006
B. G. Elmegreen & J. Palouš, eds.

Cloud formation from large-scale instabilities

Woong-Tae Kim

Department of Physics and Astronomy, FPRD, Seoul National University, Seoul 151-742,
Republic of Korea
email: wkim@astro.snu.ac.kr

Abstract. We discuss recent advances in cloud formation via gravitational instability under the action of self-gravity, magnetic fields, rotational shear, active stars, and/or stellar spiral arms. When shear is strong and the spiral arms are weak, applicable to flocculent galaxies at large, swing amplification exhibits nonlinear threshold behavior such that disks with a Toomre parameter $Q < Q_c$ experience gravitational runaway. For most realistic conditions, local models yield $Q_c \sim 1.4$, similar to the observed star formation thresholds. When shear is weak, on the other hand, as in galactic central parts or inside spiral arms, magneto-Jeans instability is very powerful to form spiral-arm substructures including gaseous spurs and giant clouds. The wiggle and Parker instabilities proposed for cloud formation appear to be suppressed by strong non-steady motions inherent in vertically-extended spiral shocks, suggesting that gravitational instability is a primary candidate for cloud formation.

Keywords. galaxies: ISM, instabilities, ISM: kinematics and dynamics, ISM: magnetic fields, method: numerical, MHD, stars: formation

1. Introduction

Most galactic star formation takes place in cold, giant molecular clouds (GMCs). Star-forming GMCs distributed along spiral arms tend to appear in groups, forming giant molecular associations with mass $\sim 10^7 - 10^8 M_\odot$ (e.g., Vogel *et al.* 1988; see also Tosaki *et al.* and Hitschfeld *et al.* in this volume for recent observational results). They are also closely associated with other spiral-arm substructures such as gaseous spurs (or feathers) and OB star complexes. Gaseous spurs are prominent in an *HST* image of M51 (Scoville & Rector 2001) as short dust lanes jutting out almost perpendicularly from main spiral arms, and as warm dust filaments in a *Spitzer Legacy* image of M51 (Kennicutt 2004). The recent analyses of *HST* archive data by La Vigne *et al.* (2006) indicate that gaseous spurs are in fact very common in grand design spirals and coincide with the density peaks of molecular gas.

Many mechanisms have been proposed for the formation of giant clouds and spurs. These include collisional agglomeration of small clouds into large clouds, the Parker instability, and gravitational instability. Although the first, stochastic coagulation model has been successful in reproducing the observed GMC mass and velocity distributions (e.g., Das & Jog 1996), the basic premise of this mechanism is doubtful because cloud collisions usually lead to disruption rather than merger (e.g., Kim *et al.* 1999). In addition, there is insufficient mass in small clouds to build a steady mass spectrum (e.g., Heyer & Terebey 1998), and it takes too long time to achieve GMC masses (e.g., Blitz & Shu 1980).

Parker (1966) showed that giant clouds can form at the magnetic valleys due to magnetic buoyancy force in a vertically stratified disk. Because the wavelengths and growth times of the most unstable Parker modes are comparable to observed GMC spacings and lifetime, the Parker instability has been favored for GMC formation (Blitz & Shu

1980). However, the results of nonlinear simulations suggest that the Parker *alone* cannot produce overdense structures like GMCs (e.g., Santillán *et al.* 2000; Kim *et al.* 2000). In fact, the Parker instability is self-limiting because of stabilizing magnetic tension forces. Neither galactic differential rotation (Kim, Ostriker, & Stone 2003) nor spiral density waves (Kim & Ostriker 2006a) helps the Parker instability much.

On the other hand, self-gravity is a long-range force and thus allows runaway growth of condensations. The observed typical mass and separation of giant clouds along spiral arms are consistent with the characteristic Jeans mass and length of galactic disks at large (e.g., Elmegreen 1987). Recent work of La Vigne *et al.* (2006) showed that the spacing of gaseous spurs in spiral galaxies is also in good agreement with that from gravitational instability operating inside spiral arms. In recent years, we have studied cloud formation via gravitational instability in a local patch of galactic disks (Kim & Ostriker 2001, 2002, 2006a, 2006b; Kim, Ostriker, & Stone 2002, 2003). Effects of magnetic fields, (both thin and thick) self-gravity, galactic differential rotation, spiral arms as well as an active stellar component have been included. In what follows, we highlight the differences among various self-gravitating instabilities and summarize the main results of our investigation. The interested reader is referred to the original work for more detailed discussion.

2. Self-gravitating mechanisms

2.1. *Axisymmetric instability*

Consider a self-gravitating, rotating, gaseous disk threaded by azimuthal magnetic fields. The disk has surface density Σ, angular velocity Ω, sound speed c_s, Alfvén speed v_A, and vertical scale height H. We concentrate on axisymmetric perturbations that do not rely on vertical motions. For local WKB modes for which the radial variations of all physical quantities (except Ω) are unimportant, the dispersion relation for axisymmetric disturbances with wavenumber k becomes

$$\omega^2 = \kappa^2 + (c_s^2 + v_A^2)k^2 - \frac{2\pi G\Sigma|k|}{1 + |k|H}, \tag{2.1}$$

where $\kappa \equiv (4\Omega^2 + d\Omega^2/d\ln R)^{1/2}$ is the epicyclic frequency (e.g., Kim, Ostriker, & Stone 2002). Note that the denominator of the gravity term in equation (2.1) accounts approximately for the geometrical dilution of self-gravity due to finite disk thickness.

For infinitesimally thin, unmagnetized disks, one can show from equation (2.1) that disks become unstable only if $Q < 1$, with the Toomre stability parameter Q defined by

$$Q \equiv \frac{\kappa c_s}{\pi G\Sigma}. \tag{2.2}$$

The range of unstable wavenumbers is $1 - (1 - Q^2)^{1/2} < 2k/k_J < 1 + (1 - Q^2)^{1/2}$, where the Jeans wavenumber $k_J \equiv 2\pi G\Sigma/c_s^2$. Sonic motions and Coriolis forces stabilize short and long wavelength perturbations, respectively. It should be noted that the usual critical value $Q_c = 1$ applies only to *axisymmetric* instability in *unmagnetized, razor-thin* disks.

Clearly, the presence of magnetic fields plays a stabilizing role, decreasing the critical value to $Q_c = (1 + 1/\beta)^{-1/2}$, with the plasma parameter $\beta \equiv c_s^2/v_A^2$. Finite disk thickness decreases the critical value, as well. For isothermal, unmagnetized, self-gravitating disks, equation (2.1) yields $Q_c = 0.65$. Galactic disks are further compressed by the external stellar gravity. In the solar neighborhood, the strength of the external gravity is comparable to the self-gravity at one scale height, and in this situation equation (2.1) gives $Q_c \sim 0.75$, 0.72, and 0.57 for $\beta = \infty$, 10, and 1 cases, respectively (Kim, Ostriker, & Stone 2002).

On the other hand, the presence of a dynamically-active stellar disk helps destabilize the system. Jog & Solomon (1984) and many other authors analyzed axisymmetric stability of two-component (gas + stars) disks, treating the stellar disk as an isothermal fluid. Using a collisionless description of the stars, Rafikov (2001) derived a dispersion relation for axisymmetric waves in the combined, razor-thin disks. Kim & Ostriker (2006b) extended Rafikov's work to allow for the effect of finite disk thickness. When the stellar parameters are similar to the solar neighborhood conditions, they found $Q_c = 1.27$ for (unrealistically) razor-thin disks, while $Q_c = 0.67$ when both disks possess realistic scale heights, implying that the stabilizing effect of finite disk thickness is considerable. Although observed star formation thresholds at $Q_{\rm th} \sim 1.4$ have often been attributed to gravitational instability in two-component disks (e.g., Martin & Kennicutt 2001), it should not be a consequence of *axisymmetric* gravitational instability.

2.2. *Swing amplification*

Since real perturbations are more likely non-axisymmetric, axisymmetric gravitational instability, albeit mathematically simple, would not be readily materialized in real disk galaxies. In more general, non-axisymmetric cases, perturbations are able to amplify either through swing amplification or magneto-Jeans instability.

Swing amplification arises due to the conspiracy among background shear, epicyclic shaking, and self-gravity (e.g., Toomre 1981). The kinematics of background shear causes the wavefronts of disturbances to rotate from leading to trailing, which occurs in the same sense as epicyclic motions. Consequently, fluid elements stay longer in wave crests, enhancing self-gravity and amplifying perturbations. Swing amplification is not a true instability but a transient mechanism, efficient only when disturbances are loosely wound. The density amplification factor is largest when the local shear rate $q \equiv -d\ln\Omega/d\ln R \sim 1$, a condition easily met at outer galactic disks, while tending to zero as q decreases.

While swing amplification in the linear theory yields an amplification factor that is a continuous function of Q, numerical simulations show that it exhibits nonlinear Q threshold behavior for gravitational runaway. That is, when $Q < Q_c$, swing amplification puts the system in a state where nonlinear interactions of swing-amplified filaments eventually cause bound cloud formation, while disks with $Q > Q_c$ remain stable with only mildly fluctuating density fields. Bound clouds that form in unstable models have a typical mass of a few $10^7 M_\odot$, similar to the characteristic Jeans mass of initial disks. For razor-thin, gas-only disks, Kim & Ostriker (2001) found $Q_c \sim 1.2 - 1.4$ for $\beta = 1 - \infty$, indicating that Q_c for nonlinear swing amplifier is insensitive to the strength of azimuthal magnetic fields. Strong density fluctuations associated with magnetorotationally-driven turbulence in a vertically stratified disk increase Q_c to 1.6 (Kim, Ostriker, & Stone 2003).

Kim & Ostriker (2006b) studied the effects on Q_c of a live stellar component as well as finite disk thickness, by following the orbits of collisionless stars using a particle-mesh method. They found that the two effects nearly cancel each other, giving $Q_c \sim 1.4$ for the stellar parameters corresponding to the solar neighborhood. This Q_c value is consistent with the recent results of Li, Mac Low & Klessen (2005) for gravitational runaway in global models. The quantitative agreement between the numerically-obtained Q_c and observationally determined thresholds for star formation suggests that nonlinear swing amplification of *non-axisymmetric* disturbances may be responsible for star formation boundaries of disk galaxies.

2.3. Magneto-Jeans instability

Now, consider non-axisymmetric perturbations in disks with no (or weak) shear. The instantaneous dispersion relation then reads

$$\omega^4 - \left[\kappa^2 + (c_s^2 + v_A^2)k^2 - \frac{2\pi G\Sigma|k|}{1+|k|H}\right]\omega^2 + v_A^2 k_y^2\left(c_s^2 k^2 - \frac{2\pi G\Sigma|k|}{1+|k|H}\right) = 0, \qquad (2.3)$$

where $k^2 = k_x^2 + k_y^2$, with k_x and k_y denoting the perturbation wavenumbers in the radial and azimuthal directions, respectively (e.g., Lynden-Bell 1966; Kim, Ostriker, & Stone 2002). Note that equation (2.3) recovers equation (2.1) when $k_y = 0$. When $k_y \neq 0$, the instability criterion is the same as the usual two-dimensional Jeans condition (modified by thick-disk gravity) in the absence of rotation and magnetic fields; that is, magnetic fields removes the stabilizing effect of galactic rotation.

In a rotating disk, a hurdle to overcome for self-gravitating modes to grow is Coriolis forces that cause epicyclic gas motions. When perturbations are non-axisymmetric, azimuthal magnetic fields exert tension forces that resist epicyclic orbits across the field lines. The constraint of potential vorticity conservation does no longer hold, and a contracting region is able to grow. This destabilizing effect of magnetic tension on non-axisymmetric perturbations is in sharp contrast to the stabilizing effect of magnetic pressure on axisymmetric modes. Since the presence of magnetic fields is essential for non-axisymmetric instability, we term this magneto-Jeans instability (MJI). The MJI grows rapidly at a rate of $\sim \Omega^{-1}$ and occurs under low shear conditions; when shear is strong, k_x increases rapidly with time and the sonic term in equation (2.3) dominates eventually. The MJI may be responsible for star burst activity toward the galactic central parts where rotation curves are almost linearly rising. Also, a spatially varying sense of shear makes spiral arms ideal places for the operation of MJI to produce spiral-arm substructures, as discussed in the next section.

3. Cloud formation inside spiral arms

The presence of stellar spiral potential perturbations not only compresses the gas and magnetic fields into spiral shocks but also changes the velocity structure significantly, causing streaming motions near the spiral arms. For isothermal spiral arms, the conservation of potential vorticity requires the Toomre Q parameter and the local shear rate q to vary as

$$Q = Q_0\left(\frac{\Sigma}{\Sigma_0}\right)^{-1/2}, \qquad q = 2 - (2 - q_0)\left(\frac{\Sigma}{\Sigma_0}\right), \qquad (3.1)$$

where Σ_0 is the mean density and Q_0 and q_0 refer to the respective values in the absence of the spiral arm forcing (e.g., Kim & Ostriker 2002). For flat rotation with $q_0 = 1$, spiral arm regions with $\Sigma/\Sigma_0 > 2$ experience shear reversal and thus are not prone to swing amplification. With low surface density, on the other hand, interarm regions achieve stronger forward shear. This spatially varying (i.e., reversed shear followed by normal shear) sense of shear caused by spiral arms maintains the overall shear rate small. With high density, strong magnetic fields, and low net shear, therefore, spiral arm regions are favorable places for the development of MJI (e.g., Elmegreen 1987, 1994).

Numerical simulations of local spiral arms indeed show that spiral arms with sufficient peak density are stable to swing amplification but unstable to MJI. Figure 1 shows evolution of surface density and magnetic fields due to MJI in a two-dimensional thick-disk model presented in Kim & Ostriker (2006a). As perturbations grow, gaseous spurs emerge nearly perpendicularly downstream from the spiral shock and become trailing in

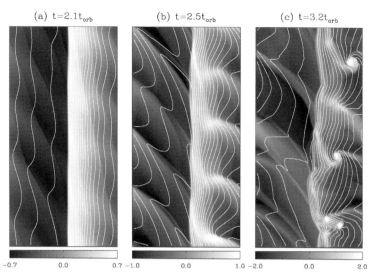

Figure 1. Snapshots of gas surface density (logarithmic gray scale) and magnetic field lines (solid curves) for a local two-dimensional spiral arm model with $Q = 1.2$, $\beta = 10$, and a spiral arm strength of 10%. This model takes allowance for finite disk thickness in self-gravity.

the interarm regions (Fig. 1b), simply reflecting the kinematics of shearing and expanding background flows off the spiral arm. Observed spurs have similar shapes. In fully three-dimensional disks, the mean separation of MJI-driven spurs is about 10 times the Jeans length at the arm peak, which is also consistent with observations (see La Vigne *et al.* 2006). When spurs grow and become sufficiently nonlinear, they experience fragmentation to form gravitationally bound condensations (Fig. 1c). These bound clouds have a mean mass of a few $10^7 M_\odot$, are magnetically supercritical, and would evolve into arm and interarm HII regions. Although magnetic fields pinch inward within the spurs, Figure 1 shows that they almost parallel the spiral arms overall, demonstrating that the material in MJI is collected along the spiral arms.

Wada & Koda (2004) showed that spiral shocks in two-dimensional disks (with the vertical dimension suppressed) are unstable to vorticity-generating wiggle instability. The nature of the wiggle instability is not well known, but numerical work suggests that it is potentially Kelvin-Helmholtz instabilities at a spiral shock; it needs spiral shocks to be quite strong; it requires neither magnetic fields nor self-gravity; and it appears to be suppressed by the equi-partition magnetic fields (Shetty & Ostriker 2006). Most importantly, the wiggle instability is absent in full three-dimensional disks (Kim & Ostriker 2006a). When the vertical dimension is explicitly included, spiral shocks exhibit vigorous non-steady motions and strong vertical shear (Kim, Kim, & Ostriker 2006), preventing the growth of coherent vortical structure that is essential for the wiggle instability. These turbulent gas flows across spiral shocks in vertically extended disks appear to suppress the Parker instability, as well.

4. Summary

Recent numerical magnetohydrodynamic simulations have investigated giant cloud formation in the presence of self-gravity, magnetic fields, galactic differential rotation, dynamically active stars, and/or passive stellar density waves, all of which are crucial for galactic gas dynamics. These works have shown that self-gravity plays a fundamental role in the formation of giant clouds and spiral-arm substructures. When shear is strong, as in outer galaxies without strong spiral arms, swing amplification is subject to nonlinear threshold behavior such that disks with $Q < Q_c$ undergo runaway collapse to form bound clouds of a few $10^7 M_\odot$, roughly the Jeans mass. For swing amplifier, the inclusion of an active stellar component nearly compensates for the stabilizing effect of finite disk thickness. For the parameters representing the solar neighborhood conditions, $Q_c \sim 1.4$, similar to the observationally-inferred thresholds for active star formation. When shear is weak, on the other hand, as in galactic central regions or inside spiral arms, gaseous spurs and giant clouds naturally form as a consequence of magneto-Jeans instability that critically relies on magnetic tension forces to resist the stabilizing Coriolis force of galaxy rotation. The MJI predicts a mean spur separation of 10 times the Jeans length at the arm peak and an average cloud mass of a few $10^7 M_\odot$, consistent with the observed spur spacings and giant cloud masses near spiral arms. Non-steady flows associated with spiral shocks in vertically stratified disks stabilize the wiggle and Parker instabilities, making them unlikely mechanisms for giant cloud formation in real disk galaxies.

Acknowledgements

I gratefully acknowledge Eve Ostriker for her stimulating advice and constructive comments. This work is supported in part by Korea Science and Engineering Foundation (KOSEF) grant R01-2004-000-10490-0.

References

Blitz, L. & Shu, F.H. 1980, *ApJ* 238, 148

Das, M. & Jog, C.J. 1996 *ApJ* 462, 309

Elmegreen, B.G. 1987, *ApJ* 312, 626

Elmegreen, B.G. 1994, *ApJ* 433, 39

Heyer, M.H. & Terebey, S. 1998, *ApJ* 502, 265

Jog, C.J. & Solomon, P.M. 1984, *ApJ* 276, 114

Kennicutt, R.C. 2004, Spitzer press release at http://www.spitzer.caltech.edu/Media/releases/ssc2004-19/ssc2004-19a.shtml

Kim, C.-G., Kim, W.-T. & Ostriker, E.C. 2006, *ApJ* 649, L13

Kim, J., Franco, J., Hong, S.S., Santillán, A. & Martos, M.A. 2000, *ApJ* 531, 873

Kim, W.-T. & Ostriker, E.C. 2001, *ApJ* 559, 70

Kim, W.-T. & Ostriker, E.C. 2002, *ApJ* 570, 132

Kim, W.-T. & Ostriker, E.C. 2006a, *ApJ* 646, 213

Kim, W.-T. & Ostriker, E.C. 2006b, *ApJ* submitted

Kim, W.-T., Ostriker, E.C. & Stone, J.M. 2002, *ApJ* 581, 1080

Kim, W.-T., Ostriker, E.C. & Stone, J.M. 2003, *ApJ* 599, 1157

Kim, W.-T., Hong, S.S., Yoon, S.-C., Lee, S.M. & Kim, J. 1999, *in Numerical Astrophysics* eds. S.M. Miyama, K. Tomisaka, & T. Hanawa (Boston: Kluwer), 111

La Vigne, M.A., Vogel, S.N. & Ostriker, E.C. 2006, *ApJ* in press; astro-ph/0606761

Li, Y., Mac Low, M.M. & Klessen, R.S. 2005, *ApJ* 620, L19

Lynden-Bell, D. 1966, *Observatory* 86, 57

Martin, C.L. & Kennicutt, R.C. 2001, *ApJ* 555, 301

Parker, E.N. 1966, *ApJ* 145, 811

Rafikov, R.R. 2001, *MNRAS* 323, 445

Santillán, A., Kim, J., Franco, J., Martos, M., Hong, S.S. & Ryu, D. 2000, *ApJ* 545, 353

Scoville, N. & Rector T. 2001, HST press release at http://oposite.stsci.edu/pubinfo/PR/2001/10/index.html

Toomre, A. 1981, in: S.M. Fall & D. Lynden-Bell (eds.), *Structure and Evolution of Normal Galaxies* (Cambridge: Cambridge Univ. Press), p. 111

Vogel, S.N., Kulkarni, S.R. & Scoville, N.Z. 1988, *Nature* 334, 402

Wada, K. & Koda, J. 2004, *MNRAS* 349, 270

Discussion

HANASZ: (1.) I don't agree with your statement that the Parker instability is irrelevant. It may be indeed irrelevant if cosmic rays are neglected. However, if cosmic rays are taken into account, the Parker instability becomes very violent and its timescale is very short. (2.) Density condensations in the Parker instability are very small only in isothermal or adiabatic approximation. If realistic cooling and heating mechanisms are taken into account, then it appears that the combined action of Parker and thermal instabilities produced high density condensations, even in a thermally stable medium (see Kisinski & Hanasz, MNRAS 2006).

KIM: Numerical studied (e.g., J. Kim *et al.* 2001) suggest that the Parker instability "alone" even with the effect of cosmic rays included cannot produce large density enhancement. It is basically because the Parker instability is not a runaway process, stabilized by magnetic tension forces. So cosmic rays are unlikely to play a major role in cloud formation inside spiral arms, although it may be important for turbulence generation. (2.) Then it will be very interesting to see how clouds for in magnetized spiral arms under realistic cooling and heating. Time scales for thermal and Parker instabilities are very different with the former longer than the latter by about two orders of magnitude. So, I wonder if the Parker instability grows in an inhomogeneous medium already produced by gas cooling and heating in a paper you mentioned.

Triggered Star Formation in a Turbulent ISM
Proceedings IAU Symposium No. 237, 2006
B. G. Elmegreen & J. Palouš, eds.

© 2007 International Astronomical Union
doi:10.1017/S174392130700172X

Turbulent structure and star formation in a stratified, supernova-driven, interstellar medium

M. K. Ryan Joung[1,2]† and Mordecai-Mark Mac Low[1,2]

[1]Department of Astronomy, Columbia University, New York, NY 10027, USA.

[2]Department of Astrophysics, American Museum of Natural History,
New York, NY 10024, USA. email: mordecai@amnh.org

Abstract. We report on a study of interstellar turbulence driven by both correlated and isolated supernova explosions. We use three-dimensional hydrodynamic models of a vertically stratified interstellar medium run with the adaptive mesh refinement code Flash at a maximum resolution of 2 pc, with a grid size of $0.5 \times 0.5 \times 10$ kpc. Cold dense clouds form even in the absence of self-gravity due to the collective action of thermal instability and supersonic turbulence. Studying these clouds, we show that it can be misleading to predict physical properties such as the star formation rate or the stellar initial mass function using numerical simulations that do not include self-gravity of the gas. Even if all the gas in turbulently Jeans unstable regions in our simulation is assumed to collapse and form stars in local freefall times, the resulting total collapse rate is significantly lower than the value consistent with the input supernova rate. The amount of mass available for collapse depends on scale, suggesting a simple translation from the density PDF to the stellar IMF may be questionable. Even though the supernova-driven turbulence does produce compressed clouds, it also opposes global collapse. The net effect of supernova-driven turbulence is to inhibit star formation globally by decreasing the amount of mass unstable to gravitational collapse.

Keywords. hydrodynamics, ISM: kinematics and dynamics, methods: numerical, turbulence

After briefly describing our numerical model of the ISM, I will concentrate on two results from our work that are most relevant to this meeting: (1) I will report on the density and velocity power spectra of the ISM driven by multiple interacting supernovae (SNe); (2) I will try to determine if SN-driven turbulence enhances or inhibits star formation (see contributions by Vázquez-Semadeni and by Clarke in this volume).

1. Model

We set up a 3D stratified atmosphere elongated in the z direction to study the vertical structure. It is a small patch of the galaxy, $(500 \text{ pc})^2$ in area, and extends from -5 to $+5$ kpc in z. We use periodic boundary conditions in the x and y directions, and outflow boundary condition at the top and bottom surfaces.

We add instantaneous, localized thermal energy that accounts for both isolated and correlated SNe. (By correlated SNe, I mean superbubbles from OB associations.) The SN rate declines exponentially as the distance from the midplane increases, with different scale heights for *type I* and *type II* SNe. The explosion rate is an input parameter, and for our fiducial model, we use the Galactic SN rate.

† Present address: Department of Astrophysical Sciences, Princeton University, Princeton, NJ 08544, USA. email: joung@astro.princeton.edu

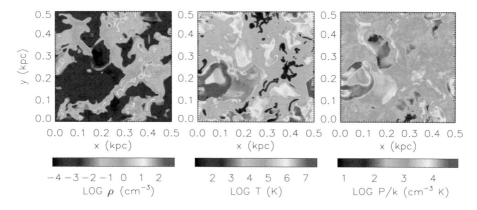

Figure 1. Cuts through the midplane of our model showing distributions of (*a*) density, (*b*) temperature, and (*c*) pressure at $t = 79.3$ Myr.

We include diffuse heating (Wolfire *et al.* 1995) and radiative cooling (Dalgarno & McCray 1972; Sutherland & Dopita 1993) terms and a static vertical gravitational field for stellar and dark halo components (Kuijken & Gilmore 1989). However, we do not include self-gravity of the gas, magnetic field, or differential rotation of the Galaxy.

Our simulations are performed using Flash (Fryxell *et al.* 2000). It is a grid-based hydrodynamics code with adaptive mesh refinement (AMR) capability. The AMR allows us to focus resolution only where it is necessary, i.e., where density and pressure gradients are large. We end up maximally resolving the region near the midplane of the galaxy, $|z| \lesssim 300$ pc, because the bulk of the gas and most of the SNe reside there. See Joung & Mac Low (2006, hereafter JM06) for more details of the model.

2. Results

Figure 1 (*a,b,c*) shows the density, temperature, and pressure distributions in the midplane. The hot regions, i.e. red and yellow regions in figure 1 (*b*), always correspond to relatively low densities. Correlated SNe produce much of the volume occupied by hot gas. Cold clouds, the red, high density regions in figure 1 (*a*), are filamentary in shape, and are surrounded by thick layers of warm gas. They have temperatures as low as 10 K, so they resemble molecular clouds observed in the ISM. These clouds form when SN blast waves interact with each other. Note that they are present even though self-gravity of the gas is not included in our model.

Both density and temperature vary by about 7 orders of magnitude, but because of the approximate inverse relationship they have, for most the volume, the thermal pressure varies only by 2 orders of magnitude. The pressure PDF shows that there is as much gas below the average pressure as above it. The variation in pressure is inconsistent with ISM models based on pressure equilibrium between phases (e.g., Field *et al.* 1969). This implies that we need a more dynamic picture (Mac Low & Klessen 2004).

Figure 2 (*a,b*) shows the density and kinetic energy power spectra from our model. The density spectrum is completely dominated by clumpy cold clouds, because they have orders of magnitude higher densities than the hot or the warm medium.

To characterize the distribution of kinetic energy, people have usually plotted the power spectrum of the velocity field. However, in a strongly compressible medium, a more dynamically meaningful quantity is the power spectrum of $\sqrt{\rho}\,v$, which I call the

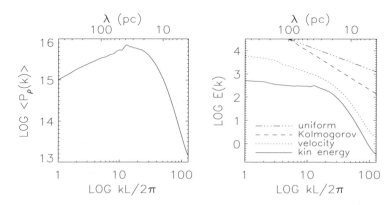

Figure 2. (*a*) Angle-averaged density power spectrum, displaying a wide peak around $kL/2\pi \approx 20$. The box size $L = 0.5$ kpc. (*b*) Kinetic energy spectrum (solid) and angle-averaged velocity power spectrum (dotted). Because of the highly intermittent density structure, the velocity power spectrum is not parallel to the kinetic energy spectrum, especially at large scales. To guide the eye, two straight lines are plotted: the Kolmogorov energy spectrum (dashed) and the spectrum containing an equal amount of energy per decade (dot-dashed).

kinetic energy spectrum. In a uniform medium, this distinction will be unnecessary. The velocity spectrum does not distinguish between high density and low density regions, and in our case has more power at large scales, so it is slightly steeper than this kinetic energy spectrum. Instead of single driving and dissipation scales and a constant slope of $-5/3$ in between, as predicted by Kolmogorov (dashed line in figure 2(*b*)), the kinetic energy spectrum of a SN-driven ISM (solid curve) shows that there is no single effective driving scale; in fact, we see energy injection over a range of scales, as expected in a medium driven by multiple interacting explosions. Despite that, though, 90% of the total kinetic energy is contained in scales shortward of 190 pc (JM06). On a side note, let me remind you that our model included only SN driving. Gravitational instability, for example, could have contributed more power on large scales, if we had included it.

Of course, what is actually observable is the column density power spectrum. We have created power spectra of the total gas column density from our ISM simulations and compared them with the two-dimensional power spectrum of the H I column density in the LMC from Elmegreen, Kim, & Staveley-Smith (2001). Although our spectrum does not have a constant slope, we find that the overall slope is intriguingly close to the value $(-8/3)$ measured from the H I observations.

In our model, cold dense clouds form directly in the turbulent flow. We may ask: what fraction of these clouds are gravitationally unstable, given the density and velocity structures in our model? Although our numerical model does not include self-gravity, it does represent other physical processes reasonably well, such as SN shock compressions or thermal instability, Hence the formation of unstable clouds, if any, must have been induced purely by SN-driven turbulence.

To do this, we took a rectangular region near the midplane, tiled it with smaller cubes, and applied a simple criterion for gravitational collapse, i.e., a modified Jeans criterion (Chandrasekhar 1951) to identify Jeans unstable boxes: $M_{box}/M_J > 1$, where the Jeans length $\lambda_J = (\pi/G\bar{\rho})^{1/2}\sigma_{tot}$ and $\sigma_{tot} \equiv (\bar{c}_s^2 + \frac{1}{3}\sigma^2)^{1/2}$. We also assume that the gas in the unstable boxes turn into stars in the local free-fall time, with some efficiency factor. We repeated this procedure for various box sizes. Then we can compare the total gravitational collapse rate and the input SN rate (which can be converted to a star formation rate using some reasonable IMF). In steady state, the two rates should be equal. In this way,

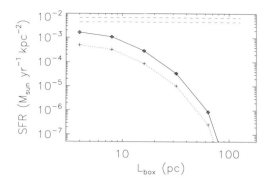

Figure 3. Predicted star formation rate from the model plotted against the subbox sizes used. The dotted line is drawn assuming that 30% of the mass in Jeans unstable regions turn into stars. The red dashed lines show the star formation rates consistent with the assumed Galactic SN rate, if 130 or 200 $M_\odot$ of stellar mass is required per SN.

we can quantitatively check the validity of our high-resolution simulations as a model for star formation (JM06).

While doing this analysis, we obtained as a by-product a simple relationship between the total velocity dispersion (σ_{tot}; thermal+turbulent) and the average gas density within each box ($\bar{\rho}$): $\sigma_{tot} \propto \bar{\rho}^{-1/2}$, which implies interestingly $\bar{\rho}\,\sigma_{tot}^2 \approx const.$ (Joung, Mac Low, & Bryan, in preparation). But due to the time limit, I will not discuss it here.

Figure 3 displays the result of our simple calculation. For boxes that are bigger than 20 pc on a side, that is, where turbulent motions are resolved, even if we assume all the gas in turbulent Jeans unstable boxes collapses and forms stars within the local freefall time, the resulting star formation rate will remain lower than the value consistent with our input SN rate (red lines). This shows that it may be misleading to use the statistics of turbulent fluctuations to predict physical properties of star-forming regions. The discrepancy arises probably from the neglect of self-gravity of the gas in our model. Hence, SN-compressions alone can indeed trigger star formation, but not all of it, maybe only ∼10% of the total star formation rate, comparable to the estimate by Mizuno (this volume).

In contrast, the initial setup of our simulation was gravitationally stable on all scales. It was, however, out of thermal equilibrium. Without extra stirring from SNe, the gas would have promptly cooled and collapsed into a thin she*et al.*ng the midplane, making *more* gas gravitationally unstable. Therefore, the net effect of SN-driven turbulence in this case was to *inhibit* collapse globally.

Acknowledgements

We would like to thank G. Bryan, I. Goldman, Y. Li, C. McKee, and J. Oishi for useful comments and stimulating discussions. M. K. R. J. was supported by an AMNH Graduate Student Fellowship for the duration of this work. M.-M. M. L. acknowledges support by NSF grants AST03-07793 and AST03-07854. The software used in this work was in part developed by the DOE-supported ASCI/Alliance Center for Astrophysical Thermonuclear Flashes at the University of Chicago. Computations were performed at the Pittsburgh Supercomputing Center.

References

Chandrasekhar S. 1951, *Proc. Royal Soc. London A* 210, 26

Dalgarno, A. & McCray, R.A. 1972, *ARA&A* 10, 375

Elmegreen, B.G., Kim, S. & Staveley-Smith, L. 2001, *ApJ* 548, 749

Field, G.B., Goldsmith, D.W. & Habing, H.J. 1969, *ApJ* 155, L49

Fryxell, B., Olson, K., Ricker, P., Timmes, F.X., Zimgale, M., *et al.* 2000, *ApJS* 131, 273

Joung, M.K.R. & Mac Low, M.-M. 2006, *ApJ* in press (JM06)

Kuijken, K. & Gilmore, G. 1989, *MNRAS* 239, 605

Mac Low, M.-M. & Klessen, R.S. 2004, *Rev. Mod. Phys.* 76, 125

Sutherland, R.S. & Dopita, M.A. 1993, *ApJS* 88, 253

Wolfire, M.G., McKee, C.F., Hollenbach, D., Tielens, A.G.G.M. & Bakes, E.L.O. 1995, *ApJ* 443, 152

Discussion

PALOUŠ: Does σ in your modified Jeans mass include bulk motions?

JOUNG: Yes. For any given subbox, I compute the rms velocity dispersion of all the cells within it. It is actually what I called σ in the expression for σ_{tot}.

ELMEGREEN: What fraction of your supernova remnants look like shells, and do you have clustered supernovae?

JOUNG: I assumed a power-law distribution for the number of supernovae in clusters, following Clarke & Oey. A small fraction of the supernova remnants do look like shells as you can see in figure 1(b), but I have not quantified the fraction.

GOLDMAN: Your kinetic energy power spectrum shows a break around 20-30 pc. Is your power spectrum steeper than the Kolmogorov spectrum on small scales?

JOUNG: The shape of the kinetic energy power spectrum below about 20 pc is determined by numerics, so the spectrum shortward of $\sim$20 pc should not be trusted.

NAKAMURA: You said that turbulence inhibits star formation, but that might change when you include magnetic field in your model.

JOUNG: Since the turbulence in our model is driven, rather than decaying, it decreases the star formation efficiency, just as Enrique has pointed out. However, I agree that adding magnetic field to the model is important. It is probably the next step one should take.

Triggered Star Formation in a Turbulent ISM
Proceedings IAU Symposium No. 237, 2006
B. G. Elmegreen & J. Palouš, eds.

© 2007 International Astronomical Union
doi:10.1017/S1743921307001731

Cold H I in turbulent eddies and galactic spiral shocks

Steven J. Gibson[1], A. Russell Taylor[2], Jeroen M. Stil[2], Christopher M. Brunt[3], Dain W. Kavars[4] and John M. Dickey[4,5]

[1]Arecibo Observatory, National Astronomy and Ionosphere Center, Arecibo, PR 00612, U.S.A.

[2]Dept. of Physics & Astronomy, University of Calgary, Calgary, Alberta T2N 1N4, Canada

[3]School of Physics, University of Exeter, Exeter, United Kingdom EX4 4QL

[4]Department of Astronomy, University of Minnesota, Minneapolis, MN 55455, U.S.A.

[5]School of Mathematics and Physics, University of Tasmania, Hobart, TAS 7001, Australia

Abstract. H I 21cm-line self-absorption (HISA) reveals the shape and distribution of cold atomic clouds in the Galactic disk. Many of these clouds lack corresponding CO emission, despite being colder than purely atomic gas in equilibrium models. HISA requires background line emission at the same velocity, hence mechanisms that can produce such backgrounds. Weak, small-scale, and widespread absorption is likely to arise from turbulent eddies, while strong, large-scale absorption appears organized in cloud complexes along spiral arm shocks. In the latter, the gas may be evolving from an atomic to a molecular state prior to star formation, which would account for the incomplete HISA-CO agreement.

Keywords. radiative transfer, surveys, ISM: clouds, ISM: evolution, ISM: kinematics and dynamics, ISM: structure, Galaxy: structure, radio lines: ISM

1. Imaging the cold atomic medium

Cold atomic gas contains a large fraction of the mass of the interstellar medium (ISM) and is a critical precursor to molecular cloud formation. However, this "cold atomic medium" is hard to map in isolation, since warmer gas is often brighter in traditional H I 21cm-line emission observations. Fortunately, with proper angular resolution, cold atomic clouds can be imaged as H I self-absorption (HISA) against warmer background H I emission (Gibson *et al.* 2000). Recent large-scale radio synthesis surveys like the Canadian and VLA Galactic plane surveys (CGPS: Taylor *et al.* 2003; VGPS: Stil *et al.* 2006) are both well suited to HISA studies.

Figure 1 compares some sample CGPS HISA to CO emission. The intricate structure of the HISA clouds is clear, as is their frequent lack of apparent CO. This lack is a puzzle unless significant H_2 is present without CO, since the low temperatures of many HISA features ($T < 50$ K) are hard to explain without molecular gas. But if these clouds are evolving rather than stable objects, then perhaps many have not yet formed enough CO to detect (e.g., Klaassen *et al.* 2005).

Using an algorithm to identify and extract HISA features in the H I data, we recently published a HISA census of the $73° \times 9°$ area covered by the first 5 years of the CGPS (Gibson *et al.* 2005). As shown in **Figure 2**, a low-level froth of weak, disorganized HISA is found throughout the regions of the CGPS where the background emission is bright enough for the HISA to be reliably detected. By contrast, stronger absorption is organized into discrete clouds and complexes.

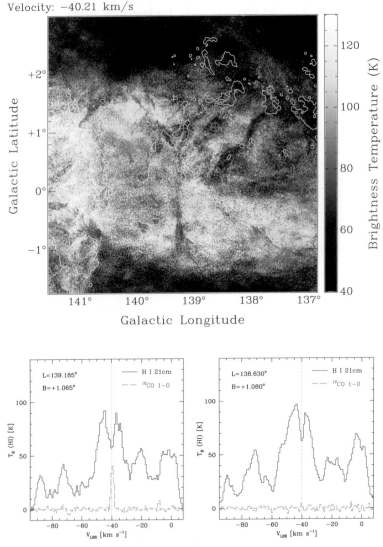

Figure 1. Sample HISA in the CGPS. The upper panel shows an H I channel map with ^{12}CO 1-0 contours from Heyer *et al.* (1998); the LSR velocity places this gas in the Perseus spiral arm some 2 kpc away. H I and CO spectra at the two marked positions are plotted in the lower panels, showing that HISA is found with and without CO, with the latter case more common in the outer Galaxy.

2. Velocity perturbation mechanisms

HISA requires background line emission at the same radial velocity as the foreground cloud. Consequently, HISA radiative transfer probes both the temperature and the velocity field of Galactic H I. The CGPS is primarily in the outer Galaxy, where pure differential rotation allows only one position along the line of sight to have a particular velocity. Since we detect HISA in the outer Galaxy, the real velocity field must be

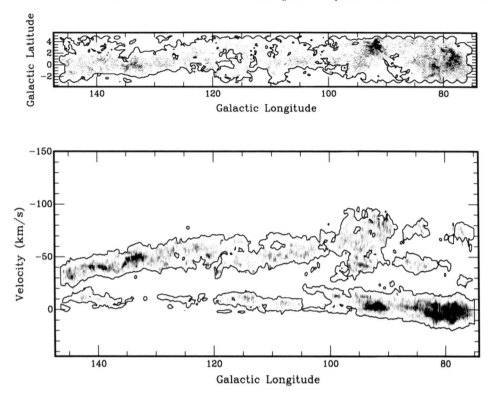

Figure 2. Longitude-latitude and longitude-velocity maps of HISA extracted from the initial $73° \times 9°$ portion of the CGPS, in which the HISA ON-OFF absorption amplitude is integrated over velocity (top) and latitude (bottom), with stronger absorption being darker. Contours mark where the H I emission background becomes too faint for reliable HISA detection.

perturbed from this simple rotation model to provide background fields to absorb against. Two known mechanisms for this are turbulence and spiral density waves.

The weak, widespread HISA in **Figure 2** can be explained as an ambient froth of cold atomic gas in the ISM made visible by turbulent eddies; perhaps the cold gas is even a product of convergent turbulent flows, as some models suggest (e.g., Vázquez-Semadeni *et al.* 2006). The stronger HISA, however, is concentrated into distinct complexes, especially along the Perseus arm near $-40\,\text{km s}^{-1}$, arguing for a more organized mechanism for these features. The very cold temperatures implied by this strong absorption could arise from gas that is forming H_2 but has not yet formed much CO.

3. Interpreting the galactic HISA distribution

Figure 3 shows a new HISA survey incorporating the VGPS and extensions to the CGPS (Gibson *et al.* 2006). This plot includes CO contours and a curve marking the maximum velocity departure from circular rotation predicted by Roberts (1972) for the Perseus spiral shock. Apart from some irregular ISM structure Roberts did not model, the strong Perseus HISA lies near the shock curve but at less extreme velocities, consistent with clouds lying just downstream of the spiral shock. Similar but fainter shock-related HISA may be seen in the Outer arm at negative velocities in the VGPS data. Both arms have a poor HISA-CO match, as would be expected for evolving gas; in this scenario,

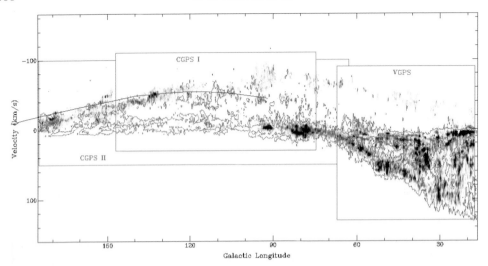

Figure 3. Longitude-velocity HISA distribution over the entire area covered by the original CGPS, the CGPS II extension, and the VGPS. Contours show molecular gas traced by Dame *et al.* (2001) ^{12}CO 1-0 emission, and the curved line marks the maximum velocity departure from Galactic rotation predicted by the Perseus arm spiral shock model of Roberts (1972).

the visible CO traces more evolved gas further downstream, where little background H I emission is left to show the remaining cold atomic gas in the CO clouds as HISA.

In the inner Galaxy, the picture is more complex. A much larger amount of HISA is seen, since even simple rotation provides near and far points on each sight line with the same velocity. This allows a much more widespread H I emission background, which may account for the stronger HISA-CO agreement and the lack of clearly-defined spiral arms. However, the latter requires a significant amount of cold interarm H I in the disk. Whether the interarm cold cloud population is related to the outer-Galaxy turbulent HISA population is still under investigation.

Acknowledgements

This work has been supported by the National Astronomy and Ionosphere Center operated by Cornell University under Cooperative Agreement with the U.S. National Science Foundation and by grants from the Natural Sciences and Engineering Research Council of Canada to the University of Calgary.

References

Dame, T. M., Hartmann, D. & Thaddeus, P. 2001, *ApJ* 547, 792

Gibson, S. J., Taylor, A. R., Dewdney, P. E. & Higgs, L. A. 2000, *ApJ* 540, 851

Gibson, S. J., Taylor, A. R., Stil, J. M., Brunt, C. M., Kavars, D. W. & Dickey, J. M. 2006, in preparation

Gibson, S. J., Taylor, A. R., Higgs, L. A., Brunt, C. M. & Dewdney, P. E. 2005, *ApJ* 626, 195

Heyer, M. H., Brunt, C., Snell, R. L., Howe, J. E., Schloerb, F. P. & Carpenter, J. M. 1998, *ApJS* 115, 241

Klaassen, P. D., Plume, R., Gibson, S. J., Taylor, A. R. & Brunt, C. M. 2005, *ApJ* 631, 1001

Roberts, W. W. 1972, *ApJ* 173, 259

Stil, J. M., Taylor, A. R., Dickey, J. M., Kavars, D. A., Martin, P. G., Rothwell, T. A., Boothroyd, A., Lockman, F. J. & McClure-Griffiths, N. M. 2006, *AJ* 132, 1158

Taylor, A. R., *et al.* 2003, *AJ* 125, 3145

Vázquez-Semadeni, E., Ryu, D., Passot, T., González, R. F. & Gazol, A. 2006, *ApJ* 643, 245

Discussion

FUKUI: An important aspect of cold HI is that we can detect enveloping lower density gas that is not detectable in CO. This part of the mostly atomic gas will be important to study cloud formation, and sub-mm CI surveys on a large scale will be coming soon to provide more knowledge of this part of the less dense gas.

GIBSON: Thank you. I'm pleased to hear of the coming CI surveys, which should be very useful for probing the state of the gas. I will only add that the cold HI does not always appear in CO envelopes; it can also be seen in cores or even throughout the cloud, depending on the object. In many cases of course, it appears without CO at all, in which case the molecular content of the cloud is unclear.

ELMEGREEN: I think the lack of CO in the outer galaxy HISA is not surprising considering the general gradient of molecular fraction with radius. This is presumably the result of a pressure gradient because low pressure regions have low density cloud envelopes and then a high column density is needed to shield molecules against photodissociation.

GIBSON: That is a very good point. However, are you saying that we should expect no H2, or merely no CO? If there are no molecules at all, then it becomes more difficult to explain some of the very low temperatures found for HISA clouds, which are typically less than 50 K and can be much less in some cases. To me the most likely explanation in this case is that the clouds are unstable, transitional objects.

Triggered Star Formation in a Turbulent ISM
Proceedings IAU Symposium No. 237, 2006
B. G. Elmegreen & J. Palouš, eds.

© 2007 International Astronomical Union
doi:10.1017/S1743921307001743

Dense gas formation triggered by spiral density wave in M31

T. Tosaki[1], Y. Shioya[2], N. Kuno[1], K. Nakanishi[1], T. Hasegawa[3], S. Matsushita[4], K. Kohno[5], R. Miura[6], Y. Tamura[6], S. K. Okumura[1] and R. Kawabe[6]

[1]Nobeyama Radio Observatory, Minamimaki, Minamisaku, Nagano, 384-1805, Japan
email: tomoka@nro.nao.ac.jp

[2]Physics Department,Graduate School of Science and Engineering,Ehime University, 2-5
Bunkyo-cho, Matsuyama, Ehime 790-8577

[3]Gunma Astronomical Observatory, Nakayama, Takayama, Agatsuma, Gunma 377-0702

[4]Institute of Astronomy and Astrophysics, Academia Sinica, P.O.Box 23-141, Taipei 106,
Taiwan, R.O.C.

[5]Institute of Astronomy, University of Tokyo, 2-21-1 Osawa,Mitaka, Tokyo 181-8588

[6]National Astronomical Observatory of Japan, 2-21-1 Osawa,Mitaka, Tokyo 181-8588

Abstract. We present the high-resolution ^{12}CO($J = 1-0$), ^{13}CO($J = 1-0$) and ^{12}CO($J = 3-2$) maps toward a GMA located on the southern arm region of M31 using Nobeyama 45 m and ASTE 10 m telescopes. The GMA consists of two velocity-components, i.e., red and blue. The blue component shows a strong and narrow peak, whereas the red one shows a weak and broad profile. The red component has a lower ^{12}CO($J = 1-0$)/^{13}CO($J = 1-0$) ratio (~ 5) than that of the blue one (~ 16), indicating that the red component is denser than the blue one. The red component could be the decelerated gas if we consider the galactic rotational velocity in this region. We suggest that the red component is "post shock" dense gas decelerated due to a spiral density wave. This could be observational evidence of dense molecular gas formation due to galactic shock by spiral density waves.

We also present results from on-going observations toward NGC 604, which is the supergiant HII region of M33, using Nobeyama 45 m and ASTE 10 m telescopes. The ratio of ^{12}CO($J = 3 - 2$) to ^{12}CO($J = 1 - 0$) ranges from 0.3 to 1.2 in NGC 604. The ^{12}CO($J = 1 - 0$) map shows the clumpy structure while ^{12}CO($J = 3 - 2$) shows a strong peak near to the central star cluster of NGC 604. The high ratio gas is distributed on the arc-like or shell-like structure along with Hα emission and HII region detected by radio continuum. These suggest that the dense gas formation and second generation star formation occur in the surrounding gas compressed by the stellar wind and/or supernova in central star cluster.

Keywords. interstellar medium, star formation, spiral arm

1. Introduction

It is well established that molecular gas in spiral arms shows very large structures, often referred as Giant Molecular Associations (GMAs; Vogel *et al.* 1988; Rand & Kulkarni 1990). Their typical sizes and masses are a few 100 pc and $\sim 10^7 M_\odot$, respectively (Rand & Kulkarni 1990; Lundgren *et al.*2004), and observational studies of GMAs in galaxies provide us with invaluable clues on the physics which governs the large scale star formation in the disk regions of galaxies (e.g., Kuno *et al.* 1995; Tosaki *et al.* 2003; Sakamoto 1996; Wong & Blitz 2002; Lundgren *et al.* 2004). However, GMAs are much larger than Giant Molecular Clouds (GMCs), a major form of molecular contents in our Galaxy, with a typical size and mass of a few 10 pc and $10^5 M_\odot$ (Scoville & Sanders 1987). Further, massive stars are formed in the densest regions of these GMCs, not in the diffuse part of the clouds. Our main interest is to understand the physical process of star formation inside Giant Molecular Associations (GMAs) along the spiral arms, i.e., physical and evolutional links among GMAs (a few 100 pc), GMCs (a few 10 pc), and dense cores ($\leqslant$ 1 pc). To address these issues, we carry out the observations of molecular clouds toward the GMA located on the spiral arm of M31 and the supergiant HII region NGC 604 of M33.

2. Dense gas formation triggered by spiral density wave in M31

We performed observations of $^{12}CO(J = 1-0)$ and $^{13}CO(J = 1-0)$ emissions toward the south bright arm region of M31 on 2003 December with the 45-m telescope at the Nobeyama Radio Observatory (NRO). The full width at a half-power beam (FWHP) was $16''$ and $17''$ at rest frequencies of $^{12}CO(J = 1 - 0)$ (115.2 GHz) and $^{13}CO(J=1-0)$ (110.2 GHz), respectively. The size of the observed region is $3' \times 4'$ (0.6 kpc $\times$ 0.8 kpc) with a grid spacing of $10.''3$. We used the 25 BEam Array Receiver System (BEARS), which can simultaneously observe twenty-five positions separated on the sky by $41''$ each (Sunada *et al.* 2000).

We also have carried out observations of $^{12}CO(J = 3-2)$ emission toward a part of the region observed with $^{12}CO(J = 1 - 0)$ and $^{13}CO(J = 1 - 0)$ emissions on 2005 August using ASTE 10-m submillimeter telescope located at the Atacama desert in Chile (Ezawa *et al.* 2004). The beam size at the $^{12}CO(J = 3 - 2)$ observation was $23''$.

The total integrated intensity maps of emission lines are presented in Fig.1. We find that the averaged $^{12}CO(J = 1-0)/^{13}CO(J = 1-0)$ and the $^{12}CO(J = 3-2)/^{12}CO(J = 1 - 0)$ ratios of the GMA in M31 are 10 and 0.3, respectively. Large Velocity Gradient calculations show that these ratios correspond to the gas density and temperature of 10^{2-3}cm^{-3} and 10 - 20 K, respectively.

The GMA consists of two distinct velocity-components, i.e., red and blue (see Fig.2). The blue component shows a strong and narrow peak, whereas the red one shows a weak and broad profile. The red component has a lower $^{12}CO/^{13}CO$ ratio (~ 5) than that of the blue one (~ 16). Because LVG calculation also shows that the gas density with the lower $^{12}CO(J = 1 - 0)/^{13}CO(J = 1 - 0)$ ratio is larger than that with high ratio at low temperature region such as galactic disk, these ratios indicate that the red component is denser than the blue one. The red component could be the decelerated gas taking into account of the galactic rotational velocity in this region. Based on these results, we suggest that the red component is "post shock" dense gas decelerated due to a spiral density wave. This could be observational evidence of dense molecular gas formation due to galactic shock by spiral density waves.

3. Dense gas and star formation triggered by first generation star formation in NGC 604

$^{12}CO(J = 3 - 2)$ observations toward NGC 604 in M33 were performed with ASTE 10-m telescope on July 2006 - August 2006. The size of the observed region is $5' \times 5'$ (0.6 kpc $\times$ 0.8 kpc) and we employed On-the-fly method.

We also performed observations of $^{12}CO(J = 1-0)$ toward NGC 604 on 2005 December - 2006 March with the 45-m telescope at NRO. We used the BEARS. The full width at a half-power (FWHP) of beam was $16''$ at rest frequency of $^{12}CO(J = 1 - 0)$ (115.2 GHz). We made a convolution of the map to $25''$, to compare with the data by ASTE 10-m telescope.

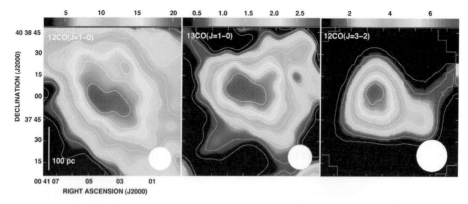

Figure 1. M31 GMA: total integrated intensity maps of ^{12}CO$(J = 1 - 0)$, ^{13}CO$(J = 1 - 0)$ and ^{12}CO$(J = 3 - 2)$. Bottom white filled circles in each panel are beam sizes, corresponding to 54 pc, 57 pc, and 73 pc, respectively, at the distance of M31 of 690 kpc.

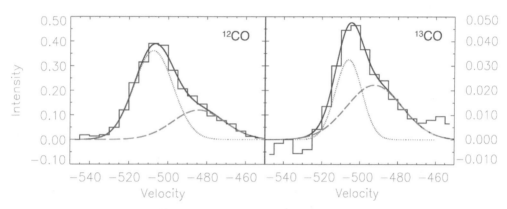

Figure 2. GMA profile of ^{12}CO$(J = 1 - 0)$ and ^{13}CO$(J = 1 - 0)$. Dashed and dotted lines indicate red and blue components. Solid line shows the sum of red and blue components.

The results are shown in Fig. 3. The emission of ^{12}CO$(J = 1 - 0)$ was distributed around NGC 604 and we found several clumpy structures in the map. A typical size of the clumps in the map are ~ 100 pc. The molecular clouds around NGC 604 are mainly located at the south of the center of NGC 604, where the central star cluster is located. The southern side corresponds to the upstream side in the NGC 604 with consideration of the galactic rotation of M33.

We found an arc-like or shell like distribution of high ^{12}CO$(J = 3-2)/^{12}$CO$(J = 1-0)$ ratio gas surrounding the central star cluster of NGC 604. The ratio in the arc-like structure is larger than 1. The Large Velocity Gradient calculation suggests that such high ratio is observed in gas with higher temperature than 60 K and higher density of 10^{3-4}cm^{-3}. In addition, there is also arc-like Hα emission around the central star cluster (Gómez de Castro *et al.* 2000), where several compact HII regions detected by radio continuum (Churchwell & Goss 1999) are embedded in them. The shape of the Hα distribution shows similarity to that of the high ratio gas distribution. Putting together these facts, the arc-like distribution of warm and dense gas together with on-going star formation is surrounding around central star cluster.

Based on these results, we propose the following scenario; First, stars were formed at the northern part of GMA as "first generation star formation". They are observed as

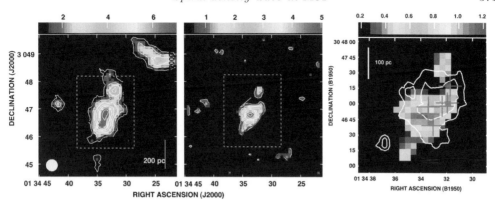

Figure 3. NGC 604: maps of total integrated intensities of $^{12}CO(J = 1 - 0)$ (left) and $^{12}CO(J = 3 - 2)$ (middle), and $^{12}CO(J = 3 - 2)/^{12}CO(J = 1 - 0)$ ratio (right), respectively. The area of the ratio map is shown as dashed squares in the integrated intensity map. The $^{12}CO(J = 1 - 0)$ maps were convolved to 25'' beam (white filled circle) which is same as that of $^{12}CO(J = 3 - 2)$ map. Crosses in the ratio map indicate the HII regions detected by radio continuum.

the central star cluster now. After that, the stellar wind and/or supernova from them compressed surrounding interstellar medium. As a result, dense gas was formed there, and such dense gas is distributed around the central star cluster and shows the high $^{12}CO(J = 3 - 2)/^{12}CO(J = 1 - 0)$ ratio. This is observed as arc-like distribution of high ratio gas. And in such dense gas, new stars are forming, and this is "second generation star formation" triggered by first generation star formation. These are seen as a radio compact HII regions now.

Acknowledgements

We would like to thank the staff of Nobeyama Radio Observatory for their kind support for our observations. The authors wish to express their deep gratitude to Dr. Shigehisa Takakuwa for kindly providing us with his LVG code. This study was financially supported by the MEXT Grant-in-Aid for Scientific Research on Priority Areas No. 15071202. The Nobeyama Radio Observatory is a branch of the National Astronomical Observatory of Japan, the National Institutes of Natural Sciences (NINS).

References

Churchwell, E. & Goss, W. M. 1999, *ApJ* 514, 188
Ezawa, H., Kawabe, R., Kohno, K. & Yamamoto, S. 2004, *SPIE* 5489, p. 763
Gómez de Castro, A. I., Sanz, L. & Beckman, J. 2000, *Ap&SS* 272, 15
Kuno, N., Nakai, N., Handa, T. & Sofue, Y. 1995, *PASJ* 47, 745
Lundgren, A. A., Wiklind, T., Olofsson, H. & Rydbeck, G. 2004, *A&A* 413, 505
Rand. R. J. & Kulkarni, S. R. 1990, *ApJ* 349, L43
Sakamoto, K. 1996, *ApJ* 471, 173
Scoville, N. Z. & Sanders, D. B. 1987, in: D. J. Hollenbach & H. A. Thronson, Jr. (eds.), *Interstellar Processes* (Dordrecht: Reidel), p. 21
Sunada, K., Yamaguchi, C., Nakai, N., Sorai, K., Okumura, S. K. & Ukita, N. 2000, *SPIE* 4015, 237
Tosaki, T, Shioya, Y., Kuno, N., Nakanishi, K. & Hasegawa, T. 2003, *PASJ* 55, 605
Vogel, S. N., Kulkarni, S. R. & Scoville, N. Z. 1988, *Nature* 334, 402
Wong, T. & Blitz, L. 2002, *ApJ* 569, 157

Discussion

BIEGING: Have you considered PDR models to explain the enhanced CO(3-2)/CP(1-0) ratios observed in M31 and NGC 604?

TOSAKI: Not yet, but we will compare our results with PDR models.

ROSOLOWSKY: When you observe with the millimeter interferometer do you see GMCs preferentially associated with one component of the line seen in the single dish? (Also, you may find the work of Maiz-Apellaniz *et al.* 2004 (AJ, 128, 1196) relevant to NGC 604 and PDRs). A flux recovery of < 50% is common for interferometer observations of extragalactic GMCs.

TOSAKI: Yes, we indeed detect two components corresponding to pre-shock and post-shock gas in Nobeyama Millimeter Array data. We will also compare our CO(3-2)/CO(1-0) results of M33 with PDR models.

Triggered Star Formation in a Turbulent ISM
Proceedings IAU Symposium No. 237, 2006
B. G. Elmegreen & J. Palouš, eds.

© 2007 International Astronomical Union
doi:10.1017/S1743921307001755

Triggered star formation in the Magellanic Clouds

Kenji Bekki[1]

[1]School of Physics, University of New South Wales, Sydney 2052, Australia
email: bekki@phys.unsw.edu.au

Abstract. We discuss how tidal interaction between the Large Magellanic Cloud (LMC), the Small Magellanic Cloud (SMC), and the Galaxy triggers galaxy-wide star formation in the Clouds for the last ~ 0.2 Gyr based on our chemodynamical simulations on the Clouds. Our simulations demonstrate that the tidal interaction induces the formation of asymmetric spiral arms with high gas densities and consequently triggers star formation within the arms in the LMC. Star formation rate in the present LMC is significantly enhanced just above the eastern edge of the LMC's stellar bar owing to the tidal interaction. The location of the enhanced star formation is very similar to the observed location of 30 Doradus, which suggests that the formation of 30 Doradus is closely associated with the last Magellanic collision about 0.2 Gyr ago. The tidal interaction can dramatically compress gas initially within the outer part of the SMC so that new stars can be formed from the gas to become intergalactic young stars in the inter-Cloud region (e.g., the Magellanic Bridge). The metallicity distribution function of the newly formed stars in the Magellanic Bridge has a peak of [Fe/H] ~ -0.8, which is significantly lower than the stellar metallicity of the SMC.

Keywords. stars: formation, ISM: abundances, galaxies: star cluster

1. Introduction

The Magellanic system composed of the LMC and the SMC is believed to be an interacting one where star formation histories of the Clouds have been strongly influenced by dynamical and hydrodynamical effects of galaxy interaction (Westerlund 1997). It is however unclear how galaxy interaction between the Clouds and the Galaxy triggers star formation in the gas disks of the Clouds. Recent observations on spatial distributions of HI (Staveley-Smith *et al.* 2003), molecular gas (Fukui *et al.* 1999), and young stars (Grebel & Brandner 1998) have provided vital information on galaxy-wide triggering mechanisms of star formation in the Clouds. By comparing numerical simulations of the Magellanic system with these observations, we here discuss (1) how the tidal interaction changes the spatial distribution of high-density gaseous regions where new stars can be formed in the LMC, (2) whether the formation of 30 Doradus is triggered by the interaction, and (3) how the interaction triggers star formation in the Magellanic Bridge (MB).

2. The last Magellanic interaction

We investigate the last 0.8 Gyr evolution of the Clouds orbiting the Galaxy based on GRAPE chemodynamical simulations of the Clouds with star formation models (Bekki *et al.* 2004; Bekki & Chiba 2005; Bekki & Chiba 2006). Since the details of the numerical methods and the initial conditions of the Clouds have been already discussed in our previous papers, we here summarize the models briefly. The total masses of the LMC and the SMC are set to be $2.0 \times 10^{10} M_{\odot}$ and $3.0 \times 10^9 M_{\odot}$, respectively. Gas particles are

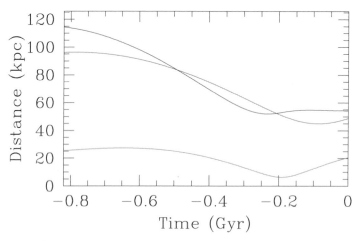

Figure 1. Time evolution of distance between the LMC and the SMC (magenta), the LMC and the Galaxy (red), and the SMC and the Galaxy (blue) for the last 0.8 Gyr. Note that the LMC-SMC distance becomes minimum (8kpc) about 0.2 Gyr ago.

assumed to be converted into new stars according to the Schmidt law with the observed threshold gas density (Kennicutt 1998). Figure 1 shows that the pericenter distance of the SMC orbit with respect to the LMC is 8 kpc about 0.2 Gyr ago. The tidal force from the LMC is therefore about 20 times stronger than that from the Galaxy for the SMC, which means that the SMC can be more strongly influenced by the LMC-SMC interaction than the SMC-Galaxy one. This LMC-SMC interaction can also significantly influence the gaseous evolution of the LMC and thus its recent star formation history.

3. Formation of 30 Doradus

Strong tidal effects of the LMC-Galaxy and the SMC-LMC interaction can induce the formation of asymmetric spiral arms with high densities of gas so that new stars can be formed in the arms. Figure 2 demonstrates that (1) the spatial distribution of young stars is quite irregular and clumpy, (2) there is a strong concentration of young stars along the stellar bar (composed of old stars), and (3) there is an interesting peak just above the eastern edge of the bar. This interesting peak of the stellar density of very young stars corresponds to the location where two asymmetric spiral arms emerge in the LMC disk. The location of the peak is very similar to the location of 30 Doradus, which suggests that the formation of 30 Doradus is closely associated with the formation of strong spiral arms due to the last Magellanic interaction about 0.2 Gyr ago. The simulated two high-density gaseous arms in eastern and southern parts of the LMC are morphologically similar to the observed gaseous arms composed of molecular clouds in the southern part of the LMC (i.e., "the molecular ridge"). This similarity suggests that the origin of the observed peculiar distributions of molecular clouds (Fukui *et al.* 1999) is due to the recent Magellanic interaction. The mean star formation rate of the LMC is increased rapidly by a factor of 5 about 0.2 Gyr ago and the rapid increase is synchronized with the enhancement of the star formation rate of the SMC in our models.

4. Star formation in the MB

The tidal interaction can also significantly change the recent star formation histories not only in the central region of the SMC's gas disk but also in its outer part, which

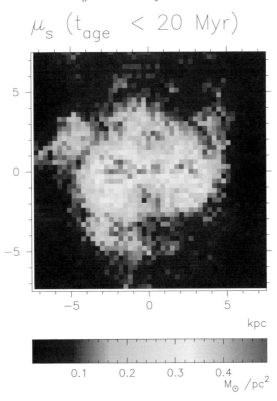

Figure 2. The projected distribution of surface mass densities of young stars with ages less than 20 Myr formed in the LMC during the LMC-SMC-Galaxy interaction.

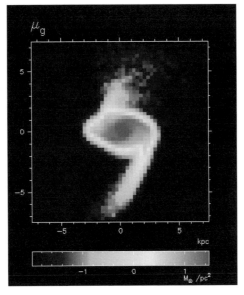

Figure 3. The projected distribution of smoothed (column) gas densities of the SMC about 0.14 Gyr ago (i.e., 60 Myr after the last Magellanic collision). The lower tidal tail with a higher gas density is the forming MB.

finally becomes the MB after the interaction. Figure 3 shows that the SMC's outer gas disk is strongly compressed by the interaction so that gas densities along the forming MB can exceed the threshold gas density of star formation (i.e., $3M_\odot$ pc^{-2}). Since the MB is formed from the outer gas, where the metallicity is significantly smaller owing to the negative metallicity gradient of the SMC, the metallicity distribution of new stars in the MB shows a peak of [Fe/H] ~ -0.8 (i.e., 0.2 dex smaller than the central stellar metallicity of the simulated SMC). About 25% of the initial gas mass of the SMC is finally distributed in the MB whereas only 0.1% of the gas mass is converted into new stars in the MB. The present model thus provides a physical explanation for the origin of the observed formation sites of new stars along the MB (Mizuno et $al.$ 2006).

5. Conclusions

The present study has suggested that tidal interaction between the Clouds and the Galaxy is closely associated with the formation of 30 Doradus, the southern molecular ridge of the LMC, and inter-Cloud stars with low metallicities. The observed asymmetric and clumpy distributions of young stars in the LMC are demonstrated to be due to the last Magellanic interaction about 0.2 Gyr ago, which forms asymmetric spiral arms with high-density gas. The synchronized burst of star formation in the Clouds about 0.2 Gyr ago (by tidal interaction) can be proved by the observed age distributions of young star clusters in the Clouds (e.g., Girardi et $al.$ 1995). The simulated distributions of star-forming regions will be compared with the latest results of the Spitzer observations on young stellar objects (YSOs) in the Clouds.

References

Bekki, K. & Chiba, M. 2005, *MNRAS* 356, 680
Bekki, K. & Chiba, M. 2006, *ApJ* (Letters), submitted
Bekki, K., Couch, W. J., Beasley, M. A., Forbes, D. A., Chiba, M. & Da Costa, G. S. 2004, *ApJ* 610, L93
Fukui, Y., et $al.$ 1999, *PASJ* 51, 745
Girardi, L., Chiosi, C., Bertelli, G. & Bressan, A. 1995, *A&A* 298, 87
Grebel, E. K. & Brandner, W. 1998, in: T. Richtler & J.M. Braun (eds.), *The Magellanic Clouds and Other Dwarf Galaxies* (Aachen, Shaker Verlag), p. 151
Kennicutt, R. C., Jr. 1998, *ARAA* 36, 189
Mizuno, N., Muller, E., Maeda, H., Kawamura, A., Minamidani, T., Onishi, T., Mizuno, A. & Fukui, Y. 2006, *ApJ* 643, L107
Staveley-Smith, L., Kim, S., Calabretta, M. R., Haynes, R. F. & Kesteven, M. J. 2003, *MNRAS* 339, 87
Westerlund, B. E. 1997, *The Magellanic Clouds* (Cambridge: Cambridge Univ.)

Discussion

DICKEY: Do you assume a dark matter halo in the LMC, or constant mass to light ratio?

BEKKI: My simulation is fully self-consistent in the sense that a galaxy is composed of a dark matter halo, stellar disk, and gas. The rotation curve of the LMC is consistent with observations.

CHU: The close encounter was 200 Myr ago, but the star formation at 30 Dor and to its south is much younger than this. A major star formation several 10^7 yrs ago was in the present-day supergiant shell LMC-4. The long ridge of cold gas to the south of 30 Dor has a hot component ($> 10^6$ K gas shown in x-ray images). How does the close encounter explain the hot gas?

BEKKI: My simulations do not include dynamics of hot gas. Therefore I cannot discuss this. This is an interesting problem.

WALBORN: What epoch does your simulation image showing a possible "proto"-30 Dor correspond to? 30 Dor is 2 Myr old and the rotation period of he LMC is comparable to the SMC interaction age. Comment: Also, as you showed, the eastern edge of the LMC leads its proper motion through the Galactic halo, and many Magellanic Irregulars have giant HII regions off the ends of bars, which are possible alternative causes of the 30 Dor regions.

BEKKI: I showed the present LMC in the figures. I agree with you on your comment.

ZINNECKER: Adding to Nolan Walborn's remarks, yet another possibility for the origin of 30 Doradus could be infall of a gas stream (~ 1 $M_\odot$/yr) from the SMC. Can you say something about this possibility from your numerical simulations?

BEKKI: As I showed in my animation, about 1% of the SMC's gas disk can be transferred into the LMC. If this SMC gas collides with the LMC's gas, and if cloud-cloud collisions can trigger star cluster formation, your idea is viable, I think.

FUKUI: concerning the hot gas which is missing in this simulation, I suggest it is not a serious discrepancy because the hot gas is much more short-lived (10^{6-7} yrs at most) than the star formation process dealt with by the author. We can add something to form the hot gas and this is not a serious short coming of the model.

BEKKI: I agree with you on this. This simulation is designed to investigate the long-term (recent 0-1 Gyr) evolution of cold gas in the LMC and the SMC, so I did not discuss the formation of the host gas. I think that if I include strong thermal feedback of Type II supernovae (from 30 Doradus regions etc), I could possibly reproduce the observed host gas.

Triggered Star Formation in a Turbulent ISM
Proceedings IAU Symposium No. 237, 2006
B. G. Elmegreen & J. Palouš, eds.

© 2007 International Astronomical Union
doi:10.1017/S1743921307001767

Turbulence, feedback, and slow star formation

Mark R. Krumholz[1]

[1]Department of Astrophysical Science, Princeton University, Peyton Hall, Ivy Lane, Princeton, NJ 08544, USA
email: krumholz@astro.princeton.edu

Abstract. One of the outstanding puzzles about star formation is why it proceeds so slowly. Giant molecular clouds convert only a few percent of their gas into stars per free-fall time, and recent observations show that this low star formation rate is essentially constant over a range of scales from individual cluster-forming molecular clumps in the Milky Way to entire starburst galaxies. This striking result is perhaps the most basic fact that any theory of star formation must explain. I argue that a model in which star formation occurs in virialized structures at a rate regulated by supersonic turbulence can explain this observation. The turbulence in turn is driven by star formation feedback, which injects energy to offset radiation from isothermal shocks and keeps star-forming structures from wandering too far from virial balance. This model is able to reproduce observational results covering a wide range of scales, from the formation times of young clusters to the extragalactic IR-HCN correlation, and makes additional quantitative predictions that will be testable in the next few years.

Keywords. turbulence, stars: formation, ISM: clouds, galaxies: ISM, galaxies: starburst

1. Introduction

Zuckerman & Evans (1974) were the first to point out perhaps the most surprising fact about star formation: it is remarkably slow. Inside the solar circle there are roughly $M_{\rm mol} \approx 10^9 \ M_\odot$ of molecular gas (Bronfman *et al.* 2000), organized into giant molecular clouds (GMCs) with typical densities of ~ 100 H atoms cm^{-3}, giving a free-fall time of about $t_{\rm ff} \approx 4$ Myr (McKee 1999). However, the star formation rate in the Milky Way is only $\sim 3 \ M_\odot$ yr^{-1} (McKee & Williams 1997), vastly less than the rate of $\sim 250 \ M_\odot$ yr^{-1} that one would expect if molecular clouds were converting their mass into stars on a free-fall time scale. More recent observations of nearby Milky Way-like galaxies (Wong & Blitz 2002) find that this factor of ~ 100 discrepancy occurs in them too. Nor is the discrepancy any smaller in systems like ULIRGs with much larger star formation rates. For example, Downes & Solomon (1998) find that Arp 220 contains roughly $2 \times 10^9 \ M_\odot$ of molecular gas with a typical free-fall time of ~ 0.5 Myr, but the observed star formation rate of $\sim 50 \ M_\odot$ yr^{-1} is a factor of 100 smaller than $M_{\rm mol}/t_{\rm ff}$.

Recently, Krumholz & Tan (2006) pointed out that objects much denser than GMCs form stars just as slowly. If one repeats the Zuckerman & Evans (1974) calculation of dividing total mass by characteristic free-fall time for any class of dense, gaseous objects (e.g. infrared dark clouds, dense molecular clumps), one again obtains a rate roughly 100 times larger than the observed star formation rate. This is true in galaxies from normal spirals to ULIRGs, and for objects with densities from ~ 100 cm^{-3}, typical of GMCs, to $\sim 10^4 - 10^5$ cm^{-3}, typical of molecular clumps forming rich star clusters. The trend may continue to even higher densities. Figure 1 summarizes the observations.

378

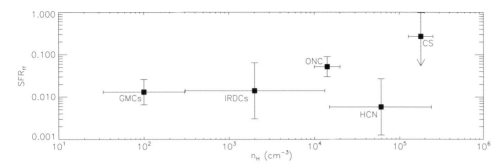

Figure 1. Fraction of mass converted into stars per free-fall time (SFR$_{\rm ff}$) versus characteristic density for various objects. We show GMCs, infrared dark clouds (IRDCs), HCN-emitting gas clumps, the Orion Nebula Cluster, and CS-emitting gas clumps. Note that the CS point is only an upper limit. The Figure is adapted from Krumholz & Tan (2006).

The apparent universality of this factor of 100 discrepancy over such an immense range of densities and galactic environments suggests it must be set by fundamental physics. In these proceedings I present a model that attempts to explain the observations as a result of the properties of supersonic turbulence, and in which that turbulence is itself a side-effect of the star formation process. This theory has two parts: a physical mechanism by which turbulence determines the factor of 100, described in § 2, and a physical mechanism for generating the turbulence, described in § 3.

2. Turbulent regulation of the star formation rate

Krumholz & McKee (2005) propose a simple model for how turbulence regulates the star formation rate, based on the premise that star formation occurs in any sub-region of a molecular cloud in which the gravitational potential energy exceeds the kinetic energy in turbulent motions. This is sufficient to determine the star formation rate in a supersonically turbulent isothermal medium, because such media are governed by two universal properties. First, they obey a linewidth-size relation, meaning that the velocity dispersion over a region of size ℓ varies as roughly $\ell^{1/2}$ (Larson 1981). Second, they show a lognormal distribution of densities (Padoan & Nordlund 2002).

These determine the star formation rate as follows: the linewidth-size relation sets the kinetic energy per unit mass in any given sub-region of a cloud, normalized to the cloud's total kinetic energy. From this, one can show that the potential energy will be larger than the kinetic energy in regions where the density exceeds a certain critical value. In turn, the density probability distribution determines what fraction of the mass is at densities larger than this critical value. Bound regions collapse on a free-fall time scale. This determines the dimensionless star formation rate SFR$_{\rm ff}$, defined as the fraction of its mass that a gas cloud turns into stars per mean-density free-fall time, in terms of two dimensionless numbers: the cloud's Mach number $\mathcal{M}$ and virial ratio $\alpha_{\rm vir}$ (roughly its ratio of kinetic to potential energy). For $\alpha_{\rm vir}$ and $\mathcal{M}$ in the range observed for real clouds, one may approximate the relationship by a powerlaw

$$\mathrm{SFR}_{\rm ff} \approx 0.073 \alpha_{\rm vir}^{-0.68} \mathcal{M}^{-0.32}. \qquad (2.1)$$

This is an extremely powerful result, and it allows numerous comparisons to observation. First, notice that, due to its very weak dependence on $\mathcal{M}$, SFR$_{\rm ff}$ is a few percent in any virialized, supersonically turbulent object, regardless of its density or environment.

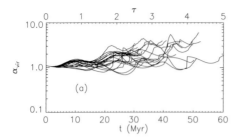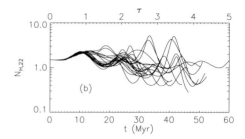

Figure 2. Virial ratio (panel a) and column density in units of 10^{22} cm^{-2} (panel b) versus time measured in years (t) and measured in cloud crossing times (τ) for a sample of GMC evolution models. Lines end when clouds are disrupted by HII regions. The Figure is adapted from Krumholz *et al.* (2006a).

This is exactly what the observations summarized in Figure 1 demand; a corollary is that this model naturally explains the extra-Galactic IR-HCN correlation (Gao & Solomon 2004; Krumholz & Tan 2006). A second prediction of this model is that achieving the star formation efficiencies of tens of percent estimated for rich star clusters requires that star formation continue for several crossing times, a result in good agreement with the observed age spreads of young clusters (Tan *et al.* 2006).

A third test of the model is to use it to predict the star formation rate in the Milky Way from equation (2.1) and the observed properties of Milky Way molecular clouds. Doing so gives a predicted star formation rate of 5.3 $M_\odot$ yr^{-1} in the Milky Way, within a factor of 2 of the observed value (Krumholz & McKee 2005). Moreover, observations of the molecular cloud populations of nearby galaxies such as M33, M64, and the LMC (see review by Blitz *et al.* 2005) are starting to reach precisions comparable to those available for the Milky Way. Repeating this calculation for these cloud populations and comparing to observed star formation rates provides a direct future observational test of this model.

If one adopts the additional hypotheses that molecular clouds in a galaxy should have masses of ~ 1 Jeans mass in the galactic disk, and should be in rough pressure balance with the rest of the ISM, then one may extend the model to predict the star formation rate in galaxies as a function of their gas surface densities, rotation rates, and molecular fractions. This prediction agrees with the data of Kennicutt (1998) as well as Kennicutt's purely empirical fit. Recent observations show that the model also agrees with the radial distribution of star formation in the Milky Way (Luna *et al.* 2006).

3. Star formation regulation of the turbulence

Having shown that turbulent regulation very naturally explains a large number of observations about star formation, we now turn to the question of the origin of the turbulence itself. This is a problem because simulations (e.g. Stone *et al.* 1998) show that, if it is not continually driven, supersonic turbulence will decay away over time scales much shorter than the observationally-estimated ~ 30 Myr lifetimes of GMCs in local group galaxies (Blitz *et al.* 2005). Since the clouds all have virial ratios ~ 1 and roughly constant column densities $\sim 10^{22}$ cm^{-2}, something must be driving the turbulence and keeping the clouds in at least approximate equilibrium.

To investigate this problem, Krumholz *et al.* (2006a) construct simple one-dimensional semi-analytic models of GMCs, the goal of which is to investigate their global energy balance. In these models one follows the evolution of GMCs using the non-equilibrium virial and energy conservation equations for a homologously-moving, evaporating cloud, including source terms describing decay of turbulence at the rates measured by simulations, and a countervailing injection of energy (and removal of mass) by star formation feedback. Both energy injection and mass loss are dominated by HII regions launched by newborn star clusters (Matzner 2002), so the models focus on them.

These models show that feedback is able to explain the observed properties of GMCs extremely well. Feedback destroys giant clouds in ~ 30 Myr, during which time they turn $5 - 10\%$ of their mass into stars, and remain turbulent, virialized, and at constant column densities. Figure 2 shows some typical examples of molecular cloud evolution, which are in good agreement with observations. In contrast, clouds with masses $\sim 10^4\ M_\odot$, like those in the solar neighborhood, survive only ~ 1 crossing time. The next step in this sort of modeling is to do full ionizing radiative-transfer MHD simulations to study feedback in a more realistic context, a project already underway (Krumholz *et al.* 2006b).

4. Summary

Feedback-driven turbulence provides a simple, natural explanation for a host of observations about star formation. The quantitative prediction of the star formation rate one derives by computing the fraction of bound mass in a turbulent, virialized object, using no physics other than the invariant properties of supersonic isothermal turbulence, matches observations of the rate of star formation in dense gas, the rate and radial distribution of star formation in the Milky Way, the Kennicutt law, and the extragalactic IR-HCN correlation. The turbulence is in turn driven by the feedback from star formation itself. The physics of the driving process explains the observed lifetimes, column densities, and virial ratios of giant molecular clouds in local group galaxies.

Acknowledgements

I thank T. A. Gardiner, C. D. Matzner, C. F. McKee, J. M. Stone, and J. C. Tan, all of whom collaborated with me on parts of this project. I acknowledge support from NASA through Hubble Fellowship grant #HSF-HF-01186 awarded by STScI, which is operated by AURA for NASA, under contract NAS 5-26555.

References

Blitz, L., Fukui, Y., Kawamura, A., Leroy, A., Mizuno, N. & Rosolowsky, E. 2005, in: B. Reipurth, D. Jewitt & K. Keil (eds.), *Protostars & Planets V*, in press, astro-ph/0602600
Bronfman, L., Casassus, S., May, J. & Nyman, L.-Å. 2000, *A&A* 358, 521
Downes, D. & Solomon, P. M. 1998, *ApJ* 507, 615
Gao, Y. & Solomon, P. M. 2004, *ApJS* 152, 63
Kennicutt, R. C. 1998, *ApJ* 498, 541
Krumholz, M. R., Matzner, C. D. & McKee, C. F. 2006a, *ApJ* in press, astro-ph/0608471
Krumholz, M. R. & McKee, C. F. 2005, *ApJ* 630, 250
Krumholz, M. R., Stone, J. M. & Gardiner, T. A. 2006b, *ApJ* submitted, astro-ph/0606539
Krumholz, M. R. & Tan, J. C. 2006, *ApJ* in press, astro-ph/0606277
Larson, R. B. 1981, *MNRAS* 194, 809
Luna, A., Bronfman, L., Carrasco, L. & May, J. 2006, *ApJ* in press, astro-ph/0512046
Matzner, C. D. 2002, *ApJ* 566, 302
McKee, C. F. 1999, in NATO ASIC Proc. 540: The Origin of Stars and Planetary Systems, 29
McKee, C. F. & Williams, J. P. 1997, *ApJ* 476, 144
Padoan, P. & Nordlund, Å. 2002, *ApJ* 576, 870
Stone, J. M., Ostriker, E. C. & Gammie, C. F. 1998, *ApJ* 508, L99
Tan, J. C., Krumholz, M. R. & McKee, C. F. 2006, *ApJ* 641, L121

Wong, T. & Blitz, L. 2002, *ApJ* 569, 157

Zuckerman, B. & Evans, N. J. 1974, *ApJ* 192, L149

Discussion

BONNELL: 1) As molecular clouds with and without star formation have similar turbulent properties, it would appear unlikely that outflows are the dominant process in driving turbulence. 2) The requirement for molecular clouds to last a few (2-4) crossing times is consistent with purely decaying turbulence, as the formation time of the cloud, as it must form from lower density gas, is easily several crossing times.

KRUMHOLZ: To question 1: first, while there are clouds without HII regions, there do not appear to be starless clouds. At 24 microns, all the GMCs in the LMC show embedded sources. Second, GMCs are born turbulent, and some very young ones will not have had much turbulent decay or formed many stars yet. This doesn't mean that feedback is unimportant, because all turbulence looks the same, so you can't easily tell whether it is "primordial" and feedback-driven. To question 2: recall that 30 Myr is the time for which $> 10^5$ M$_\odot$ of CO is visible, and CO only forms in clouds dense enough to be self-shielding. You could have 30 Myr correspond to 2 free-fall times if you first saw the cloud when its density was ~ 10 H atoms per cm^{-3}, but at such low densities there would be no CO.

WALBORN: Your theoretical picture is in reasonable agreement with the empirical morphology of massive star-forming regions. The observation that triggered second generations are always predominantly on one side of the first generation clusters indicates that the latter form near the surface of the GMC, and not from a central collapse, perhaps caused by external events. Thus, the free-fall time of the entire GMC is irrelevant. Moreover, the difference between a super star cluster (e.g. 30 Doradus) and a seeded OB association (e.g. NII and NGC604) may arise from the bound or unbound state of the progenitor cloud subvolume, as proposed by Bally & Zinnecker.

KRUMHOLZ: This probably gives me a bit too much credit, since my GMC model does not include explicit triggering. However, I do recover the observed result that there are generations of star formation modulated by feedback. I would disagree with a couple of your suggestions. First, I don't think the free-fall time of the entire GMC is irrelevant. There are dense sub-regions within GMCs that evolve faster, but if you want to know about the global evolution, you have to look at the bulk of the mass, and that gas evolves on the mean-density free-fall time scale. Second, I seriously doubt that the progenitors of OB associations are unbound. These clouds have virial ratios near unity, they are strongly centrally concentrated, and their morphologies are very round, observations that seem hard to reconcile with colliding, unbound gas. I'd say whether something winds up bound is likely determined by feedback. Systems that convert a large fraction of their mass into stars before disruption wind up bound, while systems that are disrupted while turning only a small percentage of their mass into stars wind up unbound.

CLARK: 20 Myr is a long time in massive star formation. Would you not expect there to be supernova events in the clouds by this piont, which you have not modelled?

KRUMHOLZ: We actually did examine the effects of supernovae in our paper, and decided that they are less important than HII regions. That's because if a supernova goes off outside its parent cloud, the fraction of energy deposited in the cloud is very small.

Supernovae that go off inside their parent clouds don't have a huge effect because the supernova shock loses most of its energy while still inside the star's HII region.

VAZQUEZ-SEMADENI: You seem to be taking for granted that GMCs live for ~ 30 Myr. What is your evidence for that? If you refer to Fukui's observations of GMCs in external galaxies, about 30% of these don't have stars.

KRUMHOLZ: The evidence is indeed the observations of external galaxies by Fukui *et al.* (2005), Engargiola *et al.* (2003), and others quoted in the Blitz *et al.* PPV review. Regarding the 30% starless figure, you have to be careful with what is called starless. As I said in my answer to Ian, supposedly starless GMCs have plenty of embedded stars if you observe them at 24 microns.

Triggered Star Formation in a Turbulent ISM
Proceedings IAU Symposium No. 237, 2006
B. G. Elmegreen & J. Palouš, eds.

© 2007 International Astronomical Union
doi:10.1017/S1743921307001779

Conference summary: triggered star formation in a turbulent ISM

Bruce G. Elmegreen

IBM T. J. Watson Research Center, 1101 Kitchawan Road, Yorktown Hts., NY 10598 USA
email: bge@us.ibm.com

Abstract. While the overall star formation rate in a galaxy appears to depend primarily on the gas mass and density, with the timescale for conversion of gas into stars given by the dynamical time, turbulence and explosions are still important for the process of star formation because they control the birth correlations in space and time. Most star formation appears triggered by some specific process, whether it is a galactic spiral shock, the expansion of a superbubble, the compression of a bright-rimmed globule, or some seemingly random compressive event in a supersonically turbulent flow. These processes give space and time sequences for star birth that are well observed. Many examples were given at this conference. Shocks are the link between large-scale but weak galactic processes and small-scale but strong final collapses. The rate limiting step is on the largest scale, where the dynamical time is slowest. Both gravitational instabilities and pressurized triggering seem to work on the same local dynamical time, making it difficult to tell that star formation is highly triggered when observing only galactic scales.

Keywords. turbulence, instabilities, shock waves, stars:formation, ISM: structure

1. Introduction

Instabilities, explosions, galaxy interactions, and other dynamical processes all produce relative motions that can exceed the local thermal sound speed. If the momentum transferred by these motions through a galaxy cannot be quickly damped, then they should generate turbulence. This damping condition is essentially the same as the Reynolds number condition, i.e., that forced motions lead to turbulence when the Reynolds number is large (the Reynolds number is the ratio of the mixing velocity multiplied by the mixing scale to the viscous coefficient). In galaxies, the primary motions are supersonic and much of the damping is in shocks.

We have seen many talks at this conference showing how interstellar turbulence might be generated. Each model was able to highlight certain specific processes to the exclusion of others and then study the results. This method is less confusing than observing the real ISM, which combines them all. As a result, we have a good picture for the origin of ISM motions, although the relative importance of various processes in different regions is still uncertain.

The primary question addressed at this conference is why processes like ISM turbulence and expansional motions from HII regions, supernovae, and stellar winds are relevant to star formation. These processes would seem to be disruptive rather than collective, whereas the collection of gas into dense clouds is the first step toward star formation. Observations of actual star birth, consisting of gravitational collapse to a disk and the emergence of a protostar, do not reveal much about ISM turbulence and triggering. The boundary conditions for star formation do, however. Turbulence and triggering show up in various space and time correlations surrounding the star birth event (Sect. 2). They do not seem to influence the area-average star formation rate, though.

We might guess that the star formation rate (in $M_\odot$ pc^{-3} Myr^{-1}) in some region of average density ρ scales with $\epsilon\rho(G\rho)^{1/2}$, where ϵ is the fraction of the gas that is converted into stars at the average dynamical rate (which is $[G\rho]^{1/2}$). If star formation occurs in D dynamical times, then ϵ equals $1/D$ times the overall efficiency of star formation in that region. This expression for the star formation rate should apply if the observation of a particular molecule is sensitive to a wide range of intrinsic densities, so both the clump and the interclump gas are observed. CO is like this for molecular clouds at normal pressures in normal galaxies (not dwarf Irregulars). Then the molecule traces the total mass, and the observed mass per unit volume is the average density used in this expression. If a molecule traces only dense clumps, however, then the total emitting mass of the molecule divided by the cloud volume is not the average density, but less. Also in this case, the mass above the threshold for detection increases as the average density of the cloud increases. Then the star formation rate per unit core mass would be more constant than it is for CO; it could be relatively independent of the average density detected by that molecule. This difference between diverse-density tracers and high-density tracers could partially explain the difference between the 1.5 power law for scaling with HI+CO column density (Kennicutt 1989) and the 1 power law for scaling with HCN (Gao & Solomon 2004). In addition, the HCN molecule probably traces star-forming clumps which are already evolving toward stars as fast as they can with a certain efficiency. Then the star formation rate scales only with the average density of this molecule. Lower-density tracers like CO observe the gas before it reaches this stage, so there has to be an extra factor that determines the rate at which CO turns into dense clumps; this is the role of the $(G\rho)^{1/2}$ factor for average density ρ determined from CO.

A local star formation rate per unit area equal to $A\Sigma^{1.5}$ also gives a global star formation rate per unit area proportional to the average $\langle\Sigma\rangle^{1.5}$, which may be written as $B\langle\Sigma\rangle^{1.5}$ for $B \sim A$. This can be seen from the integral over the disk of the local rate, using an exponential disk $\Sigma(r) = \Sigma_0 \exp(-r/r_D)$. The integrated rate is

$$\frac{\int_0^R A\Sigma_0^{1.5} \exp(-1.5r/r_D)\,2\pi r dr}{\pi R^2} = \frac{2Ar_D^2\sigma_0^{1.5}}{1.5^2 R^2} \tag{1.1}$$

where small terms $\exp(-1.5R/r_d)$ have been dropped in comparison to 1. The average column density is $\langle\Sigma\rangle = \int_0^R \Sigma(r)2\pi r dr/(\pi R^2) = 2r_D^2\Sigma_0/R^2$. Writing the star formation rate as $B\langle\Sigma\rangle^{1.5}$ then gives

$$B = \frac{2AR}{2^{1.5}1.5^2 r_D} = 0.3\frac{AR}{r_D} \sim A \tag{1.2}$$

because $R/r_D \sim 3-4$ for Freeman disks. Thus the global Kennicutt (1989) law is also consistent with a local relation of this type, as expected from theory (Elmegreen 2002).

The final efficiency of star formation depends on the density. It is $\sim 5\%$ for OB associations and up to $\sim 30\%$ for the cores of OB associations where clusters form. On a galactic scale, $\epsilon \sim 1.2\%$ from the rotation rate correlation in Kennicutt (1998), considering that the average density in a galaxy is about the critical tidal density. In M33, $\epsilon \sim 60\%$ if only the CO-emitting gas is used for ρ (from data in Heyer *et al.* 2004, as derived in Elmegreen 2005). The increase for the CO coefficient is because the CO-emitting gas is a small fraction of the total.

We can get pretty far in predicting the star formation rate using only the density. This suggests that star formation is controlled primarily by gas self-gravity. In that case, why should we be concerned with sequential triggering and turbulence? The answer to this question lies entirely in the observed correlations.

2. It's the correlations...

Star formation is linked to sequential triggering processes because of observed space and time sequences where young stars appear in the bright rims of OB associations, and in swept-up shells and spiral density waves. Early models of triggering were for different reasons. Öpik (1953) suggested that supernovae trigger star formation in order to explain the expansion of OB associations, while Oort (1954) suggested that HII region pressures do the same thing. There may be some component of the expansion from triggering processes, but generally the density of a giant molecular cloud is so high that supernovae and HII regions do not accelerate the molecular gas much. Stellar expansion should come from gas dispersal out of clusters, a proposal first made by Zwicky (1953), and from runaway stars. In fact, most of the large expansion observed by Blaauw (1952) is the result of a few runaway O-type stars.

The more subtle correlations that suggest triggering, namely the spatial and temporal sequences of subgroups in OB associations, were first systematically described by Blaauw (1964). Blaauw's review paper effectively began the modern study of triggered star formation, although the real beginning had to wait a decade for the discovery of molecular clouds. Today, triggering of this type is commonly revealed by molecular cloud regions that contain young stars inside swept-up layers and bright rims. A previous generation of stars made these disturbances through some combination of stellar winds, ionization, and supernovae. Many talks and posters at this conference have illustrated shells, bright rims, and globules in which star formation has been triggered by compression from an older generation of stars.

Similarly for turbulence triggering, the importance of turbulence is revealed through space and time correlations in young stars that are reminiscent of those in turbulent fluids. For example, there is hierarchical structure in young stellar groupings (Scalo 1985; Feitzinger & Galinski 1987; see Elmegreen et al. 2006 and references therein) that ranges from flocculent spiral arms on large scales to star complexes (Efremov 1995), OB associations, OB subgroups, and dense cores on small scales. There are also power-law size, luminosity, and mass distributions for stellar groupings, clusters, and star-forming clouds that reflect the scale-free conditions of turbulent flows. The optical light in galaxies also has a power-law power spectrum similar to that of the gas, partly because young stars cluster together like the gas and partly because of extinction effects (Elmegreen et al. 2003; Willett et al. 2005). Star clusters also group together with a power law autocorrelation function (Zhang et al. 2001).

Temporal correlations for turbulence triggering show up in comparisons between the duration of star formation and the size of the region: bigger regions form stars longer, in proportion to the square root of their size (Efremov & Elmegreen 1998). This is like the size-linewidth relationship if we convert the ratio between size and linewidth to a turbulent crossing time. The observation is different than the correlation expected if star formation "spreads like a disease" (Baade 1963). In that case, the duration should scale with the square of the size, as in a random walk. It is also different from that expected from shear, in which case the relation should be linear (the correlation was found in the LMC where there is no shear). The result makes sense locally: Gould's Belt size regions and star complexes ~ 300 pc in size form stars for ~ 50 Myr, OB associations ~ 30 pc in size form stars for ~ 15 My, OB subgroups and GMC cores ~ 3 pc in size form stars for ~ 5 My, and so on, with the square root dependence. The startling point about this is that a wide range of scales satisfies the correlation, which means that each scale has a duration of star formation proportional to the local crossing time of ISM turbulence on that scale. The large and small scale processes are similar in their dynamics.

3. If there is a sensitivity to $\epsilon\rho(G\rho)^{1/2}$ and Q, then how can there also be sequential & turbulence triggering?

A star formation rate proportional to $\epsilon\rho(G\rho)^{1/2}$ and the Q threshold for star formation suggest that self-gravitational instabilities initiate star formation on a galactic scale and that gravity-dominated processes, like collapse or retarded collapse, control the rate. These relations do not seem to fit in with star formation that has spatial and time correlations reminiscent of sequential triggering and turbulence triggering. How can all three processes operate at the same time? Are the triggering mechanisms insignificant compared to self-gravitational instabilities? Or do the correlated processes have the same threshold and Schmidt law?

ISM instabilities and stellar compressions all operate at pressures much less than the pressure of a final pre-collapse star-forming clump. The instabilities sensitive to Q begin on the scale of the ambient Jeans length, which is several kpc in a typical spiral galaxy disk, and primarily produce flocculent spiral arms, which shear away and disperse in ~ 100 Myr. This is a relatively mild disturbance on average. A giant shell is also relatively mild compared to the pressure of a star-forming clump. The high pressure phases of all stellar-type explosions and expansions occupy only a small fraction of the ISM volume – not enough to influence much star formation by direct compression of random clouds (Elmegreen 2004). Jenkins & Tripp (2001) have shown, for example, that most regions have pressures within only a factor of ~ 3 times the average; 1% of the mass has a pressure 30 times the average. Still, what begins as a mild ISM disturbance can end up as a network of high-pressure star-forming cores because of the further action taken by turbulence. Turbulence, and the expansional motions that lead to swept-up shells and layers, have the property that they concentrate low-pressure, high-volume energy (e.g., flocculent arms and supernova cavities) into high-pressure, low-volume forms, such as shocks, multiple shocks (converging or intersecting), and vortices. In these high-pressure forms, enhanced magnetic diffusion, enhanced thermal cooling, and enhanced kinetic energy dissipation (also via shocks) remove supportive energy and let local self-gravity take over. Then the local pressure can go up enormously as a dense core forms. Star formation soon follows.

Shocks are the link between galactic scale processes (which are big but intrinsically weak) and final star formation processes (which are small but intrinsically strong). Free-fall collapse (Hoyle 1953) is not the link because magnetic fields retard this collapse, random motions divert it, and density gradients stretch it. And while turbulence and compression help convert low density gas into high density gas, they also set up the spatial and temporal correlations that appear in young star fields without affecting the overall star formation rate much, which is still limited on the large-scale by self gravitational dynamics. The rate limiting process for star formation is on the largest scale, where the density is lowest and the timescale is longest. If intermediate scales are too slow to keep up the pace dictated by the large scale gravitational instabilities, then they will produce a bottleneck, changing the distribution of density so that more material at the bottleneck compensates for the relatively low rate there. But the overall star formation rate should not change. This means that the galactic-scale star formation rate is not controlled by the rate at which tiny solar-mass clumps collapse into stars. The rate is determined at the start of the whole process, on a large scale.

To put this another way, the star formation rates, but not the correlations, are produced by ambient disk gravity, and the correlations, but not the star formation rates are produced by the intermediate-scale processes (turbulence, shells) which channel the low-P, high-V perturbations down to high-P, low-V structures in which final collapse begins. In an analogy my former-teenagers would have understood: Q is the green light, $\epsilon\rho(G\rho)^{1/2}$ is the speed limit, and the correlations are the road patterns.

The Schmidt law and Toomre threshold also apply somewhat directly to sequential and turbulence triggering, which have similar average rates and thresholds. The triggering time for star formation in a swept-up shell is about equal to the dynamical time in the surrounding medium, $0.3\,(4\pi G\rho_0)^{-1/2}$, and the condition that collapse occurs before the shell shears away is about the same as the Toomre $Q < 1$ condition (Elmegreen, Palouš & Ehlerova 2002). Turbulence triggering also has its rate-determining step on the largest scale, and the compression time over a scale height is about the dynamical time in the disk (by definition of the scale height). Also, turbulence compression requires that the random motions be able to move for at least a scale height in order to build up enough mass to be gravitationally unstable at the ambient rms velocity dispersion. These motions are influenced by Coriolis forces, so the condition that the epicyclic radius for the random motions exceeds the gaseous scale height is again $Q < 1$.

4. The efficiency

Turbulence and stellar pressures affect the efficiency ϵ somewhat. In a turbulent star formation model, ϵ is related to the mass fraction of gas in a dense form (i.e., in the tail of the log normal pdf – Elmegreen 2002; Kravtsov 2003; Krumholz & McKee 2005; for applications to the IMF, see Padoan & Nordlund 2002). At higher Mach numbers and higher mean densities, the tail fraction exceeding a certain critical density is higher, so the efficiency should be higher. In a sequentially triggered model, ϵ is proportional to the number of generations of star formation in a cloud. Star formation, can, in principle, continue in a sequentially triggered fashion, generation after generation, until most of the gas is used up. In both cases, a lot of the microphysics is in ϵ, while the macrophysics is still galactic-scale self-gravity.

5. The even bigger picture

When star formation is regulated by the Toomre parameter Q, the gas mass tends to stay comparable to the threshold value (Zasov & Smirnova 2006) and the star formation rate is, on average, equal to the average ISM accretion rate. This is because Q self-regulates via gas consumption and perhaps massive-star feedback to remain at a value near 1 on a Gyr timescale (Fuchs & von Linden 1998; Bertin & Lodato 2001). Accretion and mass-redistribution inside a galaxy, from spiral and bar torques, for example, determine the details of the density and column density profiles.

When the Sun formed (at redshift $z = 0.45$), the mass column density from stars in the Milky Way was only half what it is today. The disk was smaller and more fragile. The star formation rate has apparently not changed much since then, aside from factor-of-5 variations either way, but star formation is still winding down universally. The "red and dead" Hubble types, i.e., the ellipticals, S0's and Sa's, are creeping toward the Milky Way's Hubble type, which is Sbc. Still we have a long way to go. The LMC stream and even the whole LMC and SMC will eventually accrete to the Milky Way. Such accretion increases Σ/Σ_{crit} locally, and then star formation proceeds at the usual dynamical rate.

References

Baade, W. 1963, in: C. Payne-Gaposchkin (ed.), *The Evolution of Stars and Galaxies* (Cambridge: Harvard University Press), Chapter 16

Bertin, G. & Lodato, G. 2001, *A&A* 370, 342

Blaauw, A. 1952, *BAN* 11, 405

Blaauw, A. 1964, *ARAA* 2, 213

Efremov, Yu. N. 1995, *AJ* 110, 2757

Efremov, Y.N. & Elmegreen, B.G. 1998, *MNRAS* 299, 588

Elmegreen, B.G. 2002, *ApJ* 577, 206

Elmegreen, B.G., Leitner, S.N., Elmegreen, D.M. & Cuillandre, J.-C. 2003, *ApJ* 593, 333

Elmegreen, B.G., Palouš, J. & Ehlerova, S. 2002, *MNRAS* 334, 693

Elmegreen, B.G. 2004, in: H.J.G.L.M. Lamers, L.J. Smith & A. Nota (eds.), *The Formation and Evolution of Massive Young Star Clusters* (ASP-CS), 322, 277

Elmegreen, B.G. 2005, in: Jose Carlos del Toro Iniesta, *et al.* (eds.), *The many scales in the Universe – JENAM 2004 Astrophysics Reviews* (Dordrecht: Kluwer), p. 99

Elmegreen, B.G., Elmegreen, D.M., Chandar, R., Whitmore, B., & Regan, M. 2006, *ApJ* 644, 879

Feitzinger, J.V. & Galinski, T. 1987, *A&A* 179, 249

Fuchs, B. & von Linden, S. 1998, *MNRAS* 294, 513

Gao, Y. & Solomon, P.M. 2004, *ApJ* 606, 271

Heyer, M.H., Corbelli, E., Schneider, S.E., & Young, J.S. 2004, *ApJ* 602, 723

Hoyle, F. 1953, *ApJ* 118, 513

Jenkins, E.B & Tripp, T.M. 2001, *ApJS* 137, 297

Kennicutt, R.C. 1989, *ApJ* 344, 685

Kravtsov, A.V. 2003, *ApJ* 590, L1

Krumholz, M.R. & McKee, C.F. 2005, *ApJ* 630, 250

Oort, J.H. 1954, *BAN* 12, 177

Opik, E.J. 1953, *Irish J. Astron* 2, 219

Padoan, P. & Nordlund, A. 2002, *ApJ* 576, 870

Scalo, J. 1985, in: D.C. Black, & M.S. Matthews (eds.), *Protostars and Planets II* (Tucson: Univ. of Arizona), p. 201

Willett, K.W., Elmegreen, B.G. & Hunter, D.A. 2005, *AJ* 129, 2186

Zasov, A.V. & Smirnova, A.A. 2006, *AstL* 31, 160

Zhang, Q., Fall, S.M. & Whitmore, B.C. 2001, *ApJ* 561, 727

Zwicky, F. 1953, *PASP* 65, 205

INTERNATIONAL ASTRONOMICAL UNION
SYMPOSIUM NO. 237

POSTERS
(in alphabetical order)

Observatory Tower (center) of the Klementinum Jesuit College, Prague

photograph by B.G.E.

Triggered Star Formation in a Turbulent ISM
Proceedings IAU Symposium No. 237, 2006
B. G. Elmegreen & J. Palouš, eds.

© 2007 International Astronomical Union
doi:10.1017/S1743921307001792

High supernova rate and enhanced star-formation triggered in M81-M82 encounter

B. Arbutina[1], D. Urošević[2] and B. Vukotić[1]

[1]Astronomical Observatory, Volgina 7, 11160 Belgrade, Serbia

[2]Department of Astronomy, University of Belgrade, Studentski trg 16, 11000 Belgrade, Serbia

email: barbutina@aob.bg.ac.yu

It is a general belief that the starburst activity of a nearby galaxy M82 was triggered in a close encounter with its massive companion M81, a few tens of million years ago. Despite the lack of supernovae observed, multiwavelength radio observations of M82 discovered a considerable number of compact supernova remnant candidates. We use these remnants to estimate the supernova rate (SNR) and the enhanced star-formation (SFR) rate in M82, and compare them with rates in normal galaxies.

Since the nature and evolutionary status of M82 remnants is controversial we will rely on the most compact objects only (Table 1). For the three unresolved sources 42.7+58.2, 44.9+61.2 and 46.6+73.8, we can estimate the time of explosion by assuming that they are all younger than 43.3+59.2, and knowing that 46.6+73.8 appeared in the 1990 image of Huang *et al.* (1994), but not in the 1981 image of Kronberg, Biermann & Schwab (1985), that 44.9+61.2 must be older than 1981 since it appeared in both images, and that 42.7+58.2 in a complex region must be older than 1990. The mean time between two successive explosions is $\tau = 12 \pm 7$ yrs. The SNR is then simply $\nu = \tau^{-1} = 0.08 \pm 0.05$ yr^{-1}, and SFR$(\mathcal{M} \geqslant 5\mathcal{M}_\odot) \approx 25\nu = 2\ \mathcal{M}_\odot$ yr^{-1}. Expressed in supernova units (SNu = number of SNe per $10^{10}L_\odot^B$ per 100 yrs), $\nu_{\mathrm{M82}} = (0.22 \pm 0.19) \times 100$ SNu, while the average core-collapse supernova rate in irregular galaxies (Cappellaro *et al.*1999) is $\nu_{irr} = 0.87 \pm 0.55$ SNu. SNR and SFR in M82 thus exceeds the rate in a corresponding non-starburst irregular galaxy of the same blue luminosity for about two orders of magnitude!

Table 1. Supernova event date estimates for M82 remnants, including SN 2004am.

Supernova	Diameter[a] D (pc)	Date t (yr)
41.9+58.0	0.5 ± 0.1	1955 ± 10^a
43.3+59.2	0.6 ± 0.1	1967 ± 5^b
44.9+61.2	< 0.5	1971 ± 10
42.7+58.2	< 0.5	1975 ± 15
41.5+59.7	< 0.5	1980 ± 1^a
46.6+73.8	< 0.5	1985 ± 5
SN 2004am	—	2004 ± 0^c

[a]Huang *et al.* (1994), [b]Beswick *et al.* (2006), [c]Singer, Pugh & Li (2004).

1

Acknowledgements

The authors acknowledge the financial support provided by the Ministry of Science and Environment of Serbia through the projects No. 146003 and 146012.

References

Beswick, R.J., Riley, J.D., Marti-Vidal, I., Pedlar, A., Muxlow, T.W.B., McDonald, A.R., Wills, K.A., Fenech, D. & Argo, M.K., 2006, *MNRAS* 369, 1221

Cappellaro, E., Evans, R. & Turatto, M., 1999, *A&A* 351, 459

Huang, Z.P., Thuan, T.X., Chevalier, R.A., Condon, J.J. & Yin, Q.F., 1994, *ApJ* 424, 114

Kronberg, P.P., Biermann, P. & Schwab, F.R., 1985, *ApJ* 291, 693

Singer, D., Pugh, H. & Li, W., 2004, *IAU Circular No. 8297*

Triggered Star Formation in a Turbulent ISM
Proceedings IAU Symposium No. 237, 2006
B. G. Elmegreen & J. Palouš, eds.

Numerical simulations for the interaction of the NGC 1333 IRAS 4A outflow and an ambient cloud

Chang Hyun Baek[1,2], Jongsoo Kim[2] and Minho Choi[2]

[1] ARCSEC, Sejong University, Seoul 143-747, KOREA
email: chbaek@kasi.re.kr

[2] Korea Astronomy and Space Science Institute, Daejeon, 305-348, KOREA
email: jskim@kasi.re.kr, minho@kasi.re.kr

NGC 1333 contains numerous young stellar objects and outflows and is a well-studied star formation region. High resolution SiO observations of the NGC 1333 IRAS 4A region showed a highly collimated outflow with a substantial deflection angle. It was also suggested by the observations that the deflection was due to the interaction of the outflow and a dense cloud core (Choi 2005a, 2005b). In order to make a detailed model of the deflected outflow, we have carried out three-dimensional hydrodynamic simulations of outflow/cloud interactions with a hydrodynamic code based on the TVD scheme. In our models, an initial outflow with number density 10 cm^{-3} and temperature 10^4 K interacts with a spherical cloud with a power-law density distribution. The cloud has a uniform temperature of 10 K and is surrounded by a homogeneous gas of density 100 cm^{-3} and temperature 10 K. The radius of the cloud is 0.02 pc, and the outflow has a sectional radius 300 AU. Through the numerical experiments, we found that the deflection angle is mainly determined by the impact parameter and the density ratio between the outflow and the impact zone of the cloud. The deflection angle is, however, not sensitive to the velocity of outflow. Using initial conditions and parameters which are particularly suitable for NGC 1333 IRAS 4A, we can reproduce its bent morphology. We therefore confirm that the northeastern deflected outflow of the NGC 1333 IRAS 4A may be colliding with a dense cloud core and deflected as a result.

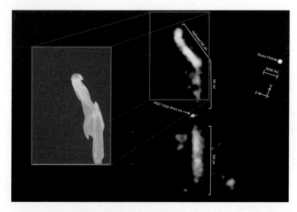

Figure 1. Comparison of deflected outflows seen in observations and simulation. The right box contains an SiO image toward the northeastern region of the NGC 1333 IRAS 4. A left box shows a map of the column density of shocked gas in our simulation. The deflection angle seen in the simulation is similar to the one seen in the observations.

Triggered Star Formation in a Turbulent ISM
Proceedings IAU Symposium No. 237, 2006
B. G. Elmegreen & J. Palouš, eds.
© 2007 International Astronomical Union
doi:10.1017/S1743921307001810

An inventory of supershells in nearby galaxies: first results from THINGS

Ioannis Bagetakos[1], Elias Brinks[1], Fabian Walter[2] and Erwin de Blok[3]

[1]Centre for Astrophysics Research, University of Hertfordshire, UK

[2]Max–Planck–Institut für Astronomie, Heidelberg, Germany

[3]Mount Stromlo Observatory, Australian National University, Canberra, Australia

Abstract. The HI Nearby Galaxy Survey (THINGS), is a 21–cm HI line survey of a sample of 34 nearby (3–10 Mpc) galaxies (Walter *et al.* 2005). The observations were carried out with the VLA and have a velocity resolution of $5 \, \mathrm{km \, s^{-1}}$ or better and an angular resolution of $7''$ which at this distance range corresponds to a linear resolution of 100–300 pc. One of the primary goals of THINGS is to look at the fine–scale structure of the Interstellar Medium (ISM) and examine how it varies as a function of Hubble type, star formation rate, galaxy mass, metallicity, etc. We present one of the first science results from this project, an inventory of HI shells in the galaxies NGC 628, NGC 3184, and NGC 6946.

Keywords. galaxies: ISM — ISM: structure — radio lines: ISM

1. Results

The morphology of the neutral ISM is greatly affected by massive stars through the combined effects of their stellar winds and, eventually, their demise as supernovae. Because massive stars tend to form in associations they will end their lives within the same, relatively short time span and within a small volume which leads to the formation of expanding bubbles of coronal gas in the ISM. These structures push the neutral gas outwards, compress it, and at least in some cases trigger secondary or induced star formation on the expanding rims. We have started the analysis of the structure of the ISM of all THINGS galaxies and some first results are given here. We find 19 well–defined HI shells in NGC 3184, 40 in NGC 6946 and 58 shells in NGC 628. The expansion velocities of these shells range from 5 to $20 \, \mathrm{km \, s^{-1}}$ in all three galaxies. The distribution of the kinematical ages peaks at about the same value (40–60 Myr). The size distribution of the holes appears to be different in the three galaxies: NGC 3184 has the largest holes (mean $\sim$1140 pc) and NGC 6946 the smallest (mean $\sim$700 pc). All three galaxies are large, late–type (Sc or Scd) galaxies. The difference in the number and size of holes detected depends on many factors, such as the inclination of the galaxy, its proximity, its (past) star formation rate, etc. We expect that an analysis of all 34 galaxies in the THINGS sample will allow us to verify if there are indeed clear trends of the properties of the HI holes (size, expansion velocity, energy requirement and kinematical age) with those of the host galaxies (mass, metallicity, SFR, and Hubble type, among others).

Reference

Walter, F., Brinks, E., de Blok, W. J. G., Thornley, M. D., & Kennicutt, R. C. 2005, *First Results from THINGS: The HI Nearby Galaxy Survey* (ASP-CS), 331, 269

Triggered Star Formation in a Turbulent ISM
Proceedings IAU Symposium No. 237, 2006
B. G. Elmegreen & J. Palouš, eds.

© 2007 International Astronomical Union
doi:10.1017/S1743921307001822

Fine–scale structure of the neutral ISM in M81

Ioannis Bagetakos[1], Elias Brinks[1], Fabian Walter[2] and Erwin de Blok[3]

[1]Centre for Astrophysics Research, University of Hertfordshire, UK

[2]Max–Planck–Institut für Astronomie, Heidelberg, Germany

[3]Mount Stromlo Observatory, Australian National University, Canberra, Australia

Abstract. We present an analysis of the fine–scale structure of the neutral ISM as traced via the 21-cm line of atomic hydrogen (HI) in the nearby galaxy M 81. The data show a stunning amount of detail in the form of 330 expanding shells and holes in the neutral ISM of M 81. A comparison with similar structures found in two other spirals and two dwarf galaxies (M 31, M 33, IC 2574 and Holmberg II) reveals that the ISM in M 81 shares a lot of similarities with the two spirals, whereas the structure of its ISM is different to that in dwarf galaxies. The sizes of the HI holes in M 81 range from 80 pc (close to the resolution limit) to 600 pc; the expansion velocities can reach 20 km s^{-1}; estimated ages are 2.5 to 35 Myrs and the energies involved range from 10^{50} to 3.5 x 10^{52} ergs. The amount of neutral gas involved is of order 10^4 to 10^6 solar masses.

Keywords. galaxies: individual (M 81) — galaxies: ISM — ISM: structure — radio lines: ISM

M 81 forms part of "The HI Nearby Galaxy Survey" (THINGS), a survey performed with the NRAO Very Large Array resulting in data with 7$''$ angular and 5 km s^{-1} or better velocity resolution (Walter *et al.* 2005). At this resolution the HI distribution shows a wealth of structure which is interpreted as being due to expanding HI shells. We conducted a search for HI shells by inspecting channel maps, analysing position–velocity cuts and looking at the integrated HI map. A total of 1212 candidate holes were detected in the HI distribution of M 81 for each of which a set of properties was determined. We selected the 330 clearest defined shells and compared their characteristics with those found in two spiral (M 31, M 33) and two dwarf galaxies (IC 2574, Holmberg II). The distributions of the size and the kinematical age of the HI shells were found to be similar for all three spirals, whereas they were clearly different from those in the dwarf galaxies, the diameter of shells in dwarf galaxies being larger than in spirals. Importantly, the amount of energy typically deposited in the ISM is the same, irrespective of galaxy type. This is ascribed to dwarf galaxies being lower mass systems which implies that for an observed velocity dispersion which is similar to that of spiral galaxies, they have a much thicker HI disk (Brinks *et al.* 2002). This means that a shell with the same energy input can grow larger in a dwarf galaxy than in a spiral. Also, the absence of shear in dwarf galaxies makes that structures in the ISM can persist longer.

References

Brinks, E., Walter, F., Ott, J. 2002 *Bloated Dwarfs: The Thickness of the HI Disks in Irregular Galaxies* (ASP-CS), 275, 57

Walter, F., Brinks, E., de Blok, W. J. G., Thornley, M. D., Kennicutt, R. C. 2005, *First Results from THINGS: The HI Nearby Galaxy Survey* (ASP-CS), 331, 269

Triggered Star Formation in a Turbulent ISM
Proceedings IAU Symposium No. 237, 2006
B. G. Elmegreen & J. Palouš, eds.

Methanol masers and massive star formation

Anna Bartkiewicz[1], Marian Szymczak[1]
and Huib Jan van Langevelde[2,3]

[1]Centre for Astronomy, Nicolaus Copernicus University, Gagarina 11, 87-100 Torun, Poland,
e-mail: annan@astro.uni.torun.pl

[2]Joint Institute for VLBI in Europe, Postbus 2, 7990 AA Dwingeloo, The Netherlands

[3]Sterrewacht Leiden, Postbus 9513, 2300 RA Leiden, The Netherlands

Abstract. It has been established that massive stars form in dense clusters, when large molecular clouds collapse. However, the high obscuration and small spatial scales make it difficult to investigate the earliest stage of high–mass protostellar objects (HMPOs). Therefore methanol masers are of special interests as they are closely associated with HMPOs and offer high spatial resolution; they probe the massive star formation environment at the unique scale of a few AU (1 mas corresponds to 5 AU at 5 kpc).

We present VLBI observations of methanol masers discovered in an unbiased survey along the Galactic plane. We compare their positions with infrared surveys. In general, the masers do not coincide with infrared objects. That implies they are already present at a very early evolution stage of HMPOs when the dense surroundings still absorb other radiation. In addition we present maps of five methanol masers towards HMPOs with milliarcsecond resolution taken with the recently extended European VLBI Network. These sources show a wide variety of morphologies, indicating they arise in different events going on in close surroundings of HMPOs, i.e. outflows, discs or shocks. One source G23.657−00.127 displays a regularly shaped ring which appears as a great laboratory for further research of a single HMPO. Proper motion studies should reveal an expansion or rotation of the maser components in 2–3 years.

Keywords. stars: formation, circumstellar matter; techniques: high angular resolution

Triggered Star Formation in a Turbulent ISM
Proceedings IAU Symposium No. 237, 2006
B. G. Elmegreen & J. Palouš, eds.

Sequential star formation in the Sh 254-258 molecular cloud: HHT maps of CO J=3-2 and 2-1 emission

J. H. Bieging[1], **W. L. Peters**[1], **B. Vila Vilaro**[2], **K. Schlottman**[1] and **C. Kulesa**[1]

[1]Steward Observatory, University of Arizona, Tucson, AZ 85721, USA
[2]National Astronomical Observatory of Japan, Tokyo
email: jbieging@as.arizona.edu

Abstract. The molecular cloud associated with the Sh 254-258 group of 5 small H II regions appears to be forming a (late)-OB association. We have mapped the associated molecular cloud in the J=2-1 line of the CO molecule over $0.75° \times 1°$, and the CO J=3-2 line toward the 2 main peaks, with the University of Arizona Heinrich Hertz Submm Telescope (HHT). We propose a scenario for sequential formation of the stars exciting the H II regions, triggered by the compression/heating of the molecular gas.

Keywords. ISM: clouds, ISM: molecules, submillimeter

The CO J=2-1 emission is shown as white contours in the left figure, overlaid on the 8 μm *MSX* image showing warm dust, and blue contours of VLA 20 cm continuum (Fich 1993 ApJS, 86, 475) showing the ionized gas. The main CO peak is directly between H II regions Sh 255 and 257. A second CO peak at top left is associated with warm dust but no ionized gas. The right figure shows the CO J=3-2 image ($10' \times 10'$, $24''$ resolution) centered on the main CO peak, with the H II region free-free emission (VLA 20 cm map) in white contours. The expanding H II regions appear to be compressing and heating the molecular gas, thereby initiating further star formation that is creating a small OB association. The CO images clearly show that the oldest H II region (Sh 254) has swept away or dissociated the molecular cloud on the right side of the ridge.

The second CO/mid-IR peak at top left is not obviously being affected by the visible H II regions. We suggest that it may be a compressed molecular core produced by colliding gas streams as seen in recent hydrodynamic simulations of cloud dynamics.

See website *mira.as.arizona.edu/~jbieging/IAUSymp237/Sh254cloud.pdf*

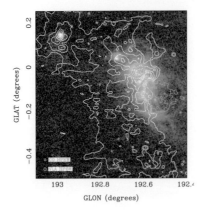

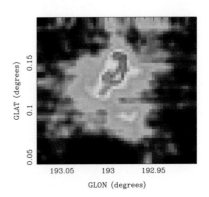

Triggered Star Formation in a Turbulent ISM
Proceedings IAU Symposium No. 237, 2006
B. G. Elmegreen & J. Palouš, eds.

Star formation thresholds derived From THINGS

F. Bigiel[1], F. Walter[1], E. de Blok[2], E. Brinks[3] and B. Madore[4]

[1]Max-Planck Institut für Astronomie, Königstuhl 17, 69117, Heidelberg, Germany
[2]Research School of Astronomy & Astrophysics, Mount Stromlo Observatory, Cotter Road, 2611, Weston, Australia
[3]Centre for Astrophysics Research, Science & Technology Research Institute, University of Hertfordshire, AL10 9AB, Hatfield, UK
[4]Observatories of the Carnegie Institution of Washington, 813 Santa Barbara St., 91101, Pasadena, USA

Abstract. We present first results from THINGS (The HI Nearby Galaxy Survey), which consists of high quality HI maps obtained with the VLA of 34 galaxies across a wide range of galaxy parameters (Hubble type, mass/luminosity). We compare the distribution of HI to the UV emission in our sample galaxies. In particular we present radial profiles of the HI (tracing the neutral interstellar medium) and UV (mainly tracing regions of recent star formation) in our sample galaxies. The azimuthally averaged HI profiles are compared to the predicted critical density above which organized large-scale star formation is believed to start (this threshold is based on the Toomre-Q parameter, which in turn is a measure for local gravitational instability).

Keywords. galaxies: ISM, radio lines: ISM, galaxies: evolution

1. Main

The HI Nearby Galaxy Survey (THINGS) provides high-resolution maps of the 21 cm line of atomic hydrogen (HI) at high spatial resolution (6") of a sample of nearby galaxies. This resolution is very well matched to the resolution of GALEX, the Galaxy Evolution Explorer, which operates in the ultraviolet regime. In a first analysis, we compare the azimuthally averaged HI column densities to the UV profiles in our sample galaxies (the deprojection parameters have been derived from a tilted ring analysis of the targets). To first order, the UV emission traces the distribution of the HI well in the centers of galaxies: The UV emission typically picks up at radii where the HI column density exceeds the critical density for star formation as given by either a constant density threshold of $\sim 7 \times 10^{20}$ cm^{-2}, or, for galaxies where rotation curves could be derived, as given by a radius dependent critical density as calculated from the Toomre-Q parameter (this parameter is related to local gravitational instabilities in the gas and can be used to derive a column density which is believed to be critical for those instabilities to collapse). Comparisons of HI column density maps with maps of UV emission show hints of faint UV emission reaching far out to large galactocentric radii, i.e. well out to the extended HI disks. Pixel–by–pixel plots of HI- and UV emission show that the UV emission is, to first order, a function of HI flux in more massive galaxies. Our unique dataset will allow us to search for trends in the relation between HI and UV as a function of environment and galaxy type.

Triggered Star Formation in a Turbulent ISM
Proceedings IAU Symposium No. 237, 2006
B. G. Elmegreen & J. Palouš, eds.

© 2007 International Astronomical Union
doi:10.1017/S174392130700186X

Intergalactic star formation around NGC 5291

Médéric Boquien[1], Pierre-Alain Duc[1], Jonathan Braine[2], Elias Brinks[3], Vassilis Charmandaris[4] and Ute Lisenfeld[5]

[1]AIM – Unité Mixte de Recherche CEA – CNRS – Université Paris VII – UMR n° 7158
CEA/Saclay; DSM/DAPNIA/Service d'Astrophysique, CEA/Saclay L'Orme des Merisiers
Bat. 709, 91191 Gif-sur-Yvette, France

[2]Observatoire de Bordeaux, UMR 5804, CNRS/INSU, B.P. 89, F-33270 Floirac, France

[3]Center for Astrophysics Research, University of Hertfordshire, College Lane, Hatfield AL10
9AB, UK

[4]Department of Physics, University of Crete, GR-71003, Heraklion, Greece

[5]Dept. de Física Teórica y del Cosmos, Universidad de Granada, Granada, Spain

Numerous instances of intergalactic star forming regions have been recently reported (see Duc *et al.* in this proceedings book). They are fueled by gaseous material expelled from parent galaxies. One spectacular example is the HI ring-like structure around the interacting system NGC 5291 (Malphrus *et al.* 1997) which hosts numerous HII regions (Duc & Mirabel 1998). In order to study how star formation proceeds in this specific environment, we have combined ultraviolet (Galex), Hα, 8 μm (Spitzer) and HI (VLA B-array) images of this system.

We have found that, qualitatively, the three star-forming indicators – ultraviolet, Hα and 8 μm bands – have a very similar morphology. The normalised infrared emission at 8.0 μm which was previously shown to be dominated by PAH bands (Higdon *et al.* 2006) is comparable to the integrated emission of dwarves of the same metallicity and to the emission of individual HII regions in spirals. The 8.0 μm emission in the intergalactic environment is therefore an estimator of the star formation rate (SFR) as reliable as for spirals. There is a clear excess of ultraviolet emission compared to individual HII regions in spirals, i.e. the $[8.0]/[NUV]$ and $[H\alpha]/[NUV]$ flux ratios are on average very low although there are some large variations from one region to the other. While the scatter of $[8.0]/[NUV]$ is largely due to large spatial variations of the dust extinction, as traced by the HI column density, the scatter of $[H\alpha]/[NUV]$ ratio is best explained by age effects. A model of the evolution $[H\alpha]/[NUV]$ with time favours young but already fading quasi-instantaneous starbursts. The total SFR measured in the intergalactic medium surrounding NGC 5291 is up to 1.3 M$_\odot$ yr^{-1}, a value typical for spirals, assuming the standard SFR calibrations are valid.

References

Boquien, M. S. *et al.*, to be submitted to *A&A*
Duc, P.-A. & Mirabel, I. 1998, *A&A* 333, 813
Higdon, S. *et al.* 2006, *ApJ* 640, 768
Malphrus, B. *et al.* 1997, *AJ* 114, 1427

Triggered Star Formation in a Turbulent ISM
Proceedings IAU Symposium No. 237, 2006
B. G. Elmegreen & J. Palouš, eds.

Comparing the structure of three dark globules

L. Campeggio, B. M. T. Maiolo, F. Strafella and D. Elia

Physics Department, University of Lecce, 73100 Lecce, Italy,
email: campeggio@le.infn.it; maiolo@le.infn.it; strafella@le.infn.it; eliad@le.infn.it

We present a comparative study of three dark globules CB52, CB107, and DC267.4-07.5. By means of an accurate photometry, near-IR two-colour diagrams were derived for the stellar backgrounds and used to determine the colour excesses of the reddened stars (Campeggio *et al.* 2004). By assuming a normal interstellar reddening law (Rieke & Lebofsky 1985), the visual extinction can be obtained as $A_V = 15.87E(H - K)$.

We sampled the three images with a spatial grid, computing, for each box, the mean extinction A_V and the relative dispersion σ_{A_V}. It is noteworthy that the extinction maps, obtained by means of this information, closely follow the shapes of the globules with values increasing from the boundary to the innermost dark regions. To analyze the structural properties of these globules we investigated the extinction dispersion σ_{A_V} as a function of A_V (Lada *et al.* 1994; Padoan *et al.* 1997), pointing out three different behaviours: parabolic (increasing up to a maximum and then decreasing), linear and fanlike for CB52, CB107, and DC267.4-07.5 respectively. For the globules DC267.4-07.5 and CB107, the scatter plots can be fitted with a linear function, while, for comparison purpose, we adopted for CB52 the tangent slope to the fitting parabola in the point $A_V=2$. To give a physical interpretation of these trends we also developed a code to simulate clouds. In this model a given volume of space is filled with a collection of small spheres (clumps) randomly distributed and characterized by different positions, central densities, and density profiles. Our simulations are completed by considering artificial stellar backgrounds, characterized by density, magnitude and colour distributions similar to those actually observed in the unreddened regions of the observed fields. We applied to these synthetic backgrounds the same procedures used for the analysis of the observations. Our results point out that such a kind of simulation can account for the different behaviours of the σ_{A_V} vs. A_V scatter plots observed in clouds with different internal structures (Maiolo *et al.* 2006). We also extended this analysis to a range of angular scales to study the non-homogeneity of these clouds for different spatial scales. We note that in each case the slope decreases with increasing angular resolution, pointing out that the globules, particularly CB107 and DC267.4-07.5, appear more structured toward the larger scales. Similar evidence was given by Lada *et al.* 1999 for the cloud IC 5146. In conclusion, our investigation suggests different internal structures for the studied clouds.

References

Campeggio, L., Strafella, F., Elia, D., Maiolo, B., Cecchi-Pestellini, C., Aiello, S., & Pezzuto, S. 2004, *ApJ* 616, 319
Lada, C.J., Lada, E.A., Clemens, D.P., & Bally, J. 1994, *ApJ* 429, 694
Lada, C.J., Alves, J.F. & Lada, E.A. 1999, *ApJ* 512, 250
Maiolo, B., Strafella, F., Campeggio, L., & Elia, D. 2006, (in preparation)
Padoan, P., Jones, B.J.T. & Nordlund, Å. 1997, *ApJ* 474, 730
Rieke, G.H. & Lebofsky, M.J. 1985, *ApJ* 288, 618

Triggered Star Formation in a Turbulent ISM
Proceedings IAU Symposium No. 237, 2006
B. G. Elmegreen & J. Palouš, eds.

© 2007 International Astronomical Union
doi:10.1017/S1743921307001883

A multifrequency study of the active star forming region NGC 6357

C. E. Cappa[1,2], R. H. Barbá[3], M. Arnal[1,2], N. Duronea[2], E. Fernández Lajús[1], W. M. Goss[4] and J. Vasquez[2]

[1]Facultad de Ciencias Astronómicas y Geofísicas, UNLP, Argentina
[2]Instituto Argentino de Radioastronomía, Argentina
[3]Departamento de Física, Universidad de La Serena, Chile
[4]NRAO, Socorro, USA

To investigate the interaction of the massive stars with the gas and dust in the active star forming region NGC 6357, located in the Sagittarius spiral arm at a distance of 1.7-2.6 kpc (Massey *et al.* 2001), we analyzed the distribution of the neutral and ionized gas, and that of the dust, based on Hα, [O III] and [S II] images obtained with the Curtis-Schmidt telescope at CTIO, radio continuum observations at 1.465 MHz obtained with the Very Large Array (NRAO) in the DnC configuration (synthesized beam = 38"), H I data from the Parkes survey (angular resolution = 15'), CO(1-0) observations obtained with the Nanten radiotelescope at Las Campanas Observatory (angular resolution = 2.7'), and IR images in the four MSX bands (angular resolution = 18.3").

NGC 6357 consists of a low excitation ionized envelope, 50' in diameter, H II regions in different evolutionary stages (Felli *et al.* 1990), molecular clouds, OB stars (most of them belonging to the open cluster Pismis 24), and IR sources (Persi *et al.* 1986).

The [S II]/Hα and [O III]/Hα line ratios confirm the low excitation of the ionized envelope. Although the H I emission distribution is dominated by absorption due to radio continuum sources, the optical filaments to the E and W are seen surrounded by neutral gas with velocities in the range –12 to +2 km s^{-1}. The CO emission distribution shows molecular gas associated with the region with velocities in the range –7 to +1 km s^{-1}.

A detailed analysis of the region reveals interstellar bubbles and photodissociation regions created by the massive stars in Pi 24 and by undisclosed ionizing sources.

Keywords. ISM: H II regions - ISM: individual (NGC 6357) - ISM: bubbles

Acknowledgements

CEC would like to thank financial support from the IAU and the Check Academy of Sciences, which facilitated her participation in the Symposium. RHB acknowledges financial support from FONFECYT (Chile) Programs No. 1050052 and 7050215. This work was partially financed by CONICET of Argentina under project PIP 5886/05 and Universidad Nacional de La Plata, Argentina, under project 11/G072. The National Radio Astronomy Observatory is a facility of the National Science Foundation operated under cooperative agreement by Associated Universities, Inc.

References

Felli, M., Persi, P., Roth, M., *et al.* 1990, *A&A* 232, 477
Massey, P., DeGioia-Eastwood, K. & Waterhouse, E. 2001, *AJ* 121, 1050
Persi, P., Ferrari-Toniolo, M., Roth, M. & Tapia, M. 1986, *A&A* 170, 97

Triggered Star Formation in a Turbulent ISM
Proceedings IAU Symposium No. 237, 2006
B. G. Elmegreen & J. Palouš, eds.

© 2007 International Astronomical Union
doi:10.1017/S1743921307001895

Triggered massive star formation in the LMC HII complex N44

C.-H. R. Chen[1], Y.-H. Chu[1], R. A. Gruendl[1] and F. Heitsch[2]

[1]Dept. of Astronomy, Univ. of Illinois, USA; [2]Dept. of Astronomy, Univ. of Michigan, USA

Abstract. We have used *Spitzer* IRAC and MIPS observations of N44 to identify young stellar objects (YSOs). Sixty YSO candidates with masses $\gtrsim 4\ M_\odot$ are identified. We have compared the distribution of YSOs with those of the ionized gas, molecular clouds, and HI gas to study the properties of star formation.

Keywords. stars: formation, stars: pre–main-sequence, Magellanic Clouds, HII regions

---

The distribution of YSO candidates relative to the ionized gas, molecular clouds (Fukui *et al.* 2001), and HI gas (Kim *et al.* 2003) is shown in the figure below. All YSOs are projected within molecular clouds. The majority are found near the peaks of molecular clouds; those off molecular peaks are in HI peaks. About 3/4 of the YSOs are in the two southern clouds associated with massive stars and ionized gas, and the other 1/4 in the northern cloud without much ionized gas, indicating that these YSOs are the first-generation massive stars in that cloud. The star formation history and stellar energy feedback may be responsible for the larger velocity dispersions (Mizuno *et al.* 2001) in the two southern clouds.

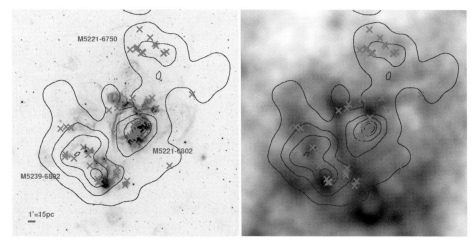

Figure 1. Left: Hα image of N44 marked with YSOs (X), CO contours, and molecular clouds (designation from Mizuno *et al.* 2001). Right: ATCA+Parkes HI image of N44 (grey scale) marked with YSOs (X) and CO contours.

References

Fukui, Y., Mizuno, N., Yamaguchi, R., Mizuno, A., & Onishi, T. 2001, *PASJ* 53, L41
Kim, S., *et al.* 2003, *ApJS* 148, 473
Mizuno, N., *et al.* 2001, *PASJ* 53, 971

Triggered Star Formation in a Turbulent ISM
Proceedings IAU Symposium No. 237, 2006
B. G. Elmegreen & J. Palouš, eds.

Star formation activity in the cluster spiral NGC 4254

Krzysztof T. Chyży[1], Rainer Beck[2] and Stanisław Ryś[1]

[1]Astronomical Observatory, Jagiellonian University, Kraków, Poland
[2]Max-Planck-Institut für Radioastronomie, Bonn, Germany

Abstract. With sensitive radio observations of a Virgo Cluster spiral NGC 4254 we are able to decompose thermal and synchrotron radio emission and, based on the thermal component, construct the SFR distribution within the galaxy, unaffected by dust extinction. The mean SFR per unit area is several times higher than in other galaxies of similar Hubble type. Contrary to other Virgo spirals the SFR distribution is not spatially truncated, in concordance with the observed weak HI deficiency. We propose that the SFR enhancement and the observed disturbed morphology of this galaxy can be attributed to tidal triggering by another nearby cluster member.

The star formation rate (SFR) is a sensitive indicator of environmental processes that influence cluster-embedded galaxies and their evolutionary state. Multifrequency radio data are especially useful to derive and study the dust-unaffected SFR within galaxies. We investigated the SFR of the Virgo Cluster spiral NGC 4254 for which we performed high-sensitivity radio polarimetric observations at 8.44 GHz, 4.86 GHz, and 1.43 GHz using the VLA and Effelsberg telescopes. With our data we are able to decompose the thermal and nonthermal (synchrotron) radio emission over the whole galaxy by least-square fitting of the thermal fraction to the observed radio emission, assuming a constant synchrotron spectral index. The obtained radio thermal distribution is used to predict the extinction-free luminosities in the Hα line, from the beam-independent regions within NGC 4254.

The radio-based predicted Hα luminosities are on average larger than the measured from the observed Hα emission, as expected. The estimated internal extinction seems to be dependent on the region luminosity and its mean value for the whole galaxy is about 0.8 in magnitude. According to our findings the production of stars in the most vivid regions in spiral arms of NGC 4254 is about 20 times higher than in more quiet interarm regions. The mean SFR of the whole NGC 4254 is several times larger than in nearby normal galaxies of similar Hubble type.

The derived enhanced SFR in NGC 4254 is in contradiction to the main trend of reduced SFR in Virgo cluster galaxies compared to isolated spirals. The confrontation of radio thermal emission of NGC 4254 with the optical (R) image does not reveal any reduced thermal emission, even in the southern part, which is most likely to be exposed to the cluster ram-pressure action. This seems to be in concordance with the weak HI deficiency observed in this galaxy. We propose that the enhanced SFR in NGC 4254 is due to tidal triggering by another nearby cluster member. The interaction is able not only to enhance star formation (as observed in some other Virgo Cluster spirals, too) but also to amplify the magnetic field and enhance the radio emission. It also explains the observed disturbed morphology of this galaxy in the optical and radio spectral ranges.

Acknowledgments. This work was supported by the Polish Ministry of Science and Higher Education, grant 2693/H03/2006/31. Based on data observed with the 100-m telescope of the MPIfR at Effelsberg and with the VLA of the NRAO.

Triggered Star Formation in a Turbulent ISM
Proceedings IAU Symposium No. 237, 2006
B. G. Elmegreen & J. Palouš, eds.

© 2007 International Astronomical Union
doi:10.1017/S1743921307001913

The Parkes methanol multibeam survey

R. J. Cohen[1], J. L. Caswell[2], K. Brooks[2], M. G. Burton[3],
A. Chrysostomou[4], J. Cox[9], P. J. Diamond[1], S. Ellingsen[5],
G. A. Fuller[6], M. D. Gray[6], J. A. Green[1], M. G. Hoare[7],
M. R. W. Masheder[8], N. McClure-Griffiths[2], M. Pestalozzi[4],
C. Phillips[2], M. Thompson[4], M. Voronkov[2], A. Walsh[3],
D. Ward-Thompson[9], D. Wong-McSweeney[6] and J. A. Yates[10]

[1]Jodrell Bank Observatory, The University of Manchester; [2]Australia Telescope National
Facility; [3]University of New South Wales; [4]University of Hertfordshire; [5]University of Cardiff;
[6]University of Tasmania; [7]The University of Manchester; [8]University of Leeds; [9]Bristol
University; [10]University College, London.

Abstract. A new 7-beam methanol multibeam receiver was successfully commissioned at Parkes
Observatory in January 2006, and has begun surveying the Milky Way for newly forming massive
stars, that are pinpointed by strong methanol maser emission at 6.7 GHz. The receiver was jointly
constructed by Jodrell Bank Observatory and the Australia Telescope National Facility for use
on the Parkes and Lovell Telescopes. The whole galactic plane is being surveyed within latitudes
$\pm 2°$, with a velocity resolution of 0.1 km s^{-1} and a 5-σ sensitivity of $\sim$0.7 Jy. Altogether 200
days of observing will be required.

Keywords. masers, surveys, stars: formation, ISM: molecules, radio lines: ISM

The Parkes Methanol Multibeam Survey is the most sensitive survey yet undertaken for
massive young stars in the Galaxy, and will provide the first unbiased catalogue of these
objects. The first 26 days of observations have yielded 377 methanol sources, of which 150
are new discoveries. Most of the new sources are weak ($\leqslant$4 Jy). The radial velocities show
that the survey is detecting masers on the far side of the Galaxy, outside the solar circle.
The most distant source found so far is 13.7 kpc from the galactic centre and 19.6 kpc
from the Sun. Accurate (0.1 arcsecond) positions for the masers are being measured
with the ATCA and with MERLIN, and cross-correlated against other galactic surveys
including MSX, IRAS, Spitzer/GLIMPSE and UKIRT/UKIDSS, to identify counterparts
at other wavelengths. The Parkes survey is expected to take 100 days, after which the
receiver will be moved to Jodrell Bank to complete the survey of the northern galactic
plane.

The receiver covers the frequency range 6.0–6.7 GHz, delivering two 300 MHz IF bands
in dual circular polarization. The telescope is scanned in galactic longitude, with data
read out every 5 seconds. Spectra cover a 4 MHz band, corresponding to a velocity range
of 180 km s^{-1} (2048 channels, 0.09 km s^{-1} resolution). Regions towards the galactic
centre are scanned more than once with different velocity settings to cover all likely radial
velocities for the maser emission. A second frequency band, centred on the 6035 MHz
line of OH. is observed in parallel with the methanol. Detection using the multibeam
provides a position accurate to $\sim 20''$, with follow-up interferometry able to increase the
precision to $\sim 0.1''$.

The survey webpage has further information and updates:

http://www.manchester.ac.uk/jodrellbank/research/methanol

Triggered Star Formation in a Turbulent ISM
Proceedings IAU Symposium No. 237, 2006
B. G. Elmegreen & J. Palouš, eds.

© 2007 International Astronomical Union
doi:10.1017/S1743921307001925

The Mopra DQS survey of the G333 region

M. R. Cunningham[1], I. Bains[1], N. Lo[1,2], T. Wong[1,2], M. G. Burton[1], P. A. Jones[1] and the DQS Team

[1]School of Physics, University of New South Wales, NSW 2052, Australia
email: Maria.Cunningham@unsw.edu.au
[2]Australia Telescope National Facility, PO Box 76, Epping NSW 1710, Australia

Keywords. ISM: clouds, turbulence, ISM: molecules, radio lines: ISM

Any successful model of star formation must be able to explain the low star forming efficiency of molecular clouds in our Galaxy. If the collapse of gas is regulated only by gravity, then the star formation rate should be orders of magnitude larger than the 1 $M_\odot$ per year within our galaxy. The standard model invokes magnetic fields to slow down the rate of collapse, but does not explain star formation in cluster mode, or the lack of observed variations in the chemistry of molecular clouds if they are long-lived entities.

Turbulent models invoke turbulence to regulate star formation, but require continuous injection of energy into the ISM to counter the rapid decay of turbulence observed in numerical simulations (Stone, Gammie & Ostriker 1998). The sources and their relative importance are as yet unclear but probably include outflows due to massive star formation, expanding supernova bubbles and large-scale galactic flows of gas.

We are using the Mopra telescope to survey a 1.5 × 1 degree region around G333.6-0.2 (the DQS) in a number of molecules tracing different densities, to examine the relationship between turbulence and star formation. A multi-line dataset will also allow us to look at the relationship between interstellar chemistry and turbulence.

During 2004, we observed ^{13}CO, and in Bains *et al.* (2006) we present analysis of the structure using clumpfind and comparison to the 1.2-mm continuum. During 2005, we observed C^{18}O and CS in the dense regions. A new digital filterbank (MOPS) has recently been installed that allows up to 8 different simultaneous 138 MHz bands of 4096 channels over 8 GHz in the zoom mode. We observed CS and C^{34}S in 2005 simultaneously with a (2 band) prototype of this system, and in the 2006 Winter season we are observing 8 bands including HC$_3$N, HNC, HCN, HCO$^+$, C$_2$H, H^{13}CN and H^{13}CO$^+$ and expect to observe another 8 bands by the end of the season.

Comparing the distribution of molecular transitions that trace gas of different densities can help in constraining the turbulent driving scales and strengths (Ballesteros-Paredes & Mac Low 2002). Strongly driven turbulence (i.e. greater energy injection into the ISM) leads to larger density fluctuations about the mean density than weakly driven turbulence.

We are using a variety of methods to characterise and compare the spatial and velocity structure in the DQS region: Power spectra (Jones *et al.* 2006, these proceedings); Gaussclumps (Stutzki & Guesten 1990; Kramer *et al.* 1998); Delta variance (Stutzki *et al.* 1998; Bensch, Stutzki, & Ossenkopf 2001); Velocity channel analysis (VCA) (Lazarian & Pogosyan 2000); Cross correlation of emission from different tracers.

References

Bains, I., *et al.*, 2006, *MNRAS* 367, 1609
Ballesteros-Paredes, J. & Mac Low, M-M. 2002, *ApJ* 570, 734
Bensch F., Stutzki J. & Ossenkopf V. 2001, *A&A* 366, 636
Kramer, C., Stutzki, J., Rohrig, R. & Corneliussen, U. 1998, *A&A* 329, 249
Lazarian, A. & Pogosyan, D. 2000, *ApJ* 537, 720
Stone, J.M., Gammie, C.F. & Ostriker, E.C., 1998, *ApJ* 508, L99
Stutzki, J. & Guesten, R. 1990, *ApJ* 356, 513
Stutzki, J., Bensch, F., Heithausen, A., Ossenkopf, V., & Zielinsky, M. 1998, *A&A* 336, 697

Triggered Star Formation in a Turbulent ISM
Proceedings IAU Symposium No. 237, 2006
B. G. Elmegreen & J. Palouš, eds.

© 2007 International Astronomical Union
doi:10.1017/S1743921307001937

Studies of the Perseus Region.I.

Srabani Datta

Department of Physics & Astronomy, University of Manchester,
Sackville Street, Manchester M60 1QD, U.K.
email: Srabani.Datta@manchester.ac.uk

Abstract. Studies of molecular clouds have shown that they evolve from turbulent gas and dust to form coherent, dense and connected structures. We have conducted a multi-wavelength study of one such molecular cloud, the Perseus star-forming region, which includes Barnard 1 (B1), Barnard 3(B3), Barnard 5 (B5), NGC 1333, IC 348, L1455 and L1448. The data obtained using the Infrared Array Camera (IRAC), Multiple Imaging Photometer (MIPS), the Sub-mm Common User Bolometer Array(SCUBA) and the 2 Micron All Sky Survey (2MASS)provides information about the geometric structure of the dust and gas covering large areas around young stellar objects (YSO), dust temperatures, effect of turbulence and processes of molecule formation and their relevance in the chemical and physical evolution of the cloud. This paper presents our first results.

Keywords. molecular cloud ,star formation, turbulence

The Perseus molecular cloud, located at a distance of about 350 pc, contains star-forming regions, molecular outflows, masers, dense gas and dust. Our analysis of this region (table 1) consists of the following- measuring the extinction in the K band (2.15 μm) as it is the most sensitive to embedded young stars; measuring the K band luminosity function (KLF); looking for significant asymmetric structure. The K band extinction is measured by an algorithm which calculates an average H-K colour from stars embedded in the molecular cloud taken from 2MASS survey. Then SCUBA maps are used to compare against this value and improve upon it. The KLF is measured using K band apparent histograms. We also look for new sources from the IRAC (3.6, 4.5, 5.8 and 8 μm) and MIPS 24 μm maps.

Acknowledgements

The author wishes to thank those who helped, especially Dr. G.A. Fuller, Prof. A. Zijlstra, Dr. S. Dalla, Dr. S. Williams and Dr. M. Matsuura at the Department of Physics & Astronomy, University of Manchester (UMIST).

Table 1. Objects in Perseus with their distances.Coordinates taken from Simbad(J2000). The program ID (PID) of the IRAC and MIPS observations from Spitzer taken for this study are given in Column 5.

Object	RA	Dec	Distance(pc)	PID
Barnard 1	3h33m16.3s	31d7m51.0s	330	178
Barnard 3	3h40m 0.0s	31d 58m 0.0s	-	178
Barnard 5	3h47m38.3s	32d52m43.0s	350	178
IC 348	3h44m34.0s	32d9m48.0s	300	178,6,58,36
L1455	3h28m4.0s	30d10m24.0s	350	178
L1448	3h24m20.6s	31d20m47.0s	300	178
NGC1333	3h29m2.0s	31d20m54.0s	300	58,178,6

Triggered Star Formation in a Turbulent ISM
Proceedings IAU Symposium No. 237, 2006
B. G. Elmegreen & J. Palouš, eds.

Evidence for triggered star formation in the Carina Flare supershell

Joanne Dawson, A. Kawamura, N. Mizuno, T. Onishi and Y. Fukui

Department of Astrophysics, Nagoya University, Chikusa-ku, Nagoya, 464-8602, Japan.
Email: joanne@a.phys.nagoya-u.ac.jp

Abstract. Theory predicts the triggered formation of molecular clouds stars through the fragmentation and collapse of swept-up ambient gas. Yet the majority of Galactic HI shells show no more than a scattering of small molecular clouds. The Carina Flare supershell (Fukui *et al.* 1999) is a rare example of an HI shell with a striking molecular component. Here we present the large-scale morphology of the molecular and atomic gas and the location of YSO candidates. A detailed look at two molecular clumps in the shell walls reveals active, intermediate mass star forming regions at various stages of early evolution.

Keywords. ISM: bubbles, ISM: structure, stars: formation.

New HI data from the Parkes SGPS reveals the Carina Flare to be a galactic chimney. The molecular and atomic components show good correspondence in position and velocity. Morphological features include molecular clouds with cometary HI tails and star-forming molecular clumps in HI filaments. IRAS YSO candidates are scattered across the region and their correlation with loop-like structures of the supershell provides evidence for star formation occurring preferentially in the shell walls. Two star-forming molecular clouds have been examined in detail using ^{12}CO, ^{13}CO and C^{18}O(J=1-0) observations made with the NANTEN and SEST telescopes, combined with IR and optical survey data. We find evidence for ongoing intermediate mass star formation in both. We suggest that the Carina Flare is a promising candidate for the triggered formation of both stars and their parent molecular clouds.

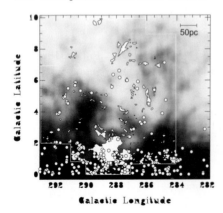

Figure 1. Parkes SGPS HI 21cm (greyscale), NANTEN ^{12}CO(J=1-0) (contours) & IRAS YSO candidates (circles) selected from the point source catalogue. All emission is integrated over the velocity range -35 < v_{lsr} < -3kms^{-1}. The white box marks the region observed in ^{12}CO(J=1-0). Carina Flare basic parameters: distance $\sim$ 2.6$\pm$0.4kpc; expansion velocity $\sim$ 8kms^{-1}; size $\sim$ 250 × 350pc; atomic mass $\sim$ 2×10^{5}M$_\odot$; molecular mass $\sim$ [1−3] × 10^{5}M$_\odot$.

Reference

Fukui, Y., Onishi, T., Kawamura, A., Tachihara, K., Yamaguchi, R., Mizuno, A., & Ogawa, H. 1999, *PASJ* 751, 51

Triggered Star Formation in a Turbulent ISM
Proceedings IAU Symposium No. 237, 2006
B. G. Elmegreen & J. Palouš, eds.

© 2007 International Astronomical Union
doi:10.1017/S1743921307001950

New insights into the nature of mid-infrared emission associated with massive star formation: disks and outflow

James M. De Buizer

Gemini Observatory, Casilla 603, La Serena, Chile
email: jdebuizer@gemini.edu

Abstract. Recent observations in the mid-infrared (5-25 μm) of massive young stellar sources have yielded a surprising result: many show evidence of mid-infrared emission from outflows and jets. These observations correlate well with other larger-scale outflow indicators and their geometries, such as what is seen in shock-excited H_2 and CO emission. In some cases these mid-infrared observations identify the local maser emission as outflow or jet related.

Keywords. ISM: jets and outflows, infrared: ISM, circumstellar matter, stars: early-type

1. Introduction

Thanks to the increase of facility-class mid-infrared imagers on large aperture telescopes (8-10m), we are achieving high spatial resolutions in the mid-infrared (∼0.25-0.60") that allow us to see the detailed morphologies of the mid-infrared environments of massive young stellar sources. Mid-infrared (MIR) emission is usually attributed to circumstellar disk dust emission, however recent observations are showing circumstellar structure that is interpreted as arising from the directly heated dust on the walls of outflow cavities or from material caught up in outflow. An outflow cavity can be created by a molecular outflow or jet punching a hole in the dense clump of obscuring material surrounding a young stellar source. Direct associations of outflows and jets with individual massive stars are few. Outflows from massive stars may demonstrate that they form by accretion processes similar to low mass stars.

There have been two recent publications showing evidence of MIR emission from jets and outflows from massive young stellar objects. The clearest example comes from the observations of G35.20-0.74 by De Buizer (2006a), which show MIR emission from the jet cavity previously detected at other wavelengths. The larger-scale MIR structure of NGC 7538 IRS 1 observed by De Buizer & Minier (2005) also shows what is believed to be emission from an outflow. The third and newest addition to this list is IRAS 20126+4104. Using Michelle on Gemini North, high spatial resolution images revealed that the MIR emission at the center of this region is distributed in a double-lobed structure similar to the 2 μm emission (Sridharan *et al.* 2005). De Buizer (2006b) argues that the MIR, near-IR, and maser emission are likely coming from the outflow cavities centered on the central radio continuum source at the heart of this region.

References

De Buizer, J. M. 2006a, *ApJ* 642, L57
De Buizer, J. M. 2006b, *ApJ* submitted
De Buizer, J. M. & Minier, V. 2005, *ApJ* 628, L151
Sridharan, T. K., Williams, S. J. & Fuller, G. A. 2005, *ApJ* 631, L73

Triggered Star Formation in a Turbulent ISM
Proceedings IAU Symposium No. 237, 2006
B. G. Elmegreen & J. Palouš, eds.

From young massive star clusters to old globulars: long-term survival chances

Richard de Grijs

Department of Physics & Astronomy, The University of Sheffield, Hicks Building, Hounsfield Road, Sheffield S3 7RH, UK

Young, massive star clusters (YMCs) are the most notable and significant end products of violent star-forming episodes triggered by galaxy collisions and close encounters. The question remains, however, whether or not at least a fraction of the compact YMCs seen in abundance in extragalactic starbursts, are potentially the progenitors of ($\gtrsim 10$ Gyr) old globular cluster (GC)-type objects. If we could settle this issue convincingly, one way or the other, the implications of such a result would have far-reaching implications for a wide range of astrophysical questions, including our understanding of the process of galaxy formation and assembly, and the process and conditions required for star (cluster) formation. Because of the lack of a statistically significant sample of YMCs in the Local Group, however, we need to resort to either statistical arguments or to the painstaking approach of case-by-case studies of individual objects in more distant galaxies.

A variety of methods have been employed to address the long-term survival issue. The most promising and most popular approach aimed at establishing whether a significant fraction of an entire *population* of YMCs (as opposed to individual objects) might survive for any significant length of time (say, in excess of a few $\times 10^9$ yr) uses the "cluster luminosity function", or its equivalent mass function (CLF, CMF), as a diagnostic tool. In essence, the long-term survival of dense YMCs depends sensitively on the low-mass section (below a few $M_\odot$) of their stellar initial mass function (IMF). Clearly, assessing the shape of the stellar IMF in unresolved extragalactic clusters is difficult, potentially ambiguous and riddled with pitfalls. Nevertheless, and despite these difficulties, an ever increasing body of observational evidence lends support to the scenario that GCs, which were once thought to be the oldest building blocks of galaxies, are still forming today.

By focusing on the shape of their *initial* mass distribution, we concluded that the CMF at the time of starbirth in the starburst galaxy M82 (de Grijs *et al.* 2003, 2005), and possibly also in the Large Magellanic Cloud (de Grijs & Anders 2006) and the Antennae interacting system (P. Anders *et al.*, in prep.), may not have been a power-law function of mass, as often assumed. Instead, we have uncovered evidence for a significant flattening of the initial CMF towards lower cluster masses, most significantly and robustly so among the M82 YMCs. This result lends strong support to evolutionary scenarios that start from initial log-normal CMFs, e.g., as proposed by Vesperini (1998). This scenario is attractive, because it is based on as few restricting assumptions as physically possible. Other popular scenarios often need to invoke significant, sometimes unrealistic assumptions on, e.g., the dynamics of the cluster populations in their host galaxies.

References

de Grijs R., Bastian N. & Lamers H.J.G.L.M. 2003, *ApJ* 583, L17
de Grijs R., Parmentier G. & Lamers H.J.G.L.M. 2005, *MNRAS* 364, 1054
de Grijs R. & Anders P. 2006, *MNRAS* 366, 295
Vesperini E. 1998, *MNRAS* 299, 1019

Triggered Star Formation in a Turbulent ISM
Proceedings IAU Symposium No. 237, 2006
B. G. Elmegreen & J. Palouš, eds.

Star formation In Bright-Rimmed Clouds: a comparison of wind-driven triggering with millimeter observations

Christopher H. De Vries[1], G. Narayanan[2] and R. L. Snell[2]

[1]Physics and Geology Department, CSU Stanislaus,
Turlock, CA 95382, USA
email: chris@physics.csustan.edu

[2]Department of Astronomy, University of Massachusetts Amherst,
Amherst, MA 01003, USA
email: gopal@astro.umass.edu, snell@astro.umass.edu

Bright-rimmed clouds (BRCs) are logical laboratories in which to study triggered star formation, however it is difficult in any single cloud to definitively show that star formation was triggered. In this study we compare the hydrodynamic models produced by Vanhala & Cameron (1998) that treat the problems of star-formation triggered by wind-driven implosion to millimeter and submillimeter molecular line observations of BRCs with embedded IRAS sources. These latter sources are derived from a catalog by Sugitani, Fukui, & Ogura (1991). In order to make an accurate comparison we implement a radiative transfer model based on the Sobolev or LVG approximation, and generate molecular line maps which can be directly compared to our observations. We observed several millimeter and submillimeter transitions of CO, $C^{18}O$, HCO^+, and $H^{13}CO^+$ using the FCRAO, SMT, CSO, and SMA observatories (De Vries, Narayanan, & Snell 2002). We compare these observations with 3 hydrodynamic models of wind-driven shock fronts interacting with pre-existing, but unbound cloud cores. In two cases these model cores are triggered to collapse under the influence of the external wind.

We find that the LVG radiative transfer method provides similar line shapes compared to the Hogerheijde & van der Tak (2000) radiative transfer model. Further, the gross morphologies of the hydrodynamic models, when run through our radiative transfer analysis, match the BRC morphologies we observe. The models produce an excitation temperature which increases with distance from the core center. This eliminates the blue-asymmetric infall signature often found in collapsing cores, and is supported by the relative rarity of infall signatures in our observed BRC sample. Closer examination reveals critical differences, including the absence of high-velocity gas streaming from the cores predicted by the hydrodynamic models. We conclude that while the wind-driven hydrodynamic model is not sufficient to explain all the features we observe, it does sufficiently explain both the overall morphology of BRCs, and the generally symmetric, broad line shapes observed within these clouds. The similarity between the hydrodynamic predictions and observations suggest that shock fronts may trigger collapse and the formation of stars within BRCs.

References

De Vries, C.H., Narayanan, G. & Snell, R.L. 2002, *ApJ* 577, 798
Hogerheijde, M.R. & van der Tak, F.F.S. 2000, *A&A* 362, 697
Sugitani, K., Fukui, Y. & Ogura, K. 1991, *ApJ* 77, 59
Vanhala, H.A.T. & Cameron, A.G.W. 1998, *ApJ* 508, 291

Triggered Star Formation in a Turbulent ISM
Proceedings IAU Symposium No. 237, 2006
B. G. Elmegreen & J. Palouš, eds.

© 2007 International Astronomical Union
doi:10.1017/S1743921307001986

The virial balance of clumps and cores in molecular clouds

Sami Dib[1], Enrique Vázquez-Semadeni[1], Jongsoo Kim[2], Andreas Burkert[3] and Mohsen Shadmehri[4]

[1]Centro de Radioastronomía y Astrofísica, UNAM, Apdo. 72-3 (Xangari), 58089 Morelia, Michoacán, Mexico
email: s.dib@astrosmo.unam.mx

[2]Korea Astronomy and Space Science Institute, 61-1, Hwaam-dong, Yuseong-gu, Daejeon 305-764, Korea

[3]University Observatory Munich, Scheinerstrasse 1, D-81679 Munich, Germany

[4]Department of Physics, School of Science, Ferdowsi University, Mashhad, Iran

Abstract. We study (i.e., Dib *et al.* 2006) the virial balance of clumps and cores (CCs) in a set of three-dimensional numerical simulations of driven, magnetohydrodynamical, isothermal molecular clouds (MCs). The simulations represent a range of magnetic field strengths in MCs from subcritical to non-magnetic regimes. We developed a clump-finding algorithm to identify CCs at different threshold levels in the simulation box, and for each object, we calculate all the terms that enter the virial theorem in its Eulerian form. We also calculate, other quantities commonly used to indicate the state of gravitational boundedness of CCs such as the Jeans number J_c, the mass-to magnetic flux ratio μ_c, and the virial parameter α_{vir}. Our results suggest that a) CCs are dynamical out-of-equilibrium structures. b) The surface energies are of the same order than their volume counterparts and thus are very important in determining the exact energy balance in CCs. c) CCs can be either in the process of being compressed by the velocity field or of being dispersed. Yet, not all CCs that have a compressive velocity field at their boundaries are necessarily gravitationally bound. d) There is no one-to-one correspondence between the state of gravitational boundedness of a CC as described by the energy balance analysis (i.e., gravity versus other energies) or as implied by the classical indicators J_c, μ_c, and α_{vir}. In general, from the energy analysis, we observe that only the inner regions of the objects (i.e., the dense cores selected at high threshold levels) are gravitationally bound, whereas J_c and α_{vir} estimates tend to show that they are more gravitationally bound at the lowest threshold levels. g) We observe, in the non-magnetic simulation, the existence of a bound core with structural and dynamical properties that resemble those of the Bok globule Barnard 68 (B68). This suggests that B68 like cores can form in a larger molecular cloud and then be confined by the warm gas of a newly formed HII region, which can heat and rarefy the gas around the core, confine it, and extend its lifetime.

Keywords. ISM: structure, kinematics and dynamics, magnetic fields, clouds, globules, individual (Barnard 68)

Acknowledgements

S. D. is supported by a UNAM postdoctoral fellowship and is grateful to the Symposium organizers, Profs Jan Palouš and Bruce Elmegreen for the financial support that allowed him to attend the conference.

Reference

Dib, S., Vázquez-Semadeni, E., Kim, J., Burkert, A., & Shadmehri, M. 2006, *ApJ* submitted, (astro-ph/0607362)

Triggered Star Formation in a Turbulent ISM
Proceedings IAU Symposium No. 237, 2006
B. G. Elmegreen & J. Palouš, eds.

© 2007 International Astronomical Union
doi:10.1017/S1743921307001998

Structural analysis of molecular cloud maps: the case of the star forming Vela-D cloud

Davide Elia[1], F. Strafella[1], F. Massi[2], M. De Luca[3,4], L. Campeggio[1] and B. M. T. Maiolo[1]

[1]Dipartimento di Fisica, Università di Lecce, CP 193, I-73100 Lecce, Italy
email: eliad@le.infn.it

[2]INAF – Osservatorio Astrofisico di Arcetri, Largo E. Fermi 5, 50125 Firenze, Italy

[4]Dipartimento di Fisica, Università degli Studi di Roma "Tor Vergata", Via della Ricerca Scientifica 1, I-00133 Roma, Italy

[3]INAF – Osservatorio Astronomico di Roma, via Frascati 33, 00040 Monteporzio Catone, Italy

Abstract. We present the preliminary results of a statistical analysis carried out on a $1° \times 1°$ CO(1-0) map of the intermediate mass star forming region Vela-D Cloud. Our goal is to determine statistical parameters suitable to quantify the structure of the observed cloud, in particular the power-law exponent of the map power spectrum. Furthermore, to help in removing the degeneracy implied in using a single parameter, we also resort to the multifractal approach.

Keywords. turbulence, ISM: clouds, structure, stars: formation

---

In last decades, several statistical tools were developed in order to characterize the structure of the molecular cloud maps obtained in the radio, submm and IR domains, and to infer on the 3D mass distribution. The comparison of the results obtained for maps which can differ in observing technique, wavelength, size, resolution, physical conditions, galactic position, fluidodynamical regime and presence of star formation can offer a way to better understand the differences observed in the star forming activity. Here we discuss the case of the SEST CO(1-0) observations of Vela-D Cloud presented and widely described in Elia, Massi, Strafella, *et al.* 2006. We applied the robust method of the Δ-variance (Stutzki, Bensch, Heithausen, *et al.* 1998) to determine the power-law exponent β of the power spectrum of the integrated intensity map from -2 to 20 km s^{-1} and of the single channel maps obtained integrating over intervals of 3 km s^{-1} starting from 2 km s^{-1}. We obtained $\beta = 2.78$, corresponding to a fractal dimension $D = 2.60$. As indicated by Stutzki, Bensch, Heithausen, *et al.* (1998), this parameter can be related with the clump mass spectrum of the investigated cloud (Elia, Massi, Strafella, *et al.* 2006), suggesting a scenario in which clumps have, on average, constant column density. Moreover, we found $\beta = 2.51, 2.87, 2.89, 2.77$ for the four channels map, respectively. As a further preliminary result, we found that the multifractal spectrum calculated as in Chappell & Scalo (2001) (but considering here also negative orders) for the above mentioned maps of Vela-D Cloud appears to be wider for the integrated intensity map than for the single channel maps, in particular for negative orders.

References

Chappell, D. & Scalo, J.M. 2001, *ApJ* 551, 712

Elia, D., Massi, F., Strafella, F., De Luca, M., Giannini, T., Lorenzetti, D., Nisini, B., Campeggio, L., & Maiolo, B.M.T. 2006, *ApJ* submitted

Stutzki, J., Bensch, A., Heithausen, A., Ossenkopf, V., & Zielinsky, M. 1998, *A&A* 336, 697

Triggered Star Formation in a Turbulent ISM
Proceedings IAU Symposium No. 237, 2006
B. G. Elmegreen & J. Palouš, eds.

© 2007 International Astronomical Union
doi:10.1017/S1743921307002001

Different evolutionary stages in S235A-B

Marcello Felli[1], Fabrizio Massi[1] and Riccardo Cesaroni[1]

[1]INAF Osservatorio Astrofisico di Arcetri, Largo Enrico Fermi 5, I-50125 Firenze, Italy
email: felli@arcetri.astro.it

Abstract. The star forming region S235A-B has been studied at high resolution with radio (IRAM Interferometer and VLA) and infrared (JCMT and Spitzer) observations. The region was mapped in HCO^+, $C^{34}S$, H_2CS, SO_2 and CH_3CN as well as in the 1.2 and 3.3 continuum, in the cm continuum at 6, 3.6, 1.3 and 0.7 cm and in the 22 GHz water maser line, in the far infrared at 450 and 850 μm and in the mid infrared from 3.6 to 8 μm. Finally, use was made of the Medicina water maser patrol, from 1987 to 2005, to study the maser variability.

Keywords. Stars: formation – ISM: individual objects: S235A-B – ISM: jets and outflows – Radio continuum: ISM – Masers

Discussion

S235A is a classical HII region, well resolved and with a very good match between radio and infrared images. To the south of it, molecular and continuum mm observations as well as infrared observations unambiguously reveal the presence of a newly formed YSO placed in between the S235A and S235B: a molecular core and an unresolved mm source are centred on a water maser, with indication of mass infall onto the core. Two bipolar outflows detected in HCO^+ and a jet originate from the same position (Felli *et al.* 1997; Felli *et al.* 2004).

No cm radio continuum emission is detected either from the compact molecular core or from the jet-like structure, suggesting emission from dust in both cases. A weak evidence is found from $C^{34}S$ observations for a molecular rotating disk perpendicular to the main bipolar outflow. The derived parameters indicate that the YSOs is an intermediate luminosity object in a very early evolutionary phase. Its main source of energy could come from gravitational infall thus making of this YSO a rare link between the earliest evolutionary phases of massive stars and low mass protostars of Class 0–I.

Two compact radio continuum sources (VLA-1 and VLA-2) and three separate maser spots were found. VLA-1 coincides with one of the maser spots and with an IR source (M1). VLA-2 lies towards S235B and represents the first radio detection from this peculiar nebula. The two other maser spots coincide with the rotating disk perpendicular to the bipolar molecular outflow. The Spitzer images reveal a red object towards the molecular core which is the most viable candidate for the embedded YSO.

The picture emerging from all these data shows the extreme complexity of a small star forming region where a more evolved region (S235A) may trigger the formation of a younger one (Felli *et al.* 2006).

References

Felli, M., Testi, L., Valdettaro, R., & Wang, J.-J. 1997, *A&A* 320, 594
Felli, M., Massi, F., Navarrini, A. *et al.* 2004, *A&A* 420, 553
Felli, M., Massi, F., Robberto, M., & Cesaroni, R. 2006, *A&A* 453, 911

Triggered Star Formation in a Turbulent ISM
Proceedings IAU Symposium No. 237, 2006
B. G. Elmegreen & J. Palouš, eds.

© 2007 International Astronomical Union
doi:10.1017/S1743921307002013

Angular power spectra of Galactic HI

D. A. Green

Astrophysics Group, Cavendish Laboratory, 19 J. J. Thomson Avenue, Cambridge, CB4 3PT,
UK
email: dag@mrao.cam.ac.uk

Abstract. As an alternative to identifying and then studying particular features seen in Galactic HI 21-cm images, studies of the angular power spectra of the emission provide a concise, statistical description of HI emission. The angular power spectra of several fields near $l = 140°$, from the Canadian Galactic Plane Survey, as observed with the DRAO Synthesis Telescope, have been analysed in this way. The derived power spectra, which typically cover angular scales from about 0·15 to 0·9 degree, are generally well-fitted by a simple power-law dependence on angular scale.

Keywords. ISM: structure, turbulence, techniques: interferometric, radio lines: ISM

Images of Galactic HI emission show a variety of features over a wide range of scales. Rather than investigate particular features identified as 'interesting' by eye – which usually represent a small subset of an observed HI data 'cube' – I have made a study of a statistical description of the whole of the HI observed in several fields, using angular power spectra.

The 21-cm HI data are taken from the Canadian Galactic Plane Survey (CGPS, see Taylor *et al.* 2003), which measures Galactic HI emission using the DRAO Synthesis Telescope. This interferometer has baselines between ≈13 and 600 mm – i.e. an angular resolution of ≈1 arcmin – and an ≈2° field of view. The method used to analyze the data is to derive an angular power spectrum for the HI emission, over a range of different angular scales, from the observed visibilities (which extends an earlier study of one field made by Green 1993). This method is in contrast to other studies of the angular power spectrum of HI emission, e.g. Dickey *et al.* 2001, which work from mosaiced images, which have been synthesised from the observed visibilities, and then back Fourier transformed. These angular spectra may be related to the underlying turbulence in the ISM. The method estimates the angular power spectrum of noise, as observed in velocity channels where there is no HI emission, and the appropriately scaled noise power spectrum is removed from the observed visibilities. The 'noise removed' visibilities are then averaged in logarithmically spaced annular bins in the *uv*-plane, which is very well sampled by the DRAO Synthesis Telescope observations.

Several fields in the Galactic plane, near $l = 140°$ which are free from obvious HII regions and SNRs have been analysed, with HI emission typically measured over angular scales from ≈ 0·15 to 0·9 degree. Generally the angular power spectra are well described by simple power laws, with power $\propto r^{-(2.5 \text{ to } 3.0)}$ (where r is radius in the *uv*-plane). However, unlike the studies of the HI in the inner Galaxy (Dickey *et al.* 2001), preliminary results do not show a clear steepening of the angular power spectra when the velocity over which the analysis is made is increased.

References

Dickey, J. M., McClure-Griffiths, N. M., Stanimirović, S., Gaensler, B. M. & Green, A. J. 2001, *ApJ* 561, 264
Green, D. A. 1993, *MNRAS* 262, 327
Taylor, A. R., *et al.* 2003, *AJ* 125, 3145

Triggered Star Formation in a Turbulent ISM
Proceedings IAU Symposium No. 237, 2006
B. G. Elmegreen & J. Palouš, eds.

© 2007 International Astronomical Union
doi:10.1017/S1743921307002025

Gemini VRI data of counterparts associated to X-ray sources in CMa R1

J. Gregorio-Hetem[1], C. V. Rodrigues[2] and T. Montmerle[3]

[1]Universidade de São Paulo, Brazil, email: jane@astro.iag.usp.br

[2]INPE, Brazil, email: claudiavr@das.inpe.br

[3]Université de Grenoble, France, email: montmerle@obs.ujf-grenoble.fr

Abstract. The molecular cloud Canis Major R1 (CMa R1) contains several embedded stellar clusters associated to a ring of nebular emission, which is an expanding shell suggested to be a supernova remnant (SNR) inducing the star formation in this region (Herbst & Assousa 1977, Comerón *et al.* 1998). However, there are alternatives to the SNR hypothesis, since the shell-like structure could be produced by strong stellar winds or an evolving HII region, as suggested by Reynolds & Ogden (1978), Blitz (1980), and Pyatunina & Taraskin (1986), for example. Two main challenges have motivated us to investigate this interesting region: (i) to conduct a stellar population study, from 7 to 0.4 solar masses, and (ii) to verify the evolutionary status of embedded cluster members. This contribution is dedicated to report VRI data obtained with Gemini South telescope in the direction of six X-ray sources that are probably unresolved. The results reveal several faint candidates that could be multiple counterparts of X-ray emitters detected by ROSAT as single sources (Gregorio-Hetem, Montmerle & Marciotto 2003). These fields have not been observed in more recent X-ray surveys. The V-R and R-I colours were estimated for the objects associated with the position of the X-ray emission, aiming to distinguish between field stars and members of the cloud. For each ROSAT source, it has been detected the following number of candidates, which we suggest to be stellar groups: src15 has 7 possible optical counterparts (86% of them are NIR sources); src17 has 14 counterparts (71% are NIR sources); src37 has 11 (73% NIR); src42 has 16 (56% NIR); src44 has 10 (80% NIR); and src55 has 6 (67% NIR). Investigating the evolutionary scenario of the embedded stellar clusters associated to X-ray emitters, which are probably very young, is a unique opportunity to better understand the star formation process in CMa R1 and to test SNR models, verifying the hypothesis of induced star formation in this region.

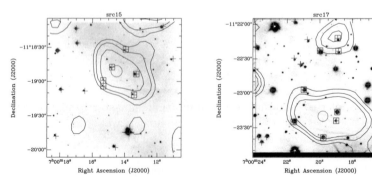

Figure 1. Optical image showing the contours of the X-Ray emission in the fields src15 and src17. The 2MASS catalogue was inspected searching for near-infrared (NIR) counterparts (crosses) related to the optical candidates (squares). Colour-magnitude diagrams have been constructed to evaluate the evolutionary status of the stellar groups.

Keywords. ISM: CMa clouds, Stars: pre-main sequence, X-rays: stars

Triggered Star Formation in a Turbulent ISM
Proceedings IAU Symposium No. 237, 2006
B. G. Elmegreen & J. Palouš, eds.

© 2007 International Astronomical Union
doi:10.1017/S1743921307002037

MHD simulations of supernova driven ISM turbulence

Oliver Gressel[1] and Udo Ziegler

MHD group, Astrophysikalisches Institut Potsdam
An der Sternwarte 16, 14482 Potsdam, Germany
[1]email: ogressel@aip.de

Abstract. Large-scale magnetic fields, that can be observed in numerous galaxies, are most likely the outcome of a dynamic process, a so-called dynamo. The favoured mechanisms for driving such a process in the ISM are supernovae(SNe) and/or magneto-rotational instability(MRI). In this work we simulate the dynamic evolution of the turbulent ISM utilising a three-dimensional MHD model.

Keywords. turbulence, ISM: kinematics and dynamics, magnetic fields, methods: numerical

---

Model description: Adopting the early models of Korpi *et al.* (1999) our spatial domain covers a box of $0.5^2 \times 2.0$ kpc^3 at a resolution of ~ 4 pc. The adiabatic equation of state is supplemented by a parameterised heating- and cooling-function allowing for thermal instability(TI). The update due to heating and cooling is implemented implicitly using a Patankar-type discretisation. Turbulence is driven by SNe which are modelled as local injections of thermal energy. SN-rates are adopted for typical cited values. We make a distinction between typeI and typeII SNe. Latter are clustered by the (artificial) constraint that the density at the explosion site be above average, former are spatially uncorrelated. The initial setup includes a differentially rotating background (with shearing BCs in radial direction) as well as vertical stratification. The initial density- and pressure-profiles are numerically integrated ensuring hydrostatic equilibrium with respect to the equation of state given by the radiative equilibrium. Including z-dependent heating rates this leads to a considerable deviation from usual isothermal initial models.

Preliminary Results: The amplification of the turbulent magnetic field is found to be independent of seed field amplitudes. Typical e-folding times for the magnetic energy are $\sim 1/10$ of the galactic period. Obtained velocity dispersions are ~ 25 km s^{-1} and thus at the upper limit of observed values.

Implications: The primary focus of this work is on the galactic dynamo and the generation of large-scale magnetic fields. As a secondary target we are interested in general properties of the ISM that are of importance for star formation(SF). To date only few simulations of gravo-turbulent SF do include magnetic fields or SN-feedback. Our model, however, lacks self-gravity. Peak densities are $\sim 10^2$ cm^{-3}, well below typical values for molecular clouds.

Acknowledgements

This work was supported by DFG under grant ZI 717/2-2.

Reference

Korpi, M., Brandenburg, A., Shukurov, A., Tuominen, I. & Nordlund, Å. 1999, *ApJ* 514, L99

Triggered Star Formation in a Turbulent ISM
Proceedings IAU Symposium No. 237, 2006
B. G. Elmegreen & J. Palouš, eds.

Bright knots along spiral arms in disk galaxies

P. Grosbøl[1] and H. Dottori[2]

[1]European Southern Observatory, Karl-Schwarzschild-Str. 2, 85748 Garching, DE
email: pgrosbol@eso.org

[2]Instituto de Física, Univ. Federal do Rio Grande do Sul, Av. Bento Gonçalves 9500,
91501-970 Porto Alegre, RS, BR
email: dottori.voy@terra.com.br

Abstract. Many spiral galaxies show bright knots along their arms on high resolution K-band images. Spectroscopy of such knots suggests that they are very young stellar clusters which formation was triggered by a large-scale front associated to a density wave. We have studied a sample of around 80 disk galaxies (with $i < 65°$) for which deep K-band maps with a resolution of $<1''$ are available and present preliminary statistics of such bright knots.

Keywords. galaxies: spiral, galaxies: star clusters, infrared: galaxies

1. Conclusions

Images of spiral galaxies in the near-infrared (NIR) K-band appear much smoother than those in visual bands since NIR maps are dominated by older stellar populations and are less affected by dust attenuation. Never the less, bright knots along spiral arms on K-band images are frequent (Grosbøl & Patsis 1998). Such knots were studied by Patsis *et al.* (2001) using narrow band filters and by Grosbøl *et al.* (2006) using K-band spectroscopy. These studies suggest that the knots are very young stellar clusters which formation is triggered by a front associated to a density wave. Investigating such clusters on K-band maps has the advantage that they are easy to identify and more reliable statistics can be obtained due to the low attenuation by dust.

Deep NIR K-band maps of 83 spiral galaxies were studied and bright knots along their spiral arms were identified. A preliminary analysis found that:
- more than half of the spirals show bright knots along their arms in the K-band,
- frequency of knots in a galaxy correlates with HI mass but not with spiral type,
- knots are very well aligned along the spiral arms over a significant radial range, and
- galaxies with strong spiral perturbations in K also show many knots

A detailed study of such young stellar clusters may be used to constrain their star formation history and thereby improve the understanding of how density waves may trigger formation of massive stellar clusters. If accurate ages can be determined for such clusters, one may estimate the location of their formation relative to the density wave and derive parameters such as pattern speed.

References

Grosbøl, P. & Patsis, P.A. 1998, *A&A* 336, 840
Patsis, P.A., Héraudeau, P. & Grosbøl, P. 2001, *A&A* 370, 875
Grosbøl, P., Dottori, H. & Gredel, R. 2006, *A&A* 453, L25

Triggered Star Formation in a Turbulent ISM
Proceedings IAU Symposium No. 237, 2006
B. G. Elmegreen & J. Palouš, eds.

© 2007 International Astronomical Union
doi:10.1017/S1743921307002050

Young stellar clusters triggered by a density wave in NGC 2997

P. Grosbøl[1], H. Dottori[2] and R. Gredel[3]

[1]European Southern Observatory, Karl-Schwarzschild-Str. 2, 85748 Garching, DE
email: pgrosbol@eso.org

[2]Instituto de Física, Univ. Federal do Rio Grande do Sul, Av. Bento Gonçalves 9500,
91501-970 Porto Alegre, RS, BR
email: dottori.voy@terra.com.br

[3]Max-Planck Institut für Astronomie, Königstuhl 17, 69117 Heidelberg, DE
email: gredel@mpia-hd.mpg.de

Abstract. Bright knots along the arms of grand-design spiral galaxies are frequently seen on near-infrared K-band images. To investigate their nature, low resolution K-band spectra of a string of knots in the southern arm of the grand design, spiral galaxy NGC 2997 were obtained with ISAAC/VLT. Most of the knots show strong Brγ emission while some have H_2 and HeI emission. A few knots show indications of CO absorption. Their spectra and absolute K magnitudes exceeding -12 mag suggest them to be very compact, young stellar clusters with masses up to 5×10^4 M$_\odot$. The knots' azimuthal distance from the K-band spiral correlates well with their Brγ strength, indicating that they are located inside the co-rotation of the density wave, which triggered them through a large-scale, star-forming front. These relative azimuthal distances suggest an age spread of more than 1.6 Myr, which is incompatible with standard models for an instantaneous star burst. This indicates a more complex star-formation history, such as several bursts or continuous formation.

Keywords. galaxies: spiral, galaxies: star clusters, infrared: galaxies, techniques: spectroscopic

1. Conclusions

Several K-band emitting knots, aligned along the southern arm of NGC 2997, were observed by Grosbøl *et al.* (2006). The analysis of the low-resolution, near-infrared spectra gave the following main results:

- their absolute K magnitudes exceed -12 mag,
- K-band emission spectra are similar to cocoon-enshrouded star forming regions,
- age range is 7-10 Myr and masses are up to 5×10^4 M$_\odot$,
- sizes and masses are similar to that of W49A in our Galaxy,
- alignment with spiral pattern along almost 4 kpc is striking, and
- locations of the knots are consistent with being inside co-rotation of spiral pattern.

This suggest a well-synchronized mechanism of piling up the material of which they were formed and for the triggering of star formation similar to the scenario of density wave propagation. Agreement between large-scale, synchronously triggered star formation and knot age differences requires several bursts or continuum star formation instead of a single burst within the knots.

Reference

Grosbøl, P., Dottori, H. & Gredel, R. 2006, *A&A* 453, L25

Triggered Star Formation in a Turbulent ISM
Proceedings IAU Symposium No. 237, 2006
B. G. Elmegreen & J. Palouš, eds.

© 2007 International Astronomical Union
doi:10.1017/S1743921307002062

A method for detecting preferred scale sizes

Maiken Gustafsson[1]†, Jean-Louis Lemaire[2] and David Field[1]

[1]Department of Physics and Astronomy, University of Aarhus, 8000 Aarhus C, Denmark
email: maikeng@phys.au.dk, dfield@phys.au.dk

[2]Observatoire de Paris & Université de Cergy-Pontoise, LERMA & UMR 8112 du CNRS,
92195 Meudon, France

Abstract. Molecular clouds are usually thought to be dominated by turbulence where the structures are inherently self-similar and lack characteristic scale. However self-similarity must break down at scales associated with star formation which imposes a characteristic scale. The turbulence may be driven by energy injection at some larger scale which also imposes characteristic scale. In order to understand the evolution of molecular clouds it is important to identify the departures from self-similarity associated with the scales of self-gravity and the driving of turbulence.

We describe a method based on structure functions for determining whether a region of gas, such as a molecular cloud, is fractal or contains structure with characteristic scale sizes (Gustafsson, Lemaire & Field 2006). Using artificial data containing structure it is shown that derivatives of higher order structure functions provide a powerful way to detect the presence of characteristic scales should any be present and to estimate the size of such structures. The method is easy to implement and compared with other techniques such as Fourier transform or histogram techniques (Blitz & Williams), the method appears both more sensitive to characteristic scales and easier to interpret.

The method is applied to observations of hot H_2 in the Kleinman-Low nebula, north of the Trapezium stars in the Orion Molecular Cloud, including both brightness and velocity data (Gustafsson, Kristensen, Clénet, *et al.* 2003). It is found that the density structure, represented by H_2 emission brightness in the K-band (2-2.5μm), exhibits mean characteristic sizes of 110, 550, 1700 and 2700 AU. The velocity data show the presence of structure at 140, 1500 and 3500 AU. These scales are respectively disk scales (140 AU) and outflow scales (>1000 AU), the latter being associated with (re-)injection of energy.

Keywords. techniques: image processing, ISM: structure, ISM: kinematics and dynamics

References

Gustafsson, M., Kristensen, L.E., Clénet, Y., Field, D., Lemaire, J.L., Pineau des Forêts, G., Rouan, D., & Le Coarer, E. 2003, *A&A* 411, 437
Gustafsson, M., Lemaire, J.L. & Field, D. 2006, *A&A* 456, 171
Blitz, L. & Williams, J.P. 1997, *ApJ* 488, L145

---

† Present address: Max-Planck-Institute for Astronomy, Königstuhl 17, 69117 Heidelberg, Germany

Triggered Star Formation in a Turbulent ISM
Proceedings IAU Symposium No. 237, 2006
B. G. Elmegreen & J. Palouš, eds.

HST emission line images of the Orion HII region: proper motions and possible variability

Leonel Gutiérrez[1,4], Corrado Giammanco[2] and John E. Beckman[1,3]

[1]Instituto de Astrofśica de Canarias, La Laguna, Tenerife, Spain. email: jeb@iac.es

[2]Physikalisches Institut der Universität Bern, Switzerland. corrado.giammanco@space.unibe.ch

[3]Consejo Superior de Investigaciones Científicas, Spain.

[4]Universidad Nacional Autónoma de México, Ensenada, México. leonel@astrosen.unam.mx

Using HST emission line images of the Orion Nebula, separated by 7 years in epoch, we have obtained evidence of localized temporal variability of both density and temperature during this period. Applying a digital filter to reduce high frequency noise, we used images in $H\alpha$ and [OIII] to quantify separately the variability in these two parameters. We detected fractional temperature variations of order 0.4% on scales of 2×10^{-2} pc. The same images yielded proper motion information; using cross-correlation to optimize the accuracy of the differential measurements we produced velocity field maps across the nebula, with vectors ranging up to $\sim$130 km s^{-1} across the line of sight. It is notable that in zones of rapid proper motion we find by far the largest density variations, as would be expected. It is much easier to quantify the temperature variations, on the other hand, in zones with low or zero detectable proper motion (see the other figure here), though these temperature variations appear across the whole face of the nebula.

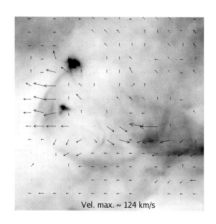
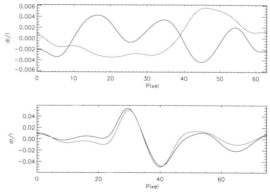

Left: Proper motion velocity vector map superposed on a portion of the $H\alpha$ image. The map presented covers an area of 0.18 x 0.18 parsec. *Right - Upper panel*: Scan across a kinematically quiescent field. The changes in surface brightness in $H\alpha$ (red, thick line) and [OIII] (black, thin line) are essentially in antiphase, as predicted for changes produced by localized temperature variations between the two epochs with only small density changes. *Lower panel*: Scan across a part of the nebula where rapid movement is detected. The changes in $H\alpha$ and [OIII] surface brightness are in phase as predicted for zones affected by density variations. We can use plots of these kinds to separate the changes in the two variables.

References

Corrado, G., Beckman, J.E. & Gutiérrez, L. 2006, in preparation
O'Dell, C.R., Peimbert, M. & Peimbert, A. 2003, *ApJ* 125, 2590
Peimbert, M. 1967, *ApJ* 150, 825

Triggered Star Formation in a Turbulent ISM
Proceedings IAU Symposium No. 237, 2006
B. G. Elmegreen & J. Palouš, eds.

The high latitude low mass star forming region Cometary Globule 12: two compact cores and a C18O hot spot

L. K. Haikala[1], M. Juvela[1], J. Harju[1], K. Lehtinen[1], K. Mattila[1], M. Olberg[2] and M. Dumke[3]

[1]Observatory, 00014 University of Helsinki Finland

[2]Onsala Space Observatory, S439 00 Onsala, Sweden

[3]European Southern Observatory, Casilla 19001, Santiago, Chile

Abstract. Cometary globule CG 12 lies at the distance of 630 pc more than 200 pc above the Galactic plane. The cloud's structure could be due to the passage of a supernova blast wave. Curiously, the cometary tail points at the galactic plane which would put the putative supernova even farther above the Galactic plane than the globule. The globule contains a low/intermediate mass stellar cluster with at least 9 members (Williams *et al.* 1977). The head of CG 12 has been observed using NIR imaging (NTT SOFI), mm continuum (SEST SIMBA) and sub mm (APEX) and mm (SEST) spectroscopy (Haikala & Olberg 2006, Haikala *et al.* 2006). The molecular material is distributed in a North-South 10' long elongated lane with two compact maxima separated by 3'. Strong $C^{18}O$ (3-2), (2-1) and (1-0) emission is detected in both maxima and both have an associated compact 1.2 mm continuum source. The Northern core, CG 12 N, is cold and is possibly still pre-stellar. A dense and compact core is observed in DCO^+ and CS emission in the direction of the Southern core, CG 12 S. A remarkable $C^{18}O$ hot spot was detected in CG 12 S. This is the first detection of such a compact, warm object in a low mass star forming region. The hot spot can be modelled with a 60" to 80" diameter (~0.2 pc) hot (80 K $\lesssim T_{ex} \lesssim$ 100 K) 1.6 solar mass clump (Haikala *et al.* 2006). The hot spot lies at the edge of a dense cloud core and on the axis of a highly collimated bipolar molecular outflow (White 1993). The driving source of the outflow is most probably embedded in the dense core. NIR imaging reveals a bright cone like feature with a faint counter cone in the centre of CG 12 S. The size of the CG 12 compact head, 1.1 pc by 1.8 pc, and the $C^{18}O$ mass larger than 100 solar masses are comparable to those of other nearby low/intermediate mass star formation regions.

References

Haikala, L.K., Juvela, M., Harju, J., Lehtinen, K., Mattila, K., & Dumke, M., 2006, *A&A* 454, L71
Haikala, L.K. & Olberg, M., 2006, *A&A* in press
White G. 1993, *A&A* 274, L33
Williams, P.M., Brand, P.W.J.L., Longmore, A.J., & Hawarden, T.G. 1977, *MNRAS* 181, 179

Triggered Star Formation in a Turbulent ISM
Proceedings IAU Symposium No. 237, 2006
B. G. Elmegreen & J. Palouš, eds.

© 2007 International Astronomical Union
doi:10.1017/S1743921307002098

Diffuse X-ray emission from the Carina Nebula observed with Suzaku

K. Hamaguchi[1,2], R. Petre[1] and the Suzaku η Carina team

[1]NASA Goddard Space Flight Center, Greenbelt, MD 20771, USA
email: kenji@milkyway.gsfc.nasa.gov

[2]Universities Space Research Association, 10211 Wincopin Circle, Suite 500, Columbia,
MD 21044-3432, USA

Abstract. The Carina Nebula possesses the brightest soft diffuse X-ray emission among the Galactic giant HII regions, but the origin has not been known yet. The XIS1 back-illuminated CCD camera onboard the Suzaku X-ray observatory has the best spectral resolution for extended soft sources so far, and is therefore capable of measuring these key emission lines in the soft diffuse plasma. Suzaku observed the Carina nebula on 2005 Aug. 29. The XIS1 spectra of the Carina nebula clearly showed spatial variations in emission line strengths. In the south, the spectrum showed strong L-shell lines of iron ions and K-shell lines of silicon ions, while in the north these lines were much weaker. Fitting the spectra with an absorbed thin-thermal plasma model with $kT \sim 0.2$, 0.6 keV and $N_H \sim 1-2 \times 10^{21}$ cm^{-2} showed that the silicon and iron abundance is about $2-3$ times higher in the south than in the north. Because of its large size (~ 40 pc), the diffuse emission in the Carina nebula might have been produced by an old supernova, or a super shell produced by multiple supernovae.

Keywords. ISM: abundances, supernova remnants, X-rays: ISM

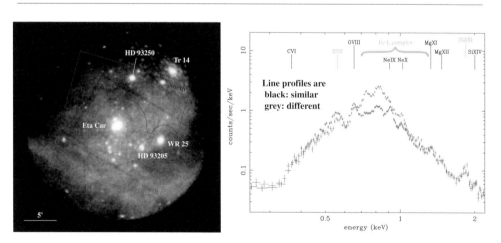

Figure 1. *Left*: *XMM-Newton* image with the source regions defined. *Right*: *Suzaku* normalized spectra of the north (lower curve) and south. See details for Hamaguchi *et al.* (2006).

Acknowledgements

K. H. is financially supported by the US *Chandra* grant No. GO3-4008A.

Reference

Hamaguchi *et al.* 2006, *PASJ* submitted

Triggered Star Formation in a Turbulent ISM
Proceedings IAU Symposium No. 237, 2006
B. G. Elmegreen & J. Palouš, eds.

Discovery of large-scale masers in W3(OH)

L. Harvey-Smith[1] and R. J. Cohen[2]

[1]Joint Institute for VLBI in Europe, Postbus 2, 7990 AA, The Netherlands. harvey@jive.nl

[2]Jodrell Bank Observatory, Macclesfield, Cheshire, SK11 9DL, UK. rjc@jb.man.ac.uk

Abstract. We report a vast filament of hydroxyl and methanol maser emission surrounding the ultra-compact HII region W3(OH). The filament stretches 3100 AU and has a linear velocity gradient. By studying the velocity structure, line profiles and extended methanol maser structures we believe we have located the position of the central star and detected around it a circumstellar disc with a large velocity gradient of 47 km s^{-1} arcsec^{-1}.

Keywords. masers, (ISM:) HII regions, ISM: molecules, radio lines: ISM, stars: formation

1. Introduction

Despite playing a vital and dominant role in feedback processes in the interstellar medium, massive stars are still not understood in terms of their formation processes. Hydroxyl and methanol masers are bright tracers of the early stages of massive star-formation, allowing us to study these objects at milliarcsecond resolution.

2. Discovery of an extended maser filament in W3(OH)

Using MERLIN we observed the OH and methanol masers around the ucHII region W3(OH). For the first time, we recovered all the single-dish flux in both species. This allowed us to discover a vast (3100 AU) filament of masing gas stretching north-south across the western face of W3(OH) (Harvey-Smith & Cohen 2005). The filament has a linear velocity gradient along its length, which is seen in both methanol and hydroxyl molecules (Harvey-Smith & Cohen 2006). The extended methanol and OH masers tend to avoid each other on small (0.1 arcsecond) scales, but together trace a continuous path across the face of the ucHII region.

3. A possible circumstellar disc in W3(OH)

Within the extended maser filament we find a small region with distinct physical properties. This region contains the brightest 6.7-GHz and 12.2-GHz methanol masers, the 13.4-GHz OH maser peak, the brightest radio continuum emission and the strongest magnetic field in W3(OH). At this location we have discovered a small 6.7-GHz methanol maser filament with broad line profiles and a linear velocity gradient of 4990 km s^{-1} pc^{-1}. This is consistent with an edge-on circumstellar disc around a central star with a mass >13 M$_\odot$. We suggest that this broadline feature houses the central star or stars powering the ultra-compact HII region.

References

Harvey-Smith, L. & Cohen, R.J. 2005, *MNRAS* 356, 637

Harvey-Smith, L. & Cohen, R.J. 2006, *MNRAS* 371, 1550

Triggered Star Formation in a Turbulent ISM
Proceedings IAU Symposium No. 237, 2006
B. G. Elmegreen & J. Palouš, eds.

IRSF/SIRIUS near-infrared survey of the Magellanic Clouds: triggered star formation in N11 in the Large Magellanic Cloud

H. Hatano[1], R. Kadowaki[1], D. Kato[1], S. Sato[1] and the IRSF/SIRIUS group

[1]Department of Astrophysics, Nagoya University, Chikusa-ku, Nagoya 464-8602, Japan
email: hattan@z.phys.nagoya-u.ac.jp

Abstract. A near-infrared survey of the Magellanic Clouds has been carried out with IRSF/SIRIUS. As a part of the results, we present a study of triggered star formation in N11 in the LMC.

We have completed a near-infrared (J, H, and Ks bands) survey of the Magellanic Clouds with IRSF/SIRIUS, covering a total area of about 40 square degree of the Large Magellanic Cloud (LMC) and 15 square degrees of the Small Magellanic Cloud (SMC). The data of the survey allow us to detect OB and Herbig Ae/Be (HAEBE) stars (down to $\sim 3\ M_\odot$) with a limiting magnitude of K ~ 17 mag, and to investigate the mechanism underlying star formation in the Magellanic Clouds.

We have explored star formation in N11 in the LMC. A total of 559 OB and 127 HAEBE star candidates were selected based on their near-infrared colors and magnitudes. Spatial correlations of the OB and HAEBE star candidates with CO clouds and Hα emission suggest that the birth of the young stellar populations in peripheral molecular clouds was triggered by an expanding shell blown by LH9 (see Fig. 1). This star formation activity has been propagated radially.

Figure 1. N11 is the second largest HII region in the LMC after 30 Dor. Distributions of Hα emission (inverted gray scale; Mac Low *et al.* 1998), ^{12}CO (1-0) integrated emission (contours; Israel *et al.* 2003), and the HAEBE star candidates (circles) are shown. The solid lines indicate our observed area. Hα emission nebulae and filaments, and CO clouds form shell structures surrounding the central OB association LH9. Many HAEBE star candidates are distributed around the periphery of LH9, and associated with the molecular cloud shell.

References

Mac Low, M. Mark., Chang, T. H., Chu, Y. H., Points, S. D., Smith, R. C. & Wakker, B. P. 1998, *ApJ* 493, 260
Israel, F. P., de Graauw, Th., Johansson, L. E. B., Booth, R. S., Boulanger, F., Garay, G., Kutner, M. L., Lequeux, J., Nyman, L.-A. & Rubio, M. 2003, *A&A* 401, 99

Triggered Star Formation in a Turbulent ISM
Proceedings IAU Symposium No. 237, 2006
B. G. Elmegreen & J. Palouš, eds.

© 2007 International Astronomical Union
doi:10.1017/S1743921307002128

A (Sub)Millimeter survey of massive star-forming regions identified by the ISOPHOT Serendipity Survey (ISOSS)

Martin Hennemann, Stephan M. Birkmann, Oliver Krause and Dietrich Lemke

Max-Planck-Institut fuer Astronomie, Koenigstuhl 17, Heidelberg, Germany
email: hennemann@mpia.de

Abstract. A sample of potential massive starforming regions identified at 170 m by ISO was observed in the submillimeter and millimeter regime. These observations allow us to infer physical properties of the molecular cloud cores. Two sources are presented in detail: ISOSS J23053+5953 and J183640221 show viable candidates for massive protocluster cores. Our analysis shows very low temperatures and low levels of turbulence of the major mass fraction in the molecular cloud cores besides active star formation at an early evolutionary stage. These conditions seem similar to the low mass case and may precede phases of luminous infrared emission observed towards young massive protostars.

Keywords. stars: formation

ISOSS J23053+5953: This object is placed at a distance of $\sim$3.5 kpc and coincides with IRAS 23032+5937. The SCUBA observations at 450 μm and 850 μm reveal two compact cloud cores separated by $\sim$17$''$ and a much more extended emission component. Only about 20% of the submillimeter flux is due to the two cores, and we deduce a large mass fraction of cold gas and dust in an extended envelope. This finding is supported by our molecular line studies in Methanol, Formaldehyde, $C^{18}O$, HCO^+ and the modelling of the thermal dust emission. **Conclusions:** The by far largest mass fraction of ISOSS J23053+5953 has a low temperature of $\sim$15 K and signatures for large scale infall are detected. It can be considered as a protocluster object probably prior to the formation of massive stars. The analysis of interferometric observations performed with the PdBI (IRAM) will clarify the state of fragmentation.

ISOSS J18364$-$0221: The distance is $\sim$2.2 kpc and the SCUBA maps at 450 μm and 850 μm show two components. The eastern seems extended and is probably composed of at least two subcores. It is cold (T $\sim$ 16.5 K) and compact (R $\sim$ 0.2 pc) with an estimated mass of 75 ± 30 M$_\odot$. The western component shows lower temperatures (T $\sim$ 12 K) and is extended (R $\sim$ 0.5 pc) with an estimated mass of 280 ± 75 M$_\odot$. We derive a gas kinetic temperature of $T_K = 11.6 \pm 1.5$ K, while the small line widths suggest a low level of turbulence. **Conclusions:** The submillimeter cores in ISOSS J18364-0221 show low temperatures, low levels of turbulence and signs of large scale infall as well as outflows driven by currently formed objects. They can be considered as massive protostellar objects representing the earliest phases of star formation.

Acknowledgements

Based on observations with ISO and Spitzer, at CAHA, with the IRAM 30 m, with the JCMT, SMT and Effelsberg 100 m. We thank J. Steinacker (MPIA) for providing the fitting routine.

Triggered Star Formation in a Turbulent ISM
Proceedings IAU Symposium No. 237, 2006
B. G. Elmegreen & J. Palouš, eds.

Non-symmetrical protoplanetary disks destroyed by UV photoevaporation

A. Hetem Jr.[1] and J. Gregorio-Hetem[2]

[1]Fundação Santo André, Brazil, email: annibal.hetem.jr@usa.net
[2]Universidade de São Paulo, Brazil, email: jane@astro.iag.usp.br

Abstract. We have developed geometric disk models to study the circumstellar geometries by fitting the spectral energy distribution (SED) of T Tauri and Herbig Ae/Be stars. The simulations provide means to recognize the signatures of different disk structures, including the effects due to external UV photoevaporation.

Following Chiang & Goldreich (1997) and Dullemond *et al.* (2001), we used hydrostatic, radiative equilibrium models for passive, reprocessing flared disks. The grains in the surface of the disk are directly exposed to the radiation from the star and the interior of the disk is heated by diffusion from the surface. Adopting this two-layers disk structure, our disk model was improved in order to optimize the parameters estimated by using a calculation technique based on genetic algorithms presented by Bentley & Corne (2002).

In the present work, we apply the code to model the SED of protoplanetary disks, which have being destroyed by photoevaporation due to the presence of ionizing OB stars, as the example of Trapezium region in the Orion Nebula. We compare geometric disk characteristics and physical conditions evaluated by our method to those obtained to the "proplyds" studied by Scally & Clarke (2001), Robberto *et al.* (2002) and Smith *et al.* (2005), among others. We also conclude that the parameter estimation by genetic algorithms assures accurate and efficient calculations.

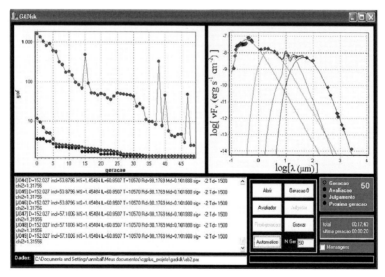

Figure 1. Fitting procedure sample session: the first panel shows the evolution of a solution; as the generation evolves, a best set of parameters is achieved. The right panel presents the SED for the best solution.

Keywords. ISM: Orion clouds, Stars: pre-main sequence, Method: genetic algorithms.

Triggered Star Formation in a Turbulent ISM
Proceedings IAU Symposium No. 237, 2006
B. G. Elmegreen & J. Palouš, eds.

ASTE Submillimeter observations of a YSO condensation in Cederblad 110

Masaaki Hiramatsu[1], Kazuhisa Kamegai[2], Takahiro Hayakawa[3], Ken'ichi Tatematsu[3], Toshikazu Onishi[4], Akira Mizuno[5] and Tetsuo Hasegawa[3]

[1]Department of Astronomy, University of Tokyo, Bunkyo, Tokyo 113-0033, Japan, email: hiramatsu.masaaki@nao.ac.jp

[2]Institute of Astronomy, The University of Tokyo, Mitaka, Tokyo 181-0015, Japan
[3]National Astronomical Observatory of Japan, Mitaka, Tokyo 181-8588, Japan

[4]Department of Astrophysics, Nagoya University, Chikusa, Nagoya 464-8602, Japan
[5]Solar-Terrestrial Environment Laboratory, Nagoya University, Toyokawa, Aichi, 442-8507, Japan

1. Observations of Cederblad 110 region

Outflow-cloud interaction is an important issue in discussions about star formation in clusters because it could generate turbulence and restrain star formation activities, as well as it causes outflow-triggered star formation.

Chamaeleon I molecular cloud complex is one of near ($D = 160$ pc) low-mass star formation regions in the southern sky. In the center of the cloud, a reflection nebula Cederblad 110 and a YSO condensation which contains 3 class I, 2 class II, 1 class III sources and a millimeter-continuum source.

We observed this YSO condensation in 2004 and 2005 with ASTE, a 10 m submillimeter telescope in Atacama. Observed lines are CO ($J = 3 - 2$), HCO$^+$ ($J = 4 - 3$) and H^{13}CO$^+$ ($J = 4 - 3$). The beamsize was $22''$ and the spectral resolution was 125 kHz, which corresponds 0.11 km s^{-1} in 345 GHz.

2. Results

Our HCO$^+$ ($J = 4 - 3$) map reveals a dense molecular clump with an extent of about 0.1 pc, which is a complex of three envelopes associated with class I sources Ced110 IRS4 and IRS11, and a millimeter-continuum source Cha-MMS1. The other two class I sources in this region, IRS6 and NIR89, are located outside the clump and have no conspicuous HCO$^+$ emission. H^{13}CO$^+$ ($J = 4 - 3$) observations toward IRS4 and MMS1 show the HCO$^+$ abundance are $X(\text{HCO}^+) = 3.3 \times 10^{-9}$ and 3.0×10^{-10}, respectively.

Bipolar outflows from IRS4 and IRS6 are detected in our CO ($J = 3 - 2$) observations. We could not detect any outflows from Cha-MMS1. This source could be a very young protostar which has no signs of outflow activity yet. The outflow from IRS4 seems to collide with Cha-MMS1. The outflow has momentum about 5 times larger than the momentum of gas in MMS1. This means the motion of the gas in MMS1 is easily affected when the outflow from IRS4 collides with the source. The time for the shock of collision to arrive at the center of MMS1 is about the same as the dynamical timescale of the IRS4 outflow, which shows induced star formation process has not triggered yet, or just triggered.

Triggered Star Formation in a Turbulent ISM
Proceedings IAU Symposium No. 237, 2006
B. G. Elmegreen & J. Palouš, eds.
© 2007 International Astronomical Union
doi:10.1017/S1743921307002153

Giant Molecular Associations in M51

M. Hitschfeld[1], C. Kramer[1], K. Schuster[2], S. Garcia-Burillo[3] and J. Stutzki[1]

[1]KOSMA, Universität zu Köln, Germany; [2] IRAM, Grenoble, France; [3] Centro Astronomico de Yebes, Guadalajara, Spain;

Abstract. We present a ^{12}CO 2-1 map of M51 (Schuster *et al.* 2006) at 11" resolution observed with HERA at the IRAM-30m telescope. The map covers the companion galaxy NGC5195 as well as the south-western arm out to 12 kpc. Using the IRAM-30m data and the clump finding procedure GAUSSCLUMPS (Stutzki *et al.* 1990), we obtain the masses, positions, peak temperatures and more intrinsic properties as i.e. deconvolved sizes of Giant Molecular Associations (GMAs) in M51 (Hitschfeld *et al.* 2007, in prep.).

1. Results

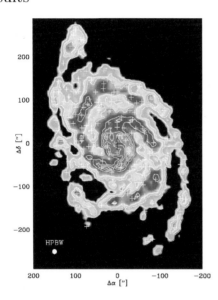

Fig. 1 Map of ^{12}CO 2–1 integrated intensities [Kkms^{-1}] showing M51 and its companion galaxy NGC 5195 in the northeast. The image has a resolution of 11″ and is constructed from a masked moment calculation. The center positions of the 155 clumps produced by GAUSSCLUMPS are indicated by white crosses. The mass range for the fitted clumps is $4.9\,10^5 M_\odot$ to $1.2\,10^8 M_\odot$.

The HERA map of ^{12}CO 2–1 (Fig. 1) is the first CO map of M51 encompassing the companion galaxy as well as the south-western arm out to radii of ~ 12 kpc in a homogeneously sampled data set at linear scales of down to 450 pc. We presented a detailed study of the distribution of molecular gas, radial averages of molecular and atomic gas densities, local Schmidt law and gravitational stability in M51 in Schuster *et al.*(2006). We decompose the ^{12}CO2-1 emission into three-dimensional Gaussian-shaped clumps using GAUSSCLUMPS and obtain i.e. positions, velocities and deconvolved sizes of the clumps.

References

Schuster, K.-F., Kramer, C., Hitschfeld, M., Garcia-Burillo, S., & Mookerjea, B. 2006, *A&A* submitted

Stutzki, J. & Güsten, R. 1990, *ApJ* 356, 513

Triggered Star Formation in a Turbulent ISM
Proceedings IAU Symposium No. 237, 2006
B. G. Elmegreen & J. Palouš, eds.

© 2007 International Astronomical Union
doi:10.1017/S1743921307002165

Galaxy open clusters and associations: study of stellar population

Alisher S. Hojaev[1]†

[1]Ulugh Beg Astronomical Institute, Center for Space Research, Uzbek Academy of Sciences,
Tashkent, Uzbekistan   email: ash[at]astrin.uzsci.net

Abstract. Some results of Galaxy star clusters and associations observation are presented.

Keywords. open clusters and associations: NGC 6823, NGC 7801, King1, King 13, King18, King20, Berkeley 55, IC 4996; HII regions: NGC 6820, stars: pre–main-sequence; techniques: photometric, spectroscopic, image processing

The star clusters and associations, especially the compact ones with real diameters up to ten arcmins, are suitable targets to search for light variability and to carry out a simultaneous CCD-photometry for all their member stars. In close collaboration with colleagues from IoA/NCU (Taiwan) at Maidanak observatory (Uzbekistan) which is notable for quite nice seeing conditions (see, for example, Frogel, 2002) we have observed the young open cluster NGC 6823 embedded in a bright HII nebula NGC 6820 (Hojaev *et al.* 2003). The cluster itself is in the core of OB association Vul OB1. 8 new and 43 suspected PMS stars of small and intermediate masses (TTS and HAeBeS) have been found. The 2MASS NIR data was used to identify young stars by the criteria described in Lee & Chen (2002). For all 8 new PMS stars the spectra recorded with 2.16 m telescope of Beijing Astrophysical observatory showed a strong Hα line emission. These PMS stars most probably form a new T-association which is closely connected with Vul OB1 association. The narrow-band images of 1×1 sq.degree area centered on NGC 6823 core were obtained with the 60/90 Schmidt telescope of the Beijing Astrophysical observatory (in the framework of BATC collaboration) with a large-format CCD camera in t band (near Hα), i band ($\lambda6660\mathring{A}$) and o band ($\lambda9100\mathring{A}$) and show the complex structure of the entire nebula NGC 6820. The o÷o-i CMD has been plotted and analysed for cluster and association stars. Afterwards 7 other compact open clusters in the Milky Way (NGC 7801, King1, King 13, King18, King20, Berkeley 55, IC 4996) were monitored for stellar variability in 2003. A homogeneous photometry has been made for NGC 7801, King 13, King20. The resulted time-series master catalogues have been prepared and analysed for stars in each of these clusters. A few interesting variables have been discovered and dozens were suspected for variability to the moment in these clusters for the first time.

References

Frogel, J.A. 2002, Image Quality at Selected Astronomical Observatories - V3.0 *A Memo prepared by Jay A. Frogel for the SNAP project at the Lawrence Berkeley National Laboratory 1 February 2002,* p. 15
Hojaev, A.S., Chen, W.P. & Lee, H.T. 2003, *A&A Transactions* 22, 799
Lee. H.T. & Chen. W.P. 2002, *Proceed. 8th IAU Asian-Pacific Regional Meeting,* p. 101

† Present address: UBAI, Astronomicheskaya 33, Tashkent 700052, UZB

Triggered Star Formation in a Turbulent ISM
Proceedings IAU Symposium No. 237, 2006
B. G. Elmegreen & J. Palouš, eds.

© 2007 International Astronomical Union
doi:10.1017/S1743921307002177

Turbulence in the G333 molecular cloud

**P. A. Jones[1], M. R. Cunningham[1], I. Bains[1], E. Muller[2],
T. Wong[1,2], M. G. Burton[1] and the DQS Team**

[1]School of Physics, University of New South Wales, NSW 2052, Australia
email: Paul.Jones@csiro.au
[2]Australia Telescope National Facility, PO Box 76, Epping NSW 1710, Australia

Keywords. ISM: clouds, turbulence, ISM: molecules

We are studying the molecular clouds in the region around G333.6-0.2 in a number of 3-mm transitions from different molecular species, to probe, among other things, the turbulent properties. The observations are being made by on-the-fly mapping with the 22-m diameter single-dish Mopra radio telescope. See Bains *et al.* (2006) and Cunningham *et al.* (2006 in these proceedings) for more details. During 2004 and 2005 we obtained ^{13}CO (1 - 0), $C^{18}O$, CS (2 - 1) and $C^{34}S$ data. Using the different molecular tracers gives complementary information about the gas density structure, due to the different critical densities, and different isotopomers allows correction for optical depth effects.

One of the simplest, and most commonly used, statistical analysis techniques is the spatial power spectrum (SPS). This is the power obtained by the Fourier Transform of the image, as a function of spatial frequency, and typically fits quite well to a power law. The slope of this power law (for $C^{18}O$ in one sub-area) is shown in Fig. 1, as a function of velocity. We note that the slope varies between -3 and -4 across the spectral line (and the slope is -1 for "red" noise away from the line emission, due to the correlated noise in the on-the-fly mapping process.)

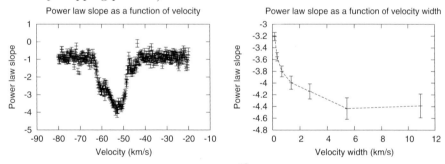

Figure 1. (a) Spatial power spectrum (SPS) of $C^{18}O$, and (b) The VCA power spectrum slope (for the 64 channels centred on the line emission) as a function of velocity width for $C^{18}O$.

For each small element of volume in the molecular cloud, the turbulence leads to a different density and velocity. Although we have only a 3-dimensional data cube (2-D column density as a function of velocity), rather than the full 6 dimensions of density and velocity, we can statistically determine the turbulent properties using the expectation that the 3 directions are equivalent. The slope of the SPS is expected to change as we go from "thin" velocity channels, where part of the structure seen is due to velocity structure, to "thick" velocity channels (Lazarian & Pogosyan 2000). In Fig 1 we show the velocity component analysis (VCA), where we can see the change in slope of the SPS as we aggregate the channels into groups of different velocity width.

References

Bains I., *et al.* 2006, *MNRAS* 367, 1609
Lazarian A. & Pogosyan D. 2000, *ApJ* 537, 720

Triggered Star Formation in a Turbulent ISM
Proceedings IAU Symposium No. 237, 2006
B. G. Elmegreen & J. Paloš, eds.

High-resolution mapping of interstellar clouds with near-infrared scattered light

M. Juvela[1], V.-M. Pelkonen[1], P. Padoan[2]
and K. Mattila[1]

[1]Helsinki University Observatory, FI-00014, Finland
[2]Department of Physics, University of California, San Diego

Abstract. We examine the intensity of scattered near-infrared (NIR) light in the case of interstellar clouds illuminated by the normal interstellar radiation field. We have developed a way to convert the observed surface brightness into estimates of the column density and have estimated the accuracy of the new method. The NIR intensities can be converted into reliable estimates of the column density in regions with A_V up to almost 20 magnitudes. The errors can be further reduced with detailed radiative transfer modelling and by using the lower resolution information that is provided by the colour excess data of background stars. Therefore, NIR scattered light is a promising new way to map quiescent interstellar clouds at a high, even sub-arcsecond resolution.

Keywords. ISM: clouds, ISM: structure, infrared: ISM, scattering

---

Current wide-field near-infrared (NIR) instruments make it possible to map the scattered light over large cloud areas (e.g., Foster & Goodman 2006). Below $A_V \sim 10$ magnitudes the surface brightness should be directly proportional to the column density. At higher column densities the scattered intensity shows signs of saturation that starts at the shortest wavelengths. The resulting changes in the intensity ratios between the NIR bands can be used to estimate the level of saturation. We have developed a way to convert observed near-infrared surface brightness values into estimates of column density (Padoan *et al.* 2006). The reliability of this new method has been examined using simulated observations (Juvela *et al.* 2006). The studies are based on three-dimensional cloud models obtained from simulations of magnetohydrodynamic turbulence. Maps of near-infrared scattered light are obtained for the model clouds with radiative transfer calculations, and are converted back into column density using the proposed method. We find that NIR intensities can be converted into reliable estimates of the column density in regions with A_V up to almost 20 magnitudes. The errors can be further reduced with detailed radiative transfer modelling and, in particular, by using the lower resolution information that is available through the colour excess data of background stars. The NIR scattered light provides a promising new method for the mapping of quiescent interstellar clouds. The main advantage is the high, even sub-arcsecond resolution that can be reached with current instruments. The comparison of surface brightness and extinction data can also serve as a sensitive indicator of changes in dust properties and radiation field within the clouds.

References

Foster, J. & Goodman, A. 2006, *ApJ* 636, L105
Juvela, M., Padoan, P., Pelkonen, V.-M. & Mattila, K. 2006, *A&A* (in press)
Padoan, P., Juvela, M. & Pelkonen, V.-M. 2006, *ApJ* 636, L101

Triggered Star Formation in a Turbulent ISM
Proceedings IAU Symposium No. 237, 2006
B. G. Elmegreen & J. Palouš, eds.

The distribution of early-type stars in the Mon-CMa-Pup-Vel region of the Milky Way

N. T. Kaltcheva

Department of Physics and Astronomy, University of Wisconsin Oshkosh, 800 Algoma Blvd.,
Oshkosh, WI 54901, USA
email: kaltchev@uwosh.edu

Abstract. An overall $uvby\beta$ photometric survey of the structure of the star-forming fields located toward the Monoceros-Canis Major-Puppis-Vela region of the Milky Way is presented.

Keywords. stars: distances, open clusters and associations: general, Galaxy: structure

1. Introduction

Unlike the external galaxies where the star-forming fields are generally evident from direct imaging, in our own Galaxy the spiral arms are strung out along the line of sight, leading to the superposition and mixing of different star-forming complexes in the sky. Thus, the study of the structure of the star-forming fields of the MW is grounded in distance determinations of young stellar tracers. At present, very few Galactic star-forming regions benefit from numerous stellar studies, capable to provide sufficient knowledge about the causal relationship between different stellar populations. The precise mapping of the Mon-CMa-Pup-Vel field presented here provides the basis for a better understanding of how the massive stars dominate the structure and evolution of their environments. The far-reaching goal is to correlate the locations of the optical spiral tracers and the interstellar material.

2. Discussion

In this context, the quantity, quality and completeness of the observational data and their interpretation play a critical role in gaining a satisfactory degree of understanding. The present study is based on $uvby\beta$ photometry of O- and B-type stars. This photometric system is arguably better suited to the study of individual stars and their groupings in terms of stellar luminosity than any other photometric system in wide use. It provides accurate stellar distances allowing not only to delineate the general Galactic spiral structure, but also to follow arm splitting and branching. Based on a complete magnitude-limited sample of O-B9 stars, the stellar content, the distribution of the interstellar absorption, the spatial stellar distribution, and metalicity and age variations across the field are analyzed. A homogeneous distance scale is established, implying that the prominent young structures studied are closer to the Sun than was previously thought (Kaltcheva & Hilditch, 2000, Kaltcheva 2006).

References

Kaltcheva, N.T. & Hilditch, R. 2000, *MNRAS* 312, 753
Kaltcheva, N.T. 2006, *AN* submitted

Triggered Star Formation in a Turbulent ISM
Proceedings IAU Symposium No. 237, 2006
B. G. Elmegreen & J. Palouš, eds.

Submillimeter-wave observations of outflow and envelope around the low mass protostar IRAS 13036-7644

Kazuhisa Kamegai[1], Masaaki Hiramatsu[2], Takahiro Hayakawa[3], Ken'ichi Tatematsu[3], Tetsuo Hasegawa[3], Toshikazu Onishi[4] and Akira Mizuno[5]

[1]Institute of Astronomy, The University of Tokyo, 2-21-1 Osawa, Mitaka, Tokyo 181-0015, Japan, email: kamegai@ioa.s.u-tokyo.ac.jp

[2]Department of Astronomy, The University of Tokyo, Bunkyo, Tokyo 113-0033, Japan

[3]National Astronomical Observatory of Japan, 2-21-1 Osawa, Mitaka, Tokyo 181-8588, Japan

[4]Department of Astrophysics, Nagoya University, Chikusa, Nagoya 464-8602, Japan

[5]Solar-Terrestrial Environment Laboratory, Nagoya University, Toyokawa, Aichi, 442-8507, Japan

1. Introduction and Motivation

The interaction between molecular outflow from a protostar and ambient molecular cloud would play an important role in dissipating circumstellar envelope, changing chemical composition, and triggering next generation star formation. In order to investigate the interaction in submillimeter wavelength, we have made line observations toward the low mass protostar IRAS 13036-7644 (Class 0/I) in the Cha II dark cloud. Although millimeter observations found CO outflow and evidence of mass infall toward the protostar (e.g. Lehtinen 1997), no submillimeter observation has been reported so far.

2. Observations and Results

The observations were carried out with ASTE 10 m telescope at Atacama, Chile in June 2005. The CO $J = 3 - 2$ (345 GHz) and HCO$^+$ $J = 4 - 3$ (356 GHz) lines were mapped in a $10' \times 6'$ area including the protostar position with a grid of $10''$ around the protostar and $20''$ in the other region. The CO $J = 3 - 2$ map shows a pair of molecular outflows which extend 0.12 pc westward and 0.08 pc eastward around the protostar. The directions of red and blue lobes are consistent with the outflow in the CO $J = 1 - 0$ line. A dense circumstellar envelope traced by the HCO$^+$ $J = 4 - 3$ line is distributed around a 0.08 pc region centered at the protostar. It is interesting that both lines have opposite velocity gradient across the protostar. Furthermore the HCO$^+$ channel map seems to show expanding motion. These velocity structures can be interpreted as an expanding circumstellar envelope which is entrained outward by the molecular outflow.

The optically thin H^{13}CO$^+$ $J = 4 - 3$ (346 GHz) line has been also detected successfully at the protostar position. The mass of circumstellar medium is estimated to be 4.8×10^{-2} $M_\odot$ from the H^{13}CO$^+$ intensity, which is consistent with the previous result that IRAS $13036 - 7644$ is in a relatively late evolutionary stage of Class 0 protostar.

Reference

Lehtinen, K. 1997, *A&A* 317, L5

Triggered Star Formation in a Turbulent ISM
Proceedings IAU Symposium No. 237, 2006
B. G. Elmegreen & J. Palouš, eds.

© 2007 International Astronomical Union
doi:10.1017/S1743921307002219

3D Spectroscopy of the Blue Compact Dwarf Galaxies IZw18 and IIZw70

C. Kehrig[1,2], J. M. Vílchez[1], S. F. Sanchez[3], L. Lopez-Martin[4], E. Telles[2] and D. Martin-Gordón[1]

[1]Instituto de Astrofísica de Andalucía (CSIC)
Camino Bajo de Huetor, Apartado 3004, 18080 Granada, Spain
email: kehrig@iaa.es, jvm@iaa.es

[2]Observatório Nacional, Rua José Cristino, 77, E-20.921-400, Rio de Janeiro, Brazil
[3]Calar Alto Observatory (CAHA), Almería E-040004, Spain
[4]Instituto de Astrofísica de Canarias, c/ Vía Lctea s/n, 38200, La Laguna, Tenerife, Spain

Abstract. Blue compact dwarf (BCD) galaxies are low-metallicity objects undergoing violent star formation (Searle & Sargent 1972). We present our ongoing work on integral field spectroscopy (IFS) of the two prototypical BCD galaxies IIZw70 and IZw18 (Papaderos *et al.* 2002). Two-dimensional spectroscopy allows us to collect simultaneously the spectra of many different regions of an extended object, combining photometry and spectroscopy in the same data set. The great advantage of using IFS for the investigation of galaxies is that it allows us to obtain data on the galaxies' positions, velocity fields and star-forming properties all in one data cube. The observations were taken with the instruments INTEGRAL/WYFFOS at the WHT (ORM, La Palma) and PMAS installed on the 3.5m telescope at the CAHA, covering a spectral range from λ 3600 to 6800 Å. The data are mainly used to study the ionized gas and stellar clusters. Our main goal is to investigate the presence of spatial variations in ionization structure indicators, physical conditions and gaseous metal abundances, in these galaxies (Vílchez & Iglesias-Páramo 1998). We also study the kinematical properties of the ionized gas, as well as systematic variations of ionization structure and physical-chemical parameters as a function of the surface-brightness in Hα emission. Maps of the relevant emission lines are presented.

Keywords. ISM: ionization structure, ISM: HII regions, Galaxies: abundances, Galaxies: dwarf

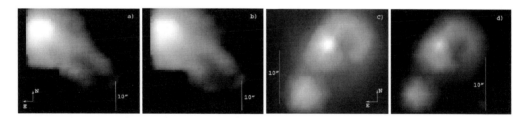

Figure 1. Maps in emission lines Hα λ6563Å and [O III] λ5007Å for IIZw70 (a,b) and IZw18 (c,d), respectively. The emission line images have subtracted the continuum using images using adjacent wavelengths obtained with PMAS at CAHA and INTEGRAL/WYFFOS at WHT. North is up and east is left.

References

Papaderos, P., Izotov, Y.I., Thuan, T.X., *et al.* 2002, *A&A* 393, 461
Searle, L. & Sargent, W.L.W. 1972, *ApJ* 173, 25
Vílchez, J.M. & Iglesias-Páramo, J. 1998, *ApJ* 508, 248

Triggered Star Formation in a Turbulent ISM
Proceedings IAU Symposium No. 237, 2006
B. G. Elmegreen & J. Palouš, eds.

H I clouds in the Large Magellanic Cloud

S. Kim[1]

[1] Astronomy & Space Science Department, Sejong University, Seoul, South Korea
email: sek@sejong.ac.kr

Abstract. We discuss the results of H I survey of the LMC and a catalog of H I clouds.

1. Analysis of H I observations

A 21 cm neutral hydrogen interferometric survey of the Large Magellanic Cloud (LMC) (Kim *et al.* 2003) combined with the Parkes multi-beam H I single-dish survey (Staveley-Smith *et al.* 2003) clearly shows that the H I gas is distributed in the form of clumps or clouds. These features are also well demonstrated in the ATCA survey alone (Kim *et al.* 1998). A cloud or a clump can be identified as an object composed of all pixels in longitude, latitude, and velocity that are simply connected and that lie above some threshold intensity. Ideally, one would like to define clouds with a zero threshold intensity. However, low threshold intensities are impractical in view of the noise level in the spectra and more importantly because of the blending of adjacent clouds which often occur in crowded regions. On the other hand, with too high a threshold intensity, regions are severely truncated, and it is impossible to obtain a reliable estimate of the sizes and velocity dispersions, thus the related parameters of the clouds identified (Scoville *et al.* 1987). Identification of the clouds are conducted within IRAF using a modified code of Lee *et al.* (1997)'s. The H I clouds and clumps have been identified and cataloged with a brightness temperature threshold (T_b) from a $21-$cm neutral hydrogen gas survey of the LMC. The catalog of H I cloud candidates and the power law relationship between the sizes and the velocity dispersions of the H I cloud candidates proves that the identified H I cloud candidates follow Larson's linewidth-size relation with a slope of 0.55 ± 0.07. The close match of the clouds and line of virial equilibrium indicates that self$-$gravity is important in the dynamics of H I clouds even though most of clouds are not gravitationally bound.

Acknowledgements

SK was supported in part by the Korea Science and Engineering Foundation (KOSEF), under a cooperative agreement with the Astrophysical Research Center of the Structure and Evolution of the Cosmos (ARCSEC). SK thanks to HI ATCA+Parkes project team members and collaborators who helped this project.

References

Kim, S., Staveley-Smith, L., Dopita, M., Freeman, K.C., Sault, R.J., Kesteven, M.J., & McConnell, D. 1998, *ApJ* 503, 674

Kim, S., Staveley-Smith, L., Dopita, M., Sault, R.J., Freeman, K.C., Lee, Y.U., & Chu, Y.-H. 2003, *ApJS* 148, 473

Larson, R.B. 1981, *MNRAS* 194, 809

Lee, Y., Jung, J. & Kim, H. 1997, *PKAS* 12, 185

Scoville, N.Z., Yun, M.S., Sanders, D.B., Clemens, D.P., & Waller, W.H. 1987, *ApJS* 63, 821

Staveley-Smith, L., Kim, S., Calabretta, M. R., Haynes, R. F., & Kesteven, M. J. 2003, *MNRAS* 339, 87

Triggered Star Formation in a Turbulent ISM
Proceedings IAU Symposium No. 237, 2006
B. G. Elmegreen & J. Palouš, eds.

The dependence of the IMF on the density-temperature relation of pre-stellar gas

S. Kitsionas[1], A. P. Whitworth[2], R. S. Klessen[3] and A.-K. Jappsen[1]

[1]Astrophysikalisches Institut Potsdam, An der Sternwarte 16, D-14482 Potsdam, Germany
email: skitsionas@aip.de

[2]School of Physics and Astronomy, Cardiff University, P.O. Box 913, CF24 3YB Cardiff, U.K.

[3]Institut fuer Theoretische Astrophysik, Universitaet Heidelberg, Albert-Ueberle-Strasse 2,
D-69120 Heidelberg, Germany

Abstract. It has been recently shown by several authors that fragmentation of pre-stellar gas (i.e. at densities from 10^4 to 10^{10} particles cm^{-3} and temperatures of order 10-30 K) depends on the gas thermodynamics much more than it was anticipated in earlier studies, in which only an isothermal behaviour has been assumed for the gas. Here we review the results of a number of numerical hydrodynamic simulations (e.g. Li *et al.* 2003, Jappsen *et al.* 2005, Bonnell *et al.* 2006) in which departure from isothermality has been attempted by employing a polytropic equation of state (eos) with exponent different from unity. In particular, in these studies it has been shown that the dominant fragmentation scale of pre-stellar gas, and hence the peak of the initial mass function (IMF), depends on a polytropic exponent that changes value, from below to above unity, at a critical density (Larson 2005). Furthermore, this piecewise polytropic eos depends on the gas metallicity and fundamental constants. Therefore, the peak of the IMF depends, in turn, also on the gas metallicity and fundamental constants rather than on initial conditions, as it has been previously suggested (e.g. Larson 1995). Hence, we are for the first time in a position to infer theoretically the notion of a universal IMF (at least for its low-mass end).

We also present two test cases in which a non-isothermal eos has been used in the context of smoothed particle hydrodynamic (SPH) numerical simulations. In the first case star formation is triggered by means of low-mass clump collisions. These calculations have shown that clump collisions can be a relatively efficient mechanism for the formation of solar-mass protostars and their lower-mass companions (efficiency greater or of order 20-25%; Kitsionas & Whitworth 2006). We have also found that in such collisions protostars form mainly by fragmentation of dense filaments along which it is likely that pairs of protostars capture each other in close binaries surrounded by circumbinary discs. In the second case, the use of a polytropic eos with a varying exponent appropriate for the metallicity of starburst regions (Spaans & Silk 2000, 2005) is shown to be sufficient to obtain a top heavy IMF similar to that observed e.g. in the Galactic centre (Klessen, Spaans & Jappsen 2006). These are preliminary results in the direction of revisiting earlier isothermal calculations that were resolving all densities up to the opacity limit for fragmentation (e.g. Bate *et al.* 2002ab, 2003), this time also taking into account the thermal properties of the gas in the density range between 10^4 and 10^{10} particles cm^{-3}. The next step would be to include self-consistent radiation transport in the calculations, the first attempts for which are already in the making (e.g. Whitehouse & Bate 2004).

Keywords. equation of state, hydrodynamics, method: numerical, stars: formation, stars: mass function

Triggered Star Formation in a Turbulent ISM
Proceedings IAU Symposium No. 237, 2006
B. G. Elmegreen & J. Palouš, eds.

© 2007 International Astronomical Union
doi:10.1017/S1743921307002244

ASTE observations of dense molecular gas in galaxies

K. Kohno[1], K. Muraoka[1], K. Nakanishi[2], T. Tosaki[2], N. Kuno[2],
R. Miura[1,2], T. Sawada[2], K. Sorai[3], T. Okuda[1], K. Kamegai[1],
K. Tanaka[1], A. Endo[1,2], B. Hatsukade[1], H. Ezawa[2], S. Sakamoto[2],
J. Cortes[2,4], N. Yamaguchi[2], H. Matsuo[2] and R. Kawabe[2]

[1]University of Tokyo
email: kkohno@ioa.s.u-tokyo.ac.jp
[2]National Astronomical Observatory of Japan
[3]Hokkaido University
[4]University of Chile

Abstract. Atacama Submillimeter Telescope Experiment (ASTE) is a joint project between Japan and Chile for installing and operating a 10 m high precision telescope in the Atacama Desert in order to explore the southern sky through the submillimeter wavelength. We have achieved an accuracy of 19 μm (rms) for the main reflector surface and a stable radio pointing accuracy of about 2 arcsec (rms). A 350 GHz cartridge type SIS mixer receiver achieves good performance with a typical system noise temperature of $150 \sim 250$ K in DSB and a main beam efficiency of $0.6 \sim 0.7$ during winter nights.

A large scale CO(3-2) imaging survey of nearby galaxies using ASTE is now in progress. One of our goals is to compare our wide area CO(3-2) images with existing CO(1-0) data as well as distributions of massive star formation tracers (i.e., Hα and radio continuum emission) in order to understand the physical mechanism which controls the global star formation properties such as star formation efficiency. Initial CO(3-2) maps of some sample galaxies (M 83, NGC 604 in M 33, NGC 1672, & NGC 7130) are reported.

Keywords. galaxies: ISM, galaxies: starburst, submillimeter, telescopes

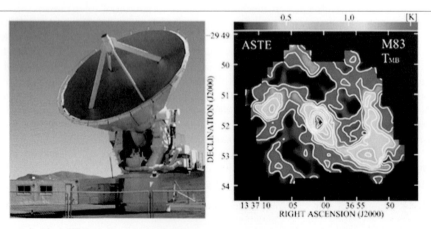

Figure 1. (left) ASTE observatory at Pampa la Bola (4860 m) in the Atacama desert, Chile. (right) A CO(3-2) peak temperature map of M83 taken with ASTE (Muraoka *et al.* 2006).

References

Muraoka, K., *et al.* 2006, *PASJ* submitted
Tosaki, T., *et al.* 2006, this volume

Triggered Star Formation in a Turbulent ISM
Proceedings IAU Symposium No. 237, 2006
B. G. Elmegreen & J. Palouš, eds.

© 2007 International Astronomical Union
doi:10.1017/S1743921307002256

Three-dimensional MHD simulations of magnetized molecular cloud fragmentation with turbulence and ion-neutral friction

T. Kudoh[1] and S. Basu[2]

[1] National Astronomical Observatory of Japan, Mitaka, Tokyo 181-8588, Japan
email: kudoh@th.nao.ac.jp

[2]Department of Physics and Astronomy, University of Western Ontario,
London, Ontario N6A 3K7, Canada
email: basu@astro.uwo.ca

Abstract. We perform a 3D-MHD simulation of a self-gravitating isothermal gas layer that is initially penetrated by a uniform magnetic field. The strength of the initial magnetic field is such that the cloud is slightly subcritical. In this system, we input random supersonic turbulence initially. Ion-neutral friction is also introduced in the magnetized gas so that the magnetic diffusion allows gas to go across the magnetic field and form self-gravitating cores. We found that self-gravitating cores are formed in the dense region enhanced by the shock waves if ion-neutral friction is introduced. The time scale of core formation is on the order of the 10^6 years, which is faster than the usual magnetic diffusion time (10^7 years) estimated from the initial condition. Our result is consistent with the results of 2D-MHD simulations by Li & Nakamura (2004).

Keywords. MHD, turbulence, methods: numerical, stars: formation, ISM: clouds, ISM: magnetic fields

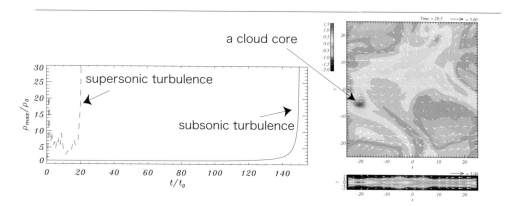

Figure 1. LEFT: Time evolution of maximum density. The dashed line shows the case when the initial perturbation is supersonic (3 times sound velocity). The solid line shows the case when the initial perturbation is subsonic (0.1 times sound velocity).The unit of time is $t_0 \sim 2 \times 10^5$ years. RIGHT: Density contour at the final stage for the case of initial supersonic turbulence. White arrows show velocity vectors normalized by sound velocity. Upper panel shows the cross section at $z = 0$. Lower panel shows the cross section at $y = -5.9$.

Acknowledgements

Numerical computations were carried out mainly on VPP5000 at the Center for Computational Astrophysics of the National Astronomical Observatory of Japan.

Triggered Star Formation in a Turbulent ISM
Proceedings IAU Symposium No. 237, 2006
B. G. Elmegreen & J. Palouš, eds.

© 2007 International Astronomical Union
doi:10.1017/S1743921307002268

Statistics of initial velocities of open clusters

J. R. D. Lépine[1], W. S. Dias[2] and Yu. Mishurov[3]

[1]IAG, Universidade de São Paulo, Brazil; email: jacques@iagusp.usp.br

[2]UNIFEI – Universidade Federal de Itajubá, MG, Brazil

[3]Rostov State University, Russia

Abstract. Galactic clusters have well organized directions of their velocities at the instant of birth measured with respect to the local circular velocity. Preferential directions survive more than 100 Myrs. This can only be explained by star formation triggered by spiral arms.

Keywords. Open clusters, star formation, spiral arms

We made use of our large database of galactic clusters, which contains distances, proper motions and radial velocities, to determine the initial velocities of these objects (direction in the galactic plane, and amplitude). In a previous work (Dias *et al.* 2005) we showed that the birth of open clusters occurs in the spiral arms. By integrating backwards the galactic orbits of the clusters for a time equal to their age, we retrieved the birthplaces as a function of time and we determined the rotation speed of the spiral pattern. Now we use the same method to retrieve the initial velocities, and we measure the angle of the initial velocity perturbation with respect to the direction of circular motion. We find that the clusters are not born with random velocities, but with velocities that are organized in a few preferential directions with respect to the spiral arms. The existence of preferential initial directions allows us to directly observe the epicyclic frequency by plotting the orientation angle of the residual velocity (after the removal of the normal circular velocity) as a function of age (left side figure). The right side figure shows the histogram of initial directions for clusters in a narrow range of galactic radius $0.95R_0 < R < 1.05R_0$. Our results show that a preferential direction of birth velocity can survive for times longer than 100 Myr. This can be explained by star-formation in spiral shock waves, but it excludes some other star formation mechanisms such as star formation induced by supernovae.

References

Dias, W.S. & Lepine, J.R.D. 2005, *ApJ* 629, 825

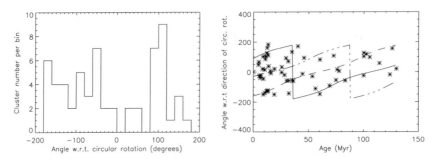

Figure 1. Histogram of initial directions (right), and direction as a function of age (left).

Triggered Star Formation in a Turbulent ISM
Proceedings IAU Symposium No. 237, 2006
B. G. Elmegreen & J. Palouš, eds.

Triggered cluster formation in the RMC

Jin Zeng Li[1] and Michael D. Smith[2]

[1]National Astronomical Observatory, Chinese Academy of Sciences, Beijing 100012, China

[2]Centre for Astrophysics & Planetary Science, University of Kent, Canterbury CT2 7NH, UK

Abstract. A comprehensive study of clustered star formation in the Rosette Molecular Complex was carried out based on archived data from the 2 Micron All Sky Survey. We presented strong evidence that triggered formation of embedded clusters and stellar aggregates took place in the working shells of the Rosette Nebula, a spectacular HII region excavated by the dozens of OB stars of the emerging massive cluster NGC 2244. Surprisingly, we have identified, within the confines of NGC 2244, a distinct congregation of young stellar objects showing prominent NIR excess that forms an arc like structure in appearance. Its location right to the south-east of the center of the main cluster and its strange morphology indicate most likely an origin from a former working shell of the HII region. This relic arc and the large, fragmented working surface layer of Rosette with the ambient cloud roughly show a concentric origin in morphology. This implies also a common origin of the clusters or stellar aggregates in association. The formation of massive star clusters was evidenced further into the heart of the molecular complex, and structured clustering star formation seemed to have taken place toward the south-east edge of the complex.

Keywords. stars: formation, ISM: clouds, ISM: structure

---

1. The Formation of New Generation OB clusters in the RMC

The surface density distribution of the 2MASS sources toward the RMC indicate distinctively the existence of two outstanding clusters, one corresponds to the emerging OB cluster NGC 2244, the other is a newly hatched massive cluster in the densest ridge of the RMC that is located ~ 20 pc to the southeast boundary of the Rosette Nebula.

2. Modes of Star Formation in the RMC

The spatial distribution of the sources indicating apparent excessive emission in the NIR toward the RMC shows the coexistence of distributed or isolated star formation and clustered star formation. However, when we restrict to only sources with higher H-K color, the distribution of the excessive emission sources appear to be congregated to multi-shell structures surrounding the main cluster. This definitely suggests an externally triggered origin of the shell clusters or stellar aggregates. Prominent new generation OB clusters were identified further into the molecular complex in the densest ridge. However, star formation toward the southeast edge of the RMC appears to be in a highly structured mode. A tree model was introduced to interpret the structured clustering of star formation that follows most likely tracks of the decay of macro-turbulence, in which the forming OB clusters are located at the root of tree pattern.

Acknowledgements

We acknowledge funding from the National Natural Science Foundation of China through grant O611081001. This work is partially supported by PPARC, UK.

Triggered Star Formation in a Turbulent ISM
Proceedings IAU Symposium No. 237, 2006
B. G. Elmegreen & J. Palouš, eds.

© 2007 International Astronomical Union
doi:10.1017/S1743921307002281

Southern IRDCs seen with Spitzer/MIPS

**H. Linz[1], Ra. Klein[2], L. Looney[3], Th. Henning[1], B. Stecklum[4]
and L.-Å. Nyman[5]**

[1]MPIA Heidelberg, Germany   [2]MPE Garching, Germany & UC Berkeley, USA
[3]University of Illinois, Urbana–Champagne, USA   [4]TLS Tautenburg, Germany   [5]ESO, Chile

Abstract. Infrared dark clouds (IRDCs) are generally assumed to be a promising hunting ground for tracing very early stages of massive star formation. Observations with Spitzer are a viable tool to probe their interiors that are still dominated by strong dust extinction even at 8 μm. With Spitzer/MIPS, we have observed several IRDCs at 24 and 70 micron. We generally find weak 24 micron sources within the IRDCs. However, at 70 micron these sources remain weak and thus indicate lower luminosities at the current state of evolution. Indications for internal substructures exist, separating regions with compact IR sources from even more dark regions.

Keywords. infrared: ISM, stars: formation, ISM: individual (G316.719+0.073)

Based on examining MSX 8.3 μm images to identify high–contrast IRDCs in the southern hemisphere, several candidates have been mapped with the SIMBA 1.2-mm bolometer at the SEST in 2001 and 2003. The vast majority of the mapped IRDCs shows clear coincidence of 1.2-mm emission with the often filamentary mid–infrared extinction structures. The IRDC shown in Fig. 1 is still surprisingly dark in the northern tip even at 24 μm, but is associated with a small cluster of deeply embedded MIR sources, comprising fluxes of 5 – 75 mJy. Only one weak source with $\sim$ 0.7 Jy remains at 70 μm. Located at roughly 3 kpc distance, the IRDC mass is 400 – 950 $M_\odot$ (depending on the extinction law), derived from the MIR extinction. Complementary, the mm emission can be translated to a total mass of 561 $M_\odot$ (assuming $T_{\rm Dust} = 20$ K, $\kappa_{1.2\rm mm} = 1$ cm^2 g^{-1}). In the future, high spatial resolution mm observations have priority, a task for the ATCA 3-mm system.

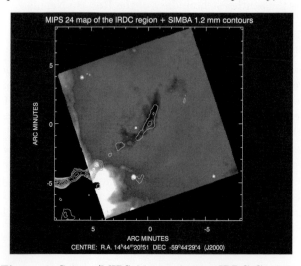
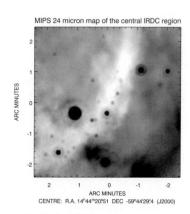

Figure 1. Spitzer/MIPS 24 μm view on IRDC G316.719+0.073. Left: With 1.2-mm contours which also trace the more evolved massive star–forming complex to the south–east. Right: Inverse grey–scale map exclusively of the IRDC, emphasising the embedded 24 micron sources.

Triggering Star Formation in a Turbulent ISM
Proceedings IAU Symposium No. 237, 2006
B. G. Elmegreen & J. Palouš, eds.

Dependence of radio halos on underlying star formation activity and galaxy mass

U. Lisenfeld[1,2], M. Dahlem[3] and J. Rossa[4]

[1]Dept. Física Teórica y del Cosmos, Universidad de Granada, Spain
email: ute@ugr.es

[2]Instituto de Astrofísica de Andalucía (IAA/CSIC), Apdo. 3004, 18080 Granada, Spain

[3]CSIRO/ATNF, Paul Wild Observatory, Locked Bag 194, Narrabri NSW 2390, Australia

[4]Department of Astronomy, University of Florida, 211 Bryant Space Science Center, P.O. Box 112055, Gainesville, FL 32611-2055, U.S.A.

Abstract. We investigate the relation between the existence and size of radio halos, which are believed to be created by star formation (SF) related energy input into the interstellar medium, and other galaxy properties, most importantly star formation activity and galaxy mass. Based on radio continuum and Hα observations of a sample of seven galaxies we find a direct, linear correlation of the radial extent of gaseous halos on the size of the actively star-forming parts of the galaxy disks. Data of a larger sample of 22 galaxies indicate that the threshold energy input rate into the disk ISM per unit surface area for the creation of a gaseous halo depends on the mass surface density of the galaxy, in the sense that a higher (lower) threshold has to be surpassed for galaxies with a higher (lower) surface density. Because of the good prediction of the existence of a radio halo from these two parameters, we conclude that they are important, albeit not the only contributors. The compactness of the SF-related energy input is also found to be a relevant factor. Galaxies with relatively compact SF distributions are more likely to have gaseous halos than others with more widespread SF activity. These results quantify the so-called "break-out" condition for matter to escape from galaxy disks, as used in all current models of the interstellar medium.
More details can be found Dahlem, Lisenfeld & Rossa, 2006, A&A 457, 121.

Keywords. ISM: general – galaxies: spirals – galaxies: evolution – galaxies: halos – galaxies: starburst – radio continuum: galaxies

Triggered Star Formation in a Turbulent ISM
Proceedings IAU Symposium No. 237, 2006
B. G. Elmegreen & J. Palouš, eds.

© 2007 International Astronomical Union
doi:10.1017/S174392130700230X

New nearby young star cluster candidates within 200 pc

Eric E. Mamajek†

Harvard-Smithsonian Center for Astrophysics, 60 Garden St., MS-42, Cambridge,
MA, 02138, USA
email: emamajek@cfa.harvard.edu

Abstract. I briefly describe two new young star cluster candidates found within 200 pc of the Sun, associated with the 4th-magnitude stars μ Oph and 32 Ori. The μ Oph group (d $\simeq$ 170 pc) has a space motion and age ($\sim$120 ± 25 Myr) suspiciously similar to the Pleiades, but lies in the opposite side of the sky behind $\sim$0.9 mag of visual extinction in Ophiuchus. The 32 Ori group is a nearby ($d \simeq$ 90 pc) loose aggregate of $\sim$25-Myr-old post-T Tauri stars co-moving with the massive binary 32 Ori (B5V+B7V) in northern Orion. The 32 Ori group accounts for part of the population of "isolated" Li-rich RASS pre-MS stars in northern Orion.

Keywords. Galaxy: open clusters and associations, solar neighborhood, stars: pre–main-sequence

1. Summary

Given the order of magnitude difference between the formation rate of embedded clusters ($\sim$4 Myr^{-1} kpc^{-2}) and open clusters ($\sim$0.3 Myr^{-1} kpc^{-2}; Lada & Lada 2003, ARA&A, 41, 57), and the recent identification of several new nearby young stellar aggregates at surprisingly close distances (d < 200 pc; e.g. η Cha, TW Hya groups; Zuckerman & Song 2004, ARA&A, 42, 685), one should not be surprised to find additional poor, young stellar groups within a few hundred pc of the Sun.

The μ Oph group (Mamajek 2) is described in detail in Mamajek (2006; AJ, in press, astro-ph/0609064). The group was first noticed as a swarm of 9 B- and A-type systems co-moving with the bright ($V = 4.6^m$) B8 giant μ Oph. μ Oph has three bright common proper motion companions ($\mu_{\alpha*}$, μ_δ = –12, –21 mas yr^{-1}) in close proximity (within 10'; $\sim$0.5 pc projected) which appear to constitute the "nucleus", including HD 160037 (A0V), HD 160038 (B9V), and HD 159874 (B9IV/V). The proper motions, parallax data, and color-magnitude diagram positions for the nine systems are consistent with having a distance of $\sim$170 pc and age of 125 ± 25 Myr. The heliocentric space motion (U, V, W = –12, –24, –4 km s^{-1}) and age are close to that of the Pleiades, α Per, and AB Dor groups, and show the cluster to be unassociated with the Gould Belt system (<60 Myr).

The 32 Ori group (Mamajek 3) was noticed as a group of X-ray-bright late-type stars from the *ROSAT All-Sky Survey* (e.g. Alcalá *et al.* 2000, A&A, 353, 186) with similar proper motions ($\mu_{\alpha*}$, μ_δ = +8, –33 mas yr^{-1}) and RVs (+18 km s^{-1}), co-moving with the nearby (d $\simeq$ 90 pc) massive binary 32 Ori (B5V+B7V). The $\sim$25-Myr-old group is defined by 32 Ori, RX J0520.0+0612, RX J0520.5+0616, RX J0523.7+0652, and a half dozen other young systems. The space motion of the new group is U, V, W = (–12, –19, –9 km s^{-1}), which is somewhat similar to that for the ill-defined Cas-Tau association in the same region, however Cas-Tau is claimed to be older ($\sim$50 Myr) and more distant ($\sim$125-300 pc; de Zeeuw *et al.* 1999, AJ, 117, 354). The group is clearly in the foreground of, and completely unrelated to, the Ori OB1 complex ($d \sim$ 400 pc).

† Clay Postdoctoral Fellow, Smithsonian Astrophysical Observatory

Triggered Star Formation in a Turbulent ISM
Proceedings IAU Symposium No. 237, 2006
B. G. Elmegreen & J. Palouš, eds.

© 2007 International Astronomical Union
doi:10.1017/S1743921307002311

Superdiffusion in molecular clouds

Gábor Marschalkó

Department of Astronomy, Eötvös University, Budapest
email: G.Marschalko@astro.elte.hu

Abstract. Turbulence has a significant influence on the chemistry of molecular clouds. Do molecular abundances alter if we use superdiffusion instead of simple diffusion? After our first simplified calculations it seems there is a notable difference.

Keywords. turbulence, diffusion, superdiffusion, ISM:clouds, ISM:abundances

---

Several (pseudo-)time-dependent models have been developed after the initial steady state gas-phase chemistry models to determine the fractional abundances in molecular clouds. Our calculations were based mostly on diffusivities and timescales of Xie, Allen & Langer (1995), and we studied what differences occur if we use superdiffusion instead of simple diffusion. After Petrovay (1999) we can speak about a 'scale-dependent diffusivity': $D(r) = K'^2 = K^{1/\zeta} r^{2-1/\zeta}$. Thus using superdiffusion we can take into account eddies with different sizes, namely the eddies exceeding the separation r do not contribute to the further separation of fluid parcels at separation r.

First we studied a simplified model neglecting the chemical processes and approximating the source term by means of the diffusionless solution based on relaxation timescales of the above mentioned paper. Our cloud was spherically symmetric with constant density of H_2. Thus the equation of diffusion: $\frac{\partial n}{\partial t} = D\nabla^2 n + \frac{(n_0 - n)}{\tau_c}$, where n_0 is the diffusionless solution of number density of a tracer and τ_c is the characteristic time. Transforming this partial differential equation into the Fourier space one can get the stationary solution solving a simple algebraic equation. In the superdiffusive case we assumed Kolmogorov spectra ($\zeta = \frac{3}{2}$) and scaled the diffusivity according to the formula: $D = D_0 (\frac{k}{k_0})^{-\frac{4}{3}}$, where k is the wavenumber and k_0 corresponds the correlation length of the cloud. Dropping the assumption of constant distribution of H_2 one may have difficulties with integral transform because of the product of operators and functions. Therefore introducing the function $F = -n_t \nabla(\frac{n}{n_t})$ we get the following diffusion equation: $\frac{\partial n}{\partial t} = D\nabla F + \frac{(n_0 - n)}{\tau_c}$, and using an iterative method the density distribution of a tracer can be solved.

We can conclude that there is a significant difference between the diffusive and superdiffusive case. We plan to refine the approximation of the source term, taking into account that this diffusive process is more complicated than a simple mixing between the inner parts and the outer layers of the cloud.

Acknowledgements

I am very grateful to Kristóf Petrovay who drew my attention to turbulence and came to my assistance during my work. This research was supported by the Hungarian Science Research Fund (OTKA) under grant no. T043741.

References

Petrovay, K. 1999, *Space Sci. Revs* 95, 9
Xie, T., Allen, M. & Langer W.D. 1995, *ApJ* 440, 674

Triggered Star Formation in a Turbulent ISM
Proceedings IAU Symposium No. 237, 2006
B. G. Elmegreen & J. Palouš, eds.

© 2007 International Astronomical Union
doi:10.1017/S1743921307002323

Stellar formation in Hi interstellar bubbles around massive stars

M. C. Martín[1], G. A. Romero[1] and C. E. Cappa[1,2]

[1]Instituto Argentino de Radioastronomía, Argentina

[2]Facultad de Ciencias Astronómicas y Geofísicas, UNLP, Argentina

Stellar winds from O and WR stars transfer large amounts of mechanical energy and momentum into the interstellar medium. They sweep up and compress the interstellar material, creating interstellar bubbles. These structures are detected as optical ring nebulae, as thermal radio continuum sources, as infrared shells, as neutral gas voids and expanding shells in the Hi line emission distribution, and as molecular shells.

Stellar formation may be induced in the compressed and dense regions around the ionized bubbles. Among the processes proposed for the onset of stellar formation, the "collect and collapse" process (Elmegreen & Lada 1977) may work in these neutral structures. Observational evidence of this process includes the presence of dense and neutral gas layers surrounding the ionized regions and high density clumps within the neutral shells.

To investigate if stellar forming processes are going on in the outer neutral shells of interstellar bubbles, we analyzed the correlation of the observed neutral gas structures with stellar formation indicators. The selected regions are the neutral gas structures associated with Anon(WR 23), RCW 52, G307.27+0.27, and RCW 78 (Cappa *et al.* 2005a,2005b, Martín *et al.* 2007) linked to WR 23 (WC6), LSS 1887 (O8V), S 221 (O6), and WR 55 (WN7), respectively.

To search for young stellar objects (YSO) in the selected regions we used the 2MASS All-Sky Catalog of Point Sources (Cutri *et al.* 2003), the IRAS Point Source Catalogue, and the MSX6C Infrared Point Source Catalog (Egan *et al.* 2003).

YSO candidates were detected towards the four selected regions. Most of the YSO candidates are projected onto areas showing Hi gas and some of them can be related to the neutral shells. Almost no YSO candidates were found projected onto the Hi cavities. A more complete study is necessary to verify if the scenario of "collect and collapse" is going on in the neutral shells around the nebulae.

Keywords. ISM: bubbles - ISM: individual (Anon[WR 23], RCW 52, RCW 78) - stars: formation

Acknowledgements

C.E.C. would like to thank financial support from the IAU and the Academy of Sciences of the Czech Republic, which facilitated her participation in the Symposium. This work was partially financed by CONICET of Argentina under project PIP 5886/05 and Universidad Nacional de La Plata, under project 11/G072.

References

Cappa, C., Rubio, M., Martín, M.C., & McClure-Griffiths, N.M. 2005a, *ASPC* 344, 179
Cappa, C., Niemela, V., Martín, M.C., & McClure-Griffiths, N.M. 2005b, *A&A* 436, 155
Cutri, R.M. *et al.* 2003, *IPAC, University of Massachusetts*
Egan, M.P. *et al.* 2003, *AFRL-VS-TR-2003-1589*
Elmegreen, B.G. & Lada, C.J. 1977, *ApJ* 214, 725
Martín, M.C., Cappa, C.E. & Testori, J.C. 2007, *in preparation*

Triggered Star Formation in a Turbulent ISM
Proceedings IAU Symposium No. 237, 2006
B. G. Elmegreen & J. Palouš, eds.

© 2007 International Astronomical Union
doi:10.1017/S1743921307002335

Triggered star formation in spiral arms

Eric E. Martínez-García[1], Rosa Amelia González-Lópezlira[1] and Gustavo Bruzual-A.[2]

[1]Centro de Radioastronomía y Astrofísica, UNAM, México
[2]Centro de Investigaciones de Astronomía, Venezuela

Abstract. We present preliminary results for six spiral galaxies from a sample of 25, where we have used the method developed by González & Graham (1996) to search for and analyze azimuthal color gradients across spiral arms. The six galaxies analyzed here are NGC 1703 (SBrb), NGC 3001 (SABrsbc), NGC 3059 (SBrsbc), NGC 3513 (SBrsc), NGC 4593 (RSBrsb), and NGC 4603 (SAsc).

NGC 1703 : We found one azimuthal color gradient in NGC 1703. Star formation was traced with the reddening free parameter Q, and the dust lane was located with the (g - J) color. This gradient lies at a distance of 1.45 kpc from the center of the galaxy. In order to get some physical parameters of the star formation processes that are taking place in the spiral arms of the galaxies in our sample, we compared the observed Q profiles with the stellar population synthesis models of Bruzual & Charlot (2003). The fitted Q model is shown in figure 1. If one assumes that stars form in the site of the shock, and that they age as they move away from this birthsite, then distance from the dust lane (at constant radius) parameterizes stellar age. In fact, stretching the model Q to the fit the data fixes the ratio between the distance and the age of the stellar population. If, in addition, the rotational velocity is known, it is possible to find the angular velocity of the spiral pattern. The spiral pattern speeds derived from the gradients (under the assumption that star formation is triggered by the density wave) yield theoretical resonance positions that are coincident with the observed spiral end points, in 3 out of 6 spirals. It would be hard to avoid the conclusion that disk dynamics and star formation are fundamentally related in these objects.

Keywords. star formation, spiral arms, density wave triggering

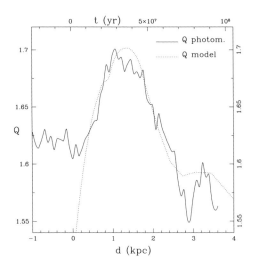

Figure 1. The fitted Q model (dotted line) compared to the observations (solid line). The zero distance indicates the location of the dust lane.

Triggered Star Formation in a Turbulent ISM
Proceedings IAU Symposium No. 237, 2006
B. G. Elmegreen & J. Palouš, eds.

© 2007 International Astronomical Union
doi:10.1017/S1743921307002347

The H_2O super maser emission of Orion KL accretion disk, bipolar outflow, shell

Leonid I. Matveyenko, V. A. Demichev, S. S. Sivakon, P. D. Diamond and D. A. Graham

Space Research Institute, Moscow, Russia, matveen@iki.rssi.ru

The H_2O super maser outbursts were observed in Orion KL in active periods 1979-1987 ($F \leqslant 8$ MJy) and 1998-1999 ($F \leqslant 4$ MJy). The line velocity was $V_{LSR} = 7.65$ km/s and line width $\Delta V \sim 0.5$ km/s. The emission was linear polarized $m \leqslant 75\%$. We studied structure of H_2O super maser region with VLBI angular resolution 0.1 mas or 0.05 AU. The emission was determined by high organized structure: a chain of bright ($T_b \sim 10^{16} K$) compact components (~ 0.05 AU), which are distributed along thin S-form structure 27×0.3 AU, $T_b \sim 10^{11}$ K, Fig. 1. The brightest components have velocities $V \sim 7.65$ km/s. The components correspond to tangential direction of the rings, velocities of which are $V \sim \Omega R$ and rotation period is ~ 180 yrs. The highly collimated bipolar outflow 9x0.7 AU and comet-like bullets were observed in the quiescent period 1995 (F = 1 kJy) and second activity period 1998-1999. The central compact (0.05 AU) bright ($T_b \sim 10^{16} K$) source is ejector of bipolar outflow, which surrounded by a torus. Compact bright features are located in the outflow, which velocities are $V \sim 10$ km/s in the beginning of activity, and $V \sim 3$ km/s in the end. The helix structure of outflow is determined by precession with period $T \sim 10$ yrs. The bullets were ejected in the first period activity. Extraordinary changing of the polarization position angle $dX/dV \sim 23°$/km/s is determined by nozzle emission.

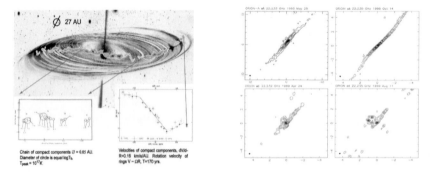

Figure 1. Accretion disk, bipolar outflow(left) Evolution of bipolar outflow(right).

Birth of a star is accompanied by the structure: torus, bipolar outflow, accretion disk and shell. The disc is divided into protoplanetary rings. Kinetic energy of the rings is transformed into bipolar outflow. Radiation and stellar wind sublimated ice and blow away H_2O molecules. The shell is amplified the structure emission of more than 3 orders of magnitude at velocity V = 7.65 km/s.

Reference

Matveyenko, L.I., Demichev, V.A., Sivakon, Diamond P.J., & Graham, D.A. 2005, *Ast.Lett.* 12, 816

Triggered Star Formation in a Turbulent ISM
Proceedings IAU Symposium No. 237, 2006
B. G. Elmegreen & J. Palouš, eds.

SUBARU near-infrared multi-color images of Class II Young Stellar Object, RNO91

Satoshi Mayama[1,2]†, Motohide Tamura[3] and Masahiko Hayashi[1]

[1]Subaru Telescope, National Astronomical Observatory of Japan, 650 North A'ohoku Place, Hilo, HI96720, USA
email: mayamast@subaru.naoj.org

[2]School of Mathematical and Physical Science, Graduate University for Advanced Studies, 2-21-1 Osawa, Mitaka, Tokyo 181-8588, JAPAN

[3]Optical and Infrared Astronomy Division, National Astronomical Observatory of Japan, 2-21-1 Osawa, Mitaka, Tokyo 181-8588, JAPAN

Abstract. RNO91 is class II source currently in a transition phase between a protostar and a main-sequence star. It is known as a source of complex molecular outflows. Previous studies suggested that RNO91 was associated with a reflection nebula, a CO outflow, shock-excited H_2 emission, and disk type structure. But the geometry of RNO91, especially its inner region, is not well confirmed yet. High resolution imaging is needed to understand the nature of RNO91 and its interaction with outflow. Thus, we conducted near-infrared imaging observations of RNO91 with the infrared camera CIAO mounted on the Subaru 8.2-m Telescope. We presented JHK band and optical images which resolved a complex asymmetrical circumstellar structure. We examined the color of RNO91 nebula and compared the geometry of the system suggested by our data with that already proposed on the basis of other studies. Our main results are as follows;
1. The K-band images show significant halo emission detected within $\sim 2''$ around the peak position while less halo emission is seen in shorter wavelength images such as J and optical. The nebula appears to become more circular and more diffuse with increasing wavelengths. The cut-off at 300AU derived from our radial surface brightness is consistent with the size of the polarization disk suggested by Scarrott, Draper & Tadhunter (1993). These consistencies indicate that this optically thick region is attributed to a disk-like structure.
2. At J and optical, several bluer knot-like structures are detected around and beyond the halo emission. These bluer knots seen in our images are comparable to the size of the envelope detected in HCO^+ emission surrounding RNO91 (Lee & Ho 2005). It is thus natural to suggest that these bluer knots are the near-infrared light scattered by an envelope structure which is disrupted by molecular outflows.
3. The pseudo-true color composite image has an appearance of arc-shaped emission extending to the north and to the east through RNO91. On the counter part of this arc-shaped structure, the nebula appears to become more extended to the southwest from the central peak position in J band and optical images. We interpret these whole structures as a bottom of bipolar cavity seen relatively edge-on opening to the north and south directions.

Keywords. stars:individual (RNO91), stars:pre–main-sequence, ISM:reflection nebulae

References

Scarrott S. M., Draper P. W. & Tadhunter C. N. 1993, *MNRAS* 262, 306
Lee C.-F. & Ho Paul T. P. 2005, *ApJ* 624, 841

† Present address: 650 North A'ohoku Place, Hilo, HI96720, U.S.A.

Triggered Star Formation in a Turbulent ISM
Proceedings IAU Symposium No. 237, 2006
B. G. Elmegreen & J. Palouš, eds.

© 2007 International Astronomical Union
doi:10.1017/S1743921307002360

Chemical structure of the massive protobinary-forming hot core, W3(H$_2$O)

Y. C. Minh[1,2] and H.-R. Chen[2,3]

[1]Korea Astronomy and Space Science Institute, Korea
[2]Academia Sinica, Institute of Astronomy and Astrophysics, Taiwan
[3]National Tsing Hua University, Taiwan
email: minh@kasi.re.kr, hchen@phys.nthu.edu.tw

Abstract. The hot molecular core, W3(H$_2$O), contains a massive protobinary system that is cocooned by dense gas (n(H$_2$) $\sim 10^7$ cm^{-3}), with about 1 arcsec (~ 2000 AU) separation of the binary system. We investigate chemical properties of the gas components around this binary system using the mm-wave transitions of complex molecules, CH$_3$CN, SiO, HNCO, and CH$_3$CH$_2$CN, observed with the BIMA array. The two protostellar objects, A and C in W3(H$_2$O), can be distinguished using chemical properties, suggesting that the source A may be younger than the source C within the time scale of less than 10^4 years. The hot core around the source C, traced by CH$_3$CH$_2$CN and the K=6 component of CH$_3$CN, seems to have more time for chemical evolution than the source A. The SiO emission in this region suggests that there was an influence from the outside of the W3(H$_2$O) and W3(OH) hot cores. The nitrogen chemistry may be more active in the later stage than the oxygen chemistry, but the chemical evolution of the protostellar envelopes may not be monotonic as previously suggested.

Keywords. ISM: abundances, ISM: clouds, ISM: molecules, ISM: individual (W3(H2O)), astrochemistry, molecular processes

A massive-star forming hot core, W3(H$_2$O), locates at the 6$''$ east of the UC HII region W3(OH). These objects are embedded in the giant molecular complex W3 at a distance of about 2 kpc. W3(H$_2$O) contains a protostellar (B0.5-0) binary system (the A and C component in continuum) in orbit with each other with a velocity difference $\Delta V_{A-C} = 2.81$ km s^{-1} (Chen *et al.* 2006, ApJ, 639, 975; and see references therein). The small angular size (a few arcsecond in diameter) of this hot core requires sub-arcsecond angular resolution to properly resolve the two protostellar objects embedded in the dense gas and dust cocoon of n(H$_2$) $\sim 10^7$ cm^{-3}.

The chemical evolutionary time scale is relatively short, which helps to understand the nature of very complicated high mass star forming cores. We model the structure of this protostellar system using the chemical tracers, CH$_3$CN, SiO, HNCO, and CH$_3$CH$_2$CN observed with the BIMA array. Chemical differences exist between A and C, probably because of the difference in their chemical evolutionary stages.

In the chemical evolution of hot cores there are, at least, two different formation periods of N-containing species, such as CH$_3$CN, for the warm extended component and the highly turbulent compact component. The fact that CH$_3$CH$_2$CN exists mainly in source C suggests that source C is more chemically evolved than source A. HNCO may trace an expanding shell around the highly turbulent gas of A and C, where YSOs are embedded. The SiO emission seems to trace the shocked gas existing outside of these hot cores.

Triggered Star Formation in a Turbulent ISM
Proceedings IAU Symposium No. 237, 2006
B. G. Elmegreen & J. Palouš, eds.

© 2007 International Astronomical Union
doi:10.1017/S1743921307002372

Triggered star formation in bright-rimmed clouds

L. K. Morgan[1]†, J. S. Urquhart[1,2], M. A. Thompson[1,3] and G. J. White[1,4,5]

[1]CAPS, The University Of Kent, Canterbury, Kent CT2 7NR
[2]University Of Leeds (UK)
[3]The University Of Hertfordshire (UK)
[4]CCLRC Rutherford Appleton Laboratory (UK)
[5]The Open University (UK)

Abstract. A sample of optically Bright-Rimmed Clouds (BRC) at the edge of HII regions has been observed at multiple wavelengths in order to investigate the possibility that star-formation is present. Such activity may be related to photoionisation induced shocks caused by the massive stars powering the HII regions.

The sample has been observed at radio, infrared and submillimetre wavelengths. Both molecular line studies and continuum observations have been made of the larger cloud structures and embedded sources within.

Radio and infrared continuum observations show the presence of ionised boundary layers coincident with the optically bright rims. These are responsible for the propagation of shocks into the clouds interiors, possibly triggering the collapse of cores into protostars.

Molecular line studies and submillimetre continuum observations show the presence of centrally condensed cores within the clouds, these cores have high densities and have submillimetre luminosities indicative of class 0 protostars. The total luminosities of the embedded sources reveal a set of forming intermediate to high-mass stars.

The identification of these regions as star-forming has important consequences for studies of triggered star-formation, not only does the high incidence of star formation in BRC suggest a high efficiency for Galactic triggered star-formation but the masses of the sources suggest a preferred process for the formation itself.

---

† Present address: Green Bank Observatory, P.O.Box 2, Green Bank, WV 24944 UNITED STATES

Triggered Star Formation in a Turbulent ISM
Proceedings IAU Symposium No. 237, 2006
B. G. Elmegreen & J. Palouš, eds.

© 2007 International Astronomical Union
doi:10.1017/S1743921307002384

Radiation driven implosion model for star formations near an H II region

K. Motoyama[1], T. Umemoto[2] and H. Shang[1,3]

[1]Theoretical Institute for Advanced Research in Astrophysics, Hsin-Chu, 300, Taiwan

[2]National Astronomical Observatory, Mikata, Tokyo 181-8588, Japan

[3]Academia Sinica Institute of Astronomy and Astrophysics, P.O. Box 23-141, 106, Taiwan

Abstract. We performed numerical simulation including UV radiation transfer, and investigated effects of radiation driven implosion on star formation processes. We also observed two bright-rimmed clouds with $C^{12}O(J=1-0)$ and $C^{13}O(J=1-0)$ in order to compare density distributions between numerical results and observational results. Density profiles of bright-rimmed clouds are consistent with those of numerical simulations. These facts insist that star formations in bright-rimmed clouds are triggered by radiation driven implosion.

Keywords. radiation driven implosion, bright-rimmed clouds

1. Introduction

Massive stars play very important roles in star formations in giant molecular clouds. Strong UV radiation of massive star ionizes surrounding gas, and creates an HII region. If pre-existing clouds are immersed by expanding HII region, then they are compressed owing to strong UV radiation. Triggered star formations due to Radiation Driven Implosion (hereafter, RDI) are suggested near HII regions (Bertoldi 1989).

Bright-rimmed clouds, which are found at the edge of relatively old HII regions, have cometary shapes. These shapes are explained well by RDI model. Bright-rimmed clouds are potential sites of triggered star formations due to RDI.

2. Method and results

We performed one dimensional and two dimensional numerical simulations. In one dimensional simulations, we assumed that a spherical core is immersed within an HII region and is exposed to diffuse radiation from surrounding ionized gas. We explored whether or not RDI enhance the accretion rate by orders of magnitude. Triggered star formations due to radiation-driven implosion increase accretion rates of protostars by 1-2 orders of magnitude compared with star formation without external trigger.

In two dimensional simulations, we assumed axisymmetry. A uniform spherical cloud is exposed to UV radiation. Our results show that the density profile is steeper at the side facing the ionized star than at the opposite side. We observed two bright-rimmed clouds and compared density profiles with those of the numerical simulation. Density profiles of bright-rimmed clouds are consistent with those of RDI model.

Reference

Bertoldi, F. 1989, *ApJ* 346, 735

Triggered Star Formation in a Turbulent ISM
Proceedings IAU Symposium No. 237, 2006
B. G. Elmegreen & J. Palouš, eds.

© 2007 International Astronomical Union
doi:10.1017/S1743921307002396

ASTE CO(3-2) observations of M 83: Correlation between CO(3-2)/CO(1-0) ratios and star formation efficiencies

K. Muraoka[1], K. Kohno[1], T. Tosaki[2], N. Kuno[2], K. Nakanishi[2], K. Sorai[3] and S. Sakamoto[4]

[1]Institute of Astronomy, The University of Tokyo, 2-21-1 Osawa, Mitaka, Tokyo, Japan
email:kmuraoka@ioa.s.u-tokyo.ac.jp

[2]Nobeyama Radio Observatory, Minamimaki, Minamisaku, Nagano, Japan

[3]Division of Physics, Grad. School of Science, Hokkaido University, Sapporo, Hokkaido, Japan

[4]National Astronomical Observatory of Japan, 2-21-1 Osawa, Mitaka, Tokyo, Japan

Abstract. We have performed $CO(J = 3 - 2)$ emission observations with the Atacama Sub-millimeter Telescope Experiment (ASTE) toward the $5' \times 5'$ (or 6.6 × 6.6 kpc at the distance $D = 4.5$ Mpc) region of the nearby barred spiral galaxy M 83. We successfully resolved the major structures, i.e., the nuclear starburst region, bar, and inner spiral arms in $CO(J = 3 - 2)$ emission at a resolution of $22''$ (or 480 pc), showing a good spatial coincidence between $CO(J = 3 - 2)$ and 6 cm continuum emissions.

From a comparison of $CO(J = 3 - 2)$ data with $CO(J = 1 - 0)$ intensities measured with Nobeyama 45-m telescope, we found that the radial profile of $CO(J = 3 - 2)/CO(J = 1 - 0)$ integrated intensity ratio $R_{3-2/1-0}$ is almost unity in the central region ($r < 0.25$ kpc), whereas it drops to a constant value, 0.6–0.7, in the disk region. The radial profile of star formation efficiencies (SFEs), determined from 6 cm radio continuum and $CO(J = 1 - 0)$ emission, shows the same trend as that of $R_{3-2/1-0}$. At the bar-end ($r \sim 2.4$ kpc), the amounts of molecular gas and the massive stars are enhanced when compared with other disk regions, whereas there is no excess of $R_{3-2/1-0}$ and SFE in that region. This means that a simple summation of the star forming regions at the bar-end and the disk cannot reproduce the nuclear starburst of M 83, implying that the spatial variation of the dense gas fraction traced by $R_{3-2/1-0}$ governs the spatial variation of SFE in M 83.

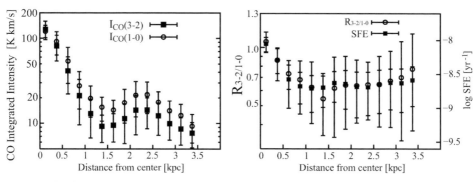

Figure 1. (left) $CO(J = 3 - 2)$ and $CO(J = 1 - 0)$ line intensities, tracing the amounts of molecular gas, as a function of the galactocentric radius of M 83. Both intensities are enhanced at the bar-end ($r \sim 2.4$ kpc). (right) The azimuthally averaged SFE and $R_{3-2/1-0}$ as a function of the galactocentric radius of M 83. No significant enhancement is visible at the bar-end, despite the fact that there is a secondary peak in the CO intensities.

Triggered Star Formation in a Turbulent ISM
Proceedings IAU Symposium No. 237, 2006
B. G. Elmegreen & J. Palouš, eds.

© 2007 International Astronomical Union
doi:10.1017/S1743921307002402

OH masers and magnetic fields in massive star-forming regions: ON1

**S. Nammahachak[1], K. Asanok[1], B. Hutawarakorn Kramer[2]†,
R. J. Cohen[3], O. Muanwong[1] and N. Gasiprong[4]**

[1]Department of Physics, Faculty of Science, Khon Kaen University,
Khon Kaen 40002 Thailand
[2]National Astronomical Research Institute of Thailand, Physics Building,
Chiang Mai University, Chang Mai 50200 Thailand
[3]University of Manchester, Jodrell Bank Observatory, Cheshire SK11 9DL United Kingdom
[4]Department of Physics, Faculty of Science, Ubon Rajathanee University,
Ubon Rajathanee 34190 Thailand

Abstract. OH masers are sensitive probes of the kinematics and physical conditions, and give unique information on the magnetic field through their polarization. Zeeman splitting of the OH lines can give the magnetic field strength and direction. Observing OH masers with MERLIN we studied the bipolar outflow in the star-forming region ON1, which hosts one of the earliest known ultra-compact (UC) HII regions. The strongest masers lie near the southern edge of the UCHII region in an elongated distribution. The maser distribution is orthogonal to the bipolar outflow seen in HCO^+, suggesting that the OH masers may be embedded in a molecular disk or torus around a young B0.3 star, most likely tracing a shock front. An isolated group of 1720-MHz masers is also seen to the East. The magnetic field deduced from Zeeman splitting of the OH maser lines shows a large-scale order, with field values ranging from -0.4 to -4.6 mG. These results add to the growing body of evidence for OH masers associated with molecular disks or tori at the centre of bipolar outflow from massive young stars, and for a significant role played by the magnetic field in generating or channeling the bipolar outflow. Further details are presented by Nammahachak *et al.* 2006.

Keywords. Masers, star-formation, OH masers, magnetic field, ON1, Onsala 1

Reference

Nammahachak, S., Asanok, K., Hutawarakorn Kramer, B., Cohen, R. J., Muanwong, O., & Gasiprong, N. 2006, *MNRAS* 371, 619

† Contact: rjc@jb.man.ac.uk, busaba@nari.or.th

Triggered Star Formation in a Turbulent ISM
Proceedings IAU Symposium No. 237, 2006
B. G. Elmegreen & J. Palouš, eds.

Subaru high-dispersion spectroscopy of Hα and [NII] 6584 Å emission in the HL Tau jet

Takayuki Nishikawa[1]†, Michihiro Takami[2] and Masahiko Hayashi[1,2]

[1]Department of Astronomical Science, The Graduate University for Advanced Studies, 650 N. A'ohoku Place, Hilo, Hawaii 96720, USA
email: nishkwtk@subaru.naoj.org

[2]Subaru Telescope, 650 N. A'ohoku Place, Hilo, Hawaii 96720, USA

Abstract. We present slit-scan observations of the Hα and [NII] 6584 Å emission lines toward the HL Tau jet with the 8.2m Subaru Telescope. HL Tau is an active young star in transitional phase from an embedded class I protostar to a class II pre-main-sequence star, and it is located in the northeastern part of the L1551 dark cloud. The slit-scan technique at high spectral resolution ($R=3.6\times10^4$) allowed for studying kinematics of individual features in unprecedented details. The Hα emission shows the main jet component ($V_{LSR} \sim -180$ km s^{-1}) and distinct lower velocity components ($|V_{LSR}| < 120$ km s^{-1}). The [NII] emission is primarily associated with the jet within 10 arcsecond from the source, and also knot B and C $\sim$30 arcsecond away from the source. These are associated with the main jet component, and absent in the lower velocity components. The velocity of Hα and [NII] emissions in the main jet component well matches each other.

Our high-resolution spectra do not show the evidence for the presence of turbulent mixing layers between the jet and surrounding gas. The lower velocity components are associated with individual knots, and explained as the lateral of bow shocks. Their line profiles suggest that shock velocity of the knots A-C is 120~130 km s^{-1} (Hartigan *et al.* 1987). The observed [NII]/Hα flux ratio markedly differ between regions: 0.1-0.7 in base of the jet; less than 0.1 in knot A; $\sim$0.2 in knot B; $\sim$0.4 in knot C; and $\sim$0.7 in knot D. Shock models predict that the [NII]/Hα flux ratio reflects the ionization of the preshock gas. This results from enhancement of N+ via the charge exchange reaction (Osterbrock 1989; Bacciotti & Eislöffel 1999). We perform more detailed comparisons between models and observations (Hartigan *et al.* 1987; Morse *et al.* 1994). The base of the jet and knot D show high [NII]/Hα flux ratios, indicating that the ambient gas surrounding the jet is considerably ionized, or the preshock density of the ambient gas is significantly low. In contrast, the knots A-C exhibit low [NII]/Hα flux ratios, indicating that the ambient gas surrounding the jet is almost neutral, or the preshock density of the ambient gas is significantly high. The [NII]/Hα flux ratio increases from knot A (< 0.1) to knot D ($\sim$0.7). This suggests that the ionization fraction of the ambient gas increases away from the source, or the preshock density of the ambient gas decreases away from the source.

Keywords. stars: individual (HL Tauri) – stars: pre-main-sequence – ISM: jet

---

References

Hartigan, P., Raymond, J. & Hartmann, L. 1987, *ApJ* 316, 323
Osterbrock, D.E. 1989, *Astrophysics of Gaseous Nebulae Active Galactic Nuclei* University Science Books, Mill Valley, CA
Bacciotti, F. & Eislöffel, J. 1999, *A&A* 342, 717
Morse, J.A., Hartigan, P., Heathcote, S., Raymond, J.C., & Cecil, G. 1994, *ApJ* 425, 738

† Present address: 650 N. A'ohoku Place, Hilo, Hawaii 96720, USA

Triggered Star Formation in a Turbulent ISM
Proceedings IAU Symposium No. 237, 2006
B. G. Elmegreen & J. Palouš, eds.

Radio observation of molecular clouds around the W5-East triggered star-forming region

Takahiro Niwa[1], Yoichi Itoh[1], Kengo Tachihara[1], Yumiko Oasa[1], Kazuyoshi Sunada[2] and Koji Sugitani[3]

[1]Graduate School of Science & Technology, Kobe University, Kobe, 657-8501, Japan
email: niwa@kobe-u.ac.jp

[2]Graduate School of Natural Sciences, Nagoya City University, Nagoya, 467-8501, Japan

[3]Nobeyama Radio Observatory, Nagano, 384-1305, Japan

1. Introduction & Motivation

It is known that most of stars are formed as clusters (Lada & Lada 2003, *ARAA* 41, L57) and clusters are formed by triggering. However, the relationships of molecular clouds' conditions and properties of formed stars by triggering is not well studied. To clarify differences between triggered and spontaneous star formation through physical properties of molecular clouds (e.g. mass, density, morphology), we observed the W5-East HII region. The W5-East HII region is located at 2 kpc and has a 10 pc extent of HII region. This region has 3 Bright Rimmed Clouds (BRCs; Sugitani *et al.* 1991, *ApJS* 77, S59), which are interface between HII regions and molecular clouds, and known as sites of triggered star formation. The molecular clouds surround the W5-East (Karr *et al.* 2003, *ApJ*, 595, 900), thus we expect molecular clouds morphology is affected by the HII region and the cloud evolution is supposed to be dominated by the expanding HII region.

2. Observation

We have carried out observation of the W5-East HII region by the Nobeyama Radio Observatory 45m telescope (HPBW = 15.6″) in ^{13}CO ($J = 1 - 0$) and C^{18}O ($J = 1 - 0$) with the observing grid spacings in ^{13}CO and C^{18}O of 40″ and 10″, respectively, and the observed areas are 0.6 deg^2 and 0.16 deg^2, respectively.

3. Main results

We identified 8 ^{13}CO molecular clouds (3 of them are associated with BRCs) and 9 C^{18}O molecular cloud cores. The masses of the clouds and cores range from 50 - 3000 $M_\odot$ and 13 - 140 $M_\odot$, respectively. The peak ^{13}CO column densities of the clouds facing the HII region are twice as large as the others. They have steep density gradients toward the HII region. We identified 18 protostellar IRAS sources and 155 2MASS sources with IR excess as YSO candidates and investigated their spatial distributions. 7 IRAS point sources are located at integrated intensity peaks of ^{13}CO, while the majority of the 2MASS sources are distributed in the front sides of BRC arcs close to the exciting star. These alignments of the YSOs and molecular clouds in the order of their ages indicate that triggered star formation occurs in the W5-East HII region. From the column densities of ^{13}CO and the spatial distribution of YSOs, we identified a new BRC candidate in the west side of the W5-East HII region.

Triggered Star Formation in a Turbulent ISM
Proceedings IAU Symposium No. 237, 2006
B. G. Elmegreen & J. Palouš, eds.
© 2007 International Astronomical Union
doi:10.1017/S1743921307002438

Dust evolution in photoevaporating protoplanetary disks

H. Nomura[1], Y. Aikawa[1], S. Inutsuka[2] and Y. Nakagawa[1]

[1]Dept. of Earth and Planetary Sciences, Kobe University, Kobe 657-8501, Japan
[2]Dept. of Physics, Graduate School of Science, Kyoto University, Kyoto 606-8502, Japan

Abstract. We construct a model of the physical structure of photoevaporating protoplanetary disks, and numerically calculate the coagulation and settling/evaporating process of dust particles in the disks. Our result show that (sub)micron-sized-dust-particles could evaporate with the gas, which leads to dispersal of infrared excess radiation from the disks.

Photoevaporation of protoplanetary disks induced by ultraviolet photons and/or X-rays from the central stars is known as one of possible mechanisms of gaseous disk dispersal. In addition, photoevaporating flow is expected to affect dust evolution that will lead to the planet formation in the disks, and observational properties of the disks.

In this work we have made a detailed model of gas density, temperature, and velocity structure of one-dimensional, steady photoevaporating flow in protoplanetary disks. By using the obtained disk structure, we compare the gravitational and the gas friction forces which affect on a dust particle in the disks to estimate a critical dust radius. The result shows that (sub)micron-sized-dust-particles could evaporate with the gas, instead of settling toward the disk midplane (Figure 1, left; Nomura & Inutsuka 2004). In addition, we calculate the dust evolution in the disks by numerically solving a coagulation equation for the dust particles (Nomura & Nakagawa 2006), taking into account the upward motion of the dust grains, which shows that (sub)micron-sized-dust-particles disappear from the disks because they can move only upward with the flow as well as coagulate very quickly near the dense disk midplane. Finally, making use of the resulting spatial and size distributions of the dust particles, we calculate spectral energy distributions (SEDs) of dust continuum emission from the disks, which suggests that the photoevaporation process could help to reduce infrared excess radiation from the disks (Figure 1, right).

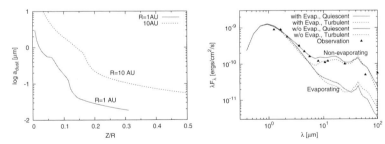

Figure 1. The critical radius of a dust particle at various positions (R, Z) in the disk (left). The resulting SEDs of evaporating (thick lines) and non-evaporating (thin lines) disks with observational median SED toward classical T Tauri stars (triangles; D'Alessio *et al.* 2006) (right).

References

D'Alessio, P., Calvet, N., Hartmann, L., Franco-Hernandez, R., & Servin, H. 2006, *ApJ* 638, 314
Nomura, H. & Inutsuka, S. 2004, in: J.-P. Beaulieu, A. Lecavelier Des Etangs & C. Terquem (eds.), *Extrasolar Planets: Today and Tomorrow* (ASP-CS) 321, 335
Nomura, H. & Nakagawa, Y. 2006, *ApJ* 640, 1099

Triggered Star Formation in a Turbulent ISM
Proceedings IAU Symposium No. 237, 2006
B. G. Elmegreen & J. Palouš, eds.

Molecular Hydrogen emission from protoplanetary disks: effects of X-ray irradiation and dust evolution

H. Nomura[1], Y. Aikawa[1], M. Tsujimoto[2], Y. Nakagawa[1] and T. J. Millar[3]

[1]Department of Earth and Planetary Sciences, Kobe University, Nada, Kobe 657-8501, Japan

[2]Department of Physics, Rikkyo University, Nishi-Ikebukuro, Toshima, Tokyo 171-8501, Japan

[3]School of Mathematics and Physics, Queen's University Belfast, Belfast BT7 1NN, Northern Ireland, UK

Abstract. We have made a detailed model of the physical structure of protoplanetary disks, taking into account X-ray and ultraviolet (UV) irradiation from a central star, as well as dust size growth and settling towards the disk midplane. Also, we calculate the level populations and line emission of molecular hydrogen from the disks, which shows that the dust evolution changes the physical properties of the disk, and then the line ratios of the molecular hydrogen emission.

Thanks to recent high resolution and high sensitivity observations, it has become possible to detect molecular hydrogen line emission from protoplanetary disks in various wavelength bands. Meanwhile, since dust particles are believed to evolve in the disks as the first step of planet formation, it is of great interest to find its observational evidence.

Our model calculations of the physical disk structure, and the level populations and line emission of molecular hydrogen from the disks (e.g., Nomura & Millar 2005) show that the level populations are controlled by the X-ray pumping if the X-ray irradiation is strong and the UV irradiation is weak. When the UV irradiation is strong, the level populations are controlled by the thermal collisions if there are enough small dust particles, or by the UV pumping process if the dust particles grow and settle. The excitation mechanism changes depending on the properties of dust grains (Figure 1) because with the decrease of small dust particles, the gas temperature drops due to the decrease in the grain photoelectric heating rate (Aikawa & Nomura 2006), and the thermal collision processes becomes less efficient. This results in a change of the line ratios of molecular hydrogen, which could be observable as evidence of the dust evolution in the disks.

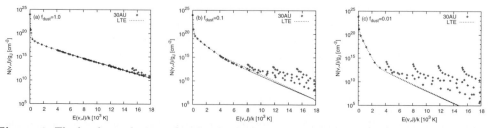

Figure 1. The level populations of molecular hydrogen, integrated vertically at the disk radius of 30AU, for various $f_{\rm dust}$, a parameter for the total surface area of dust particles per unit volume of the gas, normalized by that of the dense cloud dust model.

References

Aikawa, Y. & Nomura, H. 2006, *ApJ* 642, 1152

Nomura, H. & Millar, T.J. 2005, *A&A* 438, 923

Triggered Star Formation in a Turbulent ISM
Proceedings IAU Symposium No. 237, 2006
B. G. Elmegreen & J. Palouš, eds.

© 2007 International Astronomical Union
doi:10.1017/S1743921307002451

Photometric and spectroscopic studies of very low mass YSOs and young Brown Dwarfs in S106

Yumiko Oasa[1]

[1]Graduate School of Science and Technology, Kobe University, 1-1 Rokko-dai, Nada-ku, Kobe, 657-8501, Japan
email yummy@kobe-u.ac.jp

Keywords. star formation, young brown dwarf, near-infrared observations

1. Introduction

Young brown dwarfs have been identified in a significant population in various star forming regions. Some deep surveys have yielded less massive objects with planetary-mass (e.g.,Oasa *et al.* 1999; Lucas & Roche 2000). Nevertheless, it is not yet clear how abundant these very low-mass objects are formed. S106 is one of the nearest massive star-forming regions associated with prominent bipolar nebulae and an HII region. We have conducted near-infrared photometric and spectroscopic observations of very low-mass young stellar objects (YSOs) in the S106 region.

2. Observations and results

The photometric survey, whose limiting magnitude exceeds 20 mag in the JHK' band using the SUBARU telescope, is sensitive enough to provide a census of the stellar population down to objects below the deuterium-burning limit, a fiducial boundary between brown dwarfs and planetary-mass objects. Based on the color-color diagrams, nearly 600 embedded YSO candidates with near-infrared excesses have been identified in the area of about 25 arcmin^2. They are not uniformly distributed but are centrally concentrated. Combining the reddening-corrected luminosity of the YSO candidates with the theoretical evolutionary models (e.g. Baraffe *et al.* 2003), we suggest that there exists a substantial substellar population, including many potential isolated planetary mass objects (Oasa *et al.* 2006). The derived mass function appears to be similar to young clusters such as NGC1333 (Oasa 2003). However it is more abundant in young substellar objects compared to the mass functions obtained for other clusters such as Trapezium and IC348 (Hillenbrand & Carpenter 2000; Najita *et al.* 2000).

The spectroscopic observations of a part of above substellar YSO candidates have been subsequently carried out with SUBARU. Spectroscopy offers a means for more accurate assessment of membership and more precise measurement of the mass. We have constructed an index in the K-band to measure the strength of water. It is an indicator of low temperature although it depends on the surface gravity at cool temperatures. Using the reddening-independent water index, we confirm that some sources are cool, even if assuming they have any luminosity class. From these observations, it is considered that young brown dwarfs are formed in S106.

References

Baraffe, I., Chabrier, G., Barman, T. S., Allard, F., & Hauschildt, P. H. 2003, *A&A* 402, 701
Hillenbrand, L. A. & Carpenter, J. M. 2000, *AJ* 540, 236
Lucas, P. W. & Roche, P. F. 2000, *MNRAS* 314, 858
Najita, J. R., Tiede, G. P., & Carr, J. S. 2000, *ApJ* 541, 977
Oasa, Y., Tamura, M., & Sugitani, K. 1999, *ApJ* 526, 336
Oasa, Y. 2003, in: E. Martín (ed.), *Brown Dwarfs* (ASP-CS), p. 91
Oasa, Y. *et al.* 2006, *AJ* 131, 1608

Triggered Star Formation in a Turbulent ISM
Proceedings IAU Symposium No. 237, 2006
B. G. Elmegreen & J. Palouš, eds.

© 2007 International Astronomical Union
doi:10.1017/S1743921307002463

Luminosity functions of YSO clusters in Sh-2 255, W3 Main and NGC 7538 star forming regions

Devendra Ojha[1], Motohide Tamura[2] and SIRIUS Team

[1]Tata Institute of Fundamental Research, Homi Bhabha Road, Colaba, Mumbai 400 005, India
email: ojha@tifr.res.in

[2]National Astronomical Observatory of Japan, Mitaka, Tokyo 181-8588, Japan

Abstract. We have conducted deep near-infrared surveys of the Sh-2 255, W3 Main and NGC 7538 massive star forming regions using simultaneous observations of the JHK_s-band with the near-infrared camera SIRIUS on the UH 88-inch telescope and with SUBARU. The near-infrared surveys cover a total area of ~ 72 arcmin^2 of three regions with 10-σ limiting magnitudes of ~ 19.5, 18.4 and 17.3 in J, H and K_s-band, respectively. Based on the color-color and color-magnitude diagrams and their clustering properties, the candidate young stellar objects are identified and their luminosity functions are constructed in Sh-2 255, W3 Main and NGC 7538 star forming regions. A large number of previously unreported red sources (H-K > 2) have also been detected around these regions. We argue that these red stars are most probably pre-main-sequence stars with intrinsic color excesses. The detected young stellar objects show a clear clustering pattern in each region: the Class I-like sources are mostly clustered in molecular cloud region, while the Class II-like sources are in or around more evolved optical HII regions. We find that the slopes of the Ks-band luminosity functions of Sh-2 255, W3 Main and NGC 7538 are lower than the typical values reported for the young embedded clusters, and their stellar populations are primarily composed of low mass pre-main-sequence stars. From the slopes of the Ks-band luminosity functions, we infer that Sh-2 255, W3 Main and NGC 7538 star forming regions are rather young (age $\leqslant$ 1 Myr).

Main Results:

i) In NGC 7538, the young stellar objects in the central region are probably the result of the propagation of star forming activity from the north-western region due to the expansion of the H II region and the compression of the molecular cloud (sequential star formation). The south-eastern/southern region is independent of the above action. and presumably the star formation there is taking place in a spontaneous and gradual process.

ii) Based on the comparison of models of pre-main-sequence stars with the observed color-magnitude diagram, we find that the stellar populations in W3 Main, NGC 7538 and Sh-2 255 are primarily composed of low-mass pre-main-sequence stars.

iii) Follow-up deep JHK' imaging (J $\sim$ 22 mag at 10-σ) with CISCO/SUBARU 8.2-meter has been carried out for the search of young brown dwarfs in the cores of W3 Main and NGC 7538. The interpretation work is under way.

Triggered Star Formation in a Turbulent ISM
Proceedings IAU Symposium No. 237, 2006
B. G. Elmegreen & J. Palouš, eds.

© 2007 International Astronomical Union
doi:10.1017/S1743921307002475

The conditions for star formation at low metallicity: results from the LMC

J. M. Oliveira[1], J. Th. van Loon[1] and S. Stanimirović[2]

[1]Astrophysics Group, Keele University, UK; [2]Radio Astronomy Lab, UC Berkeley, USA

Abstract. We present our recent work on the conditions under which star formation occurs in a metal-poor environment, the Large Magellanic Cloud ($[Fe/H] \sim -0.4$). Water masers are used as beacons of the current star formation in H II regions. Comparing their location with the dust morphology imaged with the Spitzer Space Telescope, and additional Hα imaging and groundbased near-infrared observations, we conclude that the LMC environment seems favourable to sequential star formation triggered by massive star feedback (Oliveira *et al.* 2006). Good examples of this are 30 Doradus and N 113. There are also H II regions, such as N 105A, where feedback may not be responsible for the current star formation although the nature of one young stellar object (YSO) suggests that feedback may soon start making an impact. The chemistry in one YSO hints at a stronger influence from irradiation effects in a metal-poor environment where shielding by dust is suppressed (van Loon 2005).

Keywords. stars: formation; Magellanic Clouds; HII regions; circumstellar matter

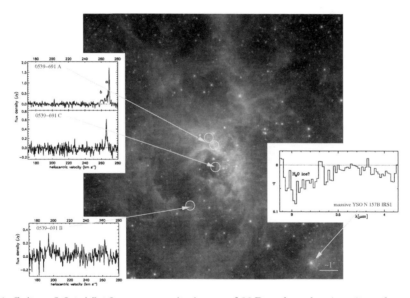

Figure 1. Spitzer $3.6 + 4.5 + 8\,\mu m$ composite image of 30 Doradus, showing sites of massive star formation. Water masers (l) pinpoint ongoing star formation; the ISAAC/VLT spectrum of the massive YSO (r) shows a hint of water ice. More details on this and other H II regions in the LMC can be found in van Loon *et al.* (2005) and Oliveira *et al.* (2006).

References

Oliveira, J.M. *et al.*, 2006, *MNRAS* in press, astro-ph/0609036
van Loon, J.Th. *et al.*, 2005, *MNRAS* 364, L71

Triggered Star Formation in a Turbulent ISM
Proceedings IAU Symposium No. 237, 2006
B. G. Elmegreen & J. Palouš, eds.

Star formation in the Eagle Nebula and NGC 6611

J. M. Oliveira, R. D. Jeffries and J. Th van Loon

Astrophysics Group, Lennard-Jones Laboratories, Keele University, UK

Abstract. We present $IZJHKL'$ photometry of the core of the cluster NGC 6611 in the Eagle Nebula. This photometry is used to constrain the Initial Mass Function (IMF) and the circumstellar disk frequency of the young stellar objects. Optical spectroscopy of 258 objects is used to confirm membership and constrain contamination as well as individual reddening estimates. Our overall aim is to assess the influence of the ionizing radiation from the massive stars on the formation and evolution of young low-mass stars and their disks. The disk frequency determined from the $JHKL'$ colour-colour diagram suggests that the ionizing radiation from the massive stars has little effect on disk evolution (Oliveira *et al.* 2005). The cluster IMF seems indistinguishable from those of quieter environments; however towards lower masses the tell-tale signs of an environmental influence are expected to become more noticeable, a question we are currently addressing with our recently acquired ultra-deep (ACS and NICMOS) HST images.

Keywords. stars: pre–main-sequence; stars: luminosity function, mass function; stars: formation; circumstellar matter

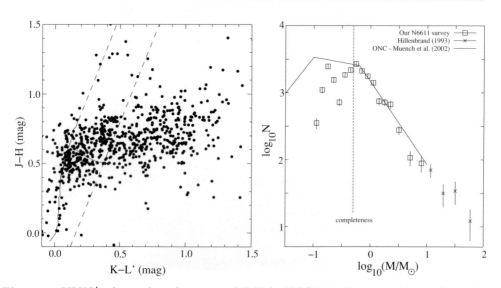

Figure 1. $JHKL'$ colour-colour diagram and IMF for NGC 6611. Down to objects of $\sim 0.5\,M_\odot$, neither the disk frequency nor the IMF show significant evidence for a strong ionizing influence from the massive O-stars in the cluster (Oliveira *et al.* 2005).

References

Hillenbrand, L.A. *et al.*, 1993, *ApJ* 106, 1906
Muench, A.A. *et al.*, 2002, *ApJ* 573, 366
Oliveira, J.M. *et al.*, 2005, *MNRAS* 358, L21

Triggered Star Formation in a Turbulent ISM
Proceedings IAU Symposium No. 237, 2006
B. G. Elmegreen & J. Palouš, eds.

© 2007 International Astronomical Union
doi:10.1017/S1743921307002499

Thermal and non thermal components of interstellar medium at sub-kiloparsec scales in galaxies

Rosita Paladino

INAF – Osservatorio Astronomico di Cagliari
email: rpaladin@ca.astro.it

Abstract. We present the result of the analysis of the point-by-point correlation between the radio continuum (RC) and CO intensities from kpc to sub-kpc scales in 22 BIMA SONG galaxies and the point-by-point correlation at sub-kpc scales between the RC, CO and 24 μm-IR emissions in 6 galaxies for which *Spitzer* images have been recently released. We found that there is no significant variation of the slope and the scatter of the correlations at this spatial resolution. All three correlations are comparably tight with scatter of less than a factor of two.

Keywords. radio continuum: galaxies; galaxies: spiral; ISM: molecules; stars: formation

One of the most intriguing relationship in astronomy is that between the thermal and non-thermal components of interstellar medium in galaxies. Although the strongest correlation observed between the far-infrared (FIR) and the radio continuum (RC) emission has been known for over two decades, the causes of the relationship are still unclear. In the standard model the FIR, RC and CO emissions are correlated because each is a tracer of massive-star formation, this conventional picture does not plausibly explain the tightness of the correlations given the entirely different processes and timescales involved.

Since we know that relativistic electrons, responsible for synchrotron radiation, diffuse from their birthplaces we would expect that the spatial correlation breaks down below the characteristic diffusion scale-length of the radiating electrons.

With the aim of determining whether the RC-CO(-FIR) correlation persists at sub-kpc scales, we studied these correlations from kpc to sub-kpc scales in galaxies selected from the BIMA SONG sample. We found that all three correlations are comparably tight up to spatial resolution of hundred parsecs. This result shows that we have not yet probed the physical scales at which the correlations break down or that there is a mechanism that compensate the electrons diffusion.

Observations of the radio spectral index help us to understand the processes involving the cosmic-ray electrons production and to determine their diffusion scales in spiral galaxies. We recently found a relation between the RC spectral index and the FIR emission in the galaxy M 51: the spectral index decreases in the spiral arms, where FIR emission is enhanced. This result indicates that electron diffusion is efficient in regions of high star formation rate, then to justify the observed correlations a mechanism compensating the leakages of the synchrotron electron should exist. A possibility is that the galaxy magnetic field is higher in molecular clouds.

Reference

Paladino, R., Murgia, M., Helfer, T. T., Wong, T., Ekers, R., Blitz, L., Gregorini, L., & Moscadelli, L. 2006, *A&A* 456, 847

Triggered Star Formation in a Turbulent ISM
Proceedings IAU Symposium No. 237, 2006
B. G. Elmegreen & J. Palouš, eds.

© 2007 International Astronomical Union
doi:10.1017/S1743921307002505

Mid-IR images of methanol masers and ultracompact HII regions

Paolo Persi[1], Mauricio Tapia[2] and Anna Rosa Marenzi[1]

[1]INAF/IASF-Roma Via fosso del Cavaliere,100, oo133 Roma-Italy
email: paolo.persi@iasf-roma.inaf.it

[2]IA-UNAM Ensenada, Mexico

Methanol masers and UCHII regions trace massive star formation sites. We have undertaken a mid-IR survey of 17 regions containing methanol masers and UCHIIs in order to locate the young stellar sources associated with them. The images were obtained from 8.7 to 18.8 μm with the mid-IR camera CID (Salas *et al.* 2003) on the 2.1m telescope of the Observatorio Astronomico Nacional at San Pedro Martir (Baja California, Mexico). The images were taken with a scale 0.55"/pix and the mean PSF was 1.5-2.0"(FWHM) close to the diffraction limit. We report as an example in Fig. 1 (left panel) our 18.8μm

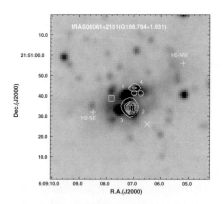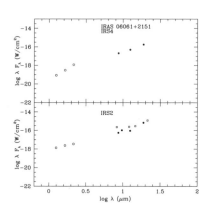

Figure 1. *Left*: Ks image of IRAS 06061+2151 with the 18.8 μm contours. *Right*: SED of the source #2 and #4 associated with the UCHIIs.

contours of IRAS 06061+2151 superimposed to the 2MASS Ks image. A young cluster of at least 4 sources has been found centered on the IRAS source (Anandarao *et al.* 2004). We have found two mid-IR sources coinciding with the source #2 and #4 of Anandarao *et al.*(2004). The source #4 is at the center of two H2 knots and a high velocity molecular outflow. The mid-IR emission from #2 is extended and coincides with the UCHII and MSX source. The methanol maser is approximately 10" south of the source #2. The SEDs of both sources are illustrated in Fig. 1 (right panel). The IR spectral indices of source #2 and #4 are α(IR)=1.9 and 2.2 respectively.

References

Anandarao, B.G., Chakraborty, A., Ojha, D.K., & Testi, L. 2004, *A&A* 421, 1045
Salas, L., Gutierrez, L., Tapia, M., *et al.* 2003, in: Instrument Design and Performance for Optical/Infrared Telescopes , *SPIE Proc.* 4941, 594

Triggered Star Formation in a Turbulent ISM
Proceedings IAU Symposium No. 237, 2006
B. G. Elmegreen & J. Palouš, eds.

© 2007 International Astronomical Union
doi:10.1017/S1743921307002517

Degree scale high resolution mapping of CO J=2-1 and 3-2 in giant molecular clouds

W. L. Peters[1], J. H. Bieging[1], C. E. Groppi[1], C. A. Kulesa[1],
C. K. Walker[1], A. S. Hedden[1] and P. S. Puetz[1]

[1]Steward Observatory, University of Arizona, Tucson, AZ 85721, USA
email: wpeters@as.arizona.edu

Abstract. We present the first results from a project to map Giant Molecular Clouds (GMCs) in the ^{12}CO J=2-1, ^{13}CO J=2-1, and ^{12}CO J=3-2 lines using the Heinrich Hertz Submillimeter Telescope (HHT) at the University of Arizona. We mapped nearly 2.5 sq. deg of W3 and 1.0 sq. deg of W51 in the J=2-1 lines. We have begun mapping in the J=3-2 line. We achieve angular resolutions of $33''$ and $24''$ in the J=2-1 and J=3-2 lines with 1.3 and 0.9 km s^{-1} resolution.

Keywords. ISM: clouds, ISM: molecules, submillimeter

W3 and W51 are GMC's with active star forming regions in the Perseus Arm and Sagittarius arms, respectively. The greatest activity in W3 takes place in the part of the cloud that has recently formed OB stars in large numbers. The radiation from these stars has ionized part of the GMC making a large H II region complex dominated by W3 Main. In other areas are found filaments and cometary clouds that have presumably been sculpted by the radiation and/or stellar winds of these OB stars.

The brightest CO emission in W51 is in the area of the cloud that also has strong infrared and radio continuum emission indicating that O stars and H II regions are located there. Another area of W51 exhibits an especially high rate of star formation possibly triggered by a spiral density wave shock.

These and other maps of the observations reveal in some detail the structures associated with the ongoing star formation in these GMC's. We will apply multiline analysis tools to derive the physical conditions in them.

Figure 1 shows the most active portion of the W3 GMC as an integrated intensity map in the ^{12}CO J=3-2 line (made with the **DesertStar** 7-beam, 345 GHz array receiver). Superimposed on it are contours of the 21 cm continuum emission which traces the associated H II regions. The green circle marks the position of OB association IC1795.

Figure 2 is a 1 square degree integrated intensity map in the ^{12}CO J=2-1 line of W51.

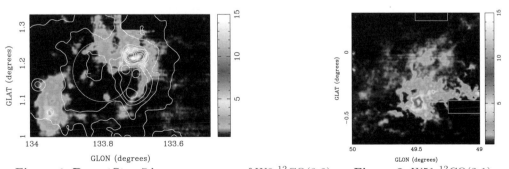

Figure 1. **DesertStar** 7-beam array map of W3 ^{12}CO(3-2)   **Figure 2.** W51 ^{12}CO(2-1)

The whole poster is available at mira.as.arizona.edu/~jbieging/IAUsymp237/GMCmaps.pdf

Triggered Star Formation in a Turbulent ISM
Proceedings IAU Symposium No. 237, 2006
B. G. Elmegreen & J. Palouš, eds.

© 2007 International Astronomical Union
doi:10.1017/S1743921307002529

Triggered star formation in the isolated cluster CB 34?

Dawn E. Peterson[1], R. A. Gutermuth[2], M. F. Skrutskie[1], S. T. Megeath[3], J. L. Pipher[4], L. E. Allen[2] and P. C. Myers[2]

[1]Department of Astronomy, University of Virginia, Charlottesville, VA 22901, USA
email: dawnp@virginia.edu, mfs4n@virginia.edu

[2]Harvard-Smithsonian Center for Astrophysics, Cambridge, MA 02138, USA
email: rgutermuth@cfa.harvard.edu, leallen@cfa.harvard.edu, pmyers@cfa.harvard.edu

[3]Department of Physics and Astronomy, University of Toledo, Toledo, OH 43606, USA
email: megeath@physics.utoledo.edu

[4]Department of Physics and Astronomy, University of Rochester, Rochester, NY 14627, USA
email: jlpipher@astro.pas.rochester.edu

Abstract. Bok globules, optically opaque small dark clouds, are classical examples of isolated star formation. However, the collapse mechanism for these cold, dense clouds of gas and dust is not well understood. Observations of Bok globules include some which appear to be starless while others harbor single stars, binaries and even small groups of forming stars. One example of a Bok globule forming a group of stars is CB 34, observed with both the IRAC and MIPS instruments as part of the Spitzer Young Cluster Survey. Based on initial analysis of 1-8 μm photometry from IRAC and the Two Micron All Sky Survey (2MASS), we identified 9 Class 0/I and 14 Class II young stellar objects within the small, $4.5' \times 4.5'$ region encompassing CB 34. This unusually high number of protostars compared with Class II sources is intriguing because it implies a high rate of star formation. Therefore we have begun a larger study of this region in order to determine why and how CB 34 started forming stars at such a high rate. Is CB 34 embedded within a larger HII region which may have triggered its collapse or does it appear to have collapsed in isolation from outside influences?

Keywords. stars: formation, stars: pre–main-sequence, ISM: clouds, ISM: globules, ISM: individual (CB 34), stars: formation

---

The immediate vicinity around CB 34 is fairly empty, with only dark cloud CB 35 nearby. To search for possible triggers, a 3° radius around CB 34 was examined. Within this region is NGC 2129, a young, 10 Myr open cluster at a distance of 2.2 kpc and with a V_r=17.5 km s^{-1} (Carraro, Chaboyer & Perencevich 2006). Kawamura *et al.* (1998) find V_{LSR}=0.6 km s^{-1} for CB 34, which rules out winds from O and B stars in NGC 2129 as triggers. Star forming region GGD 4 is nearby, at a distance of 1 kpc and with V_r=2.1 km s^{-1} (Kawamura *et al.* 1998). It is possible it is associated with CB 34, however there are not any stars in GGD 4 massive enough to trigger star formation in CB 34. Finally, the western edge of the Gemini OB 1 association is nearby. At 1.5 kpc and with V_r=3.8 km s^{-1} (Carpenter, Snell & Schloerb 1995), if there is a trigger, this one is most likely.

References

Carraro, G., Chaboyer, B. & Perencevich, J. 2006, *MNRAS* 365, 867
Kawamura, A. *et al.* 1998, *ApJS* 117, 387
Carpenter, J. M., Snell, R. L. & Schloerb, F. P. 1995, *ApJ* 445, 246

Triggered Star Formation in a Turbulent ISM
Proceedings IAU Symposium No. 237, 2006
B. G. Elmegreen & J. Palouš, eds.

Massive star formation in the outer Galaxy: S284

E. Puga[1], C. Neiner[2], S. Hony[3], A. Lenorzer[4], A.-M. Hubert[2] and L. B. F. M. Waters[5]

[1]Katholieke Universiteit Leuven, Celestijnenlaan 200D, 3001 Leuven, Belgium,
email: elena@ster.kuleuven.be

[2]GEPI, UMR 8111 du CNRS, 5 place Jules Janssen, 92195 Meudon Cedex, France,
[3]CEA Saclay, Bat.609 Orme des Merisiers, 91191 Gif-sur-Yvette, France,
[4]Instituto Astrofysico de Canarias,C/ Vía Láctea s/n, E-38200, La Laguna, Spain,
[5]Universiteit van Amsterdam, Kruislaan 403, 1098SJ Amsterdam, The Netherlands

Abstract. S284 is a diffuse HII region located in the anti-centre of the galactic disk (l=212, b=-1.3), at a distance of 5.5 kpc and it is relatively isolated. S284 harbours a cluster of stars in its centre known as Dolidze 25 (also known as C 0642+0.03). A spectroscopic study of the cluster sources (Lennon *et al.* 1990, A&A, 240, 349) revealed: a) the spectral types of these central sources and b) the low metallicity of the cluster (a factor six lower than the solar metallicity). The age of the cluster (6 Myr) has been determined through isochrone fitting of precision photometry (Turbide & Moffat, 1993, AJ, 105, 1831).
We have conducted Spitzer-IRAC observations of 0.9x1.2 degrees around S284 in the four IRAC bands (3.6, 4.5, 5.8 and 8.0 μm). These data provide us with an unprecedented perspective of the dust component toward S284. We also present complementary observations with the WFC-INT (Roque de los Muchachos) in the Hα filter. The inspection of the images reveals the existence of: i) Rings of dust that constraint the ionised gas, ii) Elephant trunks stretching toward the centre of the main HII region, namely the cluster Dolidze 25 and iii) various spatial scales (4 main shells and 22 knots).
We have calculated the dynamical age of the central HII region (S284) in a very simple pressure-driven scenario, using the strong-shock approximation. The resultant age of the main shell is 8 Myr, in good agreement with the estimated age of the central cluster (Dolidze 25). The presence of another two smaller ionised bubbles (with diameters of 3' and 5', respectively) located at the rim of this HII region suggests a later generation of high-mass star-forming regions. The color-color diagram of the IRAC photometry betrays the existence of several Class I objects, most of which are located at the rim of the most prominent shell.
Multi-wavelength photometry has been gathered to construct the Spectral Energy Distribution of the 22 blobs detected toward S284. The SEDs reveal luminosities corresponding to low- and intermediate-mass stars.
We conclude that the remarkable symmetry of the main HII bubble, the spectra of masses harboured by the different blobs around it, and most importantly, the uniform presence of elephant trunks that stretch toward the centre can be best explained by the growth of dynamical instabilities in a collected layer (Garcia-Segura & Franco, 1996, ApJ, 469, 171). In the context of this hypothesis, the low metallicity of the environment would significantly determine the length and stability over time of the elephant trunks.

Keywords. stars: formation, (ISM:) HII regions

Triggered star formation in a turbulent ISM
Proceedings IAU Symposium No. 237, 2006
B. G. Elmegreen & J. Palouš, eds.

Interaction/merger-induced starbursts in local very metal-poor dwarfs: link to the common SF in high-z young galaxies

S. A. Pustilnik[1], Ekta[2], A. Y. Kniazev[3], J. N. Chengalur[2] and L. Vanzi[4]

[1]Special Astrophysical Obs. RAS, Nizhnij Arkhyz, 369167, Russia email: sap@sao.ru

[2]National Centre of Radio Astrophysics, Pune 411007, India email: ekta@ncra.tifr.res.in

[3]South African Astronomical Obs. Cape Town 7935, South Africa email: akniazev@saao.ac.za

[4]Europian Southern Obs. Santiago, Chile email: lvanzi@eso.org

Abstract. We present a subsample of 'local' very metal-poor gas-rich galaxies that show more or less clear evidences of interactions and mergers and discuss their relevance to the study of high-redshift star-forming young galaxies.

Keywords. galaxy abundances, galaxy interactions, star formation

1. Introduction

Widespread galaxy formation from pregalactic gas took place in the first 1–3 Gyr after the Big Bang, with most of them forming in low-mass halos (i.e., M in the range of 10^7-10^{10} $M_\odot$). Observations at high redshifts ($z = 4 - 7$) are, however, mainly limited to the rare massive "tip of the iceberg" objects. Detailed studies of the properties and the evolution of more common lower-mass young galaxies will have to await the next generation mega telescopes. Their local analogs provide clues for their SF and evolution.

2. Method and Results

We have conducted a multi-wavelength (including optical/NIR morphology/photometry, HI imaging, and Hα-line kinematics) study of a sample of the local eXtremely Metal-Deficient (XMD) BCGs and find in a large fraction of them clear evidences that strong interactions or mergers with low-mass objects, provide a trigger mechanism for their observed starbursts. We present about 20 such XMD BCGs arranged in a Toomre-like sequence and also the first results of detailed studies of several individual objects.

3. Conclusions

Since both SF (through cooling rate and the IMF) and its feedback (through the massive star evolution and interaction with the ISM) depend substantially on the ISM metallicity, comprehensive multiwavelength studies of local XMD galaxy mergers, coupled with theoretical modelling, should give us substantial insight into star formation in young high-redshift galaxies.

Acknowledgements

S. A. P. would like to acknowledge the support from IAU (grant No.12330) and Czech Academy of Sciences.

Triggered Star Formation in a Turbulent ISM
Proceedings IAU Symposium No. 237, 2006
B. G. Elmegreen & J. Palouš, eds.

© 2007 International Astronomical Union
doi:10.1017/S1743921307002554

Evolutionary sequence of expanding Hydrogen shells

M. Relaño[1], J. E. Beckman[2], O. Daigle[3] and C. Carignan[3]

[1]Dpto. Física Teórica y del Cosmos, Universidad de Granada, Spain
email: mrelano@ugr.es

[2]Instituto de Astrofísica de Canarias, C. Vía Láctea s/n, 38200, La Laguna, Tenerife, Spain
email: jeb@iac.es

[3]Observatoire du mont Mégantic, LAE, Université de Montréal, C. P. 6128 succ. centre ville,
Montréal, Québec, Canada H3C 3J7
email: odaigle@ASTRO.UMontreal.CA, carignan@ASTRO.UMontreal.CA

Abstract. Giant HI shells, with diameters of hundreds of parsecs and expansion velocities of 10-20 km s^{-1} are characteristic observed features of local gas rich galaxies. Although a predictable consequence of the impact of OB associations on the ISM doubts have been raised, as OB stars are not present in the centres of the majority of these shells. Here we combine our observations of expanding ionized shells in luminous H II regions with basic dynamical models to give support to the scenario in which OB associations do produce the HI shells.

Keywords. galaxies, ISM: H II regions, kinematics and dynamics, bubbles

From Fabry-Pérot data cubes in Hα of four late-type spiral galaxies: NGC 1530, NGC 3359, NGC 5194 and NGC 6951, we extract emission line profiles of the most luminous H II regions. Symmetrically placed about the central dominant peak we find two peaks of low intensity but high velocity in the profiles of all regions with sufficient S:N. We interpret these as clear evidence for supersonic expanding shells within the regions, and use them to quantify the masses and kinetic energies of these shells (cf. Relaño & Beckman 2005). Using a basic model in which energy is supplied first by stellar winds and then by supernovae (Dyson 1981) we extrapolate the properties of these shells in time, finding results in good overall agreement with the observed properties of the HI shells, in mass and velocity ranges, after times of a few 10^{7}yr. If the formation of the OB stars occurs on timescales notably shorter than this, these stars will no longer be found at the centres of the HI shells, in agreement with observations reported by Hatzidimitriou *et al.* (2005) in the SMC.

References

Relaño, M. & Beckman, J. E. 2005, *A&A* 430, 911

Dyson, J.E. 1981 in: F. D. Kahn (ed.), *Investigating the Universe* (Dordrecht: Reidel), p. 125

Hatzidimitriou, D., Stanimirovic, S., Maragoudaki, F., Staveley-Smith, L., Dapergolas, A., & Bratsolis, E. 2005, *MNRAS* 360, 1171

Triggered Star Formation in a Turbulent ISM
Proceedings IAU Symposium No. 237, 2006
B. G. Elmegreen & J. Palouš, eds.

© 2007 International Astronomical Union
doi:10.1017/S1743921307002566

Arms pattern speed of galaxies in clusters

I. Rodrigues[1], H. Dottori[1] and D. Reichert

[1]Instituto de Física, UFRGS,cp: 15051, cep: 91501-970, Porto Alegre, Brazil
email: dottori@if.ufrgs.br

Abstract. Disks of galaxies in clusters are deeply affected by interactions. We investigate the pattern speed of galaxies in cluster in order to detect a possible dependence with cluster environmental parameters, such as galaxies density. If the perturbation of cluster galaxies is mainly produced by the interaction with the cluster ambient, the pattern speed might well depend on the history of the galaxy orbit within the cluster. Tremaine & Weinberg (1984) method is applied to 2-D H_α velocity fields, reconstructed from the isovelocity contours published by Amram *et al.* (1992), and 2-MASS K–band images (Skrutskie 2001) to obtain spirals pattern angular speed. The use of K–band images and H_α velocity maps is justified by the fact that the perturbations imprinted in H_α velocity maps are produced by the old stellar population which emits most of its energy in the near-IR. We analyzed Pegasus I cluster galaxies NGC 7593, NGC 7631 and NGC 7643 (this one shown in Figure 1). Preliminary results indicates that NGC 7593 presents a pattern speed $\Omega_p \sin(i) = 18 \pm 5 \mathrm{km\,s^{-1}\,kpc^{-1}}$,(inclination $i = 51°$) while NGC 7631 and NGC 7643 pattern speeds are $\Omega_p \sin(i) = 21 \pm 4 \mathrm{km\,s^{-1}\,kpc^{-1}}$ and $\Omega_p \sin(i) = 35 \pm 3 \mathrm{km\,s^{-1}\,kpc^{-1}}$ (inclinations $i = 64°$ and $59°$ respectively). In the three successful cases the correlation coefficients in the $\langle X \rangle$ vs $\langle V \rangle$ plots are 0.94, 0.95 and 0.98 respectively. We are presently analyzing other galaxies in the cluster.

Keywords. galaxies: spiral, galaxies: clusters

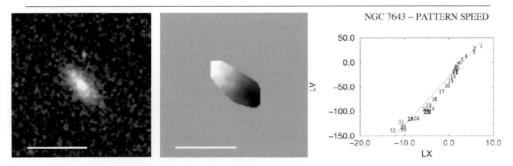

Figure 1. Left panel: 2MASS K–band image of NGC 7643. The white line is 1 arcminute long, corresponding to 16.6 kpc. Center panel: velocity map of NGC 7643. Right panel: plot $\langle X \rangle$ vs $\langle V \rangle$, from which Ω_p can be obtained.

References

Amram, P., Le Coarer, E., Marcelin, M., Balkowski, C., Sullivan, W. T., III, & Cayatte, V. 1992, *A&AS* 94, 175
Merrifield, M. R. & Kuijken, K. 1995, *MNRAS* 274, 933
Skrutskie, M. F. 2001, *BAAS* 33, 827
Tremaine, S. & Weinberg, M. D. 1984, *ApJ* 282, L5

Triggered Star Formation in a Turbulent ISM
Proceedings IAU Symposium No. 237, 2006
B. G. Elmegreen & J. Palouš, eds.

High-frequency carbon recombination line as a probe to study the environment of Ultra-compact H II regions

D. Anish Roshi

Raman Research Institute, Sadashivanagar, Bangalore 560080, India
email: anish@rri.res.in

Abstract. The far-ultra violet (6 – 13.6 eV) photons from the OB stars in Ultra-compact H II regions (UCHs) produce photo-dissociation regions (PDRs) at the interface between the ionized and the natal molecular material. In this paper, we show that carbon recombination lines (CRLs) at frequencies greater than a few GHz are detectable from these PDRs and such observations can be used to : (1) estimate the physical properties of the PDR material; (2) study the kinematics of the PDR material relative to the H II region gas; (3) constrain the magnetic fields in the vicinity of UCHs and (4) address the lifetime problem of UCHs.

Keywords. H II regions, ISM: magnetic fields, radio lines: ISM, stars: formation

---

H II regions with size less than ~ 0.1 pc are called UCHs. UCHs are associated with newly born OB stars and are embedded in dense molecular material. With the aim to detect CRLs from the interface between UCHs and their natal molecular cloud, we made a survey near 9 GHz with the Arecibo telescope toward 17 sources (Roshi *et al.* 2005a). CRLs were detected in 11 directions (65% detection rate). The detection of CRLs toward a large number of UCHs indicates that a majority of these H II regions have associated dense PDRs. We made high angular resolution (14″ to 2″), multi-frequency (6, 3.6, 2 & 0.7 cm) CRL observations toward the UCH W48A with the VLA. The C76α (near 2 cm) and C53α (near 0.7 cm) transitions were detected toward W48A (Roshi *et al.* 2005b).

Models for CRL emission toward W48A were constructed using the PDR model developed by Le Bourlot *et al.* (1993). The neutral density in the PDR inferred from our modeling is a few times 10^7 cm^{-3}. Modeling also shows that the line emission at frequencies $\lesssim 25$ GHz is dominated by stimulated emission due to the background continuum radiation arising from the UCH. This dominance of stimulated emission is made use to study the relative motion between the PDRs and the H II regions in our Arecibo data. The RMS radial velocity difference between the PDR and UCH is found to be 3.3 km s^{-1}. The observed non-thermal width of CRLs is likely to be due to Alfvén waves in the interface region. Based on this consideration we estimated magnetic field strength from the non-thermal line width. The estimated values compare well with those obtained from Zeeman observations of molecular lines. The derived physical properties and magnetic fields in the PDR indicate that the ambient pressure is high enough to pressure confine the UCHs thus increasing their lifetime.

References

Le Bourlot, J., Pineau Des Forets, G., Roueff, E., & Flower, D. R. 1993, *A&A* 267, 233
Roshi, D. A., Balser, D. S., Bania, T. M., Goss, W. M., & De Pree, C. G. 2005, *ApJ* 625, 181
Roshi, D. A., Goss, W. M., Anantharamaiah, K. R., & Jeyakumar, S. 2005, *ApJ* 626, 253

Triggered Star Formation in a Turbulent ISM
Proceedings IAU Symposium No. 237, 2006
B. G. Elmegreen & J. Palouš, eds.

© 2007 International Astronomical Union
doi:10.1017/S174392130700258X

Extraplanar gas and magnetic fields in the cluster spiral galaxy NGC 4569

S. Ryś[1], K. T. Chyży[1], M. Weżgowiec[1], M. Ehle[2] and R. Beck[3]

[1]Astronomical Observatory, Jagiellonian University, Kraków, Poland
[2]ESAC, XMM-Newton Science Operations Centre, Madrid, Spain
[3]Max-Planck-Institut für Radioastronomie, Bonn, Germany

Abstract. The Virgo Cluster spiral NGC 4569 is known for its compact starburst in the core and unusual outflow of Hα emitting gas perpendicular to the galaxy disk. Recent radio polarimetric observations with the Effelsberg telescope reveal huge magnetized outflows. Preliminary results of our XMM-Newton observations uncover not only hot gas in the disk but also an extensive X-ray envelope around it. We investigate the possibility of starburst-induced galactic outflows in various gas phases and cluster influence on the galaxy evolution.

NGC 4569 is a bright spiral (Sb) galaxy located only 0.5 Mpc from the Virgo Cluster center, known for its compact starburst in the core and a giant (8 kpc) outflow of Hα-emitting gas perpendicular to the galaxy disk. Our recent polarimetric radio continuum observations with the Effelsberg telescope at 4.85 GHz and 8.35 GHz reveal huge magnetized lobes, even extending 24 kpc from the galactic plane. This is the first time that such huge radio continuum lobes are observed in a cluster spiral galaxy.

Observing the galaxy in X-rays with XMM-Newton revealed a soft (0.2-1 keV) emission component covering the whole galaxy disk and forming an extensive X-ray envelope around it. In contrast to the radio emission the X-rays do not show similarly large extensions on both sides of the galactic disk. However, stronger X-ray emission is visible close to the disk on its western part, and corresponds to the enhanced radio and Hα emission there. The extension is broad, thus more typical for a wide-spread starburst than for a more collimated ionisation cone from an AGN. The less extended X-ray soft component is also visible to the SW direction from the disk.

The inspection of radio emission from the galaxy lobes indicates that indeed the lobes cannot be powered by an AGN but are probably caused by a nuclear starburst and superwind-type outflows which occurred ∼ 30 Myr ago. This is supported by estimates of the combined magnetic and cosmic-ray pressure inside the lobes from our radio data. The Hα spur and associated soft X-ray emission on the western part of the disk could be a recent example of such numerous events in the past.

As the galaxy is highly deficient in HI gas it probably suffered from strong ISM-ICM interactions in the form of severe gas stripping. The HI distribution and velocity field suggest that the galaxy is moving at high speed through the ICM in north-east direction. Our X-rays observations showing weaker soft emission in the northern disk and the extension to the south-west support this ram pressure scenario. The large extent of the extraplanar ionized and magnetized gas was probably enabled by the relatively weak ISM density in the heavily stripped galaxy disk.

Acknowledgements. This work was supported by the Polish Ministry of Science and Higher Education, grant 2693/H03/2006/31. Based on observations with the 100-m telescope of the MPIfR at Effelsberg and with XMM-Newton, an ESA science mission with instruments and contributions directly funded by ESA Member States and NASA.

Triggered Star Formation in a Turbulent ISM
Proceedings IAU Symposium No. 237, 2006
B. G. Elmegreen & J. Palouš, eds.

© 2007 International Astronomical Union
doi:10.1017/S1743921307002591

Physical and chemical properties of the AFGL 333 cloud

Takeshi Sakai[1]†, Tomoharu Oka[2] and Satoshi Yamamoto[2]

[1] Nobeyama, Radio Observatory, Japan
email: sakai@nro.nao.ac.jp

[2]Department of Physics, Graduate School of Science, University of Tokyo, Japan

Abstract. We have found massive clumps without any sign of active star formation in the AFGL 333 cloud. We present a study of the physical and chemical properties of the AFGL 333 cloud.

The AFGL 333 cloud is located in the W 3 giant molecular cloud (W 3 GMC). It is known that the W 3 GMC involves three star forming clouds; W 3 Main, W 3(OH), and AFGL 333 (e.g. Thronson *et al.* 1985). The three clouds exhibit different star-forming activities in spite of their similarity in size and mass. The ratio of the infrared luminosities in W 3 Main, W 3(OH) and AFGL 333 is 1.0:0.25:0.07 (Thronson *et al.* 1980), suggesting that AFGL 333 is less active than W 3 Main and W 3(OH).

We have mapped the C^0 $^3P_1$–$^3P_0$ ([CI]) and CO J=3–2 lines toward the W 3 GMC by using the Mount Fuji Submillimeter Telescope (Sakai *et al.* 2006). The [CI] emission is found to be strong in the AFGL 333 cloud, where the ^{12}CO J=3–2 emission is relatively weak. To investigate the origin of the strong [CI] emission in the AFGL 333 cloud, we have observed the AFGL 333 and W 3(OH) clouds in the CO isotopomer lines and the CCS and N_2H^+ lines with the Nobeyama Radio Observatory 45 m telescope. We have found that $N(C^0)$ linearly increases with $N(CO)$ up to A_V of 50 mag. This indicates that C^0 exists in the deep inside of the molecular clouds. The $[C^0]/[CO]$ and $[CCS]/[N_2H^+]$ ratios tend to be higher in the AFGL 333 cloud than in the W 3(OH) cloud. These results may indicate the chemical youth of the AFGL 333 cloud relative to the W 3(OH) cloud.

In the AFGL 333 cloud, we have found two massive clumps (Clump A and B) without any sign of active star formation. They are highly gravitationally bound (M_{VIR}/M_{LTE} ∼ 0.4), and the LTE mass is 2.3×10^3 $M_\odot$ and 1.4×10^3 $M_\odot$ for Clump A and Clump B, respectively. These masses are comparable to those for on-going massive star-forming clumps. We have mapped the CCS, HC_3N and N_2H^+ emissions toward the clumps, and have found that the CCS and HC_3N emissions are stronger toward Clump B than toward Clump A. There are several YSO candidates (2MASS sources with H-K>2) in Clump A, while no YSO candidate is associated with Clump B. These results suggest that Clump B is younger than Clump A and is in a very early stage of cluster formation. Therefore Clump B is a very good target to understand the initial condition of cluster formation.

References

Sakai, T., Oka, T. & Yamamoto, S. 2006, *ApJ* 649, 268
Thronson, H. A., Jr., Campbell, M. F. & Hoffmann, W. F. I. 1980, *ApJ* 239, 533
Thronson, H. A., Jr., Lada, C. J. & Hewagama, T. 1985, *ApJ* 297, 662

† Present address: Nobeyama, Minamimaki, Minamisaku, Nagano 384-1305, Japan

Triggered Star Formation in a Turbulent ISM
Proceedings IAU Symposium No. 237, 2006
B. G. Elmegreen & J. Palouš, eds.

Infrared properties of ultracompact H II regions in the Galaxy and the LMC

Marta Sewiło[1], Ed Churchwell[2], Barbara Whitney[1] and the GLIMPSE and SAGE Teams

[1]Space Science Institute, 4750 Walnut Street, Suite 205, Boulder, CO 80301, USA
email: sewilo@astro.wisc.edu

[2]University of Wisconsin - Madison, 475 N. Charter Street, Madison, WI 53706, USA

Abstract. We report the preliminary results of the study on the infrared properties of ultracompact (UC) H II regions in our Galaxy and in the Large Magellanic Cloud (LMC) based on the GLIMPSE (*the Galactic Legacy Infrared Mid-Plane Survey Extraordinaire*) and SAGE (*Surveying the Agents of a Galaxy's Evolution*) data, respectively. We found that ∼60% of the Galactic UC H II regions do not have IR counterparts. Large extinction and very strong stellar winds evacuating H II regions from dust may explain this result. The same effect is observed in the LMC. One of the goals of this research is to develop a means of identifying UC H IIs based on their mid-IR properties, e.g. positions on color-color and color-magnitude plots and/or shape of spectral energy distributions. GLIMPSE showed that bow shocks, protostellar jets/outflows, and bubbles are common phenomena in massive star formation regions (MSFRs).

Keywords. stars: formation, ISM: HII regions, infrared: stars

---

High-resolution (0.″6 pixel) GLIMPSE/IRAC image mosaics (l × b = 1.°1 × 0.°8) were used to search for the mid-IR counterparts (ionizing sources) of Galactic UC H II regions. The present sample consists of the brightest 22 UC H IIs that are not saturated in any IRAC band (3.6, 4.5, 5.8, and 8.0 μm). We found that ∼60% of the radio sources do not have an IR counterpart; however, there are always IR sources in the neighborhood. Deeply embedded UC H IIs can be hidden in the IR due to local extinction and extinction along the line of sight. It is also possible that dust is evacuated from some UC H IIs by strong stellar winds, therefore there would be no IR emission observed toward these sources. A bigger sample of UC H II regions has to be examined to draw reliable conclusions. Interestingly, our preliminary analysis of the SAGE/MIPS 24 μm data for 59 UC/Compact H II regions in the LMC showed that a similar percentage of radio sources do not have IR counterparts.

We find bow shocks in the immediate vicinity of UC H II regions. The wind-wind interactions producing bow shocks are expected to be common in regions where multiple massive stars with strong winds are forming. Bipolar jets associated with protostars stand out when the 4.5 μm band is included in MIR colored images of MSFRs. This is believed to be due to shocks that excite either the H_2 line or the CO bandheads that fall within the bandpass of the 4.5 μm band. The GLIMPSE survey has also shown that the Galactic disk has a large number of dust bubbles surrounding H II regions and stellar clusters (Churchwell *et al.* 2006, ApJ, in press). The 24 μm emission is confined to the central regions of the bubbles where 8 μm emission is weak or absent. The 8 μm emission is strong in the shells and beyond, in the PDRs of the bubbles. About 7-8% of all the bubbles found in the GLIMPSE survey show a large bubble with one or more smaller bubbles projected either on the periphery of the large bubble or within its shell. This morphology suggests triggered star formation.

Triggered Star Formation in a Turbulent ISM
Proceedings IAU Symposium No. 237, 2006
B. G. Elmegreen & J. Palouš, eds.

© 2007 International Astronomical Union
doi:10.1017/S174392130700261X

Globular clusters in NGC147, 185, and 205

M. E. Sharina[1], V. L. Afanasiev[1] and T. H. Puzia[2]

[1]Special Astrophysical Observatory RAS, N.Arkhyz, KChR, 369167, Russia
email: sme@sao.ru, vafan@sao.ru

[2]Space Telescope Science Institute, 3700 Sun Martin Drive, Baltimore, MD21218, USA
email: tpuzia@stsci.edu

Abstract. Studying the chemical compositions and color-magnitude diagrams of globular clusters in the nearby low-mass galaxies is critical to compare properties of these long-living objects situated in galaxies of different type and mass, and to establish the role of dwarf galaxies as building blocks of massive early-type and spiral galaxies. We present measurements of ages, metallicities and $[\alpha/Fe]$ ratios for 16 globular clusters (GC) in NGC147, NGC185, and NGC205 and for the central regions of the diffuse galaxy light in NGC185, and NGC205, based on measurements of absorption line indices as defined by the Lick standard system in spectra obtained with the SCORPIO multi-slit spectrograph at the 6-m telescope of the Russian Academy of Sciences. We include in our analysis high-quality HST/WFPC2 photometry of individual stars in the GGs to investigate the influence of their horizontal branch (HB) morphology on the spectroscopic analysis. The HB morphologies for our sample GCs follow the same behavior with metallicity as younger halo Galactic globular clusters. We show that it is unlikely that they bias our spectroscopic age estimates based on Balmer absorption-line indices. Almost all our sample GCs appear to be old (T > 8 Gyr) and metal-poor ([Z/H]< −1.1). We find that most of the GCs in the studied galaxies are weakly or not α-enhanced, in contrast to the population of GCs in nearby early-type galaxies, and to the halo population of GCs in M31 and Milky Way.

Keywords. galaxies: star clusters – galaxies: individual (NGC147, NGC185, NGC205)

Distributions of the GCs according to the obtained age, metallicity and $[\alpha/Fe]$ are shown in Fig.1. Figure 4 in Sharina *et al.* (2006) shows the Lick index measurements for the metal-sensitive absorption index [MgFe]' versus age-sensitive Balmer line indices H_β, $H_{\gamma A}$, and $H_{\gamma F}$, for GCs and diffuse-light fields in NGC205, 185, and 147. Almost all our sample GCs appear to be old and and metal-poor, except for the GCs Hubble V in NGC 205 ($T = 1.2 \pm 0.6$ Gyr, [Z/H]= $−0.6 \pm 0.2$), Hubble VI in NGC 205 ($T = 4 \pm 2$ Gyr, [Z/H]= $−0.8 \pm 0.2$), and FJJVII in NGC 185 ($T = 7 \pm 3$ Gyr, [Z/H]= $−0.8 \pm 0.2$).

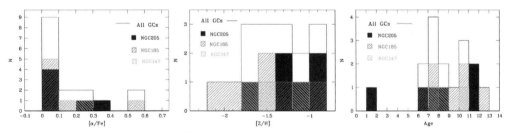

Figure 1. Age, $[Z/H]$, and $[\alpha/Fe]$ distributions of globular clusters in NGC 147, 185, and 205.

References

Sharina M.E., Afanasiev V.L. & Puzia T.H. 2006, *MNRAS* 372, 1259

Triggered Star Formation in a Turbulent ISM
Proceedings IAU Symposium No. 237, 2006
B. G. Elmegreen & J. Palouš, eds.

Formation and destruction of clouds and spurs in spiral galaxies

Rahul Shetty and E. C. Ostriker

[1]Department of Astronomy, University of Maryland, College Park, MD 20740, USA
email: shetty@astro.umd.edu, ostriker@astro.umd.edu

Abstract. We investigate the formation of clouds and substructure in spiral galaxies using high resolution global MHD simulations, including gas self gravity. Previously, local modeling by Kim & Ostriker (2002) has shown that self gravity and magnetic fields cause the growth of high density clumps in the spiral arms rather rapidly; subsequently, these clumps result in the formation of sheared, feather like structures in the interarms, known as spurs. Recently, we performed global simulations and found that gas self-gravity can cause the growth of sheared features regardless of the strength of the external spiral potential. However, a sufficiently strong spiral potential is required to produce arm clouds as well as spurs, which are the filamentary structures distinctly associated with the spiral arms, having near-perpendicular intersections with the main dust lane. We are currently performing higher resolution simulations to study the detailed properties of the clouds and spurs; we are also including a feedback mechanism, representing turbulent forcing via supernovae, to destroy the clouds. We will thus assess the role of turbulence on the clump formation rate and properties. Further, we will also follow how subsequent arm and spur morphology develops under quasi-steady conditions.

Triggered Star Formation in a Turbulent ISM
Proceedings IAU Symposium No. 237, 2006
B. G. Elmegreen & J. Palouš, eds.

© 2007 International Astronomical Union
doi:10.1017/S1743921307002633

Interaction between molecular outflows and dense gas in the cluster-forming region OMC-2/FIR4

Yoshito Shimajiri[1,3], S. Takahashi[2], S. Takakuwa[3], M. Saito[3] and R. Kawabe[3]

[1]Department of Astronomy, School of Science, The University of Tokyo, Bunkyo, Tokyo 113-0033,Japan
email:yoshito.shimajiri@nao.ac.jp

[2]Department of Astronomical Science,Graduate Univercity for Advanced Studies, Osawa 2-21-1 , Mitaka, Tokyo 181-8588 Japan

[3]ALMA-J project office, National Astronomical Observatory of Japan, Osawa 2-21-1, Mitaka, Tokyo 181-8588 Japan

Abstract. Since most stars are born as members of clusters(Lada & Lada 2003), it is important to clarified the detailed mechanism of cluster formation for comprehensive understanding of star formation. However, our current understanding of cluster formation is limited due to the followings;

(a)Cluster forming regions are located at the far distance.

(b) There are complex mixtures of outflows and dense gas in cluster forming regions.

So, we focused on the Orion Molecular Cloud 2 region (OMC-2), a famous cluster-forming region (Lada & Lada 2003) and the most nearest GMC. We observed the FIR 4 region with the Nobeyama Millimeter Array(NMA), Atacama Submillimeter Telescope Experiment (ASTE). In this region, there are 3 protostars (FIR3, FIR4, FIR5) which were identified as 1.3 mm dust continuum sources (Chini *et al.* 1997) and driving sources of mixed outflows, and FIR 4 is the most strongest source of 1.3 mm dust continuum in OMC-2. Molecular lines we adopted are a high density ($10^5 cm^{-3}$) gas tracer of $H^{13}CO^+$ (J=1-0), a molecular outflow tracer of ^{12}CO(J=1-0) and ^{12}CO(J=3-2), and SiO(J=2-1 v=0) as a tracer of shocks associated with an interaction between outflows and dense gas.

From results of the ^{12}CO(J=1-0) outflow, $H^{13}CO^+$ dense gas, and the SiO shock, the outflow from FIR 3 interacts with dense gas in the FIR 4 region. Moreover the Position-Velocity diagram along the major axis of the ^{12}CO(J=3-2) outflow shows that the ^{12}CO(J=1-0) and SiO emission exhibits a L shape (the line widths increase in the interacting region in morphology). This is an evidence of interaction between the outflows and dense gas (Takakuwa *et al.* 2003). From result of the 3 mm dust continuum, the interacted region by the molecular outflow of FIR 3 is an assemble of seven dense cores. The mass of each core is 0.1-0.8 $M_\odot$. This clumpy structure is evident only at FIR 4 in the entire OMC-2/3 region. There are possible that two cores are in the proto-stellar phase, because 3 mm dust continuum source correspond to NIR source or 3.6 cm f-f jet source. From these results, cores in the FIR 4 region may be potential source of the next-generation stars. In the other words, there is a possibility that the molecular outflow ejected from FIR 3 is triggering the cluster formation in the FIR 4 region.

Keywords. OMC-2, molecular outflows, interaction

Triggered Star Formation in a Turbulent ISM
Proceedings IAU Symposium No. 237, 2006
B. G. Elmegreen & J. Palouš, eds.

UCHII regions in the Antennae

L. Snijders[1], L. J. Kewley[2], P. P. van der Werf[1] and B. R. Brandl[1]

[1]Leiden Observatory, P.O. Box 9513, NL-2300 RA Leiden, The Netherlands
email: snijders@strw.leidenuniv.nl

[2]Institute for Astronomy, 680 Woodlawn Drive, Honolulu, Hawaii, USA

Abstract. We explore the physical characteristics of young stellar clusters in the Antennae by combining recent ground- and space-based mid-infrared observations with a newly developed set of diagnostic diagrams. Spitzer data give an overview of the star-forming regions extending over hundreds of parsecs, showing a dominant diffuse ISM component with a density of 10^2 cm^{-3} plus a small fraction of very compact material (10^6 cm^{-3}). With its higher spatial resolution VISIR gives a close-up view of the latter component. Its emission line ratios suggest that these regions are fundamentally different from local star-forming regions. Instead of having small isolated UCHII regions, as in local star-forming regions, the average density of the medium of the whole region falls in the (ultra)compact regime, exceeding 10^4 cm^{-3} over tens of parsecs.

1. Data and models

Two star-forming regions in the overlap region of the Antennae galaxies (NGC 4038/39) were observed in the mid-infrared both from the ground, with VISIR at the ESO VLT (Snijders *et al.* 2006), as from space, with Spitzer IRS.

To model the observed mid-infrared spectra we have used the latest version of Starburst99 (v5.1) combined with the photoionization code MappingsIII. We computed spectral energy distributions of stellar clusters with a Salpeter IMF between 0.1 and 100 M$_\odot$ for various metallicities, embedded in a dusty HII region. A model grid was built exploring the effect of varying densities of the ISM and intensity of the ionizing radiation (characterised by the ionization parameter U). The emission line fluxes in the output spectra are used to construct diagnostic diagrams.

2. Spitzer versus VISIR

Combining various emission line ratios from the Spitzer spectra reveals that a uniform ISM cannot explain the data. A combination of a large proportion of low density (=90%; 10^2 cm^{-3}) and a small fraction of very compact material (10^6 cm^{-3}) is required to reproduce the observations. This can be interpreted as a giant diffuse ionized region with a number of UCHII regions below Spitzers resolution limit (several hundreds of parsecs).

The observed line ratios in the VISIR spectra require a two component ISM as well, though with considerably higher densities. At least half of the matter at intermediate densities (10^4 cm^{-3}) and the remainder moderately to very dense ($10^4 - 10^6$ cm^{-3}). With its higher spatial resolution (40 – 50 pc) VISIR probes the (ultra)compact cores of star formation, which are clearly fundamentally different from local star-forming regions.

References

Snijders, L., van der Werf, P. P., Brandl, B. R., Mengel, S., Schaerer, D., & Wang, Z., 2006, *ApJ* 648, 25

Triggered Star Formation in a Turbulent ISM
Proceedings IAU Symposium No. 237, 2006
B. G. Elmegreen & J. Palouš, eds.

© 2007 International Astronomical Union
doi:10.1017/S1743921307002657

Study of photon dominated regions in IC 348

K. Sun[1], C. Kramer[1], B. Mookerjea[2], V. Ossenkopf[1,3], M. Röllig[1,4] and J. Stutzki[1]

[1]I. Physikalisches Institut, Universität zu Köln, Germany [2]Department of Astronomy, University of Maryland, USA [3]SRON National Institut for Space Research, the Netherlands [4]Argelander-Institut für Astronomie, Universität Bonn, Germany

Abstract. We present fine structure line of neutral carbon at 492 GHz ($^3P_1$–$^3P_0$, hereafter [C I] 1–0) and ^{12}CO 4–3 KOSMA observations. This data has been combined with FCRAO ^{12}CO 1–0 and ^{13}CO 1–0 data. We have used these observations to understand the emission from the photon dominated regions (PRDs) in IC 348. We confirm the anti-correlation between $N(C)/N(CO)$ and $N(H_2)$ as seen in most Galactic PDRs (Mookerjea *et al.* 2006).

1. Results

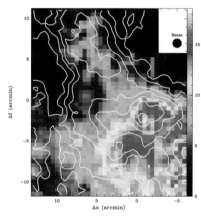

Figure 1. Integrated intensities of [C I] 1–0 emission (*color*) overlayed with contours of ^{12}CO 4–3 (*contours*) in IC 348. The center of the map is at $\alpha = 03^h 44^m 10^s$, $\delta = 32°06'$ (J2000). The intensities are integrated from V_{LSR} 2 km s^{-1} to 14 km s^{-1}. The contour levels are from 10 K km s^{-1} to 58 K km s^{-1} by a step of 8 K km s^{-1}. Both data are shown at a common resolution of 70″. The black star denotes the position of HD 281159, a B 5 type star.

IC 348 is one of the most studied young open clusters, which lies at a distance of 320 pc. The [C I] 1–0 emission peaks to the south-west of the mapped region (Fig. 1). The emission extends towards the east and the north-east. The ^{12}CO 4–3 intensity also peaks almost at the same position of the [C I] peak, but shows a second peak lying to the north of HD 281159 and elongated along the north-south direction. The features in the ^{12}CO 4–3 map agree quite well with those in ^{12}CO 3–2 map (Sun *et al.* 2006). We attribute the difference in the intensity distributions of [C I] and ^{12}CO 4–3 to the fact that ^{12}CO 4–3 traces regions of higher temperature, while [C I] traces the embedded PDR surfaces of the molecular clouds with high column density. A plot of the $N(C)/N(CO)$ ratio vs. the H_2 column density in IC 348 shows a clear anti-correlation. A linear fit to the data in logarithmic coordinates gives a slope of -0.59 (cf. Fig. 6, Mookerjea *et al.* 2006).

References

Mookerjea, B. *et al.* 2006, *A&A* 456, 235

Sun, K. *et al.* 2006, *A&A* 451, 539

Triggered Star Formation in a Turbulent ISM
Proceedings IAU Symposium No. 237, 2006
B. G. Elmegreen & J. Palouš, eds.

© 2007 International Astronomical Union
doi:10.1017/S1743921307002669

Dense core evolutions induced by shock triggering and turbulent dissipation

Kengo Tachihara[1], A. Hayashi[2], T. Onishi[2], A. Mizuno[3] and Y. Fukui[2]

[1]Graduate School of Science and Technology, Kobe University, 1-1 Rokko-dai, Nada-ku, Kobe, 657-8501, Japan
email: tatihara@kobe-u.ac.jp

[2]Department of Astrophysics, Nagoya University, Furo-cho, Chikusa-ku, Nagoya, 464-8602, Japan

[3]Solar-Terrestrial Environment Laboratory, Nagoya University, Furo-cho, Chikusa-ku, Nagoya, 464-8602, Japan

Keywords. stars: formation, ISM: clouds, ISM: evolution, ISM: molecules, ISM: structure, radio lines: ISM

1. Introduction

External shock triggering and internal turbulence play major role for the condensation of the ISM and star formation. Some evidences of shock triggering by non-isotropic compression are seen in the cloud morphologies and associated active cluster formation such as the ρ Oph and Cha I clouds. Surveys for $C^{18}O$ dense cores have shown that internal turbulence dominates the core dynamics and regulates star formation activity (Tachihara *et al.* 2002).

2. Observations and results

Based on the precedent $C^{18}O$ core surveys by the NANTEN radio telescope (Tachihara *et al.* 2002 and the references therein), the Taurus, Ophiuchus North, Lupus, and Chamaeleon clouds were surveyed for denser and more compact cores in $H^{13}CO^+$ ($J =1$–0) by the 45m telescope at the Nobeyama Radio Observatory and the SEST 15m telescope at La Silla. The results obtained in Taurus were published by Onishi *et al.* (2002). For a comparison, $H^{13}CO^+$ survey with the 45m telescope in the ρ Oph cloud by Umemoto *et al.* (2002) are compiled.

In general, one $C^{18}O$ core (typical density is $\sim 10^4$ cm^{-3}) fragments into a few $H^{13}CO^+$ cores ($\sim 10^5$ cm^{-3}) in isolated star-forming regions (SFRs), in contrast to the typical triggered cluster-forming region of the ρ Oph cloud, which consists of 57 $H^{13}CO^+$ cores. The statistics show a remarkable trend that more evolved $C^{18}O$ cores associated with $H^{13}CO^+$ cores and young stars have larger masses and smaller line widths than those without $H^{13}CO^+$ cores. This suggests that the turbulent decay is required for the dynamical relaxation of the $C^{18}O$ cores to gain more mass and then contract to form denser $H^{13}CO^+$ cores spontaneously. On the other hand, no clear trend is seen in the physical properties between star-forming and prestellar $H^{13}CO^+$ cores. Among the nearby SFRs, the $H^{13}CO^+$ cores in Taurus are larger in number and mass than in other SFRs, while the ρ Oph and Cha I clouds have larger mass fractions of the total $C^{18}O$ cores to the parental ^{13}CO clouds. We suggest that the external shock compresses the low-density part of the clouds and the internal turbulent decay leads the dense core condensations.

References

Onishi, T., Mizuno, A., Kawamura, A., Tachihara, K., & Fukui, Y. 2002, *ApJ* 575, 950
Tachihara, K., Onishi, T., Mizuno, A., & Fukui, Y. 2002, *A&A* 385, 909
Umemoto, T., Kamazaki, T., *et al.* 2002, in: S. Ikeuchi, J. Hearnshaw & T. Hanawa (eds), *8th IAU Asia-Pacific regional meeting* (Ast.Soc.Japan), Vol. II, p.229

Triggered Star Formation in a Turbulent ISM
Proceedings IAU Symposium No. 237, 2006
B. G. Elmegreen & J. Palouš, eds.

Survey observations of large-scale molecular outflows associated with intermediate-mass protostar candidates in the OMC-2/3 region

Satoko Takahashi[1], Y. Shimajiri[1], S. Takakuwa[3], M. Saito[3] and R. Kawabe[3]

[1]Department of Astronomical Science, Graduate University for Advanced Studies, National Astronomical Observatory of Japan, Osawa 2-21-1, Mitaka, Tokyo 181-8588 Japan
email:satoko.takahashi@nao.ac.jp

[2]Department of Astronomy, School of Science, University of Tokyo, Bunkyo, Tokyo 113-0033, Japan.

[3]ALMA Project Office, National Astronomical Observatory of Japan, Osawa 2-21-1, Mitaka, Tokyo 181-8588 Japan

Abstract. We have newly performed millimeter- and submillimeter-wave observations in the nearest GMC: the Orion Molecular Cloud -2/3 region (OMC-2/3). Here, we report results of our large-scale ($22' \times 14'$) outflow survey with the Atacama Submillimeter Telescope Experiment (ASTE) in the CO(3-2) emission. The OMC-2/3 region is one of the famous intermediate-mass star-forming regions and harbors several sources diagnosed as Class0 protostars (Chini *et al.* 1997). With the intensive ASTE observations, we totally identified the 8 clear, 5 probable and 6 marginal outflows in OMC-2/3. 8 clear outflows from them, MMS 2, MMS 5, MMS7, MMS9, FIR-2, FIR 3, VLA 13, and FIR 6b are associated with mm and *SPITZER* 24 μm sources. The others are more or less complicated, and two of which, VLA 13 and FIR 6, are newly identified. We found the interaction between the molecular outflows and the dust condensations at least in four regions. In addition, we confirmed the increment of the velocity width of the dense gas toward some of these condensations (i.e. at the termination of the outflow lobes). These results suggest that (i) the interaction between the outflows and the dense condensation occurs commonly in the OMC-2/3 region, (ii) the dense condensations in this region are compressed ubiquitously by these outflows and are receiving a part of the momentum from them. Particularly, one of the strongest millimeter sources, and hence protostar candidates, FIR4, is strongly compressed by a molecular outflow driven by FIR3 located at the north-east of FIR 4. These results suggest that the molecular outflows play an important role in the formation and evolution of stars and that the outflows are a driving mechanism of turbulence in the OMC-2/3 region.

Keywords. molecular outflow, OMC-2/3.

Triggered Star Formation in a Turbulent ISM
Proceedings IAU Symposium No. 237, 2006
B. G. Elmegreen & J. Palouš, eds.

© 2007 International Astronomical Union
doi:10.1017/S1743921307002682

Starbursts in isolated galaxies: burst modes in coupled star-gas systems

Christian Theis[1] and Joachim Köppen[2]

[1]Institute of Astronomy, Univ. of Vienna, Austria, email: theis@astro.univie.ac.at
[2]Observatoire Astronomique de Strasbourg, France, email: koppen@newb6.u-strasbg.fr

We studied the stability properties of isolated star forming dwarf galaxies with the aim to identify star burst modes. The impact of the stellar birth function (parametrization, IMF), the stellar feedback and the ISM model on the galactic star formation history was investigated. We focussed especially on dynamically driven star bursts induced by stellar feedback. We applied a one-zone model for a star-gas system coupled by both mass and energy transfer. Additionally, we extended the classical closed box network for **active dynamical evolution** (Theis 2004). This allows for a simple, but consistent description of the coupling between the dynamical state of a galaxy and its internal properties like star formation activity or the thermal state of the interstellar medium.

Our calculations revealed three types of repetitive star burst modes in isolated galaxies:

• mode A: initial transitory oscillations following the dynamical state: if the energy dissipation in the gas is fast, the gas temperature remains close to the equilibrium temperature for the actual density. In that case the star formation rate follows mainly the decaying virial oscillations.

• mode B: recurrent star bursts after long quiescent periods: if the energy dissipation in the gas is slow (occurs when the dissipational timescale is of the order of or longer than the dynamical time, e.g. in case of dissipation by clump-clump collisions), active periods are separated by long quiescent periods. The quiescent periods are a combination of the dynamical and the dissipative timescales. Such burst modes occur only in a small mass range. They become more pronounced in strongly dark matter dominated systems.

• mode C: cooling function induced instability: a negative temperature gradient of the cooling function leads to an oscillatory behaviour. Proper conditions are only met for low gas densities and low metallicities.

Moreover, we found that a negative thermal feedback in the star formation description leads to a destabilization, i.e. oscillatory behaviour of the system. This is in contrast with the standard closed-box models neglecting the dynamical description.

Finally, we investigated the impact of a time-dependent IMF. As an example we adopted a Weidner-Kroupa-type IMF which couples the overall star formation rate to the effective upper mass limit of the IMF, by this influencing the stellar feedback. We found that such an IMF variation has only a marginal influence on the overall star formation activity due to very efficient self-regulation in the system.

Acknowledgements

CT is grateful for financial support by the IAU.

References

Theis, Ch. 2005, in: S. Hüttemeister *et al.* (eds.), *The Evolution of Starbursts* (AIP Conf. Proc.), Vol. 783, p. 57

Triggered Star Formation in a Turbulent ISM
Proceedings IAU Symposium No. 237, 2006
B. G. Elmegreen & J. Palouš, eds.

Modeling the dust and gas temperatures near young stars

Andrea Urban[1] and Neal J. Evans II[1]

[1]Department of Astronomy, University of Texas at Austin, Austin, TX 78712, USA
email: aurban@astro.as.utexas.edu.

Abstract. As young stars form, they interact with their environment in many ways. We study the radiative interaction of a young star with its surrounding cluster environment. The change in gas temperature caused by a forming star can trigger the formation or inhibit the growth of nearby star forming cores. We calculate the gas temperature around a single star by balancing the dust-gas collisional heating, molecular cooling, and cosmic ray heating rates for a grid of models with various luminosities and density distributions. In the future, this work can be used in large-scale simulations of clustered star formation to study the effect of using a gas temperature which depends not only on density, but also on radiative environment.

Keywords. stars: formation, methods: numerical

1. Dust and Gas Temperature

In order to calculate the dust temperature, we have created a grid of models using the spherical radiative transfer code, DUSTY (Ivezic *et al.* 1999). We assume that the central star is a black body with T=10,000K and a luminosity between 10^{-2} L$_\odot$ and 10^5 L$_\odot$. The dust properties are given by OH5 dust (Ossenkopf & Henning 1994). The spherical density profile is exponentially decreasing. We use an energy rate balance code to calculate the gas temperature (Doty & Neufeld 1997). We include gas-dust collisional temperature coupling, cosmic-ray heating, and CO cooling (Young *et al.* 2004). Figure 1 shows an example model.

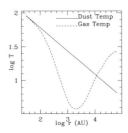

Figure 1. Model Parameters: L=10L$_\odot$, n=$10^{4.5}$cm^{-3} at 1000AU. The dust and gas are well–coupled via collisional heat transfer in the center. As the density decreases, the gas is able to cool efficiently through molecular lines. Cosmic-ray heating becomes important as the density decreases even more.

References

Doty, S. D. & Neufeld, D. A. 1997, *ApJ* 489, 122
Ivezic, Z., Nenkova, M. & Elitzur, M. 1999, astro-ph/9910475
Ossenkopf, V. & Henning, T. 1994, *A&A* 291, 943
Young, K. E., Lee, J.-E., Evans, N. J., II, Goldsmith, P. F., & Doty, S. D. 2004, *ApJ* 614, 252

Triggered Star Formation in a Turbulent ISM
Proceedings IAU Symposium No. 237, 2006
B. G. Elmegreen & J. Palouš, eds.

© 2007 International Astronomical Union
doi:10.1017/S1743921307002700

The RMS survey: radio observations of candidate massive YSOs in the southern hemisphere

J. S. Urquhart[1], A. L. Busfield[1], M. G. Hoare[1], S. L. Lumsden[1], A. J. Clarke[1], T. J. T. Moore[2], J. C. Mottram[1] and R. D. Oudmaijer[1]

[1]School of Physics and Astrophysics, University of Leeds, Leeds, LS2 9JT, UK
email: jsu@ast.leeds.ac.uk

[2]Astrophysics Research Institute, Liverpool John Moores University, Twelve Quays House,
Egerton Wharf, Birkenhead, CH41 1LD, UK

Abstract. The Red MSX Source (RMS) survey (Hoare *et al.* 2005) is a multi-wavelength programme of follow-up observations designed to distinguish between genuine massive young stellar objects (MYSOs) and other embedded or dusty objects, such as ultra compact (UC) HII regions, evolved stars and planetary nebulae (PNe). We have identified nearly 2000 MYSOs candidates by comparing the colours of MSX and 2MASS point sources to those of known MYSOs. There are several other types of embedded or dust enshrouded objects that have similar colours as MYSOs and contaminate our sample. Two sources of contamination are from UCHII regions and PNe, both of which can be identified from the radio emission emitted by their ionised nebulae. In order to identify UCHII regions and PNe that contaminate our sample we have conducted high resolution radio continuum observations at 3.6 and 6 cm of all southern MYSOs candidates ($235° < l < 350°$) using the Australia Telescope Compact Array (ATCA).

Keywords. Radio continuum: stars – Stars: formation – Stars: early-type – Stars: pre-main sequence.

1. Summary and Conclusions

Observations were made at 3.6 and 6 cm towards 826 RMS sources located within 802 fields using the ATCA (see Urquhart *et al.* 2006 for details). These observations were aimed at identifying radio loud contaminants such as UC HII regions and PNe from the relatively radio quiet MYSOs. Of the 826 RMS sources observed we have found 199 to be associated with radio emission ($\sim25\%$). More interestingly we failed to detect any radio emission towards 627 RMS sources, and therefore after eliminating one of the main sources of contamination, we are still left with a large sample of MYSOs candidates. The majority of the 199 RMS sources found to be associated with radio emission are expected to be UCHII regions. The morphologies of these sources, which are consistent with other studies of UCHII regions, relatively flat spectral indices, and their scaleheight would certainly support their identification as UCHII regions. Once combined with the results of our VLA observations we will have an unbiased sample of ~400–500 UCHII, the vast majority of which were previously known.

References

Urquhart, J. S., *et al.* 2006, astro-ph/0605738
Hoare, M. G., *et al.* 2005, in: R. Cesaroni, M. Felli, E. Churchwell, M. Walmsley, *Massive star birth: A crossroads of Astrophysics* (Cambridge: Cambridge Univ.), p. 370

Triggered Star Formation in a Turbulent ISM
Proceedings IAU Symposium No. 237, 2006
B. G. Elmegreen & J. Palouš, eds.

Triggered star formation within the bright-rimmed cloud SFO 75

J. S. Urquhart[1], M. A. Thompson[2], L. K. Morgan[3] and Glenn J. White[4,5]

[1]School of Physics and Astrophysics, University of Leeds, Leeds, LS2 9JT, UK
email: jsu@ast.leeds.ac.uk

[2]Centre for Astrophysics Research, Science and Technology Research Institute, University of Hertfordshire, College Lane, Hatfield, AL10 9AB, UK

[3]Green Bank Telescope, P.O. Box 2, Green Bank, WV 24944, USA

[4]Dept. of Physics & Astronomy, The Open University, Walton Hall, Milton Keynes, MK7 6AA, UK

[5]Space Physics Division, Space Science & Technology Division, CCLRC Rutherford Appleton Laboratory, Chilton, Didcot, Oxfordshire, OX11 0QX, UK

Abstract. We present the results of a multi-wavelength line and continuum study of the bright-rimmed cloud SFO 75 in an attempt to determine whether the ionisation front and its associated shocks, driven by the nearby O star, have triggered the formation of a new generation of stars within this bright-rimmed cloud.

Keywords. Radio continuum: stars – Stars: formation – Stars: pre-main sequence.

1. Summary and Conclusions

We present the results of CO, 1.3 cm radio continuum, mm-continuum and ammonia observations of the bright-rimmed cloud SFO 75 (Sugitani & Ogura, 1994; Thompson *et al.* 2004; Urquhart *et al.* 2006). These observations reveal the presence of an over-pressured layer of ionised gas at the surface of the cloud which is likely to be driving shocks into the cloud. This is supported by evidence of shocked gas in the surface layer of the cloud as seen in the ^{13}CO position-velocity diagram. The CO and mm-continuum data show the presence of a dense core, located slightly behind the cloud's rim, with a mass of $\sim$200–400 M$_\odot$ and a density of 3–8×10^4 cm^{-3}. From a two component fit to the IRAS and mm-continuum fluxes we obtain an total bolometric luminosity of $\sim$2.6$\times10^4$ L$_\odot$, which would support the presence of an embedded B0 ZAMS star. The ammonia observations have resolved the mm-core into two separate cores; Core A is located directly behind the rim the morphology of which correlates extremely well with that of the rim; and Core B which is located farther back from the rim. Core A is coincident with a GLIMPSE point source, and therefore, star formation is likely to be taking place. From the data we have presented it is clear that the ionisation by the nearby OB star is having a dramatic effect on the evolution of this cloud. What is not yet clear is whether this has triggered a new generation of star formation within this cloud.

References

Sugitani, K. & Ogura, K. 1994, *ApJS* 92, 163
Thompson, M. A., Urquhart, J. S. & White, G. J. 2004, *A&A* 415, 627
Urquhart, J. S., Thompson, M. A., Morgan, L. K., & White, G. J. 2006, *A&A* 450, 625

Triggered Star Formation in a Turbulent ISM
Proceedings IAU Symposium No. 237, 2006
B.G. Elmegreen & J. Palouš, eds.

A turbulence study in Dwarf Irregular galaxies

Margarita Valdez-Gutiérrez[1] and Ivânio Puerari[2]

[1]Instituto de Astronomía – Universidad Nacional Autónoma de México, campus Ensenada,
Ensenada, B. C., Mexico
email: mago@astrosen.unam.mx

[2]INAOE, Tonantzintla, Puebla, Mexico
email: puerari@inaoep.mx

Abstract. We present preliminary results of a Fourier transform power spectra analysis carried out on a sample of dwarf irregular galaxies and Local Group members. This project is intended to study turbulence and the ISM structure in irregulars as a class.

Keywords. Turbulence, ISM: structure, galaxies: dwarf, galaxies: irregular

1. Results and Discussion

Our galaxy sample is composed by six Local Group members observed in the Hα line. Observations were carried out using the 2.1m telescope at the Observatorio Astronómico Nacional at San Pedro Mártir B. C. México. The data acquisition was performed by means of the Fabry-Perot interferometer PUMA (Rosado *et al.* 1995) in direct imaging mode and covering a field of view of 10′. The Fourier transform power spectra were calculated from cuts along the galaxies major axes and tested to reproduce a noise power spectra. Power spectra slope values for four of the galaxies of our sample range from 1.3 to 1.5. This is in agreement with Willett *et al.* (2005) results for irregular galaxies with high star formation rates. On the contrary, for WLM and IC1613, their slopes are almost zero corresponding to systems with low star formation rate per unit area. These preliminary results suggest some dependence on the star formation rate. Our findings test our reduction programs and methodology with previous results reported by Willett al. (2005). The analysis of individual emission regions in the galaxies of the sample is in progress (Valdez–Gutiérrez *et al.* 2007, in preparation).

References

Willett, K.W, Elmegreen, B.G. & Hunter D.A. 2005, *AJ* 129, 2186
Rosado, M. *et al.* 1995, *Rev. Mexicana Astron. Astrofis. Ser. Conf.* 3, 263

Triggered Star Formation in a Turbulent ISM
Proceedings IAU Symposium No. 237, 2006
B. G. Elmegreen & J. Palouš, eds.

© 2007 International Astronomical Union
doi:10.1017/S1743921307002736

The history of star formation in the Galactic young open cluster NGC 6231

Mario E. van den Ancker[1]

[1]European Southern Observatory, Karl-Schwarzschild-Strasse 2, D-85748 Garching, Germany
e-mail: mvandena@eso.org

Abstract. We study the star formation history of the galactic young open cluster NGC 6231 using new, deep, wide-field $BVRI$ imaging. Contrary to previous suggestions, we do not find a lack of low-mass cluster members; our derived mass function is compatible with a Salpeter IMF. The star formation history of NGC 6231 appears to be bi-modal, with a first wave of star formation activity 3–5 Myr ago, followed by a new generation of stars forming ~ 1 Myr ago.

Keywords. Star Formation, Pre-main sequence Stars, Open clusters and associations

The star formation history of the rich open cluster NGC 6231 has been hotly debated in the literature ever since the suggestion by Eggen (1976) that a violent process must have triggered star formation in the region. In the largest photometric study of the region to date, Sung *et al.* (1998) suggested an abrupt decrease in the number of low-mass stars in NGC 6231 - giving further credibility to a scenario in which star formation in this region may have been triggered. Interestingly, Reed & Cudworth (2003), recently found that the trajectory of the globular cluster NGC 6397 intersected that of the natal cloud of NGC 6231 around five million years ago - close to the estimated age of NGC 6231 – thus providing a plausible candidate for the triggering mechanism.

New $BVRI$ images of a region of $30'$ centered on the core of NGC 6231 were obtained with the Wide Field Imager (WFI) on the ESO/MPG 2.2m telescope at La Silla, Chile. Photometry in all three broadband filters was extracted for all stars down to $V = 21$ and combined with JHK photometry from 2MASS whenever available.

A fit of isochrones to the background-corrected colour-magnitude diagram of NGC 6231 confirms its young age (~ 3 Myr). We have detected 19 intermediate-mass stars with large amounts of near-infrared excess radiation. They are most likely young stars surrounded by disks (i.e. Herbig Ae/Be and classical T Tauri stars). We do not find evidence for a lack of low-mass cluster members, as suggested by Sung *et al.*; within our completeness limit, the cluster luminosity function appears to be compatible with a Salpeter IMF. Interestingly enough, we do find strong evidence for the existence of a significant age spread within our cluster members. We conclude that if star formation in NGC 6231 is indeed triggered – either by a recent supernova or by the passage of NGC 6397 through the galactic plane – this did not simply result in one single burst of star formation activity. Rather star formation activity in this region appears to have started gradually around 5 Myr ago, reached a local maximum around 3 Myr ago, and experienced another burst of star forming activity around 1 million years ago.

References

Eggen, O.J. 1976, *QJRAS* 17, 472
Reed R.F. & Cudworth K.M. 2003, *AAS* 203, 1006
Sung H., Bessell, M.S. & Lee, S.-W. 1998, *AJ* 115, 734

Proceedings Triggered Star Formation in a Turbulent ISM
Proceedings IAU Symposium No. 237, 2006
B. G. Elmegreen, & J. Palouš, eds.

© 2007 International Astronomical Union
doi:10.1017/S1743921307002748

The generation of dense cores and substructure within them by MHD waves

S. Van Loo, S. A. E. G. Falle and T. W. Hartquist

University of Leeds, Leeds LS9 JT, UK
email: svenvl@ast.leeds.ac.uk

Abstract. By using 2D simulations, we examine the generation of dense cores and substructures by magnetosonic waves. We find that the excitation of slow-mode waves by fast-mode waves produces these high-density structures.

Keywords. MHD, stars: formation, ISM: clouds

There have been a number of 3D simulations (e.g. Vázquez-Semadeni *et al.* 2005) on the effect of MHD waves on an isothermal plasma for which the ratio β of thermal to magnetic pressure is small. These simulations show that density inhomogeneities can be formed in this way with statistical properties that are consistent with the observations. However, these calculations generally contain so many ingredients that the fundamental mechanism responsible for the clumpiness of star forming regions is unclear.

The analysis of Falle & Hartquist (2002) shows that the production of large density contrasts in a low-β plasma results when slow-mode waves are excited by the non-linear steepening of a fast-mode wave. Their 1D calculations show that this mechanism can produce transient high-density clumps if gravitational instability does not occur. We have extended this model to 2D (Van Loo, Falle & Hartquist 2006a).

Like Falle & Hartquist, we find that the formation of dense structures is associated with slow-mode waves. These slow-mode waves are excited either by the non-linear steepening of the fast-mode wave or by the interaction between the fast-mode wave and dense structures. The latter produces dense cores which reside in larger clumps. The process of producing structure by means of slow-mode waves thus works on different length-scales, but ceases, however, to be effective when β approaches unity.

This directly leads to the question of whether substructure in cold dense cores is produced by the same mechanism. Dense cores, however, may have values of β only as low as 0.1 (Ward-Thompson 2002). By examining the response of a pre-existing dense core to a fast-mode wave (Van Loo, Falle & Hartquist 2006b), we find that waves, with wavelengths shorter than the core radius, excite slow-mode waves within the core and, although β is close to unity, these slow-mode waves still produce high-density contrasts.

Acknowledgements

SVL gratefully thanks PPARC for financial support.

References

Falle, S. A. E. G. & Hartquist, T. W. 2002, *MNRAS* 329, 195
Van Loo, S., Falle, S. A. E. G. & Hartquist, T. W. 2006a, *MNRAS* 370, 975
Van Loo, S., Falle, S. A. E. G. & Hartquist, T. W. 2006b, *MNRAS*, submitted
Vázquez-Semadeni E., Kim J., Shadmehri M., & Ballesteros-Paredes J. 2005, *ApJ* 618, 344
Ward-Thompson D. 2002, *Science* 295, 76

Triggered Star Formation in a Turbulent ISM
Proceedings IAU Symposium No. 237, 2006
B. G. Elmegreen & J. Palouš, eds.

© 2007 International Astronomical Union
doi:10.1017/S174392130700275X

Photographic variability survey in the M42 region

Luiz Paulo R. Vaz[1], Gustavo H. R. A. Lima[1] and Bo Reipurth[2]

[1]Departamento de Física - ICEx - UFMG - Brazil
emails: lpv@fisica.ufmg.br, styx@fisica.ufmg.br

[2]IfA - Univ. of Hawaii - USA
email: reipurth@ifa.hawaii.edu

Abstract. In order to detect variable stars in the well known star forming region, the Orion Nebula Cluster, a series of 22 exposures taken from November 1996 to October 1998, using the ESO 100/152cm Schmidt telescope, covering a field of $5° \times 5°$ was analyzed. The films (Kodak Tech-Pan 4415 emulsions, effective spectral range from $\sim$630 nm to 690 nm) were digitized by the SuperCOSMOS machine, the measurements calibrated to the R magnitude of the USNO B1.0 catalogue and differential photometry was performed throughout the whole field. In the process, a set of 260 stars that remained constant in the 22 films and were well distributed over the field was selected and used as comparison stars for the differential photometry of all the other stars in the field. Diverse statistical studies were performed in order to characterize the type and degree of variability of the objects. The 22 films, all exposed for 30 minutes each, were stacked together at our request by the SuperCOSMOS team, producing perhaps the deepest wide field image of M42 ever taken.

This database ($>$150 000 objects, mostly stars and $\sim$2% galaxies) is going to be used as a starting point for the Variable Young Stellar Object Survey (VYSOS) project, which consists of 2 fully automated robotic telescopes of 41cm each, one installed on Mauna Loa (Hawaii, USA) and the other at Cerro Armazones (Chile), both using the Pan STARRS set of photometric filters. We discovered thousands of new variable stars within the $5° \times 5°$ region studied. We have variability statistics for all objects and are classifying the variable stars according to the variability type and amplitude. We intend to make this database available via the WEB.

Keywords. stars: variables, stars: pre-main sequence, stars: formation, astronomical data bases, surveys, open clusters and associations: individual (M42)

Acknowledgements

We are very grateful to Dr. Harvey MacGillivray of the `SuperCOSMOS` center at the Royal Observatory Edinburgh. We also thank Dr. Sue Tritton as head of the Wide Field Astronomy Unit for the allocation of time on the `SuperCOSMOS` machine. We acknowledge use of the General Catalogue of Variable Stars maintained by the Sternberg Astronomical Observatory, and thank Dr. Nikolai Samus for providing early release of data on variables in Orion. LPV and GHRAL gratefully acknowledge partial support from the Brazilian Agencies CNPq, CAPES, FAPEMIG. This publication makes use of data products from 2MASS, a joint project of the Univ. of Massachusetts and IPAC/California Institute of Technology, funded by NASA and NSF. This research has made use of the USNOFS Image and Catalogue Archive operated by the United States Naval Observatory, Flagstaff Station, of the SIMBAD database, operated at CDS, Strasbourg, France, and of NASA's Astrophysical Data System bibliographic services.

Triggered Star Formation in a Turbulent ISM
Proceedings IAU Symposium No. 237, 2006
B. G. Elmegreen & J. Palouš, eds.

© 2007 International Astronomical Union
doi:10.1017/S1743921307002761

Quiescent high mass cores in Orion region

T. Velusamy[1], D. Li[1], P. F. Goldsmith[1] and W. D. Langer[1]

[1] Jet Propulsion Laboratory, California Institute of Technology, Pasadena, CA 91109, USA

Our goal is to study relatively quiescent dense gas cores, isolated from disruptive stars, to understand the initial conditions of massive star formation. Determining their mass, size, dynamical status, and core mass distribution is a starting point to understand the mechanisms for formation, collapse, and the origin of their IMF. We obtained CSO 350 μm images of quiescent regions in Orion and detected 51 resolved or nearly resolved molecular cores with masses ranging from 0.1 $M_\odot$ to 46 $M_\odot$ (Li *et al.* 2006). The mean mass is 9.8 $M_\odot$, which is one order of magnitude higher than that of the resolved cores in low mass star forming regions, such as Taurus. Our sample includes largely thermally unstable cores, which implies that the cores are supported neither by thermal pressure nor by turbulence, and are probably supercritical. They are likely precursors of protostars. Fig. 1 shows the cores in our sample have a power law core mass function with an index α = -0.85±0.21. This mass function does not resemble the stellar IMF or turbulence cascade structure function, and it is also in contrast with core surveys done in the Ophiuchus region (Motte *et al.* 1998) and the Serpens region (Testi *et al.* 1998). We find that the differential mass function approach, while requiring more cores due to the necessity of binning, is more robust and has better defined statistical uncertainties than the cumulative mass function. Use of the cumulative mass function can erroneously suggest multiple power law indices, particularly if the core mass distribution is characterized by a power law index close to ~-1. Our results for the quiescent cores in the Orion show that the core mass function is flatter in an environment affected by ongoing high mass star formation. Thus, environmental processes likely play a role in the evolution of dense cores and the formation of stars in such regions.

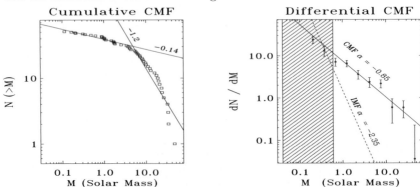

Figure 1. *(Left)* Cumulative Core Mass Function with two power laws. *(Right)* Differential Core Mass function with single power law fit. The lines represent stellar IMF (broken) and best fit to the cores (solid). The shaded area represents incomplete core sample.

This work was performed by JPL, Caltech, under contract with the NASA. Research at the Caltech CSO is supported by NSF grant AST-0229008. Di Li is a NPP Fellow.

References

Li, D., Velusamy, T., Goldsmith, P.F., & Langer, W.D. 2006, *ApJ* (in press)
Motte, F., Andre, P. & Neri, R. 1998, *A&A* 336, 150
Testi, L. & Sargent, A.I. 1998, *ApJ* 508, L91

Triggered Star Formation in a Turbulent ISM
Proceedings IAU Symposium No. 237, 2006
B. G. Elmegreen & J. Palouš, eds.

© 2007 International Astronomical Union
doi:10.1017/S1743921307002773

Integral field observations of distant cluster galaxies

D. Vergani[1,2], C. Balkowski[2], H. Flores[2], V. Cayatte[3], F. Hammer[2] S. Mei[4], and J.P. Blakeslee[5]

[1]INAF – IASF Milano, via Bassini 15, 20133 Milano, Italy
email: daniela@lambrate.inaf.it

[2]Obs. de Paris, Sec. de Meudon, GEPI, 5 place Jules Janssen, F-92195 Meudon Cedex, France
[3]Obs. de Paris, Sec. de Meudon, LUTH, 5 place Jules Janssen, F-92195 Meudon Cedex, France
[4]Dep. of Physics and Astronomy, Johns Hopkins University, Baltimore, MD 21218
[5]Dep. of Physics and Astronomy, Washington State University, Pullman, WA 99164-2814

Abstract. We have used the FLAMES multi-integral field unit system of the European Southern Observatory (VLT) centered on the cluster MS0451.6-0305 at $z = 0.5386$ to obtain the spatially resolved kinematics of the cluster members. The spectral data are supported by HST/ACS images that provide immediate morphological information of the cluster galaxies. The relevant structural parameters such as inclination, size, and orientation derived from optical high angular resolution images are compared with those derived from the kinematics. Our final goals are: 1. to derive the Tully-Fisher relation for cluster galaxies with regular kinematics. 2. to obtain the dynamical masses from resolved kinematics and stellar masses from optical images to be compared with local measurements.

Keywords. instrumentation: spectrographs, galaxies: clusters: individual (MS0451.6-0305)

1. Context

Being the largest gravitationally bound objects in the Universe, cluster galaxies are excellent tools to explore the distant Universe and to constrain cosmological models.

Integral field spectroscopic studies appear to be a pre-requisite to sample the whole velocity field of individual galaxies, and to limit uncertainties related to observations (e.g. the major-axis determination, slit misalignment) and internal processes (e.g. kinematic disturbances due to interactions). In this respect, the ability of VLT/FLAMES-IFU to provide simultaneously 15 resolved velocity fields opens the possibility to study in detail kinematics in distant clusters.

We observe the massive X-ray cluster MS0451.6-0305 at $z = 0.5386$ tracing the resolved kinematics of the [OII]3727Å emission line with VLT/FLAMES-IFU. Galaxies in the cluster have shown very different kinematic and morphological behaviors. They present anomalies at some extent, i.e. warps, out-of-plane star formation, stripping/merging events. We derived the rotation curves for eight members which represent the best data-set to perform studies on the dynamic evolution of galaxies, especially in rich environments where mechanisms of ram-pressure stripping and mergers occur more frequently.

Acknowledgements

DV acknowledges supports through a Euro3D RTN on Integral Field Spectroscopy (No. HPRN-CT-2002-00305) and the Marie Curie ERG grant (No. MERG-CT-2005-021704), funded by the European Commission. DV also acknowledges the support of the IAU Grant to attend the IAU conference.

Triggered Star Formation in a Turbulent ISM
Proceedings IAU Symposium No. 237, 2006
B.G. Elmegreen & J. Palouš, eds.

3D dust radiative transfer simulations in the inhomogeneous interstellar medium

E. Vidal Perez and M. Baes

Sterrenkundig Observatorium, Universiteit Gent, Krijgslaan 281 S9, B-9000 Gent, Belgium
email:edgardoandres.vidalperez@ugent.be

Abstract. The study of dusty discs is an important topic in astrophysics, as they seem to be abundant around different objects and are related to different phenomena. In this poster we present 3D radiative transfer simulations of T Tauri type discs with an inhomogeneous dust distribution to investigate the effect of a clumpy medium on the dust temperature distribution. Our initial results indicate that the structure of the dust temperature distribution is rather insensitive to the structure of the ISM, but nevertheless we find a clear and systematic dependence on the parameters describing the structure of the clumpiness of the dust medium.

We present detailed radiative transfer simulations of an inhomogeneous circumstellar disc around a T Tauri-type star. We create a two-phase dust medium consisting of dense clumps in a smooth interclump medium. This two-phase medium is characterized by two parameters, the filling factor ff and the density contrast C. The models presented here were performed using the SKIRT code, an efficient 3D Monte Carlo radiative transfer code. For arbitrary distributions of dust, the code computes the equilibrium dust temperature distribution and emerging spectra.

Our simulations demonstrate that the dust temperature depends systematically on the clumpiness parameters in a subtle way. As an illustration we show in Figure 1 the mean temperature of the dust medium explicitly as a function of ff for two different values of C. For a fixed density contrast, the maximum mean dust temperature is reached at intermediate filling factors. This is normal, as the medium is completely smooth in both the limits $ff = 0$ and $ff = 1$. At intermediate cases, the interclump medium is less dense and warmer, whereas the clumps are dense and cool. The maximum mean dust temperature is higher and reached at lower filling factor values if C increases.

We intend to run more simulations with a larger range of geometries, optical depths and clumpiness parameters to investigate the effects of a non-homogeneous ISM in a more systematic way.

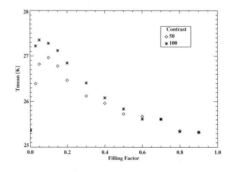

Figure 1. The mean dust temperature of our circumstellar disc for various models with different clumpiness parameters.

Triggered Star Formation in a Turbulent ISM
Proceedings IAU Symposium No. 237, 2006
B. G. Elmegreen & J. Palouš, eds.

© 2007 International Astronomical Union
doi:10.1017/S1743921307002797

Alfvén waves damping in protostellar disks providing a way to expand the action of the magneto-rotational instability

A. A. Vidotto and V. Jatenco-Pereira[1]

[1]IAG, Univ. de São Paulo, Rua do Matão 1226, São Paulo, SP 05508-900, Brazil
email: aline@astro.iag.usp.br

Abstract. In order for the magneto-rotational instability to take place, we need a sufficiently ionized disk. Here, we study, besides viscous dissipation, another heating mechanism for the disk that involves the damping of Alfvén waves due to its interaction with dust grains.

Keywords. stars: pre–main-sequence, accretion, accretion disks, MHD, turbulence, waves

---

The most accepted mechanism for redistributing angular momentum in an accretion disk is the magneto-rotational instability. This mechanism only exists if the gas in the disk is sufficiently ionized to be coupled to the magnetic field lines. Hence, a heating mechanism for the disk is needed. Besides the viscous heating mechanism often included in the models by means of the α prescription (Shakura & Sunyaev 1973), in this work we study the damping of Alfvén waves as an additional heating source. The mechanism we suggest for the damping of the waves involves dust grains, usually observed in these systems. This damping affects the Alfvén wave propagation near the dust-cyclotron frequency, since charged grains in a magnetized medium are highly coupled to the waves due to cyclotron resonances (for further details, see Vidotto & Jatenco-Pereira 2006). We assume a size grains distribution that implies different charges and thus a broad band of resonance frequencies are damped.

By assuming an initial energy flux of Alfvén waves, $\phi = v_A \langle (\delta B)^2 \rangle / (4\pi)$, where v_A is the Alfvén velocity and δB is the magnetic field perturbation, we analyse how the damping of the waves can increase the disk midplane temperature T_c. Writing $|\delta B| = fB$, we study the profile of T_c for the first 2 AU from the central star for different values of f ($f = 0$ means that the only heating mechanism is viscous dissipation). As a result we obtain that the temperature increases more significantly for the outer radii. Comparing with the temperature due to the viscous heating mechanism, if $f = 1\%$, at 0.1 AU there is not any appreciable increase in T_c and at 2 AU the increase is $\sim 9\%$. For $f = 10\%$, this scenario changes: at 0.1 AU the increase in T_c is $\sim 10\%$ and at 2 AU the temperature increases more than 2.5 times. These results show that the wave damping can be an important mechanism for disk heating.

Acknowledgements

This work is supported by FAPESP (04/13846-6) and CNPq (303459/2004-2).

References

Shakura, N.I. & Sunyaev, R.A. 1973, *A&A* 24, 337
Vidotto, A.A. & Jatenco-Pereira, V. 2006, *ApJ* 639, 416

Triggered Star Formation in a Turbulent ISM
Proceedings IAU Symposium No. 237, 2006
B. G. Elmegreen & J. Palouš, eds.

© 2007 International Astronomical Union
doi:10.1017/S1743921307002803

CONDOR observations of high mass star formation in Orion

N. H. Volgenau[1], M. C. Wiedner[1], G. Wieching[1], M. Emprechtinger[1], F. Bielau[1], U. U. Graf[1], C. E. Honingh[1], K. Jacobs[1], B. Vowinkel[1], R. Güsten[2], D. Rabanus[1], J. Stutzki[1] and F. Wyrowski[2]

[1]I. Physikalisches Institut, Universität zu Köln, 50937 Köln, Germany

[2]Max Planck Institut für Radioastronomie, 53121 Bonn, Germany

CONDOR, the **CO**, **N**$^+$, **D**euterium **O**bservations **R**eceiver, is designed to make velocity-resolved observations of the CO, [NII], and p-H$_2$D$^+$ lines in the 1.4 THz (200-240 μm) atmospheric windows. CONDOR's first light observations were made with the APEX telescope in November 2005. The CONDOR beam on APEX (at $\nu = 1.5$ THz) was expected to consist of a 4.3″ main beam and a 73″ error beam; this beam structure was verified from scans of Mars. The pointing accuracy, also determined from Mars scans, was better than 7″. The average atmospheric transmission during our Orion observations (elev$\sim 57°$) was 19 ± 4 % along the line-of-sight. A forward efficiency of $F_{eff} = 0.8$ was determined from sky dips, and observations of the Moon and Mars were used to couple the CONDOR beam to sources of different sizes ($\eta_c = 0.40$ and ~ 0.10, respectively). For more information, see Wiedner *et al.* 2006.

With CONDOR, we observed CO $J = 13 - 12$ emission from three sites of high-mass star formation in Orion (IRc2, FIR4, and NGC2024). A sample spectrum from Orion IRc2 is shown in Fig. 1. In our analysis of IRc2, we assume that all spectra from positions $< \pm 20″$ include a "spike" ($\Delta v \approx 5$ km s^{-1}) and a "hot core" component ($\Delta v \approx 35$ km s^{-1}). The optically thin spike emission arises from the interface of the Orion Ridge and the energizing M42 HII region. A simple isothermal model fit to the $J = 13 - 12$ and higher-J CO lines (e.g. Boreiko *et al.* 1989) reveals that the layer must indeed be warm ($T_{kin} \approx 620$ K), dense ($n(H_2) \approx 2 \times 10^6$ cm^{-3}), and thin ($N(CO) \approx 1.2 \times 10^{16}$ cm^{-2}). Because the Ridge has a temperature gradient, we are currently modeling the data using a PDR code. We are also analyzing the line wings to constrain the outflow properties.

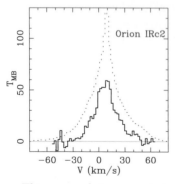

Fig. 1: CONDOR/APEX CO $J = 13 - 12$ (solid) and HHT CO $J = 7 - 6$ (dotted) spectra centered on Orion IRc2. The velocity resolution is smoothed to 2 km s^{-1}. Conversion to T_{MB} was made with $\eta_c = 0.40$ for the CONDOR data and $\eta_c = 0.54$ for the HHT data (Wilson *et al.* 2001). These coupling efficiencies are only valid for the extended, warm CO emission (the "spike" component); thus, the scale for the line wings (the "hot core" component) is underestimated.

This research is supported within SFB 494 of the Deutsche ForschungsGemeinschaft.

References

Boreiko, R.T., Betz, A.L. & Zmuidzinas, J. 1989, *ApJ* 337, 332

Wiedner, M.C., Wieching, G., Bielau, F., *et al.* 2006, *A&A* 454, L33

Wilson, T.L., Muders, D., Kramer, C., & Henkel, C. 2001, *ApJ* 557, 240

Triggered Star Formation in a Turbulent ISM
Proceedings IAU Symposium No. 237, 2006
B. G. Elmegreen & J. Palouš, eds.

© 2007 International Astronomical Union
doi:10.1017/S1743921307002815

HST ACS/HRC imaging of the intergalactic HII regions in NGC 1533

J. K. Werk[1], M. E. Putman[1], G. R. Meurer[2], E. V. Ryan-Weber[3] and M. S. Oey[1]

[1] Department of Astronomy, University of Michigan, Ann Arbor, MI, USA; jwerk@umich.edu

[2] Department of Astronomy and Physics, Johns Hopkins University, Baltimore, MD, USA

[3] Institute of Astronomy, University of Cambridge, Cambridge, CB3 OHK, UK

Abstract. Intergalactic HII regions, far from the confines of a galactic disk, represent a mode of star formation in low-density gas outside of galaxies. The figure below (left) shows an R-band continuum image of NGC 1533 from the SINGG Hα survey (Meurer *et al.* 2006) overlaid with HI contours and the location of three intergalactic HII regions discovered by Ryan-Weber *et al.* (2004). The HI contours are 1.6, 2.0, 2.4, 2.8, 3.2 and 4.0 $\times 10^{20}$ cm^{-2} and have a resolution of $\sim 1'$. ACS/HRC images of the intergalactic HII regions (right) are composites of UV, V, and I bands. The half-light radii of the clusters associated with regions 1, 2, and 5 are 24.7, 21.7, and 17.0 pc, respectively, at the distance to NGC 1533 (21 Mpc; Tonry *et al.* 2001). Assuming a Salpeter IMF with $M_{up} = 100$, Hα/UV ratios indicate a small number of ionizing O stars relative to the total number of UV-emitting O and B stars. These young (4-6 Myr), intergalactic stellar populations lend valuable insight to our understanding of the methods by which star formation is triggered and may even represent the first episodes of star formation in emerging galaxies.

References

Meurer *et al.* 2006, *ApJS* 165, 307
Ryan-Weber *et al.* 2004, *AJ* 127, 1431
Tonry *et al.* 2001, *ApJ* 421, 681

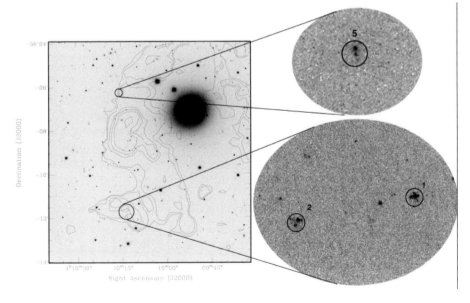

Triggered Star Formation in a Turbulent ISM
Proceedings IAU Symposium No. 237, 2006
B. G. Elmegreen & J. Palouš, eds.

© 2007 International Astronomical Union
doi:10.1017/S1743921307002827

Inferring the nature of turbulence in star-forming regions with polarimetric observations

D. Wiebe[1] and W. D. Watson[2]

[1]Institute of Astronomy of the RAS, Moscow, Russia
email: dwiebe@inasan.ru

[2]University of Illinois at Urbana-Champaign, USA
email: w-watson@uiuc.edu

Abstract. We consider the possibility to study the nature of MHD turbulence in star-forming regions with three different kinds of polarimetric data, namely, the linear polarization of starlight due to extinction by aligned dust grains, the polarized dust thermal radiation at far infrared or submillimeter wavelengths, and the linear polarization of molecular lines due to the Goldreich-Kylafis effect in the anisotropic MHD medium.

Keywords. ISM: magnetic fields, turbulence, polarization

The starlight polarimetry relates properties of the tenuous interclump medium to properties of relatively dense clumps within molecular clouds. The observed lack of contribution of dark clumps to the polarization of background stars is readily reproduced if the magnetic field structure in these clumps is less regular than in the interclump medium and the number of turbulence correlation lengths along the LOS within the clump is about 10 or more. The essential component of the model is that starlight that enters the clump is already polarized (Wiebe & Watson 2001).

Quite irregular magnetic fields in dense clumps, implied by the starlight polarimetry data, still may provide enough alignment in the dust grain ensemble, so that thermal dust emission is polarized at a level of a few per cent. We use a turbulent polarization reduction factor F to describe the influence of the irregular component of magnetic field on the observed fractional polarization. The presence of irregular magnetic field, which is 1.5 stronger than average field, decreases percentage polarization by an order of magnitude. This is compatible with most observations (Wiebe & Watson 2001).

The intrinsic anisotropy of MHD turbulence may also result in an observed polarization of thermal molecular lines. We calculate the linear polarization for the microwave lines of the CO molecule in the anisotropic MHD medium. The maximum degree of polarization occurs at a column density of 10^{15} cm^{-2} which corresponds roughly to $\tau \sim 1$. The fractional polarization reaches 8% at $C/A = 0.1$ for the strong magnetic field ($V_A/c_s = 10$). When the average magnetic field is weak, the fractional polarization also is smaller, indicating less anisotropy in the medium (Wiebe & Watson, ApJ, accepted).

Acknowledgements

D. W. is supported by the RFFI grant 04-02-16637.

References

Wiebe, D. & Watson, W.D. 2001, *ApJ* 549, L115
Wiebe, D. & Watson, W.D. 2004, *ApJ* 615, 300

Triggered Star Formation in a Turbulent ISM
Proceedings IAU Symposium No. 237, 2006
B. G. Elmegreen & J. Palouš, eds.

© 2007 International Astronomical Union
doi:10.1017/S1743921307002839

CONDOR – A heterodyne receiver at 1.25-1.5 THz

M. C. Wiedner[1], G. Wieching[1], F. Bielau[1], M. Emprechtinger[1], U. U. Graf[1], C. E. Honingh[1], K. Jacobs[1], D. Paulussen[1], K. Rettenbacher[1] and N. H. Volgenau[1]

[1]I. Physikalisches Institut, Universität zu Köln, 50937 Köln, Germany

The **CO N**$^+$ **D**euterium **O**bservations **R**eceiver (CONDOR) is a heterodyne receiver that operates between 1250–1530 GHz. Its primary goal is to observe star-forming regions in CO, N$^+$, and H$_2$D$^+$ emission.

The instrument follows the standard heterodyne design. It uses a solid state local oscillator (LO), whose signal is overlaid with that of the sky using a Martin-Puplett interferometer. The heart of the receiver is a superconducting NbTiN hot electron bolometer (HEB) (Muñoz *et al.* 2004). The bolometer has an area of $0.25 \times 2.8 \ \mu m$ and is mounted on a SiN membrane in a waveguide mixer block. To facilitate operation at remote sites, CONDOR is the first receiver that cools the HEB with a closed-cycle system. Since HEBs are particularly sensitive to temperature fluctuations as well as modulations in LO power, we use a Pulse Tube Cooler, which has less vibration than, e.g., a Gifford-McMahon cooler. In order to further minimize vibrations and temperature fluctuations, the mixer and first amplifier are mounted on a separate plate connected via flexible heat straps to the 4 K stage. CONDOR has an intermediate frequency (IF) of about 1.0–1.8 GHz. We consistently obtain receiver noise temperatures below 1800 K and minima in the spectral Allan variances at 25–35 s (see Fig. 1), which is approximately the optimum individual on-source integration time.

In November 2005, CONDOR was successfully commissioned on the 12 m Atacama Pathfinder EXperiment (APEX) telescope. Pointing observations were performed on the Moon and Mars. The first spectral line observations were obtained of CO J=13-12 emission at 1497 GHz from several sources in Orion (Wiedner *et al.* 2006).

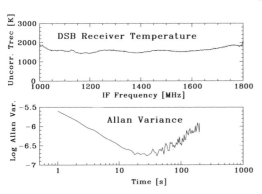

Fig. 1 Technical performance of CONDOR. *Upper panel:* DSB receiver noise temperature versus IF frequency. *Lower panel:* spectroscopic Allan variances.

Acknowledgements

This research is supported by the DFG within SFB 494.

References

Muñoz, P.P., Bedorf, S., Brandt, M., *et al.* 2004, *SPIE* 5498, 834

Wiedner, M.C., Wieching, G., Bielau, F., *et al.* 2006, *A&A* 454, L33

Triggered Star Formation in a Turbulent ISM
Proceedings IAU Symposium No. 237, 2006
B. G. Elmegreen & J. Palouš, eds.

© 2007 International Astronomical Union
doi:10.1017/S1743921307002840

Chandra and Spitzer observations of young clusters

S. J . Wolk[1], B. D. Spitzbart[1] and T. L. Bourke[1]

[1]Harvard–Smithsonian Center for Astrophysics
Cambridge, MA 02138, USA
email: swolk@cfa.harvard.edu

Abstract. The combination of spatial and spectral resolution allow us to use *Chandra* in the study regions of massive star formation which had been inaccessible even from the ground until the last decade. IRAC and MIPS data from *Spitzer* can be combined with the X–ray data to provide insight into the presence of a disk and the activity of the star. The total package allows us to better understand the evolution of the clusters. We have an ongoing program to study several young star forming clusters including distant clusters between 1-3 kpc which support O stars, RCW 38, NGC 281 and RCW 108 and well as clusters within a kpc including IRAS 20050+2720 and NGC 1579, which is a small cluster centered on the Be star LkHα101 and is of uncertain distance although the X-ray data help us refine the current distance estimates. Given the space constraints we only discuss RCW 108 below.

1. RCW 108

RCW 108 contains a deeply embedded young cluster lying in a dark cloud to the west of the young open cluster NGC 6193 (excited by two early O stars). The cluster is obscured by $A_v \sim 20$ and is at a distance of 1.3 kpc and so is more embedded and closer than RCW 38. The exciting source of the IRAS cluster is $\sim$O6-8. At 8-20 μm, the Midcourse Space Experiment Galactic Plane Survey data shows a ridge of warm dust passing through the eastern edge of the emission peak and traversing 15 minutes in a north-south ridge parallel but west of the optical ridge. Our SEST mm continuum observations show this dust ridge as well. The far infrared luminosity suggests that there is more than one significant heating source, i.e., OB stars and/or intermediate mass protostars. The colors and luminosities of the infrared sources imply an associated cluster of low mass T Tauri stars. The extended infrared nebulosity to the east of the main cluster is due to emission and not reflection, suggesting a break-out of radiation in this direction.

We detected about 350 X-ray sources in our 90 ks observation. The morphology is striking. The region is dominated by unabsorbed sources to the east. Theses are associated with the older cluster NGC 6193. The sources associated with RCW 108-IR seems to sit in the middle of a void. This is indicative of a dense cloud of neutral gas. Several specific sites of star formation are found within that cloud complex. These align with some, but not all, the 8 μm peaks in the MSX data. There are several sites within the warm dust cloud containing 1-10 stars. The extant *Spitzer* data are from a very shallow early release observation and parallel the X-ray finding of several specific sites of star formation. The overall trigger appears to be compressive with indications of involvements by both winds and photoionization. However, the direction of the star formation seems to random walk from cluster to cluster, often changing direction.

Triggered Star Formation in a Turbulent ISM
Proceedings IAU Symposium No. 237, 2006
B. G. Elmegreen & J. Palouš, eds.

© 2007 International Astronomical Union
doi:10.1017/S1743921307002852

HD simulations of super star cluster winds

R. Wünsch[1], J. Palouš[1], G. Tenorio-Tagle[2] and S. Silich[2]

[1] Astronomical Institute, Academy of Sciences of the Czech Republic, Boční II 1401, 141 31 Prague, Czech Republic

[2] Instituto Nacional de Astrofísica Optica y Electrónica, AP 51, 72000 Puebla, Mexico

Abstract. We numerically model winds driven by super star clusters (SSC) using the hydrodynamic code ZEUS with the new radiative cooling procedure. The importance of cooling on the wind dynamics depends on the properties of the central cluster: the energy and mass deposition rates L_{sc} and $\dot{M}_{sc}$, and the cluster radius R_{sc}. Low mass clusters behave adiabatically, and their winds are well described by the solution of Chevalier & Clegg (1985). However, for larger L_{sc} and $\dot{M}_{sc}$ and/or smaller R_{sc}, cooling becomes important, and the wind enters the radiative regime in which the wind temperature quickly drops to 10^4 K at a small distance away from the cluster (Silich *et al.*, 2004). There is no stationary wind solution for very energetic and compact clusters. This is expressed by the line of the critical luminosity L_{crit} shown by the left panel as a function of R_{sc}.

In the case of SSC above the threshold line, the stagnation point R_{st} appears inside the cluster. It splits the cluster volume into two parts: the outer one with $r > R_{st}$ where the wind velocity is always positive, and the inner one $r < R_{st}$ where it has a complicated time-dependent profile. The mass inserted into the outer region leaves the cluster in a form of quasi-stationary wind, while most of the mass from the inner region either accumulates there or passes the inner boundary and eventually feeds further star formation. The middle figure shows that the stagnation point R_{st} asymptotically approaches the cluster radius R_{sc} with the increasing L_{sc}.

The right figure summarises several of our calculations for a cluster with an $R_{sc} = 10$ pc. It shows the amount of the mass $\dot{M}_{out}$ outflowing from the cluster depending on L_{sc}. It can be seen that $\dot{M}_{out}$ grows with L_{sc} following the power-law fit of the simulations $\dot{M}_{out} \approx L_{sc}^{0.54}$. However, the fraction of the outflowing mass to the total mass deposited by the cluster $\dot{M}_{sc}$ decreases with L_{sc} from 100% for $L_{sc} = L_{crit}$ to several percent for $L_{sc} = 5 \times 10^{44}$ erg s^{-1}.

Keywords. galaxies: star clusters, galaxies: starburst, ISM: jets and outflows

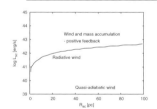 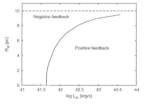 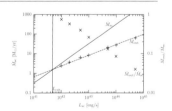

Acknowledgements

The authors gratefully acknowledge support by the Institutional Research Plan AV0Z10030501 of the Academy of Sciences of the Czech Republic, by the project LC06014 Center for Theoretical Astrophysics, and by CONACyT – México, research grant 47534-F.

References

Chevalier, R. A. & Clegg, A. W. 1985, *Nature* 317, 44
Silich, S., Tenorio-Tagle, G. & Rodríguez-González, A. 2004, *ApJ* 610, 226
Tenorio-Tagle, G., Wünsch, R., Silich, S. & Palouš, J. 2006, *in preparation*

Triggered Star Formation in a Turbulent ISM
Proceedings IAU Symposium No. 237, 2006
B. G. Elmegreen & J. Palouš, eds.

© 2007 International Astronomical Union
doi:10.1017/S1743921307002864

A sample of star forming regions triggered by cloud-cloud collision

Bei Xin[1]† and Jun-Jie Wang[1]

[1]National Astronomical Observatories, Chinese Academy of Sciences, Beijing, 100012, China.
e-mail:xp@bao.ac.cn and wangjj@bao.ac.cn

Collision between molecular clouds is considered an efficient mechanism to trigger cloud collapse to form stars. Various observations show that the process is taking place in the universe (Vallee 1995; Wang *et al.* 2004).

We conducted a large-sampling survey for the infrared sources, which are possibly regions of cloud-cloud collision, by making ^{12}CO (1-0), ^{13}CO (1-0), and C^{18}O (1-0) molecular line observations. We selected IRAS sources based on a molecular line survey in the Galaxy (Yang *et al.* 2002), and we picked up the ones whose ^{12}CO (1-0) molecular line is double-peaked or multi-peaked. Observations were made in November and December 2005 with the 13.7 m radio telescope at Delingha, the millimeter-wave radio observatory of Purple Mountain Observatory, in the J=1-0 line of ^{12}CO at 115GHz, ^{13}CO at 110 GHz and C^{18}O at 110 GHz.

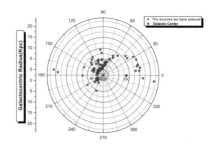

Figure 1: distribution of sources in our sample projected onto the Galactic plane.

We eliminated those whose optically thinner line peaks at the absorption part of the ^{12}CO (1-0) double line. They are possibly molecular clouds that are collapsing, and the double-peaked ^{12}CO (1-0) line is a possible sign of self-absorption (Zhou *et al.* 1993). We kept the sources whose molecular lines' features accord with the description of cloud-cloud collision, i.e. all the three molecular lines have two or more components at same velocity. Finally, we get rid of the possibility that the cloud is infalling, then built up a sample of those cold IRAS sources which are possibly cloud-cloud collision regions. A total of 214 sources were examined, and finally we got 135 of them according to cloud collision criterion. We have done some primary analysis of this data as well. Figure 1 shows the source distribution projected onto the Galactic plane. It can be seen that there are much more sources near the sun than those who are far away from us. More detailed research is carrying through now, and we hope this work can provide astronomers with useful information for their further studies.

References

Vallee, J.P. 1995, *AJ* 110, 2256
Wang, J.J., Chen, W.P., Miller, M., Qin, S.L.& Wu, Y.F. 2004, *ApJ* 614, L105
Yang, J., Jiang, Z.B., Wang, M., Ju, B.G. & Wang, H.C. 2002, *ApJS* 141, 157
Zhou, S., Evans, N.J., II, Koempe, C. & Walmsley, C.M. 1993, *ApJ* 404, 232

† Present address: 20A Datun Road, Chaoyang District, Beijing, 100012, China.

Triggered Star Formation in a Turbulent ISM
Proceedings IAU Symposium No. 237, 2006
B. G. Elmegreen & J. Palouš, eds.

© 2007 International Astronomical Union
doi:10.1017/S1743921307002876

Synthetic observations of turbulent flows in diffuse multiphase interstellar medium

Masako Yamada[1,*], H. Koyama[2], K. Omukai[1] and S. Inutsuka[3]

[1]Department of Theoretical Astronomy, National Astronomical Observatory of Japan,
Osawa2-21-1, Mitaka, Tokyo, JAPAN
*email:ymasako@th.nao.ac.jp

[2]Department of Earth and Planetary System Science, Kobe University

[3]Department of Physics, Kyoto University

Abstract. We examined observational characteristics of multi-phase turbulent flows in the diffuse interstellar medium (ISM) by calculating atomic and molecular carbon lines. Radiation field maps of C^+, C^0, and CO line emissions were generated by calculating the non-local thermodynamic equilibrium (nonLTE) level populations and high resolution hydrodynamic simulations of diffuse ISM. By analyzing synthetic line emission, we found a high ratio between the lines of high- and low-excitation energies in the diffuse multi-phase interstellar medium. Our results shows that simultaneous observations of the lines of warm- and cold-gas tracers will be useful in examining the thermal structure, and hence the origin of diffuse interstellar clouds.

Keywords. turbulence, ISM: lines and bands, radio lines: ISM, line: formation, instabilities

---

1. Observational characteristics in multi-phase turbulence

Recent studies of the multi-phase interstellar medium (ISM) have demonstrated that the origin of small scale interstellar turbulence might be attributed to the thermally unstable nature of the ISM itself. However, most of their studies focus on dynamics. We, on the other hand, tried to bring these theoretical scenarios into the observational realm by calculating the line emission field.

2. Results: line ratio as a probe of ISM models

We calculate the line ratio of high- and low- excitation energy lines in two- and one-phase model (isothermal with $T_{\rm kin}=10$K). Our results show the significantly different line ratio between two- and one- phase model irrespective of the species of emitting particles. CO line ratio $R_{(J,J-1)/1,0}$ is higher than unity up to $J \lesssim 4$ in two-phase medium, though in one-phase model $R_{(J,J-1)/1,0} \ll 1$. This trend applies to the ratio of [CI] fine structure lines as well. Our results are easily understood by level population: if kinetic temperature is higher than the transition energy ($\Delta E/k_B T_{\rm kin}$) a large number of particles that can excite the high J level enables high line ratio – this is the case of two-phase medium. On the other hand, in the cold one-phase gas $\Delta E/k_B T_{\rm kin} \gg 1$ and then line ratio is small. These results demonstrate that simultaneous observations of high- and low- excitation energy lines will reveal the existence of two-phase medium formed by thermal instability.

Acknowledgements

This research was supported in part by Grant-in-Aid by the Ministry of Education, Science, and Culture of Japan (16204012, 18026008, 15740118, 16077202, 18540238).

References

Yamada, M. *et al.* submitted to Astrophysical Journal.

Triggered Star Formation in a Turbulent ISM
Proceedings IAU Symposium No. 237, 2006
B. G. Elmegreen & J. Palouš, eds.

Large-scale CO observations of a far-infrared loop in Pegasus; detection of a large number of very small molecular clouds possibly formed via shocks

Hiroaki Yamamoto[1], Akiko Kawamura[1], Kengo Tachihara[2], Norikazu Mizuno[1], Toshikazu Onishi[1] and Yasuo Fukui[1]

[1]Department of Astrophysics, Nagoya University, Chikusa-ku, Nagoya, 464-8602, Japan

[2]Graduate School of Science and Technology, Kobe University, 1-1 Rokko-dai, Nada-ku, Kobe 657-8501, Japan
email: hiro@a.phys.nagoya-u.ac.jp

Large-scale CO observations with the millimeter/submillimeter telescope NANTEN toward a whole FIR loop-like structure whose angular extent is $\sim 20° \times 20°$ around $(l, b) \sim (109°, -45°)$ in Pegasus have been carried out in the ^{12}CO (J=1–0) at $4'$ – $8'$ grid spacing and the ^{12}CO emitting region in the ^{13}CO (J=1–0) at $2'$ grid spacing. The diameter corresponds to ~ 25 pc at a distance of 100 pc, adopted from that of the star HD886(B2IV) near the center of the loop.

The 78 ^{12}CO small clumpy clouds we detected have mass 0.04–11 $M_\odot$, of which $\sim 83\%$ have very small masses $\leqslant 1.0$ Mo. ^{13}CO emission was detected in the region where the column density of H_2 derived from ^{12}CO is $\geqslant 5 \times 10^{20}$ cm^{-2}, corresponding to $A_V \sim 1$ mag, which takes into account that of HI. The ^{13}CO clouds are far from virial equilibrium, indicating that they are not gravitationally bound, but these clouds tend to be more virialized as the mass increases.

We find no indication of star formation in these clouds in the IRAS and 2MASS point source catalogs. The very low mass clouds, $M \leqslant 1$ $M_\odot$, identified are unusual in that they have a very weak ^{12}CO peak temperature of 0.5–2.7 K and that they aggregate in a region of a few parsecs with no main massive clouds; in contrast, similar low-mass clouds $\leqslant 1$ $M_\odot$ in other regions previously observed including those at high Galactic latitude are all associated with more massive main clouds of ~ 100 $M_\odot$. The HI distribution in this area is looplike and shown in the expanding motion in the L-V diagram. The molecular clouds are distributed nearly along the HI loop but the expanding motion in CO can not be seen. Comparison with a theoretical work on molecular cloud formation suggests that the very low mass clouds may have been formed in the shocked layer through thermal instability. HD 886 may be the source of the mechanical luminosity via stellar winds to create shocks, forming the loop-like structure where the very low mass clouds are embedded.

Acknowledgements

The NANTEN project is based on a mutual agreement between Nagoya Univ. and the CIW. We appreciate the hospitality of the members of the LCO. We thank Japanese public donors and companies for the contribution to the project. This work is financially supported in part by a Grant-in-Aid for Scientific Research from the Ministry of Edu., Culture, Sports, Sci. and Tech. of Japan (No. 15071203) and from JSPS (No. 14102003, core-to-core program 17004).

Triggered Star Formation in a Turbulent ISM
Proceedings IAU Symposium No. 237, 2006
B. G. Elmegreen & J. Palouš, eds.

ⓒ 2007 International Astronomical Union
doi:10.1017/S174392130700289X

Molecular loops in the Galactic centre; evidence for magnetic floatation accelerating molecular gas

H. Yamamoto [1], Y. Fukui[1], M. Fujishita[1], K. Torii[1], N. Kudo[1], S. Nozawa[2], K. Takahashi[3,4], R. Matsumoto[5], M. Machida[3], A. Kawamura[1], Y. Yonekura[6], N. Mizuno[1], T. Onishi[1] and A. Mizuno[7]

[1]Dep. of Astrophysics, Nagoya Univ., 464-8602, Japan [2]Dep. of Science, Ibaraki Univ., 310-8512, Japan [3]NAOJ, 118-8588, Japan [4]The Graduate Univ. for Advanced Studies, 240-0193, Japan [5]Dep. of Physics, Faculty of Science, Chiba Univ., 263-8522, Japan [6]Dep. of Physics Science, Osaka Prefecture Univ., 599-8531, Japan [7]Solar-Terrestrial Environment Lab., Nagoya Univ., 464-8601, Japan E-mail:hiro@a.phys.nagoya-u.ac.jp

The new molecular image obtained by NANTEN telescope in the galactic center has revealed the existence of the two loop like structures, loop 1 and loop 2, which have never been seen before toward $l = 355°$ to $358°$. The velocities of loop 1 and loop 2 are -180 to -90 km s^{-1} and -90 and -40 km s^{-1}, respectively, and these two loops have strong velocity gradients. The foot points of the loops show a very broad linewidth of $\sim$40 to 80 km s^{-1} whose large velocity spans are characteristic of the molecular gas near the galactic center. Therefore, we classified the loops as being located in the galactic center and adopt a distance of 8.5 kpc. Then, the projected lengths of loop 1 and loop 2 were estimated as $\sim$500 and $\sim$300 pc, respectively and velocity gradients corresponds to $\sim$80 km s^{-1} per 250 pc along loop 1 and $\sim$60 km s^{-1} per 150 pc along loop 2. The heights of these loops are also estimated as $\sim$220 to $\sim$300 pc from the galactic plane, significantly higher than the typical scale height in the nuclear disk.

Each of the loops has a mass of $\sim$0.8$\times$10^5 M$_\odot$ as a lower limit by combining the ^{12}CO and ^{13}CO data and assuming local thermodynamical equilibrium at 50 K. The kinetic energy in a loop was estimated to be $\sim$0.9$\times$10^{51} erg for a velocity dispersion of 30 km s^{-1}. The velocity and the energy of loops can not be explained by a super nova. Therefore, we offer a model incorporating MHD instability to explain the formation of the two loops. From this model we found that two model loops calculated by an MHD code were a good match for the observations. We used parameters of 100 cm^{-3} in gas number density and 150 μm in magnetic field. Then, the Alfven speed was calculated as 24 km s^{-1}.

This model offers naturally significant heating of the warm molecular gas at the foot points. The velocity dispersion of the broad CO features corresponds to kinetic temperature higher than about 10^4 K if the shock is completely converted into thermal energy at the foot points. We suggest that the present model has the potential to be applied to the other salient broad velocity features in the galactic center and to the heating of the molecular gas at their foot points.

Acknowledgements

The NANTEN project is based on a mutual agreement between Nagoya Univ. and the CIW. We appreciate the hospitality of the members of the LCO. We thank Japanese public donors and companies for the contribution to the project. This work is financially supported in part by a Grant-in-Aid for Scientific Research from the Ministry of Edu., Culture, Sports, Sci. and Tech. of Japan (No. 15071203) and from JSPS (No. 14102003, core-to-core program 17004).

Author Index

Afanasiev, V.L. – 473
Aikawa, Y. – 455, 456
Allen, L.E. – 464
Alves, J. – **204**
André, Ph. – **132**, 160, 265
Arbutina, B. – **391**
Arnal, M. – 400
Asanok, K. – 452
Baek, C.H. – **392**
Baes, M. – 490
Bagetakos, I. – 76, **393**, **394**
Bains, I. – 404, 429
Balkowski, C. – 489
Bally, J. – **165**
Barbá, R.H. – 400
Barsony, M. – **177**
Bartkiewicz, A. – **395**
Basu, S. – 437
Beck, R. – 402, 470
Beckman, J.E. – 419, 467
Bekki, K. – **373**
Belloche, A. – 265
Beuther, H. – **148**
Bieging, J.H. – **396**, 463
Bielau, F. – 492, 495
Bigiel, F. – **397**
Birkmann, S.M. – 424
Blackman, E.G. – 172
Blakeslee, J.P. – 489
Bonnell, I.A. – **344**
Boquien, M. – 323, **398**
Boulanger, F. – **47**
Bourke, T.L. – 496
Bournaud, F. – 323
Braine, J. – 398
Brandenburg, A. – 183
Brandl, B.R. – 476
Breitschwerdt, D. – **57**
Brinks, E. – **76** 393, 394, 397, 398,
Brooks, K. – 403
Brunt, C.M. – 9, 363
Bruzual-A., G. – 445
Burkert, A. – 246, 410
Burton, M.G. – 160, 403, 404, 429
Busfield, A.L. – 482
Campeggio, L. – **399**, 411
Caplan, J. – 212
Cappa, C.E. – **400**, 444
Carignan, C. – 467
Carlson, L.R. – 199
Caswell, J.L. – 403
Cayatte, V. – 489

Cesaroni, R. – 160, 412
Charmandaris, V. – 398
Chastain, R.J. – 17
Chen, C.-H.R. – **401**
Chen, H.-R. – 448
Chen, W.P. – **278**
Chengalur, J.N. – 466
Choi, M. – 392
Chrysostomou, A. – 403
Chu, Y.H. – **192**, 401
Churchwell, E. – 472
Chyży, K.T. – **402**, 470
Clarke, A.J. – 482
Clarke, C.J. – **300**
Cohen, R.J. – **403**, 422, 452
Cortes, J. – 436
Costagliola, F. – 17
Cox, J. – 403
Crutcher, R.M. – **141**
Cunningham, A.J. – **172**
Cunningham, M.R. – **404**, 429
Dahlem, M. – 441
Daigle, O. – 467
Dale, J.E. – 300
Datta, S. – **405**
Dawson, J. – **406**
de Avillez, M.A. – 57
de Blok, E. – 76, 393, 394, 397
De Buizer, J.M. – **407**
de Grijs, R. – **408**
Deharveng, L. – **212**
De Luca, M. – 411
Demichev, V.A. – 446
De Vries, C.H. – **409**
Diamond, P.D. – 446
Diamond, P.J. – 403
Dias, W.S. – 438
Dib, S. – **410**
Dickey, J.M. – **1**, 363
Dobbs, C.L. – 344
Dottori, H. – 416, 417, 468
Duc, P.A. – **323**, 398
Dumke, M. – 420
Duronea, N. – 400
Ehle, M. – 470
Ehlerová, S. – **91**
Eislöffel, J. – 217
Ekta – 466
Elia, D. – 399, **411**
Ellingsen, S. – 403
Elmegreen, B.G. – **384**
Emprechtinger, M. – 492, 495

Endo, A. – 436
Evans, N.J., II – 481
Ezawa, H. – 436
Falgarone, E. – **24**
Falle, S.A.E.G. – 486
Felli, M. – **412**
Field, D. – 183, 418
Flores, H. – 489
Frank, A. – 172
Fujishita, M. – 501
Fukui, Y. – **31**, 101, 128, 406, 478, 500, 501
Fuller, G.A. – 403
Gallagher, J. – 199
Gao, Y. – **331**
Garcia-Burillo, S. – 427
Gasiprong, N. – 452
Giammanco, C. – 419
Gibson, S.J. – **363**
Goddi, C. – 160
Goldman, I. – **96**
Goldsmith, P.F. – 488
González-Lópezlira, R.A. – 445
Goss, W.M. – 400
Graf, U.U. – 492, 495
Graham, D.A. – 446
Gray, M.D. – 403
Gredel, R. – 417
Green, D.A. – **413**
Green, J.A. – 403
Gregorio-Hetem, J. **414**, 425
Gressel, O. – **415**
Gritschneder, M. – **246**
Groppi, C.E. – 463
Grosbøl, P. – **416**, **417**
Gruendl, R.A. – 192, 401
Gustafsson, M. – **183**, **418**
Güsten, R. – 492
Gutermuth, R.A. – 464
Gutiérrez, L. – **419**
Haikala, L.K. – **420**
Hamaguchi, K. – **421**
Hammer, F. – 489
Harju, J. – 420
Hartquist, T.W. – 486
Harvey-Smith, L. – **422**
Hasegawa, T. – 368, 426, 432
Hatano, H. – **423**
Hatsukade, B. – 436
Hawks, L. – 199
Hayakawa, T. – 426, 432
Hayashi, A. – 478
Hayashi, M. – 447, 453
Hedden, A.S. – 463
Heitsch, F. – 246, 401
Hennebelle, P. – **265**
Hennemann, M. – **424**

Henning, Th. – 440
Hetem, A., Jr. – **425**
Heyer, M.H. – **9**
Hily-Blant, P. – 24
Hiramatsu, M. – **426**, 432
Hitschfeld, M. – **427**
Hoare, M.G. – 403, 482
Hodapp, K. – 217
Hojaev, A.S. – **428**
Honingh, C.E. – 492, 495
Hony, S. – 465
Hubert, A.-M. – 465
Hutawarakorn Kramer, B. – 452
Inutsuka, S. – 455, 499
Itoh, Y. – 454
Jacobs, K. – 492, 495
Jappsen, A.-K. – 435
Jatenco-Pereira, V. – 491
Jeffries, R.D. – 460
Jenkins, E.B. – **53**
Jones, P.A. – 404, **429**
Joung, M.K.R. – **358**
Juvela, M. – 420, **430**
Kadowaki, R. – 423
Kaltcheva, N.T. – **431**
Kamegai, K. – 426, **432**, 436
Kato, D. – 423
Kavars, D.W. – 363
Kawabe, R. – 368, 436, 475, 479
Kawamura, A. – **101**, 128, 406, 500, 501
Kehrig, C. – **433**
Kennicutt, R.C., Jr. – **311**
Kewley, L.J. – 476
Kim, J. – 392, 410
Kim, S. – **434**
Kim, W.T. – **351**
Kiss, Z.T. – 124
Kitsionas, S. – **435**
Klein, Ra. – 440
Klessen, R.S. – 336, 435
Kniazev, A.Y. – 466
Kohno, K. – 368, **436**, 451
Köppen, J. – 480
Koyama, H. – 499
Kramer, C. – 427, 477
Krause, O. – 424
Kritsuk, A.G. – 283
Kroupa, P. – **230**
Krumholz, M.R. – **378**
Kudo, N. – 501
Kudoh, T. – **437**
Kulesa, C.A. – 396, 463
Kun, M. – **119**
Kuno, N. – 368, 436, 451
LaRosa, T.N. – 17
Lajús, E.F. – 400

Langer, W.D. – 488
Lee, H.T. – 278
Lefloch, B. – 212
Lehtinen, K. – 420
Lemaire, J.-L. – 183, 418
Lemke, D. – 424
Lenorzer, A. – 465
Lépine, J.R.D. – **438**
Li, D. – 488
Li, J.Z. – **439**
Li, P.S. – 283
Li, Y. – 336
Li, Z.-Y. – 306
Lima, G.H.R.A. – 487
Linz, H. – **440**
Lisenfeld, U. – 398, **441**
Lo, N. – 404
Longmore, S.N. – 160
Looney, L. – 440
Lopez-Martin, L. – 433
Lumsden, S.L. – 482
Mac Low, M.-M. **336**, 358
Machida, M. – 501
Madore, B. – 397
Magnani, L. – 17
Maiolo, B.M.T. – 399, 411
Mamajek, E.E. – **442**
Marenzi, A.R. – 462
Marschalkó, G. – **443**
Martín, M.C. – **444**
Martínez-García, E.E. – **445**
Martin-Gordón, D. – 433
Masheder, M.R.W. – 403
Massi, F. – 411, 412
Matsumoto, R. – 501
Matsuo, H. – 436
Matsushita, S. – 368
Matthews, B.C. – 155
Mattila, K. – 420, 430
Matveyenko, L.I. – **446**
Mayama, S. – **447**
McClure-Griffiths, N. – 403
Megeath, S.T. – 464
Mei, S. – 489
Meixner, M. – 199
Meurer, G.R. – 493
Millar, T.J. – 456
Minamidani, T. – 101
Minh, Y.C. – **448**
Minier, V. – **160**
Mishurov, Yu. – 438
Miura, R. – 368, 436
Mizuno, A. – 101, 128, 426, 432, 478, 501
Mizuno, N. – 101, **128**, 406, 500, 501
Mizuno, Y. – 101
Montmerle, T. – 414

Mookerjea, B. – 477
Moore, T.J.T. – 482
Morgan, L.K. – **449**, 483
Motoyama, K. – **450**
Mottram, J.C. – 482
Muñoz-Tuñón, C. – 238, 242
Muanwong, O. – 452
Muller, E. – 429
Muraoka, K. – 436, **451**
Myers, P.C. – 464
Naab, T. – 246
Nakagawa, Y. – 455, 456
Nakamura, F. – **306**
Nakanishi, K. – 368, 436, 451
Nammahachak, S. – **452**
Narayanan, G. – 409
Neiner, C. – 465
Nishikawa, T. – **453**
Niwa, T. – **454**
Nomura, H. – **455**, **456**
Nordlund, Å. – 283
Norman, M.L. – 283
Nota, A. – 199
Nozawa, S. – 501
Nyman, L.-Å. – 440
Oasa, Y. – 454, **457**
Oey, M.S. – **106**, 199, 493
Ojha, D. – **458**
Oka, T. – 471
Okuda, T. – 436
Okumura, S.K. – 368
Olberg, M. – 420
Oliveira, J.M. – **459**, **460**
Omukai, K. – 499
Onishi, T. – 101, 128, 406, 426, 432, 478, 500, 501
Ossenkopf, V. – 477
Ostriker, E.C. – 65, **70**, 474
Oudmaijer, R.D. – 482
Padoan, P. – **283**, 430
Paladino, R. – **461**
Palouš, J. – **114**, 242, 497
Pasquali, A. – 199
Paulussen, D. – 495
Pelkonen, V.M. – 430
Peretto, M. – 160
Pérez, E. – 238
Persi, P. – **462**
Pestalozzi, M. – 160, 403
Peters, W.L. – 396, **463**
Peterson, D.E. – **464**
Petre, R. – 421
Pety, J. – 24
Phillips, C. – 403
Pineau des Forêts, G. – 24
Piontek, R.A. – **65**

Pipher, J.L. – 464
Pomarès, M. – 212
Preibisch, Th. – **270**
Puerari, I. – 484
Puetz, P.S. – 463
Puga, E. – **465**
Pustilnik, S.A. – **466**
Putman, M.E. – 493
Puzia, T.H. – 473
Quillen, A. – 172
Rabanus, D. – 492
Reach, W.T. – **188**
Reichert, D. – 468
Reid, M.A. – **155**
Reipurth, B. – 487
Relaňo, M. – **467**
Rengel, M. – **217**
Rettenbacher, K. – 495
Rodrigues, C.V. – 414
Rodrigues, I. – **468**
Röllig, M. – 477
Romero, G.A. – 444
Roshi, D.A. – **469**
Rosolowsky, E.W. – **208**
Rossa, J. – 441
Rubio, M. – **40**
Ryś, S. – 402, **470**
Ryan-Weber, E.V. – 493
Sabbi, E. – **199**
Saito, M. – 475, 479
Sakai, T. – **471**
Sakamoto, S. – 436, 451
Sanchawala, K. – 278
Sanchez, S.F. – 433
Sato, S. – 423
Sawada, T. – 436
Schlottman, K. – 396
Schuster, K. – 427
Sewiło, M. – **472**
Shadmehri, M. – 410
Shang, H. – 450
Sharina, M.E. – **473**
Shetty, R. – **474**
Shimajiri, Y. – **475**, 479
Shioya, Y. – 368
Shore, S.N. – **17**
Silich, S. – 238, **242**, 497
Sirianni, M. – 199
Sivakon, S.S. – 446
Skrutskie, M.F. – 464
Smith, L.J. – 199
Smith, M.D. – 439
Snell, R.L. – 409
Snijders, L. – **476**
Sorai, K. – 436, 451
Spitzbart, B.D. – 496
Stanimirović, S. – **84**, 459

Stecklum, B. – 440
Stil, J.M. – 363
Strafella, F. – 399, 411
Struck, C. – **317**
Stützki, J. – 427, 477, 492
Sugitani, K. – 454
Sun, K. – **477**
Sunada, K. – 454
Szymczak, M. – 395
Tachihara, K. – 454, **478**, 500, 501
Takahashi, S. – 475, **479**
Takakuwa, S. – 475, 479
Takami, M. – 453
Tamura, M. – 447, 458
Tamura, Y. – 368
Tan, J.C. – **258**
Tanaka, K. – 436
Tapia, M. – 462
Tatematsu, K. – 426, 432
Taylor, A.R. – 363
Telles, E. – 238, 433
Tenorio-Tagle, G. – **238**, 242, 497
Theis, C. – **480**
Thompson, M.A. – 403, 449, 483
Torii, K. – 501
Tosaki, T. – **368**, 436, 451
Tosi, M. – 199
Tóth, L.N. – **124**
Tripp, T.M. – 53
Troland, T.H. – 141
Tsujimoto, M. – 456
Umemoto, T. – 450
Urban, A. – **481**
Urošević, D. – 391
Urquhart, J.S. – 449, **482**, **483**
Valdez-Gutiérrez, M. – **484**
van den Ancker, M.E. – **485**
van der Werf, P.P. – 476
van Langevelde, H.J. – 395
Van Loo, S. – **486**
van Loon, J. Th. – 459, 460
Vanzi, L. – 466
Vasquez, J. – 400
Vaz, L.P.R. – **487**
Vázquez-Semadeni, E. – **292**, 410
Velusamy, T. – **488**
Vergani, D. – **489**
Vidal Perez, E. – **490**
Vidotto, A.A. – **491**
Vila Vilaro, B. – 396
Vílchez, J.M. – 433
Vlajic, M. – 199
Volgenau, N.H. – **492**, 495
Voronkov, M. – 403
Vowinkel, B. – 492
Vukotić, B. – 391
Walker, C.K. – 463

Walsh, A. – 403
Walter, F. – 76, 393, 394, 397
Wang, J.-J. – 498
Ward-Thompson, D. – 403
Waters, L.B.F.M. – 465
Watson, D. – 494
Werk, J.K. – **493**
Weżgowiec, M. – 470
White, G.J. – 449, 483
Whitmore, B.C. – **222**
Whitney, B. – 472
Whitworth, A.P. – **251**, 265, 435
Wiebe, D. – **494**
Wieching, G. – 492, 495
Wiedner, M.C. – 492, **495**

Wolk, S.J. – **496**
Wong, T. – 404, 429
Wong-McSweeney, D. – 403
Wünsch, R. – **497**
Wyrowski, F. – 492
Xin, B. – **498**
Yamada, M. – **499**
Yamaguchi, N. – 436
Yamamoto, H. – **50**, **501**
Yamamoto, S. – 471
Yang, C.-C. – 192
Yates, J.A. – 403
Yonekura, Y. – 501
Zavagno, A. – 212
Ziegler, J. – 415
Zinnecker, H. – 270

Object Index

30 Doradus – 40-46, 47-52, 131, 239, 240, 373-377, 459
32 Ori group – 442
AFGL 333 – 471
AFGL 2591 – 152
Antennae – 210, 222-229, 231, 324, 408, 476
Barnard 1,3,5 – 405
Berkeley 55 – 428
Carina Flare – 130, 406
Carina Nebula – 421
Cas-Tau group – 442
CB34 – 464
CB52, 107 – 399
Centaurus A – 204
Cepheus Flare – 125
Chamaeleon I molecular cloud – 426, 478
CMa R1 – 414
Cometary Globule 12 – 420
CrA cloud – 274
DC267.4-07.5 – 399
DDO43,47,88 – 78
Eagle Nebula – 460
η Carina – 131, 278-282
G23.657-00.127 – 395
G29.96 – 149-152
G35.20-0.74 – 407
G307.27+0.27 – 444
G333.6-0.2 – 404, 429
GSH-242-03+37 – 88
GSH-277+00+36 – 4
Hen 206 – 274
HL Tau – 453
HoI – 78
HoII – 78, 93, 96-100, 109
Horsehead Nebula – 124, 246
Hot Molecular Cores – 148-154
IZw18, IIZw70 – 433
IC 10 – 78, 209
IC 348 – 367, 405, 477
IC 1396 – 188-191, 274
IC 1613 – 78
IC 2574 – 78, 274
IC 4996 – 428
IRAS 05358 – 150-151
IRAS 06061+2151 – 462
IRAS 13036-7644 – 432
IRAS 18089-1732 – 152
IRAS 20050+2720 – 496
IRAS 20126+4104 – 151, 407
IRAS 23032+5937 – 424

IRAS 23033-5951 – 155-159
IRAS 23151 – 150-152
ISOSS J23053+5953 – 424
ISOSS J18364-0221 – 424
King 1, 13, 18, 20 – 428
L1448 – 143-144, 168, 177-182, 218, 405
L1455 – 168, 218, 405
L1551 – 166, 167, 179-180, 453
Large Magellanic Cloud – 31-39, 40-46, 77-78, 84-90, 93, 101-105, 107-113, 130-131, 192-198, 209, 231, 274, 338, 360, 373-377, 380, 408, 423, 434, 459, 472
LkHα 101 – 496
LkHα 349a,c – 188
Lower Centaurus-Crux – 272
LSS 1887 – 444
M31 – 78, 93, 109, 209, 368-372
M33 – 78, 92, 93, 109, 209, 267, 316, 368-372, 380, 436
M42 – 487, 492
M51 – 311-316, 321, 351, 427, 461
M64 – 208-211, 380
M81 – 313, 391, 394
M82 – 77, 208-211, 234, 391, 408
M83 – 204, 436, 451
M101 – 78
Magellanic Bridge – 4, 84, 131, 373
MBM 16 – 18
MBM 40 – 19
Mon-CMa-Pup-Vel field – 431
μ Oph group – 442
N11 – 45, 110, 193, 423
N44 – 110, 401
N51D – 110, 196
N66 – 40-46, 199
N105A – 459
N113 – 459
NGC 147 – 473
NGC 185 – 473
NGC 205 – 473
NGC 253 – 77
NGC 281 – 496
NGC 299 – 202-203
NGC 346 – 40-46, 199-201
NGC 376 – 202
NGC 602 – 202
NGC 604 – 239-240, 368-372, 436
NGC 1333 – 119-123, 165, 167, 168, 172-176, 179, 180, 217-221, 265-269, 306-307, 392, 405, 457

NGC 1333 IRAS 4A – 179, 181, 265-269, 392
NGC 1530 – 467
NGC 1533 – 493
NGC 1569 – 78
NGC 1579 – 496
NGC 1672 – 436
NGC 1703 – 445
NGC 2024 – 492
NGC 2068 – 133, 137
NGC 2129 – 464
NGC 2207/IC 2163 – 321
NGC 2244 – 439
NGC 2403 – 80, 109
NGC 2997 – 417
NGC 3184 – 393
NGC 3359 – 467
NGC 4038/4039 – 222-229, 476
NGC 4254 – 402
NGC 4569 – 470
NGC 4666 – 77
NGC 5194 – 467
NGC 5195 – 427
NGC 5291 – 326, 398
NGC 6193 – 496
NGC 6231 – 485
NGC 6334 – 259
NGC 6357 – 400
NGC 6397 – 485
NGC 6611 – 460
NGC 6820 – 428
NGC 6823 - 428
NGC 6946 – 78-79, 393
NGC 6951 – 467
NGC 7023 – 51
NGC 7130 – 436
NGC 7469 – 319
NGC 7538 – 13-14, 407, 458
NGC 7593 – 468
NGC 7631 – 468
NGC 7643 – 468
NGC 7801 – 428
OMC1 – 169, 183-187
OMC-2/3 – 475, 479
ON1 – 452
Orion – 102, 126, 129, 132-140, 148-150, 165-171, 183-187, 218, 258-259, 280, 298, 418, 419, 425, 446, 479, 487, 488, 492, 495

Pegasus FIR loop – 500
Pipe Nebula – 205
Polaris Flare – 120
R136 – 40-46, 47-52, 131, 231
RCW 38 – 496
RCW 52 – 444
RCW 78 – 444
RCW 79 – 213-214
RCW 82 – 214
RCW 108 – 496
ρ Oph – 124, 132-140, 143, 233, 273, 478, 488
Rigel-Crutcher cloud – 4
RNO 91 – 447
Rosette Nebula – 110, 439
S106 – 457
S221 – 444
S255-S257 – 160-164, 396
S235 – 412
S284 – 465
Scorpius-Centaurus – 273
SFO 75 – 483
Sh2-104 – 213-214
Sh2-212 – 214
Sh2-217 – 214
Sh2-241 – 214-215
Sh2-255 – 458
Small Magellanic Cloud – 31-39, 40-46, 77-78, 84-90, 93, 96-100, 109-110, 131, 199-203, 209, 225, 373-377, 423, 467
Taurus – 10, 124, 137, 233, 256, 488
Trumpler 14, 16 – 278-282
TW Hydra – 274
UGC 1281 – 232
Upper Scorpius – 124, 272
Upper Centaurus-Lupus – 272
VDB 142 – 273
Vela-D cloud – 411
W3 – 110, 258, 274, 422, 448, 458, 463, 471
W5 – 454
W48A – 469
W51 – 463
WR 23 – 444
WR 55 – 444

Subject Index

autocorrelation function – 1-8, 9-16, 96-100

chemistry – 11, 21, 26, 27, 148-154, 337, 404, 443, 448, 459

cold HI – 363-367

dust – 40-46, 47-52, 124-127, 160, 204-207, 217-221, 311-316, 439, 455, 456, 490, 491, 494, 496

fractal geometry – 2, 60, 124, 300, 347, 411, 418

Herbig-Haro object – 165-171, 280, 423, 425, 485

infant mortality – 222-229, 233

interacting galaxies – 222-229, 317-322, 323-330, 466, 489

intermittency – 12, 18, 21, 24, 72, 74, 360

jets – 172-176

Local Bubble – 62, 119

magnetic field – 1-4, 9, 17, 48, 57-64, 65-69, 70-75, 84, 88, 108-109, 141-147, 258, 259, 269, 283-291, 292-299, 306-310, 351-357, 410, 415, 422, 437, 452, 469, 470, 474, 486, 491, 494, 501

Magneto-Rotational Instability – 65-69, 70-75

PAH – 47-52, 78, 160, 212-216, 312, 398, 472

power spectrum – 2, 10, 96, 248, 268, 283-291, 306, 359-360, 411, 413, 429

prestellar cores – 132-140

principal component analysis – 2, 11, 404

probability distribution function – 18, 379

shells and supershells – 1-8, 31-39, 47-52, 76-83, 84-90, 91-95, 96-100, 101-105, 106-113, 114-118, 119-123, 124-127, 128-131, 192-198, 212-216, 217-221, 238-241, 251-257, 368, 384-389, 393, 406, 414, 439, 444, 446, 465, 467, 472

spectral correlation function – 2

spiral shock – 74, 344-350, 351-357, 363, 438

structure function – 1-8, 9-16, 96-100, 183-187

superbubbles – 57, 76-83, 106-113, 128-131, 165-171, 192-198, 240, 270-277, 278-282

supernovae – 57-64, 106-113, 119-123, 124-127, 165-171, 192-198, 242-245, 270-277, 358-362, 391, 414, 415, 420, 421

swing amplification – 353

thermal instability – 21, 60, 72, 296, 500

tidal dwarf galaxies – 323-330

tiny scale structure – 4

Toomre Q parameter – 70-75, 336-343, 351-357, 384-389, 397

Toomre sequence – 222, 318, 466

velocity component analysis – 429

velocity channel analysis – 11

velocity-size relation – 116, 293, 347, 379, 434